Sedimentology

By the same author

A dynamic stratigraphy of the British Isles
(with R. Anderton, P. H. Bridges and B. W. Sellwood)

Sedimentology

Process and Product

M. R. Leeder

Department of Earth Sciences, University of Leeds

London
GEORGE ALLEN & UNWIN
Boston Sydney

**George Allen & Unwin (Publishers) Ltd,
40 Museum Street, London WC1A 1LU, UK**

George Allen & Unwin (Publishers) Ltd,
Park Lane, Hemel Hempstead, Herts HP2 4TE, UK

Allen & Unwin Inc.,
9 Winchester Terrace, Winchester, Mass 01890, USA

George Allen & Unwin Australia Pty Ltd,
8 Napier Street, North Sydney, NSW 2060, Australia

First published in 1982

British Library Cataloguing in Publication Data

Leeder, M. R.
Sedimentology
1. Sedimentology
I. Title
551.3'04 QE471
ISBN 0-04-551053-9
ISBN 0-04-551054-7 Pbk

Set in 9 on 11 point Times by Pintail Studios Ltd, Ringwood, Hampshire, and printed in Great Britain
by Mackays of Chatham

To Kate

The gods had condemned Sisyphus to ceaselessly rolling a rock to the top of a mountain, whence the stone would fall back of its own weight. They had thought with some reason that there is no more dreadful punishment than futile and hopeless labour.
. . . Each atom of that stone, each mineral flake of that night-filled mountain, in itself forms a world. The struggle itself towards the heights is enough to fill a man's heart. One must imagine Sisyphus happy.

Preface

The origin, dispersal, deposition and burial of natural sediment grains is the central concern of sedimentology. The subject is truly interdisciplinary, commands the attention of Earth scientists, is of considerable interest to fluid dynamicists and civil engineers, and it finds widespread practical applications in industry.

Sedimentology may be approached from two viewpoints: a descriptive approach, as exemplified by traditional petrography and facies analysis, and a quantitative approach through the physical and chemical sciences. Both approaches are complementary and must be used in tandem if the recent remarkable progress in the field is to be sustained. This text aims to introduce such a combined approach to senior undergraduate students, graduate students and to interested professional Earth scientists. Thus the many descriptive diagrams in the text are counterbalanced by the use of basic physical and chemical reasoning through equations.

I have tried to construct a text that follows logically on from the origin of sediment grains through fluid flow, transport, deposition and diagenesis (the change from sediment to rock). The text has been written assuming that some basic previous instruction has been given in the Earth sciences and in general physics and chemistry. Certain important derivations are given in appendices. I have avoided *advanced* mathematical treatment since it is my opinion that recognition of the basic physical or chemical basis to a problem is more important to the student than the formal mathematical reduction of poorly gathered data. As T. H. Huxley has written,

> Mathematics may be compared to a mill of exquisite workmanship which grinds you stuff of any degree of fineness but, nevertheless, what you get out depends upon what you put in, and as the grandest mill in the world will not extract wheat-flour from peascod, so pages of formulae will not get a definite result out of loose data.

I feel I ought to make some comments on my choice and emphasis of subject matter in this text. The book deals essentially with *principles*. I have had particular problems in deciding on the amount of space that should be devoted to case histories of ancient sedimentary rocks from the point of view of facies analysis, basin analysis and tectonic regime. Since there are a very large number of such studies in the literature (see Reading 1978 for the most complete compendium) and since each instructor approaches this aspect from his own viewpoint I have generally restricted myself to brief discussion of a few chosen ancient examples for each sedimentary environment. I suppose, in effect, I am admitting a bias in this text towards teaching facies models based upon recent sediments. Considerations of space and economy have forced me to abandon plans for separate chapters on (a) basin analysis with respect to plate tectonic setting, (b) non-uniformitarian facies (such as banded ironstones), and (c) economic sedimentology of stratabound 'metallic' ores.

Most of all I address this text to young Earth scientists. I hope that it will encourage them to reach a true understanding of those processes that play an important role in shaping the face of our planet and in producing natural resources.

Finally a note on sources. For reasons of space (and textual fluency) I have been unable to give a full acknowledgement to every statement made. Most references given are fairly recent but this implies neither ingratitude to nor ignorance of the thousands of other researchers whose results represent the sum of our present (and still incomplete) knowledge.

M. R. Leeder
Leeds
March 1981

Acknowledgements

I wish to thank the following persons for reading part or whole of this text or its early plan and for giving me the benefit of their constructive and critical reactions: J. Bridge, P. H. Bridges, J. D. Collinson, K. A. W. Crook, R. G. Jackson, J. D. Hudson, A. D. Miall, H. G. Reading, R. Steel, G. Taylor and R. G. Walker. Naturally I must take the blame for any remaining errors and bias of interpretation. My colleagues M. R. Talbot and H. Clemmey have helped in numerous ways, especially with much-needed encouragement. Eric Daniels and David Bailey of the Leeds University Photographic Service have undertaken most of the photographic work whilst Joan Fall has typed an often difficult manuscript. Roger Jones of Allen & Unwin has guided this text from conception to birth with much friendly advice and encouragement.

I am grateful to the following copyright holders for permission to reproduce halftone figures: R. U. Cooke (8.21b,c); Cambridge University Press (5.17); Society of Economic Palaeontologists and Mineralogists (4.4, 11.2c, 28.7); Elsevier Publishing Company (8.9b, 8.21a,e,f, 9.2, 9.3, 28.6); Institute of Mining and Metallurgy (30.4a–d).

Remaining line drawings have all been redrawn and relettered to a standard format by myself, with appropriate acknowledgement made in the text to the original authors. I am grateful to over 300 authors and to the following copyright holders, who have given me permission to proceed in this way (numbers in parentheses refer to text figures):

Academic Press (21.1–3, 26.1); Figure 24.13 reproduced from *The Earth and its oceans* (A. C. Duxbury 1971) by permission of Addison-Wesley Publishing Company; American Association of Petroleum Geologists (15.19, 23.15–17, 25.8, 25.11, 25.14, 27.10, 28.2, 29.17, 30.2, 31.10); American Geological Institute (20.1, 25.13); American Geophysical Union (2.3, 18.7); *American Journal of Science* (14.4, 15.6); American Society of Civil Engineers (15.13); Edward Arnold (5.22, 17.1, 17.4); A. A. Balkema (17.8, 17.9, 17.11); Blackie (5.25–7); Blackwell Scientific (19.17, 19.18, 22.14); W. S. Broecker (2.1, 2.10); Cambridge University Press (5.19, 5.20, 8.10); Canadian Society of Petroleum Geologists (14.1, 14.3, 14.8, 15.20); Chapman and Hall (4.6, 5.13, 6.9, 8.23); J. M. Coleman (19.10–12); *Economic Geology* (30.8); Elsevier (3.4, 3.5, 9.1, 10.4, 13.1, 13.6, 15.7, 15.11, 15.12, 18.13, 19.9, 22.7, 22.8, 22.11, 25.4, 26.6, 26.8, 27.2, 27.9, 29.2); Figure 7.3 reproduced from *Physical processes in geology* (A. M. Johnson) by permission of Freeman

Cooper Inc.; R. M. Garrells (27.6); Geological Association of Canada (21.10, 21.16, 23.1, 23.28, 23.29, 26.5, 26.6); Geological Society of America (1.3–5, 5.10, 14.7, 15.5, 16.6, 16.7, 19.2, 19.3, 19.6, 19.7, 19.16, 21.11, 24.11, 28.3, 28.4, 28.9, 29.1, 29.11, 29.14, 30.6, 30.7); Geological Society of London (17.7, 26.7, 26.11); Geological Survey of Canada (31.2–4); *Geologie Mijnbouw* (11.4, 22.9, 22.10, 22.14); Geologists Association (1.8, 8.5); Gordon and Breach (25.1); P. M. Harris (23.19); Figures 22.5 and 22.6 reproduced from *Shelf sediment transport: process and pattern* (Swift *et al.*, eds), © 1972 Hutchinson Ross Publishing Company; Institute of British Geographers and G. S. Boulton (17.5); International Association of Sedimentologists (6.4, 7.5, 8.16, 8.24, 8.26, 8.27, 11.3, 12.9, 13.3, 14.6, 15.10, 16.1, 16.4, 19.13, 22.12, 22.13, 26.3, 26.4); International Glaciological Society (17.3); Figures 23.10, 23.12 and 23.13 reproduced from *Sedimentation on the modern carbonate tidal flats of NW Andros Island, Bahamas* (L. A. Hardie *et al.*, eds, 1977) by permission of Johns Hopkins University Press; H. A. Lowenstam and T. W. Donnelly (2.6); Macmillan Inc. (12.8); Figure 15.12 reproduced from *Nature: Physical Science* **237**, 75–6, by permission of Macmillan Journals Ltd; Figure 12.3 reproduced from *Statistical methods for the Earth scientist* (R. Till) by permission of Macmillan Publishers Ltd; Figure 1.2 reproduced from *Introduction to geochemistry* (Krauskopf) and Figure 30.9 from *Principles of chemical sedimentology* (Berner), both by permission of McGraw-Hill; Figures 18.10 and 18.14 reproduced from *Bores, breakers and waves* (Tricker) by permission of Mills and Boon; North Holland (5.23, 8.2); Figures 24.9, 24.12 and 25.2 reproduced from *The face of the deep* (B. C. Heezen & H. D. Hollister) by permission of Oxford University Press; Pergamon Press (2.4, 2.5, 16.4); Figure 5.15 reproduced from *Mechanics of erosion* (Carson) by permission of Pion Ltd; Figure 12.5 reproduced from *Dynamic stratigraphy* (Mathews 1974, pp. 51, 58) and Figures 18.12, 21.5 and 21.9 from *Beach processes and sedimentation* (Komar 1976, pp. 136, 274, 289), both by permission of Prentice-Hall Inc.; Figure 22.2 reproduced from *The Quaternary of the United States* (Wright & Frey, eds) by permission of Princeton University Press; Royal Geographical Society (17.6); Scientific American Inc. (22.1); Scientific Press (31.8); Society of Economic Palaeontologists and Mineralogists (1.6, 1.7, 6.1, 8.11, 8.25, 13.4, 13.5, 15.9, 15.15, 17.10, 17.13, 18.5, 19.14, 19.15, 21.7, 21.8, 21.15, 23.14, 23.21, 23.24–7, 25.3, 26.9, 26.10, 27.7, 27.8, 27.12, 28.8, 28.11–14, 29.3, 29.7, 29.12, 29.15, 29.16, 31.5); Figures 6.3 and 6.8 reproduced from a paper by Chepil in *Proc. Soil Sci. Soc. Am.* **25**, 343–5, by permission of the Soil Science Society of America; Springer-Verlag (8.18, 16.3, 21.12, 23.3, 23.4, 23.9, 23.22, 23.23, 26.2, 27.1, 28.10, 28.14, 30.1, 30.3, 31.6, 31.7); University of Chicago Press (6.10, 8.22, 12.7, 22.4, 24.3, 26.5); Van Nostrand Reinhold (3.2, 5.12); Figures 24.2 and 24.5 reproduced from *Atmosphere and ocean* (J. G. Harvey 1976), published for the Open University Press by Artemis Press, Sussex, by permission of Vision Press Ltd; Wiley (15.14); Yorkshire Geological Society (19.17).

I am also grateful to Oxford University Press and Penguin Books for their permission to utilise extracts from verses of the poets Bunting and Montale respectively and also to the Geological Society for the long quotation on page 297.

Contents

PREFACE viii

ACKNOWLEDGEMENTS ix

PART ONE THE ORIGIN OF SEDIMENT GRAINS

Theme 2

1 The origin of terriginous clastic grains

1a Introduction 3
1b The role of water in rock weathering 3
1c Oxidation, reduction and Eh–pH diagrams 4
1d Rock-forming minerals during weathering 5
1e Breakdown products and new-formed minerals 8
1f Physical weathering 8
1g Sediment yields 9
1h Clastic grains and source identification 11
1i Sourcelands, differentiation and plate tectonics 13
1j Summary 14
Further reading 14

2 The origin of calcium carbonate grains

2a Introduction 15
2b Recent marine carbonate sediments 15
2c The composition of fresh water and sea water 15
2d The major carbonate minerals 16
2e Primary carbonate precipitation 17
2f Carbonate grains of biological origin 21
2g A skeletal origin for aragonite muds? 25
2h Micrite envelopes and intraclasts 25
2i Pellets and peloids 26
2j Oöliths 26
2k Grapestones 27
2l Polygenetic origin of carbonate grains 27
2m Shallow temperate-water carbonates 27
2n $CaCo_3$ dissolution in the deep ocean 28
2o Summary 29
Further reading 29
Appendix 2.1 Staining and peel techniques 29

3 Evaporites, biogenic silica, and phosphates

3a Evaporites 30
3b Biogenic silica 31
3c Phosphates 32
3d Summary 32
Further reading 34

4 Grain properties

4a Definition and range of grain size 35
4b Grain size distributions 35
4c Characteristics of grain populations 38
4d Size parameters and distributions 39
4e Grain abrasion and breakage 40
4f Grain shape and form 41
4g Bulk properties of grain aggregates 42
4h A note on grain fabric 43
4i Summary 43
Further reading 43

PART TWO FLUID FLOW AND SEDIMENT TRANSPORT

Theme 46

5 Fluid properties and fluid motion

5a Introduction 47
5b Physical properties 47
5c Streamlines and flow visualisation 48
5d Friction, pressure changes and the energy budget 49
5e The Reynolds number 51
5f Froude number 52
5g Laminar flows 53
5h Introduction to turbulent flows 54
5i The structure of turbulent shear flows 57
5j Flow separation and secondary currents 61
5k Summary 63
Further reading 63
Appendices
5.1 Bernoulli's equation 63
5.2 Reynolds number 64
5.3 Velocity profiles of viscous channel flow 64
5.4 Derivation of the Karman–Prandtl velocity law for turbulent flow 65

6 Transport of sediment grains

6a Introduction 67
6b Grains in stationary fluids 67
6c Initiation of particle motion 68
6d Paths of grain motion 70
6e Solid transmitted stresses 73
6f Sediment transport theory 74
6g Summary 75
Further reading 75
Appendix 6.1 Stokes' law of settling 75

7 Sediment gravity flows

7a Introduction 76
7b Grain flows 77

7c Debris flows 77
7d Liquefied flows 78
7e Turbidity flows 78
7f Deposits of sediment gravity flows 81
7g Summary 81
Further reading 81
Appendices
7.1 Dispersive pressure and grain flow 82
7.2 A note on autosuspension in turbidity currents 82

PART THREE BEDFORMS AND SEDIMENTARY STRUCTURES

Theme 84

8 Bedforms and structures in granular sediments

8a Bedforms and structures formed by unidirectional water flows 85
8b Further notes on bedform phase diagrams 93
8c Bedforms and structures formed by water waves 94
8d Coarse/fine laminations and graded bedding 95
8e Bedforms and structures formed by air flows 97
8f Bedform 'lag' effects 101
8g Summary 102
Further reading 102
Appendix 8.1
Notes on bedform theory for water flows 102

9 Bedforms caused by erosion of cohesive sediment

9a Water erosion of cohesive beds 103
9b Erosion by 'tools' 105
9c Summary 105
Further reading 105

10 Biogenic and organo-sedimentary structures

10a Stromatolites 106
10b Trace fossils and deposition rates 109
10c Summary 110
Further reading 110

11 Soft sediment deformation structures

11a Reduction of sediment strength 111
11b Liquefaction and water escape structures 112
11c Liquefaction and current drag structures 112
11d Diapirism and differential loading structures 114
11e Slides, growth faults and slumps 115
11f Desiccation and syneresis shrinkage structures 115
11g Summary 115
Further reading 115

PART FOUR ENVIRONMENTAL AND FACIES ANALYSIS

Theme 118

12 Environmental and facies analysis

12a Scope and philosophy 119
12b Depositional systems and facies 119
12c Succession, preservation and analysis 122
12d Subsidence, uplift and deposition 123
12e Transgression, regression and diachronism 125
12f Palaeocurrents 127
12g The Holocene 128
12h Basin analysis and plate tectonics 128
12i Summary 129
Further reading 129
Appendix 12.1
Vector statistics in palaeocurrent analysis 129

PART FIVE CONTINENTAL ENVIRONMENTS AND FACIES ANALYSIS

Theme 132

13 Deserts

13a Introduction 133
13b Physical processes and erg formation 133
13c Modern desert facies 134
13d Ancient desert facies 135
13e Summary 137
Further reading 137

14 Alluvial fans

14a Introduction 138
14b Physical processes 138
14c Modern facies 140
14d Ancient alluvial fan facies 141
14e Summary 141
Further reading 141

15 River plains

15a Introduction 142
15b Physical processes 142
15c Modern river plain facies 146
15d Ancient river plain facies 151
15e Summary 153
Further reading 153
Appendices
15.1 The helical flow cell 153
15.2 Palaeohydraulics 154

16 Lakes

16a Introduction 155
16b Physical and chemical processes 155
16c Modern lake facies 157
16d Ancient lake facies 158
16e Summary 160
Further reading 160

17 Glacial environments

17a Introduction 161
17b Physical processes 161
17c Pleistocene and modern glacial facies 164
17d Ancient glacial facies 167
17e Summary 167
Further reading 167

PART SIX COASTAL AND SHELF ENVIRONMENTS AND FACIES ANALYSIS

Theme 170

18 Physical processes of coast and shelf

18a Introduction 171
18b Wind-generated waves 171
18c Tides and tidal waves 176
18d Summary 179
Further reading 179
Appendix 18.1 Deep-water wave theory 179

19 Deltas

19a Introduction 182
19b Physical processes 182
19c Modern deltaic facies 185
19d Ancient deltaic facies 188
19e Summary 191
Further reading 191

20 Estuaries

20a Introduction 192
20b Estuarine dynamics 192
20c Modern estuarine facies 193
20d Ancient estuarine facies 194
20e Summary 194
Further reading 194

21 'Linear' clastic shorelines

21a Introduction 195
21b Physical processes 195
21c Recent facies of linear clastic shorelines 198
21d Ancient clastic shoreline facies 200
21e Summary 201
Further reading 201

22 Clastic shelves

22a Introduction 202
22b Shelf dynamics 202
22c Recent shelf facies 205
22d Ancient clastic shelf facies 207
22e Summary 209
Further reading 210

23 Carbonate–evaporite shorelines, shelves and basins

23a Introduction 211
23b Arid tidal flats and sabkhas 212
23c Humid tidal flats and marshes 215
23d Lagoons and bays 217
23e Tidal delta and spillover oölite sands 219
23f Open carbonate shelves 221
23g Platform margin reefs and buildups 222
23h Platform margin slopes and basins 223
23i Sub-aqueous evaporites 226
23j Summary 228
Further reading 228

PART SEVEN OCEANIC ENVIRONMENTS AND FACIES ANALYSIS

Theme 230

24 Oceanic processes

24a Introduction 231
24b Physical processes 231
24c Chemical and biochemical processes 232
24d Surface currents and circulation 233
24e Structure, deep currents and circulation 235
24f Slumps, debris flows and turbidity currents 237
24g Palaeo-oceanography 238
24h Summary 239
Further reading 239

25 Clastic oceanic environments

25a Introduction 240
25b Continental slopes and rises of passive margins 241
25c Submarine fans and cones 242
25d Abyssal plains 245
25e Trenches and fore-arc basins of active margins 246
25f Summary 249
Further reading 249

26 Pelagic oceanic sediments

26a Sediment types 250
26b Oceanic facies successions 251
26c Anoxic oceans and oceanic events 252

26d Hypersaline oceans 253
26e Continental outcrops of ancient facies 254
26f Summary 256
Further reading 256

PART EIGHT DIAGENESIS: SEDIMENT INTO ROCK

Theme 258

27 Diagenesis: general considerations

27a Definitions 259
27b Subsurface pressure and temperature 259
27c Petrography in diagenetic studies 260
27d Stable isotopes in diagenetic studies 260
27e Eh–pH phase diagrams in diagenetic studies 262
27f Compaction and fluid migration 264
27g Pressure solution 268
27h Diagenetic realms 268
27i Summary 269
Further reading 269

28 Terrigenous clastic sediments

28a Introduction 270
28b Marine mud diagenesis 270
28c Non-marine mud diagenesis 273
28d Classification of mudrocks 273
28e Near-surface sand diagenesis 275
28f Subsurface sand diagenesis 278
28g Secondary porosity and sandstone diagenesis 281
28h Classification 282
Further reading 284

29 Carbonate sediments

29a Introduction 285
29b Early meteoric diagenesis 285
29c Early marine diagenesis 289
29d Subsurface diagenesis by formation waters 293
29e Summary of limestone diagenesis 296
29f Models for dolomitisation 297
29g Classification 301
29h Summary 302
Further reading 302

30 Evaporites, silica, iron and manganese

30a Evaporites 303
30b Silica diagenesis 305
30c Iron minerals 308
30d Manganese 309
30e Summary 310
Further reading 310

31 Hydrocarbons

31a Introduction 311
31b Coal composition and rank 311
31c Coal-forming environments 312
31d Oil and gas – organic matter, source rock and diagenesis 314
31e Oil and gas migration 316
31f Oil and gas traps and reservoir studies 316
31g Tar sands 318
31h Oil shales 319
31i Summary 319
Further reading 319

REFERENCES 320

INDEX 339

List of tables

1.1 The 'weatherability' series for the common igneous silicate minerals. 6
1.2 The Reiche weathering potentials index (WPI) for some common silicate minerals. 6
1.3 (a) Weathering equations written with rock-forming silicate minerals as reactants with aqueous phases. (b) Gibbs free energy values for weathering reactions 1–10 in Table 1.3a. 7
2.1 Composition of average sea and river waters. 17
2.2 Solubility products (K) and ion activity products (IAPs) for calcite, dolomite and aragonite in sea water at 25 °C. 17
3.1 Chemical composition and selected values of IAP (ion activity product) and K (solubility product) for halite, gypsum and anhydrite in seawater solutions. 30
3.2 The oceanic silica cycle, with the magnitudes of dissolved silica in 10^{13} g SiO_2 a^{-1}. 33
4.1 Summary of the Udden–Wentworth size classification for sediment grains. 36
4.2 Three useful definitions of grain size. 37
4.3 Range of applicability of different techniques of size analysis. 37
4.4 Details of a sieve analysis of a medium, well sorted, positive skewed aeolian sand. 37
4.5 Sorting and skewness values for graphically obtained statistics expressed as verbal descriptive summaries. 39
5.1 Densities and molecular viscosities of some natural and artificial 'fluids'. 47
12.1 Summary (not exhaustive) of the environments of deposition on the Earth's surface. 120
12.2 (a) Number of upward facies transitions using the data of Figure 12.3. (b) Upward-transition probability matrix using the data of Table 12.2a. 123
12.3 To show that transgressions and regressions result from the interaction between deposition rate and relative sea level changes. 127
16.1 Physical processes affecting lake water dynamics. 155
19.1 Factors affecting delta regime, morphology and facies. 182
19.2 To show river discharge:wave-power ratios for seven major deltas. 184
22.1 Components of the shelf current velocity field. 203
23.1 Some differences between carbonate and siliciclastic sediments. 211
25.1 Some statistics for the Amazon, Bengal and Mississippi submarine fans. 244
27.1 Major constituents of chemical analyses of some saline 'connate' formation waters, with oceanic water for comparison. 267
28.1 Diagenetic zones for marine mud successions. 271
28.2 Links between sediment mineralogy and burial rate in marine muds. 274
28.3 Schematic ternary diagram to illustrate the nomenclature and composition of mudrocks. 275
29.1 Classification of carbonate rocks according to depositional texture. 301
29.2 Classification of carbonate rocks according to the nature and proportions of lime mud matrix and pore-filling spar cement. 302
31.1 Rank stages and important petrographic characteristics of coals. 312

PART ONE THE ORIGIN OF SEDIMENT GRAINS

I would have chosen to feel myself rough and elemental
like the pebbles that you roll;
gnawed through by the salt;
splinter beyond time, a witness to
a cold will that does not fail.

From: I would have chosen *(Montale)*

Plate 1 Calcareous algae of the genus *Penicillus* with holdfasts. Vast numbers of these algae live in the shallow lagoons of Florida and the Bahamas. Upon death the delicate organic tissues decay and release tiny aragonite needles of the plant 'skeleton' onto the lagoonal floor. Some proportion accumulates as aragonitic mud indistinguishable in external form from precipitated aragonite (coll. R. Till).

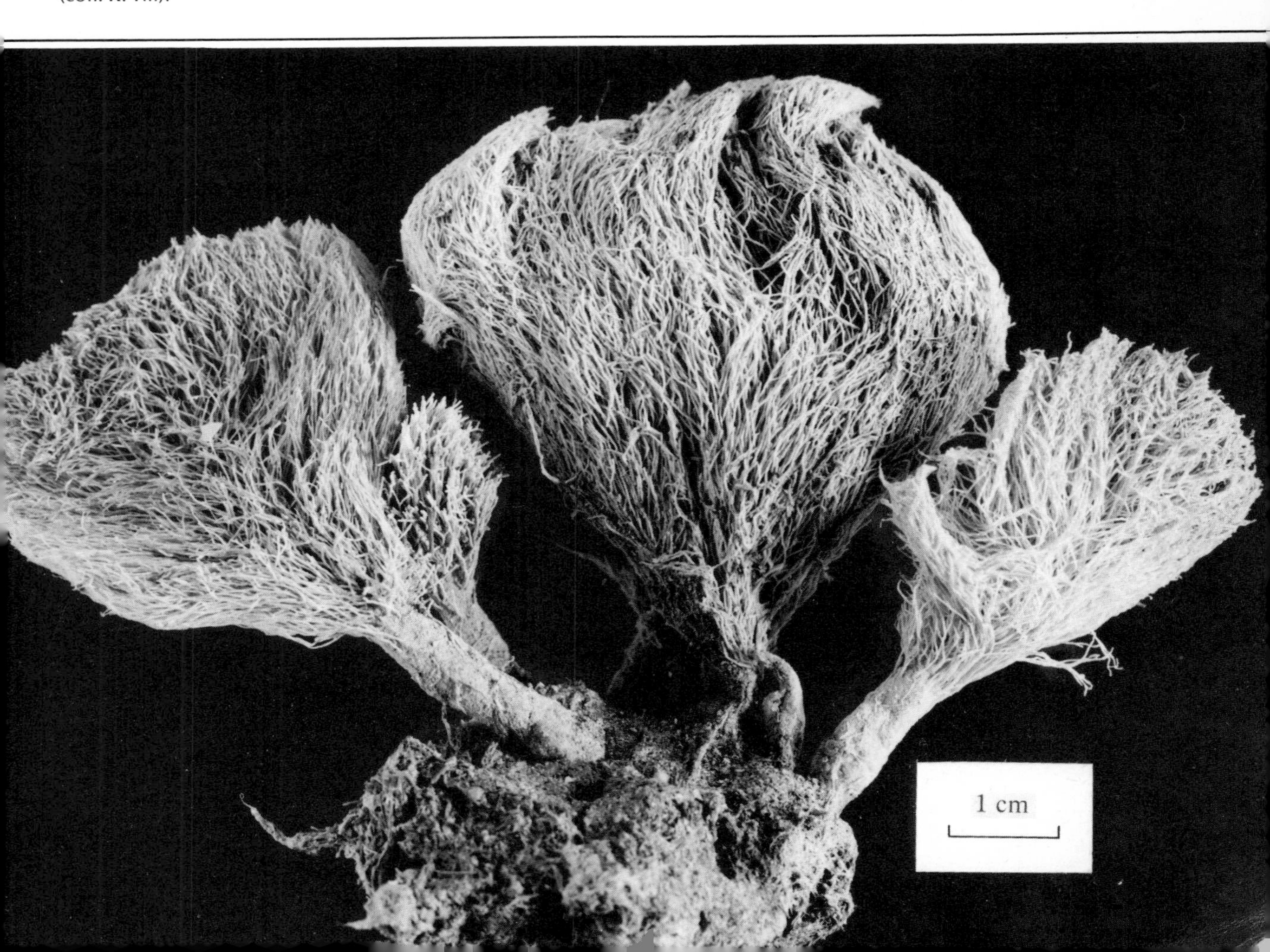

Theme

In Part 1 we examine the origin and textural characteristics of the main types of sediment grains. The grains are acted upon by fluid forces and by gravity which may ultimately cause deposition and formation of sediment layers. Broadly speaking, sediment grains originate by (a) chemical and physical weathering of pre-existing igneous, metamorphic or sedimentary rocks (giving rise to terrigenous **clastic** grains), (b) chemical precipitation of various minerals and salts (giving rise to **orthochemical** grains), and (c) biochemical precipitation of various minerals (giving rise to **biogenic** grains). The type of sediment grains found at any locality varies widely in response to climate, water chemistry, organic productivity and geomorphic location. Changes have taken place in these variables over space and in time during the Earth's history. In addition to uses in palaeogeographic reconstructions the type of grains present may provide helpful evidence towards understanding the evolution of the atmosphere, hydrosphere and biosphere.

1 The origin of terriginous clastic grains

1a Introduction

Terriginous clastic (detrital) grains are those fragments of rocks or minerals derived by physical or chemical breakdown of source rock. Every near-surface rock or surface outcrop is a potential source of such grains. Discontinuities such as joints, rock cleavage, crystal/grain boundaries and crystal cleavage are acted upon by chemical atmospheric and soil weathering and by physical weathering. Rocks are broken down into their constituent crystals or into small rock fragments, and these grains are attacked chemically to give characteristic breakdown products and solutions. Transport by gravity, wind and water (see Part 2) will further break down and change the grains, such changes being particularly effective on easily cleft or soft minerals. After deposition the particles may be further changed chemically in the diagenetic realm (see Part 8). Therefore, we may be sure at the outset that the chemical and physical nature of a deposit of grains will differ radically from that of the source rock. For example, the average feldspar content of igneous and metamorphic rocks is around 60% whereas that of sandstones is only about 12%. On the other hand, sandstone is usually greatly enriched in quartz as compared to igneous and metamorphic rocks. We shall now discuss some of the reasons why such differences should occur.

1b The role of water in rock weathering

Undoubtedly the most important feature of the surface of our planet, compared, say, to the anhydrous Moon, is the abundance of water and water vapour. Water has remarkable properties: when placed between the plates of a charged capacitor the molecules will orientate themselves, with the positive hydrogens towards the negative plate and the negative oxygens towards the positive plate. The effect is similar to the orientation of a magnet in a magnetic field. Water molecules possess poles just like a magnet and such molecules are said to act as **electric dipoles.** The strength of the dipole moment is determined by the product of the magnitude of the charges and the distance between them, measured in suitable units. Such polar molecules as water result from asymmetric covalent bonding where elements such as oxygen 'grab' more than their fair share of the available shared electrons. This gives the strongly electronegative oxygen atom a partial negative charge at the expense of a partial positive charge on the less electronegative hydrogen atom. The chemical bonds between the oxygen and the two hydrogen atoms arise from the overlap of the p-electron clouds with the s-electron clouds of the hydrogen atoms. Theoretically the bonds should be at 90° to one another and consist of a molecular orbital occupied by a pair of electrons. In practice, because of repulsion between the hydrogen atoms, the bond angles meet at an angle of just over 104°, the O–H bond length being 9.8×10^{-8} mm. When the water molecules come together *en masse* they join together by hydrogen bonding because of the polar nature of the H_2O covalent bonds just described. They form into tetrahedral groups of four by this hydrogen bonding (Fig. 1.1).

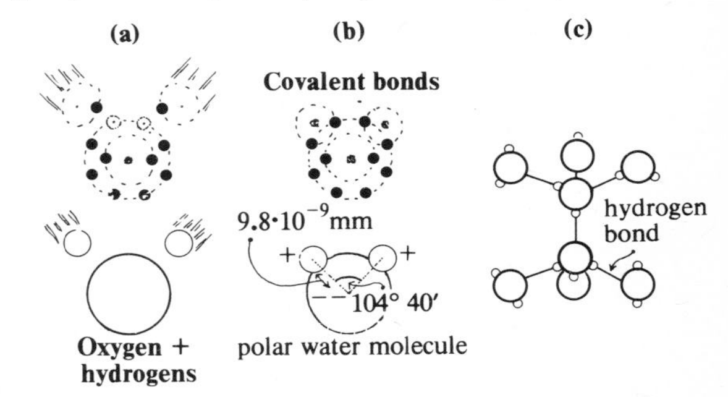

Figure 1.1 The molecular structure of water. (a & b) Formation of the polar water molecule by covalent bonding. (c) Cluster of water molecules bonded by hydrogen bonds.

Many of the distinctive properties of water are explained by the above knowledge. Thus,

(a) The very effective **solvent properties** of water upon ionically bonded compounds is due to the polar nature of the molecule. The positive and negative ends of the molecule attach themselves to the compound's negative and positive surface ions respectively, neutralising their charges so that mechanical agitation can float the compound's constituent atoms apart.

(b) The tetrahedral groupings of water molecules cause such properties as **high surface tension** and **capillarity, high melting point** and the large range of temperatures in which H_2O is a liquid phase.

(c) Decreasing temperature causes a decrease in the thermal agitation of the water molecules and an increase in the number of hydrogen bonds and hence in the cluster size of molecules bound by the hydrogen bonds. The

viscosity of water (Ch. 5) thus increases with falling temperature.

(d) **Density** increases with falling temperature due to contraction caused by decreasing molecular thermal agitation. At the same time, however, the production of more and more hydrogen bonds causes expansion. Expansion dominates below 4°C as the tetrahedral molecular groups also begin to be taken up into hexagonal ice structures and so the density decreases. Expansion continues until –22°C when ice achieves its minimum density and maximum expansive pressure. Hexagonal ice crystals have a maximum growth rate normal to the basal plane, so that ice whisker crystals growing in microcrevices of rocks can exert high stresses.

(e) Water molecules separate into the ions H^+ and OH^- at all temperatures, although normally the degree of ionisation is very small. The H^+ ion is responsible for the **acidity** of an aqueous solution. The concentration of H^+ ions is so variable that acidity is expressed as the negative logarithm of the free H^- concentration in grams per litre (pH). At room temperature there are only 10^{-7} moles per litre of hydrogen ions (with an equal number of OH^-) in pure water and a **neutral pH** (= 7) is said to exist. Larger values of pH imply alkalinity and smaller values acidity when H^+ or OH^{--} ions are provided by other reactions and reactants in aqueous solutions, e.g. H^+ ions are provided by CO_2 dissolved in rain and soilwater by the reaction

$$\begin{array}{c} H_2O + CO_2 \rightleftharpoons H_2CO_3 \\ \downarrow\uparrow \\ H^+ + HCO_3^- \end{array} \qquad (1.1)$$

and by other humic and bacterially produced acids (see also Ch. 2).

(f) The decomposition of silicate minerals is often due to **hydrolysis** in which small, highly charged H^+ ions in water displace the metallic cations in crystal lattices so that OH^{--} or HCO_3^- ions can combine with the displaced cations to form solutions or local precipitates. Hydrolysis acts along discontinuities in rocks or minerals such as joints, rock cleavage, crystal boundaries and mineral cleavage planes.

1c Oxidation, reduction and Eh–pH diagrams

The chemical elements present in primary igneous or metamorphic minerals in the weathering zone tend to reach an equilibrium with the **oxidising** or **reducing** nature of the environment. Oxidation involves the loss of electrons from an element or ion leading to an increase in positive valency or a decrease in negative valency and vice versa for reduction, e.g. in the reversible reaction below the ferrous iron in minerals such as pyroxene, olivine or hornblende may oxidise as follows:

$$\underset{\text{ferrous iron}}{Fe^{2+}} \rightleftharpoons \underset{\text{ferric iron}}{Fe^{3+}} + \underset{\text{electron}}{e^-} \qquad (1.2)$$

Dissolved oxygen in surface waters is the most important natural oxidising agent on account of its very high electronegativity. It is possible to measure the oxidation/reduction potential (redox potential) by noting the potential difference produced between an immersed inert electrode, usually platinum, and a hydrogen electrode of known potential. The redox potential for reactions, termed **Eh** for short, is compared to the arbitrary value of 0.00 mV for hydrogen in the reaction

$$2H^+ + 2e^- \rightarrow H_2 \qquad (1.3)$$

at 25°C and 1 atmosphere pressure at a concentration of 1 mole per litre (pH = 0). Negative values of Eh indicate reducing conditions, and positive values oxidising conditions with respect to the arbitrary hydrogen scale.

Most elements in the weathering zone and in the upper parts of deposited sediment columns are oxidised. Exceptions exist in waterlogged soils which are oxygen-poor and where anaerobic bacteria are abundant. Here reaction (1.2) is reversed and insoluble ferric iron is reduced to soluble ferrous iron. Reducing conditions exist below the surface aerobic zone of sub-aqueous sediment accumulations and in the so-called 'euxinic' environments of poorly ventilated marine water bodies or organic-rich brine pools.

If we know the standard oxidation/reduction potential for a particular reaction from laboratory measurements and we possess a field measurement of Eh from a particular weathering zone, then it is possible to predict the type of dissolved oxidation state for a particular ion. For example, the standard potential of the Fe^{2+}–Fe^{3+} couple of Equation 1.2 is 0.77 volts. If, for example, our field measurement shows a reading of 0.5 volts under acid conditions (pH = 2) we would be able to predict that Fe^{2+} is the stable iron phase (see Chs 27 & 30). In practice it has been found that some of the reactions that determine Eh are very slow and that the aqueous environment does not quickly come to equilibrium with the measuring electrodes. Thus field redox measurements generally give only semi-quantitative information.

A particularly informative graph may be produced by

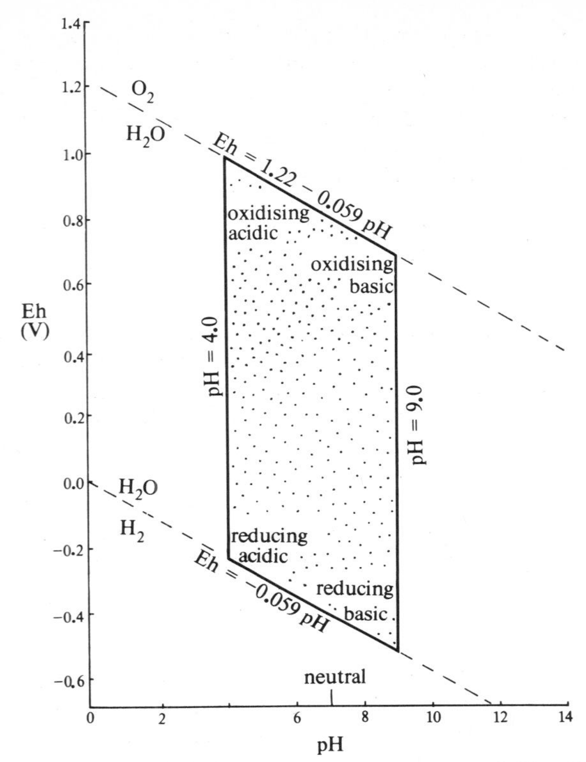

Figure 1.2 Graph to show the common range of Eh–pH values on the Earth's surface (stippled) and the stability range of water (after Krauskopf 1979).

plotting Eh and pH. The heavy box in Figure 1.2 shows the usual limits of Eh and pH found in near surface environments. The lower limit to pH of about 4 is produced by natural concentrations of CO_2 and organic acids dissolved in surface and soil waters. The upper limit to pH of about 9 is reached by waters in contact with carbonate rocks and still in contact with atmospheric CO_2. The reaction that determines the highest Eh values is

$$H_2O \rightleftharpoons \tfrac{1}{2}O_2 + 2H^+ + 2e^-; E^\circ = +1.23 \text{ volts} \quad (1.4)$$

This reaction is in fact dependent upon pH such that

$$Eh = +1.22 - 0.059 \text{ pH}$$

The reaction that determines the lowest Eh is

$$H_2 \rightleftharpoons 2H^+ + 2e^-; E^\circ = 0.00 \text{ V} \quad (1.5)$$

which is also Eh dependent, such that

$$Eh = -0.059 \text{ pH}$$

It is important to stress that the natural box in Figure 1.2 covers 'normal' conditions. Local conditions may sometimes fall well outside, e.g. oxidation of pyrite to give very acid conditions and reduction due to organic material out of contact with water. Eh–pH diagrams (Garrels & Christ 1965) may be drawn for a wide variety of reactions and reactants and are a valuable aid in interpreting both weathering, depositional and diagenetic chemical reactions (see Ch. 27). They do *not*, however, tell us anything about the rate of such reactions or about the attainment, or otherwise, of equilibrium.

1d Rock-forming minerals during weathering

We now arrive at perhaps the most important topic concerning the formation of clastic particles. Even the most elementary acquaintance with the three main groups of rocks – igneous, metamorphic and sedimentary – will soon convince the observer that the range of abundant mineral types in clastic sedimentary rocks is usually very much more restricted than that in igneous and metamorphic rocks. Why should this be?

Let us begin by considering the results of Wahlstrom's study (1948) of the mineral types left in an Upper Palaeozoic weathering zone developed upon a granodiorite from Boulder, Colorado (USA). Figure 1.3 summarises the mineralogical and chemical changes observed with depth in the weathering profile as it is traced downwards into unaltered parent granodiorite. Note the stability of quartz and, to a lesser extent, microcline; the instability of plagioclase feldspar, biotite and hornblende; and the formation of 'new' minerals such as kaolinite, montmorillonite and illite. If we crudely correlate increased depth with decreasing degree of weathering, then hornblende disappears first, followed by plagioclase, biotite and microcline, in that order. The chemical analyses of major oxides show enrichment in Al_2O_3, Fe_2O_3 and K_2O, and depletion of SiO_2, FeO, CaO, Na_2O as we go down the weathering profile.

These results are worth some thought. Some minerals were evidently more stable than others in the weathering process. Some minerals were newly formed. Some oxides suffered a net removal whilst others increased in amount. We shall concentrate on the stability problem in this section.

Numerous weathering studies have generally confirmed Goldich's (1938) original postulate that a 'weatherability' series for the common igneous minerals could be defined (Table 1.1). Notice that this series is strikingly similar to the Bowen reaction series in igneous petrology. It is usually baldly stated that the less stable minerals in the zone of weathering are the higher-temperature minerals

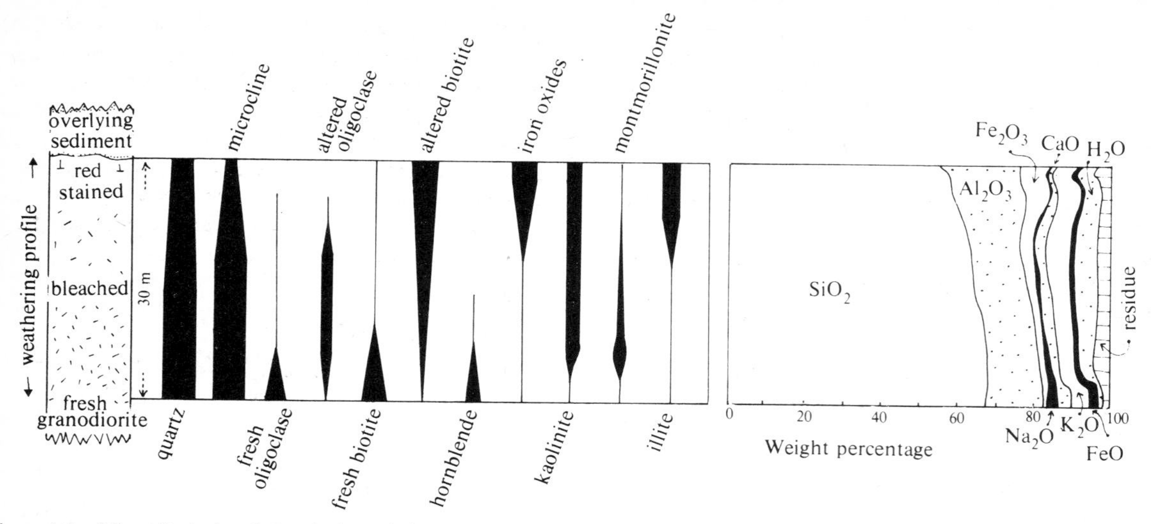

Figure 1.3 Mineralogical and chemical trends in a weathering profile of Mississippian age developed on a granodiorite near Boulder, Colorado, USA (after Wahlstrom 1948).

which, under Earth surface conditions, are most removed from the original temperatures of formation. Such statements do not, in fact, answer our original question: 'Why are some minerals more stable?'. Similarly, no answer is provided by the Reiche weathering potentials index (WPI) for rocks and minerals. This is the mole percent ratio of the sum of the alkalis and alkaline earths to the total moles present, i.e.

$$\frac{100 \times \text{moles}\,(K_2O + Na_2O + CaO + MgO - H_2O)}{\text{moles}\,(SiO_2 + Al_2O_3 + K_2O + Na_2O + CaO + MgO - H_2O)} \tag{1.6}$$

Table 1.1 The 'weatherability' series for the common igneous silicate minerals. The series is the reverse of Bowen's reaction series for mineral crystallisation from silicate melts (after Goldich 1938).

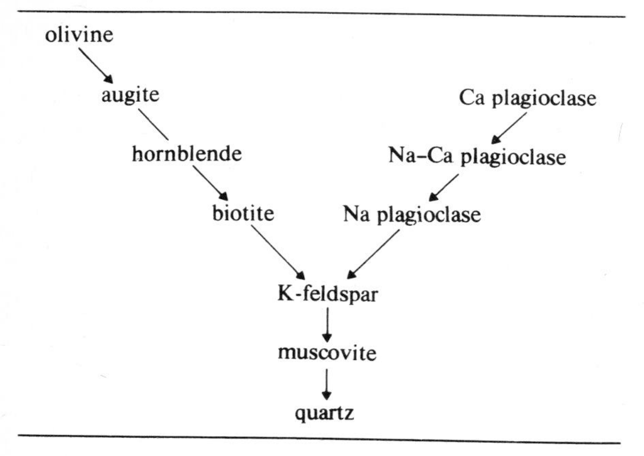

Minerals or rocks of low stability have a high WPI and vice versa (Table 1.2). The WPI is a rough guide to relative stability based upon the ease of weathering of the alkali and alkaline earth elements. Other than this, it has little explanatory value.

One notable attempt to explain relative mineral stability in terms of crystal structural properties was that of Keller (1954). He postulated that the order of formation of certain silicate minerals from magma was due in part to the relative bond strengths of the various cations with oxygen. Calculation of the total bond strengths between cations and oxygen in the minerals of the Bowen reaction series shows a fair degree of support for this proposal. It is also apparent that minerals with relatively few Si–O bonds are unstable by comparison with those with many Si–O bonds because the high strength of the bond tends

Table 1.2 The Reiche weathering potentials index (WPI) for some common silicate minerals (after Carroll 1970).

Mineral	*WPI mean*	*WPI range*
olivine	54	44–65
augite	39	21–46
hornblende	36	21–63
biotite	22	7–32
labradorite	20	18–20
andesine	14	
oligoclase	15	
albite	13	
muscovite	10	
quartz	1	

to hold the crystal structure together regardless of the removal of other weak cation–oxygen bonds. For example, the Mg–O bond is relatively strong, but forsterite (Mg_2SiO_4) is the most unstable of the igneous minerals because of the paucity of strong Si–O bonds. The most resistant minerals, therefore, are those composed solely of interlocking silica tetrahedra. When other ions are present joining up the silica tetrahedra in various ways, then the mineral is less stable because the ions are susceptible to neutralisation by dipolar water molecules.

The approach of Keller leads us towards the fundamental nature of **thermodynamics** as an explanatory framework in which to assess mineral stability. Very basically we can say that the likelihood of a particular reaction occurring in preference to another reaction may be decided by reference to the change in **free energy** of the reactions. The standard free energy change of a reaction is the sum of free energies of formation (ΔG_f°) of all the reaction products minus the sum of the free energies of the reactants. When calculated free energy changes are negative, reactions will proceed spontaneously. The greater the negative value of the change in free energy, the more tendency there should be to react.

Curtis (1976) has tackled the weathering stability problem from a thermodynamic viewpoint, with encouraging results. In order to assess stability we must write specific weathering equations with the 'primary' igneous and metamorphic minerals as reactants (Table 1.3). Experimental data on standard free energies of formation for both potential reactants and products are then assembled and the calculated free energy changes (ΔG_f°) tabulated. For example

$$\underset{\text{anorthite}}{CaAl_2Si_2O_8} + \underset{\text{in solution}}{2H^+} + \underset{\text{water}}{H_2O} \rightarrow \underset{\text{kaolinite}}{Al_2Si_2O_5(OH)_4} + \underset{\text{in solution}}{Ca^{2+}} \quad (1.7)$$

Thus

$$\begin{aligned}\Delta G_f^\circ &= (\Delta G_f^\circ Al_2Si_2O_5(OH)_4 + \Delta G_f^\circ Ca^{2+}) \\ &\quad - (\Delta G_f^\circ CaAl_2Si_2O_8 + \Delta G_f^\circ 2H^+ + \Delta G_f^\circ H_2O) \\ &= (-904 + -132.2) - (-955.6 + 0 + -56.7) \\ &= -1036.2 + 1012.3 \\ \Delta G_f^\circ &= -23.9 \text{ kcal mol}^{-1}\end{aligned}$$

Table 1.3 (a) Weathering equations written with rock-forming silicate minerals as reactants with aqueous phases (after Curtis 1976).

1. $Fe_2SiO_4 + \frac{1}{2}O_2 \rightarrow Fe_2O_3 + SiO_2$
2. $Mg_2SiO_4 + 4H^+ \rightarrow 2Mg^{2+} + 2H_2O + SiO_2$
3. $MgSiO_3 + 2H^+ \rightarrow Mg^{2+} + H_2O + SiO_2$
4. $CaMg(SiO_3)_2 + 4H^+ \rightarrow Mg^{2+} + Ca^{2+} + 2H_2O + 2SiO_2$
5. $Mg_7Si_8O_{22}(OH)_2 + 14H^+ \rightarrow 7Mg^{2+} + 8H_2O + 8SiO_2$
6. $Ca_2Mg_5Si_8O_{22}(OH)_2 + 14H^+ \rightarrow 5Mg^{2+} + 2Ca^{2+} + 8H_2O + 8SiO_2$
7. $CaAl_2Si_2O_8 + 2H^+ + H_2O \rightarrow Al_2Si_2O_5 + Ca^{2+}$
8. $2NaAlSi_3O_8 + 2H^+ + H_2O \rightarrow Al_2Si_2O_5(OH)_4 + 4SiO_2 + 2Na^+$
9. $2KAlSi_3O_8 + 2H^+ + H_2O \rightarrow Al_2Si_2O_5(OH)_4 + 4SiO_2 + 2K^+$
10. $2KAl_3Si_3O_{10}(OH)_2 + 2H^+ + 3H_2O \rightarrow 2K^+ + 3Al_2Si_2O_5(OH)_4$

(b) Gibbs free energy values for weathering reactions 1–10 in Table 1.3a (after Curtis 1976).

Mineral	*ΔG_f° kcal mol^{-1}*	*ΔG_f° kcal g atom^{-1}*
1. olivine (fayalite)	−52.7	−6.58
2. olivine (forsterite)	−44.0	−4.00
3. pyroxene (clinoenstatite)	−20.9	−2.98
4. pyroxene (diopside)	−38.1	−2.72
5. amphibole (anthophyllite)	−137.2	−2.49
6. amphibole (tremolite)	−123.2	−2.24
7. Ca-feldspar (anorthite)	−23.9	−1.32
8. Na-feldspar (albite)	−23.1	−0.75
9. K-feldspar (microcline)	−17.3	−0.32
10. mica (muscovite)	−17.3	−0.32

The result states that anorthite will react with hydrogen ions in aqueous solutions to form the clay mineral kaolinite plus calcium ions. The negative free energy change indicates that the reaction will proceed spontaneously. Similar procedures for ten other silicates with igneous affinities yield the results shown in Table 1.3. The results in kcal mol^{-1} must then be changed to kcal gram atom values by dividing by the number of product atoms for each reaction in order to facilitate comparisons between different chemical equations. This correction arises because the overall ΔG_f° values are affected by the sizes of the molecular formulae as written, whereas comparative studies require the different amounts of energy liberated for a given number of product atoms. The final results (Table 1.3b) show a good correspondence with the kinds of results found from field studies such as those quoted above. It should be stressed that all the reactions considered involve complete oxidation, as would be expected in most well drained, but wet, weathering horizons.

The results of this section would lead us to suppose that quartz, muscovite and orthoclase should dominate amongst clastic mineral components derived by erosion from weathered igneous and metamorphic terrains. However, experience teaches us that, although quartz is by far the most abundant clastic particle, the abundance of the other primary minerals is highly variable, depending upon a number of factors including climate and type of weathering, sourceland abundance, hardness, original grain size, rapidity of sedimentation, and so on. There are also important new-formed minerals produced by weathering; we shall look at these in the next section.

1e Breakdown products and new-formed minerals

The oxidation and hydrolysis weathering reactions usually lead to the liberation of the alkali- and alkali-earth elements (Ca, K, Na, Mg) in solution as hydrated ions, with silica and aluminium silicates as byproducts. Ferrous iron is oxidised to the stable, insoluble ferric form. Although the reactions listed in Table 1.3a are thermodynamically correct, the main source of natural hydrogen ions is dissolved CO_2.

$$H_2O + CO_2 \rightleftharpoons H_2CO_3 \rightleftharpoons H^+ + HCO_3^- \qquad (1.8)$$

so that the liberated alkali- and alkali-earth elements quickly form soluble carbonate or bicarbonate ions. For example

$$\underset{\text{orthoclase}}{2KAlSi_3O_8} + \underset{\text{carbonic acid}}{H_2CO_3} + \underset{\text{water}}{H_2O} \rightarrow$$

$$\underset{\text{potassium carbonate}}{K_2CO_3} + \underset{\text{kaolinite}}{Al_2Si_2O_5(OH)_4} + \underset{\text{silica}}{4SiO_2} \qquad (1.9)$$

acid conditions

or

$$\underset{\text{biotite}}{4KMg_2Fe(OH)_2AlSi_3O_{10}} + \underset{\text{carbonic acid}}{12H_2CO_3} + \underset{\text{water}}{H_2O} \rightarrow$$

$$\underset{\text{potassium bicarbonate}}{4KHCO_3} + \underset{\text{magnesium bicarbonate}}{8Mg(HCO_3)_2} + \underset{\text{hydrated iron oxide}}{FeO_3 \cdot H_2O}$$

$$+ \underset{\text{kaolinite}}{2Al_2Si_2O_5(OH)_4} + \underset{\text{silica}}{8SiO_2} + \underset{\text{water}}{H_2O} \qquad (1.10)$$

as well as the well known reaction responsible for the chemical breakdown of limestones

$$\underset{\text{calcite}}{CaCO_3} + \underset{\text{carbonic acid}}{2H_2CO_3} \rightleftharpoons \underset{\text{calcium bicarbonate}}{Ca(HCO_3)_2} + \underset{\text{water}}{H_2O} + \underset{\text{carbon dioxide}}{CO_2} \qquad (1.11)$$

The **clay minerals** are one of the most important newly formed mineral groups in the weathering zone and they may form the bulk of the weathered residue. They may be easily transported to a depositional site where, *in the absence of post depositional changes* (Chs 27 & 28) they give valuable information about weathering conditions. **Kaolinite** is formed under humid, acid weathering conditions from the alteration of feldspar-rich rocks by reactions such as (1.9 & 1.10) above. **Illite**, a potassium–aluminium hydrated silicate, is formed by weathering of feldspars and micas under alkaline weathering conditions where leaching of mobile cations such as potassium does not occur. **Montmorillonite**, a complex sheet silicate, forms from basic igneous rocks under alkaline conditions with a deficit of K^+ ions. We shall consider the clay mineral groups further below (Ch. 1h).

1f Physical weathering

The two most significant forms of physical weathering are freeze/thaw (frost) and salt weathering. Both involve the production of stresses through crystallisation of solids from solution in tiny rock fractures. **Frost weathering** is an effect partly due to anomalous expansion and decreasing density as water freezes (Ch. 1c). The accompanying increase in volume of around 10% generates enough tensile stress in small cracks to cause most rocks to split wider. Additional stresses – as much as ten times those arising from simple expansion above – may arise from ice growth as the clusters of parallel ice-crystal needles grow normal to the freezing surface. Provided the small crack is supplied with a net input of water then the stresses arising from crystal growth are limited only by the tensile strength of water, which is drawing water molecules to the ends of the growing crystals through capillary films. Frost weathering is most effective in tiny cracks and crevices of irregular shape in temperate to sub-arctic climates where repeated thawing and freezing occur on a daily basis. It is not generally realised that a certain amount, sometimes a significant amount, of daily freeze/thaw occurs in hot deserts where winter rains and dews provide enough moisture for the process to be effective.

Salt weathering has been greatly underestimated as a weathering type. There seem to be three ways in which salt expansion may give rise to stresses that lead to rock disintegration: hydration, heating and crystal growth (Goudie *et al.* 1970, Cooke 1979). The last two mechanisms are probably most important and they occur in deserts and coastal areas of all latitudes where salts are concentrated and where dews, coastal mists, sea spray and ordinary rainfall provide the necessary liquid phases. Crystal growth stresses are particularly dependent upon rock porosity and the mechanism is most effective in porous sedimentary hosts. Salts vary in their ability to disintegrate rocks by crystal growth, sodium and magnesium sulphates being most effective. Crystal growth stresses occur in tortuous cracks under pressure. Open systems, where salts crystallise due to evaporation, cannot give rise to changes of volume and hence cannot give stresses.

It is now considered impossible for **diurnal temperature changes** in deserts to cause rock exfoliation since many experiments have failed to reproduce the effects in the laboratory. The effect is ascribed to the spalling off of partly chemically weathered rock skins, often concentrated on the shady parts of rock surfaces where the effects of surface moisture upon minerals will be greatest and act for longer periods.

Despite the above remarks concerning the efficiency of physical weathering, there seems to be no doubt that it is subordinate to chemical weathering over most of the Earth's surface.

1g Sediment yields

As we might expect, there is a large variation in the yield of solid clastic grains from different areas. Climate, vegetation cover, lithology and relief are geologically the most important controlling factors, although past and present farming and clearance activities have given rise to great modifications, particularly in low latitudes. Chemically and physically weathered detritus is liberated from solid rock outcrops at a rate dependent upon lithology and weathering efficiency, the latter being controlled mainly by rainfall magnitude. The rate of transport of the liberated detritus by natural flows into river systems is then controlled by surface slope and rainfall runoff magnitude. Sediment derived from surface slopes in a drainage basin is transported out along the river system at a rate dependent upon the stream power. (Ch. 5d).

Increasing surface runoff will tend to cause an increase in sediment yield for a drainage basin of a given size; the effect of surface vegetation will oppose this trend. Arid and semi-arid areas have little vegetation cover. The incoming of grass and forest cover as rainfall increases will tend to reduce drastically the proportion of hillslope weathered mantle removed by surface runoff. These two effects give the relationship shown in Figure 1.4 (Langbein & Schumm 1958). An increase in precipitation above about $100\,cm^{-1}$ may tend to reverse the trend caused by vegetation cover, leading to gradually increased sediment yields particularly in seasonal and monsoon climates (L. Wilson 1973). It is important to note that the main portion of Figure 1.4 is based upon small drainage basins in the central United States. Although the absolute magnitude of the sediment yield may vary in other areas, it is expected that the general trend will hold true.

For the geological past we may speculate with Schumm (1968a) that in Precambrian and early Palaeozoic times (4500–400 ma), before the development of land vegetation, the constraint upon increasing yields with increasing surface runoff was absent. Thus the whole Earth's surface would have behaved as modern semi-arid areas do today. Progressive reduction of sediment yields in hinterland areas of rainfall greater than 25 cm would have followed the establishment of coniferous forests in the early Mesozoic and, most important, the grasses in the early Cretaceous (Fig. 1.4).

Several authors have attempted to provide world maps of sediment production rates. In a notable study, Fournier (1960) constructed such a map based upon an empirical equation derived to predict sediment yields as a function of precipitation and relief, with data from nearly 80 individual drainage basins. A similar map by Strakhov (Fig. 1.5) shows maximum sediment yields in the seasonally humid tropics, with yields decreasing towards equatorial regions, where seasonal effects are lacking, and towards arid zones (both hot and cold), where rainfall is low. The effect of relief is clearly seen in the Cordilleran and Himalayan regions. Other authors show essentially similar

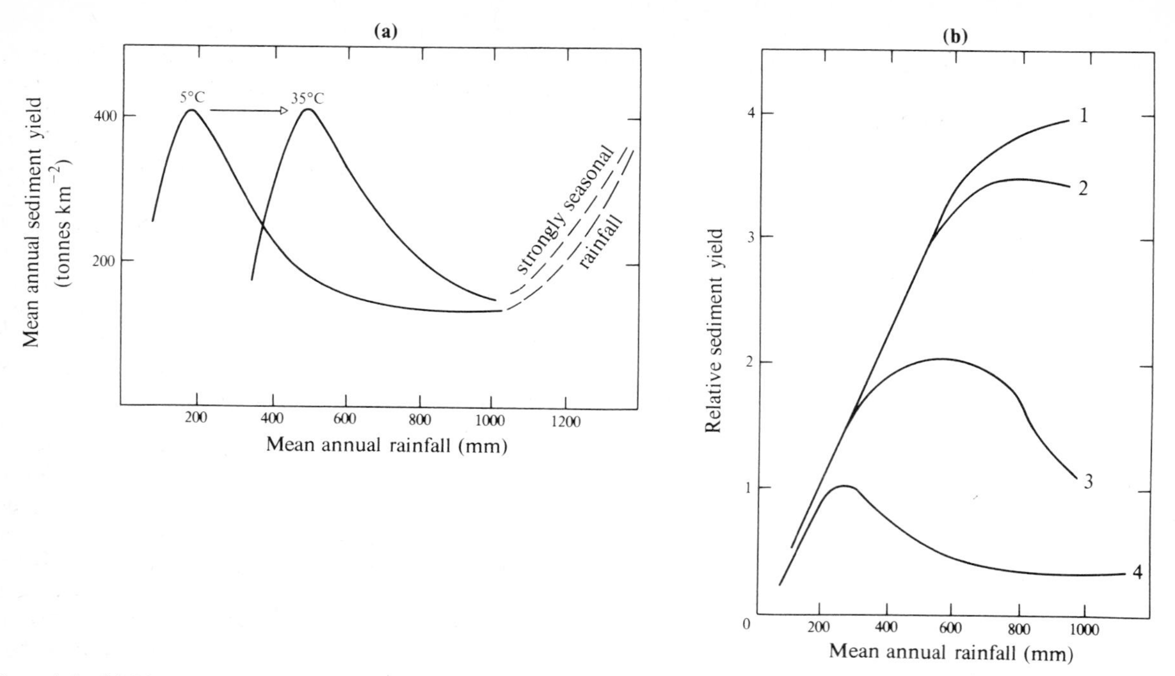

Figure 1.4 (a) The relationship between mean annual rainfall and sediment yield (after Schumm 1968a), with the generalised effects of seasonality after L. Wilson (1973). (b) Hypothetical relationships between mean annual rainfall and sediment yields: (1) before the advent of land vegetation (pre-Silurian), (2) following the appearance of primitive vegetation (Silurian to Devonian), (3) following the appearance of flowering plants and conifers (Carboniferous to Jurassic), and (4) following the appearance of grasses (post-Cretaceous) (after Schumm 1968a).

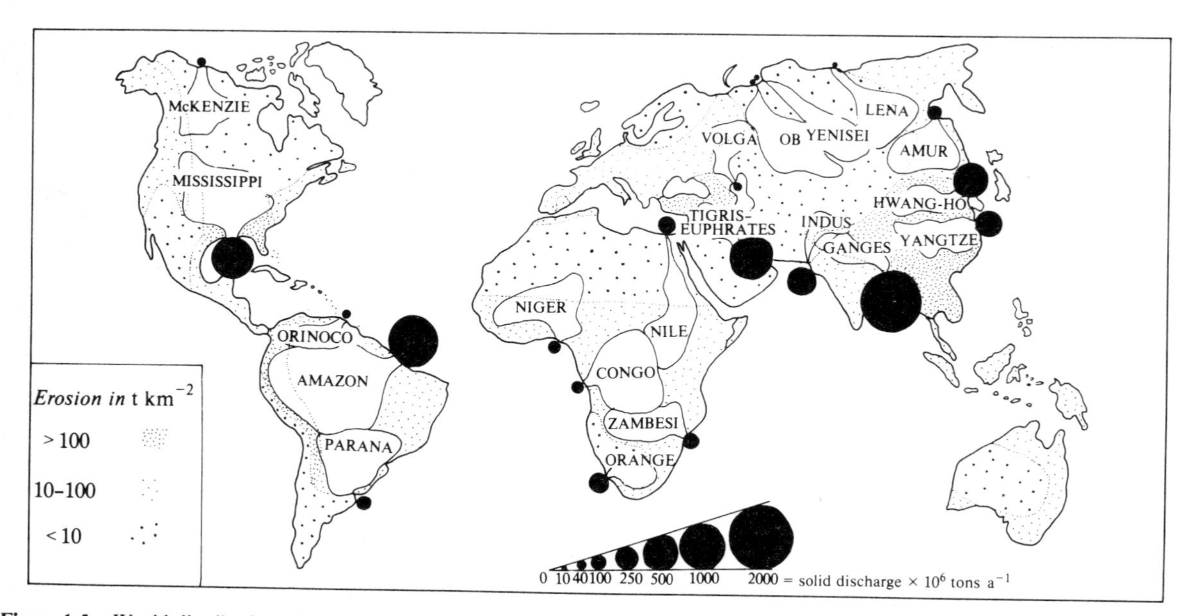

Figure 1.5 World distribution of erosion, the largest river drainage basins and the approximate magnitude of the solid load deposited at the river mouths (after Strakhov 1967 and Stoddart 1971).

trends but with more than an order of magnitude difference in the absolute values of sediment yield. This large discrepancy reflects the great difficulty of data selection and processing in this kind of study. Of particular interest are maps showing the solid sediment yields from large river drainage basins (Fig. 1.5). Note that there is in fact little correlation between drainage basin size and sediment yield. A host of other climatic, relief and lithological factors affect this relationship, including very important human effects arising from several millennia of agricultural practice.

1h Clastic grains and source identification

Quartz grains are the most abundant sand- and silt-sized (Ch. 4) grains in clastic sediments. Sections of single crystal quartz grains may show **normal** or **undulose extinction** under crossed polars. Undeformed volcanic quartz shows normal extinction, but igneous, plutonic and metamorphic quartz grains show normal or undulose extinction. Undulose extinction is due to lattice strain. Recent statistical studies (Basu *et al.* 1975) show that metamorphic quartz grains show mean extinction values >5° whereas plutonic igneous quartz grains show mean values of <5° (Fig. 1.6). Quartz grains may also be single or polycrystalline. Almost all quartz of volcanic origin is of the single crystal type. The amount of **polycrystalline quartz** is least in plutonic igneous rocks (Fig. 1.7); it increases in high-grade metamorphic rocks and is highest for low-grade metamorphic rocks. Similarly, the average number of crystal units in polycrystalline quartz is greatest in low-grade metamorphic rocks and least in high-grade and plutonic igneous rocks (Fig. 1.7). By combining these properties of quartz into a double triangular diagram and investigating the composition of quartz of known provenance from modern streams, it is possible to define a useful provenance indicator (Fig. 1.7) for first-cycle sediments (Ch. 1i).

Feldspar grains include alkali feldspars (orthoclase, microcline), perthite and the plagioclase feldspar series (albite to anorthoclase), usually in that order of abundance – reflecting the stability order of feldspars to chemical weathering. Microcline is widely distributed in both metamorphic and plutonic igneous rocks. Plagioclase compositions vary with chemical composition of plutonic igneous and metamorphic rocks. In high-gradient streams where frequent strong grain/grain impacts may be expected, twinned crystals (especially carlsbad twins) tend to break up along the twin planes and so suffer a great decrease in their abundance (Pittman 1969). Composite plagioclase twins suffer less from this effect. The low percentage of feldspar in many beach sands (compared to river sands) may reflect the tendency for feldspar grains to disintegrate by successive cleavage fractures in high energy conditions.

Rock (lithic) fragments may usefully be divided into

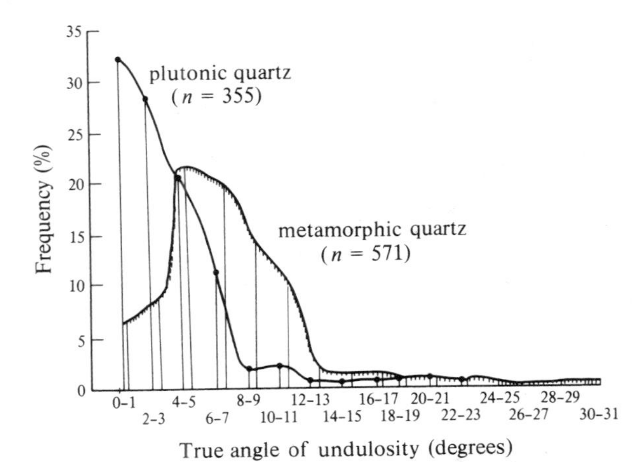

Figure 1.6 Frequency curves to show the distribution of true angles of undulosity in detrital quartz of plutonic and low-rank metamorphic parentage. Note the large degree of overlap. Values based on universal stage measurements (after Basu *et al.* 1975).

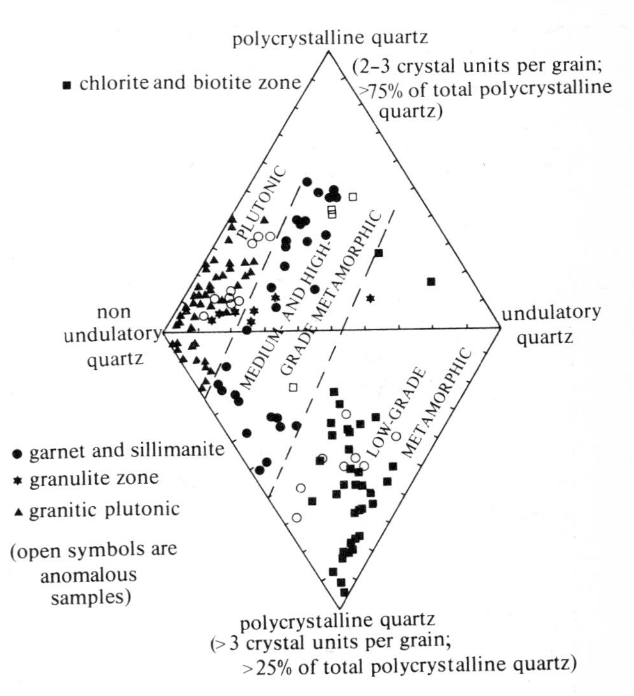

Figure 1.7 Ternary graph to show the distribution of quartz types derived from particular source rocks indicated by the symbols (after Basu *et al.* 1975).

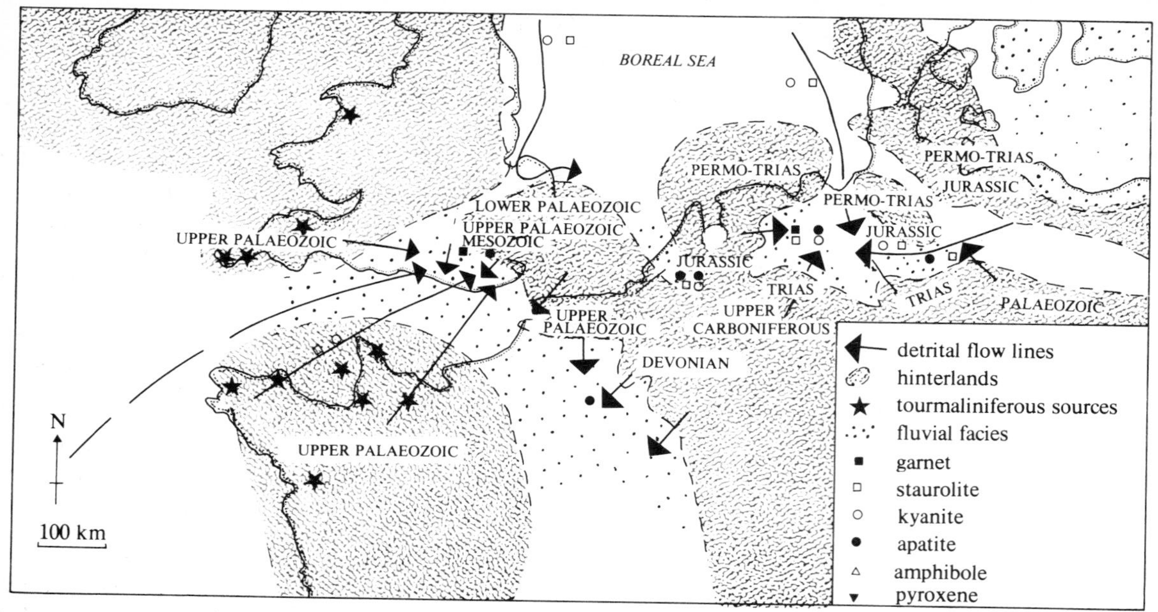

Figure 1.8 Example of the use of heavy mineral suites in the derivation of rational palaeogeographical reconstructions; Wealden (Lower Cretaceous) facies of NW Europe (after P. Allen 1967, 1972).

intraformational and exotic types. Intraformational fragments predominantly comprise the soft plastic clay fragments eroded from mud beds. Exotic rock fragments include the whole gamut of igneous, metamorphic and sedimentary rock types. The coarser plutonic, schistose and gneissose rocks commonly occur in larger, pebble-sized fragments, breaking down into their sand-size mineralogical components with continued weathering or abrasion. Important finer-grained fragments include acidic volcanics, vitric tuffs, cherts, argillites and quartzites.

Heavy minerals rarely make up more than 1% of a sediment or rock. For study they must be separated out from loose sediment or disaggregated rocks by liquids of high density (>2800 kg m^{-3}) (see Carver 1971). Opaque heavy minerals are usually most abundant and they include magnetite and ilmenite. Non-opaque forms commonly include the hard and resistant zircon, tourmaline, rutile and garnet. Abundant tourmaline serves as an indicator of boron-rich plutonic sourcelands. Recent techniques of ^{40}Ar:^{39}Ar dating applied to detrital tourmaline (P. Allen 1972) yield very valuable data on sourcelands. Garnets are most common in pelitic schists where increasing metamorphic grade is accompanied by decreasing Ca and Mg but increasing Fe and Mg in the lattice. Other metamorphic minerals encountered in heavy fractions include staurolite, kyanite and sillimanite. Although heavy mineral studies are currently 'out of fashion', the results of such studies (when presented in a suitable statistical form) combined with palaeocurrent, palaeohydraulic and facies studies may yield very interesting reconstructions of ancient drainage basins (Fig. 1.8). It should be noted, however, that both intrastratal solution and reprecipitation may take place after burial, so modifying the absolute and relative composition of the heavy mineral assemblages.

The chief **detrital clay minerals** are kaolinite, illite, montmorillonite, chlorite, and the mixed-layer clays. The basic structure of the clay mineral group comprises alternating octahedrally co-ordinated gibbsite sheets and tetrahedrally co-ordinated silica sheets (Fig. 1.9). Substitution for aluminium or silicon in both sheets occurs in many clay minerals, leading to charge deficiencies which are balanced by interlayer cations.

Kaolinite is liberated from feldspars during acidic weathering conditions by reactions such as Equations 1.9 and 1.10. The structure of kaolinite is simple, comprising alternating gibbsite and silica sheets. The minor Al for Si substitution (1 in 400) allows for no interlayer cations in the kaolinite structure. Kaolinite comprises 8–20% of the total clay minerals in modern oceanic sediments and is most abundant in sediments derived by low-latitude tropical weathering (Fig. 1.10).

Illite is closely related to the mica muscovite but differs in having more silicon and less potassium. Al^{3+} substitutes for Si^{4+} in the silica layer in the ratio 1:7. The net negative charge created by this substitution is balanced by K^+ ions which link the adjacent silica–gibbsite–silica 'sandwiches' together (Fig. 1.9). These interlayer K^+ ions

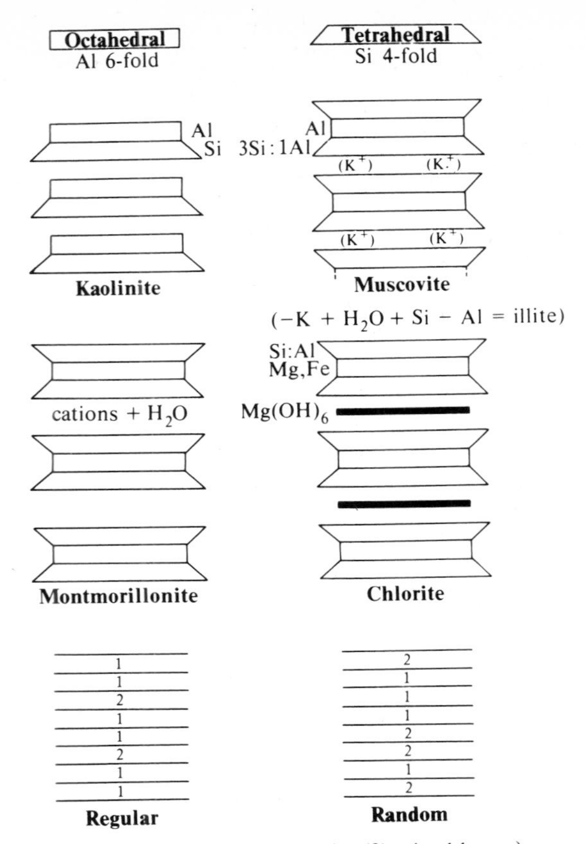

Figure 1.9 Schematic representation of the main clay mineral structures (after Pettijohn *et al.* 1972). See text for discussion.

prevent entry of H_2O and other cations so that the low exchange capacity is confined to the crystal edges. Some soils subject to leaching may contain **degraded illites** which have less K^+ than they should (their high K^+ uptake potential is made use of in agricultural practice by fertilisers). Illite comprises 26–55% of the total clay minerals in modern oceanic sediments and is most abundant in areas adjacent to temperate and semi-arid continental areas (Fig. 1.10).

Montmorillonite belongs to the smectite clay group in which Mg^{2+} substitutes for Al^{3+} in the gibbsite sheets in the ratio 1:6. There may also be other substitutions, chiefly Al^{3+} for Si^{4+} and Fe^{2+} for Al^{3+}. The net negative charge created by these substitutions is partly balanced by a small number of interlayer cations, usually Na^+ or Ca^{2+}. Water is readily adsorbed between the structural layers; 1–3 water layers may be present causing the basal spacing to range between 10–21Å. On heating to 100–200°C all smectites lose their interlayer water reversibly. All smectites show cation exchange properties, the principal cause of which is the unbalance of charge in the fundamental layers. Generally, the replacing power of ions with higher valency is greater and their replaceability is less, so that Ca^{2+} ions are the more firmly held. Montmorillonite comprises 16–53% of the total clay minerals in modern oceanic sediments and is most abundant in areas where basic or intermediate igneous rocks are being sub-aerially weathered or altered under submarine alkaline conditions; for example, material transported into the South Pacific Ocean or found along mid-oceanic ridges (Fig. 1.10).

Chlorite has a complicated structure, involving talc sheets sandwiched between silica sheets, with alternating talc 'sandwiches' being linked by brucite layers (Fig. 1.9). Minor substitution of Al^{3+} for Si^{4+} and Fe^{2+} and of Al^{3+} for Mg^{2+} occurs. The ease of oxidation of the Fe^{2+} in the brucite layer means that chlorite can occur only in weathering zones where chemical weathering is much reduced or absent. The distribution of chlorite in modern oceanic sediments thus reflects this control, so that the mineral is most common in high latitudes, particularly in glacially influenced areas (Fig. 1.10). Chlorite ranges from 10–18% of the total clay minerals in modern oceanic sediments.

Mixed-layer clay minerals are those in which different kinds of layers alternate with each other. The alternations may be regular or irregular. Of the large number of different sorts of mixed-layer minerals the most important sedimentologically are the illite–montmorillonite types.

It is briefly stressed here that all clay minerals undergo appreciable changes during diagenesis and burial so that it is misleading to imagine that the nature of a clay rock in the geological record will reflect the composition of the freshly deposited clay (Ch. 28).

1i Sourcelands, differentiation and plate tectonics

Clastic sedimentary grains derived from a metamorphic or igneous hinterland may be termed **first-cycle grains.** Those grains themselves may form a subsequent hinterland after the geological cycle of deposition, lithification (Ch. 28) and uplift turns around once more. **Second-cycle** grains would result from erosion of these hinterlands, and so on. With each cycle of deposition, diagenesis, uplift and weathering the resultant grain assemblages should become finer grained, more rounded, more quartzose (with less polycrystalline quartz) and less rich in unstable heavy minerals. Any deposit may therefore be said to become increasingly **mature** due to such processes. It is a common mistake in palaeogeographic reconstructions to

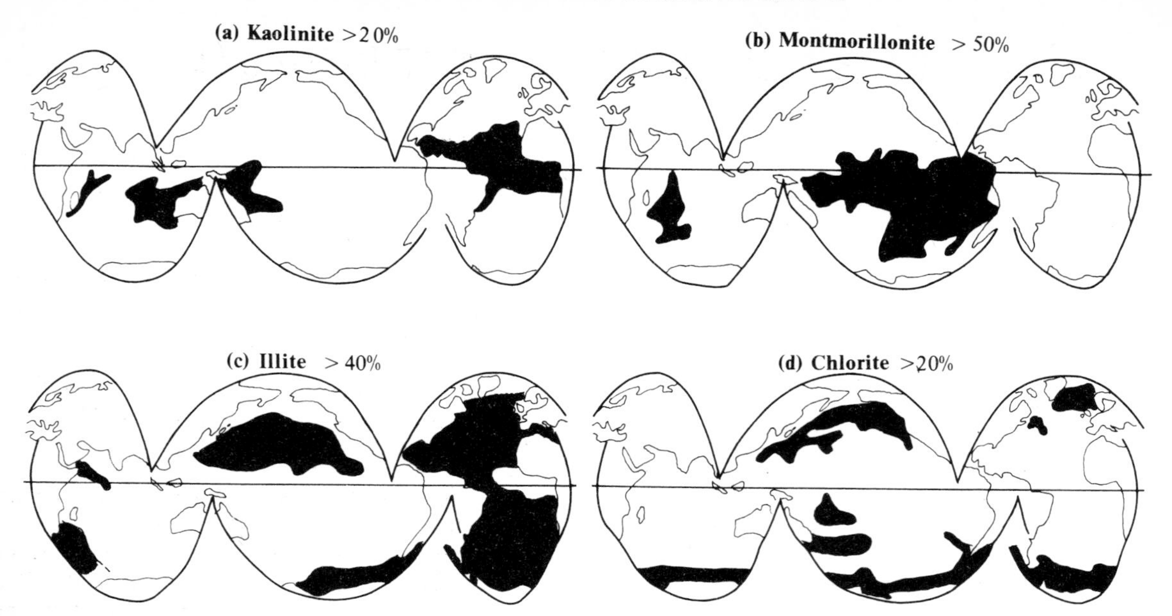

Figure 1.10 Maps to show the location of high proportions of various clay minerals in the surface sediments underlying the oceans (after Griffin *et al.* 1968). See text for discussion.

assume that clastic sediments consist of primary detritus – this is rarely the case.

Despite the above caveat and the occurrence of multicycle grains in many clastic deposits, it has become clear that the plate tectonic setting to continental margins and ocean basins can exert a strong control upon the composition of clastic deposits. Thus we may compare the stable crystalline and sedimentary 'old' hinterlands of a typical trailing-edge (Atlantic type) continental margin with the volcanically active Cordilleran hinterlands of a leading-edge (Andean type) margin. Using rigorous statistical methods Valloni and Maynard (1981) have recently subdivided deep-sea sands according to their plate tectonic setting in terms of quartz (Q), feldspar (F) and rock fragment (R) mean percentage composition. Thus sands in basins associated with: (a) trailing-edge continental margins have Q_{62} F_{26} R_{12}; (b) leading-edge continental margins, either subduction, Q_{10} F_{53} R_{31} or (c) strike-slip Q_{34} F_{39} R_{27}; (d) back-arc basins, Q_{20} F_{29} R_{51}; and (e) fore-arc basins, Q_8 F_{17} R_{75}. The amount and type of rock fragments and the type of feldspar are the most important discriminators between environments.

1j Summary

Clastic mineral grains and rock fragments are liberated from igneous, metamorphic and pre-existing sedimentary rocks by the oxidation and hydrolysis reactions of chemical weathering and by physical weathering. The mineral composition and volume of clastic grains derived from sourcelands are complex functions of hinterland geology, relief, climate, drainage basin size and vegetation cover. Clastic grains provide some evidence of provenance, but complications arise from reworking during successive geological cycles and diagenetic alterations (see Ch. 28).

Further reading

A helpful elementary treatment of physical chemistry of relevance in this and subsequent chapters is given by Krauskopf (1979). A more advanced approach to specific problems over the whole field of sedimentary geochemistry is given by Berner (1971).

Aids to the identification of mineral and rock grains are to be found in most mineralogical and petrological texts, including Kerr (1959), Pettijohn (1975) and Pettijohn *et al.* (1972). Grim (1968) is an essential reference on clay mineralogy. Techniques in petrography are dealt with by Carver (1971).

An illuminating discussion of provenance, multicyclicity, maturity and numerous other topics is to be found in Folk's incomparable book (1974a), although the tectonic concepts used are not linked to modern plate margin dynamics.

2 The origin of calcium carbonate grains

2a Introduction

In contrast to the detrital or clastic grains discussed in the previous chapter, calcium carbonate grains are usually formed within the basin of deposition and are usually marine in origin. It is impossible to understand the origin and significance of carbonate particles adequately without some knowledge of the physical chemistry of $CaCO_3$ in solution. Therefore, after briefly outlining the distribution of recent marine carbonate sediments, we shall discuss the chemical composition of sea water, the main carbonate minerals, and some of the problems involved in $CaCO_3$ precipitation.

2b Recent marine carbonate sediments

Figure 2.1 shows that marine carbonate sediments are very widely distributed. Three groups may be usefully distinguished:

(a) *Oceanic carbonates of biogenic pelagic origin.* These are widely distributed in the oceans (with the exception of the North Pacific, Arctic and Antarctic) where they are closely associated with the mid-ocean ridge system and areas of upwelling (Chs 24 & 26). These calcareous oozes accumulate slowly in maximum water depths of 3.5–5 km.

(b) *Shelf carbonates of subtropical and tropical origin.* Several of the more thoroughly investigated areas are noted on the map. Carbonates of biogenic origin dominate, but inorganically precipitated $CaCO_3$ is important locally.

(c) *Shelf carbonates of temperate origin.* These are more widespread than is sometimes thought and are wholly of biogenic origin.

In addition to these marine-occurrences many freshwater and hypersaline lakes contain carbonate particles.

Very broadly, the distribution of carbonate sediments shows the adverse effect on $CaCO_3$ concentration of major depocentres of clastic particles and the effects of dissolution of $CaCO_3$ in the deep ocean.

2c The composition of fresh water and sea water

A comparison between the compositions of 'average' river and 'average' sea water reveals some interesting facts (Table 2.1).

(a) Sea water contains about 300 times more dissolved solids than does fresh water.

(b) Seawater cations and anions in decreasing orders of abundance are Na^+, Mg^{2+}, Ca^{2+}, K^+ and Cl^-, SO_4^{2-}, HCO_3^-. Freshwater cations and anions in decreasing orders of abundance are Ca^{2+}, Na^+, Mg^{2+} and HCO_3^-, SO_4^{2-}, Cl^-.

Thus sea water is not simply 'concentrated' river water, because of the different proportions of solids. Sea water must therefore be derived by both evaporation *and* chemical/biological differentiation of river water. For example, Ca^{2+} is proportionately very much more abundant in river water than in sea water. This may reflect the relative ease with which Ca^{2+} is removed from sea water by biological and, to a lesser extent, chemical precipitation as compared with the three other major cations: Na^+, K^+ and Mg^{2+}.

This brings us on to a speculative topic of major interest: whether or not the oceans have undergone any chemical change with time. It may be argued that primaeval sea water may have been closer to present-day river waters in their proportionate divisions of the dissolved solids because the major chemical-differentiation process acting at present is the biological removal of $CaCO_3$ in skeletal hard parts. This removal has been accelerating since early Cambrian times and it probably reached its maximum rate of change only in Mesozoic times with the evolution of very abundant pelagic calcareous microfaunas and floras. It is therefore possible that the oceanic Mg:Ca ratio has gradually been increasing through geologic time (Sandberg 1975), perhaps from as low as 0.25 (fresh water) to its present value of around 5. Such a process would have had profound mineralogical effects upon carbonate sediments since, as we shall see, a high Mg:Ca ratio in sea water tends to encourage the precipitation of aragonite instead of calcite.

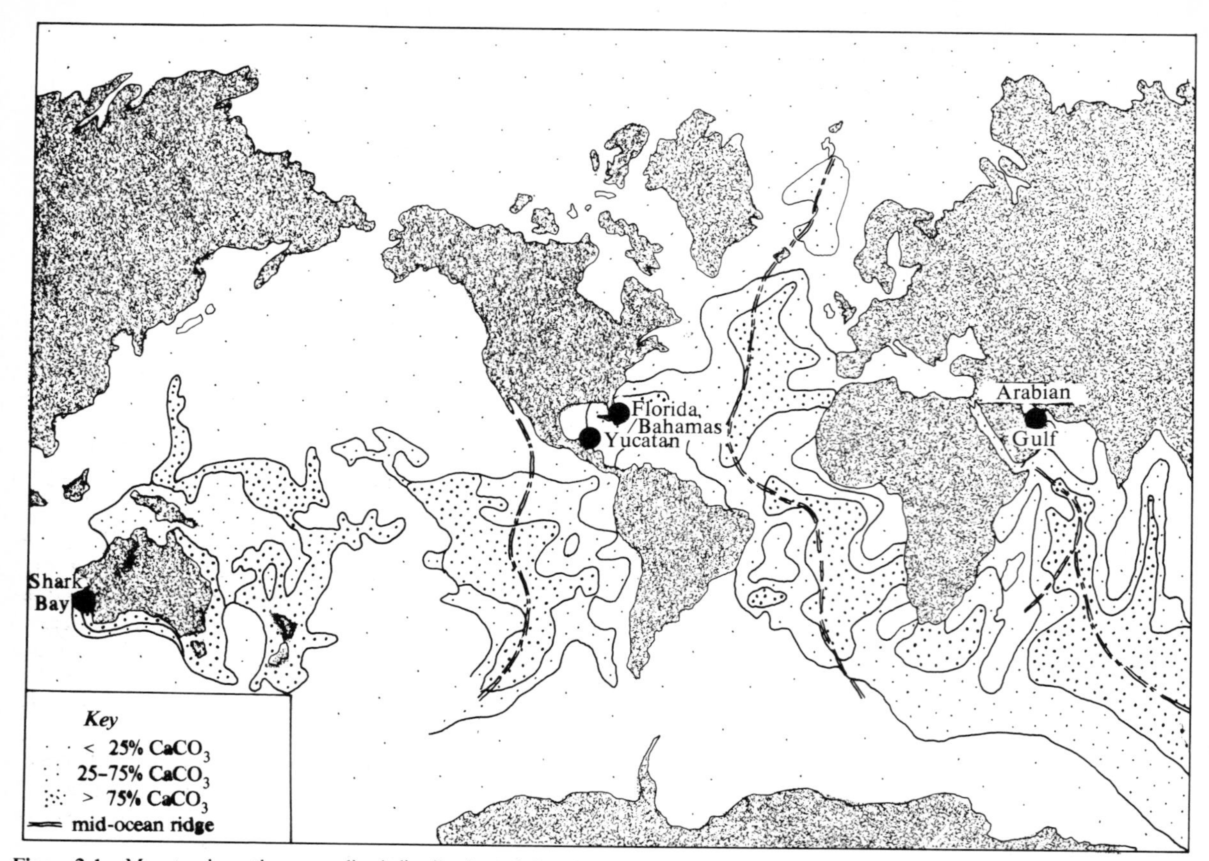

Figure 2.1 Map to show the generalised distribution of $CaCO_3$ in sediments. Note the strong association of $CaCO_3$-rich sediments with the flanks and crests of the mid-ocean ridge system, as well as the better known shallow-water carbonate platform type occurrences (after Broecker 1974 and sources acknowledged therein).

The remarkable properties of sea water may be summarised partly by noting that it is a well buffered solution of high ionic strength. By 'well buffered' we mean that the solution contains compounds that exert a strong control over pH variations. Surface sea water has a pH of 7.8–8.3. We shall discuss some of the chemical reactions that cause buffering in a later section of this chapter. Let us briefly consider **ionic strength.** The solubility of calcium carbonate in *pure water* may be calculated from thermodynamic data. It is found, however, that the solubility is very much greater in a solution like sea water where other abundant ionic species are present. These ionic species together with polar H_2O molecules tend to cluster around the oppositely charged ions of Ca^{2+} and CO_3^- and prevent the ions coming together to precipitate, i.e. the solubility increases. The greater the charge on the 'pollutant' ions, the greater the effect. Thus solutions like sea water are said to have a high ionic strength due to the formation of **complex ions** and **ion pairs**. Sea water has an ionic strength of about 0.7 whereas fresh water has values around 0.002. Because of this difference in ionic strengths, the solubility of $CaCO_3$ in sea water is much higher than in fresh water (see Krauskopf 1979 for a good discussion).

2d The major carbonate minerals

Calcium carbonate exists as the mineral polymorphs **calcite** and **aragonite**. Both may form as inorganic precipitates or as biological secretions in the hard parts of numerous organisms. Aragonite does not usually precipitate from fresh water. Properties of the two minerals are listed in Table 2.2. Aragonite is metastable under Earth surface conditions, being a high-pressure equilibrium carbonate, as found in blueschist metamorphic facies. Many of the problems of carbonate diagenesis revolve around the timing and chemical constraints upon aragonite inversion to calcite and aragonite dissolution (Ch. 29).

An important consequence of the charge similarity and

Table 2.1 Composition of average sea and river waters (after Sverdrup *et al.* 1942, Livingstone 1963, Garrels & Thompson 1962).

Ion	*Concentration (moles)*				*Sea water/ River water*
	Sea water	*Order*	*River water*	*Order*	
Na^+	0.47	(2)	2.7×10^{-4}	(4)	1740
K^+	1.0×10^{-2}	(5)	5.9×10^{-5}	(7)	170
Ca^{2+}	1.0×10^{-2}	(5)	3.8×10^{-4}	(2)	26
Mg^{2+}	5.4×10^{-2}	(3)	1.7×10^{-4}	(3)	318
Cl^-	0.55	(1)	2.2×10^{-4}	(4)	2500
SO_4^{2-}	3.8×10^{-2}	(4)	1.2×10^{-4}	(6)	317
HCO_3^-	1.8×10^{-3}	(6)	9.55×10^{-4}	(1)	1.9
pH	7.9		~7		
ionic strength	0.65		0.002		

ionic radius of Ca^{2+} and Mg^{2+} ions and of the structure of the calcite lattice is that Mg^{2+} may substitute extensively for Ca^{2+} in calcite. The calcite formula is therefore more correctly written as $(Ca_{1-x} Mg_x)CO_3$ where x is is commonly in the range 0.01–0.25, never exceeding about 0.4. These calcites with more than 5% $MgCO_3$ are known as **high-Mg calcites**. The Mg content of certain organic hard parts seems to be related to temperature, with greater amounts of Mg in warmer waters. Small amounts of Fe^{2+} (up to a few thousand ppm) may also substitute for Ca^{2+}, giving rise to **ferroan calcites** in low Eh conditions. The substitution of trace amounts of Mn^{4+} causes calcite to luminesce under the influence of cathode ray bombardment. If the flux of Mn^{4+} varied with time during calcite crystallisation, then luminescence studies reveal tell-tale growth zones which may often be mapped out in stratigraphic sections.

The aragonite lattice cannot take up Mg^{2+} when it is an inorganically precipitated phase, although certain skeletal aragonites in corals may contain about 0.001% Mg^{2+}. Strontium may be taken up in the aragonite lattice to a maximum concentration of about 10 000 ppm. The crystal habit of aragonite in precipitated and some biogenic phases (calcareous algae) is usually of the fibrous type, whilst calcite may be fibrous rhombic or 'dog's tooth' in morphology.

The double carbonate dolomite, $CaMg(CO_3)_2$, is mainly a diagenetic mineral (Ch. 29). Ferrous iron may substitute for Mg^{2+} in the solid solution series dolomite–ankerite $(CaMg_{0.75}Fe_{0.25})(CO_3)_2$. The pure calcium–iron carbonate is unknown in nature. The dolomite lattice is highly ordered, with alternating layers of cations and CO_3^{2-} groups, in which the cation layers are alternatively Ca^{2+} and Mg^{2+} (see Fig. 29.13).

Recent shallow-water tropical and subtropical calcium carbonate deposits are predominantly composed of aragonite and high-Mg calcite, whilst temperate shallow carbonates contain dominantly calcite.

Recognition of carbonate minerals is greatly aided by staining techniques in thin section and peels. Calcite Fe-calcite, high-Mg calcite, aragonite, dolomite and Fe-dolomite may all be distinguished in this way (Appendix 2.1).

Table 2.2 Solubility products (K) and ion activity products (IAPs) for calcite, dolomite and aragonite in sea water at 25°C. Solubility products for calcite and aragonite after Berner (1971), for dolomite after Hsü (1966). IAP for $CaCO_3$ is an approximate mean for surface sea water at 25°C (Berner 1971). IAP for dolomite after Hsü (1966). Note that surface sea water is supersaturated with respect to all the chief carbonate minerals.

Mineral	*K*	*IAP*	*IAP/K*
calcite	4.0×10^{-9}	1.35×10^{-8}	3.4
aragonite	6.3×10^{-9}	$(CaCO_3)$	2.1
dolomite	1.0×10^{-17}	1.0×10^{-15}	~100

2e Primary carbonate precipitation

Surface sea water is distinctly supersaturated with respect to aragonite, calcite and dolomite (Table 2.2), yet precipitation of inorganic $CaCO_3$ is confined to a few subtropical and tropical locations where it is quantitatively unimportant compared with biogenic sources of $CaCO_3$. Even though sea water is supersaturated with respect to aragonite, calcite and dolomite, only the former mineral precipitates directly from sea water (*excluding the realm of diagenetic pore waters*).

The precipitation of $CaCO_3$ is controlled by the following reaction

$$H_2O \text{ (water)} + CO_2 \text{ (carbon dioxide)}$$

$$\Downarrow\Uparrow$$

$$CaCO_3 \text{ (calcium carbonate)} + H_2CO_3 \text{ (carbonic acid)} \rightleftharpoons Ca^{2+} \text{ (calcium ion)} + 2HCO_3^- \text{ (bicarbonate ion)} \quad (2.1)$$

so that it is favoured by processes that decrease the amount of CO_2 (i.e. decrease partial pressure: pCO_2) available to the solution, such as warming or organic photosynthesis. Most of the CO_2 in sea water is held in the HCO_3^- anion due to the reactions,

$$CO_2 + H_2O \rightleftharpoons H_2CO_3 \quad (2.2)$$

$$H_2CO_3 \rightleftharpoons H^+ + HCO_3^- \quad (2.3)$$

$$H^+ + CO_3^{2-} \rightleftharpoons HCO_3^- \quad (2.4)$$

Increasing alkalinity causes the CO_3^{2-} content to increase by the reaction

$$HCO_3^- + OH^- \rightleftharpoons CO_3^{2-} + H_2O \quad (2.5)$$

From Equations 2.3 and 2.4 we see that carbonic acid ionises in steps so that the **ionisation constant**, K, for each reaction may be written as

$$\frac{[H^+][HCO_3^-]}{[H_2CO_3]} = K_1 = 10^{-6.4} \quad (2.6)$$

$$\frac{[H^+][CO_3^{--}]}{[HCO_3^-]} = K_2 = 10^{-10.3} \quad (2.7)$$

Reaction 2.3 thus provides the stronger acid solution, but the acid is still rather weak. Sea water at normal pH of 8.1–8.3 thus contains both HCO_3^- and a plentiful supply of un-ionised H_2CO_3 in addition to a small amount of CO_3^{--} (Fig. 2.2). These first two components are the main buffering components of sea water. Attempts to increase the acidity of sea water are countered by the forward reaction of 2.4, and the backward reaction of 2.3. In addition the solid $CaCO_3$ in contact with sea water will act as a buffer by the reaction

$$CaCO_3 + H^+ \rightleftharpoons Ca^{2+} + HCO_3^- \quad (2.8)$$

i.e. more carbonate is dissolved.

Attempts to increase the alkalinity of sea water are similarly foiled by the forward reaction of (2.5) and by the precipitation of $CaCO_3$ by

$$Ca^{2+} + HCO_3^- + OH^- \rightleftharpoons CaCO_3 + H_2O \quad (2.9)$$

Thus we see why sea water is such a well buffered solution. By way of contrast, land waters with their low alkalinity are very poorly buffered.

Calcium carbonate deposits, held by some authors to be of inorganically precipitated origin, occur in several areas, notably the Bahamas, Persian Gulf and Dead Sea. In each case the mineral form is aragonite in the characteristic needle-like crystal habit with crystal sizes of a few microns. In the Dead Sea, events of mass precipitation have been correlated with the appearance of the ghostly **whitings**: large patches of aragonite suspensions that appear suddenly in the surface waters. The Dead Sea whitings (Neev & Emery 1967) are characterised by an immediate decrease in HCO_3^- in the water mass, indicating $CaCO_3$ precipitation by the reverse reaction of (2.1) above. Chemical data are not available for the Persian Gulf occurrences, but the size and nature of the whitings leave little room for an alternative explanation to inorganic precipitation. Increase in CO_2 uptake during periodic diatom 'blooms' has been advanced as the cause for these whitings. A small problem here though is the complete lack of preserved aragonite in the bottom sediments of the offshore Persian Gulf (Ch. 23).

The changing chemistry of cold Atlantic water as it passes on to the broad shallow Bahamas Banks, where it is warmed up, is well illustrated in Figure 2.3. Note the increasing salinity and decreasing carbonate concentration (indicative of precipitation) from the Florida Straits over the Great Bahama Banks. Note also the occurrence of a whiting in exactly the right position by these chemical trends. The rate of loss of $CaCO_3$ is obtained by dividing the $CaCO_3$ deficit by the mean residence time on the Banks for a particular water sample. Mean residence was originally calculated (Broecker & Takahashi 1966) knowing the degree of incorporation of atomic-bomb-produced ^{14}C from the atmosphere. Mean precipitation rates thus obtained are around 50 mg

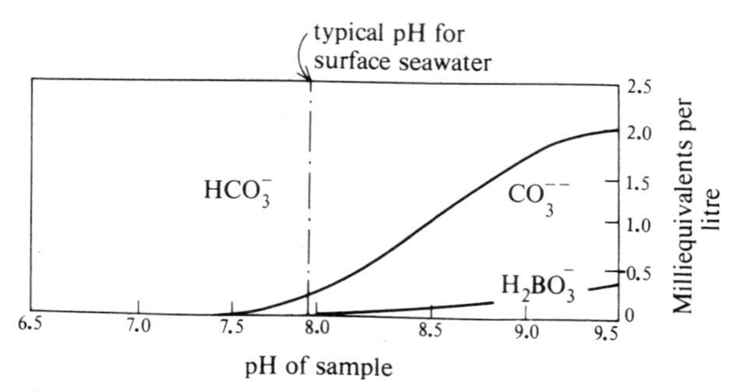

Figure 2.2 Variation of alkalinity components of sea water with pH. See reactions (2.2–2.5) in text (after Cloud 1962).

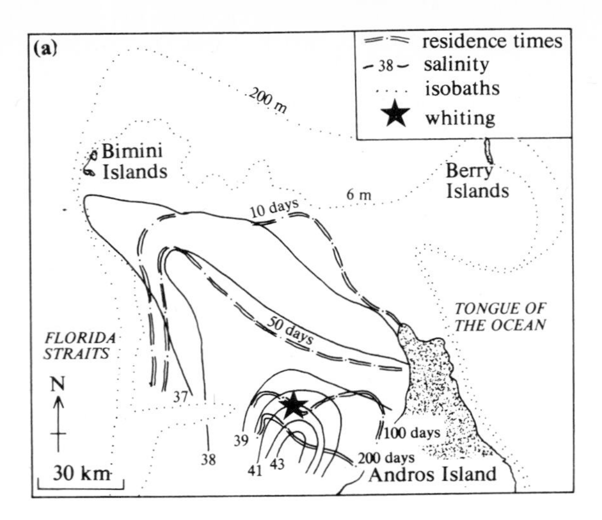

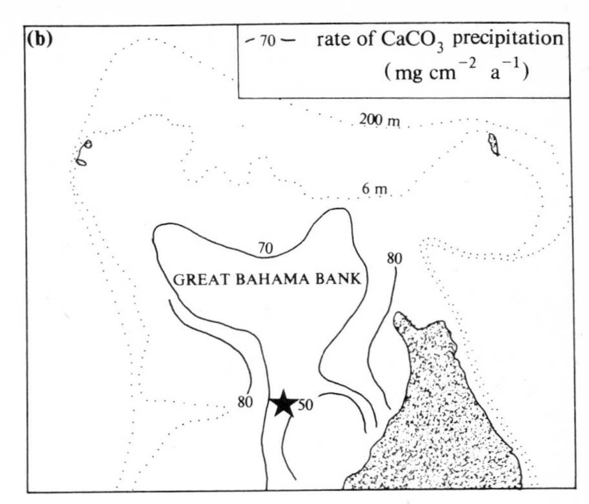

Figure 2.3 (a & b) Maps to show mean residence time of ocean water on the Bahama Bank platform, typical values of salinity and calculated rates of $CaCo_3$ precipitation. (c) The relationship between activity product and water residence time. The fall from open ocean supersaturation to approximate saturation with increasing residence time is strongly indicative of inorganic precipitation. (All after Broecker & Takahashi 1966.)

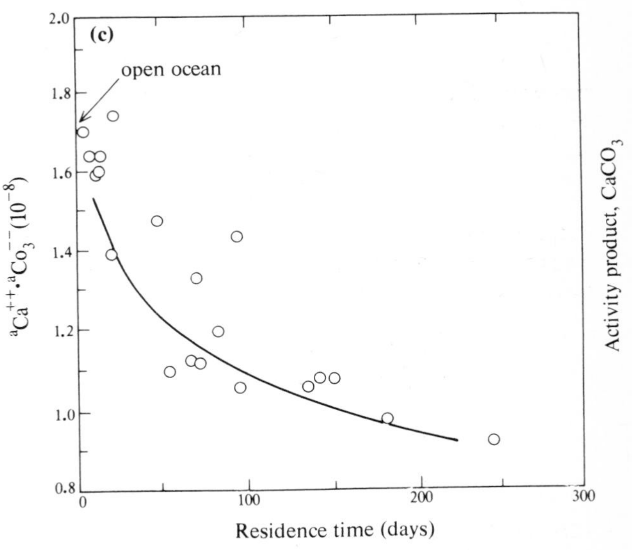

$CaCO_3\ cm^{-2}\ a^{-1}$. These gross chemical trends have been thought to indicate some degree of inorganic contribution of $CaCO_3$ in the area, perhaps as much as 50%. Problems arise, however, since the ^{14}C content of aragonite collected in whitings was inconsistent with a precipitated origin from the ^{14}C-enriched surface waters. The ^{14}C content of whiting aragonite was identical to that found in aragonites on the lagoon floors. The whitings in the area must therefore largely reflect periodic episodes of **resuspension**. As we shall discuss in Chapter 2g, most of the aragonite in Bahamian lagoons is now thought to be of organic origin.

It was noted above that although sea water is distinctly supersaturated with respect to aragonite, calcite and dolomite, only the first mineral seems to precipitate directly from sea water. Why should this be so?

Numerous experimental and theoretical studies indicate that Mg^{2+} ions inhibit the growth of calcite. Very careful laboratory precipitation studies (Berner 1975, Berner *et al.* 1978) show (Fig. 2.4):

(a) Changing pCO_2 (for fixed degrees of supersaturation) has little effect upon the rate of calcite or aragonite precipitation.
(b) Dissolved Mg^{2+} in sea water has no effect upon the seeded precipitation of aragonite.
(c) Dissolved Mg^{2+} severely retards calcite precipitation.
(d) Calcite precipitated from sea water on *pure* calcite seeds contains 7–10% $MgCO_3$ as an overgrowth of high-Mg calcite.
(e) In Mg^{2+}-deficient 'sea water', with <5% of the normal Mg content, Mg^{2+} does not appreciably retard the seeded precipitation of calcite. Low-Mg calcite may thus be stable in freshwater regimes.

Two hypotheses may be advanced for these effects of Mg^{2+} upon calcite precipitation. First, the Mg^{2+} ions may act as a surface 'poison' which inhibits crystallisation. This may be due to the fact that Mg^{2+} ions are more firmly hydrated by polar water molecules than are Ca^{2+} ions. More thermodynamic work must be done to dehydrate these Mg^{2+} ions than to dehydrate Ca^{2+} for the growth of aragonite lattices (Lippman 1973, Bathurst

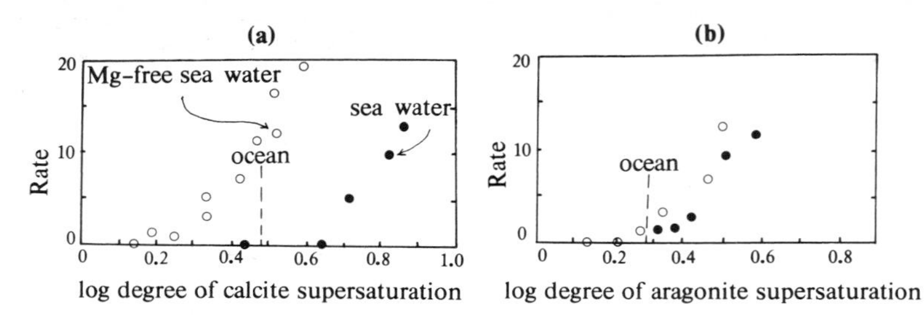

Figure 2.4 Rate of seeded precipitation (in arbitrary units) versus degree of supersaturation for (a) calcite and (b) aragonite in sea water and Mg-free 'sea water'. All rates initial. $-\log pCO_2 = 1.51$. $T = 25°C$. (After Berner 1975.) Note the evidence for retardation of calcite precipitation in sea water.

1968). Mg^{2+} has no effect upon the formation of aragonite lattices for structural reasons. Thus calcite growth is obstructed by the adsorption of firmly hydrated Mg^{2+} ions. No such adsorption occurs on aragonite crystals.

The second hypothesis states that uptake of Mg^{2+} ions into a calcite lattice, causing the growth of high-Mg calcite, causes a marked increase in the solubility of the Mg calcite over that of aragonite; therefore the latter polymorph is the preferred crystallisation product. Pure calcite is actually less soluble than aragonite (Table 2.2). The effect of increasing Mg content upon the free energies of formation of calcite, and hence upon solubility, is shown in Figure 2.5.

It may be argued that if the Mg:Ca ratio of sea water (Ch. 2c) in the geological past was much less than that at present then only calcite would have precipitated since the

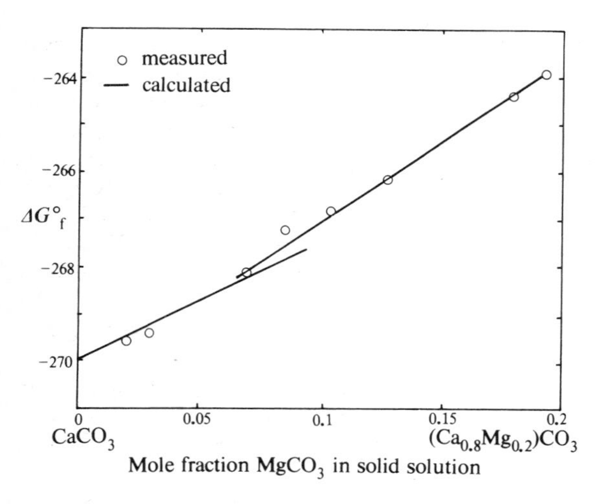

Figure 2.5 Standard free energy of formation of magnesian calcite (ΔG°_f) versus mole fraction of $MgCO_3$ in solid solution. The increase in solubility due to Mg-uptake makes the minimum degree of supersaturation necessary to precipitate calcite from sea water much higher than that which exists in the open ocean. (After Plummer & Mackenzie 1974, Berner 1975.)

above Mg-inhibiting mechanism could not occur. Although evidence is difficult to come by, the textures of some ancient, pre-Cretaceous, oöliths (Ch. 29) has recently tempted some authors to favour this hypothesis (Sandberg 1975).

The lack of dolomite precipitation in sea water must also result from an inhibition mechanism, since from the known composition of sea water dolomite should be the first mineral to precipitate (Hsü 1966; see Ch. 29). The very high degree of ordering in the dolomite lattice seems to result in extremely slow crystal nucleation and growth rates. Attempts at laboratory precipitation of dolomite result in the formation of more poorly ordered and metastable magnesian calcites of dolomite composition, known as **protodolomites**. A surface poisoning effect, when hydrated Mg^{2+} ions surround the Ca^{2+} growth planes of tiny dolomite nucleii, may impede growth much as postulated for the inhibition of calcite growth above.

To conclude this brief discussion of the reasons for the predominance of aragonite precipitation from sea water, it should be stressed that both calcite and dolomite can form in the *diagenetic* pore-water environment for various reasons to be dealt with later in the book (Ch. 29). The discussion above is restricted to *free precipitation from sea water*.

A final point concerns the reason why much surface sea water should be supersaturated with respect to aragonite, whilst inorganic precipitation of that mineral is comparatively rarely observed. The effect now seems to be due to the influence of organic compounds found in humic and fulvic acids or the phosphates that coat all potential particulate seed nucleii in the ocean with an organic phosphate monolayer, thus preventing crystallisation at normal seawater concentrations (Berner *et al.* 1978). These organic monolayers comprise organo-carbonate associations between carbonate and organic compounds such as amino-acids, fatty acids and fatty alcohols. With this control in mind, how precipitation is achieved in such areas as the Bahamas, the Persian Gulf and the Dead Sea is not certain at present. Similar sheaths of organic compounds coat aragonite within skeletal carbonates, thus protecting it from dissolution. If oxidised during diagenesis, the protective organic coats no longer exist and therefore aragonite is free to dissolve.

2f Carbonate grains of biological origin

The majority of carbonate particles are derived from the calcareous hard parts of invertebrates. A local biotic community will thus give rise to a characteristic **death assemblage** of calcareous debris. The resemblance between this death assemblage and the original faunal community will obviously reflect the degree of post-mortem physical, chemical and biological destruction and redistribution. The ease of identification of the calcareous hard parts depends upon the degree of physical and chemical breakdown. In particular, the smaller the particle the less precisely one can place it in a zoological classification. If the sediment is partly lithified or if it has become a rock, then thin-section analysis techniques must be used. Many particles are difficult to identify in this way because under thin section they show many different shapes. Consider the number of possible shapes that a thick-walled cylinder (analogous to a crinoid ossicle) might show if randomly sectioned; it is no wonder that many particles may be classified only to the order level of their appropriate phylum. Particular problems arise if the shell was originally aragonite, since diagenetic dissolution and calcite reprecipitation may totally destroy the original shell structure. However, it should be stressed that, after practice and with persistence, it is possible to do wonders with thin-section faunal identification. Carboniferous biostratigraphical zonation is, for example, increasingly dependent upon *species* identification of foraminiferida in thin section!

Organisms may secrete their calcareous hard parts as either aragonite, low-Mg calcite or high-Mg calcite. Some have multimineralic shells. The reasons why particular groups secrete certain minerals are not known. The differing chemical and mineralogical composition of carbonate hard parts clearly gives rise to a corresponding chemical/mineralogical **preservation potential.**

The nature of skeletal carbonate particles formed in a specific environment (e.g. reef, intertidal flat, shelf sand) has obviously changed through geological time in response to evolution. Thus a Silurian reef faunal and floral assemblage is distinct from a modern reef assemblage (Ch. 23) although the adaptations and exterior form of certain organisms may seem to have changed remarkably little. One practical consequence of faunal and community evolution is thus the vast number of different types of carbonate particle that the petrologist must deal

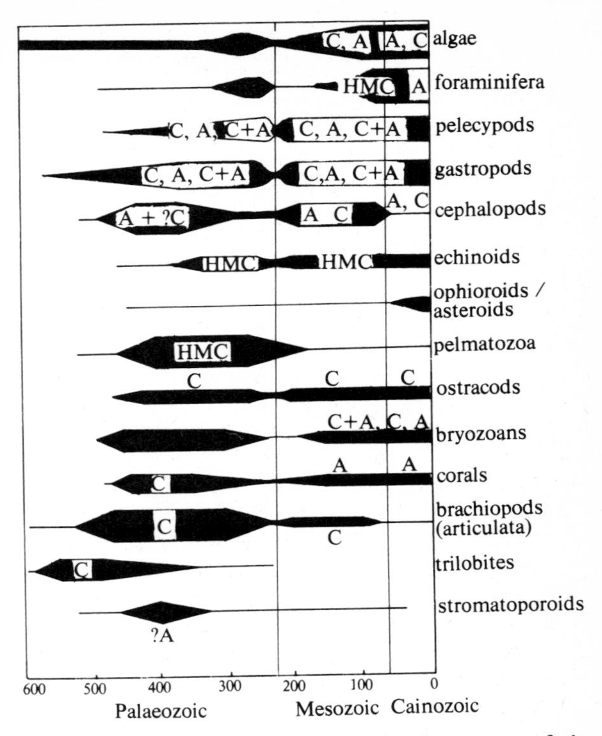

Figure 2.6 The distribution and relative importance of the various calcareous animal and plant phyla through geological time, the width of the bars representing relative importance as sediment contributors. C – calcite; HMC – high Mg-calcite; A – aragonite (after Lowenstam 1963).

with. In the following pages no attempt is made to catalogue all these types. Figure 2.6 summarises the major mineralogical composition and relative geological abundance of the commonest carbonate-secreting organisms. Figures 2.7 to 2.9 show a selection of the common grain types encountered in modern and ancient carbonate sediments. We shall now just briefly list a number of points concerning certain particle types; greater detail may be found in texts dealing specifically with such features (Horowitz & Potter 1971, Majewske 1969, Scholle 1978). Clearly, faunal identification should not be attempted without reference to standard palaeontology texts.

Mollusc shells comprise complex alternations of aragonite and calcite layers arranged in various geometrical ways with several characteristic microstructural types (see Bathurst 1975).

Brachiopod shells comprise wholly calcitic thin primary and thick secondary layers composed of fibres. A third inner prismatic layer may occur in some groups. Canals crossing the secondary layer may also occur in some groups. Normal to the shell wall, tissue-filled canals filled with calcite cement in fossils are termed

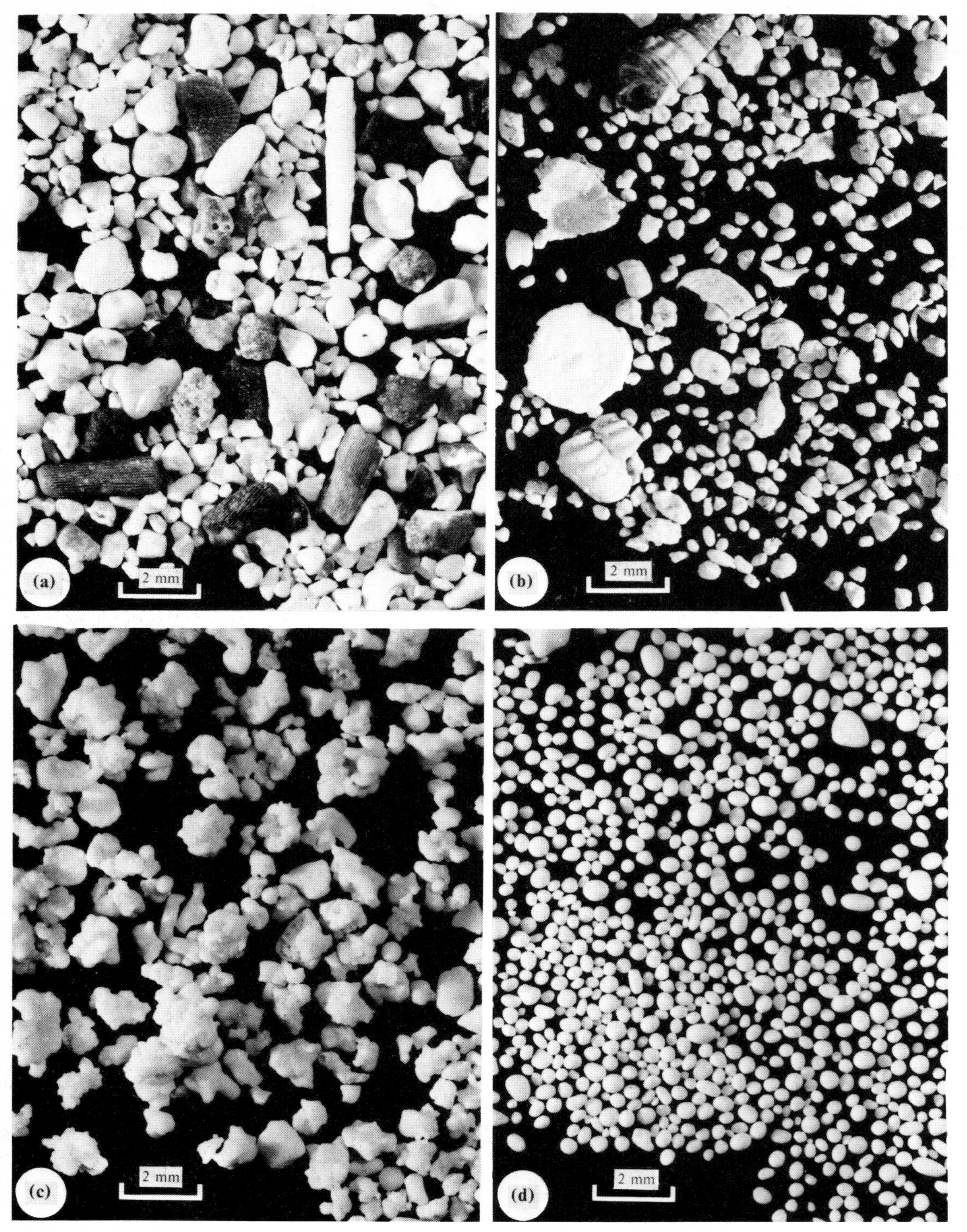

Figure 2.7 (a) Skeletal carbonate grains, comprising abraded molluscan, foraminifera, echinoid and algal fragments. (b) Skeletal–pelletal carbonate sand. Note ovoid pellets, abraded forams and gastropod. (c) Grapestone facies with aggregates composed of micritised skeletal grains and pellets. (d) Oölitic facies; note good sorting and high polish on individual oöids. All samples dredged from contemporary environments in the Bimini area, Bahama Banks (courtesy of R. Till).

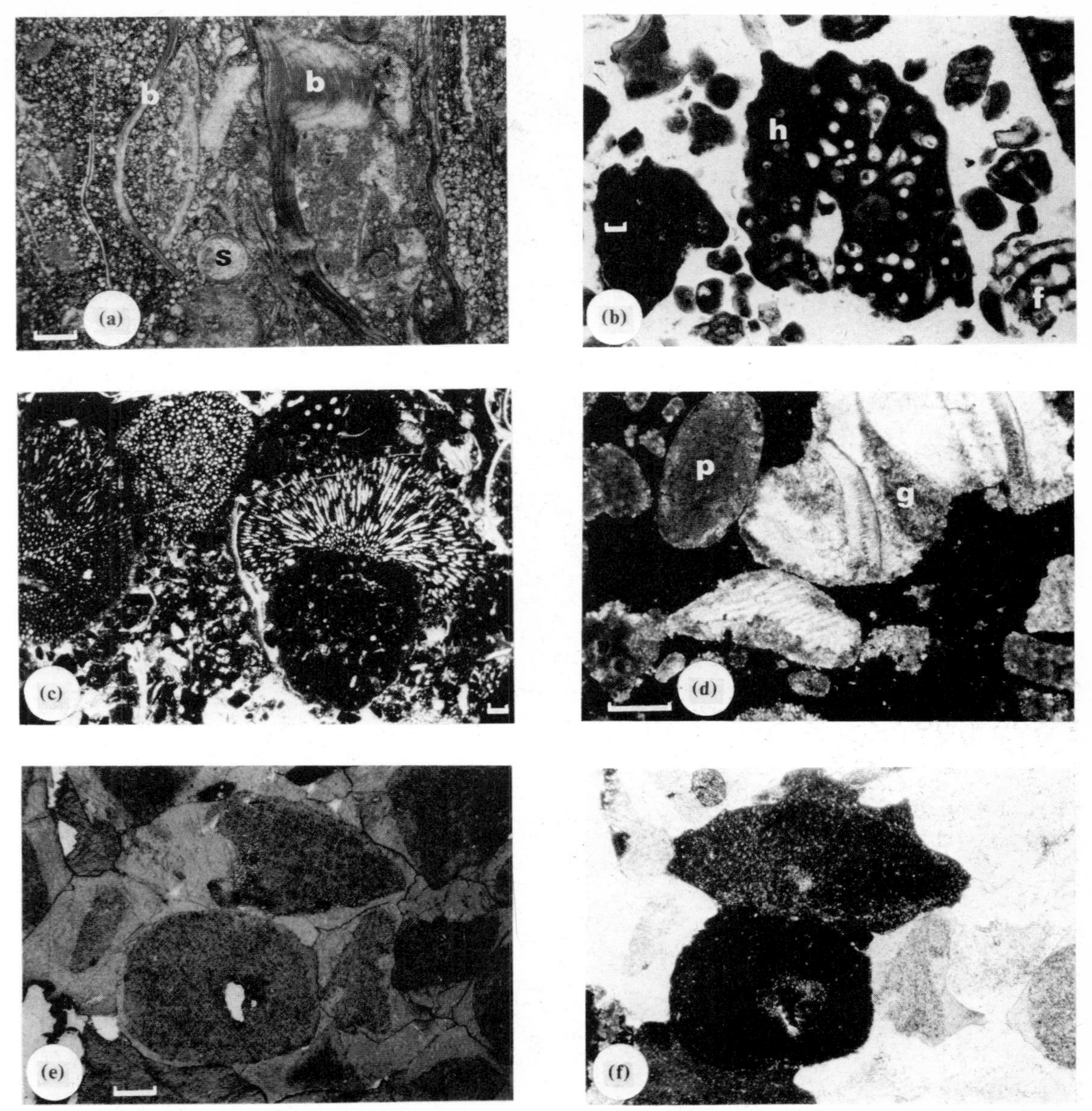

Figure 2.8 Carbonate grains in thin section. (a) Calcitic brachiopod shell debris (b) and spines (s). Dinantian of SW Scotland. (b) Large fragment of the alga *Halimeda* (*H*) with pellets, amorphous lumps and foram debris (f). Recent carbonate sands of Bimini lagoon, Bahamas (coll. R. Till). (c) Encrusting calcareous filamentous algae of the genus *Garwoodia*. Dinantian of Northumberland basin, England. (d) Micritised molluscan skeletal debris in grapestone grain (g) and ovoid structureless pellet (p) (crossed nicols). Recent carbonate sands of Bimini, Bahamas. (e & f) Views of echinoderm debris (in plane polarised light and under crossed nicols respectively) to show characteristic 'punctate' structure of calcite-infilled pores and the behaviour of individual fragments as single crystals of calcite. Note also the syntaxial rim cement (Ch. 29) in optical continuity with individual grains. Middle Jurassic of the Mendips, England. All scale bars are 100 μm.

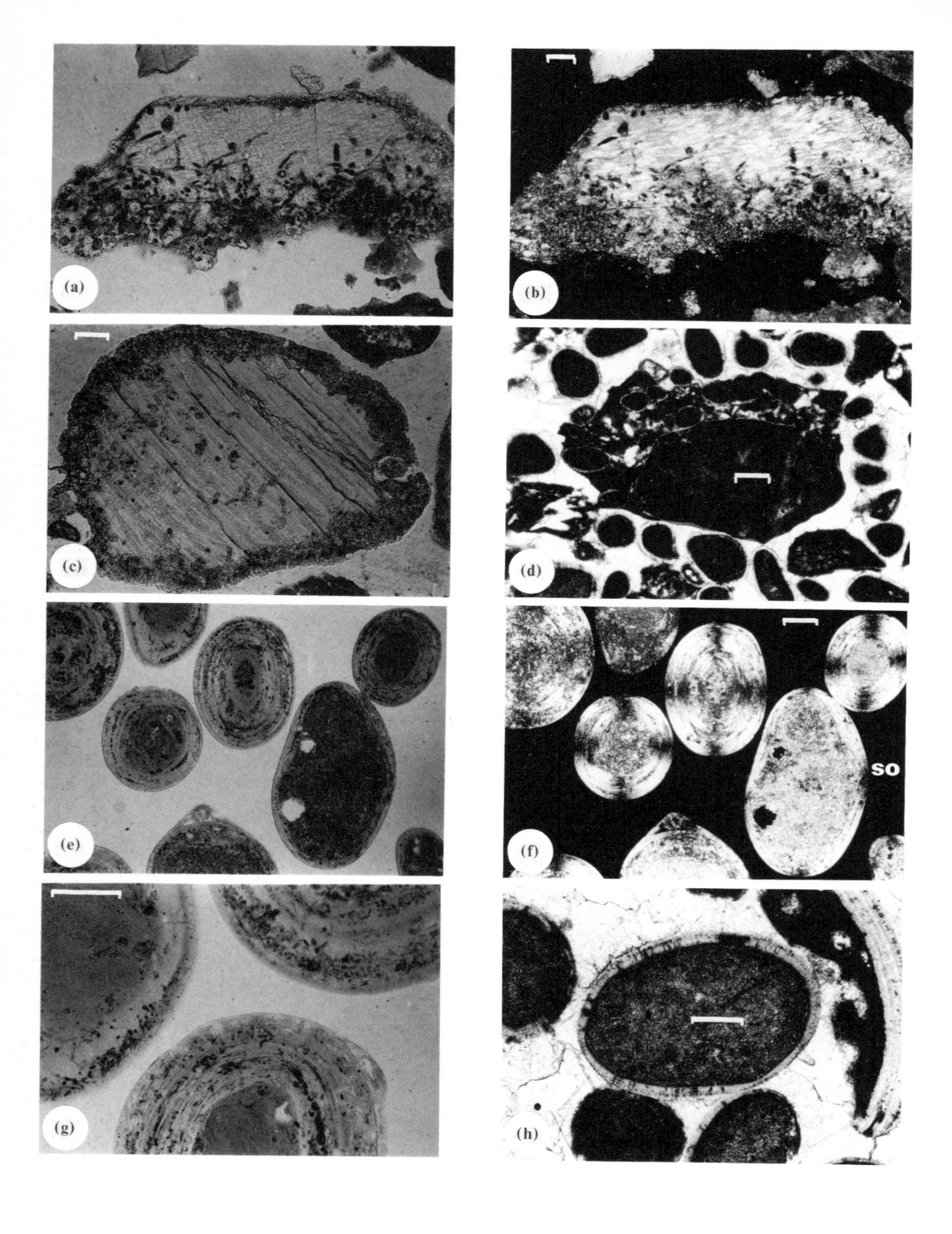

(a)
(b)
(c)
(d)
(e)
(f)
SO
(g)
(h)

endopunctae. Vertical cone-like deflections of secondary-layer fibres are **pseudopunctae**. Spines occur abundantly on some genera (Fig. 2.8a) and in certain orientations the careless observer may initially confuse them with oöliths in thin section.

Coral skeletons, including epitheca, tabulae, septae, etc., are made up of tiny calcitic (tabulata, rugosa) or aragonitic (hexacorals) fibres with various orientations.

Foraminiferal tests (Fig. 2.7a) are usually calcitic and they show agglutinating, microgranular, porcellanous, radial fibrous, spiculate and monocrystalline structures. These wall morphologies are, in fact, the primary basis for foraminferal classification at the sub-order level.

Echinoderm debris comprises fenestral calcite elements and has a very characteristic porous structure with individual particles behaving as single calcite crystals under crossed polars. Diagenetic infilling of pores by calcite cement still preserves this single crystal behaviour so that echinoderm debris in limestones may, with practice, be recognised at a glance under crossed polars (Fig. 2.8).

Calcareous benthonic algae show diverse structures. Erect frond-like forms such as the green alga *Halimeda* and the red alga *Lithothamnion*, the latter most important in temperate carbonates, break down initially into gravel-sized segments, the tubular internal structure of the former being especially characteristic (Fig. 2.8b). Many algae show a cellular internal structure. Calcification may occur within the individual cells, in the cell wall or outside the cell wall. After death, codiacean algae (e.g. *Penicillus* (Plate 1, p. 1)) may break down completely into needles of aragonite that may contribute much of the aragonite mud found in low-energy shallow environments in tropical or subtropical areas. Filamentous calcareous algae such as *Girvanella*, *Garwoodia* (Fig. 2.8c) and *Ortonella* are important encrusting agents in many ancient limestones, particularly Upper Palaeozoic and Mesozoic.

Calcareous planktonic algae are abundantly represented in many pelagic deposits from the Jurassic onwards. Originally, in life, the tiny circular to oval discs known as coccoliths surrounded a single algal cell. The calcitic coccolith plates are typically 2–20 μm in maximum diameter.

2g A skeletal origin for aragonite muds?

Regardless of the controversial chemical results concerning aragonite precipitation it is apparent that the net accumulation of tiny aragonite needles from the breakdown of benthonic calcareous algae can contribute significantly to known aragonite deposition rates in many areas of the Bahamas–Florida carbonate province. The organic origin of the aragonite is strongly supported, according to some authors, by oxygen isotope results (but see discussion in Bathurst 1975). Standing crops of *Penicillus*, *Halimeda* and other mud producers range from 0–30 plants per year per square metre in typical Floridan lagoons (Stockman *et al.* 1967). It is also known that the extensive 'aragonite' muds west of Andros Island (Great Bahamas Bank) contain a relatively large proportion of high-Mg calcite that suggests a source from the breakdown of red algae for some proportion of these muds. Earlier studies showed that *in situ* algal breakdown could account only for about 5% of deposited aragonite muds, but recent studies in Florida and the Bahamas lagoons (Stockman *et al.* 1967) suggest an overproduction of algal sediment and that the loss of aragonite as suspended plumes and its redeposition elsewhere are important features to take into account when estimating sediment fluxes (Neumann & Land 1975).

2h Micrite envelopes and intraclasts

Microscopic examination of skeletal fragments collected from quiet-water carbonate environments often shows that a dark rind surrounds their exteriors (Fig. 2.9a–c). Closer investigation shows that the rind is of variable thickness and intensity around different fragments. Individual tube-like cavities may be seen in the less advanced rinds and on the inner margins of the thicker rinds. The impression gained is that the rinds are formed by the coalescence of these tubes. The tubes may be empty or they may be filled with very fine-grained aragonite or high-Mg calcite. Fine-grained mud-size carbonate is termed micrite, but the term is usually restricted

Figure 2.9 Carbonate grains in thin section. (a & b) Molluscan fragment with well developed algal borings viewed in plane polarised light (a) and under crossed nicols (b). (c) Molluscan fragment with well developed micrite envelope formed by coalesced and infilled algal borings. (d) Compound intraclast comprising large amorphous lump and micrite-cemented peloids. Dinantian of Northumberland basin, England. (e & f) Modern oöids viewed in plane polarised light (e) and under crossed nicols (f). Note superficial oöid with pellet nucleus (SO) and pseudo-uniaxial crosses. (g) Close-up of oöid to show fine laminations and darker areas of organic mucilage and algal borings. (g) Ancient superficial oöid with pelletal nucleus and radial fabric in the oöid cortex. Dinantian of Northumberland basin, England. All scale bars 100 μm. Specimens (a, b, c, e, f & g) from modern carbonate environments around Bimini lagoon, Bahamas (coll. R. Till).

to calcitic micrite in other contexts (Ch. 29). Dissolution of the shell-fragment carbonate with dilute acid yields a gelatinous residue in which various types of blue–green algae may be identified. The tubes have thus been postulated to be the result of boring blue–green algae (fungal bores also exist) which rapidly infest all carbonate particles after deposition (Bathurst 1966). The reason for carbonate precipitation in the bores is poorly known but may be connected with the local high pH in the tube following CO_2 uptake by adjacent photosynthesising algae or by bacterial action on the organic residue left in vacated tubes. Algal bores found in temperate water carbonates do not show any carbonate precipitation (Gunatilaka 1976) even though the sea water may be supersaturated with respect to $CaCO_3$.

Micrite envelopes produced by **coatings** of filamentous endolithic algae are described by Kobluk and Risk (1977). Rapid precipitation of low-Mg calcite occurs on and within dead filaments which project from the substrate into the sea. Coalescence of the dead calcified filaments could produce an algal micrite envelope that is accreted on to a grain periphery previously bored and micritised by Bathurst's mechanism described above.

By examining further material it soon becomes apparent that the **micrite envelopes** formed by boring algae may extend inwards to include the whole shell fragment. An **amorphous lump** particle is thus produced, with no tell-tale shell structure left to testify to its original shell fragment origin (Bathurst 1966). A high proportion of such angular lumps make up the sediment in certain Bahamian and Persian Gulf lagoons. Such particles must, in fact, be defined as belonging to the class of carbonate particles known as **intraclasts**. The class is usually defined as carbonate particles that have been reworked into the basin of deposition (Folk 1962). Micritic amorphous lumps formed by micritisation are *not* true intraclasts but *must* be included in the term since they are indistinguishable from true reworked intraclasts from other sources. Intraclasts are thus a highly diverse group of particles. They include reworked beach-rock fragments (Ch. 29), hardground debris (Ch. 29) grapestones (Ch. 2k), older lithified carbonate particles and amorphous lumps of skeletal origin. The reader should note, however, that the slightest trace of a skeletal remnant structure within the lump is enough to classify the particle as skeletal debris. Careful observation and patient thought should also be applied to the firm identification of a carbonate rind as a micrite evelope – could the rind be an algal coating? How could one tell the difference?

2i Pellets and peloids

In quiet-water lagoons with aragonitic mud substrates, large numbers of molluscs (chiefly gastropods and bivalves) worms and crustaceans are continually ingesting the organic-rich muds to feed upon the nutrients. The mud is excreted as ovoid faecal pellets ranging from 0.1 mm to 3 mm long and 0.05 mm to 1 mm broad. Length:breadth ratios are in the range 1.5–3. In section the pellets are dark, fine-grained and structureless (Figs 2.7b & 2.8d). They are rich in organic matter and are soft when fresh. Older pellets are fairly hard, probably caused by rapid interstitial carbonate precipitation. Winnowing of the aragonite mud may cause a sand-grade deposit of pellets to be produced, as in lagoons of the Persian Gulf (Ch. 23).

Faecal pellets in ancient carbonate rocks may be very difficult to tell apart from abraded amorphous lumps produced by micritisation of shell fragments and other processes, including complete micritisation of oöliths. For this reason it is best to use the term **peloid** for any structureless ovoid micrite particle unless the genesis may be deduced by other observations.

2j Oöliths

No other carbonate particle has received greater attention from sedimentologists than the humble oölith. There are literally hundreds of papers on various aspects of oölite genesis and diagenesis. We shall concentrate on recent oöid genesis in this section and postpone the involved discussion of diagenesis to a later chapter (Ch. 29). Oöliths are spherical to slightly ovoid, well rounded carbonate particles possessing a detrital nucleus and a concentrically laminated cortex (Figs 2.7d & 2.9 e & f) of fine-grained aragonite or high-Mg calcite often with many thin organic mucilaginous algal layers. They occur in sandwave and dune complexes in areas of strong tidal currents (Ch. 23) or as littoral or shallow sublittoral beach deposits. Oöid diameters range from 0.1 mm to 1.5 mm. When the cortex is very thin, comprising one or two thin aragonite laminae, the oöid is said to be a **superficial** form. Usually the individual aragonite laminae are up to about 10 μm thick. Many sections through unaltered aragonite oöids show a pseudo-uniaxial cross under crossed polars, which gives a negative figure. This means that the predominant alignment of aragonite crystals must be with their *c*-axes tangential to the oöid surface. This is confirmed by X-ray and electron microscope analysis, the latter technique revealing that the aragonite crystals are in the form of rod-like particles 1–2 μm long with flattened end terminations (e.g. Loreau & Purser 1973). Some oöliths, notably those from the Great Salt Lake (USA) (Kahle 1974, Sandberg 1975, Halley 1977) exhibit spectacular radial aragonite microstructure and also certain unorientated lamellae which may be rich in clay minerals. In the

Persian Gulf, oöliths from high-energy environments have a well developed concentric arrangement of individual aragonite rods, whereas those from more sheltered areas have a radial orientation with looser packing (Loreau & Purser 1973).

Regarding the vexed question of oölith genesis we may be sure that high-energy environments somehow encourage tangential aragonite growth whilst low-energy environments (Salt Lake, Laguna Madre, Persian Gulf sheltered environments) encourage radial growth. Successful laboratory precipitation of alternating oölitic carbonate and organic laminae around spherules may occur where the concentration of organic material as humates is high and where organic membranes may form around suitable nucleii (Suess & Futterer 1972, Ferguson *et al.*, 1978, Davies *et al.* 1978). Experimentally produced carbonate laminae show radial aragonite growth under 'quiet' laboratory conditions and tangential fabrics after agitation. It is possible that aragonite crystals seed out from the oöid organic matrix to grow radially outwards, as in the growth of any uninterrupted crystal fabric from a solid surface. Such radial fabrics would be modified by turbulence and periodic abrasion into tangential fabrics in higher-energy environments. Sorby's original snowball analogue stated that aragonite needles simply stuck on to the rolling exterior of an oölith. The lack of evidence for a sticky cohesive matrix on the smooth polished oöid outer surfaces has made this 'snowball' mechanism difficult to accept.

The source of the aragonite needles remains a great problem since it is difficult to imagine such tiny crystals occurring in the bedload layer of such high-energy water bodies as tidal flows over oölite shoals. One possible solution to the problem might be the recognition that the oöliths spend much of their time stranded in the inside of moving bedforms such as sandwaves, dunes and ripples (Davies *et al.* 1978). Organic coats may develop here, causing radial aragonite laminae to begin to seed out from the pore waters between the oölite grains. Re-emergence of the grains into a turbulent bedload layer would then cause mechanical alignment of the aragonite rods into a tangential arrangement or would encourage tangential growth. Oöliths in quiet waters would remain with a radial structure. Periodic entrainment and burial also explain the development of successive concentric layers.

By way of contrast to the above approach we may note Deelman's (1978) successful formation of Bahamian-type tangentially orientated oöliths from laboratory experiments with bicarbonate solutions. Intermittent stirring caused aragonite needles to precipitate around nucleii by a process akin to collision breeding. Further agitation caused a high polish of the oöid surfaces and caused the aragonite needles to become well rounded at their edges. The tangential aragonite crystals were held together by 'normal' bonding forces. These important results match well with the observed characters of natural oöids and of their physical environment of formation. The mucilaginous laminae present in natural oöids were not observed in the experimental examples, the solutions being simple bicarbonate ones. Doubtless these must form as the oöids undergo periods of 'rest' and become infested with blue–green algal slime.

2k Grapestones

Large areas of the Bahamas Banks are covered with composite particles that comprise cemented aggregates of grains (shell fragments, oöids, pellets) resembling microscopic bunches of grapes (Fig. 2.7c). Intense micritisation by infilled algal bores usually obscures any original internal structure of the particles. It is thought that the cementation, by micritic aragonite, and the intense micritisation, reflect a mode of growth within a subtidal blue–green algal mat (Ch. 23) that stabilises the substrate in the face of tidal and wave currents. Periodic mat rip-up during storms then yields grapestone aggregates. Grapestones are thus clearly a type of intraclast, as defined previously.

2l Polygenetic origin of carbonate grains

We have now considered the five main types of recent carbonate grains: aragonitic needles, skeletal fragments, oöliths, peloids and intraclasts. The last four grain types are often termed **allochems** in contrast to the (supposed) inorganic chemical precipitation of aragonite by orthochemical mechanisms. Aragonite needles are analogous to clastic mud composed of clay minerals, whilst the allochems provide the particles of silt, sand and gravel size (Ch. 4). We have seen above how aragonite needles may be of precipitated or algal origin, and how bored and micritised skeletal grains become amorphous lumps. Abraded amorphous lumps of skeletal origin, micritised oöliths and faecal pellets may be indistinguishable and must be grouped together as peloids in many cases. Our conclusion is simple: examples of a given particle type may have quite distinct origins, i.e. the particles are **polygenetic**.

2m Shallow, temperate-water carbonates

Temperate-water carbonates lack the oölith, intraclast and pellet allochems and the contribution from green algae and corals that characterises warm-water carbonates (Lees 1975). Algal bores in temperate carbonates appear to remain unfilled; therefore micrite envelopes should be uncommon, a point that has some relevance to the preservation of skeletal aragonite in such carbonates. The absence of the above allochems and of pore-infilling, together with widespread shell dissolution (Alexanderson 1976) presumably reflects a lower degree of seawater supersaturation or even the occurrence of undersaturation in temperate waters.

2n $CaCo_3$ dissolution in the deep ocean

The world map of the distribution of $CaCO_3$ deposits (Fig. 2.1) shows a remarkable coincidence between the crests and flanks of mid-ocean ridge systems and the occurrence of pelagic calcareous oozes formed by coccoliths and forams. Detailed mapping shows that carbonate sediments are rare below a water depth of about 5 km in the Atlantic and about 3.5 km in the Pacific. This distribution depends only partly upon the unusual property of $CaCO_3$ that it is more soluble in cold than warm water. The temperature of oceanic water decreases very rapidly to about 5°C at around 1000 m and thereafter more slowly to reach a minimum of around 2°C. More important is the increase of pressure with water depth which causes increased pCO_2 and thus low pH. Both of these effects cause the oversaturation of sea water with respect to $CaCO_3$ to decrease with increasing depth. Aragonite, the more soluble polymorph, becomes undersaturated long before calcite, being undersaturated at about 500 m in the Pacific ocean and 2 km in the Atlantic ocean (Fig. 2.10). Thus pelagic organisms with aragonite skeletons, like the **pteropods** (a group of tiny gastropods), are much more prone to dissolution than are the calcitic coccoliths. Calcite becomes undersaturated at depths of between 400 m and 3500 m in the Pacific and between 4000 m and 5000 m in the Atlantic (Fig. 2.10). These zones where $CaCO_3$ approaches undersaturation for aragonite or calcite are known as **carbonate saturation depths** (CSD). Whether or not calcitic organisms can survive as debris below the CSD will be a complex function of the rate of fall of the grain through the water column versus the rate of $CaCO_3$ dissolution. The rate of dissolution thus depends partly on skeletal size. Solution inhibition by adsorbed molecules may also be important (see below).

The variation of CSD between oceans is largely a function of pH. Pacific ocean water contains much larger quantities of organic debris than does Atlantic ocean water. Oxidation of this debris produces additional H^+ ions, some of which are used up in the buffer reactions (2.3) and (2.4) noted previously (p. 18); the small remainder not balanced in this way increase the deep-water acidity. Calcium carbonate is thus dissolved and is less abundant in Pacific than in Atlantic sediments. It has been calculated that about 80% of the $CaCO_3$ produced

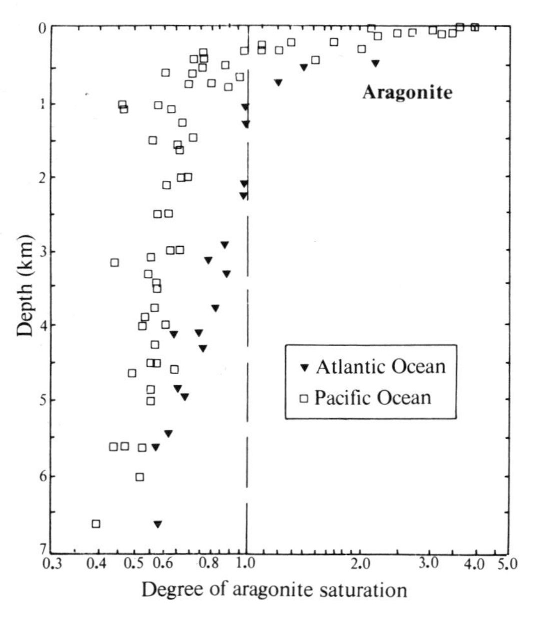

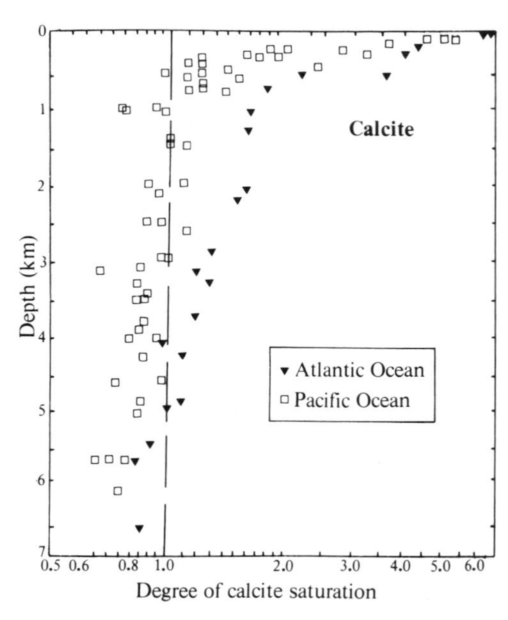

Figure 2.10 The degree of aragonite and calcite saturation as a function of depth in the Atlantic and Pacific Oceans (after Broecker 1974). Discussion in text.

in the warm sunlit areas of ocean waters by planktonic organisms is destroyed by dissolution.

In addition to defining a CSD we may also define critical depths where calcareous bottom sediments in the oceans are affected by dissolution. Depths at which calcareous shells show a rapid acceleration in dissolution, as determined by microscopic examination, are termed **lysoclines** (Berger 1971). Depths where $CaCO_3$ disappears from bottom sediments are termed **carbonate compensation depths** (CCD). Both lysoclines and CCDs fall below the carbonate saturation depth and therefore lie within the zone of undersaturation. $CaCO_3$ dissolution is greatly accelerated at a critical degree of undersaturation, the value of which depends upon the organic and phosphate concentration of the deep sea water. Adsorbed phosphate and organic ions act as inhibitors to dissolution (Chave & Suess 1970) and it is not until these are removed in the deep ocean that dissolution can proceed rapidly (Morse & Berner 1972, Takahishi 1975, Berner 1976).

2o Summary

Surface subtropical sea water is supersaturated with respect to calcite, aragonite and dolomite, but kinetic factors intervene to restrict carbonate precipitation. Most shallow-water tropical carbonate particles are thus directly or indirectly of biological origin and are made up of varying proportions of calcite, high-Mg calcite and aragonite. Carbonate dissolution in the deep sea occurs at great depths in response to a rapid decrease in adsorbed organic compounds with depth in the general zone of $CaCO_3$ undersaturation; thus again carbonate equilibria are dominated by kinetic effects.

Further reading

Useful elementary accounts of the physical chemistry of sea water and of carbonate reactions in sea water are to be found in Broecker (1974), Krauskopf (1979) and Bathurst (1975). A more advanced approach is given by Berner (1971). Bathurst (1975) also gives the most useful descriptions and illustrations of the major carbonate grain types. A more detailed account of invertebrate skeletal composition and structures is given by Horowitz and Potter (1971) and Majewske (1969), whilst Scholle (1978) provides an illustrated guide in colour to carbonate rock constituents, textures, cements and porosities.

Appendix 2.1 Staining and peel techniques

Calcite, Fe-calcite, dolomite and Fe-dolomite may be recognised by the following technique used upon either uncovered thin sections or polished blocks prepared for acetate peel making.

Dissolve 1 g alizarin red-S and 5 g of potassium ferricyanide in 1 l of 0.2% HCl (made up with 998 ml *distilled* water and 2 ml conc. HCl). *N.B. The working life of this stain is about one day only.* Etch uncovered thin section or polished block in 2% HCl (made up with 98 ml *distilled* water and 2 ml conc. HCl), for 20 s. Wash immediately in distilled water. Immerse section or block in staining solution for 4 min. Remove and *gently* wash with distilled water. Allow thin section to dry before cover slip is glued on. Many workers find it useful to stain only half of a thin section since the stain may overprint and obscure some fine petrographic detail. Calcite stains pink by the above method with the colour change pink–mauve–purple–blue indicating increased Fe^{2+} concentrations up to about 5% (see details in Lindholm & Finkelman 1972). Dolomite does not take a stain (beware of confusion with quartz!) and Fe-dolomite stains turquoise.

High-Mg calcite in recent carbonates may be preferentially stained using a solution of 500 ml distilled water, 0.5 g Clayton yellow, 4 g NaOH and 2 g EDTA together with a fixer of 20% NaOH. The sample is etched for 20 s in 5% acetic acid, dried and immersed in the staining solution for 20 min. After air drying, the samples are immersed in the fixer for 30 s prior to final drying and covering. High-Mg calcite stains red to light pink.

Aragonite in recent carbonates may be stained with Feigl's solution. Add 1 g Ag_2SO_4 to a solution of 11.8 g $MnSO_4 \cdot 7H_2O$ in 100 ml distilled water. Boil. Cool. Filter the suspension and add two drops dil. NaOH. Filter off the precipitate after 1–2 h and keep solution in dark bottles. Aragonite stains black whilst calcite and dolomite are unaffected.

Peels are taken from *damp* stained surfaces or from *dry* surfaces on unstained polished blocks. Immerse block surface (polished to at least 600-grade carborundum) in acetone, remove the drenched block and gently roll on to the surface a sheet of thin acetate film, starting from one corner of the block. Tilting of the block encourages excess acetone to run off prior to acetate fixing, but do not let the acetone on the block surface dry off prior to acetate emplacement. Set block aside for 10 min and then gently unpeel the acetate. A perfect reproduction of the limestone (and its stain) is then obtained. The beginner is advised to practise the method. Once the technique is mastered, a very rapid production rate can be achieved. Keep the finished peels pressed between glass plates fixed around the margins with masking tape. Initial pressing under a load will stop the peel curling.

3 Evaporites, biogenic silica, and phosphates

3a Evaporites

Evaporitic salts are precipitated from natural waters upon evaporation. The main evaporite mineral phases are listed in Table 3.1. In this section we shall concentrate upon marine evaporites precipitated from standing brine bodies of marine origin. Further considerations of diagenetic evaporites and evaporite facies can be found in Chapters 16, 23, 26 and 30.

Normal sea water is undersaturated with respect to all evaporitic salts (Table 3.1). Note that halite is much more undersaturated than gypsum or anhydrite so that any evaporation will cause gypsum to precipitate before halite. The most undersaturated salts are the complex series of K-salts and these represent the final precipitates from highly concentrated brines. Figure 3.1 shows schematically the sequence of salts that might precipitate as sea water is progressively concentrated by evaporation. Figure 3.2 shows that the ideal sequence of salts precipitated from sea water differs somewhat from the actual sequences recorded in evaporite deposits in the geological record. The latter show increased proportions of $CaSO_4$ and decreased proportions of Na–Mg sulphates compared with laboratory studies. Mg-depletion occurs by a combination of dolomitisation and clay mineral precipitation. Important metasomatic effects by percolating bitterns arise during the last stages of brine concentration. Also, influxes of sea water may cause dissolution and reprecipitation effects.

Anhydrite has never been observed to precipitate directly from sea water. The equilibrium constant for the reaction

$$\underset{\text{gypsum}}{CaSO_4 \cdot 2H_2O} \rightleftharpoons \underset{\text{anhydrite}}{CaSO_4} + 2H_2O \qquad (3.1)$$

is given by the activity of water, a^2H_2O, and enables stability fields for gypsum and anhydrite to be plotted from experimental results (Fig. 3.3). In highly saline brines, gypsum is still precipitated, but as a metastable phase which may subsequently alter to anhydrite. This process occurs in sabkha evaporites (Chs 23 & 30) but primary precipitation is also suspected here from textural evidence. As discussed in Chapter 30 gypsum buried to depths greater than about 1 km becomes unstable and transforms to anhydrite.

It has long been recognised that simple evaporative concentration of sea water is insufficient to produce the great thicknesses of evaporite salts observed in the geological record. For example, it may be calculated that complete evaporation of the world oceans will yield a mean thickness of only 60 m of evaporites. Some ancient evaporite successions of great areal extent may reach over 1 km in thickness. Major periods of evaporite production must have drastic short-term effects upon levels of sea-

Table 3.1 Chemical composition and selected values of IAP (ion activity product) and K (solubility product) for halite, gypsum and anhydrite in seawater solutions (after Berner 1971).

Mineral	*Formula*	*IAP*	*K*
halite	$NaCl$	0.12	38
gypsum	$CaSO_4 \cdot 2H_2O$	4.6×10^{-6}	2.5×10^{-5}
anhydrite	$CaSO_4$	4.6×10^{-6}	4.2×10^{-5}
sylvite	KCl		
carnallite	$KMgCl_3 \cdot 6H_2O$		
polyhalite	$K_2MgCa_2(SO_4)_4 \cdot 2H_2O$		

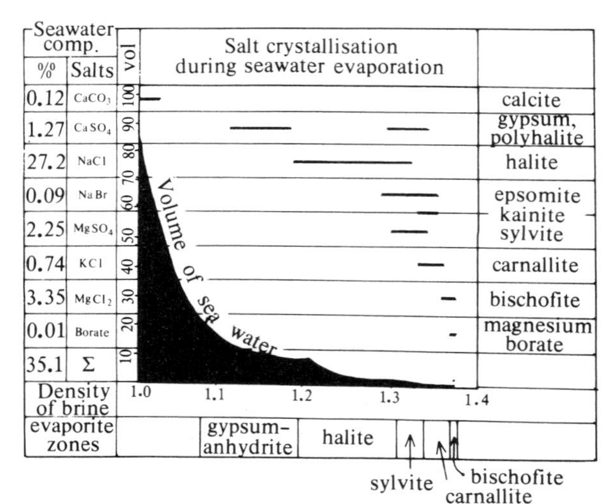

Figure 3.1 To show the effects of seawater evaporation upon brine volume, density and type of salt precipitate (after Valyashko 1972).

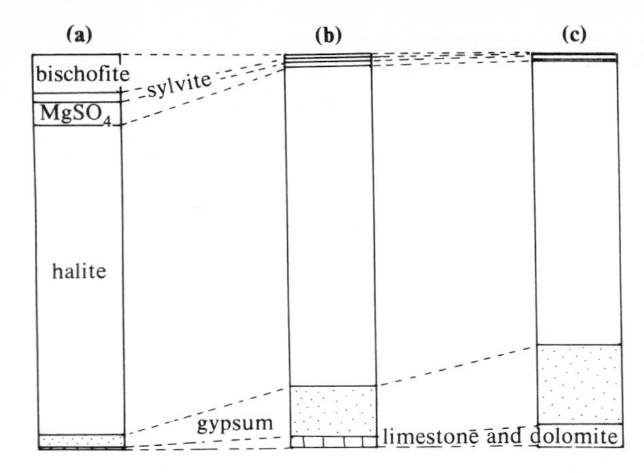

Figure 3.2 Comparative precipitation profiles from (a) experimental evaporation of sea water, (b) from a Zechstein evaporite sequence in Germany and (c) from the average of numerous other marine evaporite sequences (after Borchert & Muir 1964).

water salinity before a steady state is once more established. Thus the total volume of salts in the world's oceans is about 2.2×10^7 km^3 (Borchert & Muir 1964). The Permian Zechstein evaporites of NW Europe have a total volume of about 2.4×10^6 km^3, some 10% of the oceanic reservoir!

The simplest model for sub-aqueous evaporite formation is the **shallow-water barred basin** (Fig. 23.28) where evaporation proceeds in semi-isolation, with replenishment of sea water over a restrictive entrance sill or obstruction. Brines are thus progressively concentrated, and progressive crystallisation of the various salts then proceeds. Return flow of brines to the ocean reservoir may occur at depth so that basin brines may stay at a particular concentration for long periods. In this way abnormally thick sequences of sulphates or chlorides may accumulate. Cycles of evaporites that approximate reasonably closely to the 'ideal' cycle (Fig. 3.2a) result when basinal brines are evaporated to completion. Lateral changes in evaporite composition are to be expected in barred basins since incoming sea water will precipitate first gypsum and then halite as it spreads over the sill towards the basin shallows. Sub-aqueous evaporite facies are further discussed in Chapter 23.

One further point concerns the presence of small-scale evaporite rhythms on a millimetre to centimetre scale observed in many ancient evaporites. These may comprise alternating (a) dolomite–anhydrite + clastic clay (Fig. 3.4), (b) clay–dolomite–anhydrite + halite and (c) halite + sylvite–carnallite.

The clay interlaminae in types (a) and (b) are interpreted as a clastic influx during rainy seasons. The inflowing waters 'freshen' the brine body and this, together with

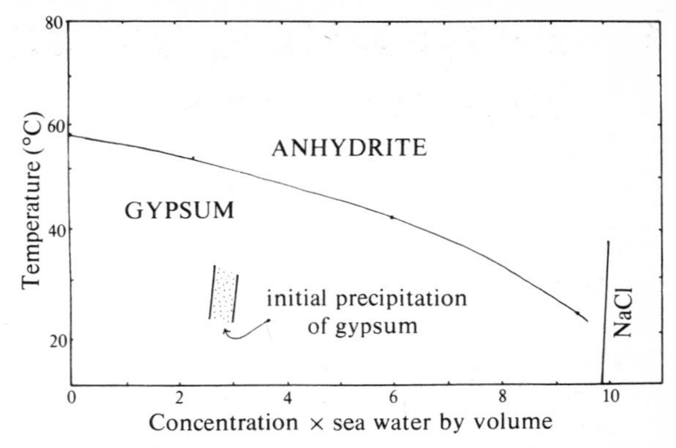

Figure 3.3 Phase diagram to show the stability fields of gypsum and anhydrite with respect to brine temperature and seawater concentration (data of Hardie 1967, as recalculated in terms of concentrations (from aH_2O) by Blatt *et al.* 1980).

cooler air temperatures, causes either cessation of evaporite precipitation or precipitation of a less undersaturated phase. The laminae thus represent annual varves by this interpretation (Richter-Bernberg 1955). The very rapid deposition rates in modern evaporating brine bodies (10–50 mm a^{-1} for halite) are far in excess of any reasonable subsidence rate and they encourage the belief that the thick accumulations of salts in ancient basins must have been precipitated in initially relatively deep brine bodies ($\pm$1 km for the Permian Zechstein).

3b Biogenic silica

Primary siliceous grains are provided to the ocean floor upon the death of the radiolarians and diatoms that live

Figure 3.4 Sub-aqueous evaporite precipitates: varved anhydrite/dolomite clay from the late Jurassic Hith Formation of Saudi Arabia. Note reworked varved clast at top of specimen.

in great numbers in the photic zones of the ocean and utilise silica in their tests. Sea water is vastly undersaturated with respect to amorphous silica. Thus for the reaction

$$\underset{\text{amorphous silica}}{SiO_2} + 2H_2O \rightarrow \underset{\text{silicic acid}}{H_4SiO_4} \qquad (3.2)$$

$K = aH_4SiO_4 = 2 \times 10^{-3}$ and the ion activity product (IAP) is 2×10^{-4} to 1×10^{-6}. Most dissolved silicon occurs in the H_4SiO_4 form, further ionisation to $H_3SiO_4^-$ being very limited ($K = 10^{-9.9}$). Despite this great undersaturation, siliceous plankton extract much silica from sea water and construct their skeletons of the amorphous opal-A form. This biogenic extraction, combined with partial dissolution of skeletal material at depth, causes a rapid increase of dissolved silica with depth in the oceans.

The main areas of siliceous oozes on the ocean floors correspond to areas of high productivity where ocean current divergence or upwelling causes fertile deep waters to rise into the warm photic zones (Fig. 3.5; see also Chs 24 & 26). An outline of the silica cycle in the oceans is shown in Table 3.2 and the reader is referred to Chapter 30 for a discussion of silica diagenesis.

3c Phosphates

Phosphates form only a minority of sedimentary rocks, the average P_2O_5 content being only 0.15%. Concentrations of phosphorus in sea water average about 0.07 ppm. Despite these low concentrations phosphorus (P) is a very important element, being an essential component of all living cells. Increased use of fertilisers and exploitation of natural phosphate reserves have inevitably focused sedimentological attention on the origins of phosphate rock.

Phosphorus occurs as phosphate ions in sea water. The warm surface waters of the ocean contain only 0.003 ppm of P because this is the region of maximum photosynthetic activity and is thus a zone of active P uptake by plankton. The deeper cold waters may contain 0.1 ppm P and they define a zone of regeneration where P is returned to solution as organic excrement and where dead organisms are dissolved and oxidised. The most favourable conditions for phosphate precipitation occur on shallow continental shelves (e.g. SW Africa) and oceanic plateaux where upwelling of deep ocean water occurs in response to Ekman transport (Chs 24 & 26). These nutrient-rich waters encourage a fantastic plankton productivity which adds to the phosphate reservoir as the cold upwelling waters move across the sloping shelf or plateau surface. Phosphate is precipitated as a **fluorapatite** phase or as a replacement after $CaCO_3$ in response to a shoreward increase in temperature, pH and salinity (Gulbrandsen 1969). Phosphate precipitation is prevented in deep oceanic water masses because of high pCO_2, despite the fact that these waters are approximately saturated with respect to calcium phosphate. The above model for phosphate precipitation, (due to Kazakov) is based upon a number of lines of chemical evidence and information on the location of pre-Recent phosphate deposits. However, it should be stressed that phosphate precipitation does not seem to be common off modern upwelling coasts. Thus, for example, the extensive phosphates off SW Africa are largely Quaternary and Tertiary relics. The reasons for this paucity of modern phosphate production are poorly understood.

Riggs (1979) has divided phosphate rocks into orthochemical and allochemical groups. Orthochemical phosphates, partly of diagenetic origin, are basically phosphate muds formed *in situ* by physicochemical or biochemical mechanisms. Periods of mud precipitation may coincide with mass mortality events involving siliceous microplankton since many phosphates have a close link with diatomite deposits. Precipitation close to or on the sea floor may be aided by bacteria whose cells may occur in great numbers within the muds. The phosphate muds are frequently pelleted and bioturbated by a burrowing infauna. Allochemical phosphates comprise pelletal, intraclast, skeletal and oölitic particles formed by reworking of semi-lithified phosphate muds or by calcium phosphate replacement of calcitic allochems.

The efficiency of the latter replacement process is also illustrated by the guano deposits of oceanic islands in which phosphate from seabird excrement extensively replaces limestone (Braithwaite 1968).

There is a close link between phosphate occurrence and the complex clay mineral **glauconite** (Ch. 30), both tending to occur at horizons of reduced sedimentation. Many pelletal phosphates have a high glauconite content (Fuller 1979).

3d Summary

Evaporites form as salts which are precipitated from sea water during evaporation. Anhydrite is not usually precipitated from brine solutions, gypsum coming out instead. Major evaporite rhythms record the progressive increase in salinity of a brine body with time. Thick deposits of a single salt indicate that the brine body is in dynamic equilibrium with an ocean-water source. Minor varve-like

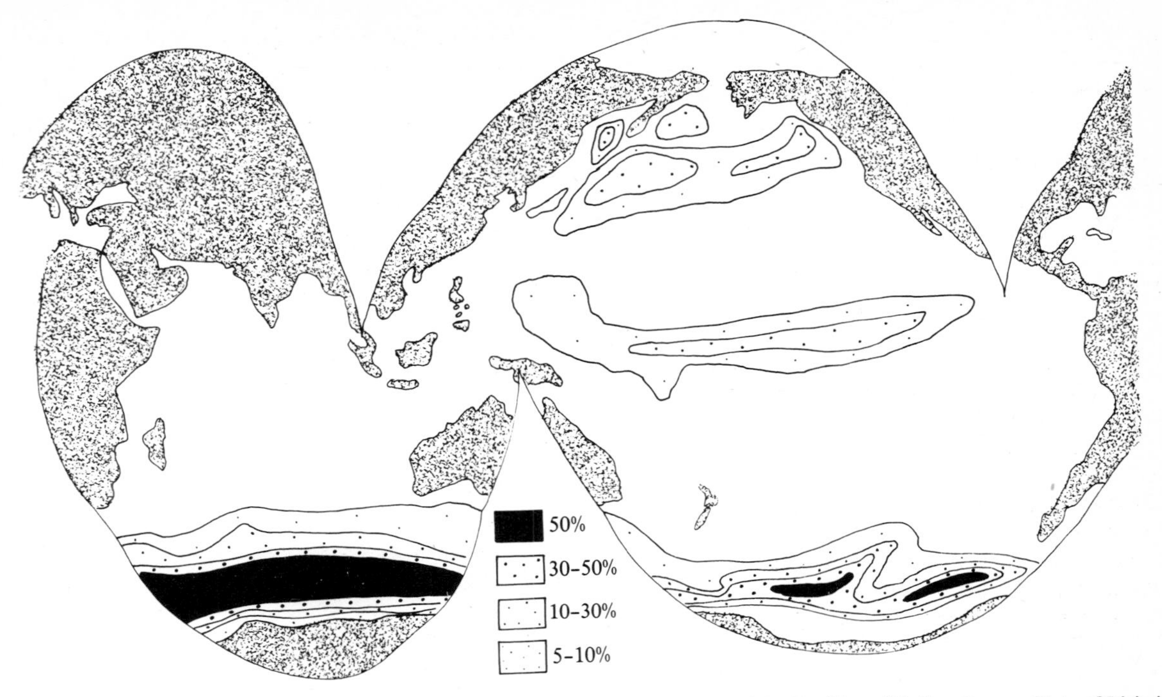

Figure 3.5 Distribution and concentration of biogenic opal in the surface sediments of the Pacific and Indian Oceans (data of Lisitzin 1967 as modified by Calvert 1974). Siliceous-rich oozes on the sea floor are strongly correlated with areas of high surface productivity (see Chs 24 & 26).

Table 3.2 The oceanic silica cycle, with the magnitudes of dissolved silica in 10^{13} g SiO_2 a^{-1} (from Heath 1974, as modified by Riech & Von Rad 1979).

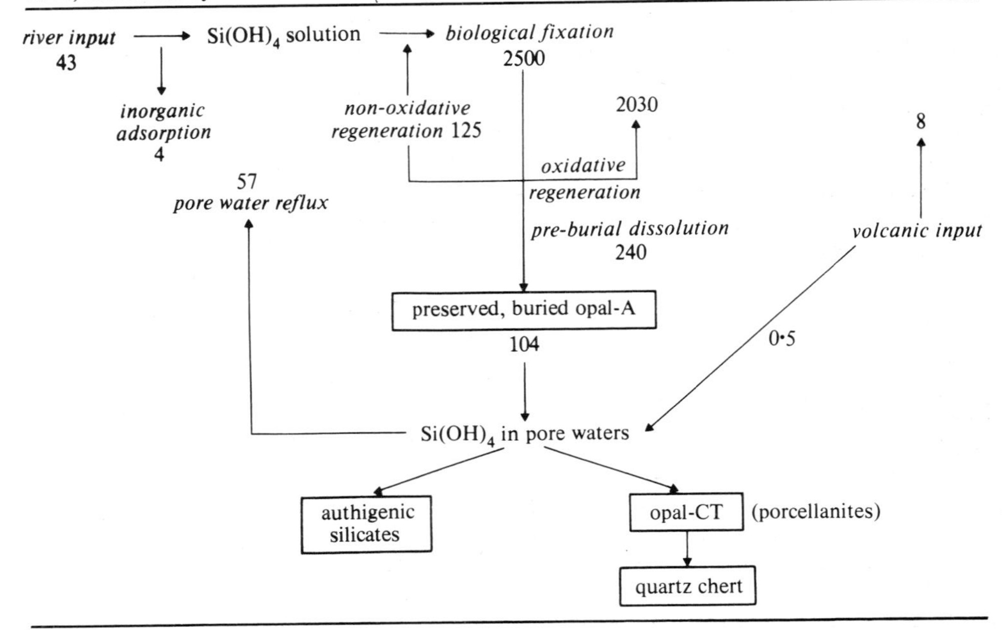

interlaminations of salts may record seasonal fluctuations in brine salinity.

Silica is provided to the ocean floors by the dead plankton that utilise amorphous opal-A in their tests.

Phosphate precipitation occurs in areas of oceanic upwelling where P-rich deep waters pass onto a shelf and where calcium phosphate precipitation and replacement of $CaCO_3$ occurs.

Further reading

Berner (1971) and Krauskopf (1979) provide good introductory accounts of evaporite genesis. Silica dynamics in the oceans are discussed by Heath (1974) and Calvert (1974). Several recent papers on phosphates and their genesis are to be found in Volume 74 of *Economic Geology* (1979).

4 Grain properties

4a Definition and range of grain size

Sediment grains range in size from minute specks of wind-blown dust to gigantic boulders (Table 4.1). Note that size is usually expressed as a linear dimension. No further difficulty would arise if all grains approximated to spheres, but unfortunately this is not the case. There are a number of alternative linear measures of grain size which take grain shape into account (Table 4.2), one of the most useful being the **volume diameter**, defined as the diameter of a sphere having the same volume as the grain in question. In the following account, grains will be treated as spheres or spheroids for the sake of simplicity and space. However, the reader should remember that many grains, particularly fragmented bioclastic debris, present grave problems in rational particle size analysis because of their shapes.

It should be noted that grain mass varies as the cube of the radius for spheroids. Thus, a 10 mm diameter sphere is five times 'larger' than a 2 mm sphere in terms of mean diameter, but $5^3/1^3$ (= 125) times 'larger' in terms of mass. The mass consideration is clearly important because it represents the resistance to movement (inertial mass) that must be overcome before grain transport occurs.

A widely adopted scale of grain size suitable for most clastic and some carbonate particles is the Udden–Wentworth scale in which the various grades are separated by factors of two about a grain size centre of 1.0 mm (Table 4.1). This scale easily adapts to the phi logarithmic transformation (Ch. 4b).

A number of alternative methods of size measurement exist, the choice of method being largely dependent upon the physical state of the grain aggregates and the size and shape of individual grains (Table 4.3). Further details may be found in specialised texts. (It should be noted that grain size measurements obtained from thin-section measurements must be 'corrected' in order that comparisons can be made with sieved samples; see Folk 1974a, Harrell & Eriksson 1979.)

4b Grain size distributions

Every sediment sample shows a range of grain size (Table 4.4). This variation must be characterised statistically so that samples may be compared and interpreted. It is therefore necessary to plot, in some way, frequency of occurrence against a measure of grain size.

The simplest plot is that of the **histogram**, where the area of each vertical bar represents the weight percentage of grains present in a given grain diameter interval (Fig. 4.1). The simple histogram has the advantage that the whole distribution may be seen at a glance, but also the disadvantage that it implies discontinuous variation at the breaks between classes. Clearly, the smooth **frequency curve** is more suitable. Errors arise, however, when reading off data from the frequency curve because of the low gradients at either end of the distribution.

It is also necessary to overcome the problem of the wide range of grain sizes that may occur in a sample; for example, a pebbly sand may have grains ranging over three magnitudes in diameter. Plotting points of this range on ordinary arithmetic graph paper is difficult; logarithmic graph paper is clearly called for. Alternatively, the millimetre Wentworth scale may be transformed into a logarithmic scale which can then be plotted on ordinary graph paper (Krumbein 1934). Note in this respect that the Wentworth scale class divisions form a series: 8, 4, 2, 1, $\frac{1}{2}$, $\frac{1}{4}$, $\frac{1}{8}$ mm, etc. This is a geometric series which may be manipulated into class divisions of equal width by taking logarithms. Unfortunately $\log_{10}$ equivalents of the Wentworth class limits come out not as integers but as fractions. Krumbein proposed that $\log_2$ should be used instead so that our series above, becomes: 2^3, 2^2, 2^1, 2^0, 2^{-1}, 2^{-2}, 2^{-3} mm. Taking negative logarithms gets over the assignation of the common sand grades to negative exponents. The exponent is now known as the phi (ϕ) grain size measure. Thus $\phi = -\log_2$ mm.

For our series above, phi units are −3, −2, −1, 0, 1, 2, 3. Since ϕ units are meant to be dimensionless, it is more correct to state that

$$\phi = -\log_2 \frac{d}{d_0}$$

where d_0 is the 'standard' grain diameter of 1 mm.

A further point arises from probability arguments. It has long been argued that since the frequency curve with a logarithmic abscissa approximates to a normal distribution, then a lognormal probability function should apply

Table 4.1 Summary of the Udden–Wentworth size classification for sediment grains (after Pettijohn *et al.* 1972). This grade scale is now in almost universal use amongst sedimentologists. Estimation of grain size in the field is aided by small samples of the main classes stuck on perspex.

	US Standard sieve mesh	Millimeters		Phi (ϕ) units	Wentworth size class
GRAVEL	*Use wire*	4096		−12	
	squares	1024		−10	boulder
		256	256	− 8	
		64	64	− 6	cobble
		16		− 4	pebble
	5	4	4	− 2	
	6	3.36		− 1.75	
	7	2.83		− 1.5	granule
	8	2.38		− 1.25	
	10	2.00	2	− 1.0	
SAND	12	1.68		− 0.75	
	14	1.41		− 0.5	very coarse sand
	16	1.19		− 0.25	
	18	1.00	1	0.0	
	20	0.84		0.25	
	25	0.71		0.5	coarse sand
	30	0.59		0.75	
	35	0.50	1/2	1.0	
	40	0.42		1.25	
	45	0.35		1.5	medium sand
	50	0.30		1.75	
	60	0.25	1/4	2.0	
	70	0.210		2.25	
	80	0.177		2.5	fine sand
	100	0.149		2.75	
	120	0.125	1/8	3.0	
	140	0.105		3.25	
	170	0.088		3.5	very fine sand
	200	0.074		3.75	
	230	0.0625	1/16	4.0	
SILT	270	0.053		4.25	
	325	0.044		4.5	coarse silt
		0.037		4.75	
		0.031	1/32	5.0	
		0.0156	1/64	6.0	medium silt
	Use	0.0078	1/128	7.0	fine silt
	pipette	0.0039	1/256	8.0	very fine silt
CLAY	*or*	0.0020		9.0	
	hydro-	0.00098		10.0	clay
	meter	0.00049		11.0	
		0.00024		12.0	
		0.00012		13.0	
		0.00006		14.0	

Table 4.2 Three useful definitions of grain size (after T. Allen 1968). Sieve diameter is the most commonly used measure but the other measures may be useful for non-quartz or quartz grains of varying shapes and density.

Name	*Definition*
sieve diameter	width of minimum square sieve aperture through which grain passes (note: a grain passing through the overlying sieve and settling on an underlying finer sieve has a sieve diameter intermediate between the two sieve values).
volume diameter	diameter of sphere having the same volume as the grain in question.
free-fall diameter	diameter of a sphere having the same density and fall velocity (Ch. 6) as the grain in question in a given fluid.

Table 4.3 Range of applicability of different techniques of size analysis (partly after Pettijohn *et al.* 1972). Indurated rocks must first be disaggregated by use of H_2O_2 or by gentle grinding in a pestle (see Carver 1971). Rocks that resist all attempts to disaggregate peacefully must be sectioned and the grains measured directly (a) through an eye-piece graticule attached to a petrological microscope or (b) on a 'shadowmaster' screen.

Sediment	*Technique*
gravel	direct measurement (callipers); sieving
sand	sieving; sedimentation tube; Coulter counter (see McCave & Jarvis 1973)
silt	sieving (coarse); sedimentation tube; Coulter counter; pipette
clay	pipette; electron microscope; Coulter counter

Table 4.4 Details of a sieve analysis of a medium, well sorted, positive skewed aeolian sand (after Bagnold 1954b). This analysis is used as a basis for the various graphical presentations of size distributions shown in Figure 4.1.

Sieve mesh (UK)	*Aperture (mm)*	ϕ	*Wt (%)*	*Cumulative wt (%)*
12	1.58	−0.65	0.005	0.005
16	1.17	−0.23	0.043	0.048
20	0.915	+0.13	0.338	0.386
24	0.755	+0.40	1.855	2.241
30	0.592	+0.75	14.120	16.361
40	0.414	+1.13	51.776	68.137
50	0.318	+1.65	20.300	88.437
60	0.261	+1.92	6.080	94.517
80	0.191	+2.40	3.860	98.377
100	0.114	+3.13	1.105	99.482
150	0.099	+3.33	0.404	99.886
200	0.073	+3.79	0.082	99.968
300	0.054	+4.21	0.024	99.992

to grain size distributions. A graphical test for lognormalcy is to plot cumulative percentage on a specially constructed probability distribution ordinate, with a phi scale on the arithmetic abscissa. Any lognormal distribution will now plot as a straight line. It is found, however, that most grain size distributions are only approximately lognormal.

Most workers plot their grain size data on a probability cumulative percentage ordinate and a phi scale arithmetic abscissa (Fig. 4.1) with the implicit assumption that a Gaussian error or probability distribution applies. Three features of grain size distributions complicate and contradict this assumption.

First, many authors have suggested that grain size distributions may comprise a number of straight line segments (Fig. 6.10) rather than a single straight line (Visher 1969, Middleton 1976). *Each* straight line segment is interpreted either as a truncated 'normal' distribution or as part of an overlapping system of normal distributions. Such segmented distributions have been ascribed to hydraulic sorting, in which case certain segments represent bedload and suspended load fractions (see Ch. 6), or to abrasion or to the provenance of the grains.

Secondly, it should be realised that most samples taken for grain size analysis include very many distinct depositional episodes. In other words the grain size analyses reflect **bulk sorting** rather than **depositional** or **transport** sorting (Emery 1978). Recent detailed size-frequency analysis of single sand laminae shows that such distributions are neither normal nor lognormal (Grace *et al.*, 1978). Most distributions are coarse-tail truncated and show that the truncation points of segmented distributions noted above cannot have a simple genetic explanation based upon bedload and suspended load processes (Ch. 6). The fact that many cumulative frequency distributions in the literature closely approach lognormality suggests that bulk samples, comprising many laminae populations, are just smoothed out aggregates of individually variable single-laminae distributions.

Thirdly, well sorted and very accurately sieved samples of wind-blown sands show that the extreme grades, of particular importance in the transport process, are usually present in greater proportions than probability would suggest (Fig. 4.1). It has recently been shown that such grain size distributions are **log-hyperbolic** and comprise a mixture of normal distributions (Bagnold 1968, Bagnold & Barndorff-Neilsen 1980).

We may note finally that materials such as scree, which have not undergone transport, do depart considerably from lognormality and show a special kind of probability distribution known as **Rosin's Law**, which plots as a straight line on specially constructed Rosin's Law probability graph paper.

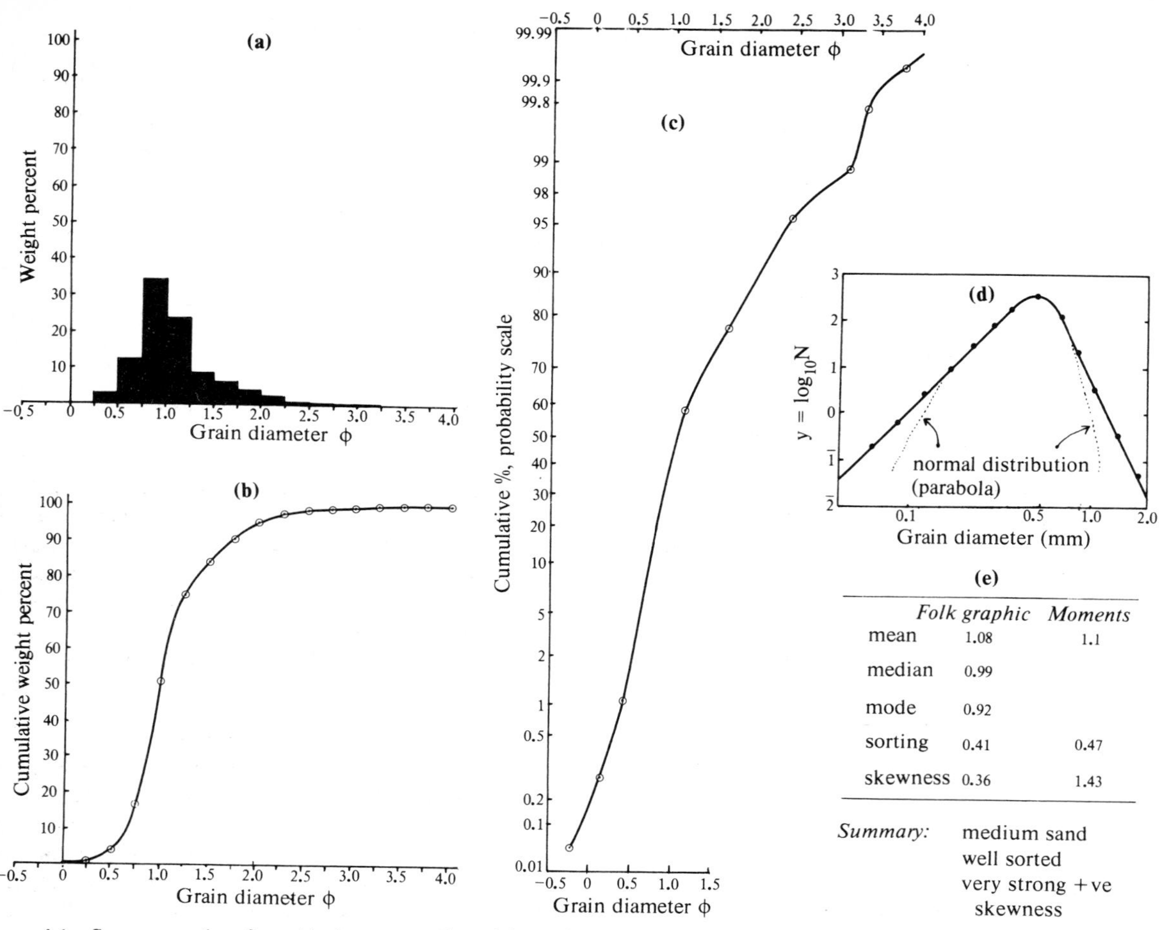

	Folk graphic	*Moments*
mean	1.08	1.1
median	0.99	
mode	0.92	
sorting	0.41	0.47
skewness	0.36	1.43

Summary: medium sand
well sorted
very strong +ve skewness

Figure 4.1 Some examples of graphical representation of the grain size data given in Table 4.3. (a) histogram; (b) cumulative plot on arithmetic ordinate; (c) cumulative plot on probability ordinate; (d) logarithmic ordinate scale where N = percentage weight/diameter interval for the sand retained between successive sieves. Note that both (c) and (d) disprove the notion that grain size distributions follow the normal distribution curve. Indeed the curve of (d) is hyperbolic rather than parabolic (see Bagnold & Barndorff-Nielson 1980). (e) shows a comparison of the Folk graphic grain population statistics with those supplied by computer moment analysis. Note how the graphical method considerably underestimates skewness.

4c Characteristics of grain populations

In addition to characterising a grain size distribution by one or more of the graphical techniques described above it is also necessary to define such features as the average grain size and the scatter about this average. The **mode** is the value of the most commonly occurring particle size, corresponding to the highest point of a frequency curve or the steepest part of a cumulative frequency curve. The **median** divides the normal frequency curve into two equal parts and corresponds to the 50% mark on the cumulative frequency curve. The **mean** (μ) is defined as the sum of measurements divided by the number of measurements; in formal statistical notation

$$\mu = \sum_{i=1}^{n} x_i/n \qquad (4.1)$$

where x_i is the value of the ith measurement and Σ means 'add together all the n values of x from 1 to n'. The mean grain size is a much superior estimator of the whole distribution than either the median or the mode. Symmetrical frequency curves have the useful property of showing a unique average value which is both mode, median *and* mean (Fig. 4.2). For **skewed** or asymmetric curves there is no such value and the three estimators are all different (Fig. 4.2). The spread of values about the mean is described by the **standard deviation**, called the **sorting** in sedimentology, which is the square root of the variance

$$\sigma^2 = \sum_{i=1}^{n} (x_i - \mu)^2/n \qquad (4.2)$$

where σ^2 = variance and σ = standard deviation.

The larger the spread about the mean, the larger the standard deviation (Table 4.5). It should be noted that one standard deviation about the mean includes 68.3% of a normal distribution, two standard deviations include 95.5%, and three standard deviations include 99.7%.

The two formulae for mean and standard deviation above are called, respectively, the first and second **moments**. Both are somewhat laborious to calculate by hand from raw grain size data but are easily dealt with by the computer. For rapid approximations to the mean and standard deviation, graphical techniques have been devised whereby selected values of size are read off from the cumulative curve (see Folk 1974a). These simple formulae refer only to cumulative/phi plots of grain size (Fig. 4.1) and give

$$\mu_z = \frac{(\phi 16 + 50 + 84)}{3} = \text{graphic mean grain size} \qquad (4.3)$$

$$\sigma_i = \frac{\phi 84 - 16}{4} + \frac{\phi 95 - 5}{6.6} = \text{inclusive graphic standard deviation} \qquad (4.4)$$

where $\phi 16$, 50, etc., are the values of phi at the various percentage levels of the cumulative frequency ordinate.

A further useful character of a distribution is the skewness, calculated as the third moment

$$sk = \sum_{i=1}^{n} (x_i - \mu)^3/n \qquad (4.5)$$

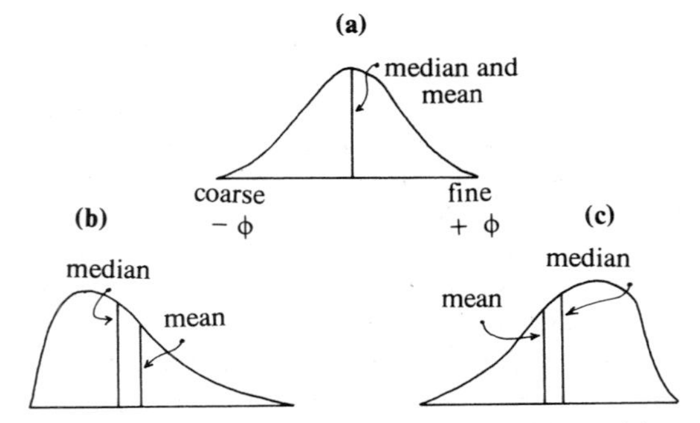

Figure 4.2 To illustrate skewness of distributions. (a) symmetrical distribution, (b) positive skewness, (c) negative skewness.

or by approximate graphical methods as

$$sk_i = \frac{\phi 16 + \phi 84 - 2\phi 50}{2(\phi 84 - \phi 16)} + \frac{\phi 5 + \phi 95 - 2\phi 50}{2(\phi 95 - \phi 5)} \qquad (4.6)$$

Note that the skewness measure, unlike the mean and standard deviation, is not expressed in phi units. Symmetrical curves have a skewness of zero. Those with excess fine grains have positive skewness and those with excess coarse grains have negative skewness (Table 4.5).

4d Size parameters and distributions

The mean grain size is a simple comparative indication of the weight force that must be balanced by an applied fluid stress before transport by water or wind is possible. Diagrams plotting mean diameter against some measure of flow strength are very useful and include competence diagrams (Ch. 6c) and bedform phase diagrams (Ch. 8a, b). Assuming the equal availability of a range of grain sizes, then the mean size of a particular deposit will reflect flow strength. A **graded bed** records time-varying flow strength at a point.

The sorting or standard deviation of a deposit is a measure of the degree of uniformity in the deposit produced by current action during grain transport and deposition. Selective winnowing may thus remove fines, as in the beach environment, and selective abrasion may produce a uniformly fine aggregate. Wind-blown deposits lack both fines, which are carried away into the atmosphere as dust, and coarse materials, due to the competence limit of air currents. The result is a uniform very fine

Table 4.5 Sorting and skewness values for graphically obtained statistics expressed as verbal descriptive summaries (after Folk 1974).

Standard deviation (sorting)	*Verbal description*
0–0.35ϕ	very well sorted
0.35–0.50ϕ	well sorted
0.50–0.71ϕ	moderately well sorted
0.71–1.00ϕ	moderately sorted
1.00–2.00ϕ	poorly sorted
2.00–4.00ϕ	very poorly sorted
4.00+ϕ	extremely poorly sorted
Skewness	
+1.00–+0.30	strongly fine-skewed
+0.30–+0.10	fine-skewed
+0.10––0.10	near-symmetrical
–0.10––0.30	coarse-skewed
–0.30––1.00	strongly coarse-skewed

to fine sand deposit. No such segregation takes place during deposition and transport of glacial tills, where all grades of particle from clay to boulders are present. The reader may care to think up further contrasting examples.

The skewness of a deposit sampled in bulk is rather sensitive to the type of depositional environment. Some river sands show positive skewness because of enrichment in fine silt-size particles which may be deposited as the river stage drops after a flood. Beach sands, by way of contrast, generally show negative skewness because the fine grains are selectively winnowed by constant wave action, leaving a 'tail' of coarser grains. Wind-blown sands generally show positive skewness becuse of the low efficiency of the wind in moving coarse particles which are usually left behind to form what is known as a **lag deposit**.

A simple plot of skewness against standard deviation for beach and river sands (Friedman 1961) brings out some of the above features although recent results (Sedimentation Seminar 1981) cast doubt on the generality of this approach. As seen in Figure 4.3, river sands are characterised by relatively poor sorting and positive skewness, whilst beach sands have good sorting and negative skewness. Further attempts to ascertain environments of deposition from grain size analysis have usually failed because of the overlap between fields and the lack of a statistical test for establishing the uniqueness of a particular field of occurrence. Examples include plots in which only three inferior properties of the grain size distribution (the one percentile, the median, the percentage finer than 3ϕ) are used (Passega 1964). As we shall stress later in this book, deductions about environments of deposition should be based on several approaches (e.g. facies, palaeocurrent, geometrical, grain size), not just one.

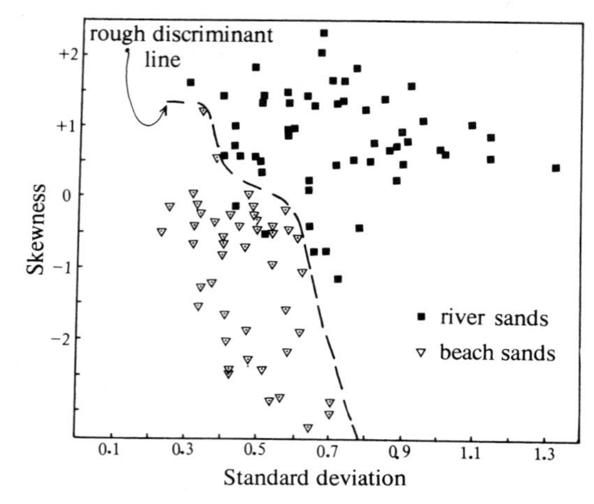

Figure 4.3 Data from modern river and beach sands plotted on to a graph of moment skewness versus moment standard deviation (sorting). Note the good separation of the two fields but also note recent adverse criticism of the method (Sedimentation Seminar 1981). Such a separation may be of use in helping to discriminate between fluvial and beach sands in the stratigraphic record (after Friedman 1961; for an application to ancient sediments see Leeder & Nami 1978).

4e Grain abrasion and breakage

Despite the importance of these processes there is little quantitative understanding of the rates and mechanisms involved. Abrasion during transport occurs by impact-induced fracture following the transfer of kinetic energy from grain to grain. As might be expected, the process is much more effective in wind than in water. In the latter case the high effective buoyancy exerted on grains and the high viscosity of the medium tend to 'cushion' impact effects. Experimental data show that wind abrasion of quartz is 100–1000 times more effective over the same distance of transport than the mechanical action of a river (Keunen 1964). In both wind and water abrasion, effects tend to zero for particles finer than about 0.05 mm.

The rounding and size reduction of cobble and large pebble grades is found to take place in significantly shorter distances in natural streams than is computed from laboratory experiments in tumbling apparatus. This is probably explicable by the observed abrasional effects of grain rocking (Schumm & Stevens 1973) at or near movement threshold, when fluid lift and shear forces act on the particle (Ch. 6). The observed downstream fining seen in many rivers may be caused by particle abrasion, breakage or hydraulic sorting, or, as seems most likely, to all three processes. The downstream fining may sometimes be expressed in a negative exponential relation, one form of which is Sternberg's 'law' which states that

$$W = W_0 \exp[-a(x - x_0)] \qquad (4.7)$$

where W is the weight of the largest particle at distance x from origin, W_0 is the weight at some reference point x_0 and (a) is a constant for a particular stream. Similar exponential 'laws' seem to fit the commonly observed downstream increase of roundness or sphericity (Ch. 4f). The form of these equations supports the experimental evidence that particle size reduction by physical abrasion becomes less and less effective for smaller particles so that, for example, quartz sand grains in water may travel 1000 km for a weight loss of less than 0.1%. Downstream grading changes in such sand-grade deposits must clearly be the result of hydraulic sorting effects. We should not be surprised at this conclusion since the stresses exchanged at contact between colliding particles are due to the rate of loss of momentum, which is largely determined by particle mass. As already mentioned, mass is a function of radius cubed so that pebbles and sand grains of 50 mm, 5 mm, and 0.5 mm radius have proportional masses of 125 000, 125 and 0.125.

Particle breakage during transport by wind or water of igneous quartz grains is greatly aided by the presence within such grains of extensive cracks (Moss *et al.* 1973) and by the occurrence (see Ch. 4f) of a number of weak cleavage planes (Wellendorf & Krinsley 1980). The cracks may be due to stresses set up within the crystals due to the alpha → beta quartz inversion that takes place as crystallised granite bodies cool below 573°C (Smalley 1966). High-temperature alpha quartz has a lower density than low-temperature beta quartz so that the inversion is associated with a volume decrease of about 1.5%. The stresses so caused are thought to both loosen the whole quartz crystal from its neighbouring crystals, enhancing 'erodeability' *and* to form internal microcrack systems within individual crystals.

4f Grain shape and form

Two aspects of grain shape are frequently confused. These are **roundness** and **sphericity**. Roundness is an estimate of the smoothness of the surface of a grain. It is thus possible to have a well rounded grain of rod-like shape. Roundness may be quantified as the ratio of the mean radius of curvature of the grain corners to the radius of the largest inscribed circle. This ratio is rather time consuming to measure for a whole population of grains. It is usual practice to estimate mean roundness using a standard set of grain images (Fig. 4.4).

Sphericity values tell one how closely a grain approximates to a sphere, in which shape the three orthogonal axes x, y and z are equal. Various formulae have been proposed but the best is the **maximum projection sphericity**, ψp, as introduced by Sneed and Folk (1958). This states that $\psi p = (s^2/li)^{\frac{1}{3}}$, where l, i and s are the long, intermediate and short dimensions respectively. The formula takes account of the settling behaviour of a grain in a fluid since it compares the grain's maximum projection area to that of the projection area of a sphere of equal volume. In thin section, sphericity is assessed as **elongation**, being the ratio of width to length of a grain.

Grain *form* involves the relation between the axes l, i and s. Grains may be classified with respect to equidimensional, rod-like or disc-like end members on a ternary form diagram (Sneed & Folk 1958).

In recent years much study has been made of the microscopic relief on sand grain surfaces using the electron microscope (Margolis & Krinsley 1974). It has been found that a number of fracture patterns and impact marks occur, some of which seem diagnostic of particular environments. It should be noted, however, that environmental analysis using the presence or absence of such patterns is extremely hazardous because of the frequent occurrence of reworking and inherited patterns. It is also a fact that the examination of a statistically viable number of grains, together with the estimation of percentage cover of particular patterns, is very tedious and time consuming. The electron microscope studies have shed light on two problems of great interest. First, it is now apparent that although quartz tends to fracture on a macroscopic scale, breakage along cleavage planes tends to dominate on a very small scale (Wellendorf & Krinsley 1980) to produce quartz debris of less than about 50 μ (Smalley & Moon 1973). Studies of glacial clays show that a significant portion of these deposits is composed of very fine (3 μm)

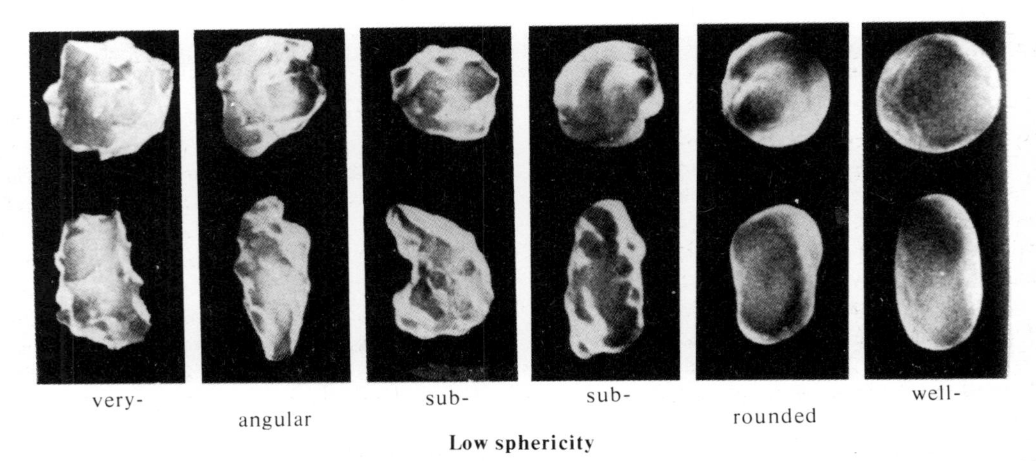

Figure 4.4 Photographic images of grains used for the visual determination of grain roundness (after Powers 1953).

tablet-shape quartz cleavage fragments. Presumably aeolian silts also contain such detritus. Secondly, it is apparent that the frictional behaviour exhibited by grains (see Ch. 7b) is due in part to the microscopic irregularities on grain surfaces in addition to the more obvious macroscopic 'corners' usually estimated as part of roundness procedures.

4g Bulk properties of grain aggregates

The accumulation of grains as a deposit inevitably leads to the development of a packing structure which determines many bulk particle properties. The **fractional volume concentration** (C) of grains within some deposit before cementation is the ratio of grain-occupied space to whole space. The amount of pore space (P), or grain-'unoccupied' space, within some volume is thus given as

$$P = 1 - C \tag{4.8}$$

Bagnold (1954) has developed the concept of linear concentration (λ) which is the ratio of particle diameter (d) to the distance between adjacent particles (s) or

$$\lambda = d/s \tag{4.9}$$

This linear concentration is related to the fractional volume concentration by

$$C = \frac{C_*}{(1/\lambda + 1)} \tag{4.10}$$

where C_* is the maximum possible concentration when $\lambda = \infty$ ($s = 0$) under conditions of rhombohedral packing (see below). C_* is 0.74 for spheres.

The packing of particles in a deposit partly determines **porosity, permeability** and **stability**. Various packing modes for sphere assemblages exist. Two simple end members are cubic and rhombohedral packing (Fig. 4.5) which have porosities of 48% and 26% respectively. These may be regarded as maximum and minimum values for equant sphaeroids. Many natural granular deposits that have undergone no cementation commonly show intermediate porosity values.

A number of factors control the packing geometry – and hence the other bulk properties – of deposited particles. Grain shape is probably of greatest importance. Consider, for example, the differences produced by the irregular packing of equidimensional spheres, cubes and plates. Very high porosities may be produced during the deposition of shell particles and 'holes' may be preserved if early cementation (Ch. 29) prevents compaction. Many

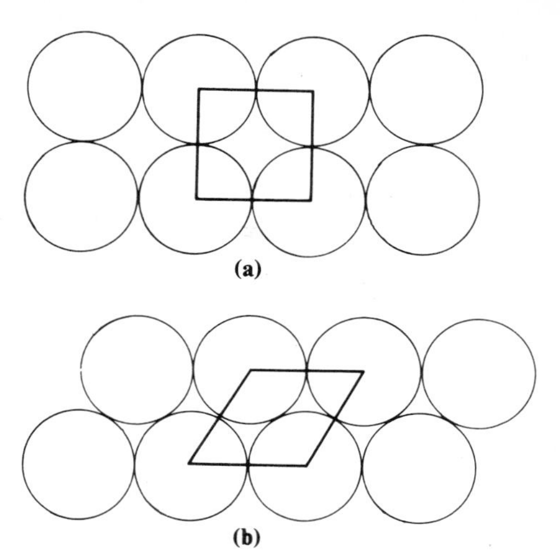

Figure 4.5 Vertical sections through piles of equidimensional spheres to show (a) cubic and (b) rhombohedral packing modes. The two modes give rise to theoretical maximum and minimum values of porosity.

freshly deposited clays show very high (up to 90%) initial porosity due to the network structures of clay mineral platelets produced by flocculation (Chs 9 & 11), but it is usual for compaction to eliminate this porosity.

The rate of deposition affects packing as follows (Gray 1968). High deposition rates lead to particle/particle collisions and interference at the (upwards) moving depositional interface. These effects prevent individual particles taking up optimum positions to ensure good, regular rhombohedral packing, a process that can occur at low deposition rates. Local cubic co-ordination and the presence of voids thus give rise to increased porosities (Fig. 4.6). Such particle interference effects will be even more marked in flaky particles since many vertical flakes may exist in the deposited mass causing development of high porosity. Increasing particle fall velocity tends to decrease porosity (Fig. 4.6) and encourage close packing since particle/bed impacts transfer kinetic energy to underlying layers causing particle shear jostling that reduces pore space. In face, vibration-induced resettling of deposited particle aggregates has many industrial and domestic applications serving to optimise available storage or carrier space.

The porosity of natural sands with similar packing is of course independent of grain size, but it will vary according to sorting (Beard & Weyl 1973). Experiments reveal a 25% difference in porosity between well sorted sands and very poorly sorted sands of the same mean grain size.

Porosity and permeability are fundamentally different parameters. As we have seen above, porosity is the fractional space between solid particles in a given volume. Permeability, by way of contrast, gives the rate at which

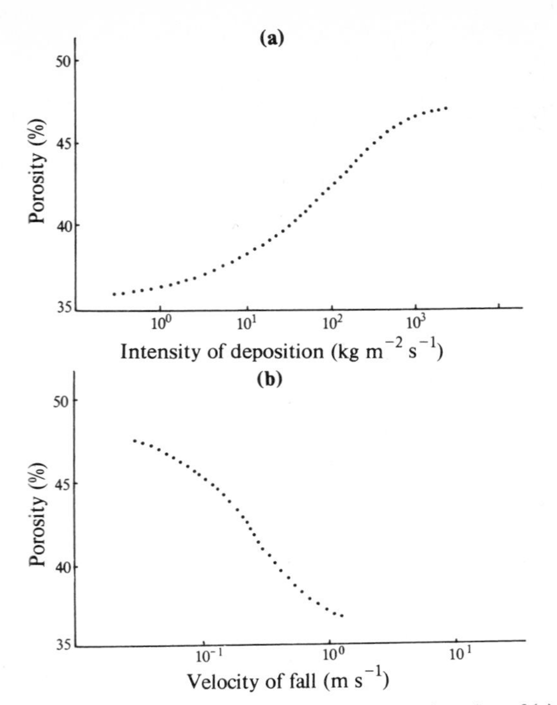

Figure 4.6 The porosity of sand aggregates as a function of (a) the rate of deposition and (b) the velocity of fall of the individual particles (after Gray 1968).

fluids stored within pores actually can move through the pore system. Clearly, there may be no relationship at all between the two. A good comparison may be made by analogy with a house where the rooms can be compared to pore space and the connecting doors and corridors can be compared to permeability. The degree of interconnectedness of pores is clearly of great economic importance in hydrocarbon and water reservoirs. Many diagenetic processes (chiefly compaction and authigenic mineral growth) tend to reduce pore interconnections (see Chs 27–31).

4h A note on grain fabric

Fabric studies reveal the ways in which grains are orientated in an aggregate. Clearly, spherical grains can show no preferred orientation. Non-equant grains may show no preferred orientation. In both cases the fabric is said to be **isotropic. Anisotropic** fabrics usually result from grain transport or depositional processes which cause grains to be aligned in some preferred direction (see Johansson 1976). Current-produced **imbrication** produced in gentle bedload transport (grains rolling or sliding) is perhaps the commonest anisotropic fabric wherein grains lie with their long *a* axis normal to flow but with the *b–c* plane inclined at a small angle (<20°) upstream. The grains therefore lie snugly in their most stable position. During stronger flows grains tends to saltate and take up *a*-axis parallel to flow orientation. Preferred orientations depend critically upon both mode of transport and mode of deposition. Elongate grains in grain flows and debris flows (Ch. 7) also take up orientations with their long axis parallel to flow and their *b–c* plane dipping upstream. Clasts in glacial tills (Ch. 17) also show *a*-axis parallelism with flow but with varying proportions of *a*-axes normal to flow. Isotropic fabrics usually result if deposition is very rapid, as in the deposits of densely charged turbidity currents (Ch. 7).

4i Summary

Modern computer processing of grain size data facilitates rapid computation of moment mean grain size, standard deviation, skewness and other statistical properties. These properties of grain size distributions are fundamental descriptive parameters and they find wide usage in sedimentological studies. Much use is made of the logarithmic grain size notation, ϕ. Interpretation of grain size distributions is a controversial subject, particularly with respect to the existence and explanation of so-called straight-line segments. Recent hydraulic interpretations and studies of single-lamina grain size distributions demonstrate that the problem is one of separating bulk sorting from transport sorting. Quartz grain abrasion and breakage during transport is greatly facilitated by the occurrence of weak cleavage planes and cracks. The bulk properties of grain aggregates, such as porosity, packing and permeability, are controlled by a number of variables, including grain shape, rate of deposition and sorting.

Further reading

Folk's book (1974a) is again recommended as a stimulating, clear, non-mathematical introduction to grain size and shape studies. Till (1974) gives a clear statistical introduction. Techniques for analysis are described by Carver (1971). The straight-line segment controversy is set out by Visher (1969) and Middleton (1976). Bagnold's own approach is set out non-mathematically in his book (1954b) and in rigorous statistical terms in his paper with the statistician Barndorff-Nielsen (1980). Gray (1968) contains much useful data on the bulk properties of grain aggregates. Beard and Weyl (1973) contains a number of photographs of sorting variations in sands that may be useful in visually assessing sorting without recourse to detailed quantitative analysis.

PART TWO FLUID FLOW AND SEDIMENT TRANSPORT

Before I had studied Zen for thirty years, I saw mountains as mountains and waters as waters. When I arrived at a more intimate knowledge, I came to the point where I saw that mountains are not mountains, and waters are not waters. But now that I have got its very substance I am at rest. For it's just that I see mountains once again as mountains, and waters once again as waters.

A saying of Ch'ing-yuan

Plate 2 Intense sediment transport over an exposed intertidal sand bar during strong winds. Note the density and thickness of the bedload layer (partly hiding the student's feet) and the well developed 'windrows' of sand which serve to visualise the powerful secondary flow vortices in the wind boundary layer (Wells, N. Norfolk, England).

Theme

Once formed, sediment grains become available for incorporation into the various flow systems that exist on the Earth's surface and under the sea. In this part of the book we examine the basic principles of unidirectional fluid flow with particular reference to the ability of flows to transport granular materials. Any moving fluid transmits shear and lift forces to its irregular solid boundaries. These forces do useful work in transporting grains or eroding sediment beds. We are not so much interested in the detailed mathematical derivations of pure fluid behaviour as in the dynamics of grain–fluid systems from the point of view of general physics. Although of interest in its own right this field of loose-boundary hydraulics provides an essential basis for the scientific study of sedimentary structures and of sedimentary facies. Further aspects of natural flows and specific environments (deserts, rivers, tides, waves) are dealt with in the appropriate chapters of Part 5.

5 Fluid properties and fluid motion

5a Introduction

In whichever climatic regime we live, everyday observations make us aware of the complexities of fluid movement: gusting winds with eddies picked out by the irregular paths of suspended leaves; the majestic sight of a turbulent, sediment-laden river in flood; flash floods and sheet flows issuing from desert wadis; the inward spiral of a plains dust devil or tornado.

In this chapter we shall examine some of the basic properties of moving fluids. The level of treatment will assume only *an elementary* knowledge of mechanics. The simplest derivation of certain fundamental equations is sketched out in the appendices at the end of this chapter. At all times the reader should seek to realise the significance of basic fluid properties and relationships to the whole problem of transport in the field of sedimentology. More advanced texts on fluid dynamics are usually 'pure' and they deal with the mathematical development of the fundamental equations of flow as applied to the fluid alone. The difficulties of obtaining solutions to these partial differential equations leave much room for experimental research in fluid mechanics. Application of fluid dynamics to natural flows transporting solid matter inevitably involves a large degree of mathematical simplification.

Let us first be aware that the term **fluid** includes a large number of substances, some natural examples being the liquids water, blood and crude oil, and the gas, air. All of these substances share the common property of a fluid that *the smallest applied external force causes the fluid to change shape continuously as long as the force is applied.* Thus a fluid may preserve its shape only if it is constrained by bounding surfaces. The same strictures apply to a gas such as air, with the added property that a gas will fill completely all the space made available to it. It follows that a gas, unlike a liquid, cannot have a surface of its own.

Secondly, let us choose to ignore the molecular-scale processes acting in fluids and assume that the physical properties we shall discuss below apply to some volume of fluid that contains a very large number of individual molecules. By formally adopting this **continuum hypothesis,** something that most of us do intuitively anyway, we can treat properties such as density, temperature, viscosity and velocity as if they were average properties applicable to small lumps or particles of fluid. This simplification is best illustrated by considering air flow. We know that in a gas the individual molecules are moving at very high speeds randomly with respect to one another. Nevertheless we may be sure that a small portion of the air can flow in some quite specific direction at a given rate. Since we may readily measure this velocity **vector** we assume by the continuum hypothesis that our measurement is an average velocity at a point.

5b Physical properties

Density (ρ) is mass per unit volume (ML^{-3}) and should not be confused with either **specific gravity** (the ratio of fluid density to that of water) which is numerically equal to ρ only in the now discarded cgs system of units, or **specific weight** (the weight force per unit volume). Values of the density of some natural fluids are shown in Table 5.1. Gases such as air have low densities and are highly compressible over wide ranges of volume. The low density is caused by the comparatively small number of molecules per unit volume and the high compressibility by the large average distance between molecules. Density decreases with increasing temperature. The much higher density of liquids as compared to gases reflects the much larger number of molecules per unit volume. For similar reasons the compressibility of liquids is much lower than gases. Most liquids show a density increase with decreasing temperature, apart from the well known exception of water below 4°C (Ch. 1).

Viscosity (μ) is the measure of resistance of a substance to a change in shape taking place at finite speeds, e.g.

Table 5.1 Densities and molecular viscosities of some natural and artificial 'fluids'.

Substance	*Density (kg m^{-3})*	*Molecular viscosity (N s m^{-2})*
air	1.3	1.78×10^{-5} (20°C)
water	1000	1.00×10^{-3} (20°C)
glycerol	1262	1.50×10^{0} (20°C)
fluidised sand	(variable)	$\approx 1.00 \times 10^{0}$
debris flows*	1500–2600**	$\approx 1.00 \times 10^{2}$–$1.00 \times 10^{3}$
basalt magma*	2700	3.00×10^{2}–3.00×10^{3}
upper mantle	3300	$\approx 1.00 \times 10^{20}$

* indicates non-Newtonian behaviour; ** indicates that density and viscosity vary according to water content.

while stirring a substance in a bowl the motion of a spoon is opposed by the viscosity of the medium. Viscosity is a force per unit area per unit velocity gradient ($ML^{-1}T^{-1}$). We are concerned here only with pure fluids, but it is worth noting that the addition of dissolved or fine suspensions of material can greatly increase the viscosity of the now impure fluid volume. Such a problem was investigated theoretically by Albert Einstein for very dilute suspensions in 1905. There are important sedimentary consequences of this result (Chs 6 & 8).

Values for the molecular viscosity of some natural fluids and gases are shown in Table 5.1. Gases have low viscosities since the constituent molecules can move along distances without hindrance from other molecules, so there is little resistance to motion. The higher viscosities of liquids arise because of the difficulty of molecules in moving fast relative to one another because of cohesion and mutual hindrance. The viscosity of all liquids decreases with an increase in temperature and this is why it is essential to state the corresponding temperature when quoting viscosity values.

So far we have treated the molecular viscosity as a constant (at constant temperature), i.e. the viscosity is not influenced by the rate of shear. Thus the rate at which the liquid is stirred (temperature kept constant) does not affect the resistance to shear. Such fluids are termed **Newtonian** (Fig. 5.1) and in view of the complications arising from variable viscosity fluids we are fortunate that water is Newtonian in its behaviour. A degree of caution must still be advised because when turbulent eddies occur during flow (Ch. 5) an additional resistance to flow arises through the **eddy 'viscosity'** which is not constant for a given fluid and temperature.

Non-Newtonian fluids, studied in **rheology**, show variable μ with shear or strain rate. Many natural water-saturated muds have this property and are highly important in processes of slumping, sliding and avalanching. Blood, cream, and household products such as emulsion paint and mayonnaise are non-Newtonian in behaviour. Rapid stirring of a can full of emulsion paint or shaking of a bottle of mayonnaise cause shearing motions which change the molecular binding forces and make both substances flow very easily. The effect is fully reversible. Similarly, the actions of an earthquake shock wave may liquefy a mass of water-saturated, uncompacted muds and cause downslope flowage. Non-Newtonian liquids usually show a high molecular weight and a readiness to form intermolecular bonds. An applied stress must be of sufficient magnitude to assist the molecular kinetic energy to break bonds at a rate to maintain the flow. Molecules of high molecular weight are very irregular so that many bonds must break before movement occurs. Flow tends to line the molecules up and so at some point in time after the application of sufficient stress the rate of flow will increase as the viscosity decreases.

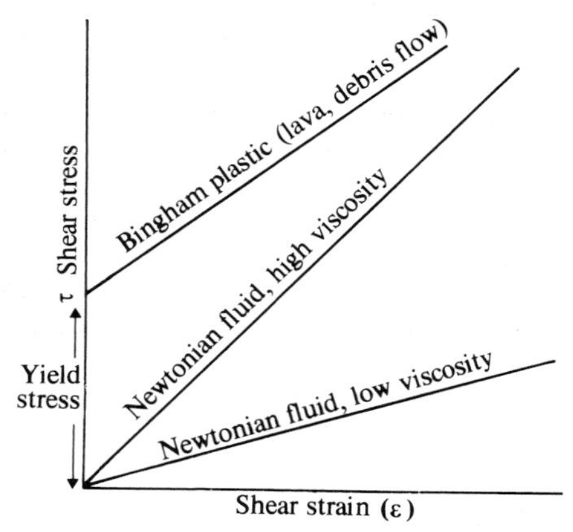

Figure 5.1 Sketch graph to illustrate 'Newtonian' and 'Bingham' behaviour of substances under shear.

Plastic substances to which an initial or 'yield' stress must be applied before strain occurs may also show a constant viscosity, when they are termed ideal or **Bingham plastics** (Fig. 5.1). Those with a variable viscosity dependent upon shear stress are called pseudoplastics or **thixotropic** substances (see Ch. 8).

Finally, it should be noted that the ratio of the molecular viscosity to the density of a fluid is often used in fluid analysis. This ratio $\mu:\rho = \nu$, is termed the **kinematic viscosity** and has the dimensions L^2T^{-1}.

5c Streamlines and flow visualisation

If we could follow the paths of numerous fluid elements with time in a flow, then we could map out the paths of the elements to give a complete picture of the flow as it passed over obstacles such as sediment grains and ripples or around river bends. A fluid flow may be 'mapped' by its **streamlines**, which may be defined as imaginary lines drawn in the fluid so that tangents drawn to them are in the direction of flow. Streamlines are usually curved but they cannot cross since at the point of interaction the fluid would then have two velocities, which is not possible.

As we can see from Figure 5.2 the quantity of fluid passing per unit time across a line between two streamlines is the same as the quantity passing any other lines between the same two streamlines. This **continuity** argument follows because no fluid can pass across the streamlines. Therefore if streamlines diverge and hence the cross-sectional area between them increases, as in the flaring mouth of a delta channel, it follows that the velocity must decrease. The converse applies to converging streamlines, as in flow over a ripple-like bedform. With reference to

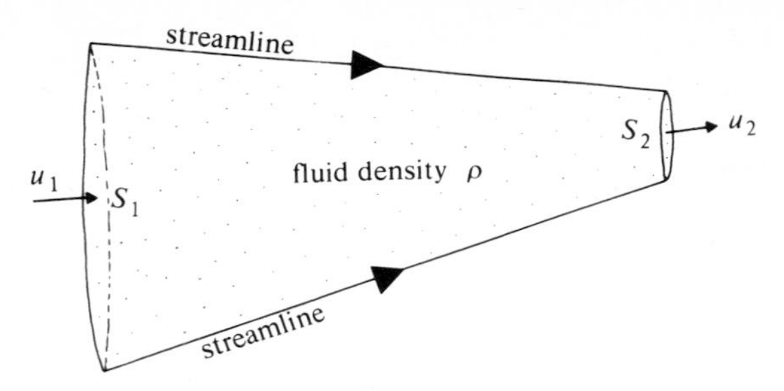

Figure 5.2 Conic streamtube to illustrate continuity of flow between converging streamlines. In a steady flow the mass of fluid passing through any cross section of area is constant, i.e. $\rho u_1 S_1 = \rho u_2 S_2$. Thus as $S_1 > S_2$, $u_2 > u_1$. ρ – fluid density; S – cross-sectional area; u – velocity.

Figure 5.2 we see that the rate at which mass enters the volume is $\rho u_1 S_1$ and the rate at which mass leaves the volume is $\rho u_2 S_2$. If the mass is constant then $\rho u_1 S_1 = \rho u_2 S_2$. Those readers with some higher mathematics will recognise that this equality is a form of the continuity equation div $\bar{u} = 0$.

Steady flows are those in which, at any point in the fluid, the mean velocity, pressure, density and temperature of the fluid remain the same and do not change with time. If the conditions change, then the flow is unsteady. **Uniform flows** are those in which the velocity is constant along the direction of flow, whilst **non-uniform** flows such as those over ripple or dune bedforms show a variation of velocity along the flow.

Streamline patterns may be computed or drawn from photographs. In flow visualisation studies, small particles (smoke, dye, powders, gas bubbles) are introduced at a point. **Particle paths** result if the dye is introduced instantaneously and then continuously observed or photographed over a long exposure. **Streaklines** are seen if the dye is introduced continuously and observed or photographed instantaneously. In steady flows streamlines, particle paths and streaklines are all identical but may be drawn with respect to either a stationary or a moving observer (Fig. 5.3).

As we shall see below, the streamline that occurs a very small distance away from the rigid boundary to a flow is very important; it is called a **skin friction line**. These lines may be drawn from dissolution scour pits in plaster of paris models (Fig. 8.5) and provide vital information concerning flow paths over bedform features (Ch. 8).

5d Friction, pressure changes and the energy budget

For ease of mathematical development in the field of **hydrodynamics** it is often assumed that fluids are **inviscid** or **ideal**. This means that the fluid is assumed to be a hypothetical incompressible substance which offers no resistance to flow, i.e. $\mu = 0$; it is thus *frictionless*. The formulation of the mathematical laws of hydrodynamics has been greatly aided by these simplifications, but, not surprisingly, there are many important fluid effects in sedimentology that cannot be explained or predicted by inviscid theory.

When a *real* fluid moves, frictional effects cause the production of zones of flow retardation next to the stationary boundaries to the flow (see Figs 5.7–9). The zones of retardation, discovered by the German physicist Prandtl in 1904, are known as **boundary layers.** The boundary layer is a zone where there is a velocity gradient and where viscous forces give rise to shear stresses. Close to the wall or flow boundary there is a layer of liquid molecules attached: the adsorbed layer. As the liquid flows, the velocity tends to zero at the wall because the adsorbed layer refuses to move. This viscous retardation gradually dies out away from the wall until at some point in the flow, termed the **free stream**, there is no velocity gradient and hence no stress.

When a fluid is at rest the only forces acting on any of its elements are *static*, due to the weight of the fluid. When it is in motion there must be, by Newton's Second Law, other forces acting upon it in addition to the static weight forces. We shall see below that the forces causing motion are due to unequal pressure on the fluid elements. When the fluid is moving with constant velocity, forces arising from viscosity and turbulence act in the direction opposing motion and these frictional forces exactly balance the pressure forces. Thus in steady flow the loss of energy due to friction must be accompanied by a fall in pressure.

Let us now consider the energy budget in an ideal moving fluid. The **potential energy** of a fluid element is the energy of position and may be thought of as the work stored in the element above some reference plane. If the element is of mass m, then the downward force on it is mg Newtons and the work needed to lift it through a distance h to its position above a reference plane is mgh Joules. The **kinetic energy** of a fluid element is the energy stored in the element when in motion and is equal to the work done in arriving at that state of motion from rest. If an element of mass m is subject to a uniform acceleration a from rest to reach velocity u in distance s, its acceleration is given by

$$u^2 = 2as \qquad (5.1)$$

and the accelerating force is

$$f = ma = \frac{mu^2}{2s} \qquad (5.2)$$

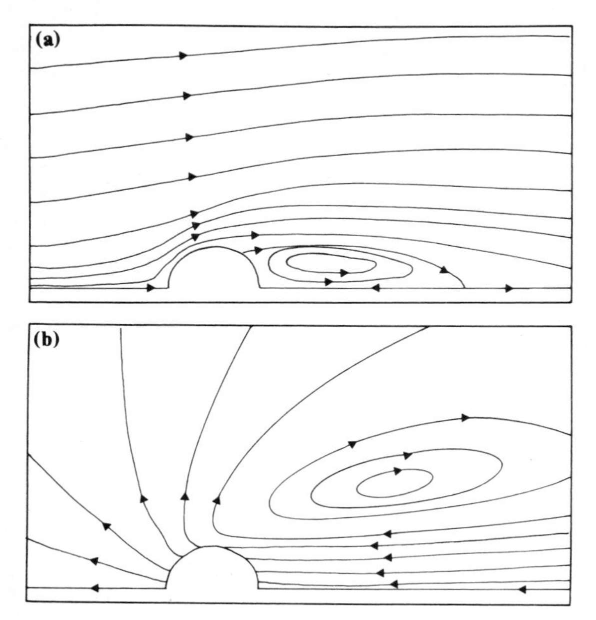

Figure 5.3 Contrasting streamline patterns produced by (a) flow past a stationary circular cylinder (only half of flow field shown) and (b) movement of the circular cylinder through the fluid, the fluid far from the cylinder being at rest. Note that in (a) the streamlines are also particle paths and streaklines since the flow pattern is steady. The closed streamlines behind the cylinder define a region of laminar separation. (see Ch. 5j). The pattern in (b) is an instantaneous one and on a figure of infinite extent all the streamlines would be closed. The streamlines here are not identical to particle paths or streaklines. (After Tritton 1977.)

The work done on the element is force times distance or

$$w = \frac{mu^2}{2s}\ s = \tfrac{1}{2}mu^2 \qquad (5.3)$$

which is the formula for kinetic energy. The law of conservation of energy says that the sum of potential and kinetic energies is constant. Thus the potential energy of a falling mass is being constantly converted into kinetic energy.

A third type of energy in a fluid element is **pressure energy** (p) which enables work to be done by building up the pressure of the fluid, as in compressed air systems.

The law of the conservation of energy as applied to fluid flow is expressed in the famous **Bernoulli's equation**

$$\underset{\text{kinetic energy per unit volume}}{\tfrac{1}{2}\rho u^2} + \underset{\text{potential energy per unit volume}}{\rho g h} + \underset{\text{total fluid pressure}}{p}$$

$$= \text{total energy} = \text{constant along a streamline} \qquad (5.4)$$

Bernoulli's equation is derived in Appendix 1 to this chapter. A number of *very* important sedimentological consequences arise from Bernoulli's equation, concerning pressure and velocity changes. We shall examine these in Chapter 5j and Chapter 6, but perhaps you have already spotted one important consequence (*Clue:* What happens if the flow speeds up along a streamline?). Bernoulli's equation is also the basis for the measurement of point velocities in fluids by an instrument called the **pitot-static tube**.

In fact, Bernoulli's equation as stated above is only strictly correct for our old friend the 'ideal' fluid. As mentioned above, in real fluid flow there is a continuous loss of energy due to friction. Therefore the total energy is not constant but will decrease in the direction of flow as a result of this energy dissipation.

We now examine some useful expressions for the 'bulk' behaviour of flows in pipes and channels. Consider the pressure difference h between the two points l apart on a pipe of diameter d. By the d'Arcy–Weisbach equation

$$h = \frac{4flu^2}{d2g} \qquad (5.5)$$

where u is mean flow velocity and f is a friction coefficient, strongly influenced by the nature of the pipe walls, that must be determined experimentally. Now consider an open channel (Fig. 5.4) with steady uniform flow, slope $\sin\alpha$, length l, and average depth h. Let ρ be water density. The downslope shear component of the water, τ, acting on unit area of the bed is

$$\tau_0 = \rho g h \sin\alpha \qquad (5.6)$$

Equation 5.6 is the tractive stress equation first due to du Boys. The d'Arcy–Weisbach and du Boys equations can be combined to give

$$\tau_0 = \frac{f\rho u^2}{8} \qquad (5.7)$$

and

$$u = \sqrt{\frac{8g}{f}} \cdot \sqrt{Rs} \quad \text{(Chezy equation)} \qquad (5.8)$$

where $8g/f$ is the Chezy coefficient and R is the hydraulic radius, usually approximated as the mean depth in wide natural channels.

The equations above are very useful in determining mean flow parameters for flows. Friction factors for fully turbulent flows depend upon the Reynolds number and the relative roughness of the containing walls (see Ch. 5e & h). We shall see in Chapter 6 that different bedforms have radically different values for f. We may distinguish at this stage between the friction caused by bed grains (**particle drag**) from the friction of bedforms such as ripples and dunes (**form drag**).

In everyday language we may refer to flows as being fast or slow, strong or weak. In fact, there are a number of alternative parameters that can be used to express flow magnitude in a more exact way. We have already come across mean velocity and boundary shear stress. The product of the first two parameters gives us the **stream power**, ω, available to unit bed area of fluid (Bagnold 1963, 1966b):

$$\omega = \bar{u}\tau_0 \quad \text{(units: MT}^{-3}\text{)} \qquad (5.9)$$

In a channel the available power supply Ω to unit length is the time rate of liberation in kinetic form of the liquid's potential energy as it descends the gravity slope S. Thus

$$\Omega = \rho g Q S \qquad (5.10)$$

where Q is the whole discharge of the stream. The mean available power supply ω to the column fluid over unit bed area is thus

$$\omega = \frac{\Omega}{\text{flow width}} = \frac{\rho g Q S}{\text{flow width}} = \rho g d S \bar{u} = \tau \bar{u} \qquad (5.11)$$

with the definition of τ given in Equation 5.6. The concept of available fluid power is an important one since Bagnold has made extensive use of it in his sediment transport theory (Ch. 6).

5e The Reynolds number

In 1883 Osborne Reynolds measured the pressure drop over a length of pipe through which tap water was passed at various speeds (Fig. 5.5). As we have seen before, loss of pressure is due to frictional losses as the real fluid moves through a system. Reynolds found that the pressure loss per unit pipe length increased with velocity but that at a certain point the losses began to increase more quickly. Below that point, or rather transition region, the graph is of a straight line such that

$$\Delta p = k_1 u \qquad (5.12)$$

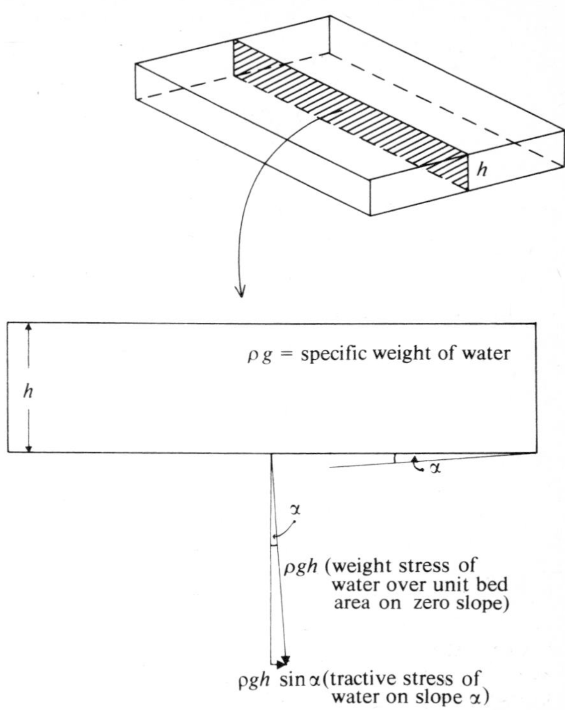

Figure 5.4 Derivation of the mean 'tractive' bed shear stress due to a uniform steady flow of water down a slope. ρ – water density.

where Δp is the pressure loss and k_1 is a constant. Above the transition region

$$\Delta p = k_2 u^n \qquad (5.13)$$

when n is between 1.75 and 2.0 and k_2 is another constant.

Deducing that the flowing water was changing its flow pattern, Reynolds confirmed this by introducing a dye streak into a steady flow of water through a transparent tube (Fig. 5.6). At low velocities the dye streak extended down the tube as a straight line and the flow is known as **laminar** or viscous flow. With increased velocity the dye streak was dispersed in **eddies** and eventually coloured the whole flow: this is **turbulent** flow.

The fundamental difference in flow types between the two flow regimes is one of the most important results in the whole field of fluid dynamics. Repetition of the pipe experiments with different viscosity fluids and different pipe diameters showed that the critical velocity for the onset of turbulence was not the same for each experiment. By applying the technique of **dimensional analysis** (see Middleton & Southard 1978 for an introduction), Reynolds found that the change from laminar to turbulent flow occurred at a fixed values of the quantity, defined as

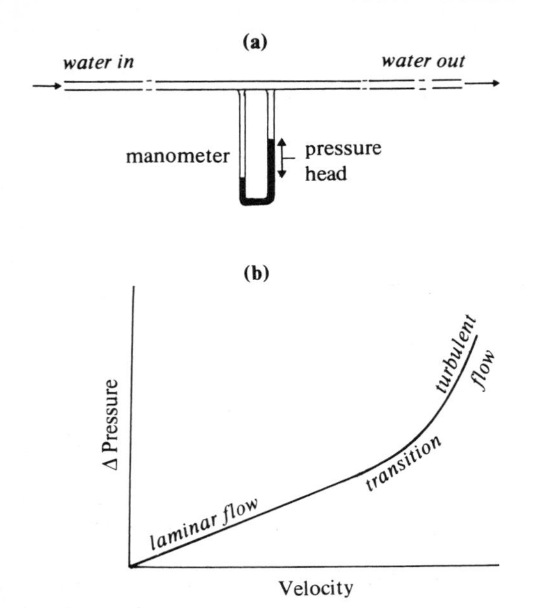

Figure 5.5 Diagrams to show (a) Reynolds apparatus and (b) results for pressure losses in pipes as a function of flow velocity.

$$Re = \frac{\rho d u}{\mu} = \frac{\bar{u} d}{\nu}, \qquad (5.14)$$

where $\bar{u}$ is mean flow velocity, ρ, μ are fluid density and viscosity ($\nu = \mu/\rho$), and d = internal diameter of pipe. Re has become known as the **Reynolds number** in honour of its discoverer.

We may think of the Reynolds number (which is dimensionless) as a ratio of two forces acting on the fluid. *Viscous forces* will resist deformation of the fluid: the greater the molecular viscosity the greater the resistance. *Inertial forces* will represent the resistance of the fluid mass to accelerations. The Reynolds number may be derived from first principles in this way, as shown in Appendix 5.2. When viscous forces dominate, as say in the flow of glycerine or syrup, then Re is small and the flow laminar. When inertial forces dominate, as in the atmospheric flow of air and most water flows in rivers, then Re will be large and the flow turbulent. For flows in pipes and channels the critical region for the laminar/turbulent transition lies between 500 and 2000. We should be careful, however, in any identification of laminar flow with high-viscosity liquids alone. The Reynolds criterion is clearly dependent upon four parameters of flow, not just one. Thus a very low density or very low velocity of flow has the same effect on Reynolds number as a very high viscosity. For example, bodies moving through air usually show turbulent effects, but a very tiny body such as a dust particle or flea might show viscous characteristics. Thus as Shapiro (1961) in his classic introductory text states '. . . it is more meaningful to speak of a *very viscous situation* than a very viscous fluid'. Flow systems with identical Reynolds numbers are said to be **dynamically similar**, a feature made use of in many modelling experiments.

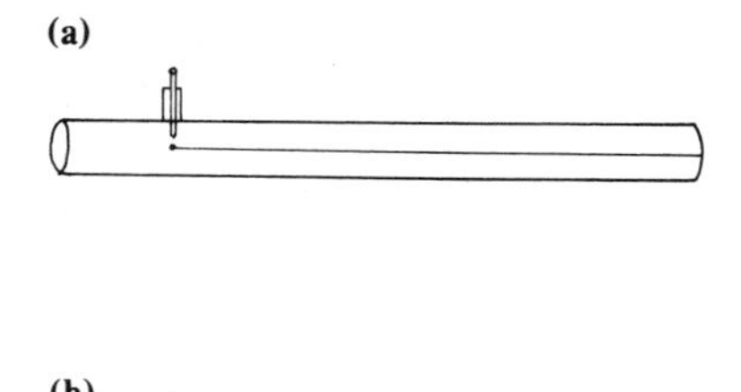

Figure 5.6 Reynolds' visualisation experiments with injected dye and glass tubes that established the existence of (a) laminar and (b) turbulent flow behaviour (see also Fig. 5.5).

Some comment is also necessary on the length scale in the Reynolds number criterion. For a pipe this is the diameter, but for a river channel or other free-surface liquid flow the mean flow depth is the correct scale. For a wind the length scale presents more problems. In an air tunnel this may be taken as the height of the tunnel, but in atmospheric flows the boundary layer thickness must be chosen. For a *grain* Reynolds number the mean diameter is used as a length scale.

5f Froude number

In Section 5e and Appendix 5.2 the dimensionless Reynolds number is defined and derived. A second relevant dimensionless number is the **Froude number**, applicable to laminar or turbulent flows having a free surface or interface such that gravity forces play an important role in causing flow. The Froude number is given by the ratio

$$\frac{\text{inertia force}}{\text{gravity force}} \quad \frac{\rho l^2 u^2}{\rho l^3 g} = \frac{u^2}{lg} = \frac{u}{\sqrt{lg}}$$

where ρ = fluid density, l is a characteristic length and u is a characteristic velocity. The ratio is named after William Froude, a pioneer naval architect who first introduced it.

Another way of appreciating the significance of the Froude number is to recognise that the expression $\sqrt{lg}$ is the velocity (celerity) of a small gravity wave in still shallow water (see Appendix 18.1). The Froude number is thus the ratio of flow velocity to the velocity of a small wave created in the flow. When the Froude number is less than unity, then the wave velocity is greater than the flow velocity, i.e. waves from a pebble thrown into a flow can travel upstream. Such a flow is said to be **tranquil**. For a Froude number of greater than unity the flow is said to be **rapid**.

Densimetric Froude numbers may be defined for density currents (Chs 7 & 19) as

$$u\bigg/\left(lg\,\frac{\Delta\rho}{\rho}\right)^{\frac{1}{2}}$$

where $\Delta\rho$ is the density difference layers and ρ is mean density.

5g Laminar flows

In natural flows, laminar behaviour is not as common as turbulent behaviour. Flows of both ice and mud-supported debris are laminar, though neither substance is Newtonian. In the wider geological context lava flows and magma or mantle convective flow are both laminar.

Figure 5.7 shows idealised laminar flow of a Newtonian fluid such as water past a lower fixed boundary. As noted above, molecular adhesion causes the fluid in contact with the boundary to remain at rest, but successive overlying 'layers' of fluid will slide relative to the layer beneath at a rate determined by the magnitude of the molecular viscosity. A velocity gradient results. Consider a rectangular element of fluid (abcd) lying between streamlines. In unit time this element will deform to the parallelogram (a′b′c′d′). The shear deformation occurs due to flow since the velocity of (a) and (b) exceeds that of (c) and (d) by an amount δu. Thus the viscous shear strain or velocity gradient

$$=\gamma=\frac{\delta u}{\delta y}=\frac{du}{dy}$$

and the viscous shear stress (τ) is the molecular viscosity times the viscous shear strain

$$\tau=\mu\,\frac{du}{dy} \tag{5.15}$$

This simple relationship was first proposed by Newton and, as noted previously, fluids in which viscosity does not change with increasing strain are known as Newtonian fluids.

We have now arrived at the point where some discussion of velocity distribution and boundary layer thickness is necessary. As already mentioned boundary layers arise

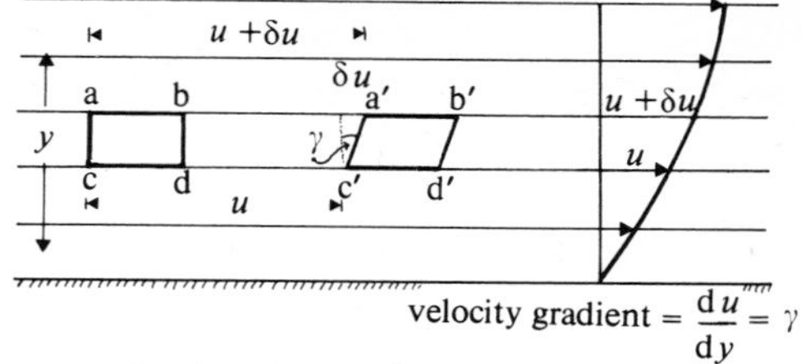

Figure 5.7 Laminar flow of fluid over a fixed bed. Molecular adhesion causes the fluid at the bed to remain stationary, with successive overlying layers of fluid sliding relative to those layers beneath at a rate dependent upon the fluid viscosity. A boundary layer therefore exists. (Explanation in text.)

from frictional retardation against some stationary solid or liquid surface. In fact the curve of velocity against height approaches the perpendicular to the boundary asymptotically. This might suggest that the boundary layer occupies the whole flow depth. In practice the thickness of the boundary layer is defined by the distance from the solid boundary where the velocity becomes equal to 99% of the free stream velocity. Let us consider the velocity distribution in a laminar flow between two walls, as in a channel (Fig. 5.8). If we measure point velocities across such a flow then a characteristic parabolic curve results. This parabolic curve may be exactly predicted by simple theory when the pressure and viscous forces acting on the fluid are balanced (Appendix 5.3).

Consider the growth of a boundary layer as a laminar flow passes over a thin flat plate orientated parallel to flow (Fig. 5.9). As the flowing fluid reaches the plate surface, friction will cause a viscous boundary layer to develop. With increasing distance from the leading edge, the larger surface area slows more fluid down and the boundary layer will grow to some equilibrium thickness. Knowledge of the growth of the boundary layer in terms of distance from the leading edge of the plate gives rise to

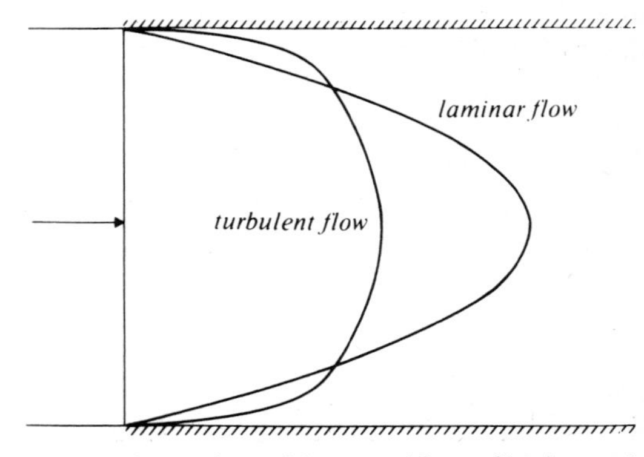

Figure 5.8 Comparison of the general form of laminar and turbulent velocity profiles as observed from above in a flow channel.

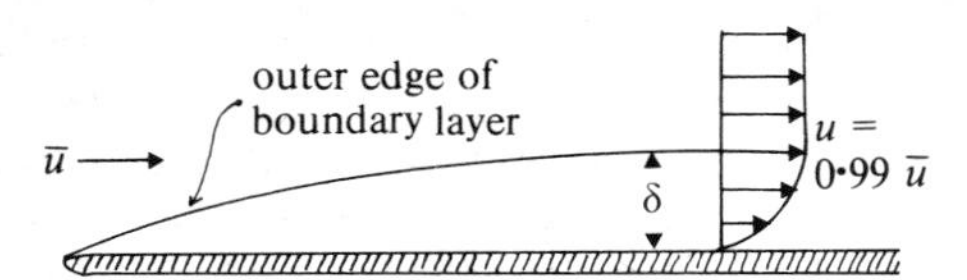

Figure 5.9 Growth of a boundary layer as fluid passes over a rigid plate. The free stream region begins at $u = 0.99\bar{u}$.

expressions for the fluid shear or drag exerted by the fluid on the surface of the plate.

The properties of boundary layers, more especially turbulent boundary layers (Ch. 5h) have a number of very important consequences in sedimentology. A particularly relevant point concerning the distribution of velocity in laminar flows comes when we consider the flow of a non-Newtonian fluid such as a mud-supported debris flow (Ch. 7) or a lava flow. The local velocity, u, at a point in a laminar flow may be related by the mean velocity, $\bar{u}$, at distance y from the boundary in a flow of thickness $2Y$ by

$$u = \bar{u}2\left[1 - \left(\frac{y}{Y}\right)^2\right] \tag{5.16}$$

yielding the parabolic shape noted previously. For a non-Newtonian fluid

$$u = \bar{u}\left(\frac{3n + 1}{n + 1}\right)\left[1 - \left(\frac{y}{Y}\right)^{n + 1/n}\right] \tag{5.17}$$

where n (<1) is a parameter that measures the degree of non-Newtonian behaviour. Fluids are Newtonian when $n = 1$. As shown in Figure 5.10, decreasing n gives rise to plug-like profiles and, in the extreme limiting case as $n = 0$, the velocity is uniform across the whole flow.

The shape of the plug-like profile for $n = \frac{1}{3}$ suggests that non-Newtonian flow will consist of well defined solid plugs of material which ride above a shearing boundary layer in contact with a solid surface. The velocity gradients, and hence the strain rates, are much higher in the boundary layers of laminar non-Newtonian fluids when compared to laminar Newtonian fluids, other things being equal. Plug flow gives rise to a number of features of sedimentological interest (Ch. 7).

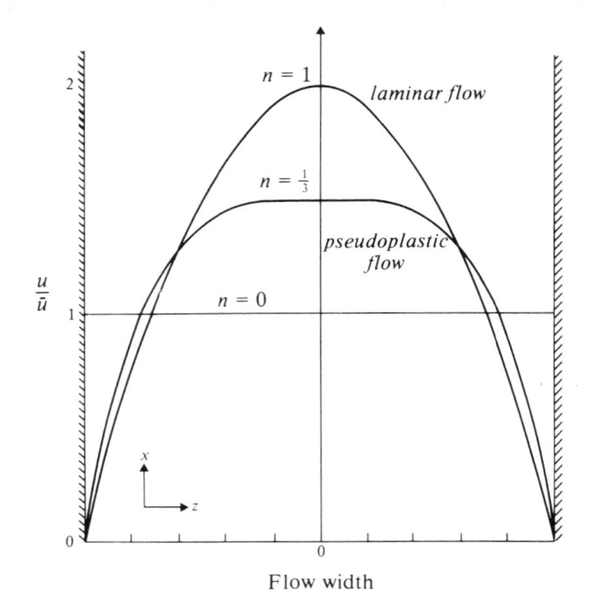

Figure 5.10 Velocity profiles of the laminar flow of a Newtonian fluid ($n = 1$; parabolic profile), a pseudoplastic non-Newtonian fluid ($n = \frac{1}{3}$; plug profile) and the limiting profile of $n = 0$ (after Komar 1972). Discussion in text. u – local velocity; $\bar{u}$ – mean velocity.

5h Introduction to turbulent flows

Turbulent flows dominate Earth surface transport processes – most wind and water flows are turbulent. Yet analysis of turbulent effects presents a formidable mathematical and physical challenge. It is one of the great (unfinished) achievements of twentieth-century fluid physics that many of the complexities of turbulent motion in shear flows have been made clear by ingenious experiment and the bold application of physical/mathematical hypotheses. Much of the necessity for understanding turbulence has come from the field of aeronautics and it is perhaps no coincidence that 'modern' fluid dynamical analysis (Reynolds, Prandtl) started around the date of man's first few uncertain attempts at controlled flight. Eighty years later man was able to photograph the effects of turbulent atmospheric flows on Earth while standing on the Moon! Let us try in the remainder of this and in the next chapter to extract the basic physical ideas and significance of turbulent effects for sedimentological studies.

Insertion of a sensitive flow-measuring device into a turbulent flow will result in a fluctuating record of fluid velocity with time (Fig. 5.11). The mean flow velocity, $\bar{u}$, plus the instantaneous deviation u' from the mean will equal the instantaneous velocity at that point. In symbols

$$u = \bar{u} + u' \tag{5.18}$$

Thus in turbulent flows we can only talk about a characteristic time–mean-flow velocity.

The use of devices such as electromagnetic flow meters, which can measure instantaneous velocity fluctuations along the three Cartesian co-ordinates, reveals that turbulent 'eddies' are three-dimensional and, therefore, at any point in time

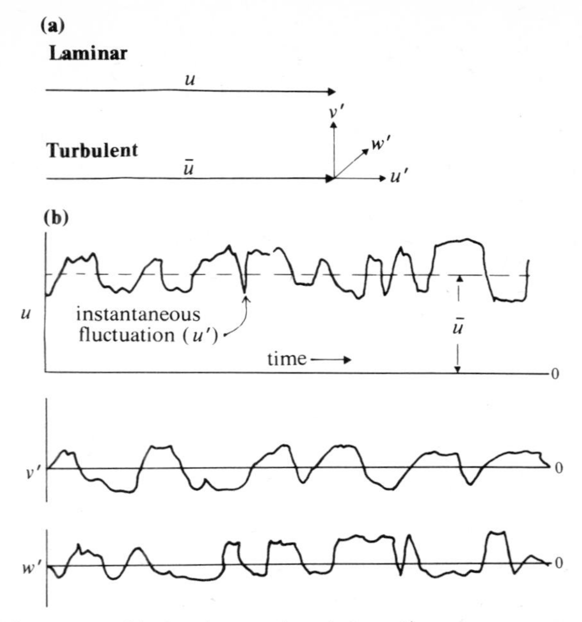

Figure 5.11 (a) Laminar and turbulent flow expressed as vectors. Steady laminar flow at a point is represented by a single vector of length u. Steady turbulent flow of the same magnitude is represented by a mean flow vector $\bar{u}$, with the addition of fluctuating velocities u', v' and w', which average out to zero over a sufficiently long period. (b) An alternative way of representing steady turbulent flow is to plot velocity at a point against time, all three components of motion u', v' and w' being shown. Components u', v' and w' average to zero over a long period.

$u = \bar{u} + u'$ measured along x-axis, parallel to flow

$v = \bar{v} + v'$ measured along y-axis, normal to xz plane

$w = \bar{w} + w'$ measured along z-axis

where $\bar{v}$ and $\bar{w}$ are usually small in comparison with $\bar{u}$ for most shear flows of sedimentological interest. By definition, the mean values of u', v' and w' taken over a long time interval are zero.

There is a very important consequence of the three-dimensional nature of random turbulent fluctuations. Local velocity gradients are set up in the flow *in addition* to those viscous effects that dominate the laminar flows discussed previously. These local velocity gradients cause local stresses which work against the mean velocity gradient to remove energy from the flow. The action of these local stresses provides turbulent energy which is ultimately dissipated by the action of viscosity on the turbulent fluctuations. Thus we now have an explanation of the greatly increased energy losses in Reynolds' pipe flow experiment for turbulent flow as compared with laminar flow.

The magnitude of the local stresses, termed **Reynolds stresses**, will obviously depend upon the magnitude of the instantaneous velocity fluctuations. Reynolds stresses have the form $\rho\overline{u'v'}$ or $\rho\overline{u'w'}$ or $\rho u'^2$ and must be regarded as additive to the viscous stresses due to fluid viscosity. Thus Newton's law for laminar flows

$$\tau_1 = \mu \frac{du}{dy} \tag{5.19}$$

is replaced in turbulent flows by

$$\tau_t = (\mu + \eta) \frac{du}{dy} \tag{5.20}$$

where η is an additional term, called the **eddy viscosity**, which takes into account the resistance to shear due to eddy motions of fluid masses. Unlike μ, η is a variable quantity dependent upon the size and velocity components of the eddies. It is usually much larger than μ. Equation 5.20 may alternatively be written as

$$\tau_t = \mu \frac{du}{dy} - \rho\overline{u'v'} \tag{5.21}$$

for the two-dimensional flow case, where the viscous stress term is negligible.

A very interesting pattern emerges if we measure the magnitude of the turbulent velocity fluctuations across a flow. It is found that these are at a maximum quite close to the boundary, but as the boundary is approached viscous stresses predominate (Fig. 5.12). As we shall see below (Chapter 5i) this feature may be explained by assuming a quite definite structure to turbulent flow.

Let us now turn to the distribution of mean downstream flow velocity with height in turbulent flows over smooth surfaces. Measurements of point velocities in a steady flow (we may talk about a steady turbulent flow if $\bar{u}$ is constant over some time interval) show that, close to the lower boundary, there is an approximately linear increase of velocity with height, but then a slower increase through a transition zone, in the form of some power function. The lowest zone of linear increase has been named the laminar sub-layer, but since it is not truly laminar it is better termed the **viscous sub-layer**. In this thin layer the stress is transmitted to the surface wholly by molecular viscosity forces. Above the viscous sub-layer the velocity is proportional to the logarithm of height, as can be shown by replotting our experimental data on semi-logarithmic graph paper, yielding a straight line (Fig. 5.13). It can be seen that the velocity distribution curve intersects the ordinate above the surface,

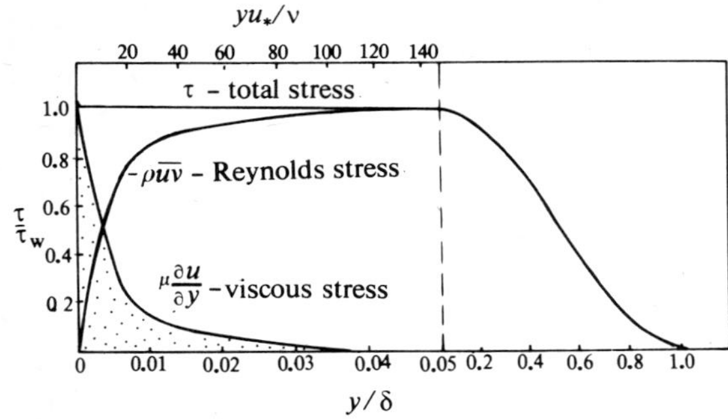

Figure 5.12 Distributions of the total stress (τ), Reynolds stress ($-\rho uv$) and viscous stress $\mu(\partial u/\partial y)$ across a turbulent boundary layer ($Re = 7 \times 10^4$). Note thirty-fold change in abscissa scale at $y/\delta = 0.05$. The only stress exerted directly on the bed is a viscous one but away from the bed turbulence generates an increasingly large Reynolds stress contribution (after Tritton 1977). τ_w – wall stress; δ – flow depth; y – height from bed; u_* – shear velocity; ν – viscosity. Viscous sub-layer extends to $yu_*/\nu = 11.5$.

implying that the fluid seems to be stationary at a small distance above that surface. This may be explained by the presence of the viscous sub-layer, since the logarithmic curve actually passes into the linear curve of velocity against height in that region, which then passes into the zero ordinate at zero height. In flows passing over rough surfaces composed of coarse sand grains, the height of the intersection of the velocity curve with the ordinate is unchanged with increased velocity. If the rough surface is immobile, then the height of the focus of all the different velocity curves (k; Fig. 5.13) is approximately equal to 1/30th of the diameter of the roughness elements.

We must now define and explain the significance of the quantity u_*, known as the **shear** or **drag velocity**. u_* has the dimensions of a velocity and is defined as

$$u_* = \sqrt{\frac{\tau_0}{\rho}} \tag{5.22}$$

where τ_0 = fluid shear stress on boundary and ρ = fluid density.

u_* is directly proportional to the rate of increase of fluid velocity with log-height and is therefore proportional to the slope of the velocity distribution curve on Figure 5.13. The proportionality constant in the lowest 10–20% of the boundary layer is 5.75. Thus u_* is a velocity, the magnitude of which is a measure of the fluid velocity gradient.

In Figure 5.13 the tangent of the velocity ray is AC/CB. u_* is then equal to (AC/CB)/5.75. Choosing points A and B so that the height of A is ten times that of B, log-height CB is then $\log_{10}10 - \log_{10}1 = 1$, so that in this case $u_* = \text{AC}/5.75$. Since the shearing stress of the fluid on the solid surface is given by

$$\tau = \rho u_*^2 \tag{5.23}$$

then the velocity difference between any two levels of which one is ten times the height of the other gives $5.75u_*$ and hence τ.

Also with reference to Figure 5.13, if u_* and the roughness constant k are known, we may find the velocity u at any height z as follows. Let the velocity u be DE. By proportion, DE = AC/CB. EO, given above as $5.75\,u_*$ EO.

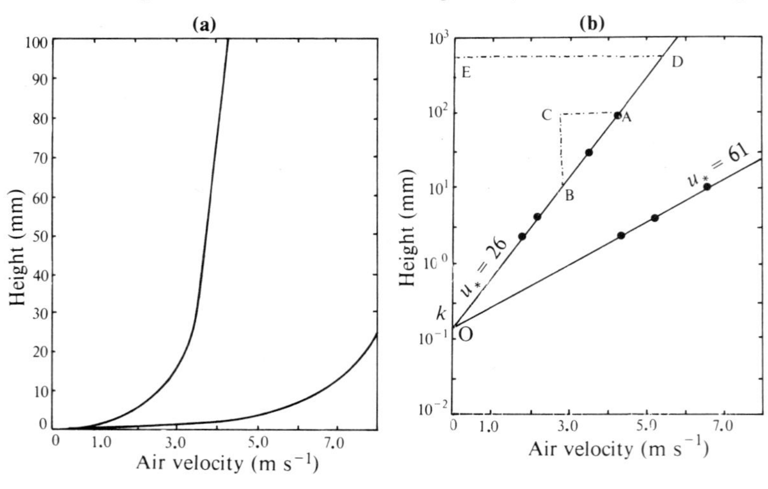

Figure 5.13 Plots of mean velocity versus height above a bed for two air flows. (a) Shows the velocity profiles on a linear ordinate scale. Owing to their flatness close to the bed the curves are awkward ones from which to derive information. (b) Shows the same profiles plotted on to a logarithmic ordinate scale from which graphical deductions may be made more easily. Discussion in text. (After Bagnold 1954b.)

But EO is the difference ($\log z - \log k$) between the log heights at E and O. Writing $\log z - \log k$ as $\log z/k$ we have

$$u = 5.75\, u_* \log \frac{z}{k}$$

or

$$u = 5.75 \sqrt{\frac{\tau_0}{\rho}} \log \frac{z}{k} \qquad (5.24)$$

Equation 5.24 is the famous Prandtl equation for the logarithmic distribution of velocity in a turbulent flow and is applicable to the lower 10–20% of a flow above the thin viscous sub-layer. The constant 5.75 arises from the Prandtl–Karman theory of turbulent mixing lengths, where the length in question is the average distance that a mass of fluid travels before it is merged in with that of fluid at a new level. There is an outline derivation of the theory in Appendix 4.

As the turbulent flow speeds up and/or the viscosity decreases, then the viscous sub-layer should get thinner. Experimental measurements show that δ, the thickness of the sub-layer, is given by

$$\delta = 11.5\nu/u_* \qquad (5.25)$$

This brings us to an important point. Surfaces whose roughness elements of sedimentary particles are enclosed entirely within the viscous sub-layer are said to be **smooth** (Fig. 5.14). When the particles project through the sub-layer, as assumed in the development above, they shed off small eddies. The surface is then said to be **transitional** or **rough** (Fig. 5.14). The graph of Figure 5.15 shows δ in relation to flow strength for sand grains in air and water flows, plotted so that the critical u_* is that appropriate to the threshold of grain motion (see Ch. 6). In water, for example, the flow boundary ceases to behave smoothly at the threshold of motion for all grain diameters greater than about 0.6 mm. As we shall see in Chapter 8, profound sedimentological consequences follow from this fact.

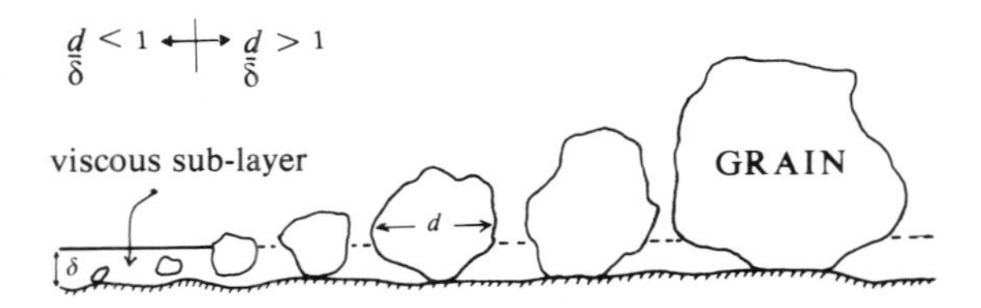

Figure 5.14 Smooth and rough boundaries. In smooth boundaries $d/\delta < 1$, for which the grain Reynolds number quantity $du_*/\nu < 5$. Transitional boundaries exist in the region $du_*/\nu = 5$ to 70. Rough boundaries correspond to d/δ values of about 14 when $du_*/\nu > 70$.

5i The structure of turbulent shear flows

The past fifteen years have seen a remarkable increase in our understanding of the structure of turbulent flows. Although the mathematical development of hypotheses for turbulent behaviour remains limited, direct observation and analysis of turbulent structure is now possible using a flow visualisation technique based upon hydrogen bubbles developed at Stanford University (Kline *et al.* 1967). Extension of this work to problems of direct sedimentological concern by workers at University College, London, (Grass 1971, Williams & Kemp 1971) has shed light on a number of interesting features.

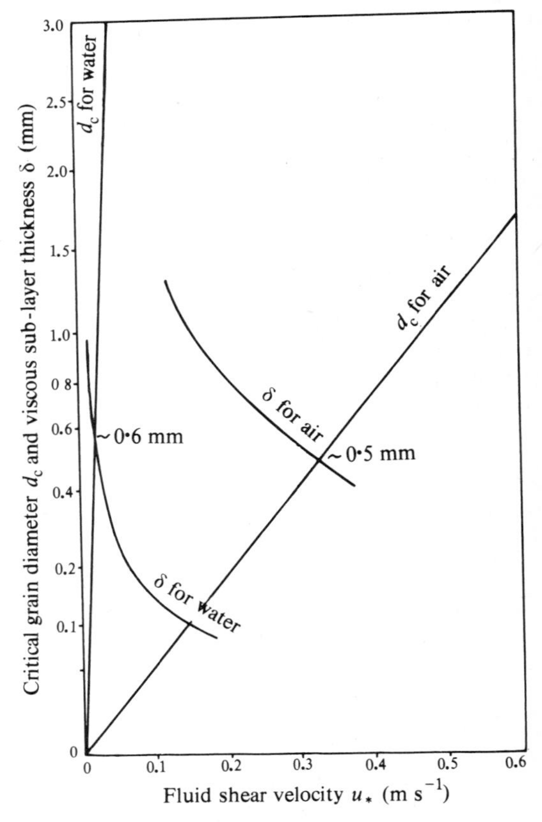

Figure 5.15 Plot of fluid shear velocity against grain diameter and viscous sub-layer thickness. Intersections of the δ and d_c curves for air (~0.5 mm) and water (~0.6 mm) define the value of grain size at which the grains will cause the disruption of a smooth boundary at the threshold of motion (after Carson 1971). See also Chapters 6 and 8.

The hydrogen bubble technique makes use of the electrolysis of water (Fig. 5.16) to produce tiny groups of H_2 bubbles from very thin platinum wires fixed either normal or parallel to the bed. A positive voltage is applied to the anode plate located on the wall of the channel and a negative voltage to the platinum wire. The platinum wire is speck insulated and a pulsed voltage input then produces blocks of minute H_2 bubbles in a very regular pattern. Intense illumination and high-speed cine camera shots then reveal the downstream fate of the bubble blocks and provide good visualisation of the flow structure and direct measurement of instantaneous velocities throughout the flow depth. Thus far, results have been obtained for mean flow velocities of $\lesssim 0.20$ m s^{-1}.

Let us first discuss the visual evidence provided by these experiments. Figure 5.17 shows views in the *xz* plane taken of bubble blocks at four depths in the flow. Each view shows a single layer of bubble blocks released at the particular height, the remainder of the platinum wire being insulated.

Close to the smooth channel bed, in the viscous sub-layer, a striking regularity of structure is evident, with downstream elongate wavy 'streaks' spaced with some regularity across the flow. Velocity measurements show that these streaks are composed of low-speed fluid. Seen from the side, in the *xy* plane (Fig. 5.18) these streaks are identified by lower than average fluid velocity and are seen to waver and oscillate within the sub-layer, intermittently 'leaping' outwards into the main turbulent boundary layer (Fig. 5.19). The outward migrations of low-velocity fluid are known as **bursts** and are associated with lower than average u velocities and positive (upward) v velocities. The low-speed streaks in the viscous sub-layer alternate across the flow with high-speed u areas associated with negative (downward) v velocities. The inward migrations of high-velocity fluid from just above the viscous sub-layer are known as **sweeps**. Graphic illustration of a sweep event is provided by Figure 5.17e where the *xz* view of the floor of the channel shows intensively illuminated 0.1 mm sand particles acting as flow tracers. The arrows show the inrush sweep phase which violently transports the sand particles forwards and outwards as it progresses down the channel floor, the particles appearing as lines because of their high velocity. Adjacent to the sweep phase the particles appear as dots, showing that the low-speed streaks are present, alternating across the flow with inrush events.

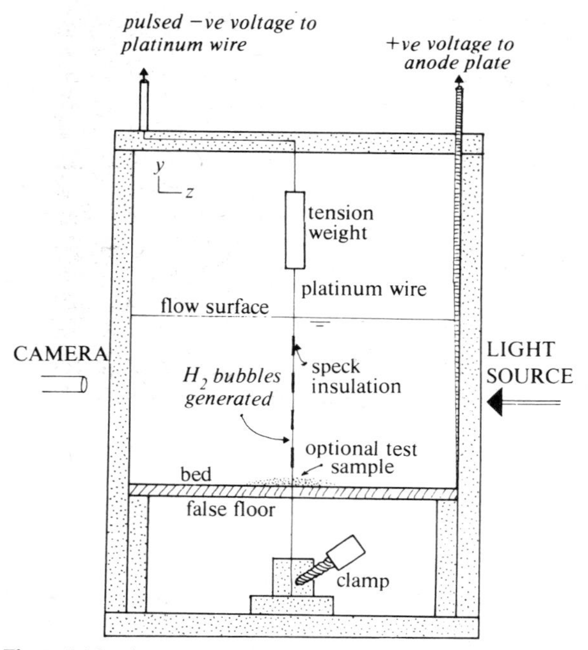

Figure 5.16 Apparatus used to generate hydrogen bubble blocks for turbulent flow visualisation studies. Flow is towards or away from the observer. (After Williams & Kemp 1971.)

Moving away from the bed, the streaky structure becomes less evident, the low speed streaks from burst events becoming 'tangled' as they rise through the boundary layer (Figs 5.17 & 5.19). Higher in the logarithmic zone the impression is of random turbulence and in the topmost zone (the so-called **wake** region) the turbulence is clearly intermittent and of larger scale than in the inner layers. Turbulent blobs may reach the surface as 'boils', causing an elevation of the water surface (Fig. 5.17).

We obviously conclude from these visual impressions that turbulent structure is spatially organised. The streak spacing, λ_s, in the viscous sublayer is given by the experimental relation

$$\lambda_s u_* / \nu \simeq 100 \tag{5.26}$$

where u_* is the shear velocity and ν the kinematic viscosity. Thus with constant ν, λ decreases as u_* increases. The rate of bursting of low speed streaks increases with u_*. It should be noted that the streak pattern is quasi-cyclic, new streaks forming and reforming constantly across the flow.

Turning now to some of the measurements made upon the bubble blocks we see in Figure 5.18 that minimum local longitudinal u velocities are directly correlated with regions of positive vertical v velocity and vice versa. These results suggest high positive contributions to local Reynolds stresses across the whole flow depth due to burst phases when migration of momentum-deficient fluid occurs. The inrush/sweep phases also give a positive contribution to Reynolds stresses but their effect is at maximum in the area close to the wall. We conclude that

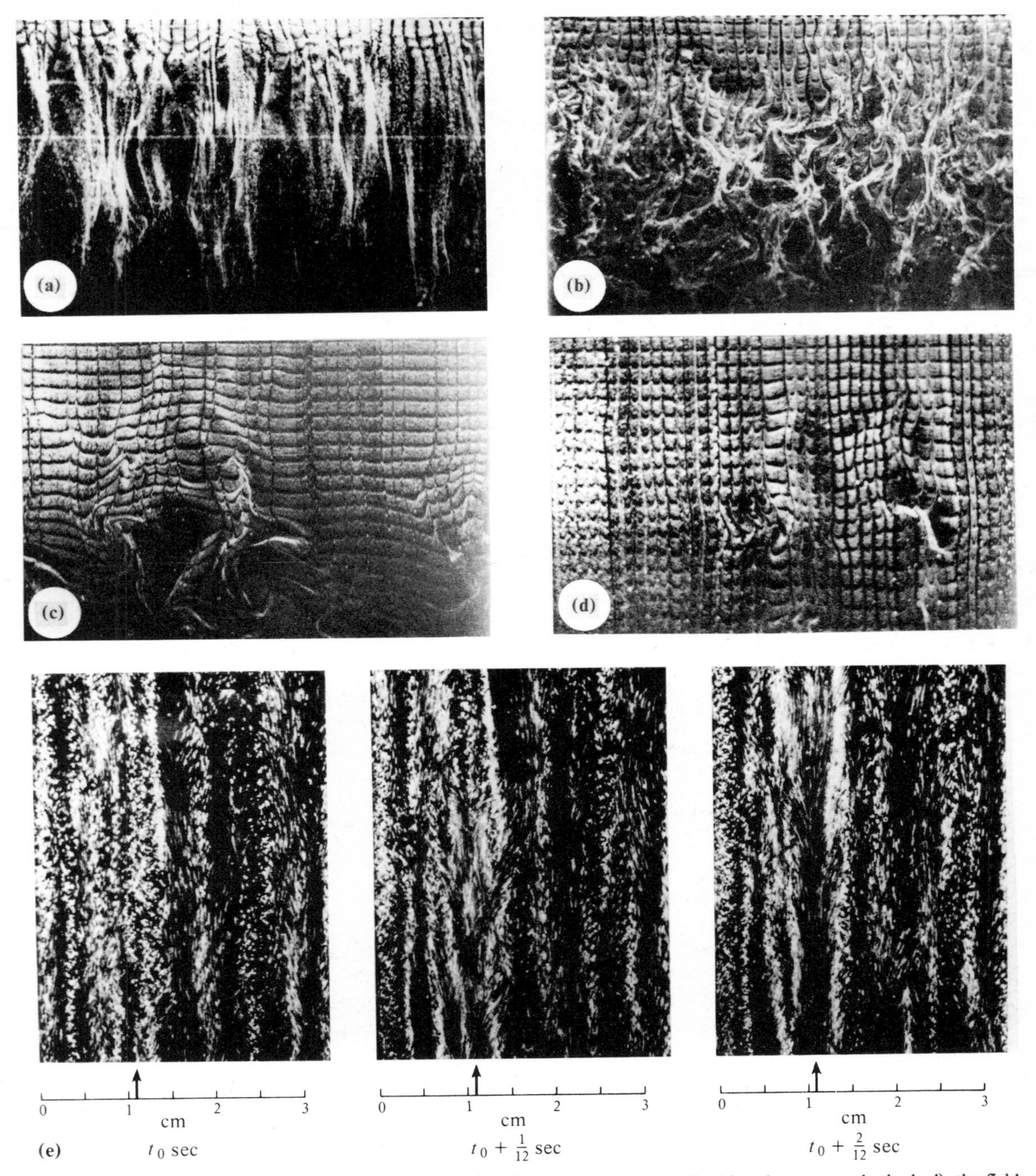

Figure 5.17 (a–d) Instantaneous photographs of H_2-bubble blocks taken from above (looking down towards the bed), the field of view being in the *xz* plane with current flow from top to bottom of each photograph. The speck-insulated platinum wire where the bubble blocks are periodically generated is at the top of each photograph. The sequence (a–d) represents successively higher positions of the wire above the bed. In (a) the streaky deformation of the bubble blocks is well shown, each streak identifying a low-speed phase of the cycle in the viscous sub-layer. In (b) the streaks have become tangled and less obvious as they pass into the logarithmic portion of the turbulent boundary layer. In (c) and (d) the blocks are mostly undisturbed in the outer regions of the flow but larger areas of macroturbulence are prominent (after Kline *et al.* 1967). (e) Viscous sub-layer structure visualised by means of 0.1 mm diameter sand moving over a smooth black boundary. The sequence of photographs, separated in time by $\frac{1}{12}$ s with a $\frac{1}{30}$ s exposure, illustrate the development of an inrush or 'sweep' event (arrowed). $u_* = 21.3$ mm s^{-1}. (After Grass 1971.)

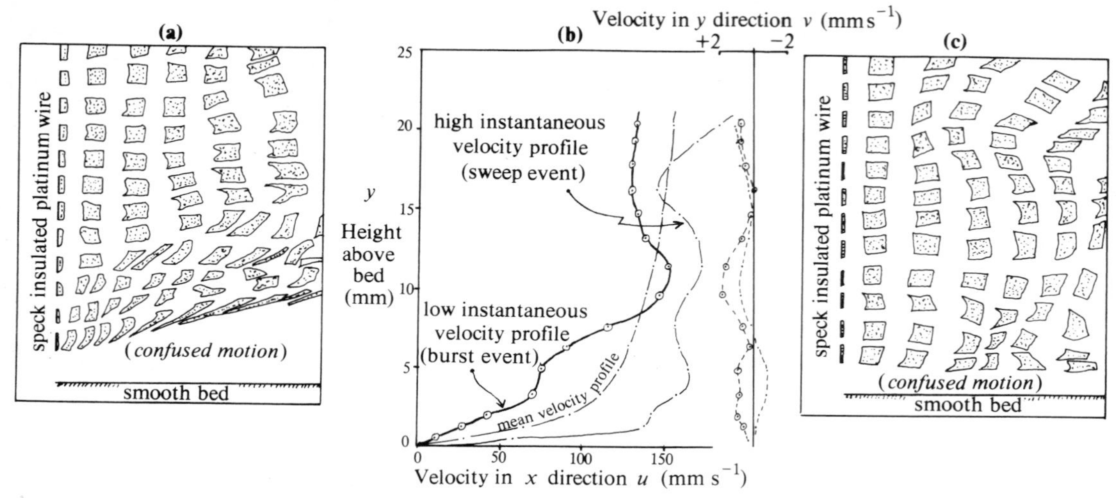

Figure 5.18 Examples of H_2-bubble flow visualisation over smooth (plywood) beds. Sketches (a) and (c) are from instantaneous photographs of bubble-block deformations associated with burst and sweep events respectively. (b) shows the x and y velocity profiles measured from such individual film frames together with a mean velocity profile measured from many film frames (after Grass 1970).

almost all (~70%) of the Reynolds stresses in turbulent flows are due to the burst/sweep processes and that the majority of these stress contributions are produced close to the wall.

Of great sedimentological interest are experimental results (Grass 1971) on burst/sweep phenomena over boundaries roughened by sediment grains glued one grain thick over the flow channel bed. 2 mm coarse sand and 9 mm pebbles provided transitional and rough boundaries respectively at flow velocities well below movement threshold for loose grains. The three boundary conditions were analysed using flows of constant Reynolds number, i.e. u, ν and h constant. Increasing boundary roughness caused increasing mean bed shear stress, as we would expect. The turbulence intensity data scale directly with u_* *independent* of roughness conditions for $h/d \gtrsim 0.2$ (Fig. 5.20), implying that beyond a certain height the intensity depends solely on boundary distance and shear stress but is independent of the conditions producing the shear stress. Closer to the bed the data separate so that with increasing boundary roughness the longitudinal intensity decreases and the vertical intensity increases (Fig. 5.20). The average Reynolds stress measurements correlate with a linear mean shear stress distribution tending towards zero at the flow free surface.

We can envisage the smooth boundary viscous sub-layer fluid and the fluid trapped between the roughness elements as 'passive' reservoirs of low-momentum fluid which is drawn on during ejection phases. Entrainment of this fluid is extremely violent in the rough boundary case, with vertical upwelling of fluid from between the roughness elements. Very significantly the streaky pattern of the viscous sub-layer observed on the smooth boundary is much less conspicuous in the transitional and rough boundary flows. Faster deceleration of sweep fluid on the rough boundary due to form drag of the grains causes the decrease of longitudinal (u) turbulent intensity and the increase of vertical (v) turbulent intensity noted above. The Reynolds stress contribution ($\overline{u'v'}$) is also increased by increased roughness close to the wall region.

We may summarise the above discussion by dividing the turbulent boundary layer into two rather distinct zones: (a) an inner zone close to the bed with its upper boundary between the transition and logarithmic region of the turbulent boundary layer, and (b) an outer zone extending up to the flow free surface. The inner zone is distinguished by

(a) being the site of most turbulence production,
(b) containing low- and high-speed fluid streaks that alternate across the flow,
(c) lift-up of low-speed streaks in areas of high local shear near the upper boundary.

The outer zone

(a) provides the source of the high-speed fluid of the sweep phase near its lower boundary that probably initiates a burst cycle,
(b) contains large vortices near the area of burst break-up which are disseminated through the outer zone and may reach the surfaces as 'boils'.

These conclusions relate directly to the origin of a range of sedimentary structures and to the mechanism of turbulent suspension (Chs 6 & 8).

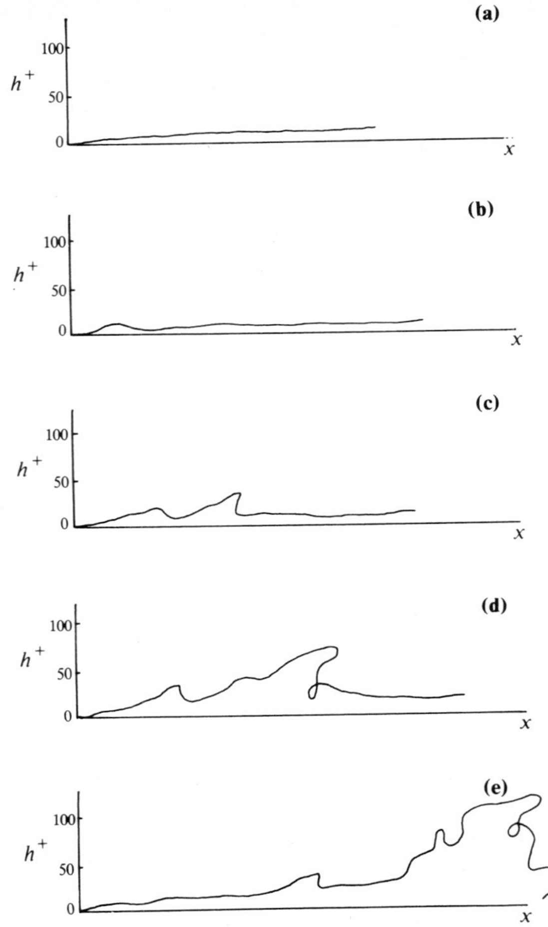

Figure 5.19 Growth of a single burst event (a–e) as illustrated by the breakup of a dye streak injected into a turbulent flow from the bed. Most of the time (a & b) the streak pattern migrates slowly downstream (x-direction) as a whole, with each streak drifting very slowly outwards. When the streak has reached a point corresponding to $h = 8$–12, it begins to oscillate (b & c) the oscillation amplifying and then terminating in an abrupt burst 'breakup' followed by contortion and stretching (c, d, e). h^+ is the dimensionless expression for the y scale hu_*/ν, where h = height from bed, u_* = fluid shear velocity, ν = kinematic viscosity. (After Kline *et al.* 1967.)

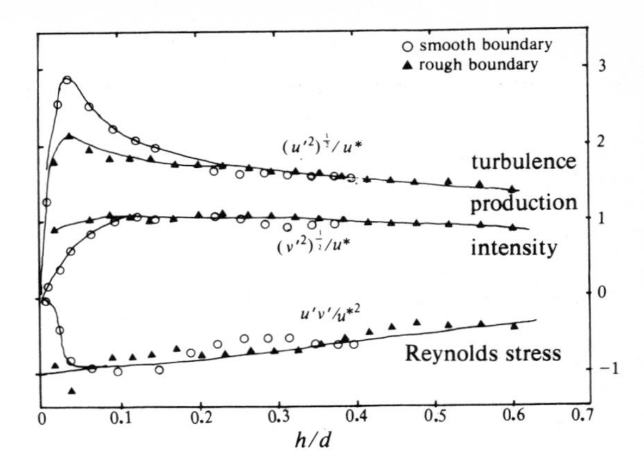

Figure 5.20 Graph to show the effect of a rough boundary upon turbulence production intensity and Reynolds stresses in turbulent flows of similar Reynolds Numbers ($Re = 6700$). Note the increase of vertical (y) turbulence intensity $(v'^2)^{\frac{1}{2}}/u_*$ and the decrease in horizontal (x) turbulence intensity $(u'^2)^{\frac{1}{2}}/u_*$ from smooth (polished plywood) to rough (9 mm pebbles) boundaries close to the bed. The turbulence intensity becomes independent of roughness for $h/d > 0.2$ (h = height above bed, d = flow depth) (after Grass 1971).

5j Flow separation and secondary currents

Consider a flow in a channel that widens downstream and is then of constant width (Fig. 5.21). In an ideal fluid, the streamlines will diverge and then become parallel. From continuity (Ch. 5c) the flow would slow down in the expanding section of the channel and again become constant downstream. Now if we think back to Bernoulli's equation (Ch. 5d) we can deduce that pressure should increase in the expanded flow section but be constant both upstream and downstream. A pressure and velocity gradient should therefore exist across the expansion such that $dp/dx > 0$; $du/dx < 0$.

In a real fluid with boundary layers the pressure gradients so produced will have greatest effect on the lower-speed fluid near to the wall in the expanding flow section. This fluid will be more easily retarded by the adverse pressure gradient. Under certain circumstances the pressure gradient will be able to push fluid close to the wall upstream, as shown in Figure 5.21. **Boundary layer separation** may take place from point S, named the **separation point**, and **boundary layer attachment** will eventually occur downstream. Within the streamline joining S and the reattachment zone will be a closed, recirculating **separation bubble**.

Boundary layer separation also occurs around spheres or cylinders placed in the flow or when a **negative step** or small defect occurs on a bed (Figs 5.22–24). In the first case there is an adverse pressure gradient behind the sphere or cylinder where the streamlines diverge and in the second case an adverse pressure gradient exists due to a sudden flow expansion. Both cases are of direct sedimentological interest (Chs 6 & 7). Of more direct practical importance is boundary layer separation on the top

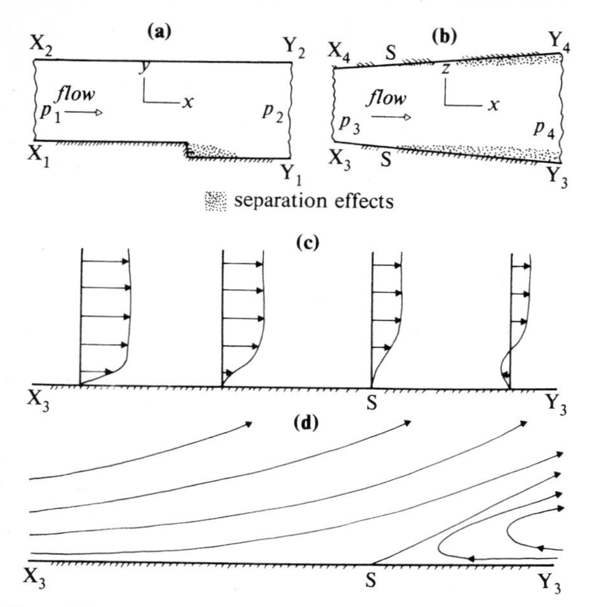

Figure 5.21 Flow separation effects of (a) a negative step approximating to a ripple crest in nature, (b) a widening channel such as a delta distributary channel in nature. (c) and (d) show schematic velocity profiles and streamlines for sections such as X_3Y_3. Discussion in text. p = pressure, S = separation point.

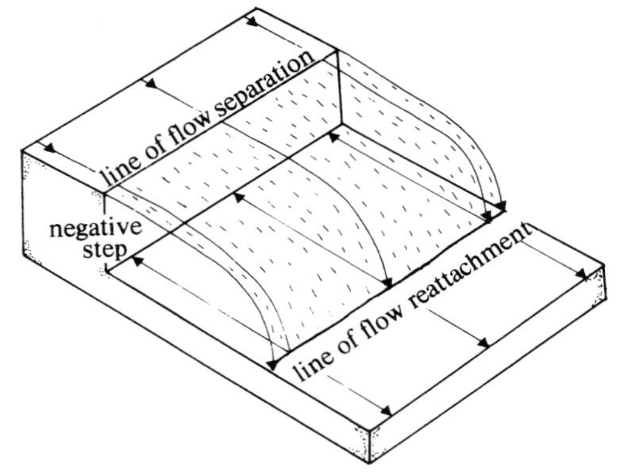

Figure 5.23 Sketch of a roller vortex due to flow separation and reattachment over a negative step approximating to a ripple lee slope in nature (after J. R. L. Allen 1968).

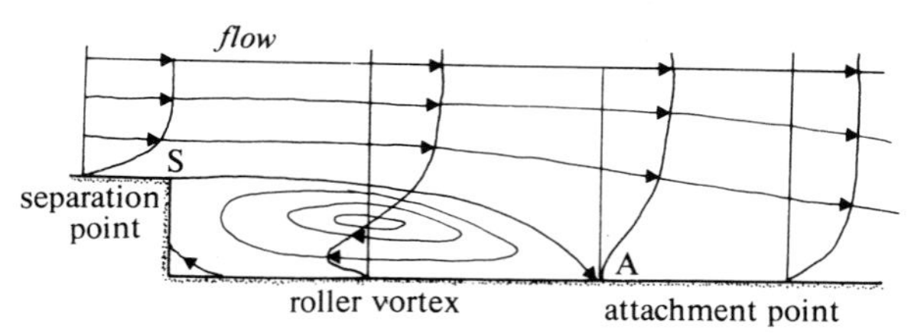

Figure 5.24 Section through roller vortex of Figure 5.23 to show time average streamlines and schematic velocity profiles.

surface of a highly inclined aerofoil which greatly reduces the lift force and causes stall.

In fact, both laminar and turbulent boundary layers can separate, the former usually separating more readily. However, in sedimentological fields we are most interested in turbulent separation. Looking a little closer at the separation bubble of a negative step, such as in the lee of a current ripple, we can define **vortex bubbles**, formed when the step is skewed up to 45° from the flow direction,

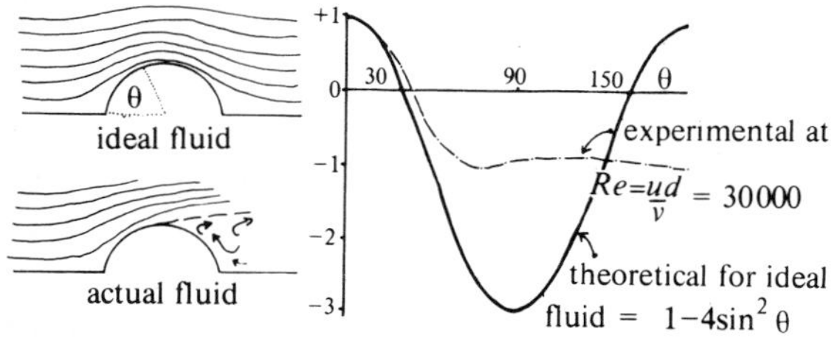

Figure 5.22 Flow round a cylinder held with its axis normal to a water stream of velocity u. Upper left shows the theoretical 'ideal' streamlines for an inviscid fluid whilst the lower left shows the actual pattern at $Re_g = 3 \times 10^4$ with separation and eddies in the wake. The graph shows the theoretical and experimental curves of pressure distribution (dimensionless ordinate) on the surface of such a cylinder. The measured pressure distribution is less than the theoretical since there is no diverging flow behind the sphere to return the fluid to its original velocity (after Francis 1969).

and **roller bubbles** which are skewed at greater than 45° to flow direction. The streamlines of a roller are closed loops and those in a vortex are helical spirals. These may be illustrated by means of skin friction lines, as in Figure 8.5. Very important effects arise at the upper junction of the bubbles with the mainstream fluid. Here we have relative motion between two flows which produces strong turbulent vortices along the unstable interface. High production of turbulent Reynolds stresses occurs along the surface to the attachment point. These may be up to three times the intensity of the stresses on the flat upstream part of the bed before the negative step (Fig. 8.4). The sedimentological moral is obvious: reattachment of a separated boundary layer will cause greatly enhanced erosion. As we shall see in Chapters 8 and 9 *this tendency for enhanced local erosion substantially explains the origin of several different bedforms and their associated sedimentary structures.*

A final type of flow instability occurs extensively in natural turbulent flows and comprises spanwise motions superimposed on downstream flow. These motions take the form of alternating pairs of spiral vortices spaced across the flow. Each vortex pair comprises alternately rotating vortices and may affect a significant portion of the whole boundary layer. Separation and attachment lines are directed parallel to the downstream mean streamlines. The spiral vortices have become known as Taylor–Görtler vortices and they occur in almost all natural turbulent flows. They are the origin of the alternating fast and slow lanes of wind-driven sand commonly observed on desert surfaces and on dry beaches, and of the widespread streaky effect observed as wind blows over water surfaces. Theoretical analysis of the Navier–Stokes equations for turbulent motion show that secondary currents are due to the imbalance of the normal Reynolds stresses $\bar{v}'^2$ and $\bar{w}'^2$, although the mathematical proof of this is rather complex (Einstein & Li 1958). Increased sediment transport or erosion occurs along the line of attachment of the secondary currents and the general transfer of particles out from the attachment skin-friction line towards the separation line. In the next chapter we shall see that several types of bedforms in both air and water seem to owe their origins to the action of secondary currents.

5k Summary

Fluid flows possess boundary layers because of viscous retardation and may be laminar or turbulent as defined by the Reynolds number. Most air and water flows of sedimentological importance are turbulent. All fluid flows cause shearing stresses to be set up at rigid flow boundaries and within the flows themselves. The product of mean bed shear stress and mean flow velocity defines the mean flow power available to the fluid over unit bed area. Turbulent flows possess a well defined structure with the site of maximum turbulence production lying close to the viscous sub-layer. In this area close to the bed, turbulent motion is manifested in the form of 'sweep' and 'burst' events. Low-velocity 'burst' fluid elements periodically rise through the flow to the surface as large-scale turbulent 'eddies'. Both laminar and turbulent flows show flow separation effects in areas of unfavourable pressure gradients such as negative steps and channel expansions.

Further reading

The mathematical complexities of fluid mechanics tend to divorce many students (and teachers) from this essential aspect of modern sedimentology. A particularly lucid and elementary introduction to fluids is to be found in Shapiro (1961). Once mastered, this short book enables one to pass on to more advanced texts that base their approach on simple Newtonian physics with only minor use of calculus (e.g. John & Habermann 1980, Massey 1979, Francis 1969, Gasiorek & Carter 1967). These titles, and many more besides, including the classic introduction by Ewald *et al.* (1933), are aimed at introducing fluids to engineering students so that the worked examples often appear irrelevant to the needs of the Earth scientist. So far as I am aware Tritton (1977) is the only text that deals with fluids from a 'geophysical' viewpoint. The text requires knowledge of higher mathematics and it deals with many aspects of geophysical flows. A fundamental reference text on turbulent flows for more advanced students is Townsend (1976). A good review of the bursting process in turbulent boundary layers is given by Offen & Kline (1975).

Appendix 5.1 Bernoulli's equation (derivation after Gasiorek & Carter 1967)

This important equation may be derived by applying Newton's Second Law to a moving fluid particle. Consider the fluid particle shown (Fig. 5.25) of cross-sectional area dA, length ds, instantaneous velocity u, density ρ and acceleration a. We are dealing with an 'ideal' fluid, so that the only forces acting on the particle are pressure (p) and gravitational (g) pull or weight force. We ignore shear stresses due to viscosity effects. Taking account of only the forces in the direction of motion, the resultant pressure force on the ends of the particle is

$$p dA - (p + dp) dA = -dp dA \tag{5.27}$$

when the direction of motion is considered positive. The component of gravity force in the direction of motion is

$$-dA ds \rho g \sin\theta = -dA ds \rho g \frac{dz}{ds} = -\rho g dA dz \tag{5.28}$$

Now the sum of these two forces is equal to mass × acceleration.

$$m = \rho ds dA \text{ and } a = \frac{du}{dt} = \frac{du}{ds}\frac{ds}{dt} = u\frac{du}{ds}$$

thus

$$ma = \rho u dA du \tag{5.29}$$

The equality becomes

$$-(dp dA) - (\rho g dA dz) = \rho u dA du \tag{5.30}$$

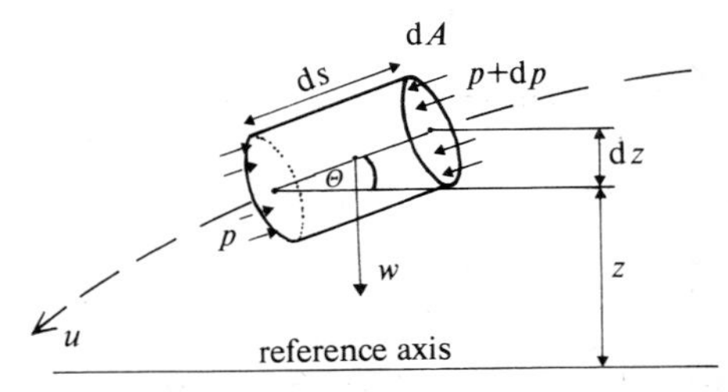

Figure 5.25 Definition sketch for derivation of Bernoulli's equation (after Gasiorek & Carter 1967). See text for symbols.

Dividing by $-\rho g dA$

$$\frac{dp}{\rho g} + dz + \frac{udu}{g} = 0 \quad (5.31)$$

By integration and assuming that in incompressible (non-gaseous) flows ρ does not vary with p

$$\frac{p}{\rho g} + z + \frac{u^2}{2g} = \text{constant} = H \quad (5.32)$$

where H is the **total head**, $p/\rho g$ is the **static pressure head**, z is the **potential head** and $u^2/2g$ is the **velocity** or **kinetic head**. Each term of the equation has dimensions of length. Multiplying through by ρg yields Equation 5.4 where the terms then express energy per unit volume.

Appendix 5.2 Reynolds number (derivation after Gasiorek & Carter 1967)

Consider a cubic fluid particle of viscosity μ, density ρ, cross-sectional area δ^2 and length δ as in the Figure 5.26. Let the upper surface of the particle move with velocity u relative to the lower surface, the velocity gradient being due to viscous forces. Motion may also cause inertial forces to be present when the particle is decelerated or accelerated.

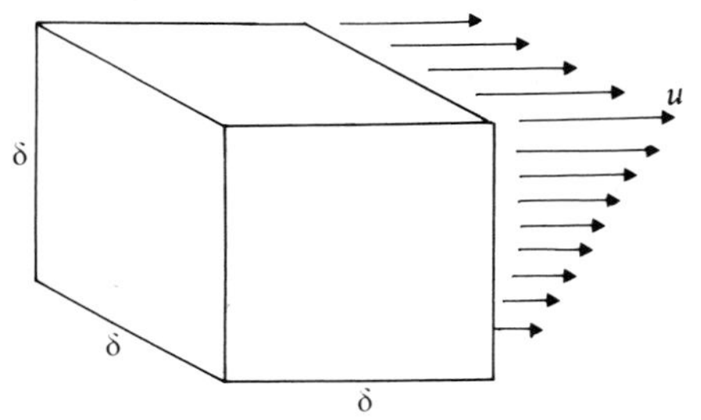

Figure 5.26 Definition sketch for derivation of Reynolds Number (after Gasiorek & Carter 1967). See text for symbols.

The **viscous force** is equal to shear stress × area, but since shear stress = $\mu(du/dy)$ by Newton's law (Eq. 5.15), in the above case

$$\text{viscous force} = \frac{\mu u \delta^2}{\delta} = \mu \delta u \quad (5.33)$$

The **inertial force** is equal to mass × acceleration by Newton's Second Law. Mass = density × volume = $\rho\delta^3$, acceleration is $u/t = u^2/\delta$, since $u = \delta/t$ and $t = \delta/u$, thus

$$\text{inertial force} = \frac{\rho\delta^3 u^2}{\delta} = \rho\delta^2 u^2$$

Arranging the two forces as a ratio

$$\frac{\text{inertial force}}{\text{viscous force}} = \frac{\rho\delta^2 u^2}{\mu u \delta} = \frac{\rho u \delta}{\mu} = \text{Reynolds number} \quad (5.35)$$

where δ is a suitable linear dimension of any flow, as discussed in the text.

Appendix 5.3 Velocity profiles of viscous channel flow (derivation after Gasiorek & Carter 1967)

Consider viscous steady flow between two parallel walls (Fig. 5.27). Let a fluid particle between the walls be of length l and unit width. Let its thickness be $2y$, y being measured from the flow centre line. Let the channel width be b. The pressure at the upstream end of the element is p_1, and at the downstream end is p_2, where $p_1 > p_2$. Since there is uniform steady flow, the pressure and viscous forces acting upon the fluid particle must be in equilibrium. Let the viscosity be μ.

The **applied force** is due to the difference in pressure forces on the ends of the particle and is equal to the pressure difference × the cross-sectional area of the particle. Thus the force is

$$p_1 l 2y - p_2 l 2y \quad \text{or} \quad (p_1 - p_2)2y \quad \text{or} \quad \Delta p 2y \quad (5.36)$$

The **resisting force** is due to shear stress τ, acting on the particle surfaces parallel to flow. Noting that we measure velocity from the centre outwards in this case, the velocity gradient being negative, we have

$$\tau 2l = -\mu \frac{du}{dy} 2l \quad (5.37)$$

For equilibrium

$$\Delta p 2y = -\mu \frac{du}{dy} 2l \quad (5.38)$$

or

$$du = -\frac{\Delta p}{\mu l} y dy \quad (5.39)$$

By integration

$$u = -\frac{\Delta p}{2\mu l} y^2 + C \quad (5.40)$$

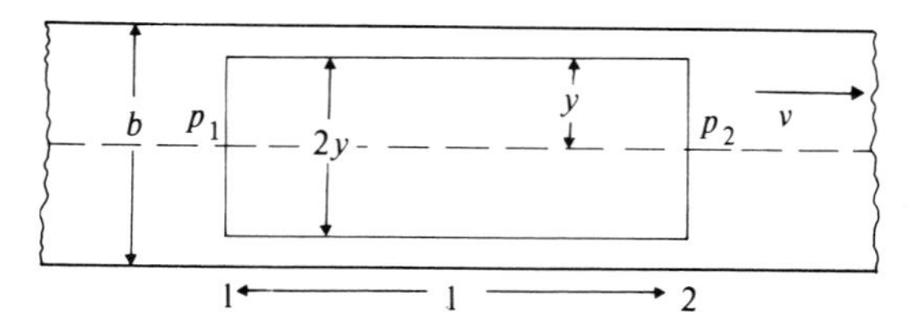

Figure 5.27 Definition sketch for derivation of the viscous flow equations (after Gasiorek & Carter 1967). See text for symbols.

which is the velocity at the surface of the element at a distance y from the centreline.

To determine C, set $y = b/2$ so that $u = 0$, then from Equation 5.40, with $u = 0$

$$C = \frac{\Delta p}{8\mu l}(b)^2 \tag{5.41}$$

Substituting for C in Eq. 5.40 yields

$$u = (b^2 - y^2)\frac{\Delta p}{8\mu l} \tag{5.42}$$

thus proving that the velocity distribution across the flow is parabolic. The maximum velocity occurs at the centreline where $y = 0$, giving,

$$u_{max} = b^2 \frac{\Delta p}{8\mu l} \quad \text{or,} \quad u = u_{max}\frac{(b - y^2)}{y} \tag{5.43}$$

The mean velocity across the flow is, without the derivation,

$$\bar{u} = b^2 \frac{\Delta p}{12\mu l} \tag{5.44}$$

and the viscous shear stress on the plate surfaces is

$$\tau_{max} = \frac{6\mu u}{b} \tag{5.45}$$

Appendix 5.4 Derivation of the Karman–Prandtl velocity law for turbulent flow (derivation after Francis 1969)

In a turbulent flow, eddy motions cause mixing of slow and fast moving fluid both across and along the mean direction of flow. With reference to the idealised rotary eddy motion in Figure 5.28, let u be the mean speed of the centre of the eddy relative to the flow boundaries and du/dy the instantaneous velocity gradient across the eddy. The speed of the upper layer is $u + \frac{1}{2}l du/dy$ and of the lower layer is $u - \frac{1}{2}l du/dy$. The eddy motions across the flow are thus $\frac{1}{2}l du/dy$ relative to the mean motion.

In the cylinders of cross-sectional area as shown on each side of the eddy, a velocity $\frac{1}{2}l du/dy$ is occurring, transferring a mass of fluid $a\rho\frac{1}{2}l du/dy$ in unit time from fast layer to slow layer and vice versa. Every unit mass of fluid changes its momentum by an amount $l du/dy$, so that the total rate of change of momentum by both cylinders is

$$2\left(a\rho\tfrac{1}{2}l\frac{du}{dy}\right)\left(l\frac{du}{dy}\right) \tag{5.46}$$

Now, forces must act to preserve the velocity gradient since if they did not the faster and slower layers would mix together to form a uniform stream at the mean velocity. Thus a shear stress

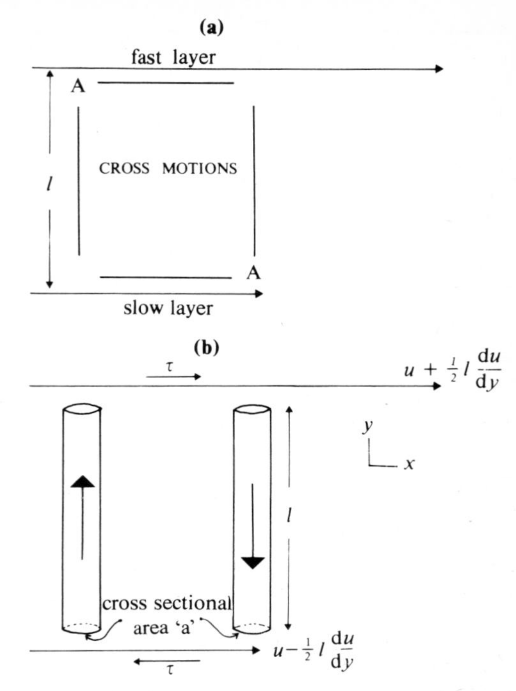

Figure 5.28 Definition sketch for derivation of the Karman–Prandtl equation (after Francis 1969). See text for symbols.

τ' must oppose the tendency of the momentum transport to eliminate the velocity gradient. Thus

$$\tau' 2a = 2\left(a\rho\tfrac{1}{2}l\frac{du}{dy}\right)\left(l\frac{du}{dy}\right) \tag{5.47}$$

or

$$\tau' = \tfrac{1}{2}\rho\left(l\frac{du}{dy}\right)^2 \tag{5.48}$$

let τ', l, and du/dy now be the mean shear stress, eddy length and velocity gradient respectively. To do this and preserve the equality we must introduce the constant k so that

$$\tau = k\tfrac{1}{2}\rho\left(l\frac{du}{dy}\right)^2 \tag{5.49}$$

Now we may solve this differential equation if we assume (a) $\tau = \tau_0$, the boundary shear stress (b) an experimental result that the quantity kl is proportional to the distance from the boundary, i.e. that eddy size is directly dependent upon depth. Writing (5.49) as

$$\left(\frac{\tau}{\rho}\right)^{\frac{1}{2}} = \left(\frac{k}{2}\right)^{\frac{1}{2}} l\frac{du}{dy} \tag{5.50}$$

and substituting the experimental fact that $(k/2)^{\frac{1}{2}}l = 0.4y$ we have

$$du = \left(\frac{\tau_0}{\rho}\right)^{\frac{1}{2}} \frac{1}{0.4y}\, dy \tag{5.51}$$

Integration yields

$$u = \left(\frac{\tau_0}{\rho}\right)^{\frac{1}{2}} \frac{1}{0.4} [\log_e y] + \text{constant} \tag{5.52}$$

Writing the constant of integration as

$$C = \left(\frac{\tau_0}{\rho}\right)^{\frac{1}{2}} \frac{1}{0.4} \log_e \frac{1}{C_1}$$

where C_1 is another constant, 5.52 becomes

$$u = \left(\frac{\tau_0}{\rho}\right)^{\frac{1}{2}} \frac{1}{0.4} \left[\log_e y + \log_e \frac{1}{C_1}\right] \tag{5.53}$$

or

$$u = 2.5 \left(\frac{\tau_0}{\rho}\right)^{\frac{1}{2}} \log_e \frac{y}{C_1} \tag{5.54}$$

In $\log_{10}$ units

$$u = 5.75 \left(\frac{\tau_0}{\rho}\right)^{\frac{1}{2}} \log_{10} \frac{y}{C_1} \tag{5.55}$$

or in the notation of the main text (see Eq. 5.24)

$$u = 5.75 \sqrt{\frac{\tau_0}{\rho}} \log_{10} \frac{z}{k} \tag{5.56}$$

It must finally be stressed that Equation 5.56 applies only to the lowest 10–20% of the boundary layer. It is sometimes referred to as the 'law of the wall'.

6 Transport of sediment grains

6a Introduction

Having established in Chapter 5 some basic principles of fluid motion we now turn to the interaction between fluid motion and sediment grains. In the 1960s Bagnold gave the name **loose-boundary hydraulics** to this field of study to distinguish it from the hitherto more common analysis of the pure fluid motion alone. Initially we can distinguish two types of sediment flow boundary: **granular-cohesionless** and **cohesive**. The first type includes all boundaries made up of discrete solid grains which are kept in contact with adjacent grains purely by gravitational effects. The second type applies most commonly to clay mineral aggregates on mud beds where the tiny clay mineral flakes are mutually attracted by electrolytic forces which may be large compared to gravitational ones (Ch. 9).

6b Grains in stationary fluids

Sediment particles falling through static water and air columns are common in nature. If we introduce a small sphere of density σ into a static liquid of density ρ, such that σ > ρ, then the sphere will initially accelerate through the fluid, the acceleration decreasing until a steady velocity known as the **terminal** or **fall velocity** (V_g) is reached. If we carried out careful experiments with a large range of spherical grain sizes, we would be able to plot up our results of fall velocity versus grain diameter as shown in Figure 6.1. These results show clearly that the fall velocity increases with increasing grain diameter but that the rate of increase gets less.

By considering viscous fluid resistance forces only, it is possible to arrive at a theoretical relation which expresses fall velocity as a function of grain and fluid properties. From the derivation given in Appendix 6.1

$$v_g = \frac{gd^2(\sigma - \rho)}{18\mu} \qquad (6.1)$$

Equation 6.1 is known as **Stokes' Law** in honour of its discoverer. Stokes' Law accurately predicts the fall velocity of particles whose particle Reynolds number, $v_g d\rho/\mu$, is less than about 0.5. This corresponds to silt-size and finer, quartz-density particles in water. Why then does this relation not apply to all grain sizes? The answer is to be found in the tendency for boundary layer separation (Ch. 5j) to occur behind the falling grain at increased rates of fall velocity, when fluid inertial effects, ignored in the Stokes derivation, cause a great increase in the drag or retarding action of fluid on a sphere. Newton derived an expression for the fall velocity when inertial forces dominate and gave

$$v_g = \sqrt{\frac{4}{3} g \frac{\sigma - \rho}{\rho}} \qquad (6.2)$$

known as the **impact formula.** In fact, Newton's formula does not give very satisfactory agreement with experimental results (Fig. 6.1), since the form of the numerical constant depends upon uncertain assumptions concerning the water impact on the falling sphere and does not take account of flow separation.

Thus far we have considered the fall of single smooth spheres through fluids. Complicating factors arise in most

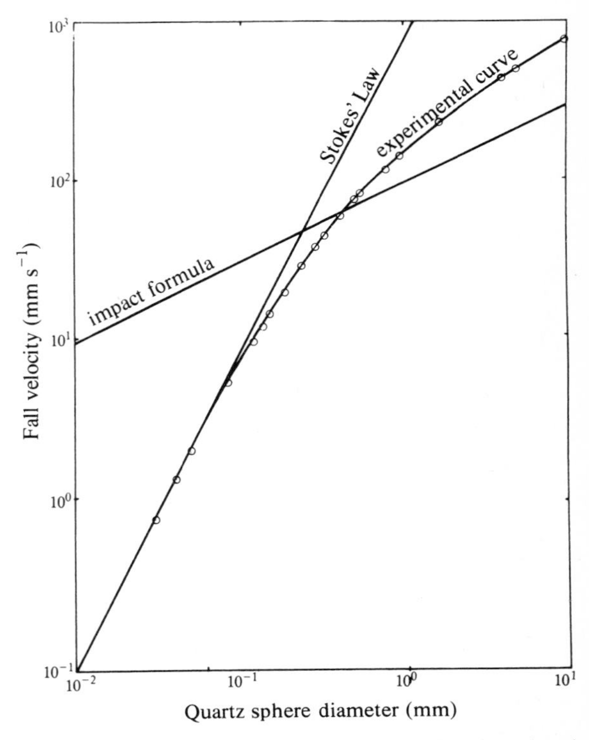

Figure 6.1 Graph to show fall velocity as a function of grain diameter for quartz spheres in water at 20°C. For comparison the predictions of fall velocity for the same system according to Stokes' Law and the impact formula are shown (after Gibbs *et al.* 1971).

natural cases. First, natural sediment grains are not perfectly rounded or spherical. Natural silt- and sand-size silicate grains are perhaps the closest in shape to spheres and may often be considered to have the same fall velocity as the equivalent diameter sphere (Table 4.2). Very accurate work would require the fall velocity for a particular grain to be determined by direct experiment. Serious problems arise when considering grains of biogenic origin or precipitates. For example, shell sands will comprise flaky particles whose descent through a water column will show to-and-fro motions normal to the mean downward velocity vector, similar to those of falling leaves. Clearly there can be no general laws or empirical expressions for such particles; direct experimental relations for each particular sample will have to be sought (Braithwaite 1973). Secondly, almost all natural examples of particle settling involve a population of particles falling as a group. Mutual particle hindrance and increased drag then result in a decrease of the grain-fall velocity relative to its velocity in a grainless fluid. It has been proposed (Richardson & Zaki 1958) that the fall velocity of a spherical particle, v_g', in a dispersion of other falling grains varies as

$$v_g' = v_g(1 - C)^n \qquad (6.3)$$

where v_g is the fall velocity of a single grain in an otherwise grainless fluid, C is the volume concentration of grains in the falling dispersion, and n is an exponent varying between 2.32 and 4.65, depending upon grain Reynolds number. This relation tells us that the fall velocity of a grain in a dispersion will be smaller than that in an otherwise grainless fluid and strongly dependent upon concentration. For fine sediment, when $n = 4.65$, and at a high value of C around 0.5, v_g' may be only 2–3% of v_g. This consideration is of great importance in understanding the settling behaviour of dense, sediment-laden flows such as turbidity currents (Ch. 7).

6c Initiation of particle motion

As fluid shear stress over a levelled plane bed of granular particles is slowly increased, there comes a critical point when grains begin to be moved downstream with the flow. Sediment transport has begun. A great deal of attention has been paid to the determination of the *critical threshold for grain movement* since it is an important practical parameter in civil engineering schemes (canals, irrigation channels, model experiments). A knowledge of the threshold value for different grain types and sizes is also of special interest to the sedimentologist.

Before we examine the results of experimental investigations into threshold values, let us look at the fluid forces acting upon bed grains (Fig. 6.2). A **drag force** due to the fluid velocity gradient at the bed is experienced by each grain. If τ_0 is the mean bed shear, derived as in Equations 5.6 and 5.7, then the mean drag per grain is given by $F_0 = \tau_0/n$, where n is the number of particles over unit bed area. A **lift force** due to the Bernoulli effect also exists. Fluid streamlines over a projecting grain will converge, the velocity will speed up and therefore to maintain the pressure equilibrium demanded by Equation 5.4 the pressure must decrease above the grain. Chepil's (1961) elegant measurements (Fig. 6.3) and later results leave no doubt that the lift force is comparable to the drag force when the grain is on the bed. The lift force rapidly dies away and the drag force rapidly increases as the grain rises from the bed.

Both fluid forces considered above will try to move bed grains. They will be resisted by the normal weight force due to the grains (Fig. 6.2). It has proved impossible to determine theoretically the critical applied shear stress necessary to move grains, despite the initial attractions of using a 'moments of force' approach. This is because there are a large number of variables involved, quite apart from the difficulty of estimating the lift force contribution. The critical conditions for the initiation of particle motion must therefore be determined experimentally. In order to have the utmost generality any plot of experimental results must be applicable to a wide range of fluids and particles. As sedimentologists we are most interested in natural mineral grains in air and water, but these systems must be treated as special cases of a more general application.

Working from first principles we might expect the critical conditions for particle motion, C_c, to be dependent upon gravity (g), grain size (d), immersed weight ($\sigma - \rho$),

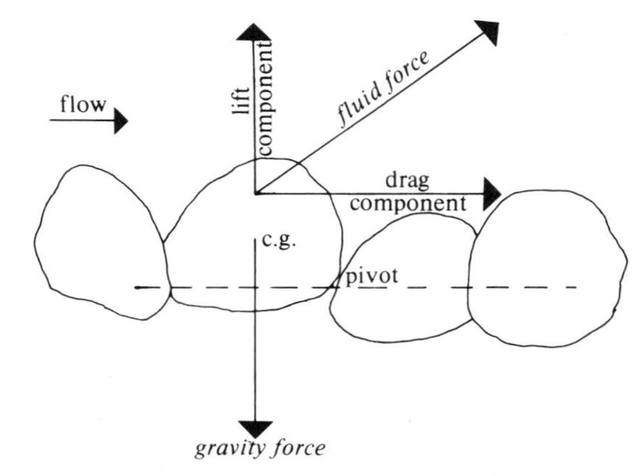

Figure 6.2 Schematic diagram to show the forces acting on a grain, resting on a bed of similar grains, subjected to fluid flow above it (after Middleton & Southard 1978).

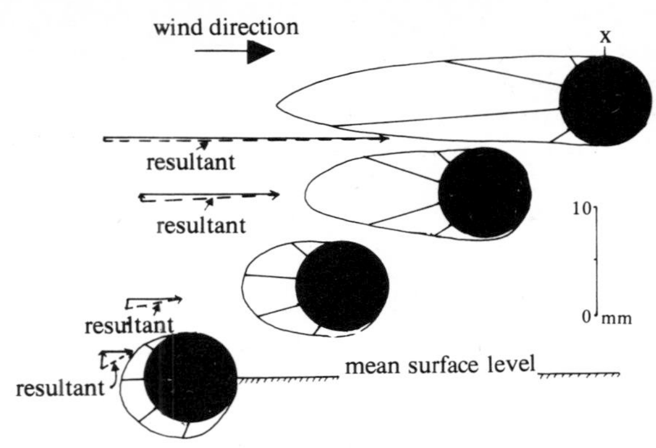

Figure 6.3 The pattern of approximate pressure differences between position x on top of a 7.5 mm sphere and other positions on the sphere at various heights in a windstream. Both lift and drag forces act on the sphere but lift decreases rapidly with height whereas drag increases because of the direct pressure of the wind. The wind velocity at 20 mm above the surface was 7.7 m s^{-1}; shear velocity was 0.98 m s^{-1}. The lengths of lines in the shaded areas outside the spheres denote the relative differences in air pressures (after Chepil 1961).

fluid kinematic viscosity (ν) and the bed shear stress (τ_0). Thus

$$C_c = f(d, g(\sigma - \rho), \nu, \tau_0) \tag{6.4}$$

Now, great generality will result if we arrange these quantities into two dimensionless groups for the purpose of plotting experimental results (see full discussion by Middleton & Southard 1978). The two groups are

$$\left(\frac{\tau_0}{gd(\sigma - \rho)}\right) = f\left(\frac{u_* d}{\nu}\right) \tag{6.5}$$

The term to the left is known as the dimensionless bed shear stress θ. The term to the right is familiar to us already (in a slightly different form) and is a grain Reynolds number, Re_g. A plot of θ versus Re_g (known as a Shields diagram) for liquids is shown in Figure 6.4c. Considerable scatter is evident, because of the many different sets of experimental conditions and the difficulty in deciding exactly when the threshold is reached. It appears that θ is nearly constant, at an average of about 0.05, for a wide range of grain diameters up to a grain Reynolds number of about 1.0. At low grain Reynolds numbers θ increases steadily to around 0.3. This increase is thought to be due to the presence of a smooth boundary to the flow, with the particles lying entirely within a viscous sublayer where the velocity gradient and the instantaneous velocity variations are less than in the lowest part of the turbulent boundary layer. The reader should note that many of the lines drawn by various authors through the

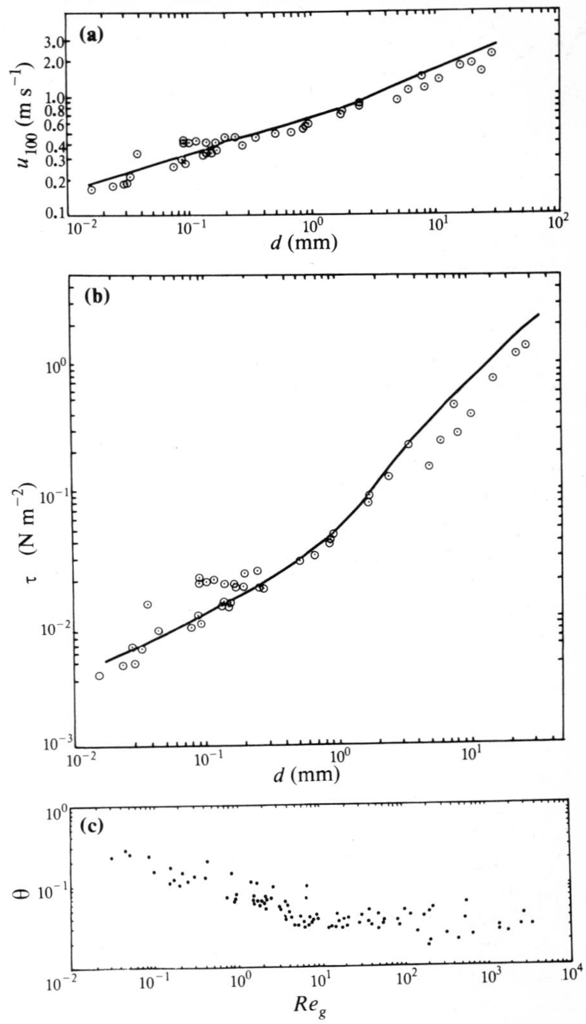

Figure 6.4 Various threshold graphs. (a) The grain diameter d versus the flow velocity 1.0 m above the bed necessary for initiation of movement of quartz density grains in water at 20°C (after Miller *et al.* 1977). (b) The grain diameter d versus the shear stress τ necessary for initiation of movement of quartz density grains in water at 20°C (after Miller *et al.* 1977). (c) The dimensionless applied shear stress θ versus grain Reynolds Number for initiation of movement of a wide variety of grain types in water at 20°C (Shields plot; after Miller *et al.* 1977).

experimental points are of doubtful statistical meaning, e.g. a dip is often shown in the line between Re_g of 100 and 10 which clearly cannot be justified by inspection of the very full data of Figure 6.4c.

Important light has been thrown on the 'threshold problem' in water by regarding the onset of grain movement as resulting from the interaction between two statistically distributed variables (Grass 1970). The first variable may be termed the initial movement characteristics of a given bed material in a fluid of given viscosity and density. Thus every grain on the bed is assumed susceptible to a local instantaneous stress and because of the random shape, weight and placement of individual grains this τ_0, termed τ_c, has a probability distribution. The second variable is the local instantaneous bed shear stress caused by burst/sweep events (Ch. 5h). These stresses have a probability distribution that depends upon $\bar{\tau}$, fluid density, and viscosity and flow boundary conditions. At the onset of grain motion the most susceptible particles (those with the lowest characteristic critical shear stress) are moved by the highest shear stresses in the shear stress distribution applied to the bed by the flow. The results of H_2 bubble experiments enable histograms to be constructed for each distribution (Fig. 6.5). These show that the critical shear stress necessary to move grains occurs when the two τ distributions overlapped by a certain constant amount. Much of the scatter on the Shields plot evidently reflects different observers deciding on different degrees of overlap.

In air flows Bagnold (1956) defined two types of threshold. At a critical value of air speed insufficient to move bed grains by fluid shear alone, grain motion could be started and propagated down wind simply by letting sand grains fall on to the bed. Other grains were bounced up into the air stream which, upon falling, caused further movement as they impacted onto the bed, and so on down wind. Grain motion ceased as soon as the introduction of artificial grains stopped. The critical wind speed necessary for this process was termed the **impact threshold**. Further increase of air speed enabled grains to be moved by the direct action of the wind at the normal or **fluid threshold**. The ability of natural sand grains in air to disturb and eject other grains after impact contrasts with the behaviour of grains in water. This is because of the viscosity contrast between air and water (Ch. 5b) which controls resistance to motion and the great effective density contrast between quartz and air (2000:1) compared to quartz and water (1.65:1). As we shall see below (Ch. 8), the types of ripples developed in the two fluids depend intimately upon these factors. For most sands in air flows the critical shear velocity is a function of the square root of the particle size.

We may finally mention that a particular fluid shear stress or shear velocity (u_*) above the threshold for motion may be expressed as a ratio with respect to the critical threshold stress or velocity (u_{*c}) for the grains in question. Thus a **transport stage** (Francis 1973) is defined as

$$\frac{u_*}{u_{*c}} \tag{6.6}$$

6d Paths of grain motion

Once the threshold for motion is exceeded, grains move downstream in three basic ways (Fig. 6.6). **Rolling** motion is simply defined as continuous grain contact with the bed and it includes the rarely observed 'sliding' motion. **Saltating** (Latin root *saltare*: to jump) motion comprises a series of ballistic hops or jumps characterised by steep-angled (>45°) ascent from the bed to a height of a few

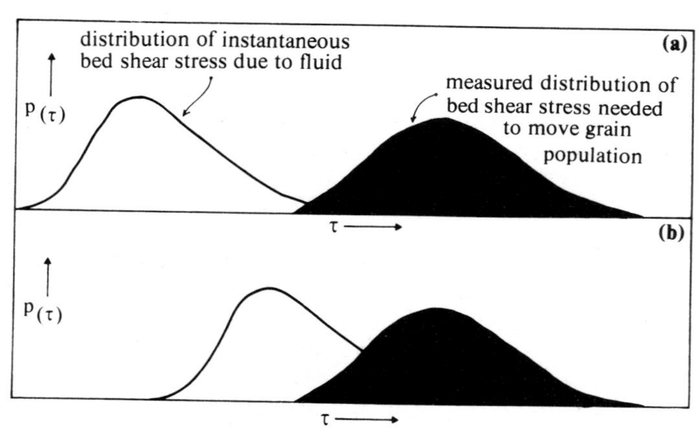

Figure 6.5 Schematic graphs to show that the threshold of grain movement must be defined by a defined degree of overlap (for details see Grass 1970) between the distribution of instantaneous applied shear stresses due to the turbulent fluid and the actual distribution of stresses needed to move the particular grain population under examination. (a) Small overlap, little motion; (b) larger overlap, general threshold exceeded. Different degrees of overlap assigned by different workers explains some of the overlap in threshold graphs such as Figure 6.3(a–c) (modified after Grass 1970).

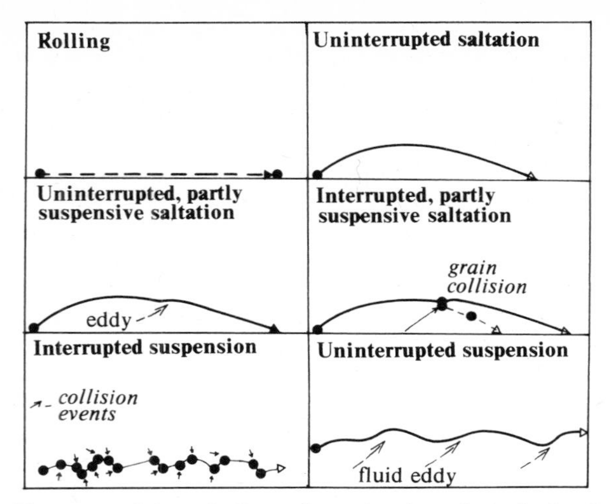

Figure 6.6 Schematic illustrations of grain paths in 'bedload' and 'suspended' load (after Leeder 1979).

grain diameters which turns into a shallow-angled (>10°) descent path back to the bed. The trajectory is *not* usually a symmetrical parabola as shown, say, by the calculation of the trajectory of an artillery projectile or arrow, in which fluid resistance is neglected. **Suspended** (literally 'held up') motion involves grains moving in generally longer and more irregular trajectories higher up from the bed than in saltation. As Bagnold (1973) points out, fully developed suspension implies that 'the excess weight of the solid is supported wholly by a random succession of upward impulses imparted by eddy currents of fluid turbulence moving upwards relative to the bed'. However, a grain may experience an upward acceleration during the descending part of a saltant trajectory as turbulence just starts to affect saltation. Although defined as suspension it is perhaps better to call such motion **incipient suspension.**

It is often difficult to follow the path of a moving sand grain, particularly in a crowd of other grains, and it is only from the application of skilful photographic techniques in recent years that we have a full idea of the different types of grain paths. Multi-exposure photographic techniques (Francis 1973, Abbott & Francis 1978) show that (a) the varying proportions of time spent in the above three modes by grains in water is a direct function of transport stage, (b) the statistics of the increase of length and height of grain saltant and incipient suspensive trajectories are direct functions of transport stage (Fig. 6.7), and (c) the mean forward speed of grains is a direct function of the mean flow velocity (Fig. 6.7).

There are some important differences in saltation mechanisms between air and water. In air the grains are much less buoyant because of the large density contrast

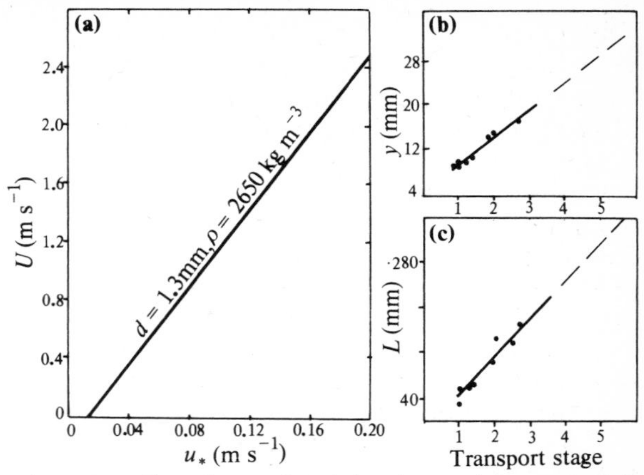

Figure 6.7 Plot of mean forward grain speed U against fluid shear velocity u_* for a 1.3 mm quartz grain (derived by Leeder 1979 from data in Abbott & Francis 1977). (b) Plot of mean maximum saltation height $\bar{y}$ against transport stage for 8.3 mm grains. (c) Plot of mean saltation length $\bar{L}$ against transport stage for 8.3 mm grains (modified after Abbott & Francis 1977). Transport stage is given by u_*/u_{*c} where u_{*c} is the critical shear velocity for initiation of motion.

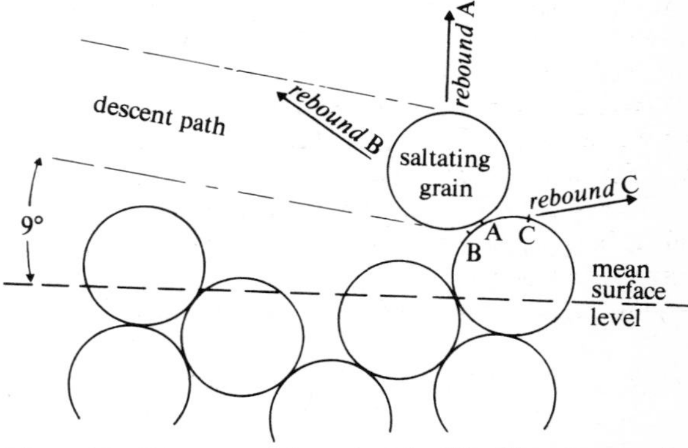

Figure 6.8 Diagram to show that the 'lift-off' trajectory of a saltating grain *in air* depends upon the collision angle of the grain with stationary bed grains (after Chepil 1961). Rebound effects are not important in sub-aqueous saltations where fluid lift forces cause the upward movement of the grains from the bed.

between fluid and solid. As noted previously (p. 70) falling grains carry sufficient momentum to dislodge and 'splash up' other grains from tiny impact craters. The high and variable initial angle of ascent is entirely due to this process (Fig. 6.8). The relatively small air resistance, due to low viscosity, helps the grains reach quite large heights (up to, say, about 500–100 grain diameters), a process aided by rebound from pebble surfaces (Fig. 6.9). In water, however, the grains are much more restrained by buoyancy effects. Impacting grains of up to fine gravel size do not disturb other bed grains and, of great

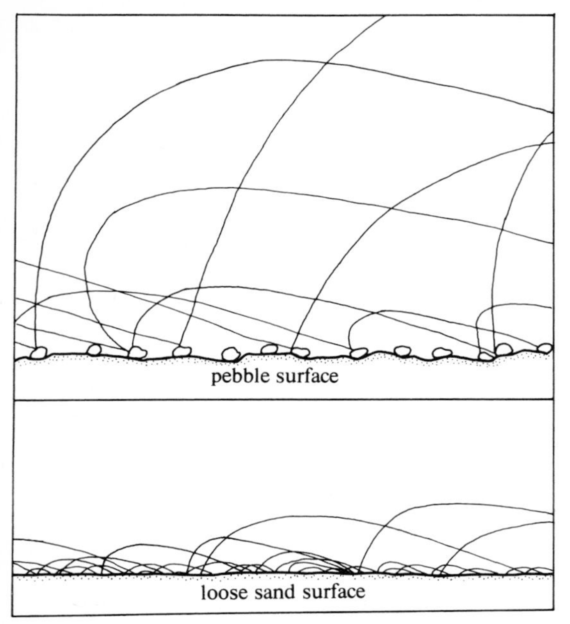

Figure 6.9 Diagram to illustrate the difference in height reached by saltating grains in air over sand and over pebble beds due to the difference in rebound 'efficiency' of the two substrates (after Bagnold 1954b).

importance, do *not* rebound from the bed immediately to begin another saltation (Abbott & Francis 1977). Linear momentum is *not* conserved by grain/bed collisions and thus saltating grains in water are rather different from bouncing balls. Multi-exposure images make this last point quite definite and it follows that the initial ascent of a saltating grain in water must be the result of a Bernoulli *lift force*, rebound effects being ruled out.

The onset of fully developed suspension may be roughly estimated by assuming that the root-mean-square values of the upward vertical velocity fluctuations close to the bed begin to exceed the fall velocity of the saltating grains, i.e.

$$v_g \gtrsim \sqrt{\bar{v}'^2} \tag{6.7}$$

Since it is known that the term involving v' reaches about 1.2 u_* at maximum we can clearly write an *approximate* criterion for full suspension as

$$v_g/u_* \lesssim 0.8 \tag{6.8}$$

This criterion successfully explains the onset of full suspension effects observed in multi-image photographic experiments. An important effect arising from this criterion is that fine bed sediments ($d < 0.1$ mm) should become suspended immediately the threshold conditions are reached.

The characteristic grain paths defined above apply only to grain transport *when grains have no effects upon each other during movement*, i.e. when simple saltant trajectories are not interrupted by grain/grain collisions or deflections. Such interference effects should be much more likely in water than in air because of the thinness of the saltation zone in the former case. As the transport stage of water flows is increased, more and more grains will be entrained and hence the grain concentration will increase. Grain/grain collisions will become certain. It can be shown by simple collision dynamics based upon kinetic theory that simple saltations will cease to exist over a transport stage of about two when the grains close to the bed will begin to move as a concentrated granular dispersion dominated by grain/grain impacts or deflections (Leeder 1979).

We have now reached a suitable point to give rational definitions of the types of transported sediment loads.

Bedload (traction load) includes rolling, saltating and collision-interrupted 'saltating' grains (the last important only in water flows). The grains comprising bedload will all transfer momentum to the stationary bed surface by solid/solid contacts and the rate of change in this momentum over unit area in unit time must equal the immersed weight of the bedload grains. Our definition of bedload is thus both positional and dynamic (Bagnold 1973).

Suspended load includes all grains kept aloft by fluid turbulence so that the weight force of the suspended grains is balanced by an upward momentum transfer from the fluid eddies. The process is more efficient in water flows than air flows.

'Washload' is a broad term used to describe the more-or-less permanently suspended clay-grade 'fines' present in water flows. **Dustload** is a more suitable equivalent term for air flows. Clay particles would fall back to the bed fairly slowly if the flow were to cease.

The transport of bedload sediment may have marked effects upon the distribution of fluid velocity with height from the bed surface. A clear example is provided by the wind case, where velocity measurements are possible within and just above the bedload zone using a very fine pitot tube (Bagnold 1954b). As the wind strength increases above the threshold for movement, there is a clear increasing retardation of the air velocity in the centre of gravity of the bedload zone (Fig. 5.13). The retardation reaches about 20% of the pure air flow velocity expected by the Karman–Prandtl equation (Ch. 5) for turbulent flow in the absence of moving solids. The departure of the measured velocity profiles from the straight line is explicable by the expected transfer of momentum from fluid to solid over the upper, accelerated, part of the saltant trajectories. A similar effect should be present in water flows (Bagnold 1973) but it is difficult to measure because of the thinness of the bedload zone in this case.

A final point in this section concerns an explanation for the discontinuities and straight line segments found when grain size distributions of water-lain particles are plotted on probability paper (Fig. 6.10; see Ch. 4). It has recently been demonstrated that the two coarsest components of these curves correspond to the bedload and intermittent suspension modes discussed above (Middleton 1976). This reasoning is based upon the criterion for suspension discussed above. The break in slope between the two populations corresponds broadly to a grain diameter at about the critical necessary for suspension by Equation 6.8. Middleton reasons that differential sorting of grains below this diameter is effected by suspension, and sorting of coarser grains occurs in the bedload or traction load.

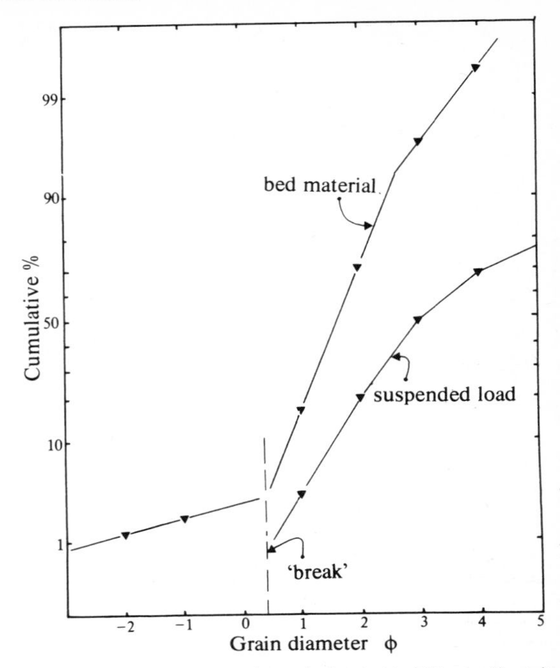

Figure 6.10 Size analyses (cumulative probability ordinate) of typical bed material and suspended sediment for flows of the Bernadillo River, New Mexico. The data are averages of all data collected in 1958. The dashed line indicates the size 'break' which is interpreted as the boundary between the traction and intermittent suspension populations. (After Middleton 1976.)

6e Solid transmitted stresses

So far we have assumed that the shearing stresses exerted on a sediment bed are entirely caused by moving fluid. As Bagnold (1954a, 1956, 1966b, 1973) recognised, this situation cannot persist once grains are in motion as bedload since impacting grains will transfer momentum from flow to the bed and give rise to additional solid transmitted stresses. For example, consider a saltating grain of mass m impacting on to a bed with velocity U and making an angle ζ with the bed surface. If elastic rebound occurs, and neglecting frictional effects, then momentum of magnitude $2mU \sin \zeta$ is transferred normal to the bed surface and momentum $2mU \cos \zeta$ is transferred tangential to the bed surface. If no rebound occurs and a period of rolling ensues after impact, then this momentum transfer is halved.

It is difficult to measure the magnitude of the solid (but see Bagnold 1955) stresses T and P due to the rate of change of tangential and normal momentum respectively in flow channel experiments. We can infer that during bedload transport the total applied shear stress to the bed must be made up of both solid and fluid components. Thus

$$\underset{\text{total applied shear stress}}{\mathcal{T}} = \underset{\text{fluid shear resistance}}{\tau_s} + \underset{\text{solid shear resistance}}{T} \qquad (6.9)$$

In an ingenious experiment using a rotating coaxial drum Bagnold (1954a) was able to measure the magnitude of T and the normal transmitted stress P. The results of the experiment reveal that there is a great increase in the shear resistance of the solid/fluid mixtures when compared to the plain fluid alone. Two regions of behaviour were definable in terms of a dimensionless number now known as the **Bagnold number**. A viscous region of behaviour at low strain rates and/or low grain concentration showed the ratio $T{:}P$ to be constant at around 0.75. In this region grain/grain effects were caused by near approaches causing a repulsion before solid collision could occur. An inertial region of behaviour at high strain rates and/or high grain concentrations showed the ratio $T{:}P$ to be constant at around 0.32. Grain/grain collisions dominated in this region.

Applying these results to natural flows, Bagnold postulated that:

(a) In air, owing to the 'chain reaction' process described above at threshold, so much fluid momentum is transferred to the saltation that virtually the whole of the applied shear stress is resisted by it. Thus $\mathcal{T} = T$ almost immediately.

(b) In water, where the density ratio of solid to fluid is low, the contribution of T gradually increased with transport stage. When one whole formerly stationary grain layer was in motion as bedload at high transport stage Bagnold argued that $T \rightarrow \mathcal{T}$.

(c) The normal stress P, also termed the **dispersive stress,** should be in equilibrium with the normal stress due to the gravity weight force of the moving bedload.

Concerning (b) it should be pointed out that at any transport stage

$$\mathcal{T} \vDash \tau_c + T \tag{6.10}$$

where τ_c is the fluid stress critical for initiation of grain movement. Calculation of T for certain experimental data (Leeder 1977) reveals that it is by no means dominant at high transport stages. These results are not yet fully understood.

(c) is the basis for Bagnold's sediment transport theory for bedload; we shall discuss this later in the chapter.

Much further experimental and theoretical work remains to be done in this important but rather difficult field. Results may be expected to shed light upon a number of fundamental problems in sediment transport and bedform theory (see also Ch. 8).

6f Sediment transport theory

The rate of sediment transport achieved by a particular flowing system depends upon a number of variables. Such rates are of great practical use in a variety of applied civil engineering and environmental disciplines. A large number of sediment transport equations have been proposed in the past 100 years, many of them being based upon measured results obtained in laboratory flow channels. They often take the general form (Allen 1978)

$$i = f(\bar{u} - \bar{u}_{crit})^n \tag{6.11}$$

or

$$i = f(\bar{\tau}_0 - \bar{\tau}_{crit})^m \tag{6.12}$$

where i is the sediment transport rate by immersed weight, $\bar{u}$ is the mean flow velocity, $\bar{u}_{crit}$ is the mean flow velocity at the threshold of sediment movement, τ_0 is the mean bed shear, τ_{crit} is the mean bed shear at movement threshold and n and m are exponents greater than 1.0. These empirical relationships simply say that the greater the flow magnitude above the threshold, the greater the transport rate. They add nothing to our understanding of the process of sediment transport and hence are unscientific. Bagnold has provided two interesting transport theories based upon simple physical principles. Let us now briefly examine these, both from the bedload viewpoint only.

Bagnold (1966b, 1973) regards flowing water as a transporting machine whereby the total power supply ω to unit bed area is (Ch. 5).

$$\omega = \rho g S h \bar{u} = \tau \bar{u} \tag{6.13}$$

Most of this power is dissipated in maintaining the flow against fluid shear resistance so that the effective power available to transport is less than this value. The sediment-transporting work rate of the fluid may be derived as follows. If the immersed weight of bedload over unit area is $m_b g$ and the mean travel velocity is $\bar{U}$ then the transport rate is $m_b g \bar{U}$. Now, the immersed normal weight stress of bedload is less than $m_b g$ by an amount $\tan \alpha$ given by the dynamic friction coefficient, so that the rate at which the fluid does transporting work per unit of bed is

$$m_b g \bar{U} \tan \alpha = i_b \tan \alpha \tag{6.14}$$

whose units are identical to those of Equation 6.13.

Equating (6.13) and (6.14) so that available power × efficiency = work rate, and introducing an efficiency factor e_b, we have Bagnold's simple expression for bedload transport rate

$$i_b = e_b \omega / \tan \alpha \tag{6.15}$$

For a given total stream power the efficiency factor e_b is inversely dependent upon the depth:grain-size ratio since in deep flows the effective velocity acting upon the bedload layer will be less than in shallow flows.

For the case of bedload transport in wind we cannot use the power analogue as in the water flow case since the height of a wind is indeterminate. Bagnold's approach (1954b) in this case is to equate the rate of applied shear by the moving air to the rate of loss of momentum by the bedload grains. Consider a grain of mass m moving from rest and striking the bed at velocity U after travelling a distance L. If all the velocity U is lost at impact then the grain may be considered to have gained momentum of magnitude mU/L per unit length of travel from the air. Now if a mass, i_b, of equal-sized grains of sand in saltation moves along a lane of unit width and passes a fixed point in unit time then the rate of loss of momentum will be $i_b\ U/L$. The rate of loss of momentum is a force so that the expression $i_b\ U/L$ is a measure of the resistance exerted on the air over unit surface area due to grain saltation. Calling the resisting or shear force τ, or, from Chapter 5, ρu_*^2, we can write

$$\tau = \rho u_*^2 = i_b U/L \tag{6.16}$$

Now it has been found that U/L is equal to g/v' where v' is the initial vertical velocity of the saltating grain. Hence

$$\frac{i_b g}{v'} = \rho u_*^2$$

or

$$i_b = \frac{\rho u_*^2}{g} v' \qquad (6.17)$$

v' should be proportional to u_* so that $v' = ku_*$ where $k = 0.8$ from experiments. Substituting in x yields

$$i_b = \frac{\rho u_*^3}{g} 0.8 \qquad (6.18)$$

Allowing for the 25% of bedload moved as creep by impact from saltating grains we arrived at the final expression

$$i_b = 1.1 \frac{\rho}{g} u_*^3 \qquad (6.19)$$

which gives reasonable agreement with experimental data.

Both the above expressions for bedload transport by wind and water take the form of cube relationships with respect to velocity. Equation (6.15) may be written as

$$i_b \propto U \cdot u_*^2 \qquad (6.20)$$

whilst Equation (6.19) may be written as

$$i_b \propto u_*^3 \qquad (6.21)$$

Thus we find sediment transport rate to be strongly affected by increasing flow velocity. Bagnold calculates, for example, that a strong wind blowing at 16 m s^{-1} (35 mph) will move as much sand in 24 hours as would be moved in three weeks by a wind blowing steadily at 8 m s^{-1} (17.5 mph). We should note, however, that the net long-term transport rate is dominated by moderate winds since these make up for their low magnitude by having high frequencies.

6g Summary

At low grain Reynolds numbers the terminal or fall velocity of a solitary spherical grain is given by Stokes' Law. The fall velocity of a grain in a group of falling grains is much reduced relative to its descent in a grainless fluid. Estimation of τ_{crit} is made difficult by the nature of the progressive overlap between an applied stress distribution and a resisting stress distribution. Great generality results if the threshold data is plotted on to a dimensionless Shields graph. Grains on a bed are transported in rolling, saltating and suspension modes once the critical threshold shear stress is exceeded. Bedload comprises all moving grains whose immersed weight force is in dynamic equilibrium with the momentum transferred to the bed by moving grains. Suspended load comprises grains kept aloft from the bedload by fluid turbulence. Sediment transport theory may be approached from a power basis (water flows) or from a momentum-extraction basis (wind).

Further reading

Bagnold's superb book (1954b) is still a good starting place for aeolian sediment transport studies. Raudkivi (1976) and Yalin (1977) deal with many advanced concepts. Bagnold's 1966b paper (revised for the bedload case in 1973 and 1977) is a fundamental starting point for all aspects of transport theory. Many introductory aspects of sediment transport are to be found in Middleton and Southard (1978).

Appendix 6.1 Derivation of Stokes' law of settling

A fundamental equation of fluid dynamics, derived as a simplification of the full Navier–Stokes equations of motion, states that there must exist in fluids a balance between the local viscous and pressure forces. Solution of this equation of 'creeping motion' for low Reynolds number flow past a sphere leads to a relationship between sphere radius a, velocity V_g, viscosity μ and surface drag D.

$$D = 6\pi\mu a V_g \qquad (6.22)$$

In terms of a dimensionless drag coefficient C_D ($= D/\frac{1}{2}\rho V_g^2 A$, where ρ is fluid density, A is the area of the sphere projected normal to the direction of motion) Equation (6.22) may be written as

$$C_D = 6\pi/Re_g = 24/Re_g \qquad (6.23)$$

where Re_e is the grain Reynolds number defined as $\rho V_g^2 a/\mu$. This dimensionless relationship is verified experimentally, *but only for grain Reynolds numbers of* $\lesssim$*0.5*.

During its steady descent through a liquid at these low values of Re_g the surface drag resisting force and the net buoyancy force acting on the grain must be balanced by the applied gravitational force acting on the grain. Thus

$$m_g g = D + m_f g$$

where m_g, m_f = grain and fluid masses, or

$$4/3\pi a^3 \rho_g g = 6\pi\mu a V_g + 4/3\pi a^3 \rho_f g$$

Rearranging and solving for V_g leads to the **Stokes' Settling Law**

$$V_g = \frac{2}{9} \frac{\rho_g - \rho_f}{\mu} g a^2 \qquad (6.24)$$

Where it can be seen that V_g is proportional to the grain radius squared for any given grain/fluid system.

7 Sediment gravity flows

7a Introduction

In Chapter 6 we saw how water and air flows transport grains. We now consider how grain aggregates may, with the aid of gravity, transport themselves with no help from the overlying stationary medium. All gravity flows involving grains must overcome the effects of friction between particles. The following four idealised flow types achieve this in different ways (Fig. 7.1).

Grain flows are characterised by grain/grain collisions between the flowing grains, as in an avalanche. No friction reduction occurs in such flows, hence they may occur only on steep sub-aerial or sub-aqueous slopes that exceed the angle of slope stability for the particular grains involved. **Debris flows** are slurry-like flows in which silt- to boulder-size grains are set in a matrix of clay-grade fines and water. The matrix has 'strength' and buoyancy to support the grains (intergranular collisions also occur) and it serves to lubricate grain irregularities so that debris flows may occur on very gentle sub-aerial and sub-aqueous slopes. **Liquefied flows** are very concentrated dispersions of grains in water which result from the collapse of loosely packed grain aggregates subjected to cyclic shock. The grains are suspended in their own upward-moving pore water and they settle downwards into a new, tighter, packing as the pore fluid is expelled upwards. Friction between grains is thus much reduced during liquefaction. **Turbidity flows** are those in which particles are kept aloft in the body of the flow by turbulent suspension. The suspended particles cause the density of the suspension to be greater than that of the ambient flow, with the result that the suspension will move down slope.

As the reader will have realised, intergradations must exist between the above flow types. It must also be stressed that the different grain support mechanisms are not limited to any one type of flow.

7b Grain flows

If a volume of dry grains in a container is tilted at ever increasing angles then at some critical angle θ_i, some of the grains will flow off the tilted grain surface as an avalanche or grain flow (Fig. 7.2). The remaining grains on the surface will be found to be resting at an angle, θ_r, some 5°–15° less than θ_i. These results are unaffected by the experiment being conducted under water. Attempting to explain this phenomenon we may first note that the downslope movement of a mass of grains must involve an expansion of the whole mass at failure. This expansion is known as **dilation** and it requires energy to be expended (Rowe 1962, Bagnold 1966a). Evidently θ_i, termed the **angle of initial yield**, must include the effects of the dilation involved in moving one grain over another on the zone of potential shear. No such effect is included in θ_r, termed the **angle of residual shear** (Allen 1970a). The value of θ_i is strongly dependent upon porosity, which is controlled by the type of particle packing, and it may vary between 40° for tightly packed natural sands and 30° for loosely packed sands. Grain shape is also an important variable.

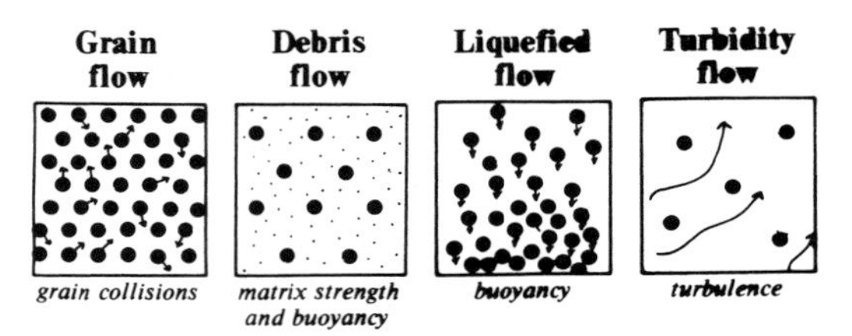

Figure 7.1 Diagram to show the four main types of sediment gravity flows. Note that 'overlaps' can occur between these four end-members and their mechanisms of deposition.

After slope failure of a grain aggregate, the resulting grain flow will consist of a multitude of grains kept aloft above the basal shear plane. Equilibrium demands that the weight stress of the grains should be resisted by an equal and opposite stress arising from the transfer of grain momentum onto the shear plane (Appendix 7.1). The latter stress is known as the **dispersive stress** (Bagnold 1954a). From an analysis of the mechanics of grain flow (Lowe 1976) we can prove that a near-parabolic velocity profile exists, with a thin surficial plug of non-shearing grains that moves passively upon the actively shearing grains just above the shear plane (Appendix 7.1). Of great importance is Lowe's further conclusion that grain flows cannot be thicker than a few centimetres for sand-size grains and therefore cannot be solely responsible for the deposition of single, thick structureless sand beds.

A final feature of grain flow deposits is the frequent occurrence of **reverse grading**. Two hypotheses have been proposed to explain this feature. The first (Bagnold 1954a) notes that the dispersive stress is greatest close to the shear plane and that large particles can exert a higher stress than small ones. Hence the larger particles move upwards through the flow to equalise the stress gradient. The second, termed **kinetic filtering**, says that small grains

Figure 7.2 Grain flows on the steep leeward side of a desert dune. Note the steep scarp to the dune crest and the wrinkled, ropey surfaces to the grain flows that indicate motion of a non-shearing plug zone over the active shear plane. Sahara, Mali (photo by I. Davidson).

simply filter through the gaps between larger colliding grains until they rest close to the shear plane (Middleton 1970). A simple test for the rival hypotheses is to shear grains of equal size but contrasting density, since the dispersive stress also depends upon grain density. It is observed that the densest grains do indeed rise to the flow surface. Although the dispersive stress hypothesis is confirmed, it still cannot be proven that kinetic filtering is absent in multisized grain flows.

7c Debris flows

Sub-aerial debris flows are common in most climatic regimes and are usually initiated after heavy rainfall. They are of particular importance in volcanic areas when the torrential rains that frequently follow eruptions lead to widespread and catastrophic flows on the volcanic slopes.

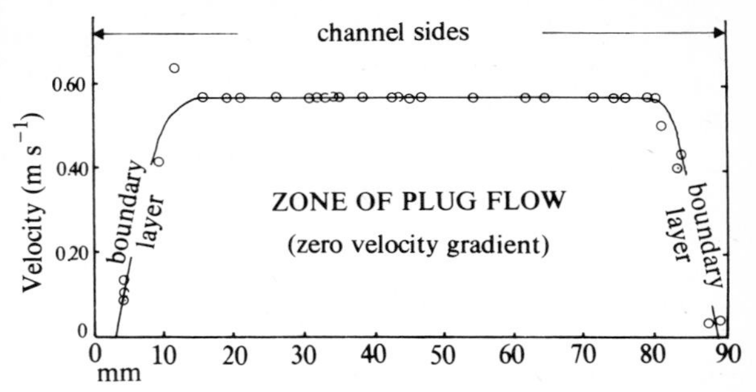

Figure 7.3 Measured velocity profile across an experimental debris flow. Flow is viewed from above. Note the well defined plug zone which is typical of Bingham-like debris flows (after Johnson 1970).

Less is known about the initiation of sub-aqueous flows, although these seem to develop from slumps (see Ch. 24), after earthquake shocks in some instances.

The 'strength' of a debris flow depends upon the matrix property of cohesion and upon the granular friction caused by particle interlocking, in addition to the resistance to shear caused by the matrix 'viscosity' (Johnson 1970). We can write the following expression for the internal shear stress of a moving debris flow:

$$T = \underbrace{c}_{\text{cohesion}} + \underbrace{F_N \tan\theta}_{\text{granular friction}} + \underbrace{\mu \, du/dy}_{\text{Newtonian shear (laminar flow only)}} \qquad (7.1)$$

and for the point at which flow begins

$$T \gg c + F_N \tan\theta_d, \qquad (7.2)$$

where c = cohesion, F_N = normal weight stress, θ_d = angle of dynamic internal friction ($\approx \theta_r$ discussed above), μ = viscosity, du/dy = velocity gradient.

Inspection of Equation (7.1) reveals that as the first two terms on the right tend to zero then the equation reduces to that for shear stress in any Newtonian fluid. We may simplify Equation (7.1) to

$$T = \underbrace{k}_{\text{yield strength}} + \underbrace{\mu_B}_{\text{Bingham viscosity}} du/dy; \; T \gg k \text{ (for flow)} \qquad (7.3)$$

Equation (7.3) is known as the Bingham plastic model for debris flow and the qualifier simply states that the yield stress must be exceeded for flow to occur. The velocity profile across a flowing Bingham substance shows a marked plug profile ($du/dy = 0$) bordered by narrow zones of very high shear strain (Fig. 7.3). The flows are usually of laminar type, but turbulent examples are known.

Grains within debris flows are supported by the strength of the matrix and by their own buoyancy within the matrix rather than by dispersive stress as in the grain flow case, although examples of the latter may occur. Matrix strength gives rise to leveed margins to debris flows. It has been widely noted that debris flows have the ability to transport coarse material such as large boulders and to flow on gentle slopes. These properties are caused by small amounts (sometimes $<1\%$ by volume) of clay/water slurry matrix. This slurry supports fine-grained particles of silt and sand size within debris flow, so reducing the normal stress between particles, and thus internal friction. Experiments investigating this last effect alone show that the friction angle θ_i remains very low as grains are added up to 60% by volume in slurries (Rodine & Johnson 1976). Above these concentrations, particle interlocking and increased normal stress between particles causes θ_i to increase rapidly. It is possible that in hot semi-arid climates, infiltrated diagenetic clay minerals (Ch. 28) may be important contributors towards the production of a suitable cohesive matrix for debris flows. As noted further below, sub-aqueous debris flows may change into turbidity flows as they pass rapidly down a slope (Fig. 7.4).

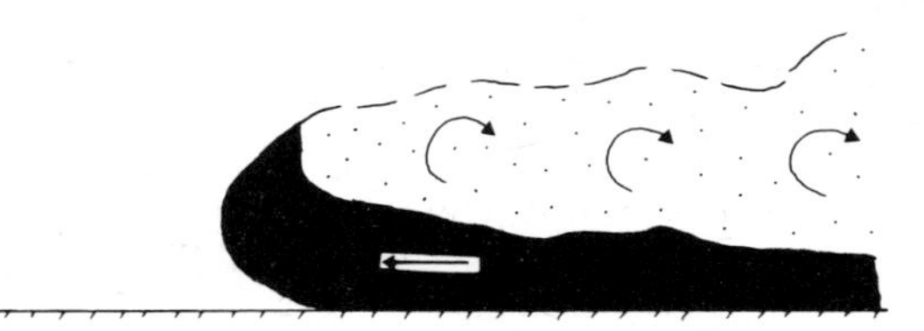

Figure 7.4 Diagrammatic profile of the front of an experimental *sub-aqueous* debris flow (black) to show the turbulent cloud (stippled) generated by flow separation in the lee of the debris flow. Such clouds probably give rise to turbidity currents as the debris flow mixes with the ambient fluid (after Hampton 1972).

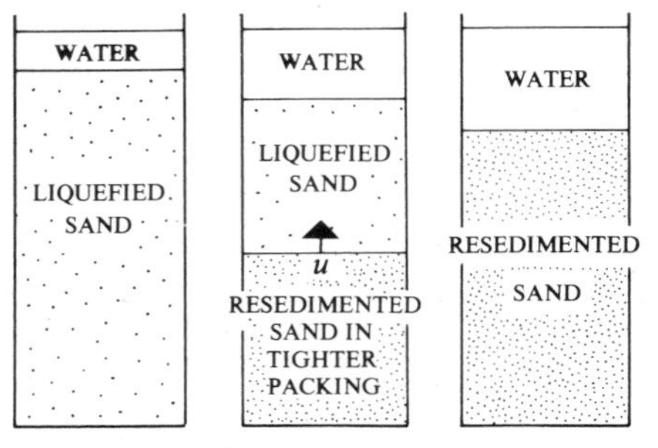

Figure 7.5 Diagram to illustrate the settling and water expulsion associated with liquefaction of subaqueous sands (after Allen & Banks 1972).

7d Liquefied flows

The process of sub-aqueous liquefaction of a sand bed is illustrated in Fig. 7.5 and discussed in more detail in Chapter 11. The loosely packed sand bed is subjected to cyclical shock which causes the grains to become momentarily suspended in their own pore fluid. The liquefied aggregate has now become a concentrated grain dispersion of negligible friction, and flow is therefore possible on very low slopes. The liquefied flow must soon begin to 'settle out' as the grains come into contact once more with their neighbours and displace pore fluid upwards as they assume a tighter packing. Experiments show that a surface of 'settled out' grains rises up through the dispersion (Fig. 7.5) at a rate determined by the fall velocity of the dispersion grains (see Eq. 6.3). The flow of a liquefied bed must always be a race against time, the flow gradually freezing upwards with time and becoming immobile.

In many liquefied beds the upward displacement of pore water is not uniform and it may be concentrated in pipes where the upward fluid escape velocity is sufficiently high to entrain grains or fine material. Dish and pillar structures result within the bed (Ch. 11) whilst sand volcanoes may form at the flow surface. The suspension and upward movement of grains by upward moving fluid is termed **fluidisation**. It is important to distinguish between fluidisation and liquefaction.

7e Turbidity flows

Using the apparatus shown in Figure 7.6, the lock gate is removed and a surge of dense fluid moves along the floor of the tank as a **density current**. In this experiment the excess density is provided by a salt solution, but in nature sediment particles achieve the same end and such density flows involving a turbulent sediment/water mixture are known as **turbidity flows**.

The experimental density flows show well developed 'head' and 'tail' regions. The 'head' is usually 1.5–2 times thicker than the tail, although the ratio approaches unity as the depth of the ambient fluid approaches the depth of the density flow. Experiments (Keulegan 1957) show that the head flows with a velocity u_h given by

$$u_h = 0.7\sqrt{\frac{\Delta\rho}{\rho}gh} \qquad (7.4)$$

where $\Delta\rho$ = density contrast between flow and ambient fluid, ρ = density of ambient fluid, h = head thickness. The constant in Equation (7.4) is insensitive to changes of bed slope.

Close examination of the head (see Fig. 7.8) shows it to

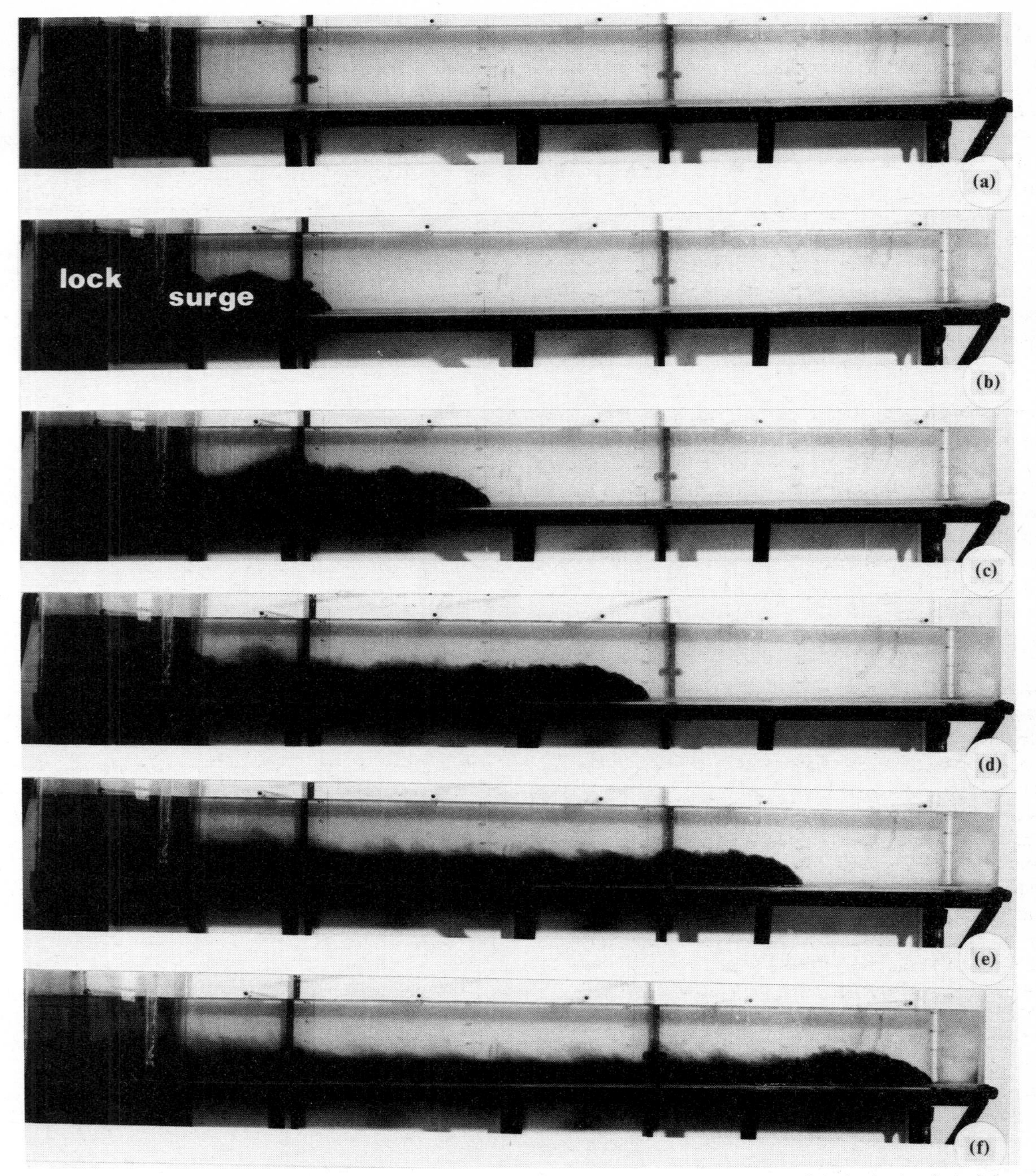

Figure 7.6 (a–f) A sequence of photographs to show the development of a laboratory density current (ink-dyed salt solution) from a surge. Note well-defined head and tail to current in f. The tank is 2.5 m long.

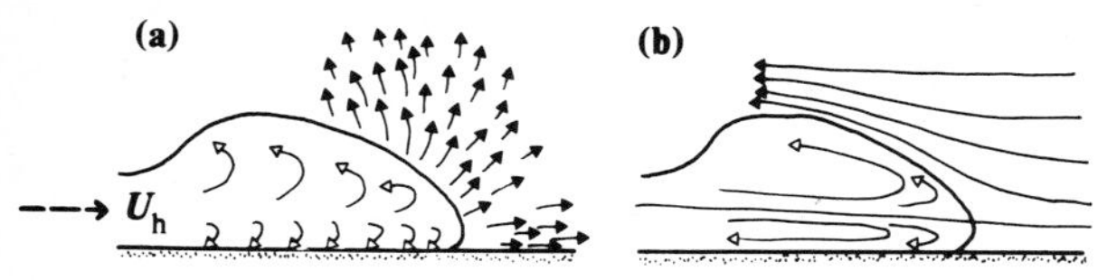

Figure 7.7 The patterns of motion within and around the head region of an advancing density current. (a) The motion relative to the ground, i.e. observer stationary; (b) the motion relative to the moving head, i.e. observer moving with same velocity as head (after Middleton 1966a, Allen 1970b).

be divided into an alternating series of trumpet-shaped clefts and bulbous lobes (Allen 1971b, Simpson 1972). Ambient fluid must clearly pass into the clefts whereupon it can mix internally within the head, causing dilution. Internally, the head region shows strongly diverging flow (Fig. 7.7) which provides the turbulence needed for sediment suspension (Middleton 1966a).

Continuous forward motion of the head at constant velocity requires a continuous transfer of denser fluid into the head from the tail in order to compensate for the fluid mixing across the front of the head region. A steady state is brought about in flows that have a near constant input of dense solution with time. Such flows might occur over a period of time as sediment-laden river water debouches into a water body and travels along the bottom as a continuous undercurrent. Experiments and theory reveal that a slope angle of only about 1° is needed to offset the energy losses caused by friction (Kersey & Hsü 1976).

Surge-like turbidity flows, however, must decelerate because the supply of denser fluid from behind the head is finite. The head thus shrinks until it is completely dissipated (but see note on autosuspension below and in Appendix 7.2). Rapid dissipation also results when a channelised flow enters into a wide reservoir (Fig. 7.8). Such effects are responsible for submarine fan formation.

The study of experimental turbidity flows using correctly scaled-down particle density reveals two categories of flow (Middleton 1966c). **Low concentration** ($C < 0.3$) flows show sediment deposition beginning only a short distance behind the head. Slow deposition from suspension is followed by extended bedload movement and then by rapid deposition from suspension. Finally there is very slow deposition of the finest sediment from the tail. The resulting flow deposits show good sorting, well developed upward-fining of grain size and suites of internal sedimentary structures recording diminishing flow strength with time. In **high concentration** ($C > 0.3$) flows the initial deposition of sediment is followed by mass shearing and formation of a liquefied sediment deposit which develops instability waves at its upper surface, causing circular shearing motions to extend deep into the bed. As the liquefied bed consolidates, in the way described above (p. 78), a plane bed surface is formed upon which the finest particles settle from the tail of the flow. The resulting flow deposit shows poor sorting, poor grading and no primary internal sedimentary structures, although liquefaction structures may be present.

Natural turbidity flows probably originate from sediment slumps caused by earthquake shocks (see Ch. 25). Liquefied and debris flows can both give rise to turbulent turbidity flows following slump episodes, but the nature of the mixing mechanism required to transform these dense, laminar flows into a turbulent sediment suspension remains obscure (Hampton 1972). It is known from experiments, however, that sub-aqueous debris flow fronts are unstable and give rise to a superimposed turbidity flow (Fig. 7.4).

It should be noted finally that a state of **autosuspension** may exist in a turbidity flow (Bagnold 1962, Pantin 1979). Here, continuous sediment suspension is possible with no internal energy loss (Appendix 7.2).

Figure 7.8 View from above of a channelised density current (ink-dyed salt solution) that has entered a wide reservoir. Submarine fans at the toes of submarine canyons form in this way as the spreading current rapidly decelerates. Note the *lobes* and *clefts* forming around the current margin. Grid squares are 50 mm × 50 mm.

7f Deposits of sediment gravity flows

A discussion of deposits has been left to the end of this chapter in order that a comparison may be made between the various flow types and their deposits (Fig. 7.9).

Grain flow deposits occur most frequently as avalanches on dune (Fig. 7.2) and ripple slip faces which give rise to small and large-scale cross stratification (Ch. 8). The laminae are well sorted with evidence of reverse grading in some cases. Internal structures are absent but individual grains may show orientations parallel to flow. The thickness of **debris flow** deposits ranges from decimetres to metres. Clasts from granule to boulder size occur in a matrix of fines. The fines are clay- and silt-grade grains which may be present in concentrations as low as 1% by volume. There is poorly developed grading, poor sorting and no preferred grain orientations. A basal zone of shearing may cause a faint fabric to be present above excavated scours 'dug out' by boulders at the flow base. Debris flows may occupy water-cut channels but cannot cut channels themselves. **Liquefied flows** are dominated by water-escape structures such as convolute laminations, fluid-escape pipes, dish structures, etc. (see Ch. 11). **Turbidity current** deposits differ according to distance from source and to their position on complex depositional surfaces such as submarine fans. Thick-bedded turbidites are coarse grained, relatively poorly graded and laminated, lack basal scour marks and may be channelised. Thin-bedded turbidites are fine grained, well graded and laminated, laterally extensive and show basal scour marks. The 'ideal' sequence of structures in turbidite beds is, from base to top: A – massive unit, B – planar laminated unit, C – small-scale cross-laminated unit, D – interlaminated silt and mud, and E – homogeneous mud and silt. The sequence records a decay in flow strength with time as different bedforms equilibriate with the decelerating current (see Ch. 8). In terms of this **Bouma sequence**, thick-bedded turbidites tend to show A and B units with subordinate C–E, and thin-bedded turbidites show the reverse. Thick-bedded turbidites result from high-concentration flows whilst thin-bedded turbidites result from deposition in low-concentration flows with a well developed bed load. *NB* The use of the terms 'proximal' and 'distal' for thick- and thin-bedded turbidites respectively must now be discontinued in the light of recent discussions by Nilsen, Walker & Normark (1980).

In conclusion it must be stressed that an intergradation exists between slumps (Ch. 11) debris flows, thick-bedded turbidites and thin-bedded turbidites. As noted previously, debris flows probably give rise to turbidity flows as more and more ambient fluid is mixed in with flowing slurry.

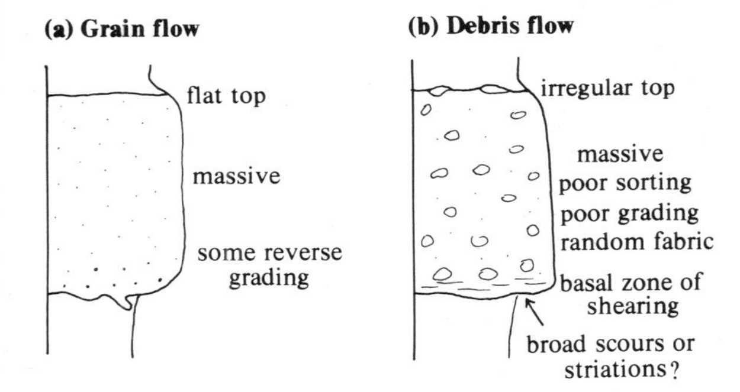

Figure 7.9 Sketch logs to show the main features of sediment gravity flows (after Middleton & Hampton 1973).

7g Summary

Aggregates of grains may flow if friction between grains is overcome. Grain flows move when the angle of initial yield is exceeded on steep slopes. Debris flows move because a fine intergranular matrix reduces friction and encourages buoyancy. Liquefied sediment moves so long as the grains are supported by an excess pore pressure. Turbidity flows move down slopes because of their excess density which is provided by suspended sediment. Since turbulence is generated by movement, further suspension is brought about which causes the excess density to be maintained against gravity; autosuspension results.

Further reading

Middleton and Hampton (1973) provide a clear introduction to gravity flows and their deposits. Johnson's book (1970) gives an advanced physical treatment of debris flow movement. Bagnold (1954a) is a fundamental reference on the mechanics of grain flow. Experimental studies of turbidity currents with much sedimentological relevance are discussed by Middleton (1966a–c). An elegant analytical model for turbidity flows, with the most up-to-date discussion of autosuspension, is given by Pantin (1979).

Appendix 7.1 Dispersive pressure and grain flow (after Bagnold 1954a, Lowe 1976)

Grain flow dynamics are best understood by reference to basic kinetic theory. Consider a vessel filled with air and closed by a perfectly sealed but movable weighted piston. At equilibrium the weighted piston is held up by air pressure: the weight force is balanced by a dispersive force due to molecular bombardment by randomly moving air molecules. In one direction there is a constant weight force and in the other there is a net balancing force arising from molecular collisions.

Now in grain flows, where collisions between grains dominate, the dispersive normal stress P must be in equilibrium with the weight stress W at each level. Bagnold (1954a) found P to vary as

$$P = \mathrm{k}\sigma\lambda^2 D^2 (\mathrm{d}U/\mathrm{d}y)^2 \cos\alpha \qquad (7.5)$$

where k is a constant, σ is grain density, λ is linear grain concentration (Ch. 4), D is grain diameter, $\mathrm{d}U/\mathrm{d}y$ is the strain rate in the flow, U is the solid velocity and α is the angle of dynamic friction ($= \theta_r$ of Ch. 7b). W is given by

$$W = \bar{C}g(\sigma - \rho)(Y - y)\cos\beta \qquad (7.6)$$

where $\bar{C}$ is the mean grain concentration by volume above point y in the flow, ρ is the fluid density, Y is the flow thickness, y is distance above the bed and β is the local bed slope.

Since $P = W$ at equilibrium we may equate (7.5) and (7.6) and solve for $\mathrm{d}U/\mathrm{d}y$. Integration with the constraint that $U = 0$ at $y = 0$ gives

$$U = \frac{2}{3}\left(\frac{\bar{C}g(\sigma - \rho)\cos\beta}{\rho \mathrm{k}\cos\alpha}\right)^{\frac{1}{2}} \frac{1}{\lambda D}\left[Y^{\frac{3}{2}} - (Y - y)^{\frac{3}{2}}\right] \qquad (7.7)$$

Solving U for various heights in the flow ($\bar{C}$, λ, α assumed constant with height) yields a velocity profile showing a thick zone of shearing overlaid by a superficial 'plug' of non-shearing flow.

Bagnold (1954a) found (see Ch. 6) that $T/P = \tan\alpha$. Thus in order for grain flows to develop $\tan\beta \gg \tan\alpha$. Since $\tan\alpha$ for quartz grains has a value of 0.5–0.6, we must conclude that grain flows are possible only on slopes steeper than about 25°–30°.

Appendix 7.2 A note on autosuspension in turbidity currents

Suspended sediment is held up by turbulence generated at the bed. In normal river flow, gravity causes fluid movement down a slope which then causes turbulence which then supports a given mass of suspended sediment. In a turbidity current the flow is caused by a suspension of sediment, giving a body of fluid excess density which causes flow down the local gravity slope. Thus a feedback effect arises whence suspension causes motion which causes turbulence which causes suspension. This is the state of autosuspension (Bagnold 1962, 1963).

In terms of flow power, autosuspension must arise when the total available power in the current ω exceeds the power ω_τ expended against bottom friction. Thus $\omega \gg \omega_\tau$. Put in terms of transporting efficiency of suspended load, the proportion of ω_τ available for suspension must exceed the power ω_N needed to support the suspended load. Thus $e_x\omega_\tau \gg \omega_N$, where e_x is an efficiency factor. Combining these two criteria for autosuspension and with various substitutions (see Pantin 1979 for full derivation), we arrive at a condition for autosuspension.

$$\frac{e_x \beta U_s}{V_g} \gg 1 \qquad (7.8)$$

where β is bed slope, U_s is the velocity of the suspended load ($\approx$ flow velocity) and V_g is the suspended grains' fall velocity. Thus autosuspension is favoured by high efficiency, slope, and flow velocity and fine sediment size.

After solving the equations of motion for a two-dimensional turbidity flow, Pantin (op. cit.) shows that turbidity currents must either decelerate and deposit their load or accelerate into the autosuspension field where deposition may occur only as the bed slope levels out. The reader is referred to Pantin's elegant paper for a full discussion of this topic.

PART THREE BEDFORMS AND SEDIMENTARY STRUCTURES

Trees show the bodily form of the wind;
Waves give vital energy to the moon.

From the Zenrinkushu

Plate 3 This weathered sandstone scarp (–10 m high) shows a variety of large-scale cross sets (some overturned foresets are visible directly under the channel base), massive beds and a well defined channel with its faintly laminated to massive fill. The sediment is thought to have originally been deposited in a braided river channel environment (Fell Sandstone, Northumberland, England)

Theme

The transport of sediment grains is often accompanied by the development of morphological features termed bedforms. Much of the loose sediment on the Earth's surface is moulded into these features which range in size from the humble and ubiquitous ripples (of diverse origins) to gigantic mountains of desert sand, termed draa. Man's longstanding fascination for these features has recently been reinforced by a reasonable level of understanding of the mechanics of their formation and migration. Of particular importance in geological studies is the role of bedforms in the generation of the sedimentary structures so commonly observed in the geological record. A true understanding of the origins and relevance of sedimentary structures can be gained only by a good grasp of the dynamics of grain–fluid systems as outlined in Part 2. Further aspects of bedforms and structures in specific environments are dealt with in Part 5.

8 Bedforms and structures in granular sediments

8a Bedforms and structures formed by unidirectional water flows

Sediment movement is accompanied by the organisation of grains into morphological elements known as **bedforms.** Experimental results from flow channels have shown that a number of bedforms exist which are stable only between certain values of flow strength. These **bedform states** occupy distinct fields on plots of flow strength against grain size for well sorted quartz sands (Fig. 8.1). Although **bedform phase diagrams** such as that shown in Figure 8.1 are in common use, we shall see below (Ch. 8.2) that use of bed shear stress as the ordinate term for flow strength has certain disadvantages. The reader is advised that there is much divergence of use amongst researchers concerning bedform nomenclature.

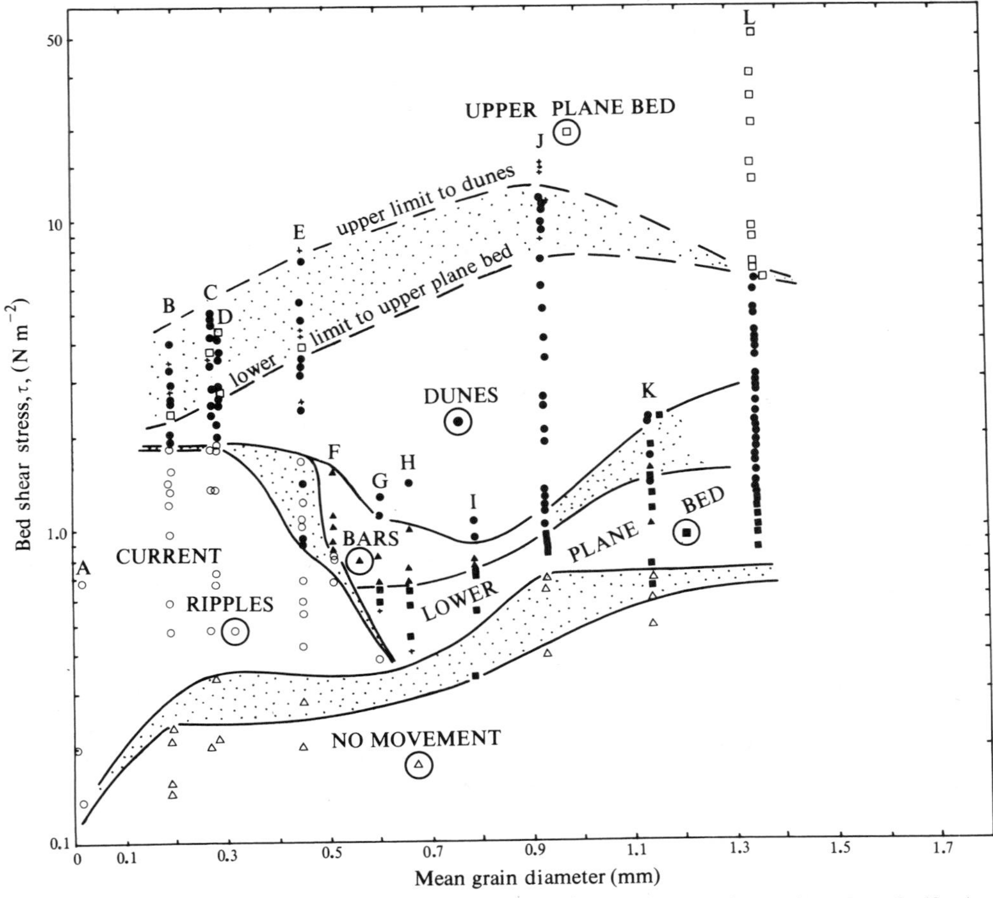

Figure 8.1 Bedform phase diagram to show the equilibrium stability fields of the various sub-aqueous bedforms developed by uniform, steady flow over granular beds in straight laboratory channels. Note the degree of overlap (stippled) between fields, some of which is controlled by differing nomenclature of individual experimenters. Transitional forms denoted by crosses. Recalculated data of Costello (1974) (F, G, H, I, K), Guy *et al.* (1966) (B, C, D, E, J), Mantz (1978) (A), Williams (1970) (L) (after Leeder 1980). τ calculated as $\rho gh \sin \theta$ (Eq. 5.6).

Certain alternative usages will be noted below in the appropriate sections. There is also some uncertainty regarding the application of flume data gained from shallow (<0.5 m) flows to deeper natural flows (see discussion in Middleton & Southard 1978). Correctly scaled down experiments currently in progress (see Southard *et al.* 1980) should provide data useful in assessing this uncertainty.

Referring to Figure 8.1 let us first discuss the various bed states developed in fine bedstock as flow strength is gradually increased.

Current ripples are the stable bedforms above the threshold for sediment movement on artificially smoothed fine sand beds at relatively low flow strengths. They may also form from *initial* bed irregularities well below this smooth bed threshold. Current ripples do not occur in sands coarser than about 0.7 mm. The ripples are roughly triangular in the xy plane parallel to flow, having a gentle sloping upstream or **stoss** side, sometimes with a prominent crestal platform, and a more steeply sloping (~30°–35°) downstream or **lee** side (see Fig. 8.3). Ripple heights range from 0.005–0.03 m, ripple wavelengths from 0.05–0.40 m and typical **ripple indices** (wavelength:height ratios) are between 10 and 40 (J. R. L. Allen 1968). There is a broad increase of ripple size with increasing bed shear stress, but size is independent of water depth. Ripple wavelengths vary with grain size, being roughly 1000 d, although the plot shows much scatter. Viewed from above, the ripple crestline may be straight, sinuous or linguoid (tongue-shaped) (Figs 8.2 & 8.3a, b). This increasing complexity occurs as flow velocity is increased for a given depth and is accompanied by increasing influence of longitudinal flow vortices (Ch. 5).

Mapping of skin-friction lines from plaster-of-paris ripple models (J. R. L. Allen 1968) shows a pattern of flow separation (Ch. 5) at ripple crests with flow reattachment down stream of the ripple trough. A captive bubble of recirculating fluid is held in the ripple lee (Figs 8.4 & 5). Grains are moved in bedload up the ripple stoss side until they fall or diffuse from the separating flow at the crest to accumulate high up on the steep ripple lee face. Periodically the accumulated grains become unstable as the angle of accumulation exceeds the angle of initial yield (Ch. 7). Small grain avalanches then occur, terminating at the toe of the lee face, and a single avalanche lamina is thus accreted on to the ripple lee. Ripple advance by lee slope deposition results in the flow attachment point shifting up the back of the downstream ripple where increased erosion occurs because of the very high turbulent stresses generated at the reattachment point. In this way the ripples constantly shift down stream, preserving their overall equilibrium shapes, at a fairly constant velocity dependent upon the magnitude of the flow strength.

Sections cut through current ripples in the xy plane parallel to flow reveal successive accreted avalanche laminae which define the sedimentary structure **small-scale cross lamination**. Sectioned normal to flow in the xz

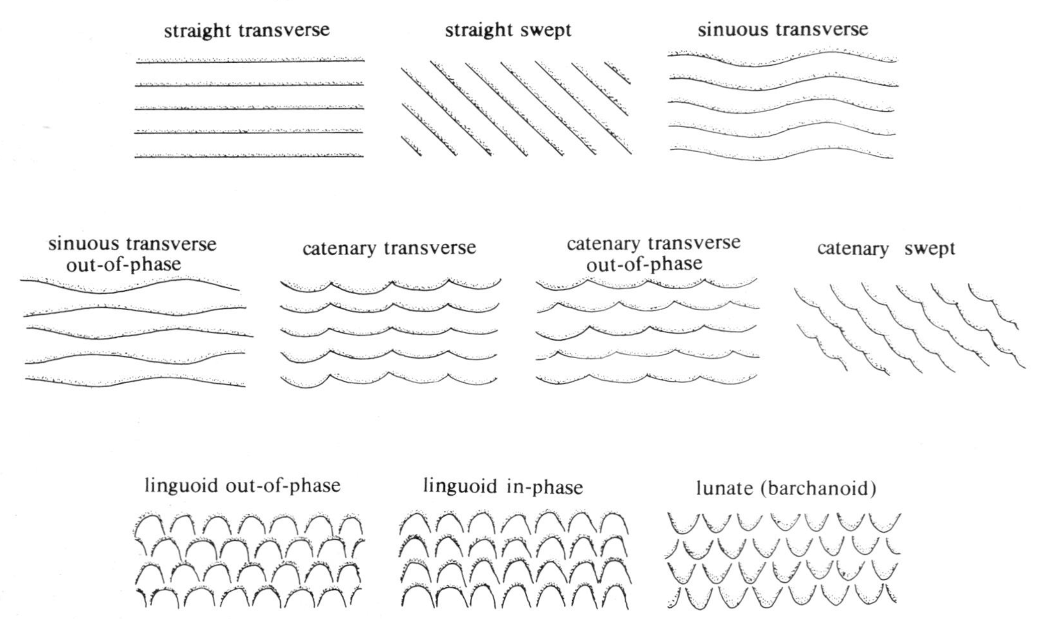

Figure 8.2 A classification of ripple and dune crest types. View from above, with flow from bottom to top in each case (after J. R. L. Allen 1968).

Figure 8.3 (a) Sinuous and (subordinate) linguoid current ripples; flow from right to left; scale bar = 0.15 m; Solway Firth, Scotland. (b) Linguoid current ripples; flow from bottom to top; packet = 0.1 m long; Severn Estuary, England. (c) Large-scale trough cross stratification, flow towards observer; hammer shaft = 0.30 m; Old Red Sandstone, Welsh Borders. (d) Downward-dipping sets of large-scale cross stratification; scale bar = 0.1 m; Fell Sandstone, Northumberland, England.

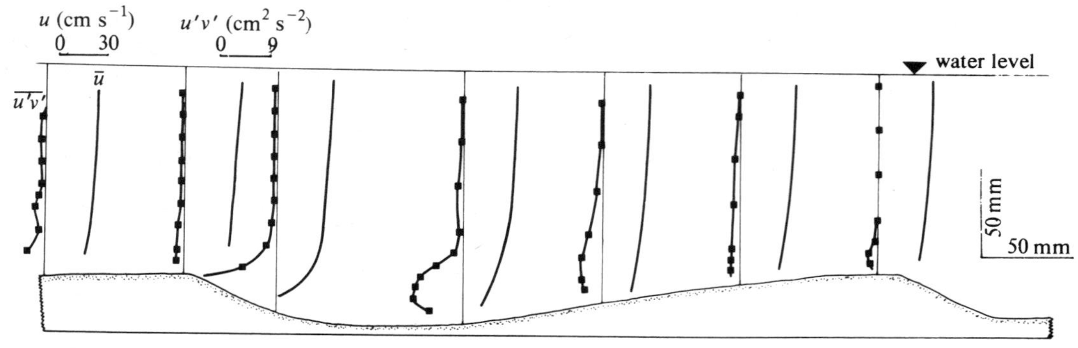

Figure 8.4 Profiles of mean velocity ($\bar{u}$) and a measure of turbulent intensity ($\overline{uv}$) as measured over a fixed and sand-coated experimental ripple model. Note the presence of the roller vortex in the ripple lee and the large amount of turbulence produced in the flow reattachment region (after Sheen 1964 as reported in Raudkivi 1976).

plane the cross laminae may be parallel and horizontal, defining **planar cross lamination**, or trough shaped, defining **trough cross lamination** (Fig. 8.6). Tabular cross laminae result from the migration of straight crested ripples whilst trough cross laminae result as sinuous to linguoid ripples migrate forward into heel-shaped scour troughs eroded by the isolated separation eddies (Fig. 8.6). If there is no net sediment deposition at a particular point, then no cross lamination can be produced other than that found within the individual ripple elements. When net deposition occurs, then a particular ripple crest must have a vertical component of motion as well as an horizontal component (Fig. 8.7). **Sets** of cross lamination may thus be formed, bounded by erosive surfaces. The thickness of the sets is directly proportional to the rate of upward movement. Internally the set boundaries are seen to 'climb' at an angle to the horizontal (Fig. 8.7), and the structure is known as **climbing-ripple cross lamination.** High angles of climb, with preservation of stoss-side laminae, indicate high rates of net deposition as in decelerating flows such as river floods or turbidity currents (J. R. L. Allen 1972).

With increasing flow strength, **dunes** result. These large bedforms (Figs 8.8 & 9) are similar to current ripples in general shape but are dynamically distinct (J. R. L. Allen 1968), as is shown by the lack of overlap between ripples and dunes on a plot of height against wavelength. Dunes do not form in sediment of coarse silt grade and finer (Fig. 8.1). Dunes are sometimes termed large-scale

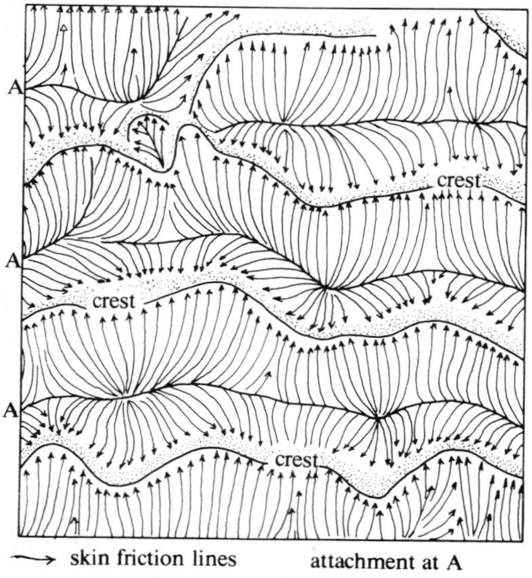

Figure 8.5 Pattern of skin friction lines for flow (mean velocity 0.22 m s⁻¹, depth 0.095 m) over a current rippled bed. Flow from bottom to top. The ripple crests mark lines of flow separation. The steep ripple avalanche faces are stippled (after J. R. L. Allen 1969a).

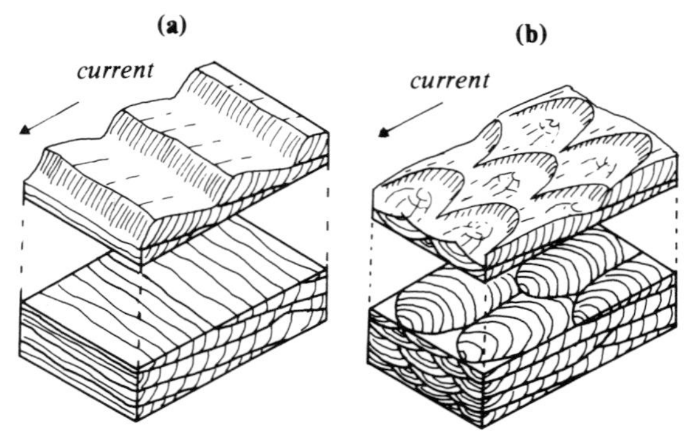

Figure 8.6 Diagrams to show that the migration of (a) straight-crested and (b) curved-crested bedforms produces planar and trough cross sets respectively (after Allen 1970b). Note that preservation of successive cross lamina sets necessarily involves a degree of net deposition. Most sets therefore 'climb' at some angle from the local bed inclination (see Fig. 8.7).

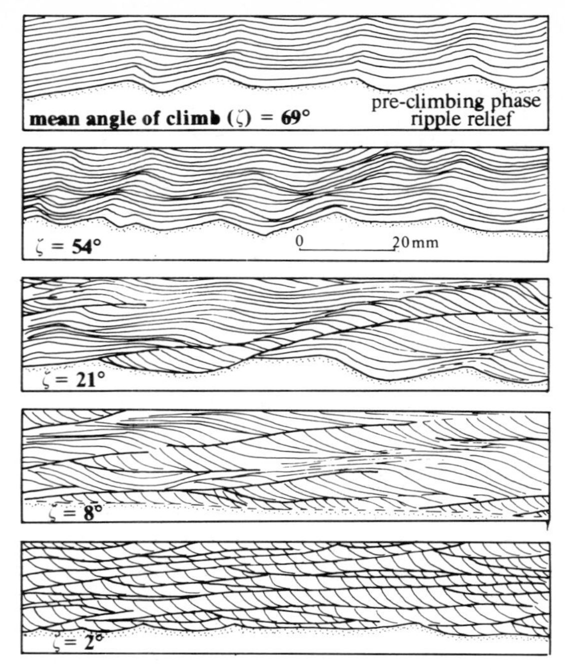

Figure 8.7 Experimentally produced climbing-ripple cross-lamination seen in vertical profile parallel with flow. The increasing angle of climb from bottom to top is caused by the increasing rate of net vertical deposition relative to the speed of advance of the ripples (after J. R. L. Allen 1972).

ripples or megaripples, but these terms do not emphasise the distinctiveness of the dune bedform sufficiently. Dune wavelengths commonly range from 0.6 m to hundreds of metres, and heights vary from 0.05 m to 10.0 m or more. Dunes commonly show a correlation of wavelength λ and height H with flow depth y, (Fig. 8.10) in contrast to current ripples. Thus from experimental and field measurements (Allen 1970b) for the range $0.1 \text{ m} > y < 100 \text{ m}$

$$\lambda = 1.16 y^{1.55} \qquad (8.1)$$

and

$$H = 0.086 y^{1.19}, \qquad (8.2)$$

although there is much scatter about the regression lines. In plan view dune crests resemble the ripple crests noted above, with the addition of lunate (barchanoid) forms. The flow pattern over dunes is similar to that over ripples, with well developed flow separation and reattachment. In the lower part of the dune stability field, ripples are commonly superimposed on the backs of dunes, introducing the concept of **bedform hierarchies**. Similarly, small dunes may be superimposed on the stoss sides of larger 'dunes'. There is debate as to whether such superimposition is an equilibrium flow effect or is the result of variations in flow strength with time (see Allen & Collinson 1974). If the first hypothesis is true, then the larger bedform should be considered distinct from the dune state. Some authors who believe this hypothesis refer to such bedforms as **sandwaves**, but this term is also used by other authors for large-scale structures in the marine environment.

Dune migration gives rise to **large-scale cross stratification** of planar or trough (Fig. 8.3c) type in a similar way to that outlined for current ripples. Tangential contacts between the individual cross laminae and the bounding set surface are encouraged by relatively weak lee separation eddies and by a high fallout rate of particles from suspension in the dune lee. **Counterflow ripples** (Boersma 1967) result when grains are swept back up the lower parts of tangential foresets by near-bed flow in the separation bubble. The regularly dipping cross sets may also often be cut by erosive surfaces (see review by Jones & McCabe 1980). These result from erosion of the dune crest and lee by falling stage or low stage flows. Such **reactivation surfaces** (Collinson 1970) are preserved within the migrating dune when normal avalanching events begin once more at rising and high flow stages. Sometimes, successive reactivation surfaces may enclose smaller-scale cross stratification in an arrangement aptly known as **downward-dipping cross stratification** (Fig. 8.3d). Here, small dune forms have migrated up the stoss side of the parent bedform *and* down the gently dipping lee side where they are preserved as cross sets (Banks 1973).

As flow strength is further increased, dunes give way to an **upper-stage plane bed** (Fig. 8.9b) associated with intense sediment transport over a virtually flat bed. The surface of the upper-stage plane bed is marked in detail by a system of low linear ridges, a few grain diameters high, aligned parallel to flow direction (Allen 1964; Fig. 8.9b). These ridges constantly shift position over the plane bed surface. The ridges, hundreds or thousands of grain diameters long, are separated by flat-bottomed hollows and this characteristic microstructure of flow-parallel ridges and hollows is known as **primary current lineation.** Primary current lineation is the direct result of the viscous sub-layer structure discussed in Chapter 5. Incoming sweeps, spaced parallel to flow, push grains aside to form the tiny grain ridges separated by broad troughs (Fig. 8.9b; see Fig. 5.17e). It is important to stress, however, that primary current lineation is not restricted to the upper stage plane bed regime but may occur on the stoss side of ripples and dunes. The lateral spacing of primary current lineation agrees with the measured spacing of low-speed streaks on an hydraulically smooth boundary

Figure 8.8 (a) A straight crested dune or bar with superimposed linguoid ripples formed during the falling stage of tidal ebb flow. (b) Strongly sinuous dunes with well developed scour pools. (Both photographs courtesy of T. Elliott & A. Gardiner.) Loughor Estuary, Swansea, Wales.

Figure 8.9 (a) View over a sinuous dune crest to show scour pool and ripple fan on back of next dune downstream. Scale = 0.5 m long. Loughor Estuary, Swansea, Wales. (b) Primary current lineations in fine sands (from Allen 1964). (c) Upper phase plane beds. Knife = 0.15 m long. St. Bees Sandstone, Cumberland. (d) Train of antidunes (wavelength ≈ 0.3 m) in fast shallow tidal channel flow. Barmouth estuary, Wales. (e) Train of upstream-breaking antidunes in tidal channel; flow left to right. Shovel handle = 20 cm long. Solway Firth, Scotland.

(Eq. 5.25). Primary current lineation progressively dies out on coarse sand beds since in such cases the spatially organised low- and high-speed streaks are progressively disrupted as the grains disrupt the viscous sub-layer and replace it with a granular dispersion (Allen & Leeder 1980).

Upper-stage plane beds give rise to an internal structure of **planar laminations** ranging between 5 and 20 grain diameters thick (Fig. 8.9c). Such laminae, which are responsible for 'flagstone' lithologies in some sandstones, must result from pulsating net deposition over the plane bed surface, but the exact relationship between this process and the burst/sweep cycle is not clear (see Bridge 1978b for one hypothesis).

There now remain two final bedform types to be discussed which do not appear on the bedform phase diagram in Figure 8.1. The first bedforms are sinusoidal forms with accompanying in-phase water waves. They are somewhat misleadingly termed antidunes (Fig. 8.9d, e). Antidunes are commonly seen in very fast, shallow flows with a Froude number (Ch. 5) of greater than about 0.8, i.e. antidunes are roughly indicative of rapid (supercritical) flow. The wavelength of antidunes is dependent chiefly upon the square of the mean flow velocity by the equation (Kennedy 1963)

$$\lambda = u^2 g/2\pi \tag{8.3}$$

Antidunes commonly occur in long trains; the wave form may be stationary or may periodically steepen, move upstream and break up in a great rush of turbulence, the process then beginning over again. This upstream migration (Fig. 8.11) gives rise to upstream dipping sets of low angle (<10°) cross lamination (Middleton 1965). These sets have a low preservation potential in the absence of net deposition since any deceleration of the flow will cause a plane bed to develop, causing destruction of the antidune laminations. Even when preserved, the laminae are very faint since the rapid upstream antidune migration is not accompanied by avalanching and, thereby, grain sorting (Ch. 7). The rare occurrence of antidunes preserved on bedding planes facilitates direct determination of palaeoflow velocity from Equation 8.3 above, if λ can be measured.

As antidune flows are further increased in velocity, **chute-and-pool** structures develop. The chutes are shallow, rapid (supercritical) flows with high slopes which end abruptly in the deeper pool where the flow is tranquil (subcritical). The rapid flow into a pool, whose upstream boundary is marked by violently breaking water, is a form of **hydraulic jump**, being a zone of marked degradation of kinetic energy into thermal energy, seen as a flow deceleration. Sediment accumulation may occur in the

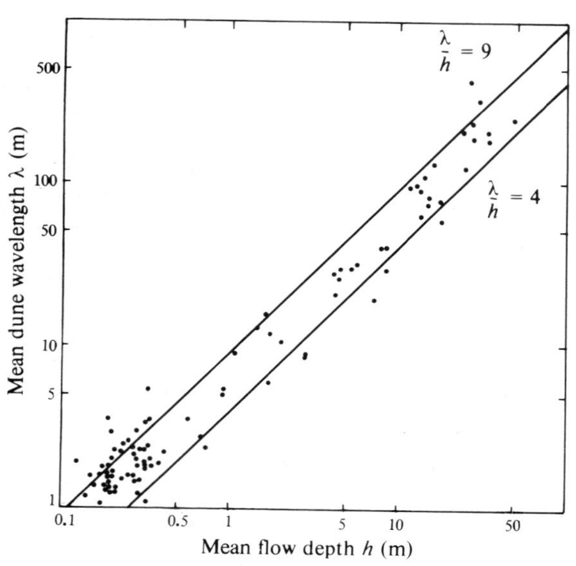

Figure 8.10 Group-mean wavelength of sub-aqueous dunes versus mean flow depth in unidirectional open channel and tidal flows. A minimum of five successive dunes was used to determine each value plotted (after Jackson 1976a).

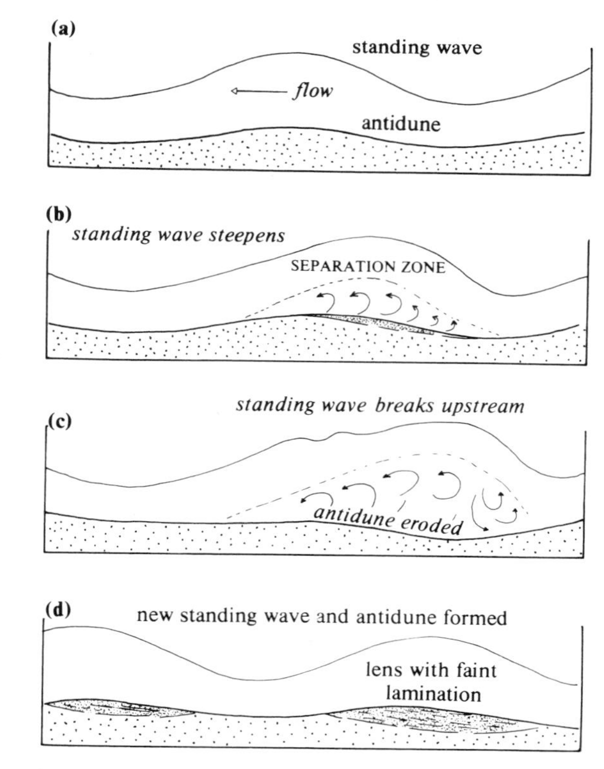

Figure 8.11 Breaking of a 'standing' wave to form faint upstream dipping internal laminations (after Middleton 1965).

relatively tranquil pool region where steeply dipping backset laminations develop, as in the upstream-breaking antidunes noted above. The only known occurrence of preserved chute-and-pool laminations is in volcanic base-surge deposits around the rims of maar-type volcanic craters (Schminke *et al.* 1975).

Referring back to Figure 8.1, let us now note the rather different succession of bedforms that develop as flow is increased over a coarse sand bedstock. As noted briefly above, ripples do not form in coarse sands above a grain diameter of about 0.7 mm. Above the movement threshold on artificially planed beds in such coarse sands an equilibrium plane bed exists instead of ripples. This is termed the **lower-stage plane bed** and it exhibits shallow scours and narrow irregular grooves 2–3 grain diameters deep over its surface (Leeder 1980). Net deposition on a lower-stage plane bed should give rise to crude planar laminations made up from shallow scours, but convincing examples have yet to be described from the sedimentary record.

As the flow is further increased, dune-like structures termed **bars** (or, by some authors, sandwaves), develop over the lower-stage plane bed. Bars display great variation in wavelength (Costello 1974). They have high wavelength:height ratios (Fig. 8.12). They show straight crests, and leeside avalanche-induced migration giving rise to planar sets of large-scale cross stratification. Bars do not show leeside scour hollows (cf. Fig. 8.8a) and the variation of wavelength along the flow direction in experimental examples indicates that the bedform repeat distance is not determined by processes acting in the weakly developed leeside separation eddy. In natural environments, bars may show superimposed dune forms. Bars do not show any strong correlation of wavelength or -height with water depth. It is thus clear from this brief discussion that bars must be considered as a bedform distinct from dunes, which they superficially resemble. It is probable that bars also occur as a separate bed configuration in medium sands since low down in the dune stability field (Fig. 8.1) the 'dunes' have higher wavelength:height ratios and have been termed 'intermediate flattened dunes' (Pratt 1973).

With further increase of flow strength in coarse sands, dunes, and then upper-stage plane beds, develop in coarse sands, with antidune development if the Froude number exceeds about 0.8.

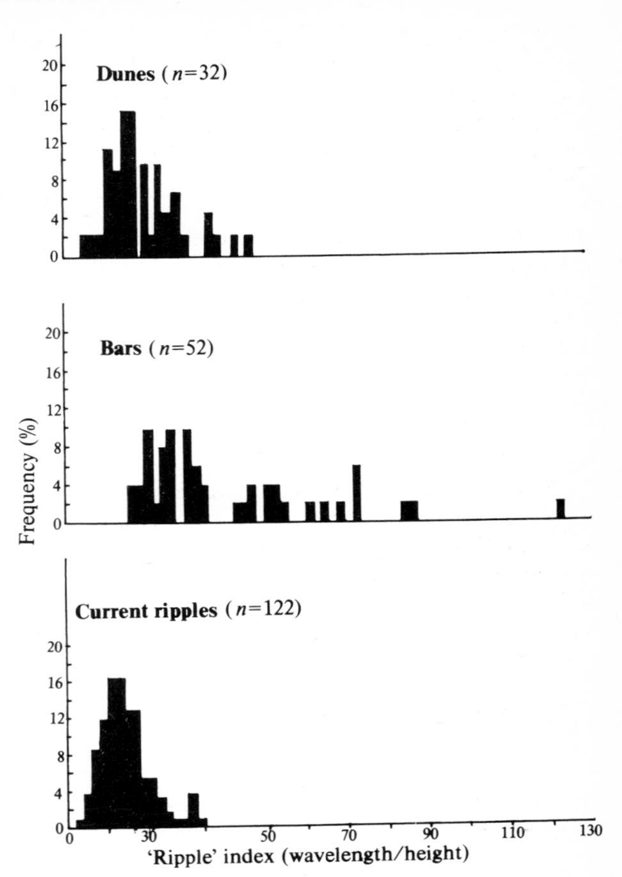

Figure 8.12 Histograms of length:height ratios ('ripple' index) for current ripples, bars and dunes formed in an experimental flow channel (after Costello 1974). Note the very broad spread of values for bars compared to dunes and current ripples.

8b Further notes on bedform phase diagrams

It is usual to separate the bedforms discussed above into two broad groups (Simons *et al.* 1965). Ripples, lower-phase plane beds, bars and dunes occur in a **lower flow regime** where flow resistance is relatively high and where water waves and large eddy bursts at the water surface are out of phase with the bed undulations. Upper plane beds, antidunes, and chute and pool structures occur in an **upper flow regime** where flow resistance is relatively low and where surface waves, where present, are in phase with any bed undulations. As we have seen, ripples and dunes both show flow separation and reattachment features. The turbulent energy generated by these processes causes a marked increase in the frictional resistance exerted by the bed on the flows. Lower flow regime bedforms commonly show friction coefficients 2–5 times those calculated for upper flow regime bedform. This is not true for lower phase flat beds where the resistance to flow arises solely from grain drag and not to bedform drag.

The comments above on friction coefficients raise an important point concerning the use of bedform phase diagrams such as Figure 8.1 in which bed shear stress or flow power is used as the measure of flow strength.

Since the applied fluid shear may be written as $\tau = \rho f \bar{u}^2/8$ (Eq. 5.7) where f is the d'Arcy–Weisbach friction coefficient, ρ is fluid density and $\bar{u}$ is mean flow velocity, we see that shear stress is a direct function of the friction coefficient. Since the friction coefficient is itself dependent on the type of bedform developed, we can see that bed shear must itself be a function of bedform type (Southard 1971). We could therefore imagine the situation where the same shear stress could be produced by slow flow over a very rough (e.g. dune) boundary or a fast flow over a very smooth (e.g. upper plane bed) flow boundary. This effect is partly responsible for the overlap of the dune and upper phase plane bed fields in Figure 8.1. This 'overlap problem' may be overcome (Southard 1971) by plotting bedform phase diagrams as mean flow velocity versus flow depth for particular grain sizes or by plotting diagrams based upon mean flow velocity versus grain size for particular flow depths (Figs 8.13 & 14).

The reasons for the existence of sub-aqueous bedforms and for their observed conditions of stability are relatively poorly understood. An introduction to this problem, that of **bedform theory**, is provided in Appendix 8.1.

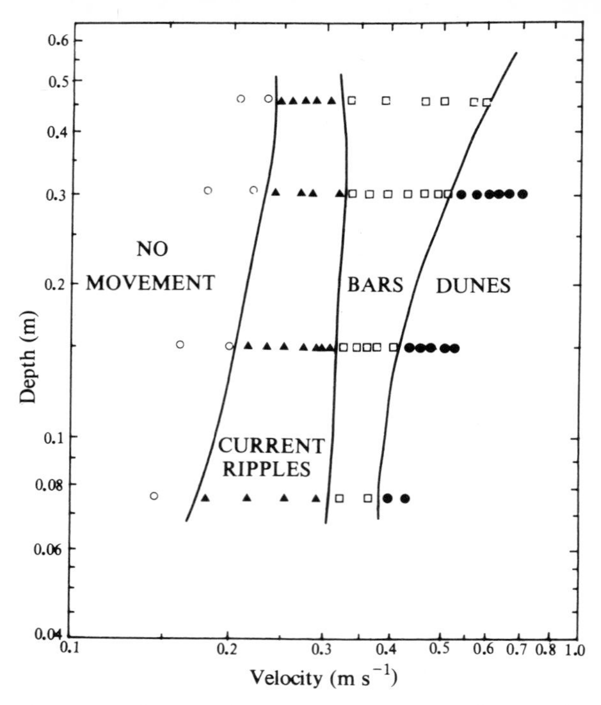

Figure 8.13 Velocity–depth bedform phase diagram for 0.49 mm sand as determined by experiment in an hydraulic flume (after Costello 1974).

8c Bedforms and structures formed by water waves (see also Ch. 18)

It is a common mistake to imagine water waves as heaps of water moving along a surface. In fact, at any fixed vertical point the water simply rises and falls (Figs 18.2 & 14); it is the wave energy that is transferred. As discussed in more detail in Chapter 18, any particular element of surface water involved in waves passing over deep water defines a circle of rotation as the wave passes by. In shallow water the circular motion becomes elliptical downwards, so that a regular to-and-fro motion occurs in response to the water waves at the water surface (Fig. 18.2). If the water were deep enough, the to-and-fro motions would die away completely. In shallow water the motions cause shearing stresses to be set up on sediment beds.

The effect of the shearing stresses set up by the to-and-fro motion on the bottom is to cause rolling grain movement at some critical wave condition on an intially plane bottom. The formation of symmetrical (oscillation) **wave-formed ripples** follows, whose crests are usually very persistent laterally but which bifurcate in a characteristic manner. The wave-formed ripples vary greatly in size since they are dependent only upon the dimensions of surface waves. They vary in wavelength between 0.009 m and 2.0 m and in height from 0.003–0.25 m, the ripple index varying between 4 and 13. Such ripples may form at depths of up to 200 m on continental shelves (Ch. 22).

Once the threshold for motion is reached, the rolling grains tend to come to rest along crests which lie parallel to the crestlines of the water waves, the grains at the ripple crests resting at the angle of residual yield. At low values of bed shear stress the ripple crests are low, with broad, flat or gently curved troughs in which no grain movement occurs. These ripples are a stable form and are termed **rolling grain ripples** (Bagnold 1946). As the bed shear stress is increased, the ripple crests reach a critical height, causing the formation of vortices (see the superb visualisation studies of Honji *et al.* 1980) on either side of the ripple crest (Fig. 8.15) during the to-and-fro bottom motion of the water (Bagnold 1946). The vortices scour sand from the ripple troughs, increase the ripple amplitude and thus markedly decrease the ripple index. The ripples so formed are termed **vortex ripples** (Bagnold 1946). These are the common symmetric wave-formed ripples seen on most beaches. Their wavelengths are given by $\lambda = 0.65 d_0$ from experiment (Miller & Komar 1980). Sections through oscillation ripples (Fig. 8.16) reveal an internal structure of chevron-like laminae accreted onto either side of the ripple crest during successive vortex motions (Fig. 8.16). Increased wave strength eventually causes all ripples to be washed out, with the establishment of a **plane bed** (Fig. 8.17).

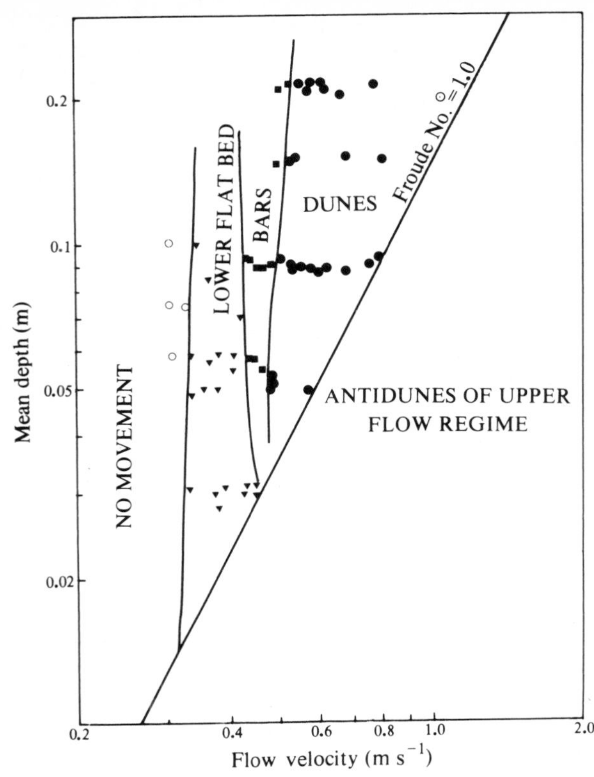

Figure 8.14 Velocity–depth bedform phase diagram for 1.14 mm sand as determined by experiment in an hydraulic flume. Note the absence of current ripples and their replacement at low flow velocities by a lower flat (plane) bedform. (After Costello 1974.)

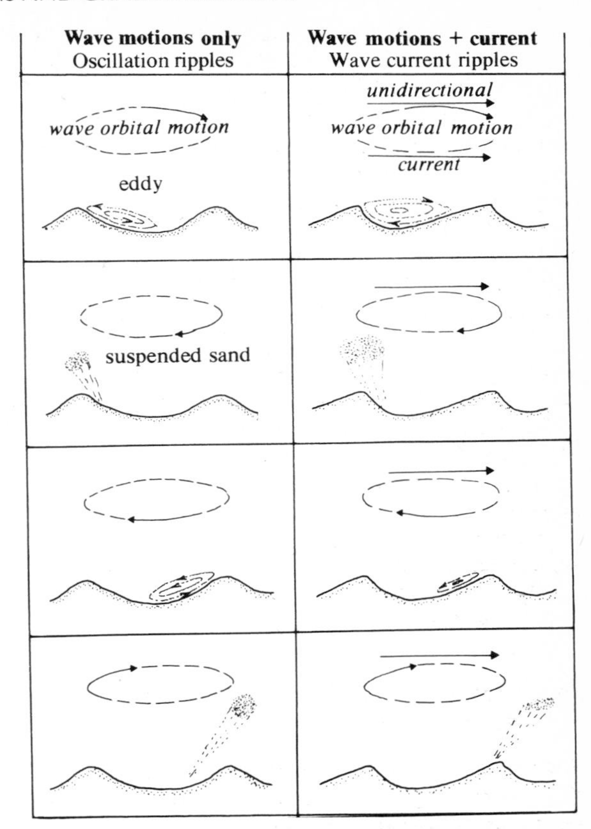

Figure 8.15 The relationship between sand transport over a rippled sand bed and the orbital motions of wave action with and without a superimposed unidirectional current (after Komar 1976 from original data of Inman & Bowen 1963).

In shallow water near the breaker zone there is a net landward transport of water superimposed upon any surface wave pattern (Chs 18 & 21). This causes the formation of asymmetric **wave current ripples** (Fig. 8.15) which give rise to small-scale cross lamination similar to that produced by current ripples. Preserved ripple forms may be distinguished by the ripple index (Reineck & Wunderlich 1968a), wave current ripples rarely showing values greater than about 15 whereas current ripples may show values up to 40, and by the fact that current ripples rarely show crest bifurcation. In environments subject to both pure oscillatory flow and periodic net landward flow a complex alternation of internal cross laminations may result (Fig. 8.16).

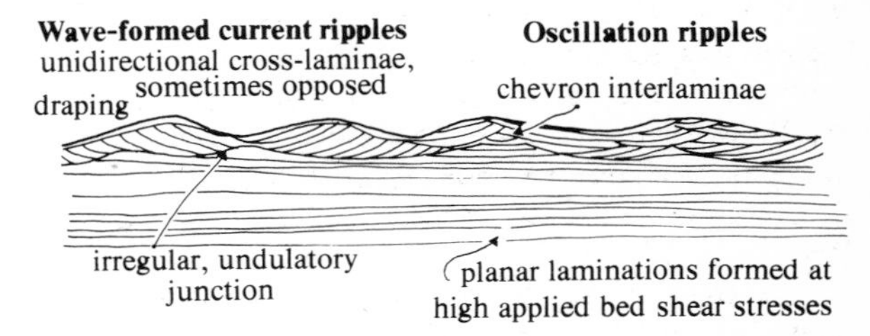

Figure 8.16 Some diagnostic internal features of wave-formed ripples (after de Raaf *et al.* 1977).

8d Coarse/fine laminations and graded bedding

Figure 8.18 shows that a complete transition exists betwen mud beds deposited from suspension fallout to sand beds deposited by ripple migration. Four types of laminations may usefully be defined: **streaky, lenticular, wavy** and **flaser**. Such laminations (Figs 8.18 & 19) are formed in environments of deposition where alternating periods of moving and slack water occur, as in tidally influenced regimes, or where the process of sediment supply is rhythmic or periodic, as in delta fronts or river floodplains. Once deposited, mud laminae have a high preservation potential because of their cohesiveness and

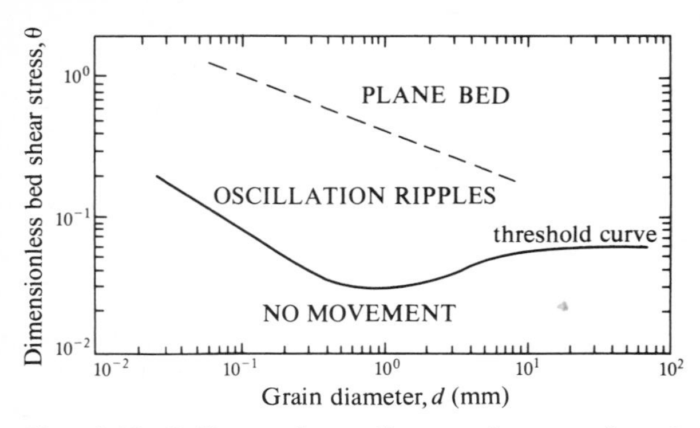

Figure 8.17 Bedform phase diagram for wave-formed oscillatory flows (after Komar 1976).

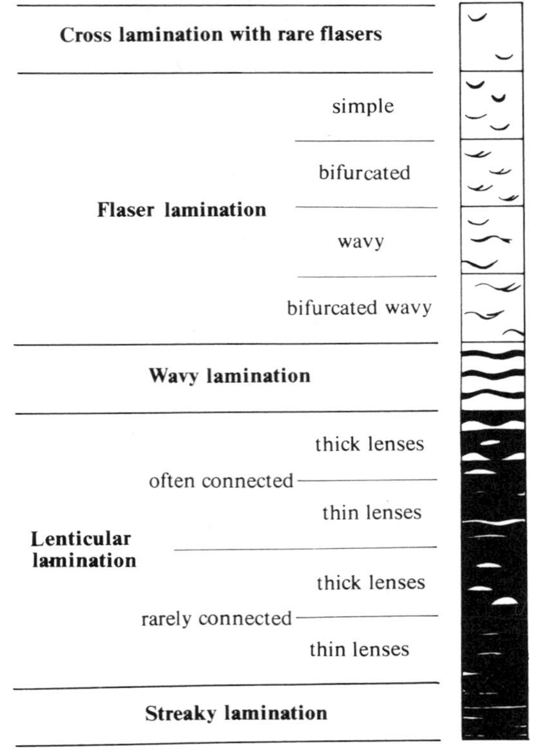

Figure 8.18 Classification of flaser and lenticular laminations. Mud is black, sand is white. (After Reineck & Singh 1973.)

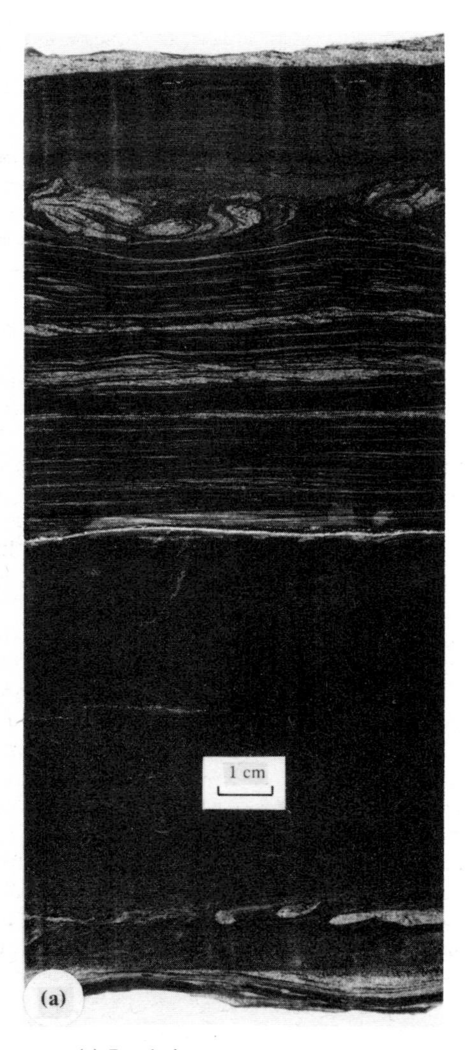

Figure 8.19 (a) Borehole core to show homogeneous muddy siltstones alternating with streaky, lenticular, wavy and contorted siltstone laminations. (b) Borehole core to show laminations and burrow mottling. Both Westphalian, Nottinghamshire coalfield, England.

because of small but significant amounts of early post-depositional compaction. Such laminae must greatly increase the chances of preservation of any bedform, such as a ripple, that they may cover over.

Normal grading from coarse upwards to fine within a deposit occurs in response to flow deceleration which causes particles with the largest mass to be deposited first. As already noted (Ch. 7b), the development of grading is rather sensitive to the concentration of particles within the decelerating flow, high concentration flows rarely showing good grading. We have also noted the occurrence of reverse grading in some grain flows (Ch. 7b).

8e Bedforms and structures formed by air flows

Desert bedforms are perhaps the most spectacular and awe-inspiring products of flow over loose sediment beds. The great thickness of the atmospheric boundary layer (up to 3 km) enables truly gigantic bedforms to exist, as in the high **draa** of the Saharan sand seas which may reach 100 m in height. As we saw previously for the water flow case, a definite hierarchy of aeolian bedforms seems to exist (Wilson 1972b). Wind tunnel experiments reveal that typically long, straight-crested ripples with only slight asymmetry and high wave-length:height ratios give way to steeper ripple forms which seem identical with some sinuous-crested sub-aqueous current ripples as the flow strength is increased (Bagnold 1954b). Ultimately the ripples are replaced by a plane bed at very high sand transport rates (transport stage $\gtrsim 3$) which gives rise to a parallel laminated sand deposit. However, wind tunnel experiments cannot hope to match the scale and diversity of the larger aeolian bedforms found in nature. Three distinct bedform groups may be defined from deserts by plotting the grain size of the coarsest 20% of the grain size distribution against bedform wavelength (Wilson 1972b). Distinct groupings exist which are termed **ripples, dunes** and **draas**. The lack of transitional forms (Fig. 8.20) shows that these forms do not represent a growth sequence analogous to the kind: child, adolescent, adult; a conclusion also reached previously (Ch. 8.1) in connection with sub-aqueous current ripples and dunes. Further, superimposition of ripples on dunes and draas, dunes on larger dunes and dunes on draas all occur.

Aeolian ripples (Figs 8.21a & 22) range in wavelength between 0.01 m & 20.0 m, and in height from a few millimetres to 1 m. Ripple indices fall between 12 and 50. In cross section parallel to wind flow, aeolian ripples show variable asymmetry, often with a short, well defined steep initial lee face at the angle of repose of sand (~30°) which passes into a gentler sloping face dipping into the ripple trough. There is often no clear internal structure of cross lamination in aeolian ripples (Fig. 8.22), unlike their sub-aqueous counterparts, since the ripples migrate less by repeated leeside avalanches than by saltation bombardment (see below). In plan view the ripple crests are frequently very persistent normal to wind flow direction, but sinuous crests are also common. Linguoid ripples, analogous to those previously described from water flows, occur in faster wind flows blowing very fine sands.

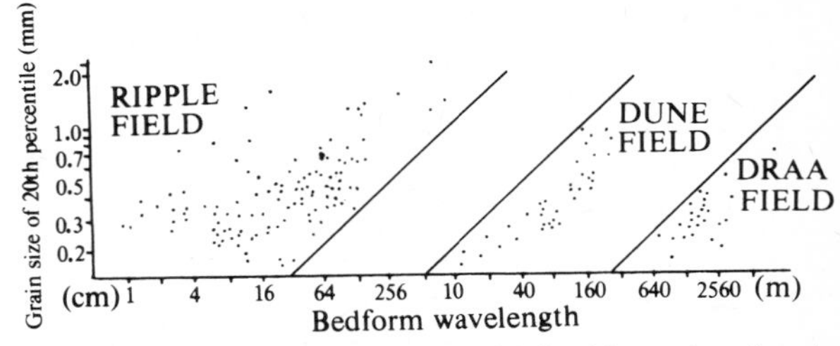

Figure 8.20 Aeolian bedform fields defined by a plot of grain size at the 20th percentile versus bedform wavelength. Field data from the Sahara (after Wilson 1972b).

Concerning the origin of aeolian ripples (Bagnold 1954b), it is important to realise that the superficial resemblance of form to sub-aqueous ripples hides a fundamental contrast in the mechanism of formation. As discussed previously, saltating grains in air are much heavier than saltating grains in water because of the greatly increased solid:fluid density ratio. Conditions on the air/bed interface are thus dominated by grain splashdown effects and not by the small-scale viscous sub-layer streaks which seem to control grain movement and ripple initiation on the water/bed interface.

Consider an air/bed interface with small hollows produced by grain splashdown effects or other mechanisms (Fig. 8.23). Since saltating grains descend at a low, roughly constant angle at the end of their saltation path, many of the upstream parts of hollows, A–B, will be sheltered from the impact effects of these grains. It should follow that many more grains will leave B than will arrive down the slope A–B. The hollow at B thus deepens. Assuming that there is a mean or characteristic saltation path length L for a reasonably well sorted sand at a particular wind strength, then at a distance downstream from A to B given by L, an area will tend to form a low negative slope which will then cause a further slope to form, and so on. Observations show that, to begin with, this process is enacted all over the initially flat bed due to the random distribution of initial bed defects. As time goes on, certain hollows will spread laterally to join up with others and the whole bed gradually evolves into a stable rippled configuration in a way similar to the coalescence and growth of the streak-produced defect mounds that give rise to sub-aqueous current ripples.

10 cm
(a)
(b)
(c)
(d)
(e)
(f)

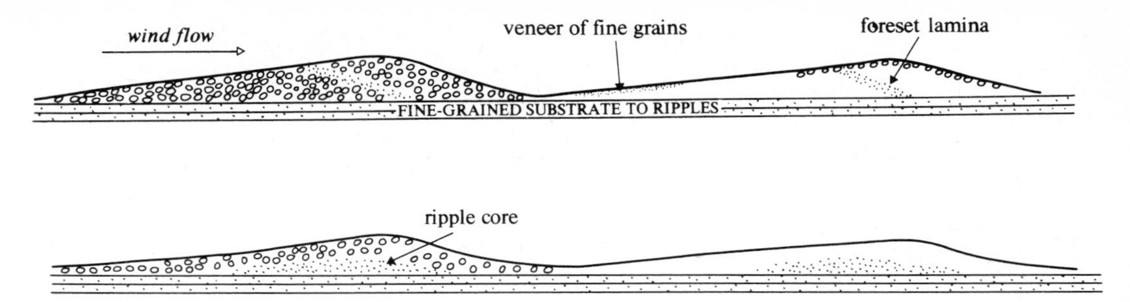

Figure 8.22 Sketches from impregnated specimens to show the internal structure of aeolian ballistic ripples. The poorly defined internal structure of the ripples may be defined by occasional fine-grained laminae deposited on the ripple lee during periods of reduced wind action (after Sharp 1963).

If the hypothesis above is correct then aeolian ripples should have a wavelength roughly equal to the characteristic saltation path. Since the length of this path increases with wind strength and grain diameter, the ripple wavelength should also follow this trend. Experimental results (Bagnold 1954b) bear out these points remarkably well (see Fig. 8.21a). Ripples produced in fine sands by the above mechanisms are known as **impact** or **ballistic** ripples. In areas of high deposition, ripples may 'climb', as in the sub-aqueous case (Ch. 8.1), producing inclined planes defining the ripple set boundaries (Hunter 1977).

There is every gradation between ballistic ripples and the large ripples called **ridges** or **granule ripples**. These large ripples are often composed of coarse sand or granules that are too large to be moved by saltation (Bagnold 1954b). They may show a crude, internal cross stratification close to the ridge crest. The clue to the origin of such large ripples comes from the fact that the sands are often bimodal, comprising distinct large and small grain sizes. The fine sands are those moved by saltation, and upon impact their kinetic energy is such that they are capable of nudging coarser grains (up to six times their diameter) along by intermittent rolling or sliding known as **creep**. Ripples produced by this process will be of greater wavelength since the saltation jump length of fine particles increases markedly with the size of the bed grains with which they collide (Ch. 6d). Disturbances on the bed will be amplified, as noted above for ballistic ripples, but the coarse-grained particles on the ripple crests can never be moved away, so that, once formed, large granule ripples tend to grow in height with time.

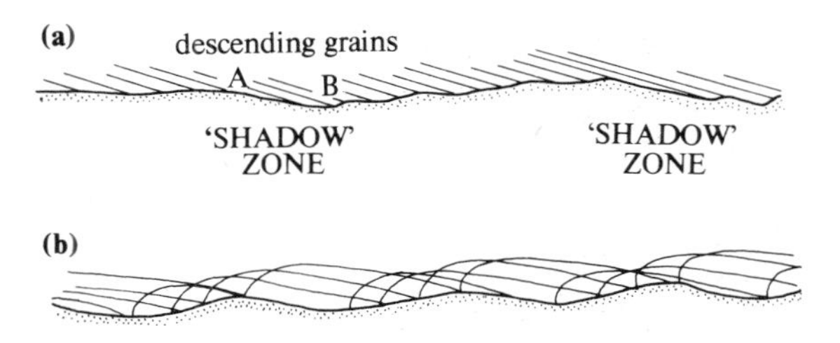

Figure 8.23 (a) Initial bed irregularities define 'shadow' zones sheltered from the impacting effects of saltating grain descending at a near constant flight angle. Deepening of the hollows formed at the bases of the shadow zone slopes initiates ballistic ripple formation. (b) The ballistic ripple wavelength coincides with the mean saltation jump lengths. (After Bagnold 1954b.)

Aeolian dunes show diverse morphologies. However, a simple classification into flow-transverse and flow-longitudinal types is possible (Wilson 1972a). Dune patterns commonly show combinations of both elements. Amongst the transverse dunes, sinuous-crested types, termed **aklé** dunes (Fig. 8.21b) are common in areas of plentiful sand supply. These show slip faces orientated normal to the local flow vectors on the dune leesides, giving rise to large internal sets of cross stratification. Frequent internal reactivation surfaces and downward-dipping intraset cross laminae result from periodic modifications of dune shape by aberrant winds. The **barchan** dune (Fig. 8.21c) is perhaps the most evocative

---

Figure 8.21 (a) Ballistic ripples illustrating grain-size control of wavelength; the smaller ripples on the left are in finer-grained sand and have been affected by a later gentle wind, at 90° to the first, which did not disturb the coarser sand on the right (from I. G. Wilson 1973). (b) Aerial view of aklé dunes (scale unknown) from Utah (from Cooke & Warren 1973). (c) Barchan dune advancing across a lag gravel pavement; La Joya, South Peru (from Cooke & Warren 1973). (d) Draa comprising superimposed aklé dunes; draa height $\approx$ 30 m; Erg Occidental, Algeria (photo I. Davidson). (e) Aerial view of barchanoid draa (*c.* 50 m high) with superimposed dunes; Sahara, Algeria (after I. G. Wilson 1972a). (f) Aerial view of 'meandering' seif dunes migrating across a lag of coarse sands; Edeyenubari, Libya (after I. G. Wilson 1972a).

and most widely known dune type. Barchans occur only in areas of reduced sand supply, with individual dunes separated from their neighbours by either solid rock floor or by immobile coarse pebbles (a **lag** deposit). Sediment transport rates are higher on the solid margins of an initially formed barchanoid sand mound where intragranular frictional effects upon saltation impact are much reduced. Barchanoid 'wings' thus result which cause the slip face in the lee of the mound to become concave downwind. The internal structure of a barchan is shown in Figure 8.24. **Domal** dunes without a prominent slip face seem to form from the degradation of barchanoid dunes during long periods of gentle wind flows. As expected, such dunes show complex internal cross-stratification patterns.

Longitudinal dunes are often referred to as **seif** dunes, individual examples of which may sometimes be traced for upwards of 200 km, as in the Simpson Desert of Australia. Dune heights range up to 50 m with typical lateral spacing between dunes of about 500 m. Dune coalescence produces Y-junctions which always fork upwind (Folk 1971). Some seifs show a sinuous plan form (Fig. 8.21f) or periodic humps. The latter forms occur when barchan dunes are subjected to winds from two directions at acute angles to each other. One barchan wing becomes elongated, later to become the nucleus of a new barchan as the wind re-establishes itself in its former mode. The resultant **beaded seif dune** (Bagnold 1954b) has its long axis orientated parallel to the resultant of the two wind azimuths. Internally, seif dunes show a bimodal pattern of large-scale cross stratification produced by avalanche accretion on either side of the crest (Fig. 8.25).

Concerning the origins of aeolian dunes there is little doubt that the presence of secondary flows is of major importance. Aklé dunes clearly indicate minimal effects of longitudinal secondary flow vortices whereas in seif dunes the dune may develop along the axis of the meeting point of pairs of oppositely rotating vortices. Finer saltating sands are thus always swept inwards in broad lanes where deposition occurs and, given sufficient sand supply, the dune form grows into equilibrium with the flow. A most important point is that, once formed, the dunes will then reinforce the secondary flow cells. The origin of the simple aklé dune is still not adequately known, although instability theories involving initial large-scale mounds of sand due to flow irregularities and stationary objects are often quoted. It is possible that the wavelength of aeolian dunes is related to the flow 'depth' (i.e. boundary layer thickness), as in sub-aqueous dunes. Although there is no supporting field evidence it is quite possible that dune wavelengths are controlled by the repeat distance of large-scale 'burst' events occupying the whole wind boundary layer, as seems to be the case for sub-aqueous dunes.

Finally we come to the giant draa bedforms (see Wilson 1972a, McKee 1978) with wavelengths of 650–4000 m, reaching up to 400 m high. Draas are composite bedform elements made up of superimposed dunes (Fig. 8.21d, e). They may be aklé or barchanoid in plan, with frequent development of the spectacular stellate (star-shaped) forms known as **rhourds** which arise from complex winds and convected air masses. Draas take long time periods to form and they will require appreciable original sand cover to provide an adequate original nucleus. Some draas show giant slip faces up to 50 m high, but many do not and show instead dune migration down the lee side at a fairly shallow angle. Further comments on the dynamics and internal structures of draas are given in Chapter 19.

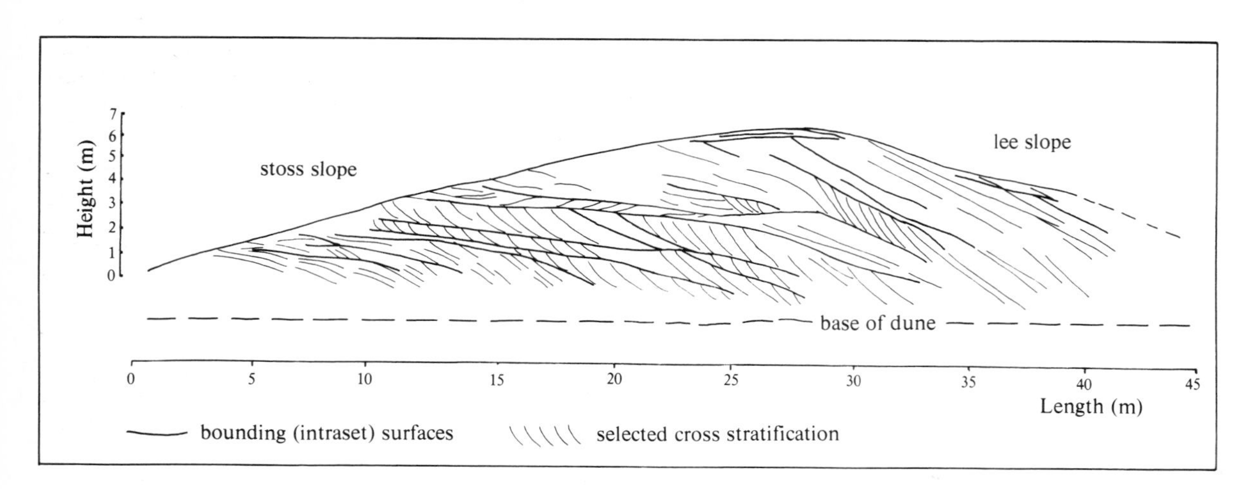

Figure 8.24 Internal structure of a barchanoid dune from White Sands, New Mexico (after McKee 1966). Note that the internal downdipping sets imply growth from a nucleus of low dunes migrating over a gently sloping sand mound.

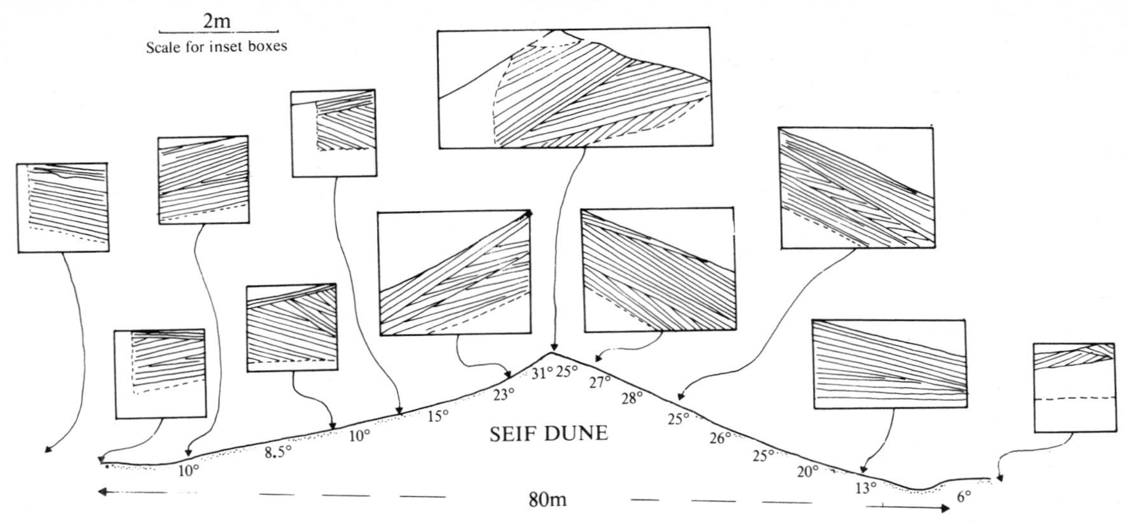

Figure 8.25 Internal structure of a seif dune from the Libyan Sahara as determined from test pits (after McKee & Tibbits 1964).

8f Bedform 'lag' effects

Many of the remarks made so far in this chapter assume the presence of an 'equilibrium' bed state adjusted to a steady flow. However, natural flows of water and wind are unsteady on a variety of scales, e.g. tidal flows for hours, and river and air flows for weeks or months (Allen 1973). Sub-aqueous dunes that form in response to a steady flow may persist for a considerable time as the flow decays into the ripple stability field. Within the dune stability field, changes in water depth during rising or falling river stage may cause changes in the dune wavelength or height to lag behind the flow. Data for the change in dune height and wavelength with respect to changing river discharge are shown in Figure 8.26. Note the clearly defined *lag* or **hysteresis** effect, the maximum dune wavelength being reached long after the peak discharge.

We may idealise our discussion with reference to Figure 8.27, in which graphs of y and z are plotted so that both variables are also functions of time. y could be dune wavelength whilst x could be discharge. If no lag exists, then a straight line plot results. Lag in Figure 8.27a reaches extreme values of 90° (cf. Fig. 8.26) whilst lag in Figure 8.27b shows intermediate values.

The wide occurrence of bedform lag in natural environments means that we should be very careful in interpreting field measurements of bedform sizes in relation to existing flow conditions. Criteria for bedform equilibrium should be investigated. One relevant example might be the equations relating dune wavelength and height to water depth (shown previously, Fig. 8.10) in which very considerable scatter was evident. Much of this scatter could have resulted from lag effects since the data was collected and assembled before the significance of lag was generally realised.

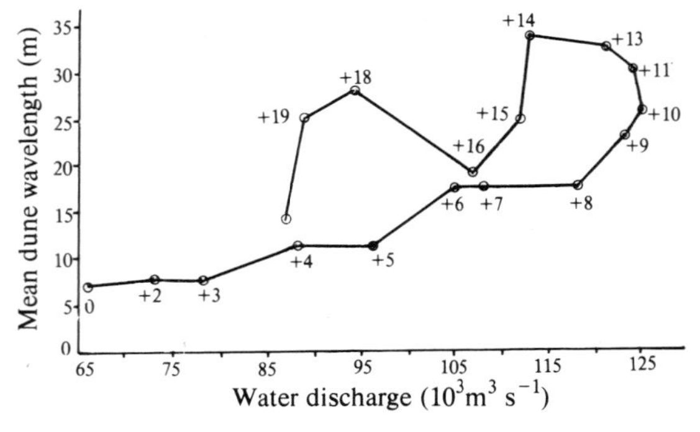

Figure 8.26 Variation of mean dune wavelength with water discharge over a period of 19 days in the Fraser River, British Columbia, Canada. Note the well developed lag effect. (After Allen 1973.)

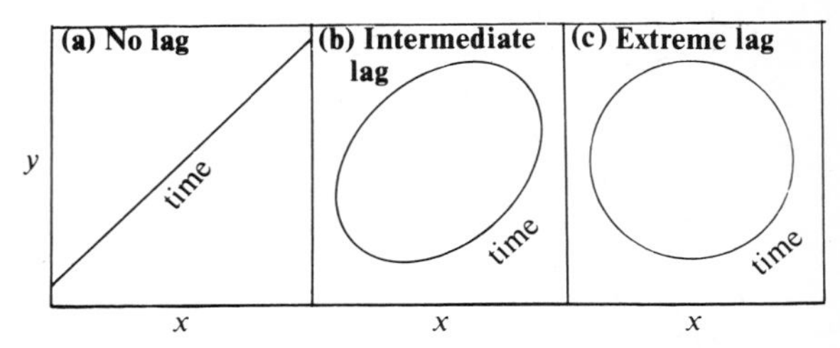

Figure 8.27 General form of the variation of a quantity y as a function of quantity x when both x and y are also functions of time. (After Allen 1973.)

8g Summary

Bedforms caused by water flows over granular beds form a definite sequence as flow strength increases. Current ripples, dunes and upper-phase plane beds form in fine-grained sediment and give rise to small-scale cross stratification, large-scale cross stratification and planar laminations respectively. Current ripples are absent in sands coarser than about 0.7 mm where the sequence of bedforms is lower-phase plane bed, bars, dunes and upper-phase plane beds. Antidunes form in all grades of sediment once a critical Froude number (≈ 0.8) is reached. Much confusion exists in the literature at the present time regarding the nomenclature of dune and bar bedforms. This confusion is compounded by the existence of bedform hierarchies and lag effects. Bedforms under water waves include a variety of oscillation ripples and plane beds. The combination of wave and current gives rise to hybrid forms. Unsteady flows and variations of wave height with time cause the formation of coarse/fine interlaminations. Bedforms caused by air flows include ballistic ripples, plane beds, dunes and draas. The latter two types may exist on a very large scale related to regional air currents and give rise to sets and complex cosets of large-scale cross stratification (see further discussion in Ch. 13). The theory of bedforms is still in its infancy and it suffers from an overabundance of data from small-scale water flume and air tunnel experiments, as distinct from data gathered in natural flows. Severe problems arise in natural environments from unsteady flow and lag effects.

Further reading

Bedforms and sedimentary structures in water flows are discussed by J. R. L. Allen (1968, 1970b). The linkage between bedforms and turbulent flow processes is discussed in stimulating papers by Williams & Kemp (1971) and Jackson (1976b) and also in the relevant chapters in Raudkivi (1976) and Yalin (1977). Bedform phase diagrams are discussed by Southard (1971). An introduction to wave theory is given in Chapter 18 of this text. Bedforms and structures in wind-blown sands are discussed by Bagnold (1954b), McKee *et al.* (1978), with superb satellite photographs and in a series of outstanding syntheses, by I. G. Wilson (1972a & b, 1973). The best introduction to the concept of bedform lag is given by Allen (1973). Many types of sedimentary structures are introduced and beautifully illustrated in Pettijohn & Potter (1964) and, more recently, by Collinson & Thompson (1982). A variety of methods for the study of sedimentary structures in both consolidated and unconsolidated sediments is given by Bouma (1969).

Appendix 8.1 Notes on bedform theory for water flows

As an introduction to this field, four simple questions are discussed:

1. Why do ripples form at close to threshold flow conditions?

The spanwise sequence of fluid bursts and sweeps in the viscous sub-layer plays an important role here (Williams & Kemp 1970). Entrainment of grains during a sweep is followed by deposition as the sweep fluid decelerates. All over a sand bed, made artificially plane prior to experiment, tiny discontinuities are formed from heaps of deposited grains 2–3 diameters high. Certain heaps now begin to influence flow structure as flow separation and reattachment occur on their lee sides. These heaps become magnified and propagate downstream since the turbulent stresses at attachment (Ch. 5) erode too much material to be held in transport. More heaps are thus formed downstream which in turn give rise to flow separation, and so on. Tiny current ripples now form, coalesce and interact all over the bed. An equilibrium ripple assemblage evolves after an hour or so. It is not known what controls the magnitude of the ripples produced at equilibrium.

2. Why don't ripples form in coarse sands?

The limit of occurrence of current ripples, at about 0.7 mm, coincides with the disappearance of smooth boundary conditions at the threshold of grain movement (Ch. 6). Ripples cannot develop on transitional to rough boundaries because of the poor development of initial streak-produced defects and to the inhibition of flow separation by the rough boundary (Leeder 1980). The initially planed bed thus remains stable.

3. Why do ripples and dunes differ in physical scale?

A popular hypothesis is that current ripples are controlled directly by flow conditions in the viscous sub-layer (see 1 and 2 above) whilst dunes are adjusted to processes acting in the whole turbulent boundary layer (Yalin 1977, Jackson 1976a). The main evidence for this hypothesis is the correlation of dune wavelength with mean flow depth (Fig. 8.10), indicating that the repeat distance of dunes is somehow adjusted to a decay length of large-scale turbulent burst events as they move from the boundary downstream throughout the whole flow depth.

4. What determines the stability of an upper-stage plane bed?

It has been argued that there is a critical concentration of grains in the bedload layer that will cause the fluid close to the bed to become markedly less turbulent, since it is known that the presence of appreciable concentrations of grains reduces turbulent intensity. Flow separation over small bed defects to cause ripple or dune formation therefore becomes impossible until the fractional grain concentration in the grain–fluid mixture falls below a critical value (about 0.1), when turbulent stresses at reattachment can begin actively to erode grains and cause the small defects to be amplified into ripples and dunes (Allen & Leeder 1980).

9 Bedforms caused by erosion of cohesive sediment

9a Water erosion of cohesive beds

Sediment beds of clay-grade material may show a cohesive strength in which particle–particle bonds between clay minerals form as a result of adsorbed water films. In addition, many freshly deposited muds are cohesive because of electrostatic attractions. Clay particles carry an electrical charge caused by isomorphous substitution of one kind of ion for another (Ch. 1). In montmorillonite Mg^{2+} may substitute for Al^{3+}, giving a net unit charge deficiency per substitution. Similarly, Al^{3+} may substitute for Si^{4+} in kaolinite. Some of the attractive forces between clay particles depend upon the existence of a small net positive charge at the edge of a clay platelet. When particles are very close together this edge charge participates in an edge-to-face linkage of electrostatic type (see Fig. 11.1). The balance between repelling effects caused by the net negative charge on particle faces and the attractive effects caused by van der Waals forces determines whether flocculation or dispersion will occur (see Ch. 11).

The very large surface area of a given volume of clay minerals, compared with the same volume of granular quartz sand, gives rise to the important effects above. Clay minerals often behave as colloids where body forces are no longer of importance. A tendency towards flocculation in colloids is caused by increasing the electrolyte concentration, ionic valency or temperature and by decreasing the dielectric constant, size of hydrated ions, or pH. Large normal stresses can be transmitted through a highly dispersed clay by long-range electrical forces so that there is no direct mineral-to-mineral contact between particles. Alternatively, in a flocculated clay the particles are effectively in contact and normal stresses are transmitted as in granular aggregates. The behaviour of natural clay aggregates usually falls somewhere in between these two states (see also Ch. 11).

As we can gather from the above discussion, the erosion of mud-grade substrates is a complex process. For example, following our discussion of flocculation, we would expect the critical erosion rate for muds to be a very sensitive function of electrolyte concentration. This is confirmed by studies in which various concentrations of $NaNO_3$ were added to the de-ionised pore waters of pure kaolinite muds. The added salt greatly increased the critical erosive stress (Raudkivi & Hutchinson 1974). Consolidation is also of great importance, leading to an increase in cohesiveness with depth so that surface erosion by flows may be followed by bed stability at depth. It is therefore impossible to generalise on values of critical erosive stress for muds without specifying mud and fluid composition and previous depositional history. In particular, experimental data pertaining to freshwater muds are hardly likely to apply to marine situations.

Once the critical erosive stress is exceeded the erosion of a perfectly flat kaolinite mud bed by water proceeds with the development of three sorts of bedform as flow strength is gradually increased (Allen 1969b). Small-scale **longitudinal grooves** and **ridges** have a typical spacing normal to flow of around 0.5–1.0 cm. Well established examples show that ridges are sharp with broad, rounded intervening furrows. Tiny striae produced by the flow enlargement of air bubble casts in the mud indicate a bottom flow structure consistent with the existence of pairs of counter-rotating flow vortices typical of the viscous sub-layer (Ch. 5). The spacing is identical to that required by Equation 5.26. At slightly higher flow velocities the longitudinal grooves change into **'meandering' grooves**, indicating a transverse instability affecting the sub-layer streaks. These grooves may cause deep corkscrew-like erosional marks to form, which may gradually develop into the characteristic spoon-shaped depressions called **flute marks** (Fig. 9.1). Flutes may also

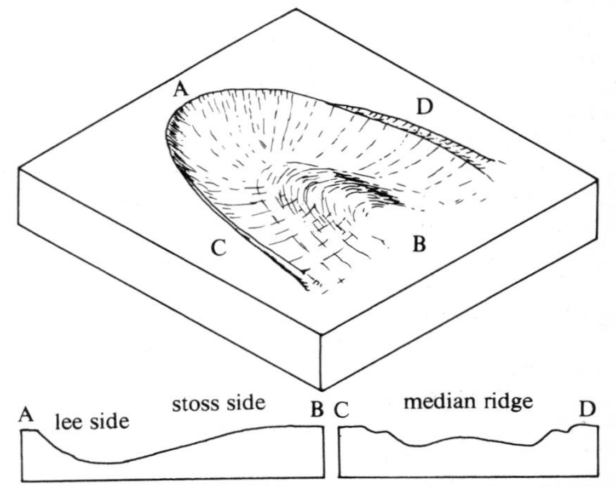

Figure 9.1 Diagram to show the morphology of an idealised flute mark cut into a cohesive substrate (after J. R. L. Allen 1971a).

commonly form from bed defects such as hollows or impact marks (Fig. 9.2). They form as a result of flow separation from the lip of the initial hollow (Fig. 9.3). Flow reattachment gives rise to high turbulent stresses (Ch. 5j) causing deepening and lengthening of the flute. In a mature flute the deepest part usually lies some distance up stream from the reattachment point in the area occupied by a captive recirculating fluid bubble. A large variety of flute shapes occur (Fig. 9.4; J. R. L. Allen 1971a).

It should be noted that each of the above bedforms may be produced by pure fluid stress alone, although most natural examples were probably aided in their formation by the pitting effects of suspended and bedload grains and by initial bed irregularities.

Laboratory tests cannot reproduce the spectacular large-scale longitudinal grooves called **gutter marks** (Whittaker 1973). These may have a spacing of a metre or more and be up to 20 cm deep. Flutes or tool marks may cover the margins of these features. Intergroove areas may be perfectly flat and show few effects of erosion. Gutter marks are probably caused by the action of large-scale secondary flow (see Ch. 5).

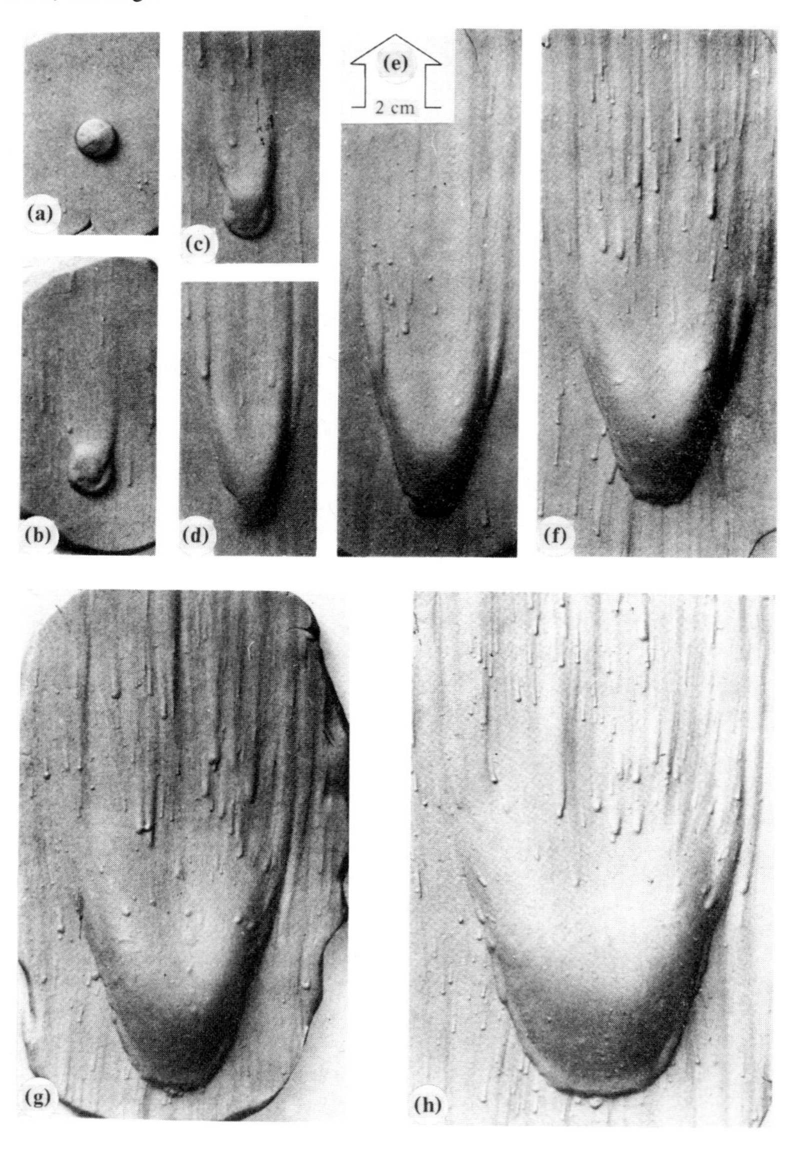

Figure 9.2 Casts to show the development (b–h) of a flute from an initial small bed defect (a). Sequence produced by water flow (mean velocity of 0.45 m s^{-1}) over a mud bed. (After J. R. L. Allen 1971a.)

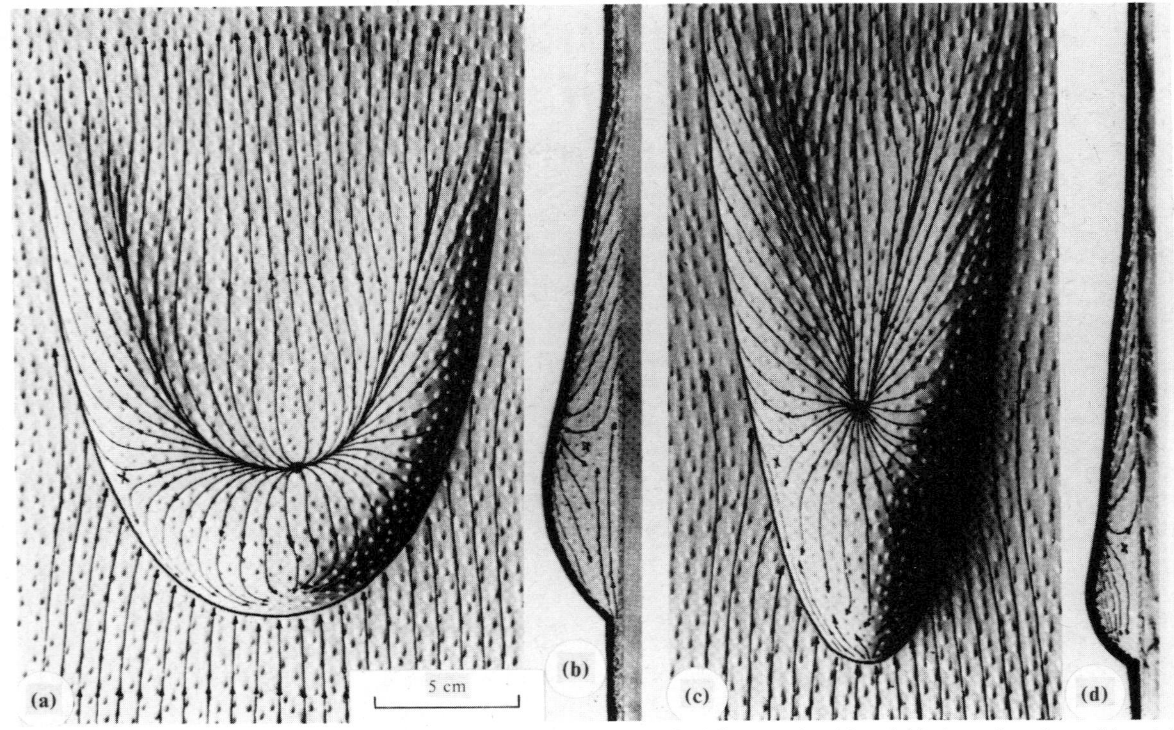

Figure 9.3 Patterns of skin-friction lines on casts of large models of idealised flute marks. (a) and (c) show plan views, (b) and (d) show side views. Note the evidence for flow separation and the line of flow reattachment. (After J. R. L. Allen 1971a.)

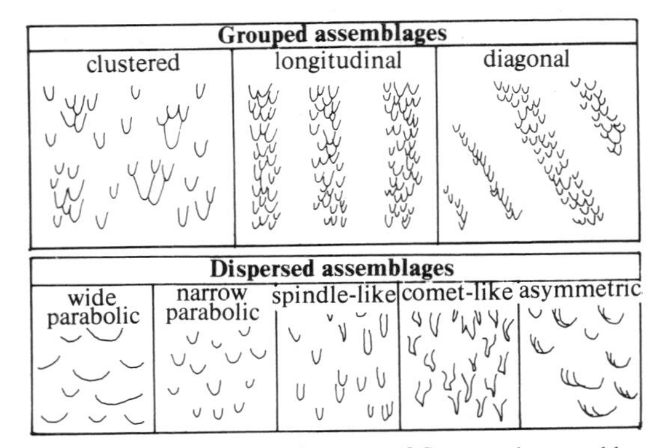

Figure 9.4 Some of the main types of flute mark assemblages and morphologies found in nature (after J. R. L. Allen 1971a).

9b Erosion by 'tools'

Detritus carried by a flow may form a great variety of impact marks on cohesive mud beds as the 'tools' bounce or are dragged over the substrate. Sometimes these marks may be recognised as caused by a particular tool, as in the case of a saltating orthocone, rolling ammonoid or dragging plant stem. Preserved tool casts may be useful palaeocurrent indicators if the original mark was slightly eroded by flow subsequent to impact, or gave rise to shear drag features such as chevron marks.

9c Summary

Bedforms in cohesive sediments are cut by the erosive effects of turbulent fluid with or without entrained sediment grains. Flutes, tool marks and longitudinal grooves of various sizes are preserved as casts after infill by later coarse-grained deposits.

Further reading

J. R. L. Allen (1971a) gives an exhaustive theoretical and experimental study of the origins and characteristics of erosive bedforms in both muds and rock (this paper includes a section on cave scallop solution). Many fine illustrations of erosive bedforms in muds are to be found in Dzulynski and Walton (1965) and in Pettijohn and Potter (1964).

10 A brief introduction to biogenic and organo-sedimentary structures

10a Stromatolites

Algal-laminated structures, termed **stromatolites**, are most prolific at the present day in very shallow waters of both marine and non-marine environments. In the marine environment they are characteristic forms in the shallow subtidal to supratidal zones of tropical to subtropical carbonate environments, although colonisation of siliciclastic shorelines has been increasingly recognised in recent years (Schwarz *et al.* 1975, Gunatilaka 1975). Stromatolite surface micromorphology varies according to the type of algae present and to ecological position in a given area (e.g. Logan *et al.* 1974, Kinsman & Park 1976), but there is no general rule when different areas are compared. Gradations of large-scale growth forms are related to degrees of wetting, frequency of exposure, and intensity of current action. Algal structures include flat, crinkle, cinder, polygonal, dome, blister, pillar and club forms (Figs. 10.1 & 10.2).

Internally, stromatolites reveal a diagnostic pattern of laminations which may closely follow the external growth surface (Fig. 10.1a). The laminations are composed of trapped or bound particulate matter, precipitated carbonate and calcified or organic-walled blue–green algal filaments (Monty 1967, Park 1977). Most modern marine stromatolites are soft and unlithified whilst many lacustrine freshwater forms are wholly or partially calcified (Monty 1967, Hardie 1977). The dominant process of sediment binding is achieved mainly by the filamentous blue–green alga *Schizothrix*, although many other genera contribute. Diurnal growth rhythms, periodic drought and storm or tidal deposition phases lead to the superposition of contrasting laminae, often of the sediment-rich/algal-rich variety.

The incorporation of sediment particles in the stromatolite may be a highly selective process. Together with the precipitation of penecontemporaneous or very early diagenetic carbonate (see notes on Bahamian algae in Ch. 23), this leads to strong contrasts between stromatolite fabrics and those of underlying or laterally equivalent detrital carbonate sediments. Early diagenetic processes, often aided by bacteria acting beneath the stromatolite surface, may lead to breakdown and compaction of algal mucilage (Park 1977) and precipitation of microcrystalline aragonite and dolomite (Dalrymple 1966, Gebelein & Hoffman 1973).

A suitably broad definition of stromatolites might be: 'Laminated structures composed of particulate sediment and/or precipitated carbonate produced by the growth, metabolism and breakdown of successive algal films or crusts' (Logan *et al.* 1964, Leeder 1975b).

Although modern stromatolites are all very shallow-water forms, it is likely that they occupied a far wider spectrum of ecological niches in the geological past, particularly in Precambrian times. This is evident from the broad conclusions of facies analysis and also from considerations of competition. Today, it may be argued, algal stromatolites are restricted to the most hostile of marine environments, the intertidal zone, because of such metazoan predators as gastropods that graze on the algal slime. No such predation would have occurred in Precambrian times and hence the algae would have been free to colonise the full range of marine environments *within the photic zone* (Garrett 1970). Another hypothesis (Clemmey, personal communication) implies that a major changeover in stromatolite types has occurred between marine and lacustrine environments. Many Precambrian marine stromatolites were calcified crustose forms with no evidence of sediment trapping or binding by gelatinous filaments. Today, however, many lacustrine and riverine stromatolites are calcified whilst their marine relatives are not. Preservation potential due to calcification has thus changed with time.

Figure 10.1 (a) Smooth, regularly laminated algal stromatolite from the lower intertidal zone of the Trucial coast. Note dark (algal-rich) and light (sediment-rich) interlaminations; white scale bar = 10 cm. (Photo R. Till.) (b) View of blister algal stromatolites from the mid–upper intertidal zone of the Trucial coast (photo by R. Till). (c) View of large-scale polygonal algal mat with raised rims and areas of blister mat growth; Trucial Coast (photo R. Till). (d) Section through smooth algal mat disrupted by polygonal shrinkage cracks. Note raised rims to polygons and periodic 'healing' (photo R. Till). (e) Lithified stromatolite columns showing seaward asymmetry, Shark Bay, W. Australia (photo P. G. Harris). (f) Lithified stromatolitic ridges in exposed, high-energy intertidal zone of Shark Bay, W. Australia. Ridges are separated by skeletal carbonate sands and show elongation parallel with the direction of wave propagation. (Photo P. G. Harris.)

(a)
(b)
(c)
(d)
(e)
(f)

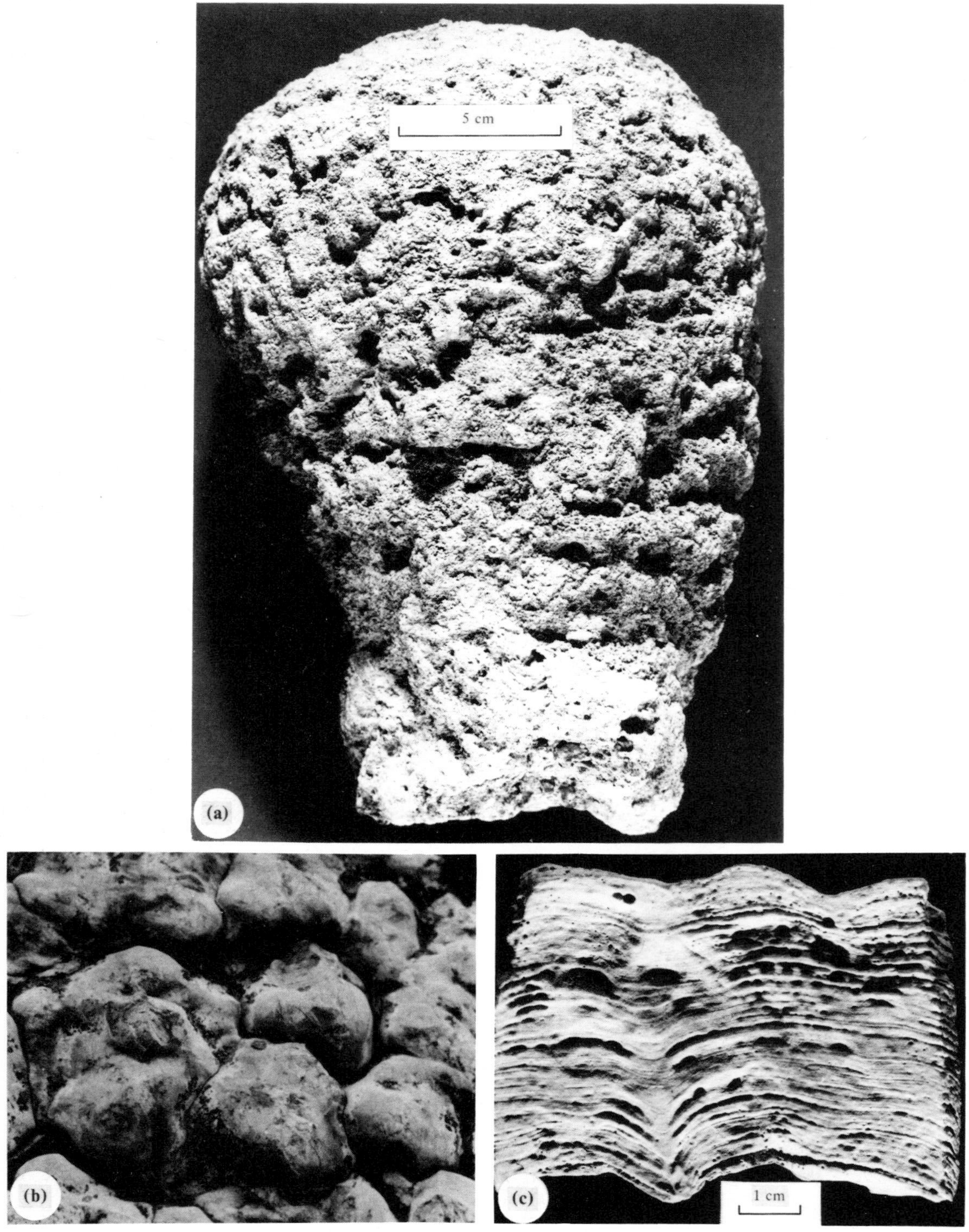

Figure 10.2 (a) Single lithified stromatolitic head with pustulose mat growth forms. Head comprises algal-bound and aragonite-cemented skeletal debris. (Recent of Shark Bay, collector E. J. Van der Graaf.) (b) Bedding plane covered with laterally linked stromatolite domes. Dinantian, Chipping Sodbury, Somerset, England. (c) Section through laterally linked domal stromatolite, Saharonim Formation (Jurassic), Ramon anticline, Israel.

10b Trace fossils and deposition rates

Trace fossils have the great advantage over normal body fossils that they are usually *in situ*. All other factors being equal, the amount of disturbance of primary sedimentary structures by burrowing activities increases as the rate of sediment deposition decreases. This is perhaps best illustrated by comparing the sparse occurrence of burrows in rapidly deposited intertidal point bar sediments (Ch. 21) compared with the abundant bioturbation seen in adjacent, more slowly deposited, tidal flat sediments. Bioturbation is easily picked out in well laminated sediments but may be underestimated in homogeneous sediments or when very substantial burrowing activity has completely homogenised a formerly laminated deposit. X-ray analysis is helpful in these latter cases.

Rapid sediment deposition may be distinguished in a deposited sequence with the aid of **escape burrows**, which are the vertical traces left as entombed animals rise back to the sediment/water interface. Poignant examples occur in offshore storm layers in the Heligoland Bight of the North Sea. Here, tiny intertidal-dwelling gastropods of the genus *Hydrobia* were swept out to sea in a storm surge and buried beneath several centimetres of sand. Box cores now reveal their corkscrew-like escape traces reaching back to the sediment/water interface (Fig. 10.3). Tragically, death soon followed since the gastropods found themselves removed from their intertidal ecological niche (Reineck *et al.* 1968).

It is a fact that sedimentologists often pay little attention to erosive episodes in the sedimentary record. The subtle interplay between successive erosive and depositional events is well shown by the up-and-down movements of trace fossils such as the U-shaped dwelling burrow *Diplocraterion yoyo* (Fig. 10.4). The concave-upwards laminae (**spreite**) formed at the base of the tube record successive positions of the burrow base and together with erosive surfaces give much information on erosive and depositional episodes (Goldring 1964).

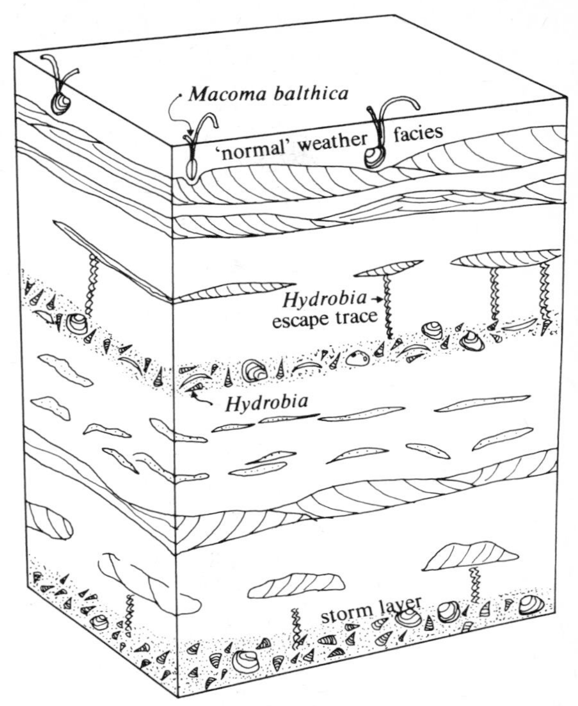

Figure 10.3 Diagram to show the mode of occurrence of the intertidal gastropod *Hydrobia* as a major constituent of storm-deposited layers located in offshore areas (after Reineck *et al.* 1968).

Identification of a trace fossil as a *boring* excavated into lithified sediment gives valuable evidence for the relative timing of lithification. Such structures are common in zones of early diagenetic cementation in limestones, termed **hardgrounds**. Truncation of borings may indicate a period of erosion of the hardground subsequent to lithification (Ch. 29).

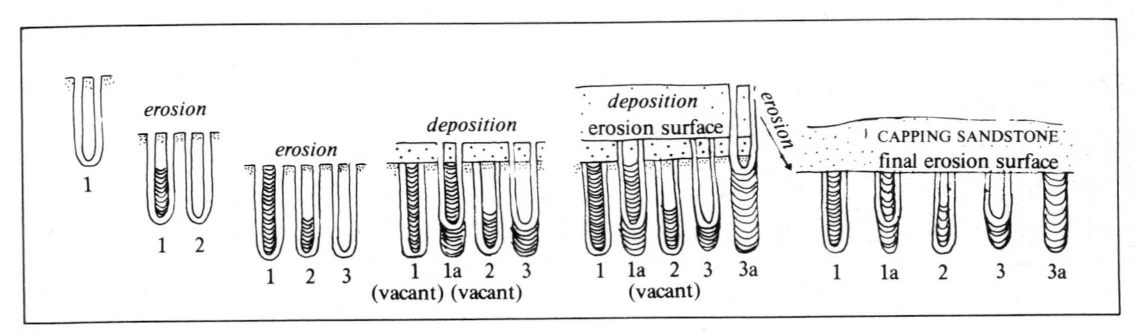

Figure 10.4 Diagram to illustrate how a colony of *Diplocraterion yoyo* might respond to periods of erosion and deposition. The final burrows and their tell-tale *spreite* give just enough information to reconstruct the previous sedimentary history. (After Goldring 1964.)

10c Summary

The growth of algal films on sediment surfaces causes characteristic internal laminations and external growth forms to develop according to ecological position, physicochemical and biochemical factors. Calcified algal-laminated structures have a particularly high preservation potential. When used in conjunction with other lines of evidence (see Ch. 23), stromatolites may yield valuable evidence for past environments of deposition.

Deposited sediment, particularly Phanerozoic marine sediment, is prone to modification by the burrowing and surface grazing activities of invertebrates. Trace fossils give valuable evidence of relative deposition rates as well as providing primary evidence for environments of deposition.

Further reading

A fundamental source for any study of stromatolites is Walter (1976). Lucid accounts of Bahamian stromatolites are in Monty (1967) and Hardie (1977). Trucial Coast stromatolites are documented by Park (1976, 1977) and Kinsman and Park (1976). The spectacular Shark Bay assemblages are discussed by Logan *et al.* (1974).

No attempt has been made in this chapter to give a full discussion of trace fossils; the reader is referred to the collections of papers in Crimes (1970), Crimes and Harper (1977) and Frey (1975). The usefulness of trace fossil assemblages in facies analysis is discussed by Seilacher (1967) and Schäfer (1972).

11 Soft sediment deformation structures

11a Reduction of sediment strength

Normally, a deposited granular sediment behaves as a 'solid' aggregate and is stable so long as the slope of the depositional surface remains below the angle of initial yield (Ch. 7). Such an aggregate is capable of withstanding an applied shear stress. By way of contrast, a *quick* condition is one where the shear strength of a sediment is zero, with the weight of the grains being borne by the interstitial pore fluid. Thus, if a water-saturated sand is subjected to ground vibrations such as cyclic earthquake shocks or to pressure fluctuations caused by water waves passing overhead (Ch. 24), it will tend to compact and decrease in volume as the grains attempt to find the closest possible packing (Seed & Lee 1966). Provided that the pore fluid cannot escape, or at least escapes slowly, an increase in the pore-water pressure results (Fig. 11.1). If the pore-water pressure increases to a level where it approaches the overburden pressure, then the sand *liquefies* (see also Ch. 7). Liquefaction in sands is possible only if the deposit is initially loosely packed. Tightly packed sands tend to expand under shear, a process known as **dilatancy** (Ch. 7). Liquefaction in sands is caused by earthquake shocks of magnitude 5.5–5.8 on the Richter scale and may occur at distances of hundreds of kilometres from the epicentre. After liquefaction, water expulsion inevitably results if the sand has a connection with the surface. As noted previously (Ch. 7) escaping water usually concentrates in vertical pipes and it may flow with sufficient velocity to squirt high in the air as the liquefied bed consolidates.

Liquefaction in clay sediment also results from cyclic stresses causing closer particle packing, but the process is more complex than in sands. Freshly deposited marine clays may often show a flocculated structure (Fig. 11.1) with many edge-to-face contacts between clay particles. This open structure is also enhanced by quartz silt particles in the clay framework (Smalley 1971). A sudden attempt to change from an open flocculated clay structure with a high porosity and water content to a denser dispersed clay structure (Fig. 11.1) is accompanied by an increase in pore pressure and the potential of liquefaction. Such fabric changes in response to shock or loading may be possible only if the flocculated structure is first acted upon by freshwater solutions which 'flush out' from the

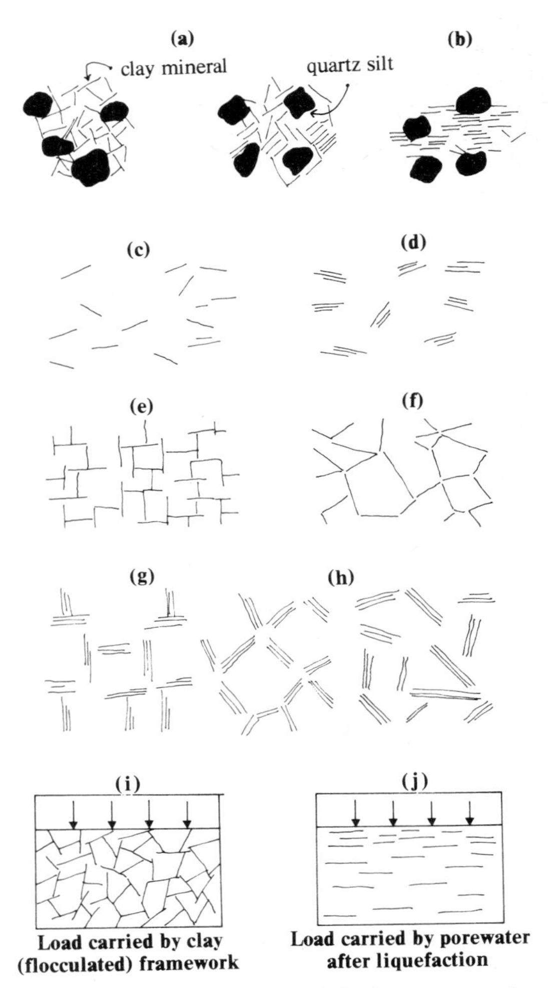

Figure 11.1 Idealised clay microfabrics in clay suspensions to show generalised (a) flocculated and (b) dispersed fabrics and detailed schematic sketches of the following structures: (c) dispersed and deflocculated, (d) aggregated but deflocculated, (e) edge-to-face flocculated but dispersed, (f) edge-to-edge flocculated, (g) edge-to-face flocculated and aggregated, (h) edge-to-edge and edge-to-face flocculated and aggregated. (Partly after van Olphen 1963.) (i) and (j) show how a load may be transferred from the clay framework to the pore waters during liquefaction.

clay the ionic species responsible for the strength of the flocculated bonds. Many notorious **quick clays** are of this type and give rise to serious civil engineering problems.

The processes of liquefaction discussed above are irreversible. Many natural, freshly deposited muds and man-made substances show a pattern of reversible liquefaction known as **thixotropy**. Thixotropic substances will not flow unless shaken or stirred. They fully regain their 'strength' after standing for a short while. Mayonnaise and emulsion paint are two everyday examples of thixotropic substances. Many **quicksands** in muddy sands show thixotropic behaviour. Some thixotropic muds lose their water on standing after liquefaction. This process is known as **syneresis** and it is accompanied by sediment contraction under the skin of escaped water (see p. 115).

11b Liquefaction and water escape structures

The upward escape of water from a consolidating fluidised bed along conduits causes structureless **de-watering pipes** (Fig. 11.2) to form. These range from a few millimetres up to a metre in height. Any pre-existing laminations in the liquefied bed serve to outline the upward drag exerted on the liquefied sand by the escaping water. Grains suspended by the violently escaping water may be transported up to the bed surface to form **sand volcanoes** with diameters of up to 1 m and cone angles of up to 16° (Gill & Keunen 1958). Sand volcanoes are preserved only in quiet sub-aqueous environments. They are best known in the rock record on the top surfaces of certain turbidite beds and sub-aqueous slumps.

Dish and pillar structures (Fig. 11.2c) are also now attributed to water escape mechanisms (Lowe & Lopicollo 1974, Lowe 1975). Dishes are thin, sub-horizontal, flat to concave-upwards clayey laminations in silt and sand units. Pillars are vertical to near-vertical cross-cutting columns and sheets of structureless sands. Both structures are clearly post-depositional since they cut primary sedimentary structures. They are thought to form as follows. During de-watering following liquefaction, less permeable horizons act as partial barriers to upward flow and they force the flow to become horizontal until upward escape is possible. As the water seeps upwards, fine sediment grains such as clay flakes are filtered out and concentrated in pore spaces. The resulting clay-enriched laminae form the dishes which may later be deformed at their margins by upward flow. Pillars form during more forceful water escape, as noted above for de-watering pipes, of which pillars are simply small-scale forms.

11c Liquefaction and current drag structures

Should liquefaction affect sub-aqueous dune bedforms, then an interesting arrangement presents itself since the flowing water now exerts a shear stress upon a liquefied bed (Allen & Banks 1972). Such a simple shear system will cause progressively less shear to be transmitted to the deeper parts of the liquefied dune (Fig. 11.3). Also, since the liquefied bed solidifies from the base upwards (Ch. 7), there will be progressively less time for shear to operate in these lower areas. The net result is to cause an initially dipping cross-stratification plane within the dune to become sheared over into a parabolic curve, defining the structure known as **overturned cross stratification**, sometimes termed omelette structure. Liquefaction does not usually destroy the internal laminations in a bed since the

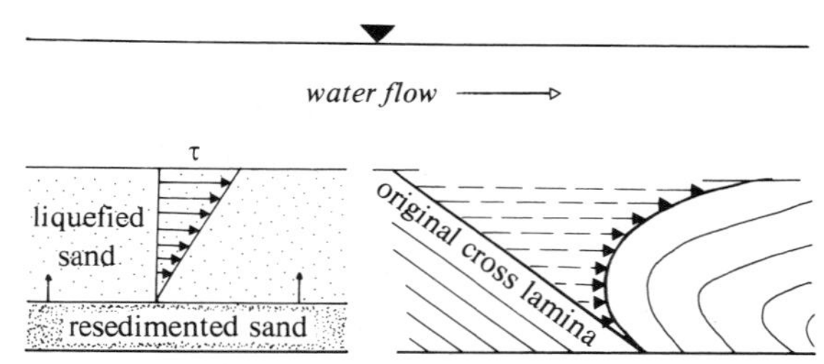

Figure 11.3 Diagram to show (left) a liquefied sand bed subjected to simple shear by water flow over its top boundary. Should the sand bed contain cross laminae (right) the shear will cause progressive overturning of the cross laminae whilst the liquefied sand consolidates upwards (modified after Allen & Banks 1972).

Figure 11.2 (a) Water escape pipe with adjacent upturned laminae. Note truncated top to pipe (arrow) which records a subsequent erosive episode (? destroying any pre-existing sand volcano). Scale = 0.2 m. Lower Carboniferous, Kirkbean, Scotland. (b) Water escape structure overlain by convolute laminations and cut by subsequent erosive episode (arrow); Upper Carboniferous, Nottinghamshire, England. (c) Dish structures defined by clay-rich laminae. Top shows increasing frequency of pillars culminating in topmost zone of structureless sand. Jackfork Gp, Okla, USA (after Lowe & Lopiccolo 1974). (d) View of load casts on base of a coarse turbiditic wacke. Note irregular, unoriented bulbous load casts and the narrow, indented flame structures bounding the casts. Upper Carboniferous, Mam Tor, Derbyshire, England. (e) Flame structure of mud intruded into base of coarse-grained proximal turbidite. Grès d'Annot, Provence, France. (f) Sandstone balls formed by periodic collapse and sinkage of silty current ripples into liquefied muds. Note normal cross laminations and parallel laminations in upper part of core. Upper Carboniferous, Nottinghamshire, England.

(a)
2 cm
(b)
(c)
2 cm
(d)
1 cm
(e)
(f)

degree of lateral particle movement is quite small. Primary depositional size contrasts, defining lamination, are thus preserved.

11d Diapirism and differential loading structures

When a light fluid lies beneath a dense one, the system is gravitationally unstable so that the lighter fluid will tend to rise above the denser fluid. Such instability, termed Rayleigh–Taylor instability, gives rise to the production of pipe- or ridge-like intrusions of the lighter fluid into the denser one. Such intrusions are known as **diapirs**. The two main sedimentary systems that give rise to diapirism are mud/sand and salt/sediment interlayers. In the former case differential loading gives rise to diapirs known as **clay lumps** which may reach the depositional surface from depths of up to 150 m. The density inversion needed to form clay diapirs disappears at depth because of compaction, causing de-watering and diagenetic bonding. No such change occurs with salt/sediment interlayers. The rate of flow of rock salt by *creep* increases with burial until a critical point is reached at which the upper surface of the salt bed deforms and amplifies by the process known as **halokinesis**. The production of one rising diapir or salt pillow often triggers an adjacent structure, and so on. One result is a fairly regular separation distance between salt diapirs in particular areas. It can be shown theoretically that for the two-dimensional case the dominant wavelength is given by

$$\lambda = \left(\frac{2\pi h_2}{2.15}\right)\left(\frac{\mu_1}{\mu_2}\right)^{\frac{1}{3}} \tag{11.1}$$

where
λ is the dominant wavelength or separation distance of diapirs,
h_2 is the thickness of original salt layer (considered much less than overburden thickness)
μ_1 is the viscosity of sedimentary overburden and μ_2 is the viscosity of salt ($\mu_1 > \mu_2$).

In some areas salt diapirs have risen vertically a distance of 5–6 km, sometimes reaching the surface to flow as salt glaciers, as in the Cambrian salt diapirs of southern Iran and the Persian Gulf. Spectacular diapirs of Permian salt are also known in the Gulf of Mexico and the North Sea.

On a smaller scale, diapiric injection of liquefied mud into sand beds gives rise to narrow **flame structures** (Fig. 11.2d, e) accompanied by broader downbulges of sand into the mud, defining **sand pillows** or **load casts**. The shape of the structures and their wavelengths are controlled by density, viscosity and layer-thickness parameters, as noted for diapirs above. Spectacular examples of deformation structures formed by basalt loading into water-saturated sands are seen in the Middle Proterozoic of Australia where the lava pillows are up to 250 m in diameter (Needham 1978).

In some instances sand pillows sink into the liquefied mud to form detached **sandstone balls** with characteristic internal deformed laminations (Fig. 11.2e & 11.4). Rarely, the lamination within detached sandstone balls records the advance of sinking current ripples over a liquefied mud patch (Fig. 11.2f, Reineck & Singh 1980).

Convolute laminations are common structures in muddy, fine-grained sands and silts. They exhibit narrow vertical upturned laminae, often truncated at the top surface, separated by broader synclinal downfolds with wavelengths of a few centimetres or decimetres. They are most commonly developed in sediments that have been

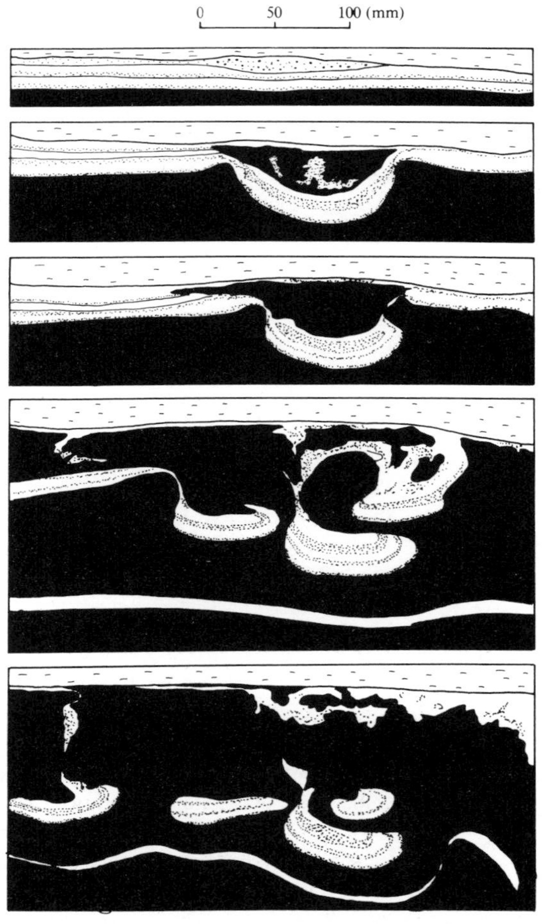

Figure 11.4 Diagram to show the experimental formation of sandstone balls by sinking of sand beds into liquefied muds (dark) (after Keunen 1965).

rapidly deposited, as witnessed by their association with climbing ripple cross laminations in distal turbidite (Ch. 15) and river flood facies. There is some evidence that the convolutions develop because of fabric readjustment following the gravitational collapse and flattening of a rippled depositional surface during liquefaction (Leppard, unpublished results). Associated water escape pipes are common.

11e Slides, growth faults and slumps

Sloping depositional surfaces, cliffs and roadcuts in cohesive mud deposits are prone to rotational failure, the slip surfaces approximating to circular arcs (Fig. 11.5). Many muddy intertidal point bar and cutbank surfaces (Ch. 14) show such **rotational slides** to perfection (Bridges & Leeder 1976). More important are the rotational slides that occur in major deltas. These slides, moving slowly along the fault planes, develop as deltaic clastic wedges prograde out over the muddy delta front deposits (Crans *et al.* 1980). The active fault is termed a **growth fault** (Figs. 11.5 & 11.6a & b). Beds on the downthrow sides of such faults are thickened since greater sediment deposition occurs in response to increased subsidence. The associated **roll-over anticlines** (Ch. 31) are important hydrocarbon traps in areas such as the Tertiary deposits of the Niger and Mississippi deltas (e.g. Weber & Daukoru 1975).

Gently dipping rotational slides on continental margins and delta fronts (see Ch. 19) give rise to large-volume sediment slumps. Some slumps are clearly triggered by cyclic earthquake shocks acting upon clay sediments of high water content in which pore waters cannot easily escape (undrained condition). The resulting bulbous-nosed slumps with pull-apart structures and imbrications may slide on their basal fault planes for large distances on gentle slopes (~3°). Such slumps may give rise to true debris flows and, ultimately, turbidity flows if the ambient waters can be mixed into the slide sediment (Ch. 7).

11f Desiccation and syneresis shrinkage structures

It is a matter of common observation that exposure of wet cohesive sediment to the atmosphere causes the formation of polygonal **desiccation cracks** as the sediment volume is reduced (Fig. 11.6c). These downward-tapering cracks, usually preserved in the rock record as casts on sandstone bases (Fig. 11.6f), exist on a variety of scales; the thicker the desiccated layer, the deeper and wider the crack systems. Deep cracks often show the plumose markings seen on rock joint surfaces (Fig. 11.6e). Desiccation cracks are rectangular on sloping surfaces such as

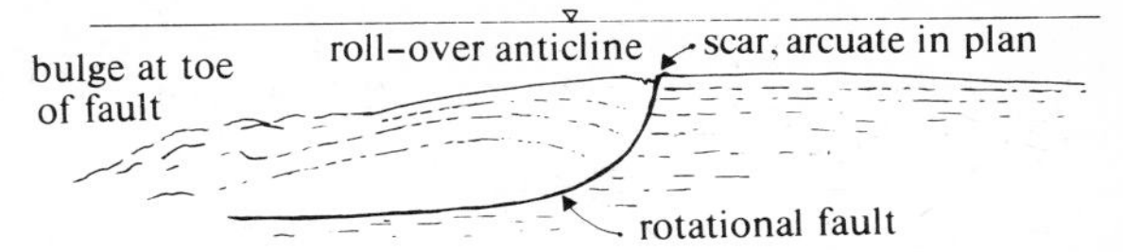

Figure 11.5 Diagram to show the formation of a rotational fault and associated features in water-saturated muds of high sensitivity.

exposed lake margins (Fig. 11.6d, Clemmey 1976). Desiccation of thin mud drapes or layers may produce **mud curls**. These have a low preservation potential, tending to become reworked to mudflake intraclasts by subsequent current action. Rarely, wind-blown sand may preserve mud curls *in situ*.

Shrinkage cracks may also form under water following pore-water expulsion of syneresis type which causes the clayey sediment to suffer a volume decrease. The **sub-aqueous shrinkage cracks** thus produced may be single, elongate 'eye-shaped' features or trilete cracks radiating from a central point with no connection to adjacent cracks (Fig. 11.6g). The single cracks often show preferred orientations (Donovan & Foster 1972). Delicate infill of sub-aqueous shrinkage cracks by sand or silt leads to preservation, but compactional effects often cause the preserved fill to be considerably deformed when seen in sections normal to the former depositional surfaces. The delicate nature of the preservation process means that sub-aqueous cracks are most common in, and characteristic of, sheltered shallow-water lacustrine environments (Clemmey 1978).

11g Summary

After deposition, sediment may be made mobile once more by the process of liquefaction caused by cyclical shocks of earthquake origin or pressure fluctuations caused by water waves. Pore-water expulsion follows liquefaction and may cause development of water escape structures. Liquefied sediment may be acted upon by currents to form overturned cross stratification or it may intrude overlying non-liquefied beds to form diapirs. Rapidly deposited muds overlain by sands on a delta front or delta slope are prone to rotational slides, slumps and growth fault development. Drying-out of wet muds produces desiccation cracks, and shrinkage of muds during syneresis produces sub-aqueous shrinkage cracks.

Further reading

Discussions of sediment strength are found in all soil mechanics texts; the reader is especially recommended to peruse Lambe and Whitman (1969) for a clear introductory account.

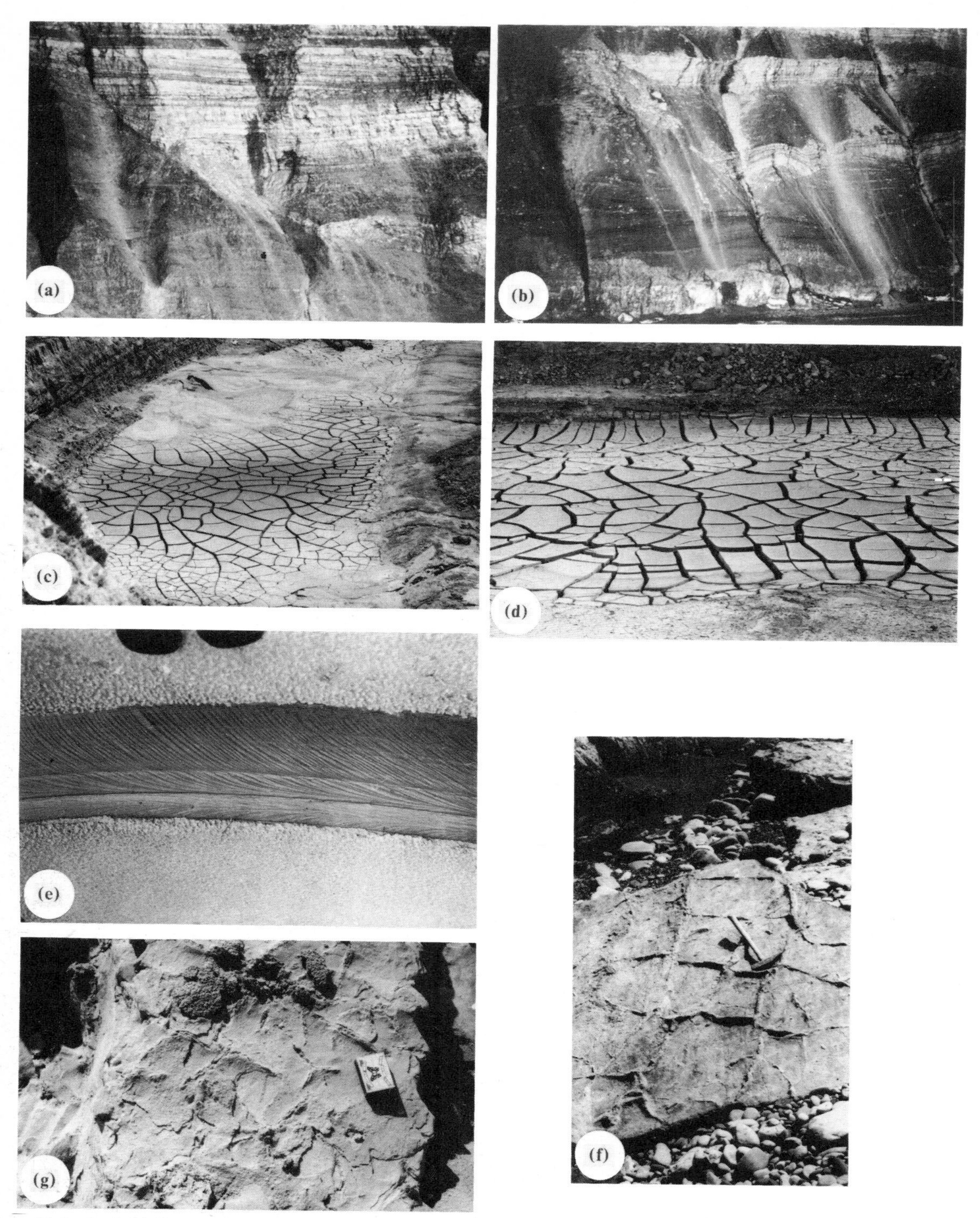

Figure 11.6 (a & b) Spectacular growth faults in Triassic sandstone and mudstone exposed in fjord cliffs in Spitsbergen. Note decrease of fault dip downwards (photos by M. Edwards; see also Edwards 1976). (c & d) Giant desiccation polygons (maximum width 2.5 m) in lacustrine fill (about 50 m across) to abandoned opencast copper workings; Kitwe, Zambia. Note rectangular cracks along sloping margins. (e) Plumose markings along desiccation crack surfaces of (c) and (d). (f) Infilled rectangular desiccation casts from Middle Jurassic floodplain facies; Scarborough, Yorkshire, England. (g) Stellate and trilete (to left of box) syneresis cracks in lacustrine facies of Ore Shale formation, Zambian Copper Belt.

PART FOUR ENVIRONMENTAL AND FACIES ANALYSIS

She opens the door of the past to me,
Its magic lights
Its heavenly heights
When forward little is to see!

From She opened the door *(Hardy)*

Plate 4 Field party amidst the folded sedimentary rocks (including Jurassic and Cretaceous limestones in the background) of the Helvetic Alps in Switzerland. Facies analysis in such terrains enables paleogeographic reconstructions to be made which help in the understanding of continental and oceanic sedimentary and tectonic history.

Theme

Sedimentary rocks found in the geological record were all deposited in certain environments of deposition. In the absence of a time machine the sedimentologist must use facies analysis and deductive reasoning to work out how and where rocks were deposited. In the past 20 years or so this aspect of sedimentary studies has changed from an obsession amongst a few geomorphologically orientated sedimentologists into one of the single most important aspects of research into ancient sediments, pure and applied. Facies analysis possesses its own elegance and combines all aspects of sedimentology in its approach.

12 General introduction to environmental and facies analysis

12a Scope and philosophy

Environmental analysis is the particular concern of geologically orientated sedimentologists. In the hands of a skilled operator the deductions that can be made from a stratigraphic sequence rival those of the celebrated Sherlock Holmes. The practice of such analysis requires a deep knowledge of sedimentary processes, a wide acquaintance with the literature dealing with modern sedimentary environments, and a degree of experience in dealing with previous problems. The particular appeal of the analysis lies in its central role, depending on insights gained from the whole of sedimentology with its ramifications and linkages with many other geological and geomorphological disciplines. Environmental analysis is of particular importance in economic studies (Ch. 31). The poor environmental analyst is dogmatic, inflexible and narrow minded, often dependent upon a simple generalised hypothesis into which he forces uncomfortable observations. Frequently he is scornful of attempts to integrate quantitative techniques into his work, with the excuse that natural processes are too 'noisy' for such exercises.

The philosophy of the present approach to environmental analysis is basically traditional in the sense that modern depositional environments provide the 'key to the past' in examining stratigraphic successions preserved in the geological record. It is, however, often forgotten that modern depositional environments can themselves only be understood by a knowledge of the appropriate physical, chemical and biological processes within these environments. This is a particularly important point when dealing with the rocks of the Precambrian, when many present day constants (e.g. g, pCO_2, pO_2) may have been appreciably different. In the following chapters we will therefore examine each depositional system by means of a fourfold approach involving:

(a) a descriptive account of the particular environment and its physiography;
(b) an analysis of the basic physics and/or chemistry of the environmental processes;
(c) the nature of the modern sedimentary sequences generated in the environment;
(d) a brief discussion of examples of environmental analysis of ancient sedimentary successions *thought* to have been deposited in the particular environment.

Regarding the last point it soon becomes obvious that there must exist an almost infinite number of stratigraphic examples that could be analysed in this way. However, such are the complexity and variety of environmental variables that almost every example provides some new slant on a depositional model. Individuals with a strong generalist tendency may find this rather depressing, but such a conclusion spurs the true addicts of environmental analysis on to even greater efforts.

12b Depositional systems and facies

Sediment deposition may occur in a wide variety of environments on the Earth's surface. The broadest possible classification of environments is into **continental, coastal, shelf** and **deep marine**. Although the geologist may wish to know only whether his rocks fall into such broad categories, further subdivisions are usually needed (Table 12.1). It is not unusual nowadays to find that environmental analysis of rocks has reached such a refined level that deposition, for example, may be deduced to have taken place in the hollows and puddles present on the surface of the point bar of a meandering river in rocks over 300 million years old!

Each depositional setting acts, or has acted in the past, as a sediment trap, preserving the products of sediment transport for posterity. Analysis of the deposited sedimentary sequence may be approached from two contrasting viewpoints.

The study of present-day processes in particular environments leads to an understanding of how fluid flow and/or chemical reactions produce a particular suite of grain types, sediment sizes, sedimentary structures and sediment geometry. Together with information gleaned from any *in situ* flora or fauna, it is then a relatively easy matter to establish a set of criteria by which the sedimentary environment may be characterised. The set of attributes thus defined enables one to erect a standard **environmental facies model** for all or part of a depositional system. The term environmental facies in this context refers to the *whole set of attributes possessed by the deposited sediment laid down in a particular environment*, e.g. tidal-delta oölite facies, intertidal mudflat facies. In such cases it is usually possible to link the production of a particular attribute directly to a direct cause.

By way of contrast to the above let us consider the

Table 12.1 Summary (not exhaustive) of the environments of deposition on the Earth's surface.

Environmental association	*Environment*	*Sub-environments or environmental variants*
Continental	desert	erg, wadi, erg apron, interdune, playa, duricrust
	alluvial fan	fanhead channel, proximal fan, distal fan, suprafan lobe
	alluvial plain and riverine cone	braided channel, meandering channel, levee, crevasse splay, floodbasin, lake
	lake	(saline, temperate stratified, tropical stratified, glacial, delta plain) lake terrace, shoreline, slope, basin, delta.
	glacial and periglacial	supraglacial (flow till), subglacial (lodgement till), intraglacial (melt out till), morainic complex, outwash fan, glacial lake, esker (also glaciomarine)
Coastal–shelf	delta	distributary channel, tidal channel, backswamp, bay, mouth bar, prodelta
	estuary	estuarine channel, marginal flats, flood tidal delta
	linear clastic shorelines	beach, nearshore, offshore, barrier, lagoon, tidal flat, tidal delta, tidal inlet, coastal aeolian dunes
	carbonate–evaporite shorelines shelfs and basins	sabkha, algal marsh, tidal flat, beach, lagoon, tidal delta, platform margin, marginal buildups, deep basin, evaporite basin.
	clastic shelf	(tide dominated, weather dominated) various tidal bedforms, sand ribbons, linear tidal ridges, sandwaves, shoal retreat massifs, buried channels, scarps (also glaciomarine)
Oceanic	passive margin	continental slope, continental rise, abyssal plain, submarine fan, submarine channel
	active margin	trench, subduction complex, perched basin, fore-arc basin, back-arc basin, fan
	oceanic pelagic	mid-ocean ridge, ridge flank, abyssal plain (hypersaline ocean, euxinic ocean)

approach of the geologist who must examine stratigraphic sequences of deposited sediment or sedimentary rock. Obviously in this case there is no *direct* evidence of the depositional environment left. Our geologist must play the role of Sherlock Holmes, beginning with a detailed examination of the rocks for all of their contained attributes such as laminations, grain size trends, sedimentary structures, etc. He will usually abstract such information onto lateral sections or onto vertical sections using a **Bouma logging** technique (Fig. 12.1). Patterns of sedimentary attributes will now be looked for so that *purely descriptive lithofacies* may be defined. The lithofacies in this case may be conveniently defined as *a body of rock with certain specified attributes that distinguish it from other rock units*, e.g. coarsening-upwards mudrock to sandstone facies and large-scale cross-stratified oölite facies. The geologically orientated sedimentologist must now compare his descriptive facies with those directly interpreted facies of Recent environments of deposition. This is where the skill and experience factors come to the fore since it is only by possessing a detailed knowledge of Recent facies that the geologist can assign realistic models to his ancient rocks. Even so it may often prove wise to assign alternative models and to avoid dogmatic conclusions as to ancient environments of deposition. It is humbling to realise that one can never actually prove that an ancient facies *was* deposited in a certain environment.

Figure 12.1 illustrates how a rational facies analysis might proceed, from Bouma Log to environmental conclusions. It should be stressed that all facies analysis is made easier by good rock exposure, particularly in three dimensions. The utmost caution should be adopted when exposure is poor since the 3-D geometry of a facies is a vital piece of information that often may prove decisive in an environmental interpretation.

The erection of a facies by an observer is always a fairly subjective process. As in systematic palaeontology or zoology there are the facies 'lumpers' or 'splitters'. A reasonable compromise between these tendencies should be searched for. A too-coarse division of facies may hide valuable trends whilst an excessive number of facies may prohibit useful generalisations. Further, there is nothing more infuriating to the reader than wading through, say, twenty different facies types in an account of a sedimentary unit. When a large number of facies definitions is unavoidable then some sort of grouping into **facies associations** may prove useful. As in our discussion of recent environmental facies and ancient lithofacies above, facies associations may be used in both contexts. For example, all the facies defined from a recent delta may be grouped into a deltaic facies association. In a rock sequence all coarsening-upwards sequences may be grouped into a coarsening-upwards facies association.

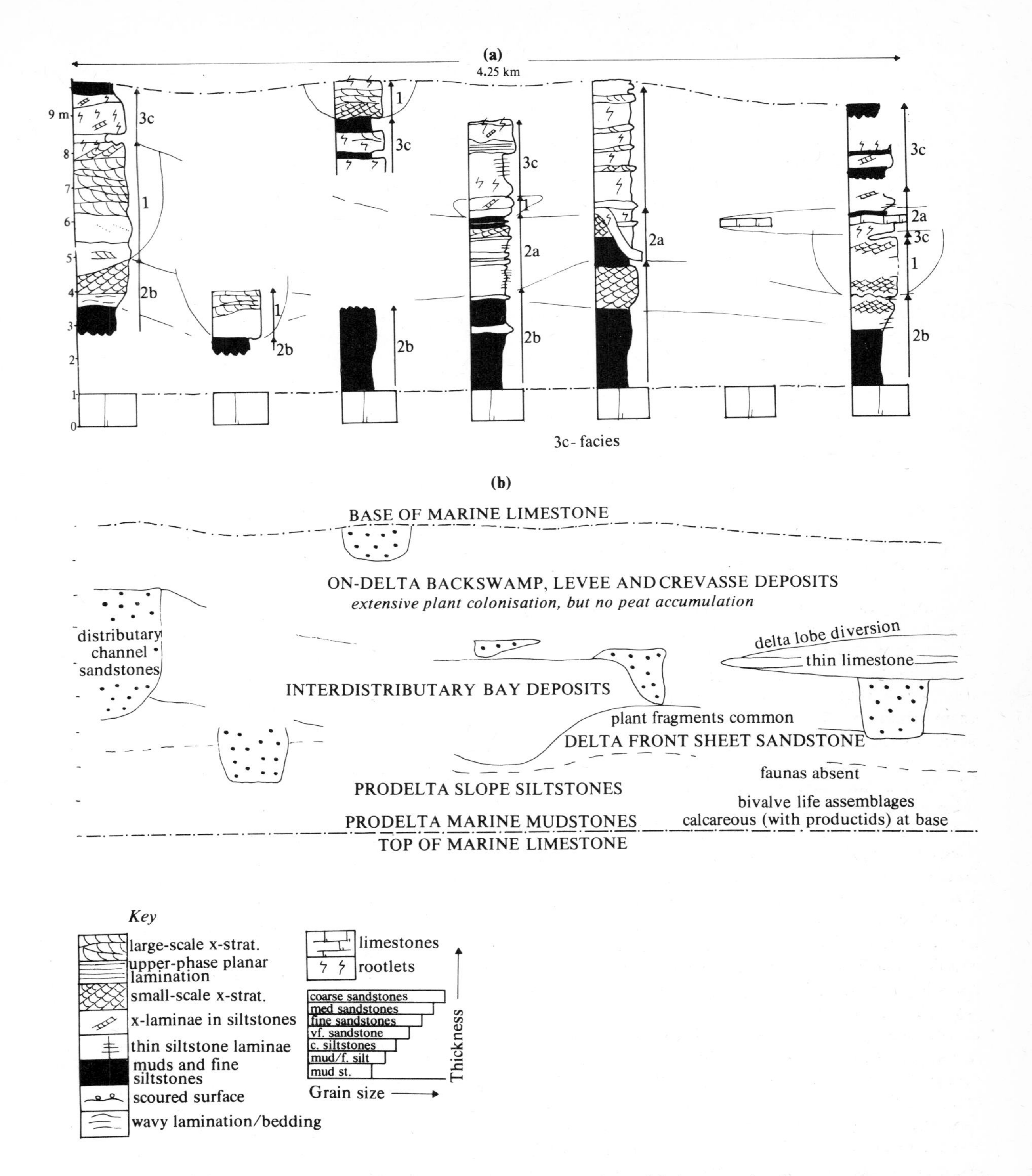

Figure 12.1 (a & b) Bouma logs and interpreted facies ribbon diagram for a deltaic succession between two transgressional limestones in the Mississippian of the Northumberland basin, England. Numbers in (a) define various facies associations: 1 – erosive-based fining-upwards association; 2 – transistional coarsening-upwards association; 3 – alternating beds. (After Leeder 1974.)

12c Succession, preservation and analysis

A particularly important aspect of facies work is the recognition of the significance of upward and lateral changes from one facies state to another, or of upward gradational or abrupt changes in grain size within a single facies. Consider, for example, the upward-coarsening sequence of facies in Figure 12.2. The upward coarsening is gradational and it implies that the increasing proximity to a sediment source, *or* to a higher-'energy' environment, was itself gradual. Such cycles usually result from a **prograding** (building out) depositional system in which a depositional hollow such as a lake or marine water body is gradually infilled by a deltaic sand influx (Fig. 12.2). Similar effects occur as high-energy coastlines prograde in areas of excess sediment supply. In low-energy coastlines, however, progradation gives rise to a fining-upwards sequence.

The examples just discussed all illustrate the point that the *vertical* succession of deposited facies may represent the *lateral* succession of environments of deposition. This conclusion is sometimes referred to as **Walther's Law**, but there are very many exceptions to the rule; it is only the simplest progradational or regradational systems that obey the 'law'. Particularly difficult problems arise when erosive-based clastic facies of channel origin occur. These channels may have been the lateral equivalents of fine-grained alluvial floodplain or deltaic backswamp sediments, or they may have resulted from a later, completely independent, process such as channel advance and incision caused by climatic or tectonic changes in the hinterland drainage basin. (e.g. Allen 1974, Leeder & Nami 1979, Haszeldene & Anderton 1980). It is often difficult to be able to tell such possibilities apart in the absence of very precise dating.

Consideration of facies with erosive bases such as channels leads us on to the concept of **preservation potential** as applied to sedimentary deposits. Once deposited, a sedimentary sequence may be partly or completely eroded away by subsequent erosional events. A good example here is fine-grained floodplain sediment laid down by periodic river floods as a channel system overtops or breaks its banks (see Ch. 15). If the channel system itself periodically migrates over its whole floodplain, then the fine flood sediment will be constantly destroyed so that the vertical sequence generated in a subsiding alluvial basin will be dominated to a greater or lesser extent (depending upon a number of variables) by channel deposits (Allen 1965a, Bridge & Leeder 1979). The only remaining evidence for the existence of floodplain deposits will be in eroded intraformational clasts associated with the erosion surfaces.

A most important control upon sediment preservation

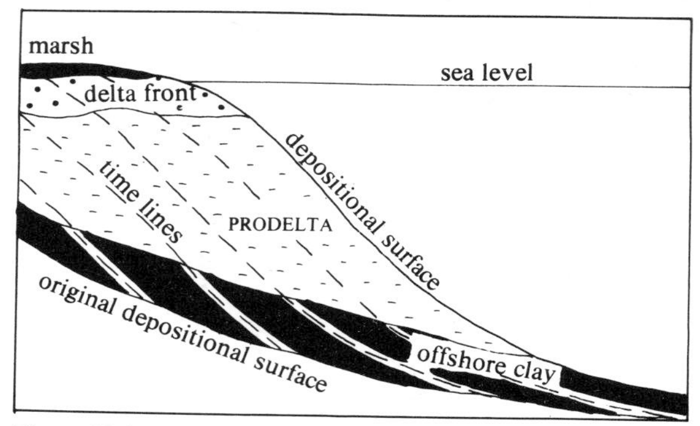

Figure 12.2 To illustrate diachronism and Walther's Law with reference to a seaward-prograding deltaic complex (after Coleman 1976).

is the rate of local or regional tectonic subsidence. Since subsidence and uplift are usually direct reflections of mantle processes, there is here an exciting direct link between sedimentary and solid earth processes.

The rigorous environmental analysis of vertical sequences of sediments or sedimentary rock may be approached from two directions. As noted previously, one may make a facies analysis *vis-à-vis* recent sedimentary environments, with subsequent attempts to explain away variations within and between facies states by reference to the environmental model. The second approach classically involves statistical analysis of beds or facies so that the frequency of any repeat patterns may be expressed as a probability. The most common (or most likely) facies changes may thus be deduced from a thick succession. Facies analysis may then be carried out on the results of the statistical analysis.

Let us consider the second approach in a little more detail, following, in a general way, the discussions of Till (1974) and Miall (1973). Consider the sequence of facies defined in the Bouma log of Figure 12.3. There is clearly some pattern about the sequence of facies. Certain facies states *tend* to follow each other but do not always do so. There is clearly a sequence of facies changes caused by events which are defined by probabilities, although each event has a random element in it. We may quantify the upward pattern of facies changes by constructing a table called an **upward-transition probability matrix** (Table 12.2). The table gives us the probability of each facies passing to the other facies and we may then know the most probable facies sequence and make our environmental analysis accordingly. In order to make sure that the probabilities are not the result of purely random features, it is useful to construct an **independent trials probability matrix** which represents the probability of any facies change occurring randomly. A test of significance such as the chi-square (χ^2) is then applied so that the

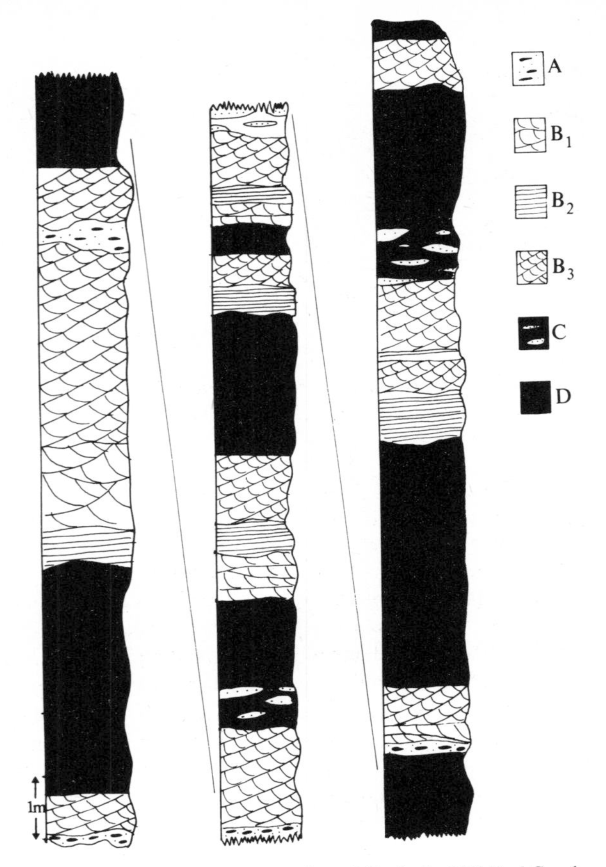

Figure 12.3 Simplified Bouma log of fluviatile Old Red Sandstone sediments from southwest Wales (see Table 12.1). Facies A – intraformational conglomerate; facies B_1 – large-scale cross-stratified sandstones; facies B_2 – planar laminated sandstones; facies B_3 – small-scale cross-stratified sandstone; facies C – interlaminated sandstone and siltsone; facies D – siltstone. (After Till 1974.)

differences between the two matrix results may be assessed for randomness. Should the upward-transition probability values be found significant, the deduction can be made that the occurrence of a given facies depends to a greater or lesser degree on the previous facies. Thus we deduce that the sedimentary process controlling the distribution of facies has a 'memory'. Such processes are termed **Markov processes**, formally defined as 'natural processes which have a random element, but also exhibit an effect, in which previous events influence, but do not rigidly control, subsequent events' (Harbaugh & Bonham-Carter 1970). Numerous sedimentary processes are Markovian, e.g. the advance of a delta into a standing water body and the progradation of a tidal flat. Markov analysis provides an indication of the **modal** (most

Table 12.2 (a) Number of upward facies transitions using the data of Figure 12.3.

	A	B_1	B_2	B_3	*C*	*D*	*Totals*
A	–	1	0	3	0	0	4
B_1	0	–	2	2	0	0	4
B_2	0	0	–	6	0	0	6
B_3	0	1	1	–	3	7	12
C	0	0	0	0	–	3	3
D	3	2	3	1	0	–	9
Totals	3	4	6	12	3	10	38

(b) Upward-transition probability matrix using the data of Table 12.2a. (After Till 1974.)

	A	B_1	B_2	B_3	*C*	*D*
A	–	0.25	0	0.75	0	0
B_1	0	–	0.5	0.5	0	0
B_2	0	0	–	1.0	0	0
B_3	0	0.08	0.08	–	0.25	0.59
C	0	0	0	0	–	1.0
D	0.33	0.23	0.33	0.11	0	–

common) **cycles** in sedimentary successions, but it cannot explain such cyclicity. Facies analysis alone can do this.

12d Subsidence, uplift and deposition

To a large degree the rate of subsidence of the Earth's crust controls the amount of preservation of particular sedimentary facies. The origins of basin subsidence are to be found in mantle and lower crustal processes which cause crustal disequilibrium. The rate of active crustal subsidence caused by the Earth's dynamic driving forces, such as mantle convection or lower crustal 'creep', must be clearly separated from the rate of subsidence caused by sediment loading and the resulting isostatic reactions. This is particularly true of 'deep' water basins ($\gg$200 m) which are being progressively infilled by a prograding sedimentary system. Some significant proportion of the final sediment thickness deposited in the basin will be caused by an isostatic response to sediment loading (Figs. 12.4 & 5).

Direct determination of present-day subsidence rates (by repeated levelling and by strain gauges) in sedimentary basins shows values in the range 0.3–2.5 mm a^{-1}. Direct measurements of present-day uplift rates fall in the range 0.2–12.6 mm a^{-1} with the higher values found in active orogenic belts and in areas subjected to glacial 'rebound' of isostatic origin. Schumm (1963b) has analysed the disparity between present rates of denudation and of tectonic uplift, and found that modern rates of uplift are about eight times greater than the average maximum rate of denudation.

The indirect determination of subsidence rates in sedimentary basins is possible only if the age and depth of deposition of a particular sedimentary facies are accurately known from sedimentary or micropalaeontological evidence (van Hinte 1978). It is dangerous practice to equate sediment thickness with subsidence unless the succession is (a) entirely of shallow-water origin, (b) deposition was continuous and (c) compaction is taken into account.

Problems similar to the above arise in the determination of sediment deposition rates. As sediment may be partly eroded within a generally depositional time interval, it becomes necessary to distinguish the **net sediment deposition rate** V', determined over a timespan, t', from the **local short-term deposition rate** V, determined over a timespan $t(t \ll t')$. Over the time interval t' the net accretion rate V' is given by T/t where T is the thickness of sediment. Ideally T should be arrived at by completely decompacting the sediment to a 'solid thickness' as outlined by Perrier and Quiblier (1974). The time, t', taken to accumulate T may be determined by radioactive decay systems or by palaeontological dating. T may be determined by direct measurement over shorter field study programmes. Numerous techniques are available to measure short-term deposition rates, a particularly useful method on river flood plains and on tidal flats being shown in Figure 12.6. One final point concerning deposition rates involves the effects of human activities upon the

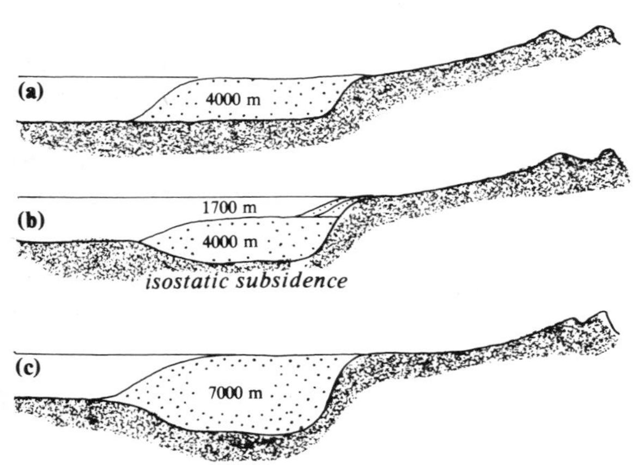

Figure 12.4 Diagram to show isostatic subsidence accompanying marine sedimentation into a basin of initial depth 4 km. For convenience the isostatic subsidence is shown occurring 'at an instant' rather than continuously as in nature. (After Matthews 1974.)

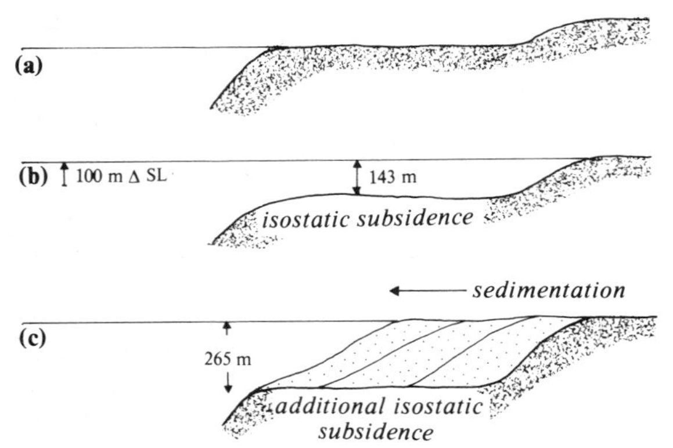

Figure 12.5 (a & b) Diagram to show how a 100 m eustatic sea-level rise may produce an extra 43 m isostatic subsidence. Sediment infill may then cause additional isostatic subsidence, as indicated in Figure 12.4. (After Matthews 1974.)

Figure 12.6 Determination of long-term deposition rates. A small cylinder of mud is extracted from the tidal flat and replaced flush to the surface by silica powder (white). The spot is marked and two years later is box-cored and sliced to reveal any erosion or deposition. (de Mowbray 1980.)

rates of deposition and erosion experienced in continental environments. Particular care has to be taken so that such effects can be minimised.

Inevitably, any discussion of erosion and deposition rates involves a decision as to the relative importance of abnormal or **catastrophic** events. Much recent work has concentrated upon the identification of deposits resulting from such events (e.g. 'storm' layers in shelf sediments, flood events in river facies) and the results have tempted some authors to assume that catastrophic events have exerted an overriding influence upon sedimentation (Ager 1973). However, stratigraphers and sedimentologists have generally ignored the conclusions of a classic study of the interrelations between magnitude and frequency in geomorphological studies by Wolman and Miller (1960). These authors point out that for many processes the rate of movement of material can be expressed as a power function of bed shear stress (see Ch. 6). The frequency distributions of these magnitudes, on the other hand, approximate to log normalcy. Therefore it follows that the amount of work performed by events – the product of frequency and rate – must attain a maximum (Fig. 12.7) and that the rarer, high-magnitude events (catastrophes) cannot be held responsible for the majority of work performed. These ideas are easily adapted to sediment deposition and erosion if it is assumed that the local deposition or erosion rate associated with an event is also proportional to the work done.

The importance of magnitude/frequency considerations in sedimentology may be brought home by restatement of this elegant metaphor by Wolman and Miller (1960):

> A dwarf, a man and a huge giant are having a woodcutting contest. Because of metabolic peculiarities, individual chopping rates are roughly inverse to their size. The dwarf works steadily and is rarely seen to rest. However, his progress is slow, for even little trees take a long time, and there are many big ones which he cannot dent with his axe. The man is a strong fellow and a hard worker, but he takes a day off every now and then. His vigorous and persistent labours are highly effective, but there are some trees that defy his best efforts. The giant is tremendously strong, but he spends most of his time sleeping. Whenever he is on the job, his actions are frequently capricious. Sometimes he throws away his axe and dashes wildly into the woods, where he breaks the trees or pulls them up by the roots. On the rare occasions when he encounters a tree too big for him, he ominously mentions his family of brothers – all bigger, and stronger, and sleepier.

We conclude that the man produces the biggest pile of chopped-up trees, i.e. moderate magnitude/frequency events are responsible for most geomorphic work and sediment deposition!

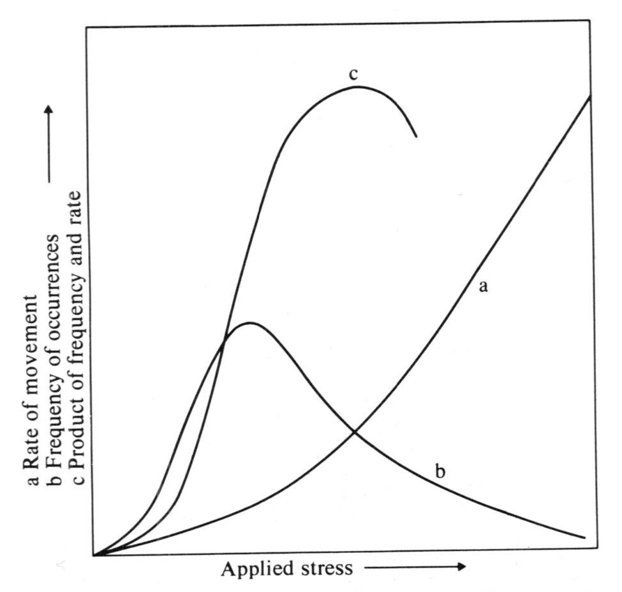

Figure 12.7 Schematic relationship between rate of movement, applied stress and frequency of stress application (after Wolman & Miller).

12e Transgression, regression and diachronism

Transgression may be simply defined as the process of migration of the shoreline of a water body in a landward direction. **Regression** is the reverse process (Curray 1964). Transgressive or regressive effects may be of local, regional or worldwide extent and may be due to a variety of causes. Shoreline movement, combined with long-term crustal subsidence, produces distinctive lateral and vertical facies changes. A particularly informative approach involves plotting the rate of sediment deposition against the rate of sea-level movement. Regression may result from falling sea level and/or high deposition rate whilst transgression may result from rising sea level and/or low deposition rate. With no net deposition or erosion and with stable sea level, the shoreline remains geographically stationary. Rising sea level usually results in transgression, but a high rate of deposition can offset this tendency and cause progradation of the shoreline. Similarly, falling sea level usually results in regression, but an excess of erosion over deposition can result in transgression of the shoreline under conditions of significant falling of sea level (Curray 1964). A classification of the various sorts of transgressions and regressions is shown in Table 12.3 and an example is given in Figure 12.8.

It is most important to realise that relative sea level changes may be produced by either rising or falling coastlands or by rising or falling sea level. The former process may be tectonic or compactional in origin and is usually of local or regional extent. The latter is termed **eustatic** if global in extent and may be caused by the growth and decay of either polar ice caps or mid-ocean ridges. In the former process the total volume of the ocean basins remains constant while the volume of water varies. In the latter the total ocean basin volume fluctuates while the water volume remains constant. Changes in eustatic sea level caused by ice sheet changes have dominated sedimentation trends during the late Tertiary and Quaternary. The rates of transgression and regression caused by ice sheet fluctuations are, by geological standards, extremely rapid. For example, in the last great transgression, the Flandrian, (caused by partial ice cap melting), the average rate of sea level rise was about 10 mm a^{-1}, far greater than most sedimentation rates.

One important consequence of the rapid Quaternary fluctuations in sea level has been that almost all sedimentary systems have not been behaving as if they were in 'steady state'. It is known that ice caps have been present at various times during the Earth's history, but they have probably never been present for *very* long periods. This means that ice-affected mid-Tertiary to Quaternary and Recent sedimentary deposits are often rather poor analogues for those sediments laid down in previous, ice-free epochs when some 'steady state' was more likely to have existed (see Ch. 12g).

In ancient sediments the case for eustasy may be empirically established if synchronous changes in depth of the sea can be followed widely and correlated around the world (see Hays & Pitman 1973). The two vital necessities for such analyses are a good idea of how sedimentary facies reflect water depths and a precise chronostratigraphic zoning scheme. Eustatic cycles in Mesozoic times have been deduced by the above methods and have been considered due to fluctuations in the volume of ocean ridges caused by variations in partial melting of the mantle causing fluctuations in spreading

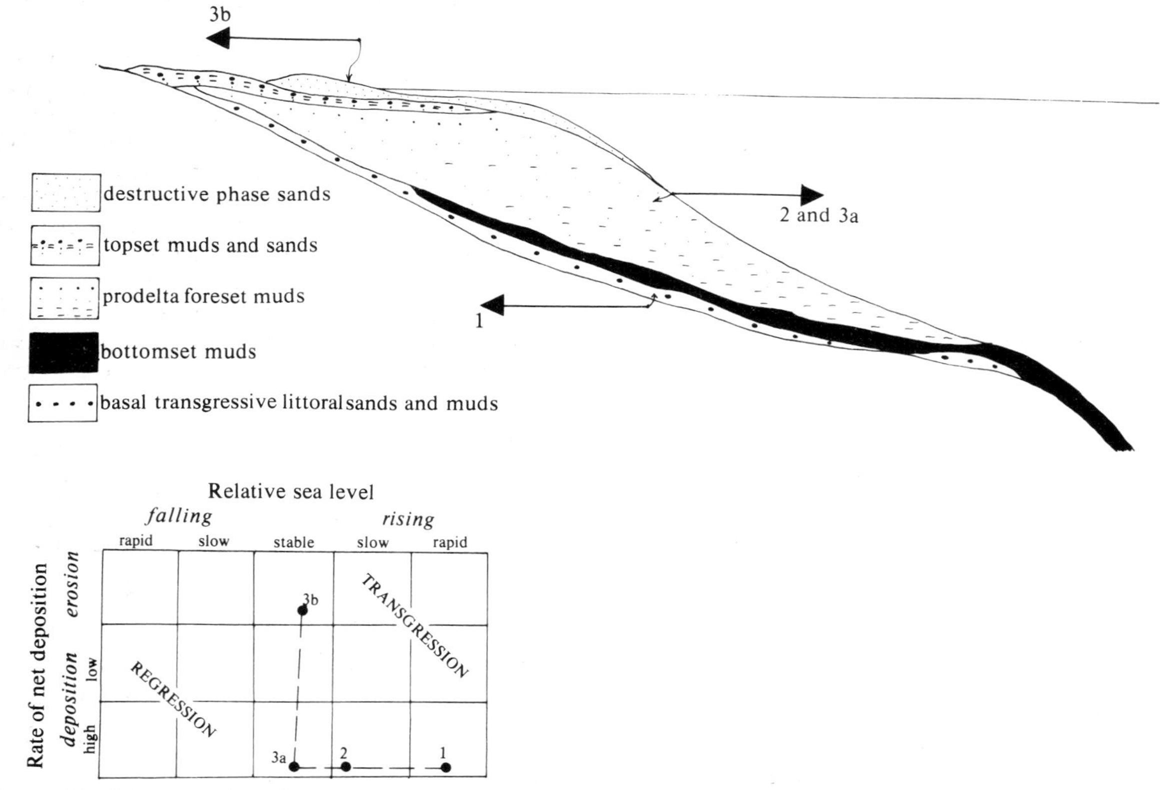

Figure 12.8 Diagrammatic section through a hypothetical subdelta of the Mississippi to show: 1 – basal marine transgression as sea level rises over a coastal plain; 2–3a – subdelta progradation; 3b – delta abandonment and compaction/erosion-induced transgression. (After Curray 1964.)

Table 12.3 To show that transgressions and regressions result from the interaction between deposition rate and relative sea level changes. (After Curray 1964.)

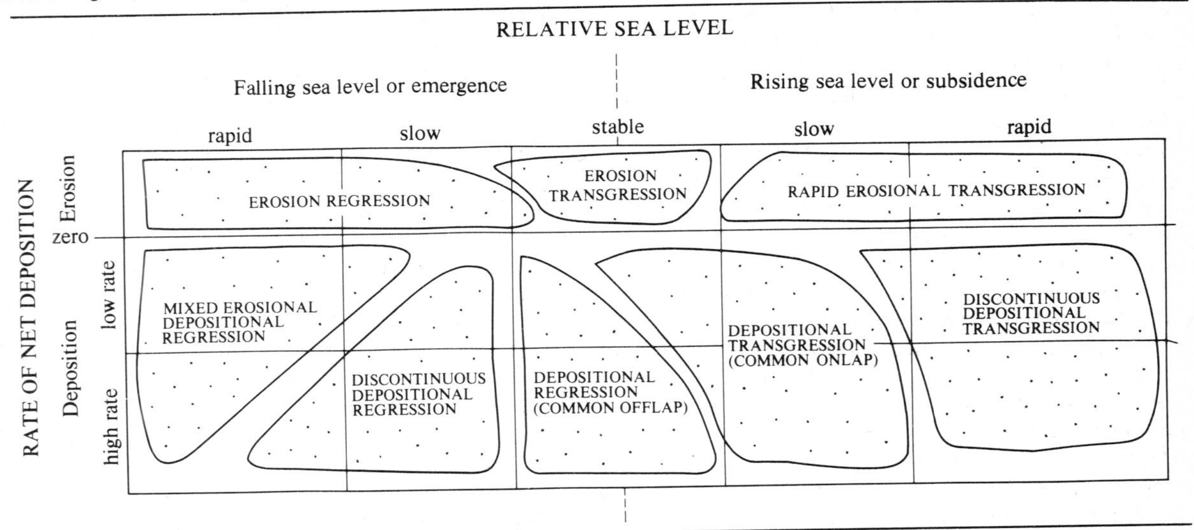

rates (Hallam 1969, Hays & Pitman 1973). It is thought that the rate of sea-level rise and fall is much less by this method, perhaps around 0.01 mm a^{-1}, than by the glacial hypothesis noted previously.

One final important consequence of transgressions and regressions is the production of **diachronous** sediment bodies. Diachronism means that a lithological unit cuts across time boundaries (e.g. see Fig. 12.2) and thus cannot be considered as a chronostratigraphic unit. The only true chronostratigraphic units composed of a single lithological unit are those built up by fall-out of sediment from a standing water body, e.g. ash layers and some deep oceanic oozes. Almost all other sedimentary units are diachronous to some degree.

12f Palaeocurrents

Many sedimentary structures may be used to give an idea of the direction of past current flow. Careful field measurement of structures in particular facies is followed by computation of outcrop **vector means** and **magnitudes** (Appendix 12.1) which give the worker valuable information upon (a) the regional palaeocurrent system and any change with time, (b) the direction of elongation of a particular facies, and (c) the location of possible hinterland regions (alluvial facies only). Three main points concerning the interpretation of palaeocurrent measurements need discussing: the nature and significance of bedform hierarchies, the significance of true flow vectors, and the construction of 'ideal' palaeocurrent models in relation to the major environments of sediment deposition.

Even the most cursory examination of bedforms on a Recent sand tidal flat or on the dried-up bed of a river will convince the reader that the variability of directional structures (Fig. 12.9) is related to the magnitude of the flow/bedform system (J. R. L. Allen 1966). Thus flow azimuths from current ripples show greater variance than for dunes because the former are produced or destroyed quickly by the falling or low-stage flows that modify the dunes which were in equilibrium with the high-stage flow. Now usually the high-stage flow will be the most important indicator of the local current trend, in the case of rivers giving the best indication of the local palaeoslope.

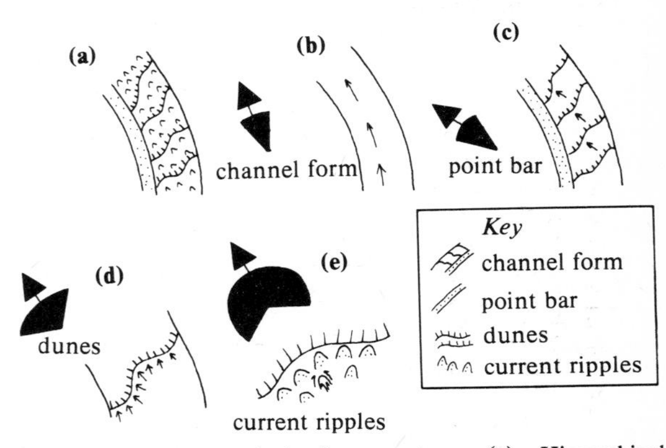

Figure 12.9 Hypothetical flow system. (a) Hierarchical organisation of bedforms; (b–d) dependence of current-directional data on rank of bedform. (After Allen 1966.)

We may conclude that the most reliable palaeocurrent indicators for regional studies are those formed by high-magnitude flows and that palaeocurrent sampling should take this into account (see Miall 1974 for a good discussion).

A related topic concerns the usage of 'vector' in palaeocurrent analysis. To the physicist a vector has both magnitude and direction; examples of vectors include velocity, force, acceleration and momentum. A measurement of the direction (azimuth) that a cross bed dips towards cannot be a vector unless a nominal value of unity is given for vector magnitude. The usage of the term 'vector analysis' is therefore misleading, as it refers solely to the resolving of an azimuth into its N–S and E–W components (Appendix 12.1). Yet it is clear that sedimentary structures are produced by fluid forces (vector quantities) having both magnitude and direction. An illuminating way of considering palaeocurrent analysis is therefore to consider the magnitude *and* direction of the fluid movement that produced the sedimentary structure. Direction may be measured in the usual way whilst magnitude may be assessed by reference to bedform phase diagrams such as those presented in Chapter 7.

The results of a palaeocurrent sampling programme cannot be interpreted in isolation from a consideration of the environments of deposition deduced for the particular facies in which the palaeocurrents were measured. For example, it would be futile to expect palaeocurrents from a tidally influenced sand facies to say anything about the ultimate sources of the sand grains. As briefly noted above, only alluvial facies can provide information regarding palaeoslopes of terrestrial areas. Further, such results will only establish the *local* palaeoslope and may give a spurious indication of sediment sourcelands if extrapolated 'up slope'. Clearly, petrographic and regional geological considerations must also be applied to such problems.

12g The Holocene

In subsequent chapters frequent mention will be made of the fact that the present-day Earth's surface and its sediment cover bear the scars of Pleistocene glacial/interglacial cycles, even in areas far removed from direct glacial action.

In effect our planet is still recovering from the last stage of glacial advance and retreat. Thus vast expanses of inactive desert now fringe such active deserts as the Sahara; the most recent fall and rise of sea level has caused tremendous changes in facies deposited on all shelves and coastal plains; and stable cratonic interiors such as that of Australia contain complex alternations of arid and humid soils and sediment facies. Additionally the remnant continental ice sheets are still a major influence upon the pattern and vigour of modern atmospheric and oceanic circulation.

The above list is by no means exhaustive and it is a stimulating exercise to try to note the many peculiarities of our Holocene environment and to ponder upon the likely contrasts between this complexity and the conditions of sedimentation during glacier-free epochs. In reading the subsequent chapters it is therefore advisable always to question the exact relevance of Pleistocene-to-Holocene sedimentary analogues for the geological past.

12h Basin analysis and plate tectonics

There is a particularly close link between plate margin morphology and the nature of basin infill sequences. During the initial stages of cratonic rifting, fault-bounded graben develop. These act as sites for lake and alluvial fan development. Lithosphere attenuation and the formation of new oceanic crust lead to marine transgression over these rift-stage continental clastics. Thick evaporative salts may accumulate in low latitudes should circulation in the new ocean be restricted. Continued seafloor spreading is accompanied by slow subsidence of the continental margin. Major drainage systems developing in the continental hinterland find an outlet to the coast and begin to deposit thick fluviodeltaic facies as the coastal plains prograde seawards. Shelf development proceeds apace because of subsidence and sedimentary accumulation. A broad shelf develops which in turn causes a regional tidal regime to become established.

In a mature ocean, pelagic marine facies accumulate over the abyssal plains and on the mid-ocean ridges. The thickness and composition of these facies reflect the surface- and deep-water dynamics of the ocean water body. Along the continental rise, various mass-flow processes transfer clastic sediment oceanwards as bottom-hugging turbidity currents, causing submarine fans to develop. When our idealised mature opening ocean eventually begins to close up, then the thick pile of coastal-plain, shelf and continental-rise sediments will begin to deform and become uplifted as a cordillera. Deep trenches develop along the line of subducting ocean lithosphere. Detritus transferred oceanwards in submarine fans may now be scraped off or 'telescoped' landwards to form an accretionary wedge complex. Massive eruption and intrusion of calc-alkaline magmas along the cordillera or island arc cause characteristic lithic-rich sands to be shed into fore-arc, back-arc or trench basins. Drainage systems debouching from the uplifted cordillera deposit fluvial (red bed) facies as molasse sequences in interior graben or exterior basins.

This brief reminder of the importance of plate margin processes in sedimentary basin analysis may be supplemented by readings on continental margin basin formation (Bott 1976), rifting margin subsidence and evaporite formation (Kinsman 1975a & b), and the contrasting nature of sequences and processes along active, passive and strike-slip plate margins (Mitchell & Reading 1969, 1978, Ballance & Reading 1980). Further reference to the importance of plate margin setting in oceanic facies models may be found in Part 7 below.

12i Summary

The analysis of facies must be approached from a number of viewpoints if it is to prove fruitful. The fundamental difference between modern and ancient facies is that in the former the depositional processes may be seen doing their work. These processes must be inferred for ancient sediments. The study of facies is enlivened by attempts to quantify such elusive variables as crustal subsidence rates, net deposition rates, the effects of infrequent catastrophic erosional and depositional events, rates of transgression and regression, and the magnitude of palaeocurrent vectors.

Further reading

Essays by two leading facies analysts are highly recommended: Walker (1978a) and Reading (1978). All students should read Ager's forceful 'born again' championing of neocatastrophism (1973). Potter and Pettijohn (1978) is a fundamental reference on palaeocurrents and basin analysis.

Appendix 12.1 Vector statistics in palaeocurrent analysis

Populations of directional measurements such as palaeocurrent azimuths follow a circular distribution that may be Gaussian. The mean of a population of such measurements cannot be arrived at by the normal method of summation and division through by the number of measurements (e.g. the mean of 350° and 10° is 180°!). Faced with a population of angular measurements, $\theta_i (i = 1, 2, 3, \ldots n)$ where θ is the azimuthal direction measured E of N, then their co-ordinates on a unit circle are

$$x_i = \cos\theta_i \quad \text{and} \quad y_i = \sin\theta_i$$

and their means are

$$\bar{x} = \sum_{i=1}^{n} \cos\theta_i / n$$

$$\bar{y} = \sum_{i=1}^{n} \sin\theta_i / n$$

This mean point may be represented in polar co-ordinates (r, θ) as

$$r = \sqrt{(x^2 + y^2)}$$

$$\cos\bar{\theta} = \bar{x}/r$$

$$\sin\bar{\theta} = \bar{y}/r$$

$\bar{\theta}$ is the mean of the palaeocurrent measurements and r is an estimate of the spread of the angular values around a unit circle; r approaches unity the closer the points are clustered. The mean angular deviation about $\bar{\theta}$ is given by

$$s = \sqrt{(2(1 - r))}$$

A worked example may be found in Till (1974), from which the above is taken. More details of directional statistics may be found in Watson (1966) and Mardia (1972).

PART FIVE CONTINENTAL ENVIRONMENTS AND FACIES ANALYSIS

I met a traveller from an antique land
Who said: Two vast and trunkless legs of stone
stand in the desert. . . . Near them, on the sand,
Half sunk, a shattered visage lies, whose frown,
And wrinkled lip, and sneer of cold command,
Tell that its sculptor, stamped on these lifeless things,
The hand that mocked them, and the heart that fed
And on the pedestal these words appear:
'My name is Ozymandias, king of kings:
Look on my works, ye Mighty, and despair!'
Nothing beside remains. Round the decay
of that colossal wreck, boundless and bare
the lone and level sands stretch far away.

Ozymandias *(Shelley)*

Plate 5 Vertical aerial photograph to show the Bozeman (centre), Helm (lower left) and (part of) Pearl meander bends of the R. Wabash near Grayville, Illinois, USA. Flow is from top to middle left; photo taken at low flow; scale approx. 1:30 000. The sub-aerial parts to the point bars are the white (sand–gravel) crescentic areas on the inside of each bend. Note the progressive colonisation by trees and shrubs of the inactive point bar top, the faint traces of scroll bar surfaces, the abandoned meander loop (right centre) and the well developed active scroll bar present on the downstream portion of Helm bend (photo by US Dept. Agriculture; for further details see Ch. 15 and Jackson 1976b).

Theme

Terriginous clastic grains are transported from drainage basin hinterlands to depositional plains or basins along river courses or down alluvial fans. Net deposition of alluvial facies occurs if there is net crustal subsidence. Thus some proportion of sediment is forever trapped beneath the coastal plain or graben. In arid areas of the world, under low-latitude high-pressure cells, detritus from upland drainage basins is reworked by wind into erg sand sheets covered with a multitude of bedform types. These larger bedforms are orientated so that they precisely reflect the regional or continental wind system. In polar arid areas of the world, sedimentation is dominated by vast continental ice sheets whose influence extends outwards via the influence of calved icebergs which drop their frozen loads onto the shelf floors and ocean bottoms. All continental environments may contain lakes, ranging from impermanent interdune playas to deep freshwater lakes in graben to proglacial lakes and deep hypersaline lakes. Lakes act as a complete sediment 'trap' whose deposits yield much information concerning Tertiary to Recent climatic and ecological changes.

13 Deserts

13a Introduction

Aeolian sedimentary processes dominate over some 30% of the continental land area enclosed by the 15 cm precipitation line (Fig. 13.1). In these semi-arid and arid zones sedimentary grains are liberated from upland areas by weathering and transferred to low-lying areas by ephemeral stream flow along wadis. The sand-, silt- and mud-grade deposits (the latter grades comprising 'dust') are then selectively taken into the local or regional wind system where they undergo sorting processes and are moulded into various aeolian bedforms (Ch. 8) which may coalesce to form sand seas or **ergs**. **Loess** sheets result in adjacent areas after deposition of the wind-blown dust. In the great trade wind deserts such as the North African Sahara and Central Australia there is a close correspondence between dominant wind flow and sand transport over the whole area. I. G. Wilson (1973) has estimated that over 99% of all 'active' aeolian sands occur in ergs having areas greater than 125 km^2.

Figure 13.1 also shows areas of 'fixed' or inactive ergs which border the active ergs. These 'fixed' ergs attest to the much expanded and more vigorous trade wind belts that existed during periods of ice sheet advance during Pleistocene glacial periods. The North African trades, for example, were active much further south during the last glaciation. (Vast areas of Pleistocene loess in temperate latitudes also indicate the vigorous nature of the planetary wind systems at this time.) The 'fixed' ergs, now stranded and stabilised by vegetation in the northern savanna fringes (Talbot & Williams 1979, Talbot 1980), attest to the long time interval needed to destroy an erg and they lead to the expectation that erg facies ought to be preserved in the stratigraphic record.

13b Physical processes and erg formation

We have discussed previously (Ch. 8) the various aeolian bedform types and their internal structures. Let us here briefly consider the origin of erg sand bodies and their relations to continental wind systems. Consider the map of North Africa in Figure 13.2. Meteorological observations make possible the computation of wind resultants during sandstorms for various stations scattered over the area. This data, together with ground and satellite

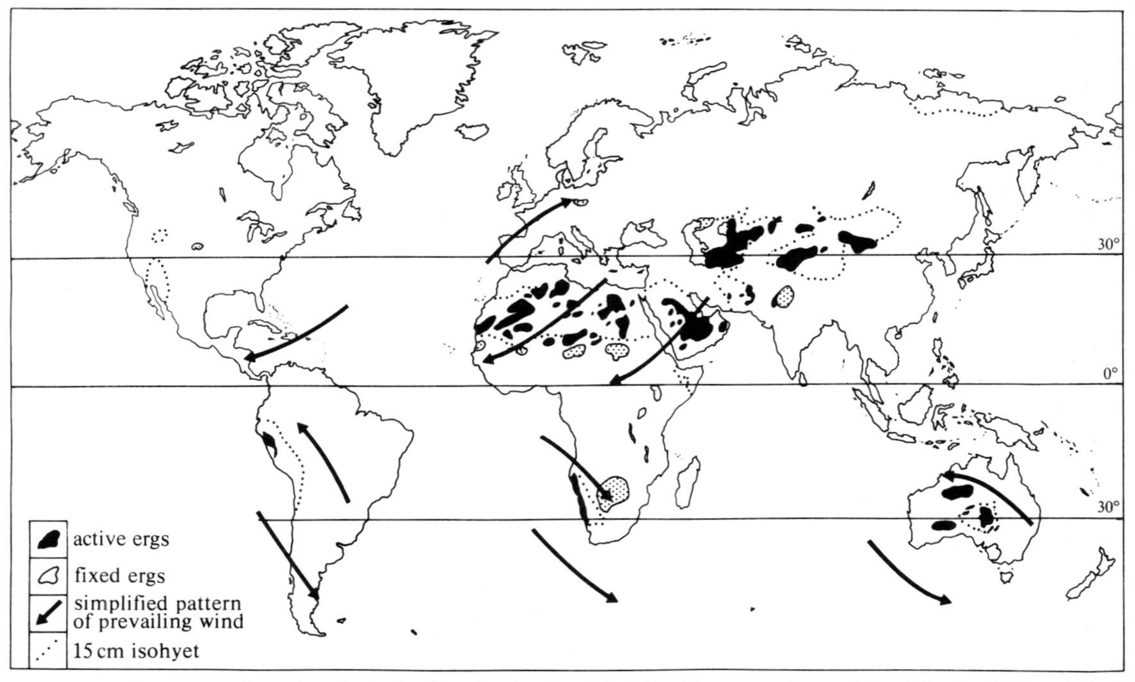

Figure 13.1 World map to show the principal active (modern) and inactive (Quaternary) ergs (after Glennie 1970 and Cooke & Warren 1973).

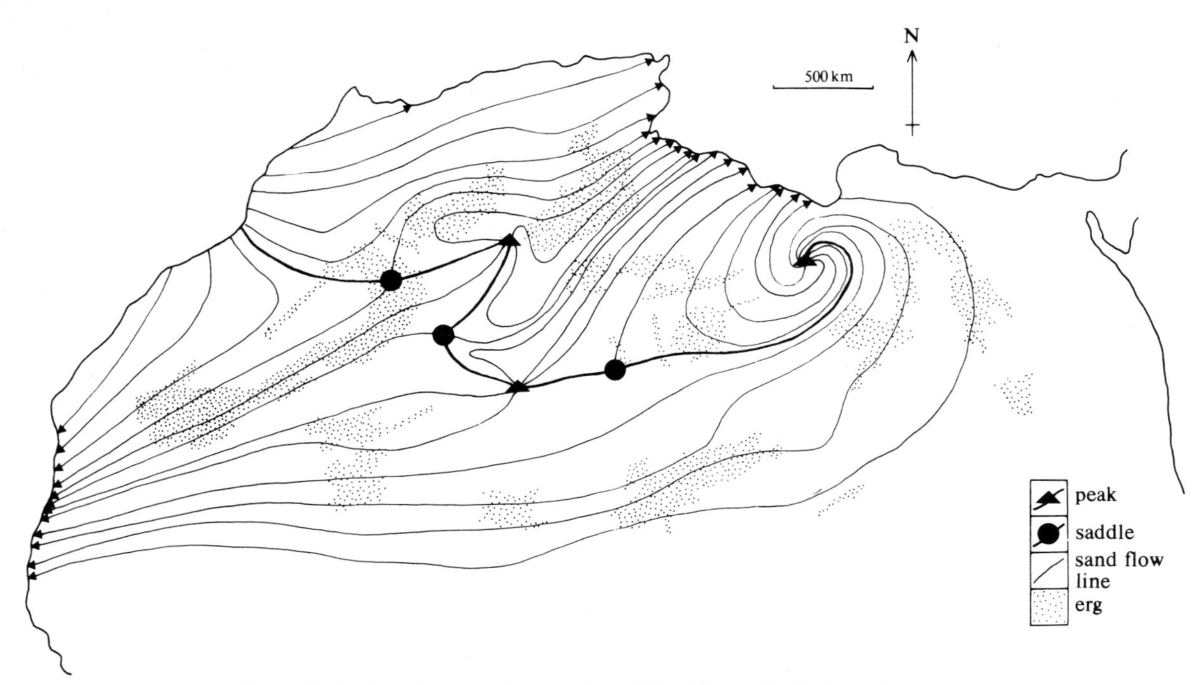

Figure 13.2 Sand flow map for the Sahara (after Wilson 1971); discussion in text.

observations on the orientation of barchanoid and longitudinal (seif) dunes and erosional lineations (**yardangs**), enables a spectacular continental sandflow distribution to be mapped out. (I. G. Wilson 1971, Mainguet & Canon 1976, Mainguet 1978, McKee 1978.) Ideally a sandflow map should show resultant directions as flowlines *and* resultant magnitudes as contours, analogous to a combined wind direction and pressure map. The information currently available is generally inadequate to achieve this aim. Sandflow maps are also analogous to drainage maps in that they show divides separating distinct 'drainage' basins: peaks in fixed high-pressure areas and saddles in between them. Unlike water drainage there is little direct relation between sand flow and topography since winds may blow up hill.

The flowlines drawn on Figure 13.2 extend from erg to erg, implying very long transport distances down wind, giving ample time for aeolian abrasion and transport processes to work. Evidence for this erg-to-erg transport is provided by satellite photographs showing linear traces of aeolian corrosion in between erg areas along sandflow lines (Mainguet 1978). Taken as a whole the map shows that all the sandflow lines arise within the desert itself, with the main clockwise circulatory cell roughly corresponding to the subtropical high-pressure zone. Note that all the sandflow lines eventually lead to the sea. In fact, a great plume of Saharan dust extends out for thousands of kilometres into the Atlantic ocean, providing a steady rain of fine silt-to-clay grade particles into the deep ocean.

The other spectacular example of trade wind orientated desert dunes occurs in the great ergs of central Australia where the migrating high-pressure cell provides the dominant control of anticlockwise sand movement along longitudinal dune systems (Brookfield 1970). The pattern of sandflow here is much simpler than in North Africa since the sandflow circulation is undisturbed by topographic obstacles.

Individual desert ergs are confined to basins, whatever their absolute height, and they terminate at any pronounced slope break. Ergs can form only when the wind is both fully charged with bedload and either decelerating or converging. They can form at sandflow centres, at saddles and in local areas controlled by topography. Deposition and deflation is controlled not only by the regional wind system, but also in a complex manner by the bedform hierarchies (Ch. 8) present on the erg (see Wilson 1971).

13c Modern desert facies

Although something is known about the internal structures of aeolian dunes (Ch. 8), virtually nothing is known about the structures within draas and erg sand bodies. The migration of dunes and draas within ergs must give rise to a hierarchy of cross-stratified sets and bounding planes (Brookfield, 1977). Within a moving draa the migration of dune bedforms should give rise to large-scale cross sets separated by truncation planes (Fig. 13.3) (termed **second order surfaces** by Brookfield 1977) that gently dip down wind or up wind. In an erg that is itself building up, and/or is in a subsiding sedimentary basin, draas themselves will overtake each other and preserve thicker cross sets between essentially horizontal truncation planes (termed **first order surfaces** by Brookfield). An alternative model for the origin of the very gently dipping to horizontal truncation planes is the **deflation hypothesis** of Stokes (1968) in which periods of dune migration and build-up, with the formation of cross sets and second order surfaces, are followed by deflation back down to the water table surface (Fig. 13.4). Erg build-up or subsidence between the events then allows preservation of a sequence below the water table. One point that goes against this idea is that water table features such as gypsum or halite crusts and cements or interdune sabkha structures (*qv*) are not found associated with the truncation planes.

An interesting study of the Great Sand Dunes erg, National Monument, Colorado (Fryberger *et al.* 1979), has shown that **low-angle sand sheet** accumulations form a transitional facies between normal aeolian dunes and non-aeolian deposits (Fig. 13.5). These deposits originate by gentle wind deceleration in the lee of small surface irregularities and contain low-angle erosion surfaces, ripple remnants and surface lag deposits.

In many ergs, areas of **interdune sabkhas** occur where the water table intersects the interdune troughs (Glennie 1970). Evaporation leads to salt precipitation as crusts and subsurface nodules. Blown sand driving across such moist areas causes build up of adhesion-rippled sands. In some areas temporary playa lakes may form in interdune depressions following extensive rainfall. Spectacular examples of interdune playas between stabilised dunes occur around Lake Chad near Timbuktoo (Talbot 1980). Many dried-up, formerly freshwater, lake beds occur within the Central Australian desert. These lakes became progressively more saline about 25 000 years ago due to climatic aridity when gypsum was deposited together with clays. Complete desiccation of the lakes led to wind reworking of the finely powdered gypsum and clay into the elliptical 'lunette' dunes now seen on the downwind lake margins. Longitudinal dunes then migrated across the lake floors (Bowler 1977). Spectacular interbedding of aeolian interdune sheets and water-laid deposits from wadis are recorded from many erg margins.

Vertical sequences of aeolian and non-aeolian facies

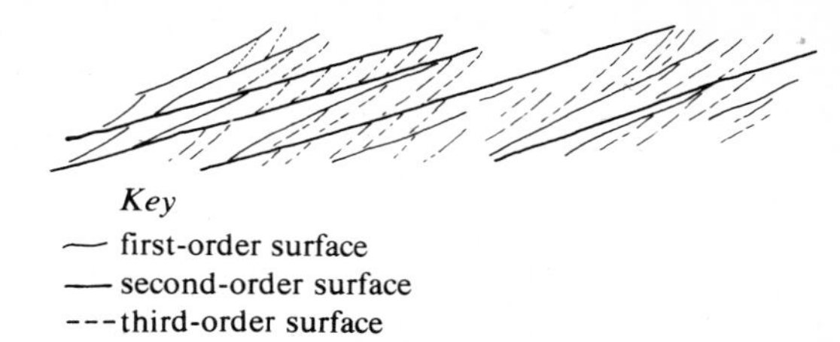

Figure 13.3 Various orders of bounding surfaces observed in a section through Permian erg deposits at Locharbriggs, Scotland (after Brookfield 1977). Tectonic dip 14° left.

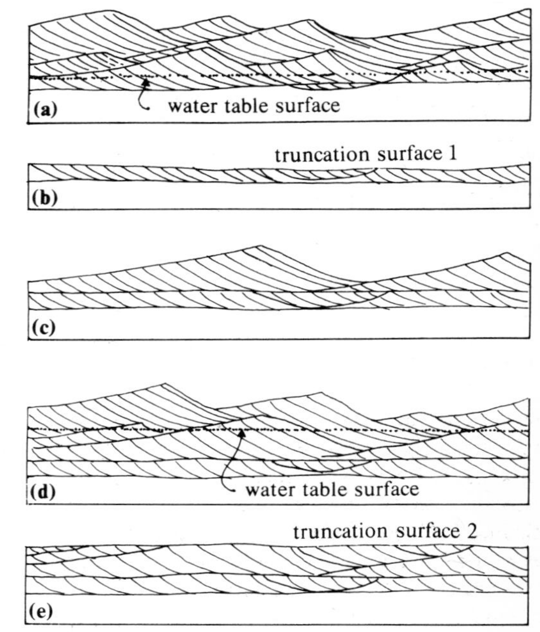

Figure 13.4 Stokes hypothesis for the formation of multiple truncation planes in aeolian sandstones. The mechanism depends upon periodic relative rises of the groundwater level (perhaps subsidence-induced) followed by erosive deflation which removes sand to the groundwater level. Deflation is followed by renewed sand build-up and so the cycle continues (after Stokes 1968).

arise in areas of inactive ergs where dunes and draas are currently being degraded by alluvial fan incision along their flanks. Dune sands are reworked by mass flow and water flow into interdune troughs, giving rise to a complex stratigraphy (Talbot & Williams 1979).

13d Ancient desert facies

The interpretation of a rock unit as aeolian is an important one and the decision should not be taken lightly on the strength of only limited evidence. Much play has

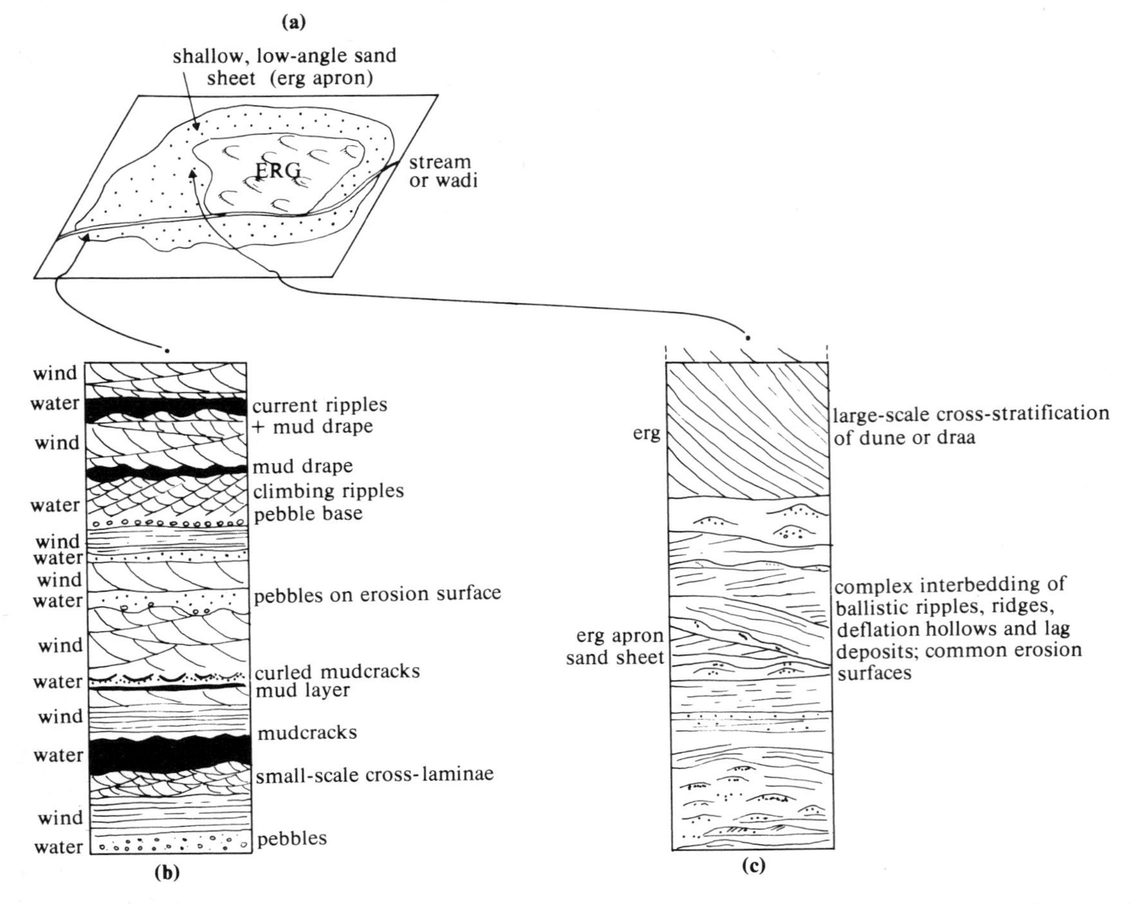

Figure 13.5 Vertical sequences liable to be generated at the margins of ergs. (b) Facies produced by alternations of water and air flow such as might be found close to a stream or wadi system. (c) Facies produced by air flow alone on the erg apron margin. (After Fryberger *et al.* 1979.)

been made recently of the importance of very large-scale cross-stratified sets, standing at or about the angle of repose of sand, in the diagnosis of wind-blown dunes. Since such sets are now known from fluvial sand bodies (McCabe 1977), it is dangerous to rely on this characteristic alone.

There have been very few detailed regional studies of suspected ancient aeolian deposits in the stratigraphic record (but see Sanderson 1974, Walker & Harms 1974, McKee 1978). One of the best known deposits is the Permian Rotliegendes sandstones of the southern North Sea basin (Fig. 13.6) which form an important regional natural gas reservoir in the British and Dutch sectors (Glennie 1972). Detailed well-log studies reveal a complex fossilised erg sand body up to 500 m thick with marginal facies of fluvial wadi sediments and a basin centre facies of lacustrine clays and playa evaporites. The cross-stratified sets (up to 5 m thick) are separated by horizontal to gently dipping truncation planes. Within each set there is generally an upward progression from sub-horizontal finely laminated to steeply inclined (20°–27°) thicker sets of laminae. This may mark the passage of a steep transverse dune avalanche face over parallel laminated interdune and dune toe deposits. The truncation surfaces may represent the second-order surfaces of Brookfield (1977). Regional palaeocurrent studies reveal that the Rotliegendes were deposited by generally northeasterly winds. Thin intercalations of wadi fluvial facies and interdune evaporites and adhesion ripples occur throughout the erg sandstone sequence. These interbeds are considerably less permeable and porous than the main aeolian sandstones and they form annoying breaks to hydrocarbon flow. The red colour of these and other ancient dune sands is thought to result

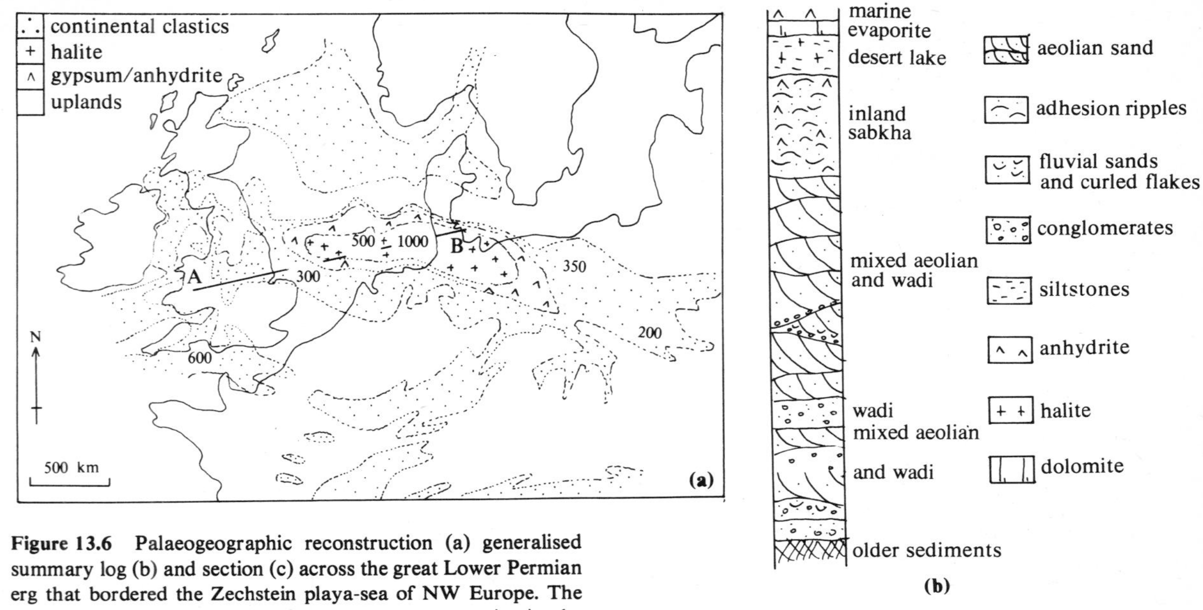

Figure 13.6 Palaeogeographic reconstruction (a) generalised summary log (b) and section (c) across the great Lower Permian erg that bordered the Zechstein playa-sea of NW Europe. The main aeolian sandstones are important gas reservoirs in the British and Dutch sectors of the North Sea. (After Glennie 1972 and Ziegler 1975.) Numbers in (a) refer to thicknesses in metres; summary log about 300 m thick.

from early diagenetic ferromagnesian mineral breakdown (Ch. 28).

Knowledge that modern ergs of continental scale are adjusted to the motions of the trade wind system encourages the hope that ancient erg deposits may be used in conjunction with continental reassemblies to reconstruct ancient wind systems. Much progress has been made in this field in recent years (e.g. Bigarella 1973).

13e Summary

The major sandy deserts of the world comprise a number of individual ergs. In any one desert, sand flow in and between ergs occurs in response to the predominant trade wind pattern. Ergs themselves are made up of a range of aeolian bedforms, from tiny ripples to giant draa. Migration of dunes and draa is thought to result in the hierarchy of bounding truncation surfaces observed within ancient aeolian sand deposits. The monotonous aspect of an erg is broken by the occurrence of interdune sabkhas, where the local water table intersects the erg surface, and by erg margin features such as ephemeral streams, and fans at wadi mouths. Ancient erg deposits are dominated by large-scale cross-stratified sandstones broken in places by waterlain facies and by thin interdune sabkha facies. These features may have important effects upon the reservoir characteristics of aeolian sandstones in hydrocarbon fields.

Further reading

Cooke and Warren (1973) is a readable summary of many features of desert geomorphology and sedimentology. Glennie's (1970) account concentrates upon depositional facies and has many fine photographs. Bigarella (1972) is also useful. McKee's (1978) recent compilation contains superb satellite photographs of erg features, many in colour, and chapters on most aspects of modern and ancient erg sedimentation. I. G. Wilson's elegant work on ergs (1971, 1973) is highly recommended but the reader is advised to consult Mainguet and Canon (1976), Mainguet (1978) and McKee (1978) for updated interpretations of French meteorological data and for the use of satellite imagery in bedform interpretations. Evocative accounts of the erg landscape and of its pull upon Western travellers are given by Bagnold (1935), Thomas (1938) and Thesiger (1964).

14 Alluvial fans

14a Introduction

Alluvial fans are localised and relatively small-scale (Fig. 14.1) accumulations of sediment formed by deposition as trunk streams emerge from upland drainage basins into some sort of lowland basin. They may form along linear mountain fronts, along the sides of major valleys, or at the margins of glacier ice. In a geological context the most important environment of fan deposition is in fault-bounded sedimentary basins where periodic fault movement enables subsidence and hence preservation of the fan sediments to occur (Fig. 14.2).

Dry fans are those in semi-arid climates where flow over the fan surface may be regarded as ephemeral and where transport by debris flows is important. **Wet fans** are those subject to perennial flow and where stream flow is the most important mechanism of transport and deposition. Some wet fans form low-gradient alluvial cones which, in contrast to dry fans, may cover very large areas, an example being the Kosi fan in Pakistan (Ch. 15).

Topographically, dry alluvial fans comprise (Fig. 14.3) (a) **rockhead valley**, (b) **fanhead canyon** or trench, (c) **fan channels,** (d) **mid-fan lobes** (e) **interlobe** and **interchannel** areas and (f) distal **fan apron**. As we shall see below, each of these areas gives rise to characteristic sedimentary facies. Wet fans show a proximal–distal change from braided coarse alluvium to finer meandering alluvium to distal flood deposits. We shall concentrate our discussion below on dry fan systems (see Chs 15 & 17 for wet fans).

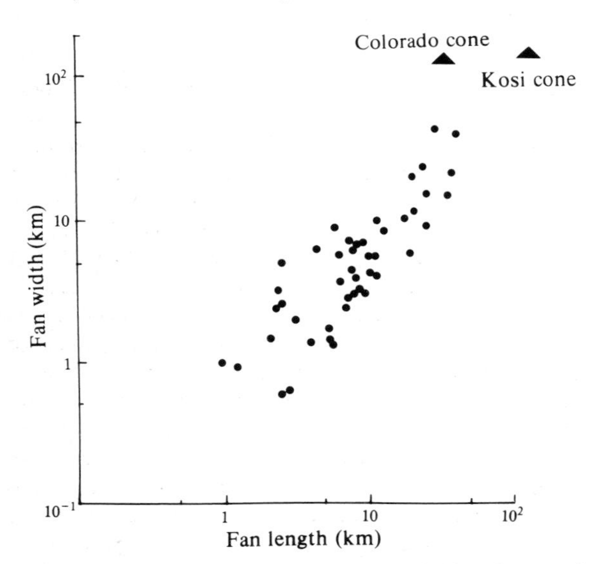

Figure 14.1 The dimensions of Recent alluvial fans from semi-arid and arid climatic zones (after Heward 1978b).

14b Physical processes

The basic form of an alluvial fan is a depositional response to expansion of confined channel flow as water leaves the rockhead valley to emerge over the fan surface. Analogies may be made with submarine fans at the ends of submarine canyons (Ch. 25) and with distributary mouth bars formed by effluent discharge at the mouths of delta distributary channels (Ch. 19). Recent work has shown that the presence or absence of a fanhead trench is one of the most important features that controls the nature and distribution of alluvial fan sediments. The fanhead trench may form in response to tectonic uplift as

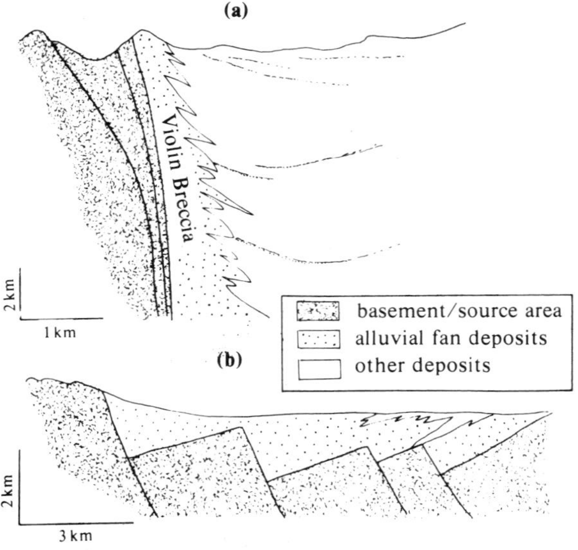

Figure 14.2 The typical association of alluvial fan facies with basin margin syndepositional faults. (a) Tertiary fan facies adjacent to a strike-slip fault in California (after Heward 1978b). (b) How a broad outcrop of fan facies may result when repeated backfaulting of tensional origin occurs. Example from the Permo-Triassic of the Scottish Hebridean basin (after Steel & Wilson 1975).

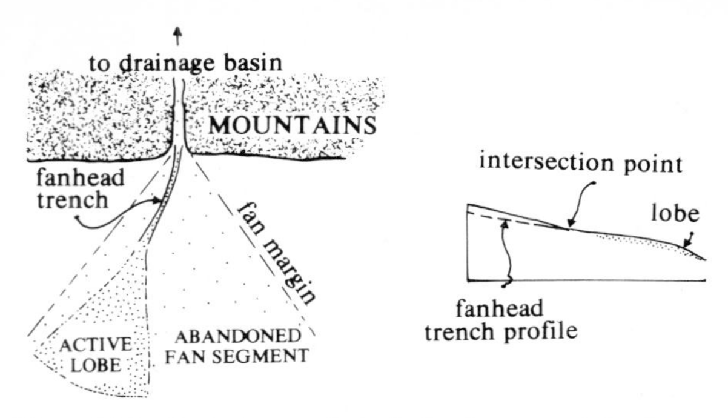

Figure 14.3 Typical morphological elements defined on modern semi-arid alluvial fans (after Heward 1978b).

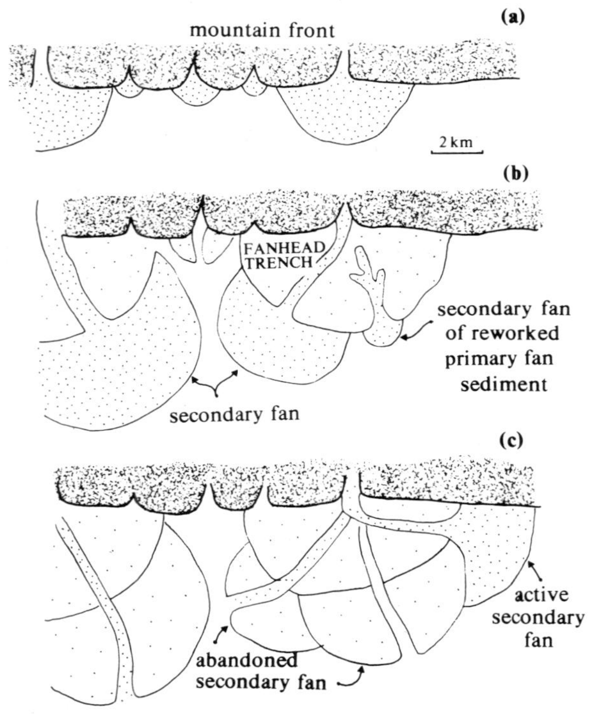

Figure 14.4 To show the sequential development of secondary alluvial fans subsequent to fanhead entrenchment and primary fan reworking (after Denny 1967).

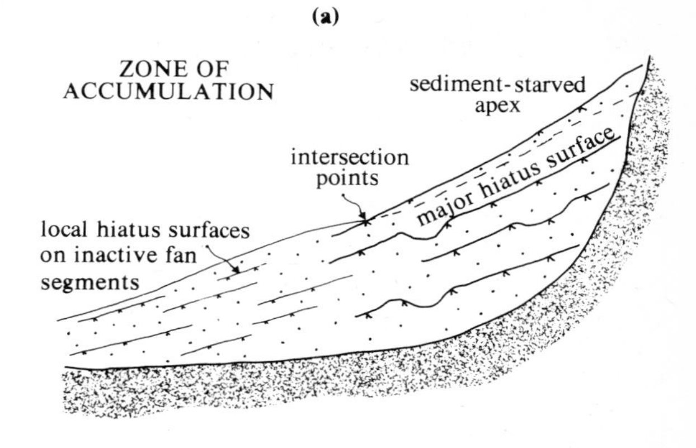

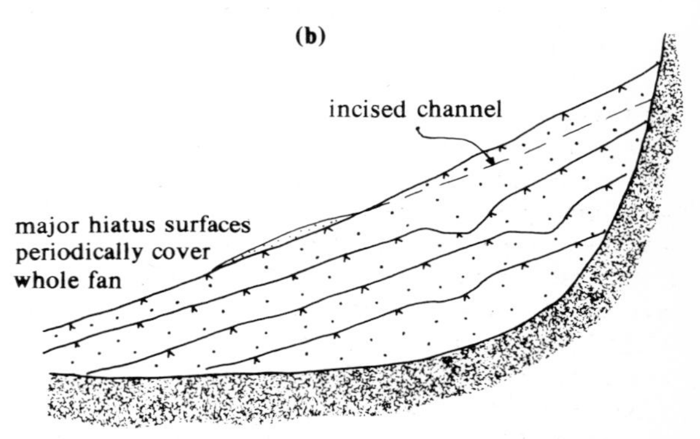

Figure 14.5 Schematic sections through alluvial fan sequences dominated by (a) threshold-type reactions such as entrenchment and changing lobe activity and (b) climatic reactions in which the whole fan surface becomes periodically inactive. (After Talbot & Williams 1979.)

a 'correction' to excessive deposition at the fan apex. Experimental studies with laboratory wet fans (Weaver, in Schumm 1977) show that with fanhead trenching the base level of the source area is lowered by incision. Alluvium in the valleys of the source area is then flushed out, with terraces appearing along the valley sides. The increased sediment load accelerates the filling of the fan trench, which in turn raises the base level of the fanhead.

Fanhead trenching, together with channel avulsion and channel plugging by debris flows, may cause and control the development of subsidiary fan lobes and of composite fans (Figs. 14.3 & 4). In many examples the site of active deposition is periodically shifted over the fan surface, leaving large areas inactive and prone to weathering and soil development. This cyclical development of alluvial fan deposition and erosion may also be caused by climatic changes, as seen in many Quaternary alluvial fans (Talbot & Williams 1978, 79). Care must therefore be taken in interpreting fan cycles (Fig. 14.5). Major depositional episodes on the lower fan may correspond to periods of entrenchment at the apex, the amount of deposition being controlled by the balance between local rainfall amount and the availability of detritus temporarily stored in the rock drainage basin. Thus some high rainfall periods may see little sediment transfer whilst others may see very substantial transfer. The 'liberation' of stored sediment in the fan drainage basin is thus seen in terms of a **geomorphic threshold** that must be exceeded before substantial deposition can occur on the lower fan (Schumm 1977).

14c Modern facies

A number of down-fan changes in the general nature of sedimentary deposits are apparent from studies of recent fans. There is usually a down-fan decrease in mean grain size, bed thickness and channel depth and a down-fan increase in sediment sorting (Beaty 1963, Bluck 1964, Bull 1972, Denny 1967). Because of the short transport distances involved in dry fans (Fig. 14.1) there is usually little discernible down-fan change of grain shape. Important changes in flow mechanism are responsible for some of these trends; thus there may be a down-fan change in dry fans from debris flow lobes to active fluvial fan channels and interchannel areas and to sheet floods on the lower fan apron. In the coarse deposits of the upper fan much of the runoff may percolate through the subsurface, depositing detrital clays in the interstices of the open gravel framework (Hooke 1967). Such gravels with infiltrated clays are termed **sieve deposits** (see also Ch. 28).

The rockhead valley and fanhead trench sub-environments contain localised accumulations of poorly sorted, angular and coarse gravels with the clasts being grain- or matrix supported. Internal stratification is poorly developed. Deposition occurs by scree fall, colluvial mass flow and in-channel debris flow. Fanhead trenches range in depth from a few metres up to tens of metres. At the intersection point of the fan trench with the general alluvial fan surface (Hooke 1967) the confined flow changes into a braiding network of shallow channels with slower flow rates. The major locus of deposition on the fan occurs downstream of this intersection point. Both debris flow deposits and water-lain deposits occur over the mid-fan area. The former comprise interdigitated sheets with non-erosive basal contacts, or occupy channels cut by water flow action (Fig. 14.6). Water-lain deposits commonly show erosive, channelled contacts (Fig. 14.6) and internal stratification related to bedload transport or bedform migration. Upward-fining trends within water-laid deposits give evidence of flow decay with time during flood events. Portions of the mid-fan area that are periodically inactive may show development of soil horizons. As the distal fan apron is approached, the braided channel networks on the surface gradually

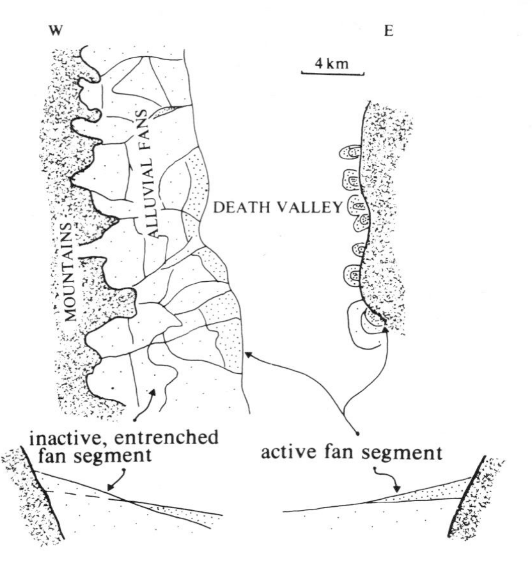

Figure 14.7 Map and schematic sections (latter not to scale) of Death Valley, California, to illustrate fan response to asymmetrical vertical crustal movements. The large western fans are heavily dissected and largely inactive because of relative uplift, and the small eastern fans are shrinking because of relative downwarp. (After Hooke 1972.)

disappear. Sheet flood deposits of great lateral extent may occur in this area and interdigitate with lake or floodplain facies developing marginal to the alluvial fan body.

Stratigraphic models for the sedimentary sequences built up over time by alluvial fans are largely based upon the hypothetical behaviour of a prograding or retrograding fan system (Heward 1978b). A prograding fan is expected to give rise to large-scale coarsening-upwards sequences as the depositional processes become increasingly proximal. Progradation may be caused by increasing intensity of basin margin faulting (Fig. 14.7) or to increasingly humid climate giving higher runoff and sediment transport rates. A retrograding fan system should give rise to a fining-upwards sequence by the reverse of the above. Initial fan growth along a new scarp or fault scarp should also give rise to a coarsening-upwards sequence, as will advance of active subsidiary fans after deep fanhead trenching.

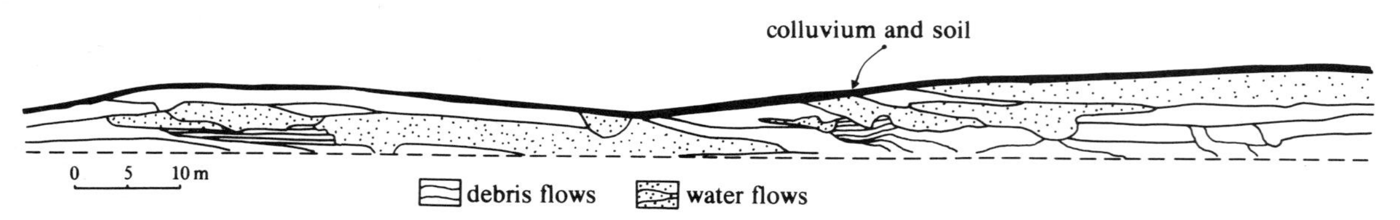

Figure 14.6 Cross section through coalesced Quaternary alluvial fans of 'Alpine' type to show the complex interbedding of water and debris flow deposits (after Wasson 1977).

14d Ancient alluvial fan facies

Progradation of small-scale alluvial fans into lacustrine environments is deduced for the thick, coarse-grained Stephanian coalfield successions of the Cantabrian Mountains, northern Spain (Fig. 14.8, Heward 1978a & b). A tropical and seasonal climate is thought to have allowed vegetation to colonise abandoned fan surfaces: coals may reach 20 m in thickness in the distal fan areas. Heward presents evidence for a general down-fan change in sedimentation mechanisms from debris flows to channelised water flows to sheet floods.

Particularly detailed studies of clast size variation and vertical facies changes related to basin margin faulting are presented by Steel (1974) and Steel and Wilson (1975) for alluvial fan facies in the New Red Sandstone of the Scottish Hebrides and by Bluck (1965) for Triassic sediments of South Wales.

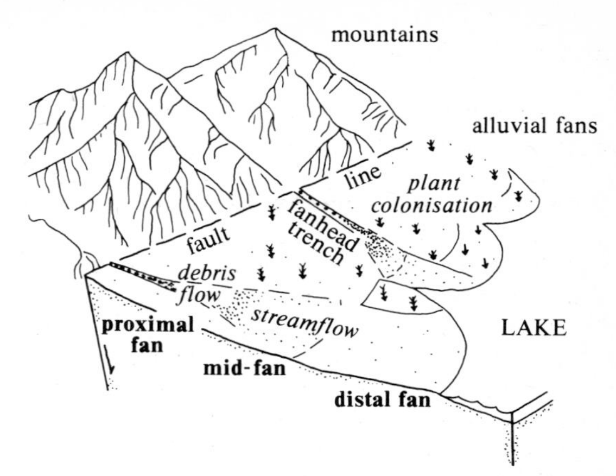

Figure 14.8 Simplified alluvial fan depositional model for the Stephanian (Pennsylvanian) coal-bearing sediments of certain northern Spanish coalfields (after Heward 1978b). Note the colonisation of abandoned fan segments by vegetation. Vertical sequences observed within the coalfields attest to periodic fan progradation and abandonment caused by the combined effects of fault movement and lobe diversions.

14f Summary

Alluvial fans are widely distributed geomorphic features with significant preservation potential along the margins of fault-bounded basins and graben. Fan morphology is largely controlled by processes of aggradation and degradation acting at the fan apex in response to periodic tectonism, climatic change and geomorphic threshold development. Depositional processes on active fan lobes are dominated by debris and stream flows. Internal stratigraphy is controlled by proximal to distal changes in depositional processes and by cyclical changes arising through the aforementioned mechanisms.

Further reading

The best review of alluvial fan sedimentation (with an extensive bibliography) is given by Heward (1978b) who synthesises an extensive literature into a number of models useful for the geologist interested in ancient fan facies. Another helpful review is by Bull (1972). Depositional mechanisms on modern fans are discussed by Hooke (1967). Debris flow mechanics have been discussed previously in this text (Ch. 7). The concept of geomorphic thresholds in relation to fan cycles is discussed by Schumm (1973, 1977).

15 River plains

15a Introduction

Rivers are the connecting link between areas of sediment production in drainage basins and areas of coastal deposition. Many of the great coastal plains of the world also lie in areas of the crust that are prone to subsidence so that the alluvium may be preserved as stratigraphic sequences. Other important environments of deposition are interior drainage basins where terminal riverine cones may build up.

The major features of the alluvial environment are **channels** and **floodbasins**, joined by a transition zone of raised **levees** and flood channel routes termed **crevasses**. Channel segments may be **braided** or **meandering** (with every intergradation between) and carry coarser sand- and gravel-grade detritus in bedforms confined within the channel banks. Flood basins receive silt- and mud-laden waters during periods of greater than bankfull discharges.

The Quaternary sea-level oscillations and climatic changes have given rise to a complex pattern of river adjustments involving incision/aggradation cycles, changes in sediment load and changes in channel morphology. Although those adjustments provide valuable clues as to the dynamics of alluvial systems, they conspire to prevent a 'normal' stratigraphic sequence being laid down in sedimentary basins such as might have been deposited during geological periods when ice sheets were absent.

15b Physical processes

Straight channels, with **thalweg** meandering between lateral bars (Fig. 15.1), are rare in nature. Laboratory studies show that such channels are quickly disturbed by a perturbation (see the useful review discussion by Callander 1978) to become meandering. Should the channel banks be made of highly cohesive muds, however, then the channel may remain straight, as in the very lowest reaches of the Mississippi River. (Fig. 15.2, Ch. 19). Laboratory and field evidence (Schumm & Khan 1971, Schumm *et al.* 1972) indicates that straight channels tend to occur on very low slopes (Fig. 15.2).

Meandering channels have always held a special fascination for scientists. Measurements of meander wavelength (λ_m) and channel width w (Leopold & Wolman 1960) show a near-linear relationship between the two variables, expressed by

$$\lambda_m = 10.9\, w^{1.01}$$

(Standard error $\approx$ 0.3 log units) (15.1)

Since channel dimensions are adjusted to carry frequently occurring water discharges it is not surprising to find that λ_m is also related to the mean annual discharge, $\bar{Q}$, by

$$\lambda_m = 106\, \bar{Q}^{0.46}$$

(Standard error $\approx$ 0.05 log units) (15.2)

where the units are imperial (Carlston 1965).

The basic dynamics of meanders (Appendix 15.1) lead to processes of erosion on the concave (outside) parts of bends and deposition on the convex (inside). Erosion of the cut bank depends upon the structure of the bank and the properties of its constituents, whether they be non-cohesive, cohesive or composite. Bank undercutting by the river leads to collapse along planes formed by soil shrinkage. Clay banks may fail along rotational slides. In *both* cases failure and collapse may tend to occur during falling river stage in response to changing pore-water levels and pore pressures.

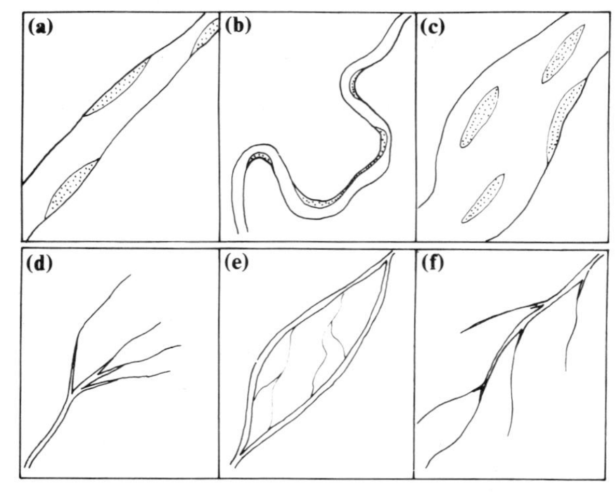

Figure 15.1 Sketches to show the three main types of river channels and channel associations: (a) straight with lateral bars, (b) sinuous with point bars, (c) braided with in-channel bars, (d) distributive, (e) anastomosing, (f) tributive.

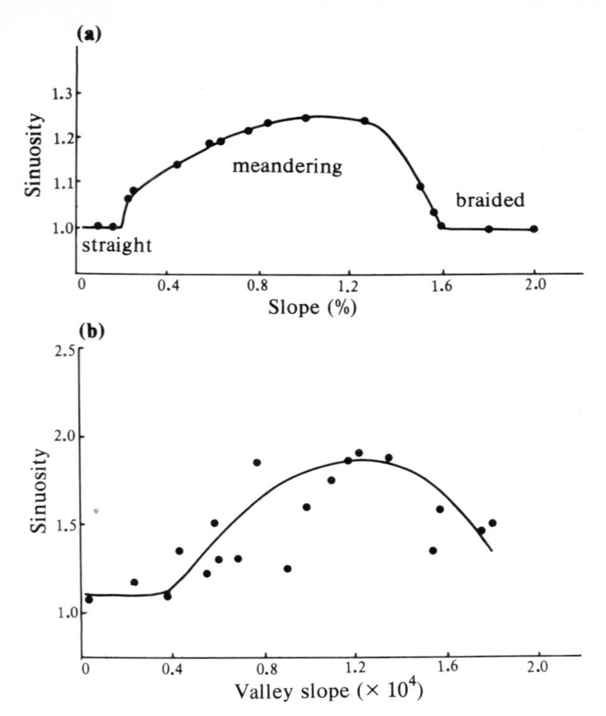

Figure 15.2 (a) Experimental relation between valley slope and sinuosity. (b) Field relation between valley slope and sinuosity for the Mississippi River between Cairo, Illinois and the Head of Passes, Louisiana (both after Schumm *et al.* 1972). Some uncertainty must arise from the form of the best-fit line in (b).

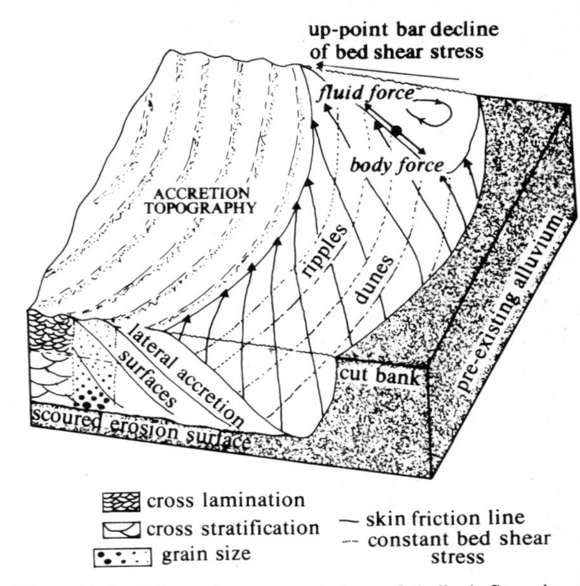

Figure 15.3 Schematic representation of helical flow in a meander bend, the balance of forces acting upon sediment grains, and lateral changes in grain size and bedforms. Dotted lines mark grain transport paths (after Allen, 1970b).

Lateral deposition occurs on the insides of bends, on point bars, as a direct result of the **helical flow cell** set up by water flow through a curved channel (Figs 15.3 & 5, Appendix 15.1). Due to the helical flow, the bed shear stress vector deviates by a small angle δ from the mean flow direction (Fig. 15.4) so that near-bed sediment grains tend to move inwards. The deviation angle is small; by theory and experiment (Rozovski 1960)

$$\tan \delta = \frac{11y}{r} \qquad (15.3)$$

where y is the local flow depth and r is the local radius of curvature of the bend. The actual path of a moving bedload grain depends upon the balance of the frictional, fluid and gravity forces acting on it. For steady, uniform flow the net rate of transverse sediment transport must be zero, or a stationary bed cannot be attained. To counterbalance the inward transport a transverse bed slope develops. This is the point bar surface where grains travel parallel to the mean flow direction, around the 'contours' of the point bar surface (Allen 1970c, Bridge 1976, 1977). The depth profile across a point bar is given by

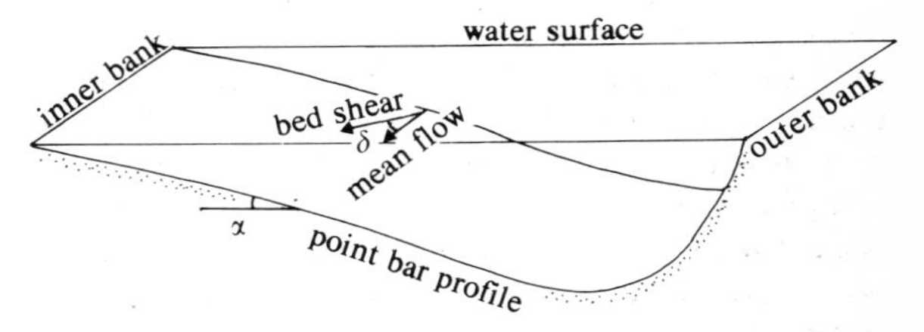

Figure 15.4 Definition diagram for flow in an open channel bend (after Bridge 1976).

$$y = y_{\max} \left(\frac{r}{r_t}\right)^{11 \tan \alpha} \qquad (15.4)$$

where r_t is the radius of curvature at the thalweg (of depth $y_{\max}$) and $\tan \alpha$ is the dynamic friction coefficient. The grain size d varies across the point bar as

$$d = \frac{3\tau}{2 \tan \alpha \Delta\sigma} \qquad (15.5)$$

where τ is the bed shear stress and $\Delta\sigma$ is the density difference of quartz and water. Thus, given a wide range of available grain sizes, d will vary directly with bed shear over the point bar, causing a general lateral fining from point bar toe to point bar top. Variation of bed shear stress also leads to the development of distinct bedform suites on the point bar surface. Current ripples, dunes and bars are the commonest bedforms observed.

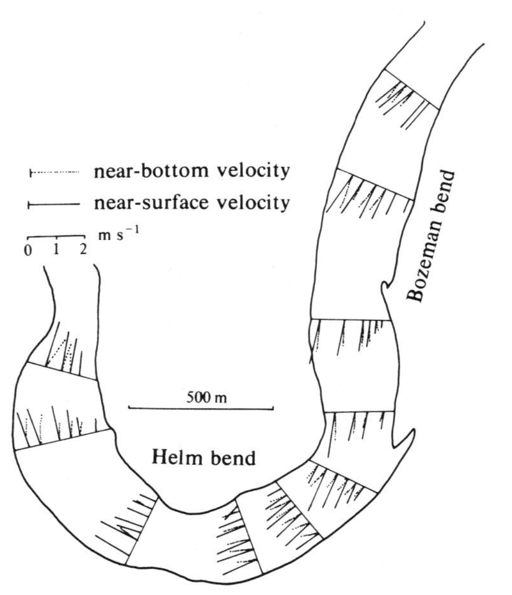

Figure 15.5 Velocity vectors measured around two bends of the River Wabash, Illinois (see Plate 5) to show surface and bottom components of the helical flow pattern (after Jackson 1975).

The migration routes of meanders (Fig. 15.6) may be studied from aerial photographs since migration leads to partial preservation of the meander form as **swale** and **ridge** surfaces. Periodic accretion of sediment on to the point bar or flow expansion down stream causes **scroll bars** to form (Hickin 1974, Nanson 1980; Plate 4) which will ultimately define a single arcuate ridge within a meander loop. Periodic cutoff leads to a renewed cycle of reach meander growth with an irregular trend (traced out by the older scroll bar ridges) towards increasingly tight bends (Hicken 1974, Fig. 15.6). Prolonged occupation of an area by a meandering river thus leads to the production of a complex **meander belt.**

The dynamics of braided rivers are more complicated than those of meandering rivers. During flood stage the relatively low sinuosity reaches contain in-channel bedforms that migrate downstream (e.g. Coleman 1969). Lowering of stage leads to bedform emersion and the 'braided' appearance becomes evident. Falling stage modifications lead to partial bedform erosion, bedform superimposition and limited lateral accretion as subsidiary reaches meander around braid 'islands' and prograde bars into deeper-water reaches covered with dune bedforms (N. D. Smith 1971, 1974). Most studies of braided rivers have taken place at low stage; hence there is only limited understanding of the cycle of erosion and deposition that occurs as stage rises and falls.

The causes of braiding or meandering behaviour remain obscure. The well known slope dependence of braiding (Fig. 15.2) cannot be the whole story since bank stability, discharge variability and sediment load characters clearly play an important role in determining whether or not a reach will meander or be of low sinuosity. Stable banks will clearly contain more cohesive sediment, such as clay or silty clay, than sand. If the river is transporting an appreciable volume of suspended fines, then the floodplain should contain dominantly fine-grained sediment. By defining a parameter M that expresses the amount of silt–clay in the channel perimeter, Schumm (1960, 1963a) was able to show a high degree of correlation in Great Plains rivers between M and both width:depth ratio ($w:y$) and sinuosity (P). Thus

$$w:y = 255\,M^{-1.08}$$

(Standard error = 0.2 log units) (15.6)

$$P = 0.94\,M^{+0.25}$$

(Standard error = 0.06 log units) (15.7)

These equations imply that changes in the nature of sediment load, and therefore of bank materials, should radically affect channel sinuosity. Confirmation of this comes from areas such as the Riverine Plain of southeastern Australia (Schumm 1968b) where a change from braided and meandering sand-bed streams to meandering suspended-load streams has occurred in response to a general decrease in runoff in the past 15 000 years. Increased temperature and humidity after the last Ice Age caused vegetation growth and substantially reduced the amount of coarse sediment liberated from drainage basins.

Perhaps the least understood aspect of river sedimentation dynamics is the out-of-channel **vertical accretion** process that occurs on levees, crevasse splays and in flood basins. Initial floodwater may break through a levee via a channel leading to a crevasse splay (Fig. 15.7) or spill into the floodbasin as a sheet flood. The deceleration of this floodwater leads to a gradual deposition of progressively finer sediment further from the channel margin. Floodbasin infill will be followed by gentle downvalley flow of the shallow sheet of floodwaters. Continued flood discharges from the channel will deepen and speed up the floodbasin flow so that a situation such as that shown in Figure 15.8 can develop, perhaps leading to widespread deposition on levees as a result of sediment settling out from periodic flow vortices at the channel margins (J. R. L. Allen 1970b).

The net effect of flooding processes in meandering

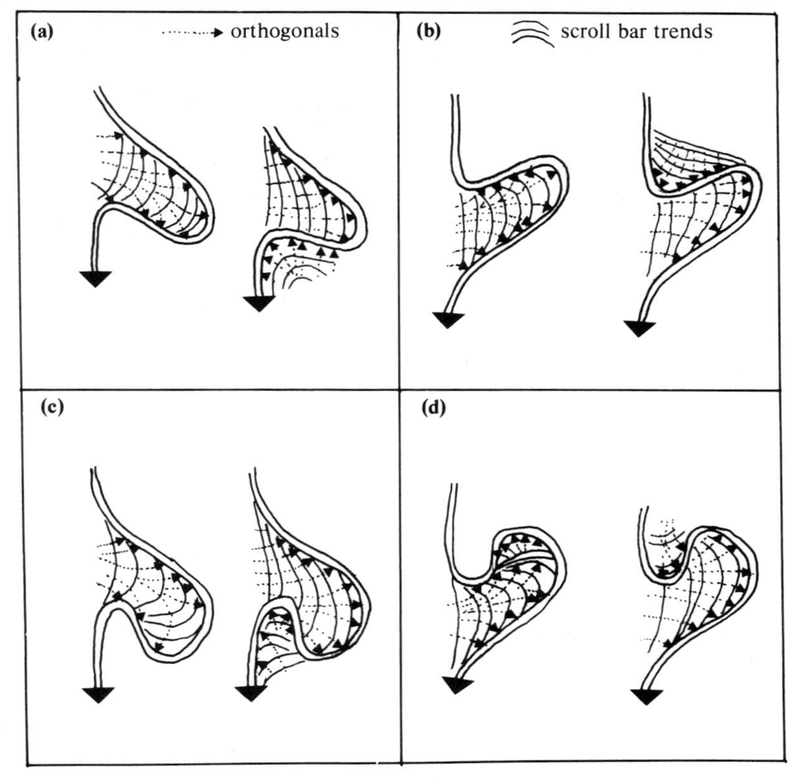

Figure 15.6 Schematic representation of meander migration routes based upon examples analysed from the Beatton River, British Columbia (after Hickin 1974).

rivers is the production of an **alluvial ridge** whose topography of levees and abandoned meander loops stands above the general floodplain level. The mean net deposition rate r at any distance z from the edge of the meander belt is given generally by

$$r = a(z + 1)^{-b} \qquad (15.8)$$

where a is the maximum net deposition rate at the edge of the channel belt, and b is an exponent which describes the rapidity with which the rate of deposition decreases with distance from the meander belt (Bridge & Leeder 1979). The constants in Equation (15.8) will vary according to factors such as climate, river size and sediment load.

Since river floods are periodic, newly deposited sediment is at once acted upon by early diagenetic processes, including soil formation. Rapid vegetation growth in humid climates leads to the formation of peat beds separated by thin partings of flood-derived muds and silts. Both soil formation and peat formation will be encouraged by slow sedimentation. Well marked horizonated soils are generally poorly developed in areas of active floodplains because of the rapidity of sediment deposition (Leeder 1975a).

A final aspect of river behaviour concerns the large scale movement of a river course, termed **avulsion**. The

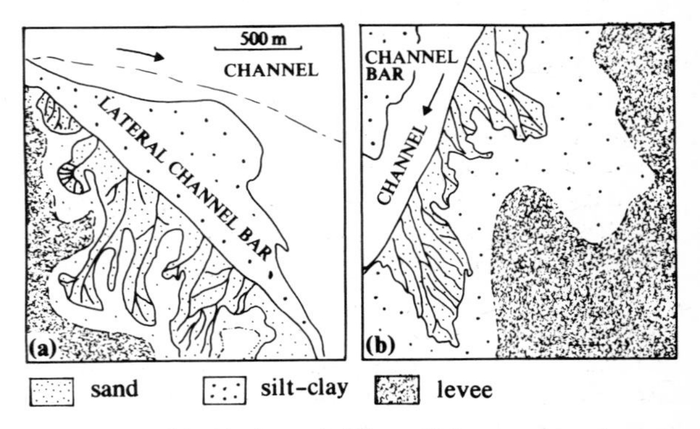

Figure 15.7 (a) Single and (b) multiple crevasse channels which breach the levees that border the River Brahmaputra, India (after Coleman 1969).

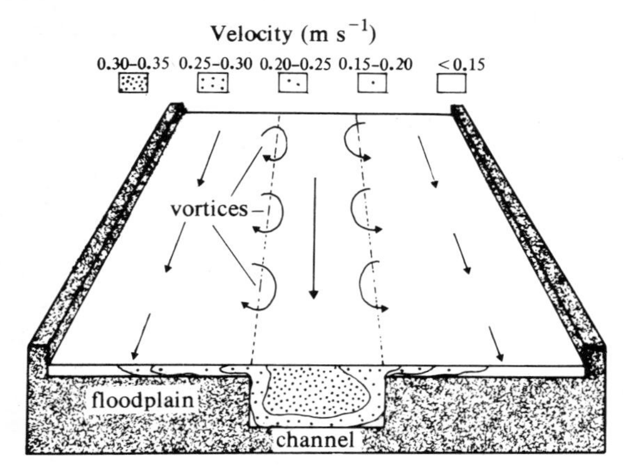

Figure 15.8 Pattern of fluid motion and velocity distribution in an experimental river (~0.45 m wide) flowing down its channel *and* over the surrounding flood basins (after Allen 1970b).

process occurs in meandering and braided rivers and is recorded by abandoned channel belts preserved on floodplains. The periodicity of avulsion appears to be of the order of 10^2–10^3 years. The diversions are actually gradual but can be considered instantaneous compared with the recurrence time of avulsion. Avulsion is more likely with greater relative elevation of the alluvial ridge since the process involves flood escape water seeking a gradient advantage over the old course. Avulsion may be related to very large-magnitude (rare) floods, but it is also known from tectonically active areas that vertical crustal movements play an important role in both initiating and controlling avulsion. Thus diversions are often established along floodplain lows because of fault subsidence in the basement.

15c Modern river plain facies

The simple subdivision of alluvial plains into low-sinuosity (braided) channel, meandering channel and floodplain hides a wide range of facies types. We shall now examine various examples to illustrate this range.

The Donjek River, Yukon (Williams & Rust 1969) is a sand–gravel bed braided river fed by glacial meltwater. Sediment grains range from clay to coarse gravel and are poorly sorted. A hierarchy of channels is present (Fig. 15.9) in response to a continuous cycle of channel migration, abandonment and fill by finer-grained sediments. The active part of the braided system lacks vegetation and shows longitudinal gravel and transverse sand bar bedforms. The higher, older channel levels are partly or completely vegetated with continuous low-strength water flow only in the main channels. These levels are infilling by a process of vertical accretion following flooding. A detailed facies model for the structure of coarse-grained, braided stream alluvium has recently been proposed by Bluck (1979).

Sand-bed braided rivers are exemplified by the rivers

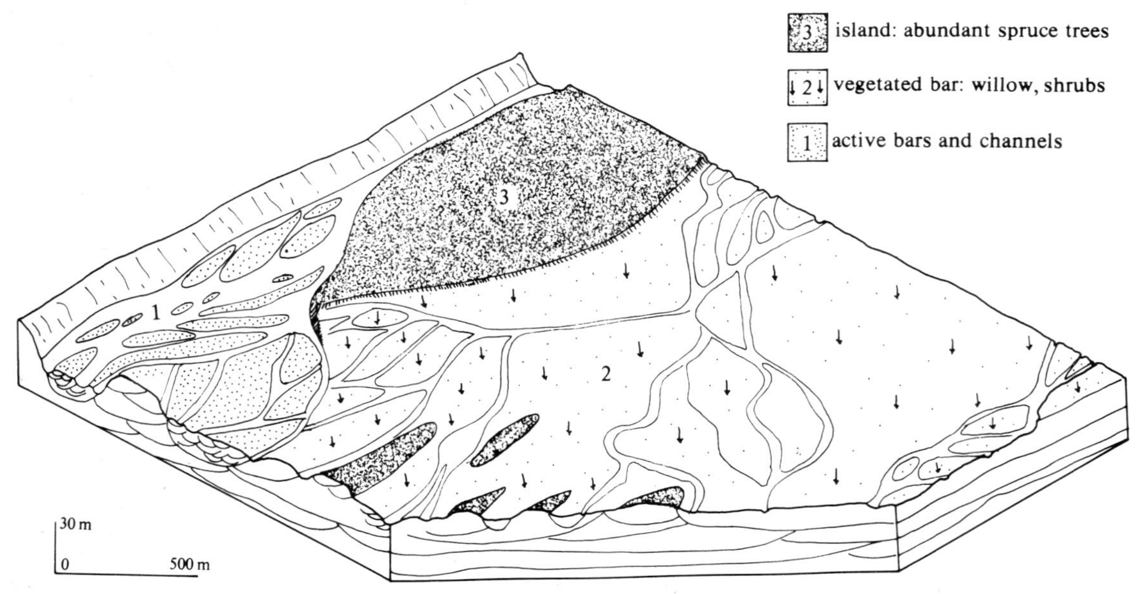

Figure 15.9 Morphology of the braided Donjek River floodplain, Alaska, to show the hierarchy of bars and reaches (after Williams & Rust 1969). The bars are mostly of longitudinal types.

South Saskatchewan (Cant & Walker 1978), Platte (N. D. Smith 1971) and Brahmaputra (Coleman 1969). Large areas of very shallow-water sand flats in the first two examples are formed by deposition around cross-channel bars which form in local flow expansions. The cross-channel bars deposit thick sets of planar cross stratification which diverge substantially from the main direction of channelised flow. Major channels flow around the sand flats. These contain sinuous crested dunes which deposit trough cross-stratification at low levels in the sequence. Vertical sequences of facies may be related to sand flat development or to channel processes (Fig. 15.10).

The braided Brahmaputra river is characterised by rapid short-term channel migrations of up to 1 km a^{-1} (Fig. 15.11). The most significant changes occur during falling stage as sediment is deposited as bars within the channel, causing a change in local flow direction and thalweg migration. As river stage rises there is a gradual growth of bedforms from small dunes (0.3–1.5 m high) to giant sandwaves (up to 13 m high) with superimposed dunes (Fig. 15.12). A complex series of vertical sequences may be generated in such a braided system because of rapid thalweg migration. Giant low-angle cross sets (caused by sandwave migration) and downward-dipping cross sets (caused by dune migration down the sandwave lee faces) should dominate these successions. The well developed flood basin sequences, with their proximal levees and crevasse splays, clearly have a very low preservation potential because of the rapidity of lateral channel movement.

The progressive combing of 'wet fan' type alluvial surfaces by braided streams is well illustrated by the Kosi River system, India (Fig. 15.13, Gole & Chitale 1966).

Detailed studies of dominantly sand-grade and mixed sand/gravel meandering rivers have been made in recent years, including the Endrick (Bluck 1971), Wabash (Jackson 1976b), the South Esk (Bridge & Jarvis 1976, 1982) and the Congaree (Levey 1978). In each of these meandering rivers fining-upwards sequences produced by lateral point bar accretion occur in the classical way (Fig. 15.3) *only* on the downstream parts of bends where spiral flow is fully developed (Figs 15.14 & 15). A significant discovery is that coarsening-upwards sequences tend to develop in the upstream part of the point bar where spiral flow is developing from the bend up stream (Fig. 15.15). The preservation potential of these coarsening-upwards cycles is limited since most meander migration is dominated by expansion, rotation and translation components (Fig. 15.16), leading to only a small portion of the point bar volume being composed of coarsening-upwards facies. Intermediate cycles of point bar sands, showing approximately constant grain size through 60% of the succession, are much commoner within meander loops. Bedforms, including sinuous dunes, sandwaves and transverse bars (terminology of Jackson 1976b) migrate

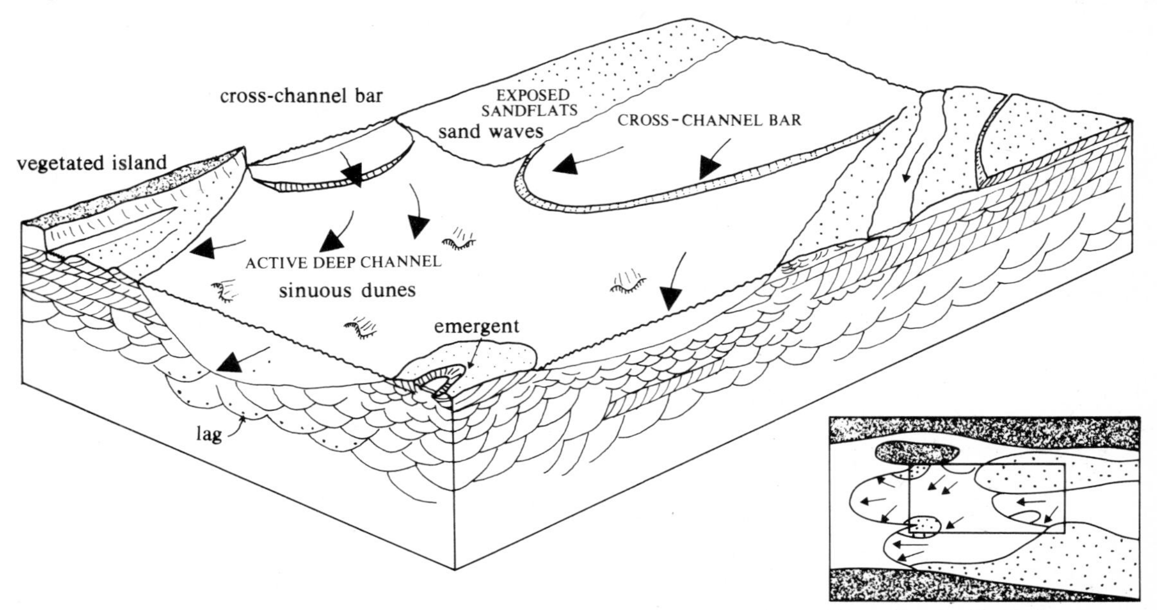

Figure 15.10 Generalised facies diagram based upon processes active in the Saskatchewan River, Canada. Vertical sequences are produced by sand flat development and by channel aggradation. The facies are dominated by trough cross-sets produced by sinuous crested dunes and by planar cross-sets produced by cross-channel bars and 'sand waves' (after Cant & Walker 1978). Note the various vertical sequences that arise within the deposited channel sand-body.

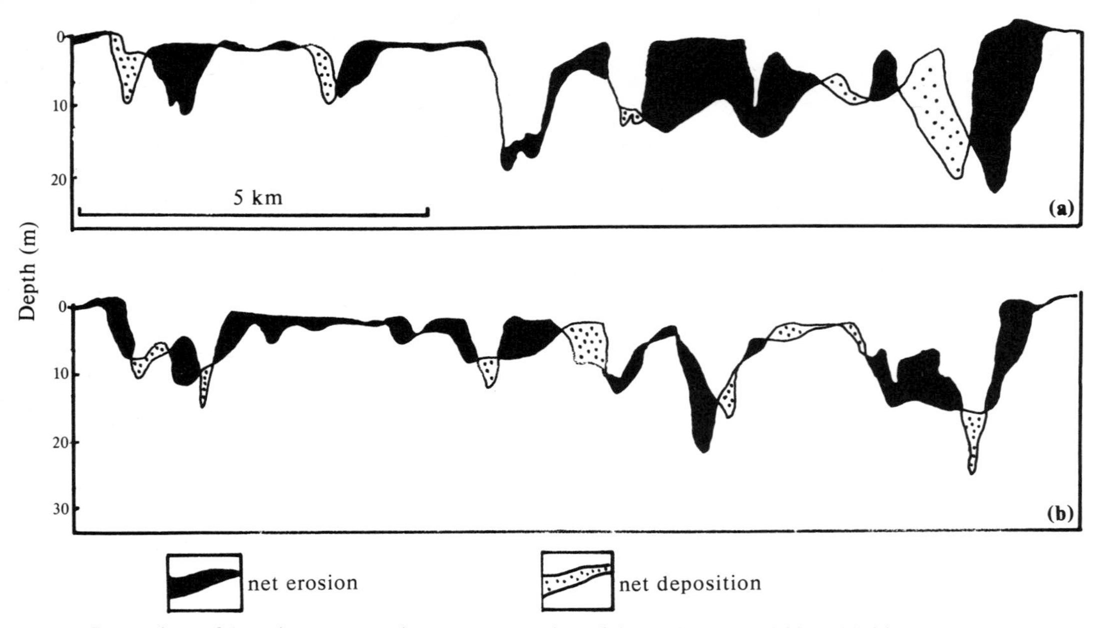

Figure 15.11 Comparison of 4 sections measured across two reaches of the Brahmaputra River with (a) a gap of 11 months and (b) a gap of 21 months. Note the preponderance of erosion. (After Coleman 1969.)

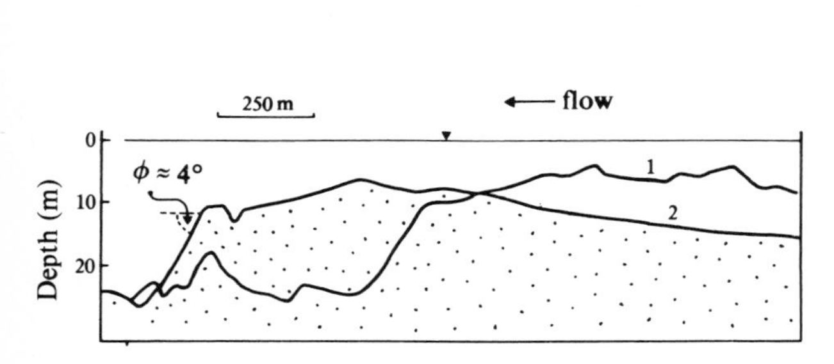

Figure 15.12 Successive fathometer traces of a large sand wave in the River Brahmaputra with a height of over 15 m. Traces measured with an interval of 24 hours (after Coleman 1969). Superimposed small dunes on the sand wave have been omitted. Note the small leeside angle to the sand wave, indicating that the bedform should be dominated internally by downward-dipping cosets of cross-stratification formed as small dunes migrate down the sand wave lee.

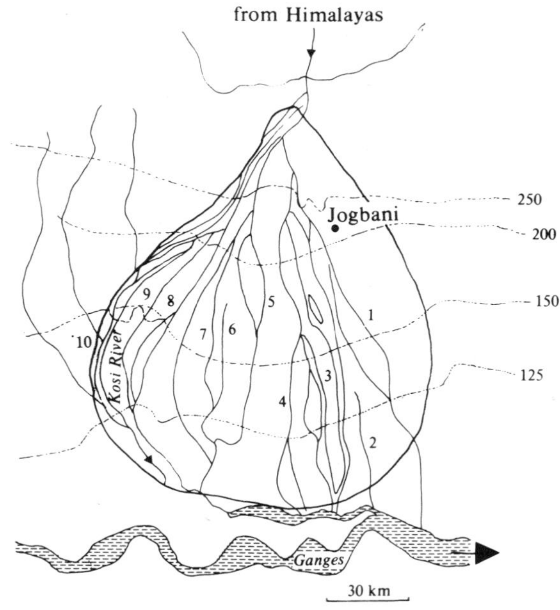

Figure 15.13 The Kosi River fan, Bihar State, India. 1–10 indicate successive positions of the channel from 1730 to the present day; note the systematic east to west combing. The dotted contour lines are given in feet. (After Gole & Chitale 1966.)

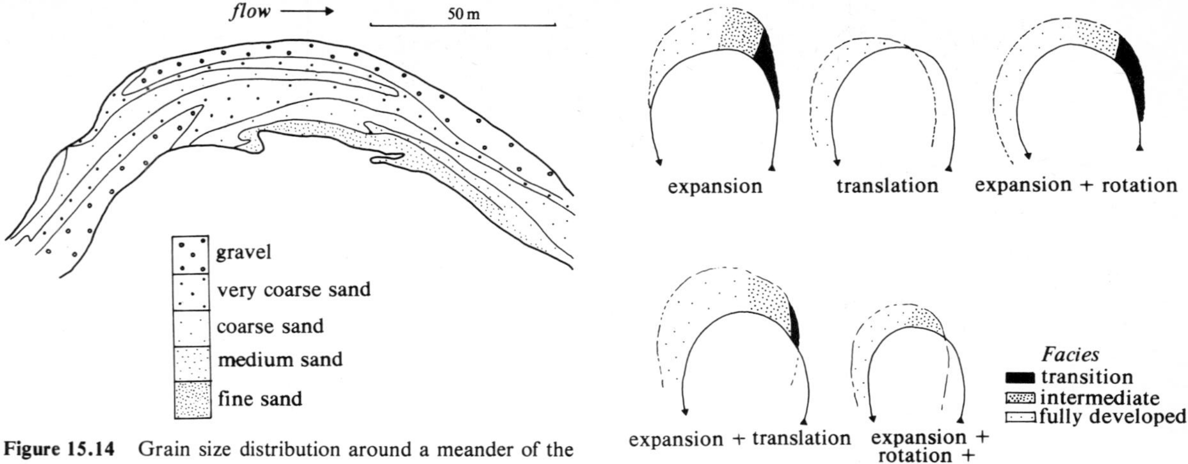

Figure 15.14 Grain size distribution around a meander of the River South Esk, Scotland, to show the round-bar fining on the inner bank, the up-bar fining in the downstream section, and the up-bar coarsening in the upstream section (after Bridge & Jarvis 1976).

Figure 15.16 Preservation of depositional facies in idealised migrations of moderately curved meander bends (see also Fig. 15.15) (after Jackson 1976b).

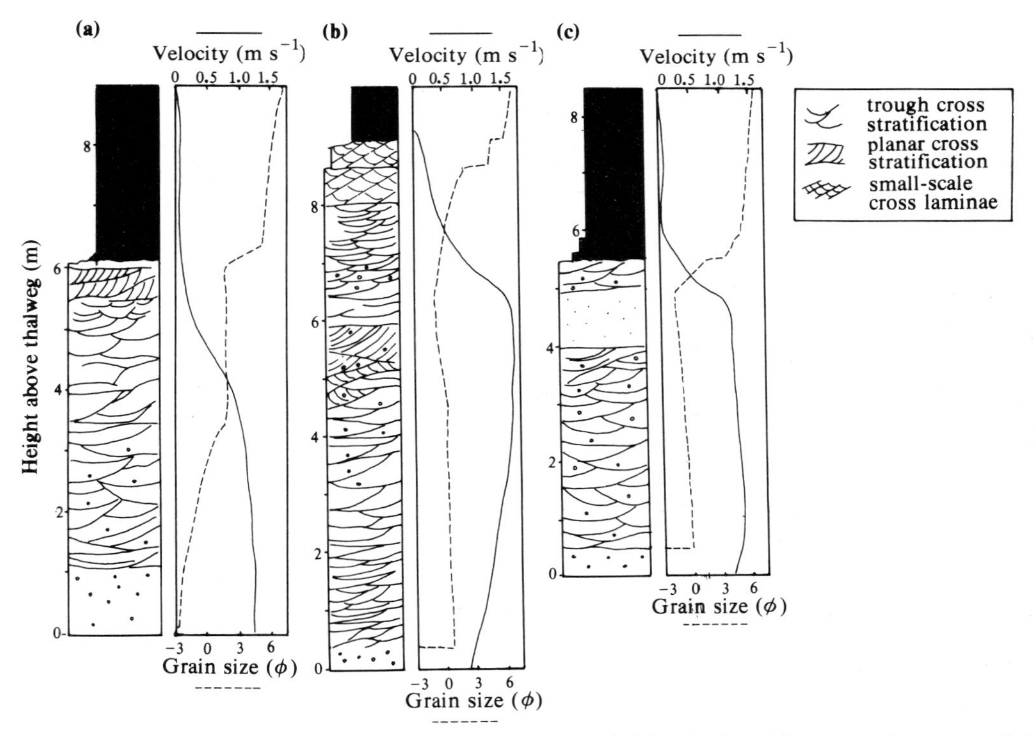

Figure 15.15 Logs to summarise vertical measured sections dug into (a) fully developed flow areas (note upward fining), (b) transitional flow areas (note slight upward coarsening), and (c) intermediate flow areas of meander bends in the River Wabash, Illinois, USA (after Jackson 1976b). See also Figure 15.16 and Plate 5.

around the point bar surfaces, giving rise to dominantly large-scale trough cross stratification in trenches dug into the point bars (Fig. 15.15). In many cases flow expansion in the downstream parts of the meander bend at high stage causes the development of scroll bars (Plate 5, p. 130) which migrate up the point bar slope, depositing large-scale planar cross-stratified sets (Jackson 1976b). Repeated accretion of these large scroll bars is responsible for the swale and ridge topography seen on many meander loop surfaces. True flow separation on the downstream portions of meanders is common in small river channels (but see Nanson 1980), being best shown in tidal flat gullies (Ch. 21). Trenches through certain Mississippi point bars show well developed, upstream-directed, cross laminations evidently produced by gentle upstream flow within a separation eddy. Flow may also be routed over the point bar upper surfaces during high stage flows, causing development of **chute bars** at the downstream end of chute channels.

As noted previously the stratigraphy of modern alluvium is dominated by the effects of Pleistocene sea-level oscillations and climatic changes. For example, sections across the Holocene alluvium of the Mississippi valley reveal a general fining-upwards trend passing from braided valley alluvial sands and gravels upwards into the relatively fine-grained alluvium of meander belt and flood basin (Fig. 15.17). These upward changes were caused by rising sea level causing aggradation and to the gradual disappearance of the continental ice sheets, causing a decrease in runoff and in coarse sediment supply.

As a final point we can briefly discuss the main controls on the architecture or stratigraphy of alluvium. The relative density of channel sand deposits and their connectedness will depend upon:

(a) the frequency of channel migration and avulsion;
(b) the rate of floodplain accretion that balances crustal subsidence;
(c) the ratio of channel belt width to floodplain width.

Computer modelling (Bridge & Leeder 1979) of channel behaviour, with due account taken of compaction and tectonic movements, overcomes some of the problems left by the Quaternary sea level and climatic oscillations noted above. The simulated alluvial successions so produced may become important tools in the interpretation of subsurface alluvium in oilfields, where porous and permeable channel deposits are set in a 'matrix' of impermeable floodbasin fines.

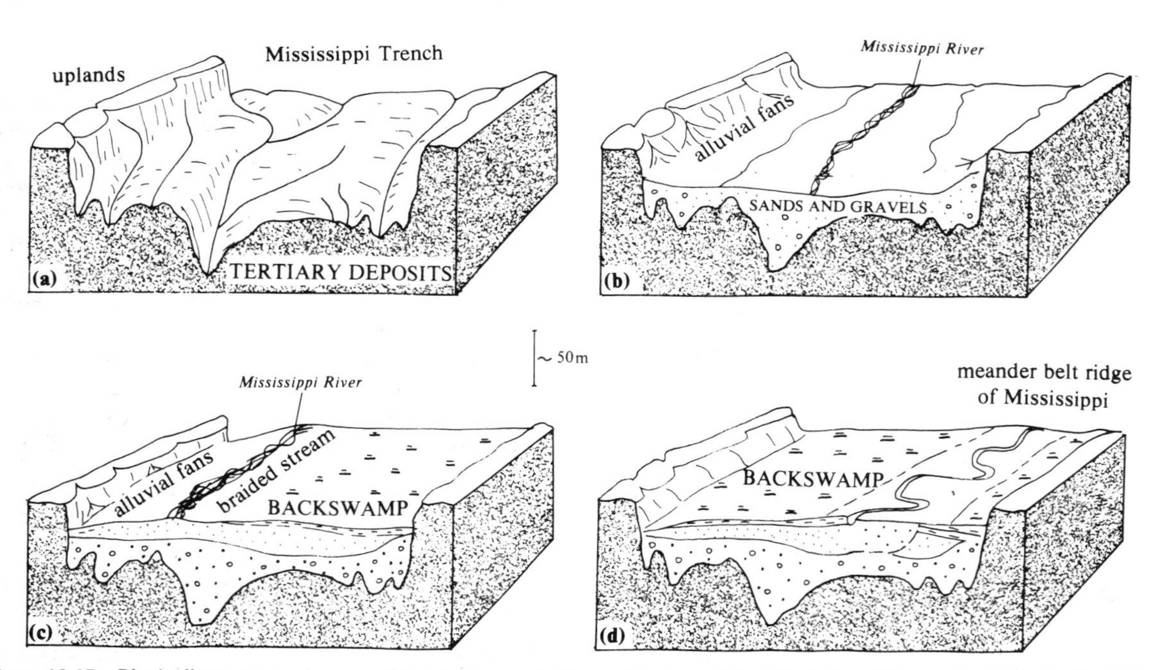

Figure 15.17 Block diagrams to show the late Pleistocene to Recent history of the Mississippi River (after Fisk 1944). (a) Late Pleistocene entrenchment at sea level minus 130 m. Braided river carries gravels to the Gulf of Mexico down a high slope. (b) Holocene aggradation at sea level minus 30 m. (c) Holocene aggradation at sea level minus 6 m. (d) The modern Mississippi meandering channel deposits sand, silts and clay. Note the final upward-fining sequence.

15d Ancient river plain facies

The identification of ancient alluvium and its division into channel and floodplain facies is the first prerequisite. The distinction between braided and meandering channel alluvium must be based on a number of criteria and is greatly facilitated by good, preferably three-dimensional, rock exposures.

In coastal exposures of the Scalby Formation, Yorkshire, England (Nami 1976, Nami & Leeder 1978), a basal regional sheet sandstone with giant trough cross sets up to 8 m thick is interpreted as a braided river deposit analogous to the sorts of successions likely to be left by a Brahmaputra-type river channel. This deposit is overlain by a perfectly preserved fluvial meander belt showing complexly interrelated meander loop deposits, (Fig. 15.18). Abundant fine-grained silty mudstones occur within the individual lateral accretion units. These horizons are frequently bioturbated and subject to soft sediment deformation. The meander belt is overlain by a thick (40 m) succession of floodbasin facies with occasional channel-deposited sandstones, mostly of meandering river origin with lateral accretion surfaces.

Lateral accretion deposits such as those of Fig. 15.18 provide *a priori* evidence of point bar deposition, but although they obviously record periodic additions of sediment onto the point bar surface (see also Ch. 21), their genesis is not entirely known (see Jackson 1978). Examples recorded in the Endrick River (Bluck 1971) occur only in the fine-grained facies at the point bar top. Good ancient analogues here are point bar sandstones from the Tertiary of the Pyrenees (Puigdefabrigas & Van Vleit 1978). Good exposures are vital for the correct identification of lateral accretion deposits. Thus in relatively poorly exposed terrain, the absence of lateral accretion units should *not* be used to dismiss a meandering origin for a sandstone facies.

By way of contrast to the above meander belt facies, the Morrison formation of New Mexico is a regional sheet sandstone. The Westwater Canyon member (Campbell 1976) comprises a sand sheet more than 100 km wide normal to the palaeocurrent direction and it is 60 m thick on average (Fig. 15.19). The sandstone body consists of coalescing fluvial channel systems that are in turn composed of still smaller channels. The channel systems are 11 km wide and 15 m deep on average whilst the individual channel deposits average 180 m wide and 4 m thick. Any one vertical section through the member is >90 % sandstone, emphasising the 'connectedness' of the various channel-deposited sand bodies and strongly suggesting deposition in the anastomosing channels and subchannels of a braided river that frequently moved laterally over its braid plain.

Let us now turn to whole-basin analysis of fluvial successions, emphasising the regional changes in alluvial facies types brought about by slope decrease and deposition. Perhaps the most spectacular example of a large-scale ancient alluvial cone complex, partly analogous to the Kosi River cone, is provided by the fill of the Devonian Hornelen basin in western Norway (Steel 1976, Steel & Aasheim 1978). This small (~2000 km^2) late-orogenic basin (Fig. 15.20) has a staggering 25 km stratigraphic fill of well exposed alluvial facies. The basin has been filled longitudinally by repeated progradation of a low-gradient alluvial cone in response to periodic basement faulting. The basin is also ringed by the coarse deposits of numerous small alluvial fan complexes (Fig. 15.20). The main basin fill is strikingly organised into about 200 basinwide coarsening-upwards cycles 100–200 m thick (see Fig. 15.20) which each pass down current from gravelly braided stream channel deposits to sandy braided channel deposits into fine-grained sheet flood and lacustrine facies (Fig. 15.20). The basin was evidently of interior type with channels petering out down slope. Such environments may be considered as a type of wet alluvial fan (see Ch. 14), although the term **terminal riverine cone** (Friend 1978) is probably more appropriate and avoids confusion with true dry fans discussed

Figure 15.18 Photograph to show an exhumed Jurassic meander loop in alternating sandstones and mudstones exposed on the foreshore north of Scarborough, England. Flow from left to right as revealed by cross laminations in the sandstone skerries. Figure (circled) indicates scale. (See Nami 1976 and Nami & Leeder 1978.)

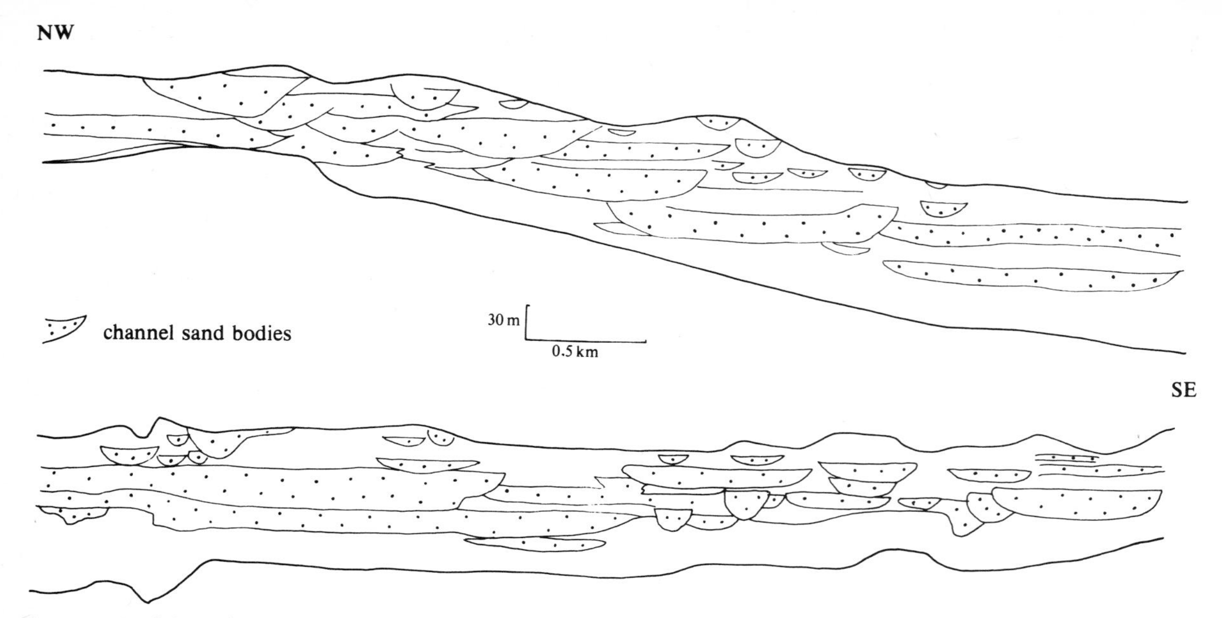

Figure 15.19 Schematic stratigraphic cross section (transverse to palaeocurrents) of the Cretaceous Westwater Canyon member to show the multistorey and time-transgressive sheet sandstone thought to have been produced by avulsing low-sinuosity channels (after Campbell 1976). Each channel sand body is a channel belt deposit made up of a number of individual channel bar deposits.

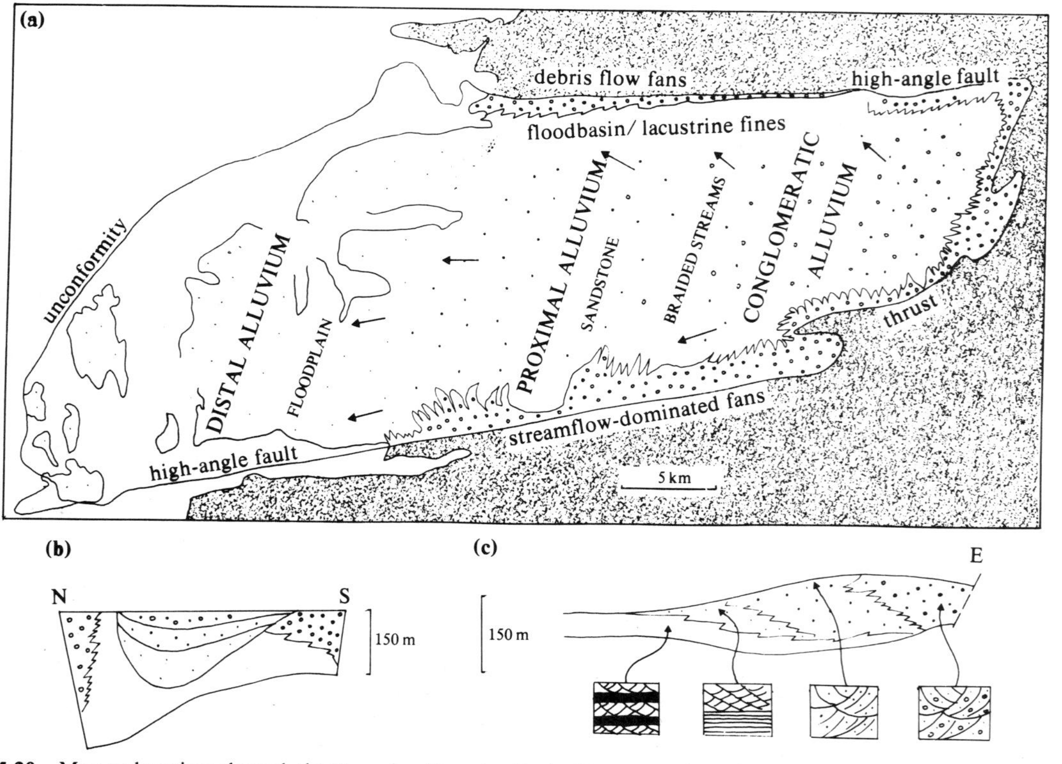

Figure 15.20 Map and sections through the Devonian Hornelen Basin, Norway to show the nature of the low angle alluvial fan with its fringing fans and down current changes in grain size and sedimentary structures. (b) and (c) refer to a single cycle of fan progradation, of which there are many present. Note the coarsening-upwards megasequences produced by fan progradation. (After Steel & Aasheim 1978.)

previously. Another such example has been discussed by Friend and Moody-Stuart (1972) from the Devonian of Spitsbergen. In a sophisticated basin analysis, based on more than a decade of field study, the authors distinguish three river systems which flowed northwards into an area of clay flats. The rivers of the eastern system were large, north- and north-northwestward-flowing, low-sinuosity rivers of braided type with abundant detrital feldspar. Rivers of the western system were small, eastward-flowing, meandering and poor in detrital feldspar. The central system flowed northwards but was similar to the western system in other respects.

15e Summary

Rivers are the natural conduits whereby detritus is transferred from the upland weathering zone into the sea. The depositional facies of rivers stand a good chance of preservation in both interior graben and in the subsiding coastal plains that border opening oceans. The channel facies of river plains comprise a variety of gravel- to sand-size sediment transported as bedforms. Sinuous meandering channels are dominated by lateral accretion processes which generate fining-upwards and coarsening-upwards deposits overlying basal scoured surfaces. Low-sinuousity braided channels are dominated by the lateral, vertical and longitudinal accretion of bars which give rise to a great variety of vertical facies sequences. Out-of-channel processes are dominated by vertical accretion of flood-derived sediment introduced from channels by decelerating overbank spills and crevasse splays. Once deposited these flood-basin fines are acted upon by soil-forming processes. Vertical successions of alluvium are generated by lateral channel migrations and avulsions combined with net crustal subsidence or progradation of alluvial cones.

Further reading

Geomorphological aspects of rivers are discussed by Leopold *et al.* (1964) and, more recently, by Schumm (1977) and Gregory (1977). Allen (1965a) is an important though somewhat dated synthesis of sedimentation processes in rivers and the mechanisms by which vertical successions may be constructed. Miall (1977) is a good review of sequence and structure in the various types of braided river alluvium. The volume edited by Miall (1978) is indispensable as the most up-to-date volume of fluvial sedimentology with a host of pertinent papers including a stimulating historical review. Callender (1978) is a useful review of the causes and effects of meandering.

Appendix 15.1 Flow around channel bends: the helical flow cell

As water flows around a bend, each water volume is acted upon by centrifugal forces which tend to throw out the volumes towards the outer concave bank. Near the surface the centrifugal force is given by wu^2/gr where w is the weight of water, u is the surface velocity and r is the radius of curvature of the bend measured outwards from a point. Note that the centrifugal force increases with increasing water depth at a point and with the square of the velocity, but decreases with increasing radius of curvature.

A water surface slope develops from the centrifugal effect, with the transverse slope inwards from the outer to inner banks. The slope dy/dz is normal to the resultant of the centrifugal force and the weight force, thus

$$\frac{dy}{dz} = \frac{wu^2/gr}{w}$$

or

$$g dz = u^2 dr/r$$

Upon integration, with the superelevation y of the water surface set at zero on the inner, convex bank, we have

$$y = \frac{u^2}{g} \log_e \frac{r_2}{r_1}$$

or

$$y = 2.3 \frac{u^2}{g} \log_{10} \frac{r_2}{r_1}$$

where r_1 is the radius of the inner bank and r_2 the radius of the outer bank. Thus we see that a small superelevation of the water surface occurs in the bend with its maximum at the outside of the bend.

For equilibrium to occur in a bend the outward-acting centrifugal force must be balanced by the inward-acting force given by the hydrostatic pressure difference arising from the transverse water slope. The spiral or **helical flow cell** (Fig. 15.3) arises because of an imbalance at depth between these forces. Thus the additional pressure caused by the centrifugal effect gradually decreases from surface to river bottom as the velocity decreases in the boundary layer. This leads to an excess of hydrostatic pressure which forces water particles to be moved inwards at these lower levels. This is compensated by an outward movement of water in the surface levels and so the helical flow cell is formed.

The essence of this theory was first proposed over 100 years ago by J. Thomson. The above derivation is taken from Leliavsky (1955).

Appendix 15.2 Palaeohydraulics

Earlier in this chapter it was pointed out that morphology and magnitude are so adjusted that channels can most efficiently transfer sediment and water. Empirical equations such as Equations 15.1, 15.2, 15.6, 15.7 may sometimes be used to reconstruct the range of magnitude of channel size and water discharge for palaeochannels.

In the simplest possible case the width of a meandering palaeochannel fill may be substituted into Equation 15.1 to derive a range of values for meander wavelength. These ranges may then be inserted into Equation 15.2 to derive broad likely ranges of mean annual discharges.

Preserved lateral accretion deposits may also be used in this way, their thickness approximating to bankfull depth (see Fig. 15.3) and their width approximating to 2/3 channel bankfull width.

More sophisticated approaches involve estimating the mean percentages of silt and clay in the channel banks and channel deposits and insertion of these values into Equations 15.6 and 15.7.

Reviews of these methods (and others) with frank appraisals of their use in reconstructing ancient river hydraulics are given by Schumm (1972), Leeder (1973), Ethridge and Schumm (1978) and Bridge (1978a). Outstanding applications of palaeohydrology to ancient deposits are given by Baker (1973, 1974).

16 Lakes

16a Introduction

There is an astonishing variety of lakes on the continental land surface, ranging from large, deep and permanent freshwater lakes (such as the North American Great Lakes, the deeper East African Rift Valley lakes and the Russian Lake Baikal) to shallow ephemeral saline lakes such as those in the western USA and in central Iran. There is a complete spectrum from deep fresh to deep saline (e.g. Dead Sea), shallow fresh (e.g. Lake Chad) and shallow saline. The major controls upon lake water dynamics (Table 16.1) include: (a) *climate*, controlling water chemistry, shoreline fluctuations, organic productivity and water temperature; (b) *water depth*, controlling lake stratification and current effectiveness; and (c) the nature and amount of *clastic sediment* and *solute* input from the lake drainage basin. Many of the world's most important lakes lie in rift valleys (East Africa, Baikal, Dead Sea) where active crustal subsidence has often encouraged the accumulation of relatively deep standing water bodies. Lake sediments in such actively subsiding regions obviously have a high preservation potential.

16b Physical and chemical processes

Water movement in lakes is controlled entirely by wind-driven waves and currents, aided in temperate lakes by seasonal density overturns (see below). The world's largest lakes are too small to exhibit a tidal oscillation, but wind-driven surface waves effectively mix the upper levels of lake water and give rise to wave currents along shallow lake margins (Csanady 1978). The size and effectiveness of lake waves depend upon the fetch of the lake winds. Observations of lake levels show that a steady wind causes a tilting of the lake water surface, the downwind part being higher than the upwind part. It can be shown (e.g. Csanady 1978) that a static equilibrium is possible if the wind stress is balanced by a surface elevation gradient of magnitude u_*/gh, where u_* is wind shear velocity and h is water depth. The effect is therefore less in deeper lakes. The surface slope may have values of between 10^{-7} (or 1 cm in 100 km) and 10^{-6} (or 10 cm in 100 km). Although these gradients are small, the sudden application or disappearance of wind stress causes lake level oscillations known as **seiches** which may further mix surface waters and cause breaking waves at the lake margin. Linear lake bodies may show reasonably fast currents (caused by direct wind stressing) in shallow marginal areas (up to 30 cm s^{-1}) with a pattern of surface eddies (Fig. 16.1) accompanied by slow compensatory flow in the lake bottom (<3 cm s^{-1}). Such surface currents might thus be able to transport very fine sands and silts as bedload and would certainly be responsible for thoroughly mixing any introduced suspended load.

Many temperate lakes show a well marked **thermal stratification** with an upper warm layer separated by a **thermocline** from deeper cold water. In spring the water of a moderately deep lake will all be at a temperature of about 4 °C. The topmost waters will be gradually warmed by solar radiation and mixed downwards by

Table 16.1 Physical processes affecting lake water dynamics (after Sly 1978)

Physical input	*Wind*	*River*	*Solar heat*	*Surface pressure**	*Gravity**
Controls	orientation, size, depth, shape and hinterland relief to lake, Coriolis force, duration	river discharge, lake shape, water temperature, sediment conc., Cor. force	latitude altitude basin depth	size of lake	
mechanisms response	shear stress waves circulation, upwelling coastal jets seiches	entrainment by jet river plumes density flows	density effects mixing, stratification, overturn, internal waves ice cover	differential pressure seiches	tidal attraction tides

*Negligible effects in all but very largest lakes.

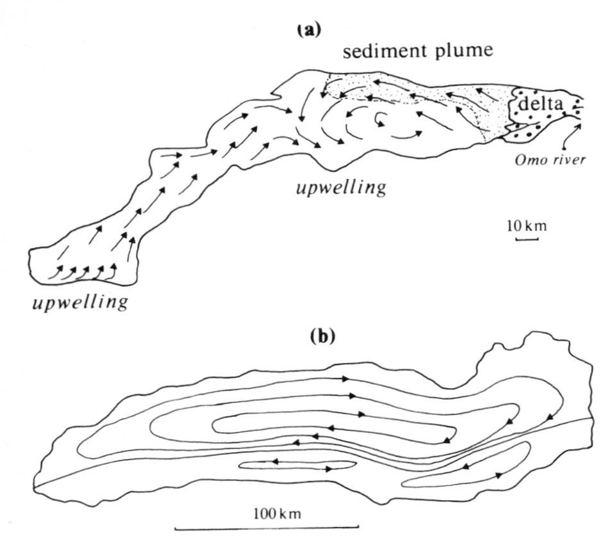

Figure 16.1 (a) Major surface currents in Lake Rudolf together with sediment plume from the River Omo. Currents caused by river inflow and southeasterly winds (after Yuretich 1979). (b) Surface currents produced by wind-driven flow in Lake Ontario (after Rao & Murty 1970).

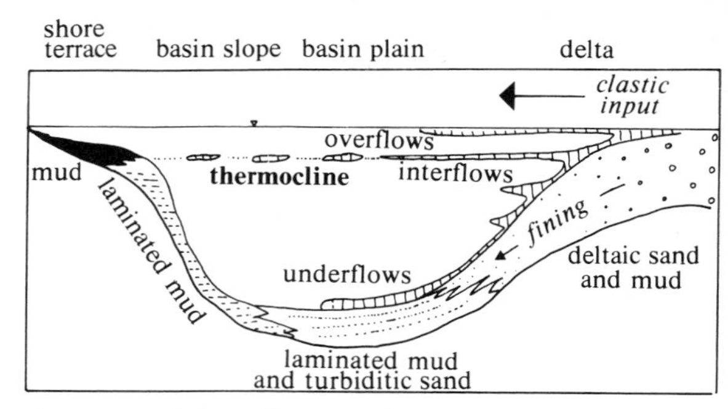

Figure 16.2 Schematic model for clastic sedimentation in a stratified temperate lake which has clastic input via a river delta (after Sturm & Matter 1978).

wind action. As heating continues the isothermal, warm surface water will become buoyant enough to resist wholesale mixing and will remain above the cold deep water, separated by the thermocline, a zone of marked temperature decrease. Most heat is thus trapped in the surface **epilimnion** until, in autumn, cooling from the water surface downwards causes density inversions and wholesale mixing of the epilimnion with the deep **hypolimnion**. Melting of winter ice causes wholesale sinking of cold surface water giving rise to the spring overturn, and the cycle begins again. In deep tropical lakes the water stratification into epilimnion and hypolimnion is permanent.

Density currents in lakes are caused by inflow of sediment-laden river waters (see Fig. 16.2). In thermally stratified lakes such as Lake Brienz, Switzerland (Sturm & Matter 1978), the density of the inflowing water may be greater than that of the lake epilimnion but less than the hypolimnion, sò that the density current moves along the thermocline as an **interflow** (Fig. 16.2). Thus, high concentrations of suspended sediment occur at this level which may then be dispersed over the lake by wind-driven circulation. Other inflows are denser than the hypolimnion and flow along the lake floor as **underflows**. The underflows bring oxygenated water into the deep hypolimnion and prevent permanent stagnation in deep lakes.

The chemistry of lake water is delicately adjusted to the input of solutes from the lake drainage basin and to the amount of evaporation. Chemical data from temperate lakes such as Lake Zürich (Kelts & Hsü 1978) show that surface waters are only slightly supersaturated during the winter and that supersaturation decreases after the spring overturn when undersaturated bottom waters are brought up to the surface and the lake water is effectively homogenised. Maximum supersaturation, with low-Mg calcite precipitation, occurs in the summer and is caused by CO_2 removal by organic phytoplankton blooms (Fig. 16.3).

A **saline lake** may be defined as one with greater than 5000 ppm of solutes (Hardie *et al.* 1978). Saline lake development occurs most commonly in closed basins, often fault bounded, with a high mountainous rim. The low basin collects runoff from the peaks but is not itself subject to rainfall. Clastic detritus is trapped on alluvial fans that fringe such basins so that only solute-laden waters and springs issue into the saline lake. Replenishment flow into the **playa** may be mainly subsurface with evaporation causing evaporite crusts to appear at the surface. A concentric arrangement of precipitates is usually seen (Fig. 16.4), with the mineral phases present being controlled by the nature of the weathering reactions in the surrounding hinterlands.

Deep permanent saline lakes such as the Dead Sea (Neev & Emery 1967) show appreciable river water influx which balances net evaporation. Most of the solute influx, however, comes from small marginal saline springs around the periphery rather than from the Jordan river itself. The lake waters are very saline (>300 000 ppm) and of an unusual Na–Mg–(Ca)–Cl type, with low sulphate and bicarbonate. The deep (~300 m) main basin is density stratified because of salinity differences. Both aragonite and gypsum are precipitating from surface waters. Natural halite precipitation does not occur today but seems to have occurred in the deep lake about 1500 years ago when the evaporation:inflow ratio was high enough to reach into the halite supersaturation field. (*NB*

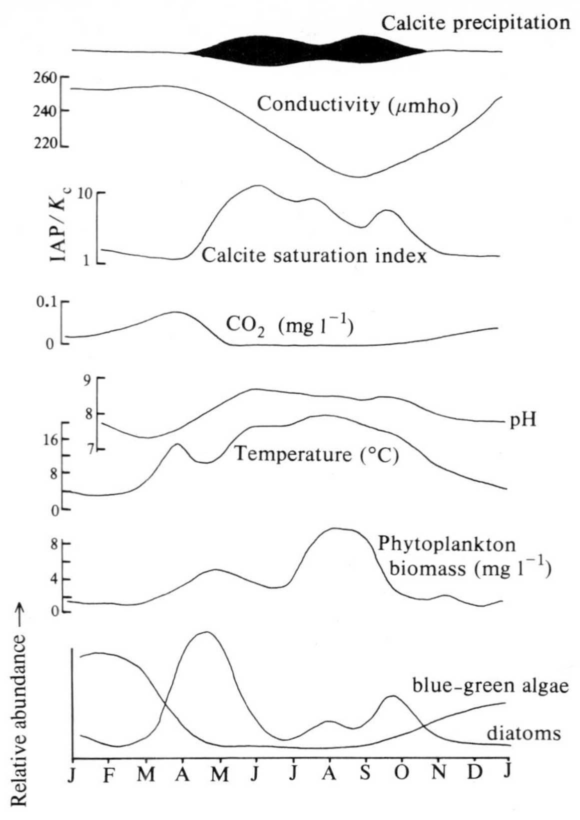

Figure 16.3 Correlation of several parameters observed in the seasonal cycle of the epilimnion water of Lake Zürich, Switzerland (after Kelts & Hsü 1978).

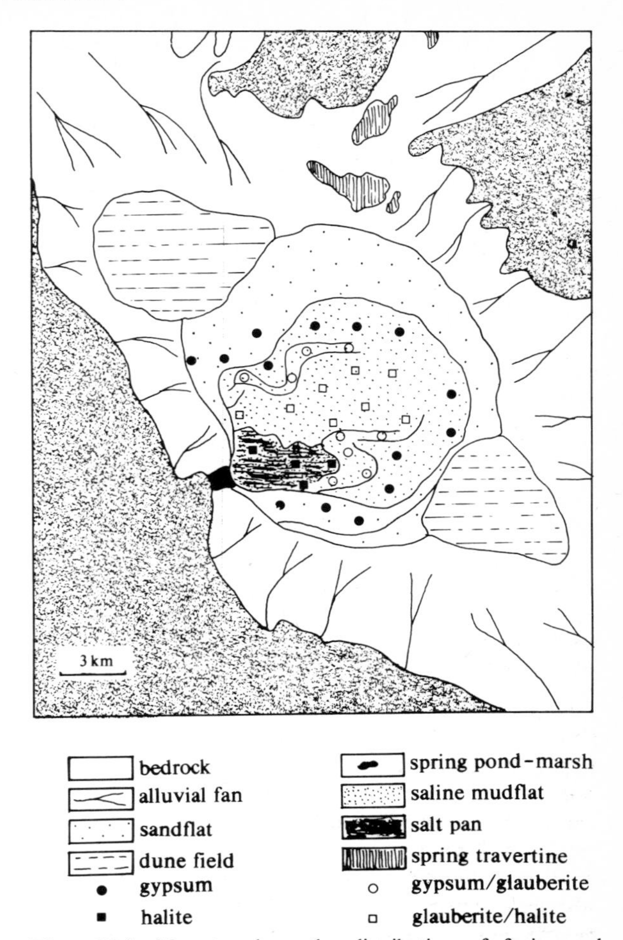

Figure 16.4 Map to show the distribution of facies and evaporite minerals in the Saline Valley playa of California. Note the roughly concentric arrangement of evaporative salts. (After Hardie 1968, Hardie *et al.* 1978.)

Recent studies (Neev 1978) indicate that the Dead Sea is now well mixed and that coarsely crystalline halite, aragonite and gypsum deposits are being precipitated and preserved on the lake floor.)

16c Modern lake facies

The facies pattern of an infilling temperate lake with thermal stratification is well illustrated by Lake Brienz, Switzerland (Sturm & Matter 1978), a 14 km long and 261 m deep lake in the Swiss Alps. Sediment deposition is entirely clastic, the detritus being introduced by rivers that enter the lake from opposite ends (Fig. 16.2). As noted above, the fluvial sediment is transported and deposited in the seasonally stratified lake by overflows, interflows and underflows, depending on the density difference between the river and lake water. High-density turbidity currents form underflows and deposit very thick ($\gtrsim$1.5 m) graded sand layers. These deposits occur only once or twice per century in response to catastrophic flooding. The underflows formed by low-density turbidity currents occur annually during periods of high river discharge and deposit centimetre-thick, faintly graded sand layers. Fine sediment introduced by overflows and interflows is mixed over the whole lake surface by circulation and it settles continuously during the summer thermal stratification to form the summer half of a varve couplet. At turnover in the autumn the remaining sediment trapped in the thermocline settles out and forms the winter half of the varve couplet. Turbidites grade laterally into thin dark laminae similar to the summer part of the varve noted above. The mechanisms of turbidite deposition and clastic summer varve formation are thus related, the two layers having a common sediment source but a differing level of introduction during periods of thermal stratification. The

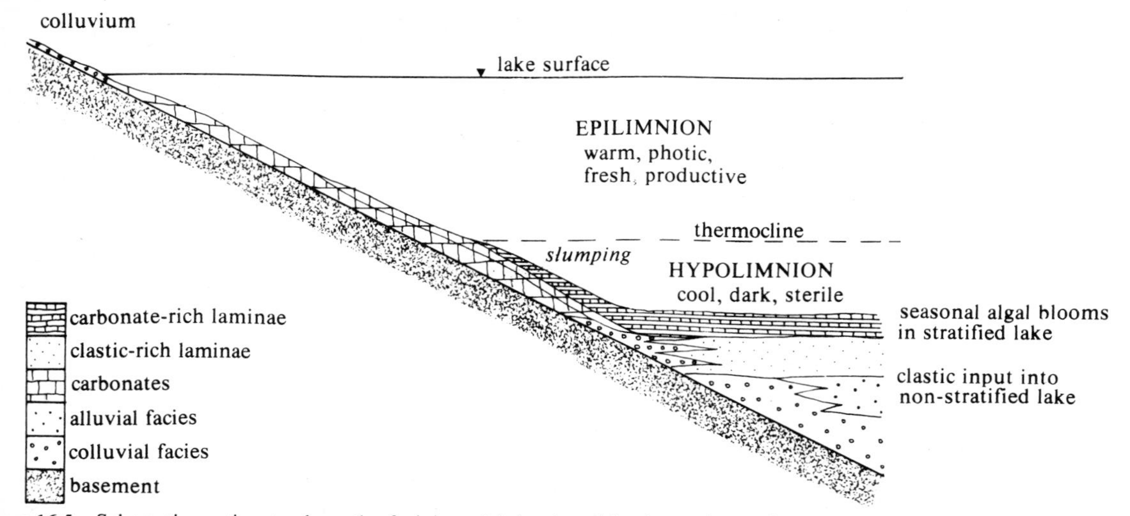

Figure 16.5 Schematic section to show the facies model developed for lacustrine sedimentation at the margins of the Devonian Orcadian Lake of northeast Scotland (after Donovan 1975). The vertical sequence (tens of metres thick) is caused by fluvial input followed by laminites produced by seasonal algal blooms as lake transgression occurred.

light-coloured winter layer is of uniform thickness over the whole basin and it forms when the lake is wholly mixed.

In contrast to the clastic facies of temperate Lake Brienz, the sediments currently being laid down in Lake Zürich, Switzerland, are mostly biogenic and chemical since flood control dams have almost stopped fluvial sediment input into the lake since about 1900. The varved sediments laid down here form in response to an annual chemical and biological cycle (Figs 16.3 & 5). The varves are present below 50 m depth but are destroyed on slopes by slow creep. Close analogies may be made between the chalky varved sediments of Lake Zürich and the Neogene lacustrine chalks penetrated by deep-sea drilling in the Black Sea (Kelts & Hsü 1978; see Ch. 26).

The varve-like sediments found in Lake Turkana, Kenya (Yuretich 1979) contain very little organic material owing to a high rate of influx of clastic sediments from the Omo River delta (Fig. 16.1) and to the fact that this relatively shallow lake (mean depth 35 m) is well mixed, oxidising and non-stratified. The fine-grained lake-bottom muds contain much montmorillonite that is believed to have neoformed in the Mg-rich interstitial waters. The mainly detrital basin fill of Lake Turkana contrasts with the organic- and chemical-rich sediments found in many other lakes in the East African Rift system. Diatoms can constitute the dominant component of many such lakes, with calcite, dolomite and siderite precipitation also common. Siderite can precipitate only if the Ca:Fe ratio is less than 20, as in Lake Kivu.

Saline lake facies are summarised in Figure 16.4, with the associated facies of alluvial origin also shown. Characteristic evaporite/clastic couplets are produced in the ephemeral and perennial lake by cycles of storm runoff from the surrounding alluvial flats, followed by evaporite precipitation. **Layered halite rock** often results, in which clastic laminae, solution or deflation surfaces separate vertically elongate crystals of halite which have nucleated from the shallow lake floor or from foundered mats of **hopper crystals** that grew on the brine surface (Shearman 1970, Arthurton 1973; see also Ch. 23).

16d Ancient lake facies

In the identification of lake facies, more than in any other single environment, emphasis must be placed on a group of attributes, with special reference to faunal and floral evidence. For example, without wider palaeogeographic evidence it may prove impossible to distinguish patchy lacustrine evaporitic facies from marine-fed playa facies located on coastal plains. Particularly severe problems arise in the interpretation of pre-Phanerozoic lacustrine facies (see Clemmey 1978).

Sediments of probable lacustrine origin make up much of the Old Red Sandstone (Middle Devonian) in the Orcadian basin of northeastern Scotland. The main Caithness Flagstone Group consists of many sedimentary cycles containing laminites (Rayner 1963, Donovan 1975, Donovan *et al.* 1974). Some of these show 0.1–1 mm laminae of alternating micritic calcite or dolomite and clastic siltstones, sometimes with phosphates. Where carbonate laminae are dominant, the

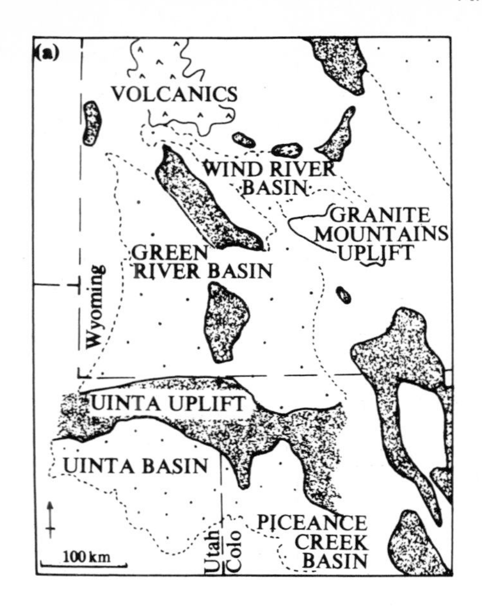

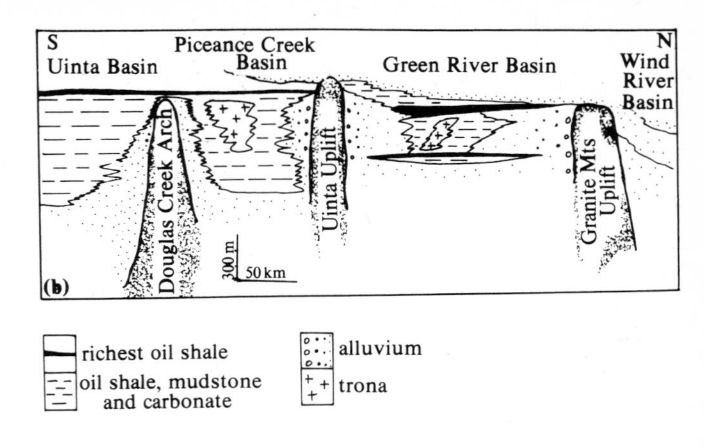

Figure 16.6 (a) Index map to show the main early Tertiary structural features of the Green River and adjacent basins. (b) Generalised lithostratigraphic scheme for the Green River Formation and coeval rocks in the area of (a). (Both after Surdam & Stanley 1980.)

clastic laminae are reduced to 0.1 mm organic streaks, as in the famous fish-bearing Achanarras Limestone. Non-carbonate laminites show alternations of siltstone and mudstone or of coarse and fine siltstones. Sub-aqueous and sub-aerial shrinkage-crack casts are abundantly present on the bases of many coarse laminae. The tops of thicker laminae show preserved symmetrical wave-formed ripples. Thicker, sharp-based and sometimes erosive sandstones cap coarsening-upwards cycles involving the above lithologies. There is a general upward trend from laminites to coarser sandstone lithologies. Palaeocurrents indicate generally western to southern derivation in the Orcadian basin.

The features noted above are interpreted as due to the interplay between a large stratified lake and streams draining into the lake margin (Fig. 16.5). The fluvial facies that increasingly dominate the succession in Caithness indicate a major fluvio-lacustrine regression with time. The lake basin was probably permanently stratified, never evaporative, and intermontane. Basinal subsidence was rapid, since over 5 km of sediment accumulated in only 10 Ma.

The Eocene Green River formation (maximum thickness 950 m) of Wyoming, Utah and Colorado (Fig. 16.6) is one of the most studied sequences of lacustrine deposits in the world and it provides an interesting contrast with the Orcadian example noted above. The formation contains the world's largest reserves of trona (Na_2CO_3) and its oil shale facies are potentially the world's largest single hydrocarbon reserve. A detailed study of the Wilkins Peak member in Wyoming (Eugster & Hardie 1975) reveals six sedimentary facies thought to have been deposited in a spectrum of lacustrine environments ranging from central playa lake to fringing alluvial fans (Fig. 16.7). *Facies 1* comprises flat pebble conglomerates of dolomite mudrock fragments thought to have been formed by strandline reworking of algally bound sediments subjected to desiccation shrinkage. *Facies 2* includes calcareous, wave rippled and cross-laminated sandstones and desiccated mudrocks of lake shoreline origin. *Facies 3* comprises desiccated mudrocks with siltstone laminae of playa mudflat origin. The oil shale lithologies of *Facies 4* comprise organic-rich dolomitic laminites and oil shale breccias. Desiccation cracks are common. The organic laminae were gelatinous, bottom dwelling, flocculant algal–fungal oozes which accumulated on a shallow, periodically exposed lake

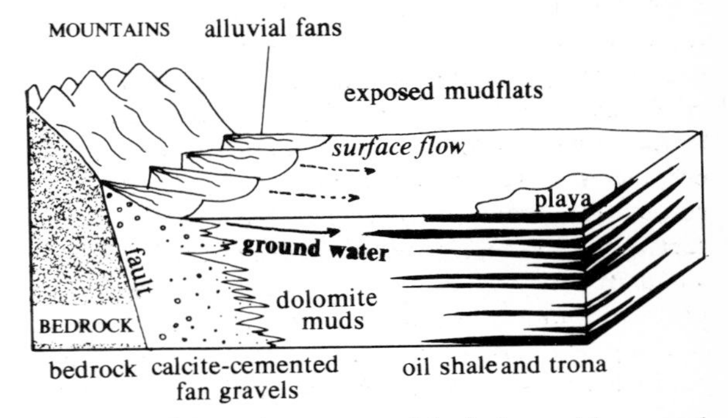

Figure 16.7 Schematic summary of the facies model proposed for the Green River Formation, Wilkins Peak member (after Eugster & Hardie 1975).

floor. High rates of organic productivity are inferred, with the playa flats of Facies 3 acting as efficient sediment traps preventing the introduction of much clastic material. The bedded trona deposits of *Facies 5* contain thin partings of dolomitic mudstone and the facies is interpreted to result from increased evaporative concentration of the playa lake that gave rise to the oil shale facies. *Facies 6* comprises immature cross-stratified sandstones with frequent channel forms. These intertongue northwards from the Uinta Mountains and are interpreted as braided alluvium sand sheets.

16e Summary

Although lakes comprise a relatively small proportion of the continental surface, the preservation potential of much lake sediment is enhanced by occurrence in tectonic depressions undergoing net crustal subsidence. Lake water and sediment dynamics are controlled by climate, water depth and the nature and amount of clastic and solute input. These factors are closely interrelated and they combine to give a diverse range of lacustrine facies models, ranging from deep, thermally stratified temperate lake models to shallow, well mixed tropical saline lake models. No purely sedimentary features are considered diagnostic in the search for ancient lake facies; rather the emphasis is on a collection of attributes combined with palaeontological features. Therefore, severe problems are likely in the identification of pre-Phanerozoic lake facies.

Further reading

Many papers of interest for both modern and ancient lake studies are to be found in the volumes edited by Matter and Tucker (1978) and Lerman (1978). A fundamental reference on the physical and chemical aspects of lakes is Hutchinson (1957).

17 Glacial environments

17a Introduction

Some 10% of the Earth's surface area is presently covered by ice which contains about 75% of the fresh water on Earth. Earlier in Quaternary times about 30% of the Earth's surface was ice covered, with vast areas of North America and Europe subjected to glacial erosion and deposition. The effects of the glaciations extended all over the globe, with the influence of ice caps upon (a) oceanic circulation (Ch. 24 & 26), (b) atmospheric circulation, (c) sea level changes, and (d) marine and terrestrial ecology. At least four previous 'Ice Ages' have been identified in the geological record: those of the early Proterozoic, late Precambrian, late Ordovician and late Palaeozoic. A discussion of the causes of glaciations is beyond the scope of this text, but in recent years it seems to have been clearly established that the periodicity of climatic changes during the Quaternary is best explained by the **Milankovitch mechanism** in which fluctuations in the Earth's orbit cause regular variations in solar radiation input to high latitudes (Hays *et al.* 1976).

The major environments of glacier ice are valley glaciers, piedmont glaciers, ice caps and ice shelves. In each environment a dynamic equilibrium between ice formation and melting is established (Fig. 17.1). Depositional products are dominated by a variety of **glacial till** facies and by the extensive **melt-out** facies of glacial outwash plains. The preservation potential of glacial environments will generally be low unless the depositional and erosional features are produced in areas of subsidence. For this reason the ice shelf environment, with its wide ring of icebergs, is probably the most likely to be preserved.

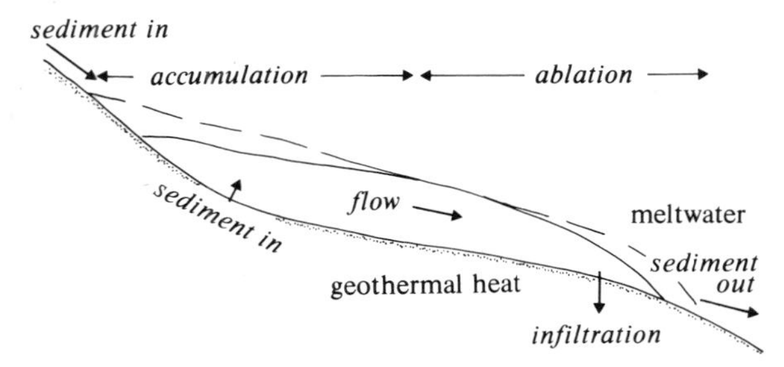

Figure 17.1 Glacier ice as an open system (after Embleton 1980).

17b Physical processes

The slow laminar flow of glacier ice (Fig. 17.2) is measured in metres per year, with values between 10 and 200 m a^{-1} for valley glaciers and 200–1400 m a^{-1} for ice caps (Paterson 1969). Ice in a glacier is composed of an aggregate of roughly equigranular crystals. Each crystal deforms internally along glide planes parallel to the basal planes of the hexagonal crystal lattices (review in Paterson 1969). The strain rate (du/dy) of creeping glacier ice is given by experiments as

$$\frac{du}{dy} = k\tau^n$$

where n is an exponent ranging between 1.5 and 3.9, k is

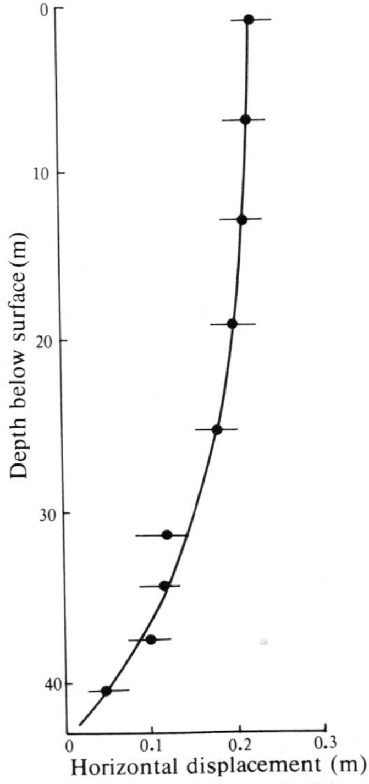

Figure 17.2 The deformation of a vertical borehole in the Saskatchewan glacier over a period of two years. Probable errors are indicated for each measurement and the line is the curve of best fit. The deformation shows a well developed boundary layer extending almost to the surface. (After Meier 1960.)

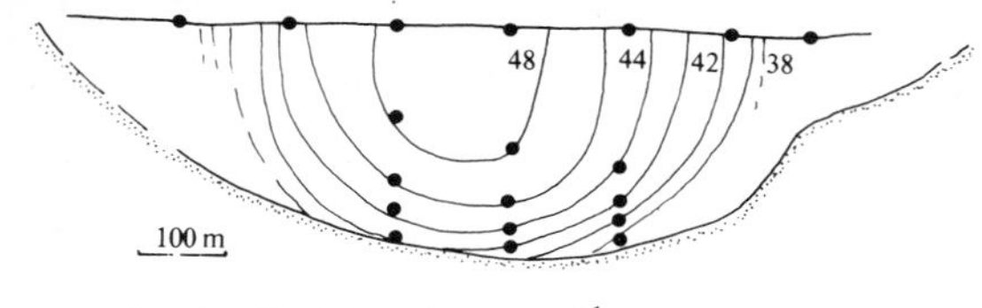

Figure 17.3 Distribution of longitudinal velocity in a cross section of the Athabasca glacier (after Raymond 1971). Note evidence for slip at the glacier base.

an experimental constant and τ is the shear stress. Field velocity profiles for ice bodies are shown in Figure 17.2 & 17.3. The shear stress arising from glacier ice sliding over a bed may be given by the tractive stress equation

$$\tau = \rho g h \sin \alpha$$

where ρ = density of ice, h = thickness of ice, α = slope.

Two fundamental types of flowing glacial ice may be identified (Boulton 1972a). **Polar ice** lies well below the freezing point at all depths (Fig. 17.4a). A condition of 'no-slip' exists at the ice/bed interface and there is a general absence of englacial or subglacial drainage. Forward motion of polar ice is therefore by internal creep alone. Glacial debris is transported within the ice, with erosion due to grinding effective only at the summits of protuberances on the bed. By way of contrast, **temperate ice** lies close to its melting point at the glacier sole, and the glacier slides over its bed (Fig. 17.4b). Up to 60% of the total movement may be caused by basal sliding. Abundant basal meltwater comes from (a) surface meltwaters let into the sole by crevasses and ice tunnels, (b) ice melted by geothermal heat, and (c) ice melted by pressure at the glacier sole. Temperate ice characteristically contains a basal zone of regelation ice, heavily charged with debris caused by pressure melting and refreezing (Fig. 17.5). Subglacial erosion is effected by both sediment-charged subglacial water and by plucking, abrasion, crushing and fracturing.

The debris found within glacial ice may come from the glacier top or sole. In the former case a variety of angular particles are supplied to the glacier margin by scree fall whence some proportion may find its way to the sole via crevasses and intraglacial tunnels. Grains are derived from bedrock in temperate ice by a plucking process around obstacles and may be inherited from preglacial sediments. Successive bands of debris within temperate glaciers record seasonal cycles of snow accumulation and debris fall from rock outcrops (Fig. 17.6). Subglacial transport processes cause frequent grain-to-grain interactions and encourage abrasion and formation of rock

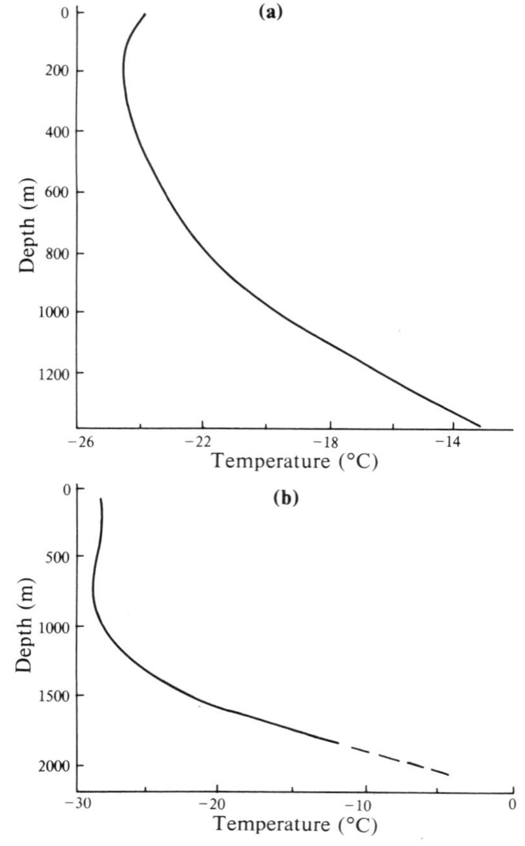

Figure 17.4 (a) Temperature profile for a polar glacier; Camp Century borehole, Greenland. (b) Temperature profile for a temperate glacier; Byrd Station borehole, Antarctica. (After Embleton 1980.)

'flour'. Thrust planes (Fig. 17.7) may arise at the glacier snout in response to compressional stresses between a slow-moving or stationary snout and the fast-moving upglacier ice. These thrusts carry subglacial debris to high positions in the ice front.

Deposition (lodgement) of glacial debris at the active ice sole will occur in response to the cohesive smearing of clay-rich sediment onto bedrock or pre-existing sediments. The process is aided by high values of pore pressure which will increase local frictional retardation. Lodgement cannot occur beneath polar glaciers because of the 'no-slip' condition. Clast orientation within lodgement till and shear foliations points to the effectiveness of the shearing/smearing process. Clasts may be orientated with their long a axis parallel to flow, as observed in many grain flows and debris flows, but with a varying proportion of imbricate intermediate b axis orientations, as seen in sediments deposited by bedload rolling mechanisms.

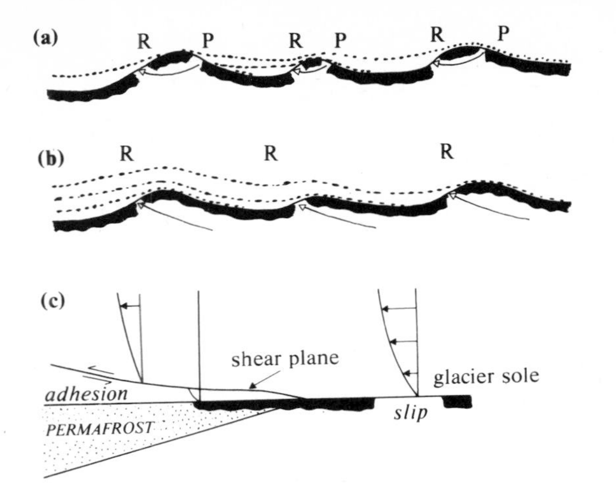

Figure 17.5 (a) The thin debris-charged regelation layer at the base of temperate glaciers. Bedrock is shaded. Pressure melting (P) occurs on the upstream flanks of protuberances whilst regelation of the meltwater so produced (arrows) occurs on the leeward flanks. (b) The net acquisition of debris-rich ice. (c) The development of thrust planes at a glacier snout as glacier slip gives way to adhesion in the permafrost zone; the resultant compression causes the thrusts and enables basal debris to be transported to a higher level. (After Boulton 1972a.)

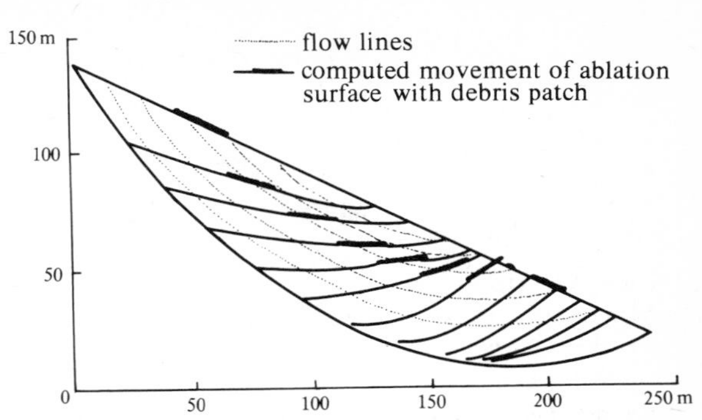

Figure 17.6 Longitudinal section of a Norwegian cirque glacier to show flow lines and successive positions of a debris patch computed for 10 year intervals (after McCall 1960).

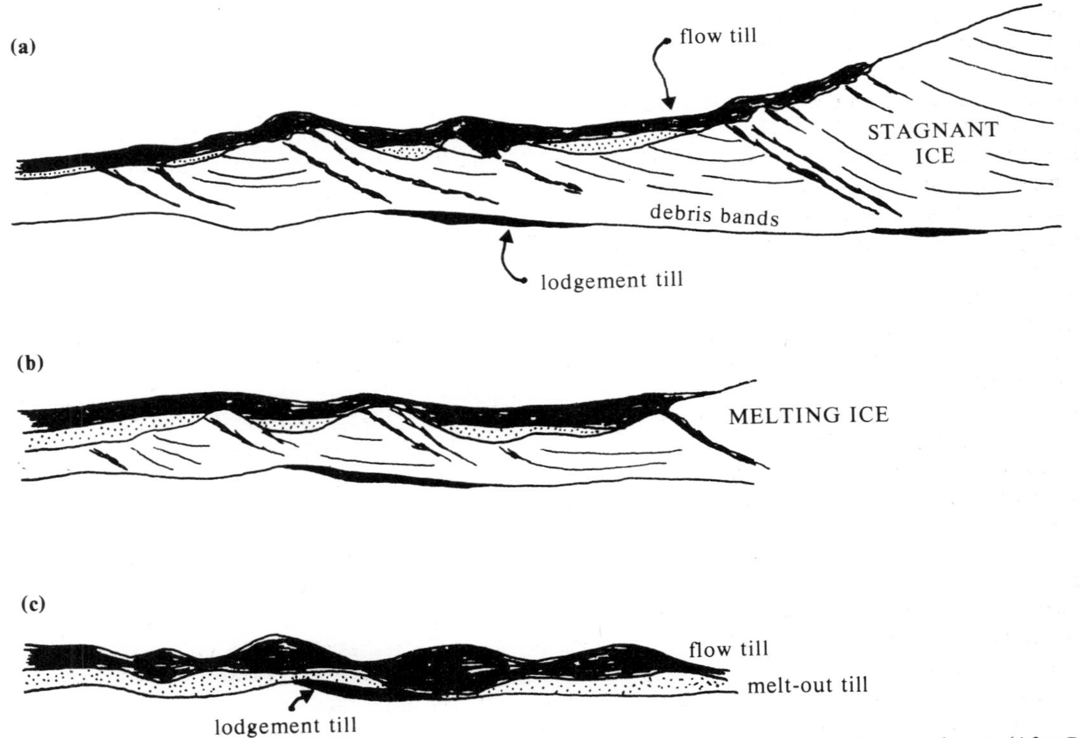

Figure 17.7 Origin of a tripartite till complex caused by flow, meltout and lodgement as a glacier snout decays. (After Boulton 1972b.)

17c Pleistocene and modern glacial facies

A broad spectrum of depositional facies exists in association with the flow and melting of glacier ice. We may usefully divide these facies into: (a) *ice-produced*, by ice flow and passive ice meltout; (b) *water-produced*, in the ice-contact zone; and (c) *water-produced*, distant from the contemporary ice terminus.

Ice-produced facies are dominated by poorly sorted, multilithologic deposits known as **tills** or boulder clays. A number of distinct sorts of till are known. As noted previously, **lodgement tills** result from deposition at the base of temperate ice by some sort of plastering effect. The till may be moulded into streamlined **drumlin** bedforms over pre-existing topographic nuclei. **Ablation** or melt-out tills result from the seasonal melting of debris-laden ice in temperate glaciers (Fig. 17.7). Such tills should show no development of internal structures or fabrics and may locally overlie lodgement tills. **Flow tills** (Fig. 17.7) result from thick supraglacial accumulations of debris derived from the melt out of debris bands at the glacier snout (Boulton 1958). These poorly sorted deposits then slump down the local gravity slope at the glacier snout by processes akin to debris flow (Ch. 7).

As shown in Figure 17.7, flow tills may come to overlie melt-out and lodgement tills to form a characteristic tripartite subdivision (Boulton 1972b). Should the glacier snout be fringed by deep proglacial lakes or be in contact with sea water, then **sub-aqueous ablation** and **flow tills** can result (Fig. 17.8) which may intertongue with lacustrine or marine facies as debris flow lobes or, more rarely, as turbidity current deposits (Dreimanis 1979, J. Shaw & Archer 1979). **Moraines** are essentially ridges produced by glacial dumping of both ice-deposited tills and water-deposited outwash. High, laterally extensive terminal moraines mark periods when forward motion is exactly balanced by melting. The structure of a typical dump moraine in a temperate valley glacier is shown in Figure 17.9.

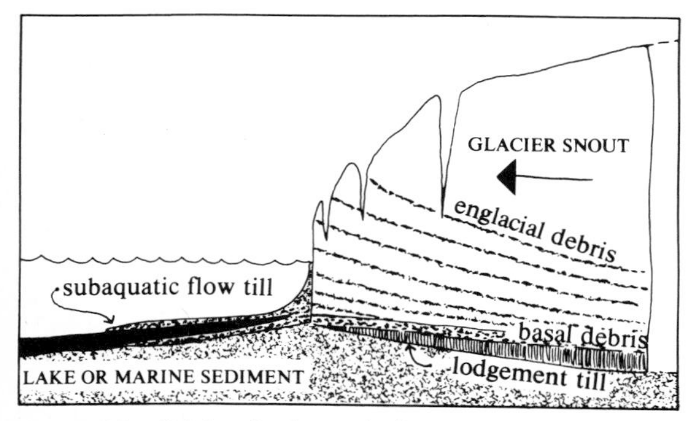

Figure 17.8 Origin of subaquatic flow till at a glacier snout bordered by a lake or sea (after Dreimanis 1979).

Water-produced facies in the ice contact zone include esker ridges, proglacial lake deposits and ice shelf deposits. An **esker** (Fig. 17.10) is a linear accumulation of stratified sands or gravels that was deposited by a stream wholly or partly confined by glacier ice. The water streams may occur at the glacier surface, within the glacier as a tunnel, or at the base of the glacier. Deposits of the first two streams are 'let down' to the local bedrock surface during ice melting and, as a result, may show marginal slump and fault structures related to this movement. Certain eskers are thought to have formed subaqueously as subglacial streams debouched into lake bodies. Most eskers are associated with temperate glaciers and are thought to have formed at or fairly close to a retreating glacier terminus. The deposits of eskers are dominated by current-produced sedimentary structures, with the frequent occurrence of climbing ripple cross laminations attesting to the rapidity of sediment deposition (e.g. J. R. L. Allen 1972, Bannerjee & McDonald 1975). Esker beads (Fig. 17.10) are thought to be caused by flow acceleration and deceleration at the front of subaqueous subglacial streams, giving rise to characteristic down-bead fining (Bannerjee & McDonald 1975). Subaqueous esker facies intertongue with marine or lacustrine facies.

Proglacial lakes are fed by seasonal outwash which may debouch into the lake to form a steeply dipping delta front of the Gilbert type (Ch. 19) dominated by coarse deposits showing downstream fining along foresets. Finer-grained material finds its way out into the lake body proper by processes of overflow, interflow or underflow, depending upon the relative density of the incoming suspended-sediment-rich water and the ambient lake waters (see Ch. 16). The seasonal melt/freeze process gives rise to **varved** deposits which may show complex internal rhythms (Ashley 1975) caused by inference between sediment supplied by overlapping density currents.

The **ice shelf** environment occurs seawards of large ice masses where the glacier ice thins and calving of icebergs occurs (Fig. 17.11). There is a general outward fining from lodgement and ablation tills to stratified marine mudstones with occasional **dropstones** from melting icebergs which penetrate the mud laminae (Fig. 17.12). Subaqueous flow tills also occur in proximal environments.

Water-produced facies distant from the ice contact zone are dominated by **braided outwash fans** (a type of wet fan; see Chs 14 & 15) where strongly seasonal meltwater flows deposit coarse-grained successions (Fig. 17.13). Down-fan trends include decreasing grain size and increasing predominance of point and lateral bars

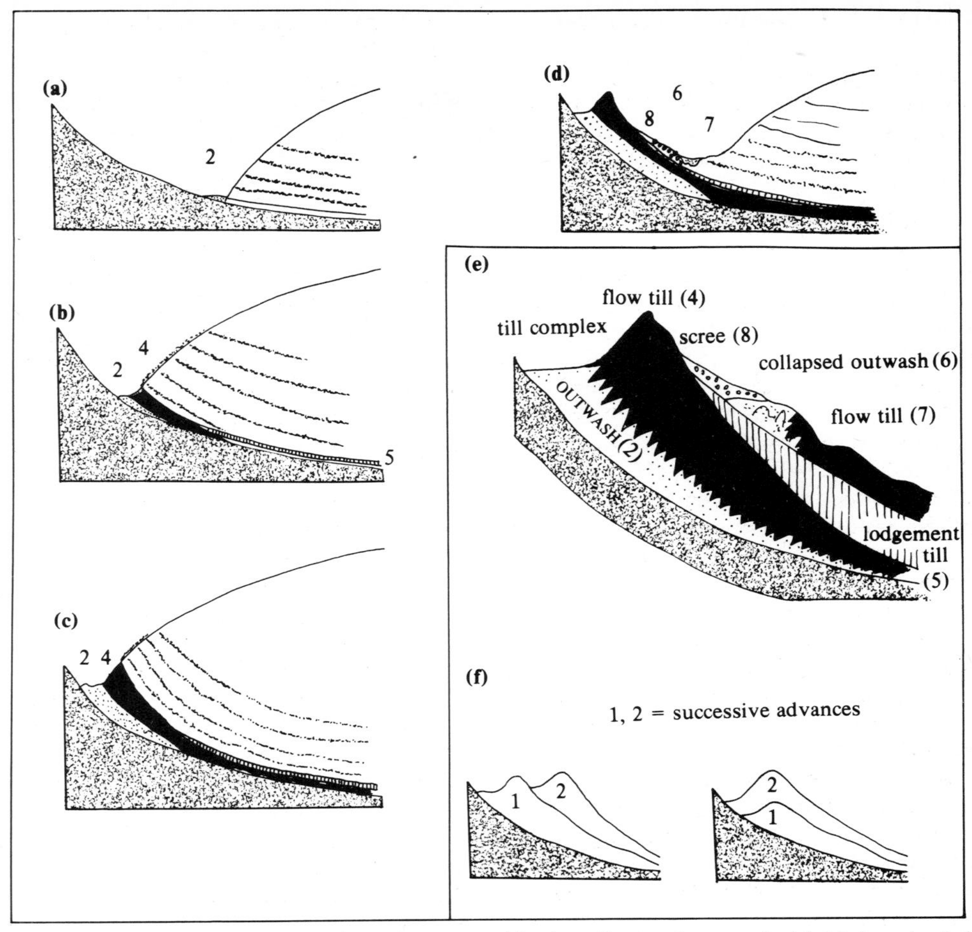

Figure 17.9 The development of morainic complexes from a combination of outwash, supraglacial debris and subglacial debris. (a–d) show a sequence of events during a single advance/retreat cycle. The morainic stratigraphy is summarised in (e). (f) shows moraines produced by successive glacier advances of contrasting extent. (All after Boulton & Eyles 1979.)

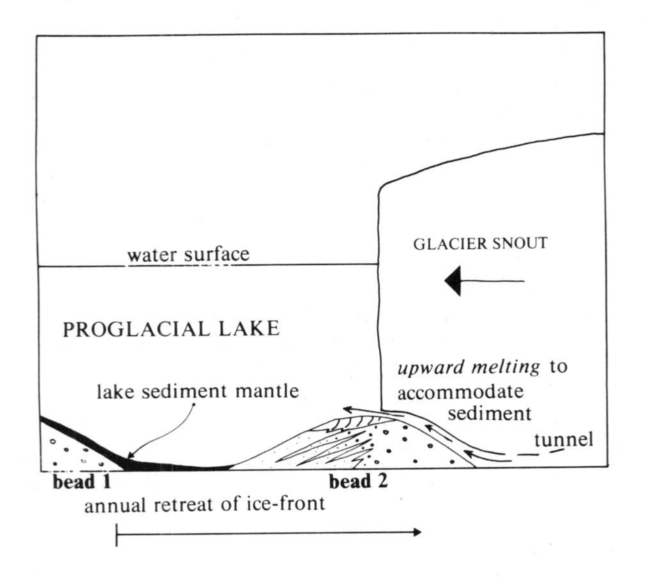

Figure 17.10 The formation of esker beads at glacier snouts as summer meltwater causes upward melting at the snout. Meltwater then decelerates from subglacial tunnels into the proglacial lake causing deposition of an esker bead. (After Bannerjee & McDonald 1975.)

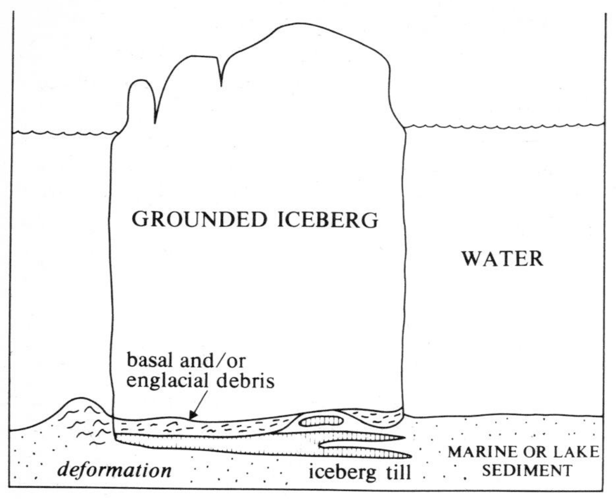

Figure 17.11 To show how waterlain till may be produced at the base of a grounded iceberg (after Dreimanis 1979).

Figure 17.12 Dropstone of granite embedded in laminated mudstones; note deformation at margins and 'puncturing' of laminae. Gowganda Formation, Upper Palaeozoic, W. Australia. Scale is 30 cm long. (Photo by Brian Jones.)

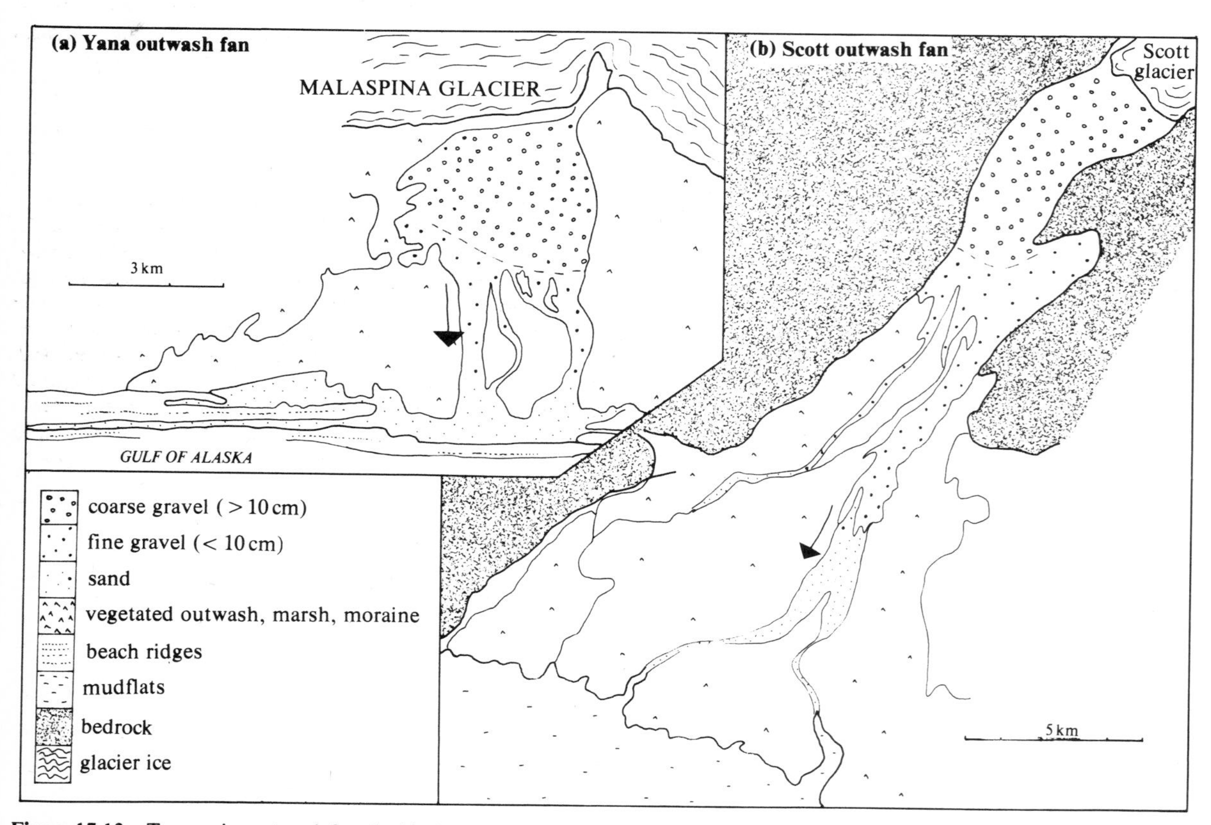

Figure 17.13 Two major outwash fans in Alaska to show major downfan changes in grain size (after Boothroyd & Ashley 1975).

(Boothroyd & Ashley 1975). Peculiar effects caused by both seasonal freezing of the fan channel beds and by melting of buried ice masses occurs on outwash fans and valley flats. In the latter case, large-scale cross beds and coarsening-upwards sequences are thought to result from the transport of outwash, as deltas prograde into local depressions formed by the melting of buried ice masses. Most shallow-water braided outwash fans do not exhibit large-scale foresets because of the very shallow water depths. Imbricate gravels are much more common in proximal areas, with particles transported and deposited in layers parallel to the surface of the bar–channel network (Rust 1975).

17d Ancient glacial facies

The identification of glacial deposits (see Harland *et al.* 1966) in the geological record should be attempted by use of all available local and regional evidence for glacier advance, retreat, melt out, and for general climatic frigidity. Deposits thought to be tills are best termed **mixtites** or **tillites** so as to emphasise their characteristics at the expense of a finalised genetic description. The broad range of modern till types should be remembered when criteria for establishing ancient tills are being searched for (e.g. Edwards 1975).

One of the oldest glacial deposits to be identified was the Port Askaig Tillite of the Scottish and Irish Dalradian Supergroup (~700 Ma b.p.). The Tillite overlies, and is overlain by, limestone and dolomite facies with stromatolitic bioherms and quartz replacements after anhydrite. The Tillite is actually made up of over 40 tillite beds separated by siltstones, dolomites and cross-stratified marine sandstones (Spencer 1971). Laminated clays often contain dropstones and are deduced to be glacial lake facies. The tillites are interpreted as the deposits of grounded ice sheets advancing onto a shallow shelf environment. The late Precambrian glaciers that deposited the Port Askaig Tillite seem to have been of worldwide extent, with both palaeomagnetic and facies evidence indicating glaciations at near-equatorial latitudes.

Other detailed studies of ancient glaciations include the Permo-Carboniferous glaciations of Gandwanaland (Wanless & Cannon 1966, Hamilton & Krinsley 1967, Lindsay 1970) and the Late Ordovician glaciations of North Africa (Beuf *et al.* 1971).

17e Summary

The preservation potential of glacial facies is generally low unless the depositional and erosional features are produced in areas of net crustal subsidence. The depositional products of glaciers may be produced by ice flow, ice melt out and meltwater processes. The most characteristic product of glaciation is till, which term includes a variety of poorly sorted mixtites deposited by lodgement, ablation and flow. Meltwater processes are dominated by fluvial deposition on giant braided outwash fans and by proglacial lake infill. At least four ancient glacial periods are identified in the geological record.

Further reading

A good up-to-date account of glacial processes is given by Embleton (1980). Many important papers are to be found in the volumes edited by Jopling and McDonald (1975) and Dreimanis (1979). Paterson (1969) is good on physical aspects of ice behaviour whilst Boulton (1972a) is an essential starting point for the application of glacial thermal regimes to depositional and erosional processes. Advanced topics in ice dynamics are dealt with in the volume edited by Colbeck (1980). A. E. Wright and Moseley (1975) is a good starting point for ways in which Quaternary glacier studies may be applied to those ancient glacial deposits found in the geological record.

PART SIX COASTAL AND SHELF ENVIRONMENTS AND FACIES ANALYSIS

Silver blades of surf
fall crisp on rustling grit,
shaping the shore as a mason
fondles and shapes his stone.

From Briggflatts *(Bunting)*

Plate 6 View seawards over intertidal flats from the salt marsh zone. Note hummocky salt marsh topography with a varied halophyte (salt-loving) flora in the foreground passing out into a flatter upper intertidal mudflat with spaced clumps of *Salicornia* and shallow pond-like depressions. In this area the salt marsh in the topmost intertidal and lowest supratidal zones is gradually extending seawards, causing a progradational regression (nr Wilhelmshaven, N. Germany).

Theme

Clastic sediment grains are introduced to the coast by way of river estuaries and deltas whose morphology and depositional processes are the result of a balance between river currents and coastal waves and tides. Although much sediment is trapped by deltaic deposition, a significant portion is reworked by wave and tide to form the deposits of linear clastic shorelines, of beaches, of tidal flats, barriers, spits and lagoons. Plumes of fine-grained sediment issuing from coastlines make their way out onto the shelf or shelf edge where deposition occurs. Shelf dynamics depend upon the complex interaction between wave and tidal currents, the latter being adjusted to shelf morphology and shape. Tide-dominated shelves show important variations in grain size and bedforms which arise from the variable strength of currents along tidal transport paths. In the absence of significant tidal currents, shelf sediment distribution is delicately adjusted to the power of storm waves whose effects decay outwards to the shelf edge.

In semi-tropical to tropical waters (and to a lesser extent in temperate waters) where clastic input is limited, a high biological production encourages the formation of biogenic carbonate grains. Locally $CaCO_3$ precipitation occurs as aragonite mud. These orthochemical and allochemical grains are then acted upon by wave and tidal currents to form carbonate beaches, tidal flats, barriers, spits and lagoons. Along the coast and on the supratidal plains (sabkhas) of many arid areas evaporite precipitation occurs in isolated water bodies and in the shallow subsurface of deposited sediment.

18 Physical processes of coast and shelf

18a Introduction

The complexity of coastal and shelf environments arises from the interplay between chemical/biogenic grains or clastic sediment introduced into the system by rivers of variable power and the marine energy spectrum available in the form of wind-generated waves and tides. Add to this the many effects of the relatively recent Holocene rise in sea level (e.g. drowned coasts, shelf relict sediment patches) and it readily becomes apparent that coastal processes and their resultant facies must be carefully studied in both field and laboratory in an interdisciplinary manner.

Some idea of the marine energy spectrum is given by Figure 18.1. Note that the term wave covers a vast range of period, more than eight orders of magnitude from tiny capillary waves (water ripples) to ocean-scale tidal waves.

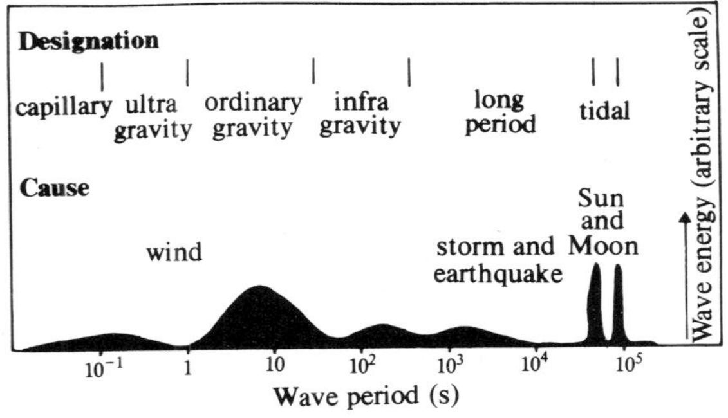

Figure 18.1 Energy density spectrum of water surface waves (after Munk 1950).

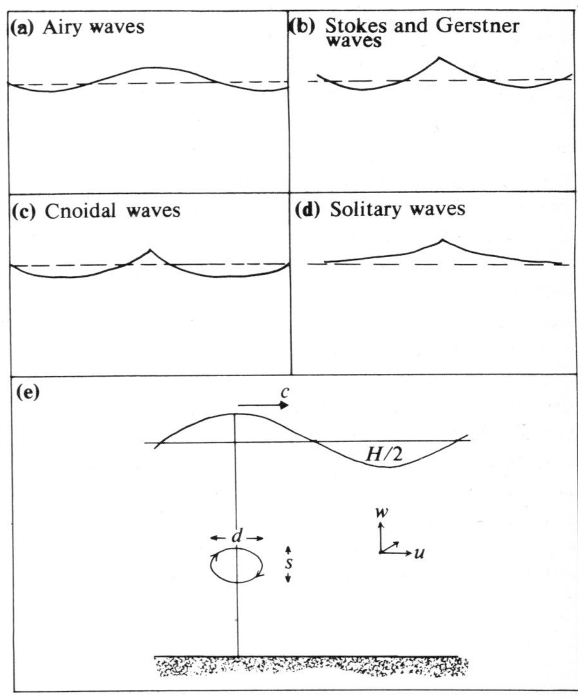

Figure 18.2 (a–d) Profiles of the major theoretical wave types. (e) In a sinusoidal Airy wave each water particle moves with a horizontal velocity u and vertical velocity ω in an elliptical orbit of major diameter d and minor diameter s.

18b Wind-generated waves

As briefly noted in Chapter 8, waves transfer energy without necessarily involving net forward water motion. Wind-generated waves are a form of surface gravity wave driven by a balance between fluid inertia and gravity. Attempts to investigate wave motion in a rigorous manner assume that the wave surface displacement may be approximated by curves of various shapes, the simplest of which is a harmonic motion used in linear (Airy) wave theory.

In such simple harmonic motion the particle displacement y is given by

$$y = a \sin \omega t \qquad (18.1)$$

where a = wave amplitude, t = time, ω = angular velocity. Since $\omega = 2\pi/T$, where T = periodic time, then

$$y = a \sin \frac{2\pi}{T} t \qquad (18.2)$$

with the wavelength of the motion given by $\lambda = cT$, where c = wave speed, and the wave speed given by $c = \lambda f$ where f = frequency = $1/T$.

Let us consider (Fig. 18.2) simple sinusoidal waves of small amplitude in water so deep that motions caused by the surface wave form cannot reach the bottom (i.e. water depth $>\lambda$). In such cases it can be demonstrated (see Appendix 18.1) that the wave speed is given by

$$c = \sqrt{g\lambda/2\pi} \qquad (18.3)$$

The coefficient $\sqrt{g/2\pi}$ is constant so with measurements in S1 units

$$c = 1.25\sqrt{\lambda} \tag{18.4}$$

and

$$T = 0.80\sqrt{\lambda} \tag{18.5}$$

We thus see that surface gravity waves in deep water are **dispersive** in the sense that their rate of forward motion is dependent upon wavelength. As developed in Appendix 18.2 the linear theory of sinusoidal waves on deep water predicts that at any fixed point the fluid *speed* caused by wave motion remains constant whilst the direction of motion rotates with angular velocity ω. The radius of these water **orbitals** decreases gradually below the surface (Fig. 18.3).

As our ideal sinusoidal waves pass into shallow water they suffer attenuation through bottom friction and induce significant horizontal motions to the bottom. Such waves are of great interest to sedimentologists. These shallow-water gravity waves move with a velocity that is proportional to the root of the water depth and is independent of wavelength or period. Thus

$$c = \sqrt{gh} \tag{18.6}$$

The wave orbits are elliptical at all depths, with increasing ellipticity towards the bottom, culminating at the bed as horizontal straight lines representing to-and-fro motion (Fig. 18.3). Under such conditions the maximum orbital horizontal velocity is given by

$$u_{\max} = \frac{H}{2h}\sqrt{gh} \tag{18.7}$$

where H = wave height.

Concerning **wave energy**, it is the rhythmic conversion of potential to kinetic energy and back again that maintains the wave motion, indeed our derivations in Appendix 18.1 are dependent upon this approach. The displacement of the wave surface from the horizontal provides potential energy that is converted into kinetic energy by the orbital motion of the water. The total wave energy per unit area is given by

$$E = \tfrac{1}{8}\rho g H^2 \tag{18.8}$$

(Note the square term in this equation and its significance for wave-power schemes from high-energy coastlines. If 1 m is a typical mean wave height, then the total world wave energy is about 4.5 to 10^{17} J. (McClellan 1965).) The **energy flux** (or wave power) is the rate of energy

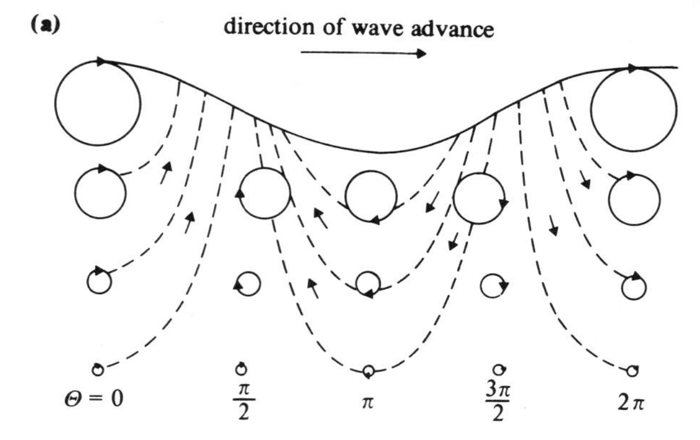

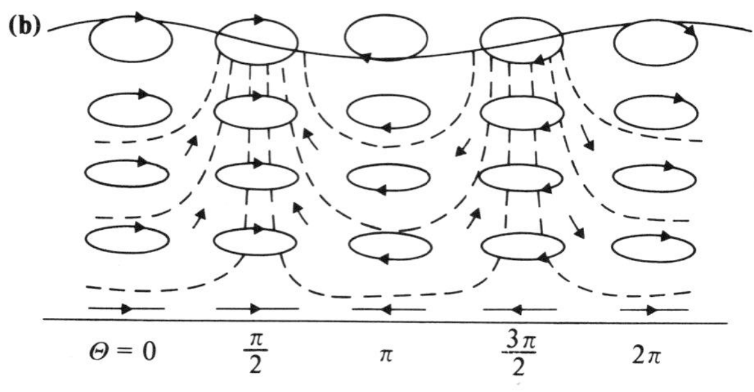

Figure 18.3 (a) Orbits, orbital velocities and streamlines for deep water waves and (b) shallow water waves. (After McLellan 1965.)

transmitted in the direction of wave propagation and is given by

$$w = Ecn = \tfrac{1}{8}\rho g H^2 cn \tag{18.9}$$

where $n = \frac{1}{2}$ in deep water and 1 in shallow water. In deep water the energy flux is related to the **wave group velocity** rather than the wave velocity and it can be shown that this group velocity is about half the wave velocity (see Tricker 1964).

The **momentum flux** (or radiation stress) is the excess flow of momentum due to the waves. The two non-zero components of momentum flux are

$$S_{xx} = E(2n - \tfrac{1}{2}) = E/2 \text{ (deep water) or } 3E/2 \text{ (shallow)} \tag{18.10}$$

and

$$S_{yy} = E(n - \tfrac{1}{2}) = 0 \text{ (deep water) or } E/2 \text{ (shallow)} \tag{18.11}$$

where the x axis is in the direction of wave advance and the y axis is parallel to the wave crest.

It should be noted that linear (Airy) wave theory (Appendix 18.1) neglects terms to the second and higher orders. These terms are included in **Stokes' theory** and they serve to enhance wave crest amplitude and decrease the trough amplitude. In fact the wave profile has become more similar to that of natural waves. Trochoidal waves (Fig. 18.2) approximate best to many natural wave profiles but the theory of trochoidal waves is complex and will not be discussed here.

Concerning Stokes' waves, one interesting departure from linear (Airy) theory is that the wave orbitals are not closed, so that a steady mass transport of water must occur. It is interesting to note that Stokes' derivation predicted, in fact, a near-bed seaward flow of water, quite the opposite to that detected by experiment and field measurements (Bagnold 1940). This disagreement between theory and practice was caused by the non-inclusion of viscous terms in the derivation. Once such terms are included (Longuet-Higgins 1953) good agreement between theory and practice occurs. The shoreward velocity resulting from Stokes' waves is now known to be

$$\bar{u}^1 = \frac{5}{4}\left(\frac{\pi H}{\lambda}\right)^2 c \frac{1}{\left[\sinh\left(\frac{2\pi}{\lambda}h\right)\right]^2} \qquad (18.12)$$

Stokes' theory can also be used to shed light on the wave-breaking phenomena. Waves will break when the water velocity at the crest is equal to the wave speed. This occurs as the apical angle of the wave reaches a value of about 120°. In deep water the tendency towards breaking may be expressed in terms of a limiting wave steepness given by

$$\frac{H}{\lambda} \simeq \frac{1}{7} \qquad (18.13)$$

Let us now return to the 'real' world of wave behaviour. Controversy still exists concerning the physics of wave generation, though pressure fluctuations in the wind boundary layer and wave sheltering effects upon wind flows must both play an important role. Storms generate a spectrum of waves in a generating area. As these waves propagate outwards with the wind, the longer waves travel faster (Eq. 18.3) so that a sorting effect operates. Even so, the generation of waves occurs at different times in different areas so that all wave-recording stations record a range of wave sizes over any period of time. Statistics such as **significant wave height** and **period** refer to the mean height of the highest one-third of waves measured over some time interval. Apropos of our previous discussion on wave energy, it should be realised that the total energy accumulated over a portion of coast is obtained by integration of the wave spectrum over the entire frequency range.

As in several other modes of wave motion, superimposition of wave forms leads to positive or negative amplification or neutralisation depending upon the wavelengths involved. Ample opportunities for wave interference occur with deep-water waves since speed of travel is directly proportional to wavelength (Eq. 18.3). Groups of gravity waves in deep water are separated by low-amplitude neutralisation areas which travel with only half the velocity of the waves themselves. Small-amplitude waves passing from neutralisation areas grow in amplitude as they pass through a group. After passing through the midpoint of the group they decrease in amplitude away from it. The energy transported by waves thus travels with the group velocity rather than the wave velocity (Eq. 18.9).

As the typical sinusoidal 'swell' of the deep ocean passes over the continental shelf towards coastlines, the waves undergo a transformation as they react to the bottom at values of between about $\frac{1}{4}$–$\frac{1}{2}$ their deep-water wavelength. Wave speed and length decrease whilst height increases. Peaked crests and flat troughs develop until oversteepening causes wave breakage. Breaking waves may be divided into spilling, plunging and surging types (Fig. 18.4). Important **wave refraction** effects occur, the direction of wave propagation tending to become increasingly normal to the submarine contours as the water shallows (Fig. 18.5).

Important effects act upon waves on beaches in response to the steepness of the beach face (Huntly & Bowen 1975). Steep beaches possess a narrow surf zone in which the waves steepen rapidly and show high orbital velocities. Wave collapse is dominated by the plunging mechanism and there is much interaction on the breaking waves by backwash from a previous wave-collapse cycle. A **rip cell circulation** (*q.v.*) may also be present. Gentle beaches show a wide surf zone in which the waves steepen slowly, show low orbital velocities, and surge up the beach with very minor backwash effects. Rip cells are not associated with such beaches; steady longshore currents exist instead.

The nearshore current system includes components other than those to-and-fro motions caused by shoaling waves. The first example is a remarkable system of cell circulation comprising **rip currents** and **longshore currents** (Shepard & Inman 1950) (Fig. 18.6). The narrow zones of rip currents make up the powerful 'undertow' on many beaches and are potentially hazardous to swimmers because of their high velocities (up to $2\ \mathrm{m\,s^{-1}}$). Rip currents arise because of variations in **wave set-up** (Longuet-Higgins & Stewart 1964) along steep beaches.

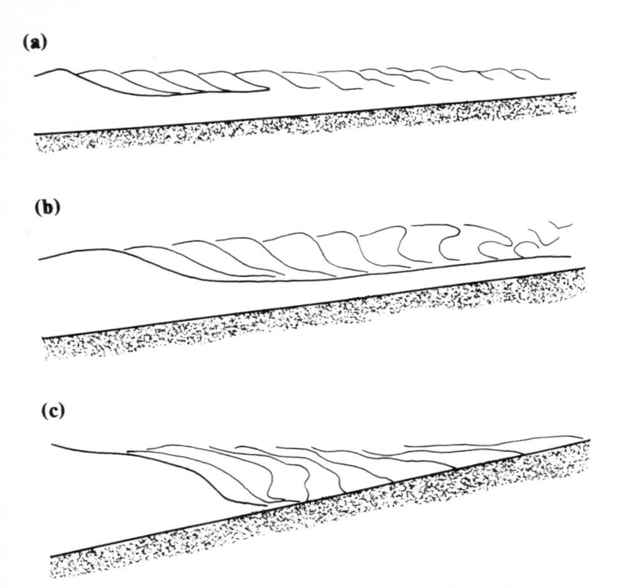

Figure 18.4 Three types of breaking waves. Note that every gradation occurs between these types. (a) In spilling waves the crest steepens and the wave becomes asymmetric until the unstable crest cascades down as bubbles and foam. (b) In plunging waves the shoreward wave face curls over and impacts upon the beach. (c) In surging waves the shoreward wave face steepens as if to plunge, but the wave then moves as a surge up the beach causing the wave face to disappear. (After Galvin 1968.)

Wave set-up is the small (cm–dm) rise of *mean* water level above still water level caused by the presence of shallow-water waves (Fig. 18.7). It originates from a portion of the momentum flux S_{xx} (Eq. 18.10) which is balanced close inshore by the pressure gradient due to the sloping water surface. Now, in the breaker zone the set-up is greater shoreward of large breaking waves than smaller waves so that a longshore pressure gradient causes longshore currents to move from areas of high breaking waves to low breaking waves (Bowen 1969, Bowen & Inman 1969). These currents turn seawards where set-up is lowest and where the currents converge (Fig. 18.6).

This analysis of cell circulation depends upon some mechanism (or mechanisms) that can produce variations in wave height parallel to the shore in the breaker zone. Wave refraction is one such mechanism and some rip current cells are closely related to offshore variations in topography (Fig. 18.8). Since rip cells also exist on long straight beaches with little variation in offshore topography, another mechanism must also act to provide lateral variations in wave height. This is thought to be the standing **edge waves** (Huntly & Bowen 1973) which form by breaking wave/backwash interactions on relatively steep beaches. The addition of the incoming waves to the edge waves gives marked longshore variations in breaker height, the summed height being greatest where the two wave systems are in phase.

The second group of nearshore currents are those produced by oblique wave attack upon the shoreline. (It should be noted that such currents may be superimposed upon the rip cells described previously.) Such currents are caused by S_{xy}, the flux towards the shoreline (x direction) of momentum directed parallel to the shoreline (y direction). Thus

$$S_{xy} = En \sin \alpha \cos \alpha \qquad (18.14)$$

where α is the angle between wave crest and shore. Longuet-Higgins (1970) derives the longshore velocity component, u_l, as

$$u_l = \left[\frac{5\pi}{8} \frac{\tan \beta}{Cf}\right] u_{max} \sin \alpha \qquad (18.15)$$

where $\tan \beta$ = beach slope, Cf = drag coefficient. However, Komar and Inman (1970) found no dependence of $\bar{u}_l$ on slope and Komar (1971, 1975) suggests $\tan \beta / Cf$ is approximately constant so that Equation 18.15 becomes

$$\bar{u}_l = 2.7 u_{max} \sin \alpha \cos \alpha \qquad (18.16)$$

in good agreement with field measurements (Fig. 18.9) at mid-surf.

Concerning sediment transport in the nearshore zone, Bagnold (1963) has related sand transport by immersed weight i_b to the wave energy flux (power) ω, such that $i_b = k\omega$, where k is a constant that approximates to 0.28 from empirical data. Under purely to-and-fro water motion, the wave power simply acts to suspend sediment grains with no net transport. Net transport occurs only in the presence of a shoreward-directed residual current u' caused by shoaling waves or to a longshore current l (or combinations thereof). In such cases the net transport rate i, is given by

$$i' = k\omega \frac{u'}{u_{max}} \qquad (18.17)$$

which for longshore transport yields

$$i_l = kECn \cos \alpha \frac{\bar{u}_l}{u_{max}} \qquad (18.18)$$

in good agreement with field measurements (Komar & Inman 1970).

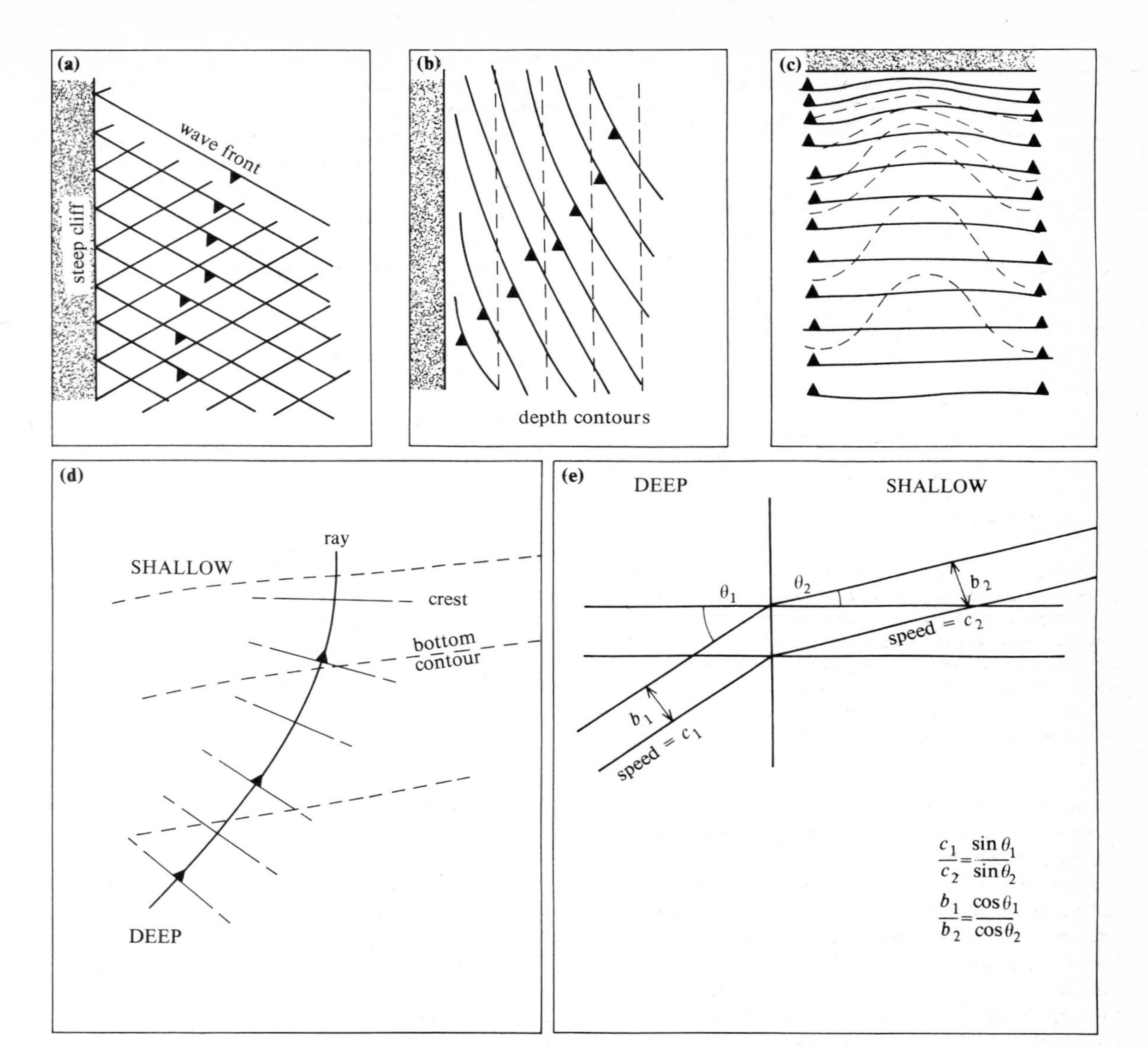

Figure 18.5 Aspects of wave refraction and reflections as waves approach a coastline. (a) Wave reflection from a steep cliff. (b) Wave refraction in shallowing water. (*NB*. Remember that wave speed decreases as water shallows.) (c) Refraction over a sea-floor hollow or drowned valley. (a–c after Allen 1970b.) (d) Definition of rays, wave crests and bottom contours. (e) Schematic refraction effects in waves. (d–e after Collins 1976.)

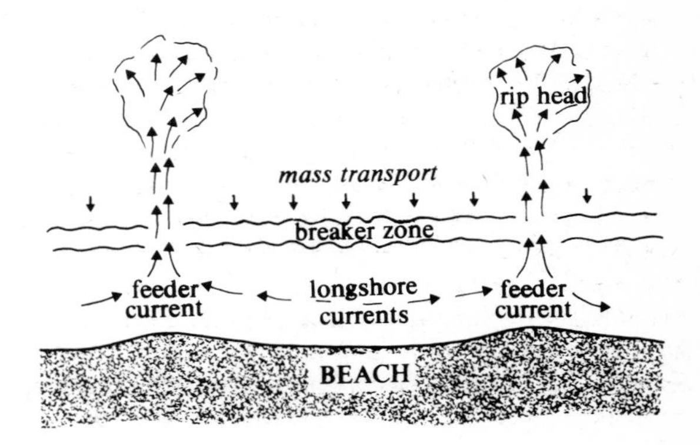

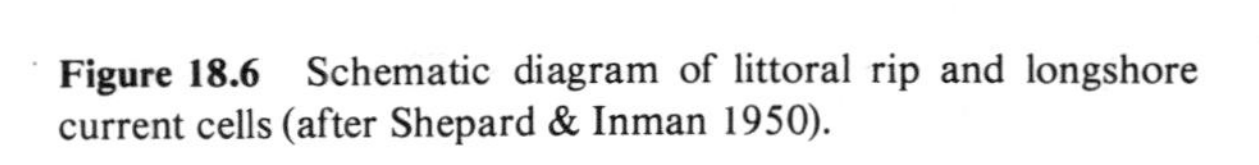

Figure 18.6 Schematic diagram of littoral rip and longshore current cells (after Shepard & Inman 1950).

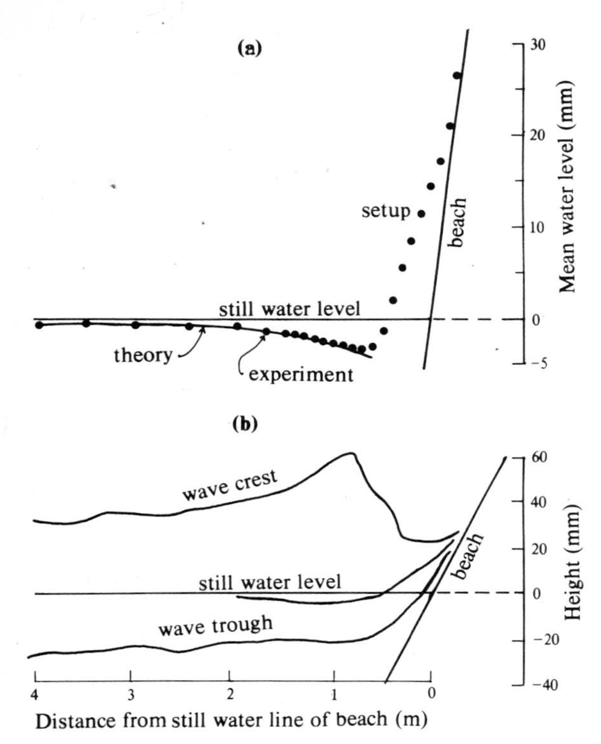

Figure 18.7 (a) Departures of mean water level from still water level. Wave set-up and set-down are produced by the radiation stress (momentum flux) of incoming waves in a laboratory wave tank and artificial beach. (b) Experimental wave heights in relation to the set-down and set-up effect. (After Bowen *et al.* 1968.)

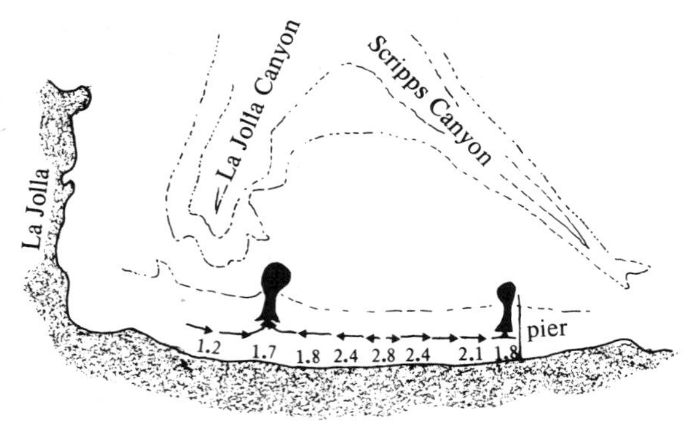

Figure 18.8 Rip cells off the Californian coast. Here, longshore variations in wave height are produced by coast refraction over offshore submarine canyons. Wave heights are indicated along the shoreline in metres. (After Shepard & Inman 1950.)

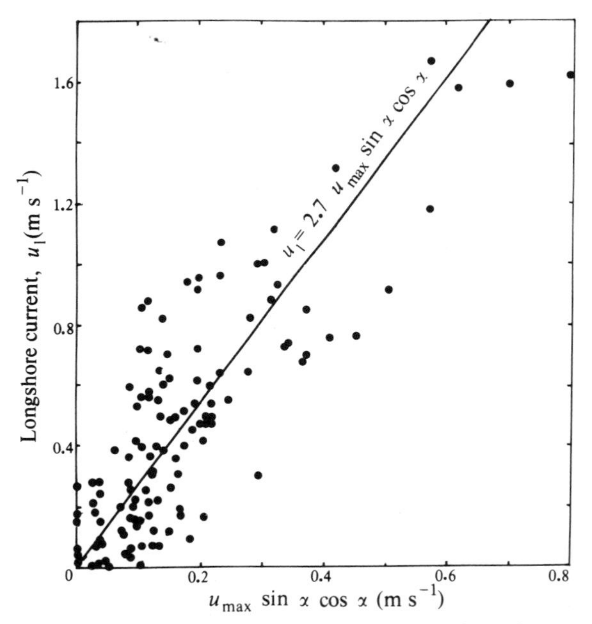

Figure 18.9 Graphical representation of Equation 18.16 to show broad agreement with field measurements (after Komar & Inman 1970).

It must be stressed that the *long-term* longshore transport of sediment depends upon the summed effects of all wave systems that impact upon a coastline. Such long-term transport vectors cancel out seasonal effects and are visually impressed upon the mind of the casual coastal visitor by piling-up of sediment against groynes, jetties and breakwaters around coastlines and by the migration directions of coastal spits.

A final point in this section concerns a few general remarks on wave-produced bedforms. We have already discussed the moulding effect that purely oscillatory and combined oscillatory–current flows have upon loose beds of sediment grains (Ch. 8). All gradations exist between purely oscillatory and combined oscillatory–current flows in the nearshore and beach environments, but in general we might expect an increasing degree of flow asymmetry, and hence bedform asymmetry, as the water depth decreases towards the shoreline. In fact large seasonal variations in the degree of asymmetry are to be expected, causing complex interrelations to be preserved in the sedimentary structures of the nearshore and beach zone.

18c Tides and tidal waves

The periodic rise and fall of sea level visible around coastlines has long fascinated both scientist and laymen alike. Newton was the first person to explain tides successfully as a 'spin-off' from his famous gravitational law

$$F = \frac{Gm_E m_M}{r^2} \qquad (18.19)$$

where F is the force of attraction, G is the universal gravitational constant, m_E and m_M are the masses of the Earth and Moon respectively and r is the distance between the Earth and Moon. Now each water particle in the oceans undergoes a centripetal acceleration of equal strength and direction as the Earth–Moon system revolves in space. Since F varies as the square of r, and since water particles vary in their distance from the Moon, a net residual force exists, given by the difference between F and the constant centripetal acceleration (Fig. 18.10). As shown in Figure 18.10 the tidal rhythm is partly caused by the tangential component of acceleration, A_t, so that approximately

$$A_t = \frac{3}{2} g \frac{m_M}{m_E} \frac{R_E^3}{r^3} \sin 2\theta \qquad (18.20)$$

The force is small but unopposed until the pressure gradient of the sloping tidal bulge balances it exactly. The so-called *equilibrium* tide produced by A_t causes two areas of high water, one directly under the Moon and another on the opposite side of the Earth. As the Earth rotates, any point on its surface will pass under the tidal bulges twice, giving two periods of high water and two periods of low water every 24 hours (a semi-diurnal tide). From Equation 18.20 we see that the magnitude of the tide-producing force varies as 2θ so that it is zero immediately underneath the Moon and also normal to this position. The maximum tidal amplitude is predicted half way between.

The effect of the Sun's large mass upon Earth tides is over-compensated by its huge distance from the Earth so that the Sun exerts about 50% of the *total* tide-raising force. Important effects arise when the Sun and Moon act together on the oceans to raise extreme high tides (**spring tides**) and act in opposition on the oceans to raise extreme low tides (**neap tides**) in a two-weekly rhythm. Variations in these tides come about because of the eccentricity of the lunar orbit, the very highest springs forming when the perigree falls at either new or full Moon, and because of variations in the Earth–Sun distance in summer and winter. A number of further complications arise because of the declination of the Earth's axis at 23.5° to the plane of orbit about the Sun. Declination with respect to both Moon and Sun introduces an inequality into the tides so that high tides will be successively larger and smaller Other longer-term periodic components of Earth–Moon–Sun behaviour also affect the tides.

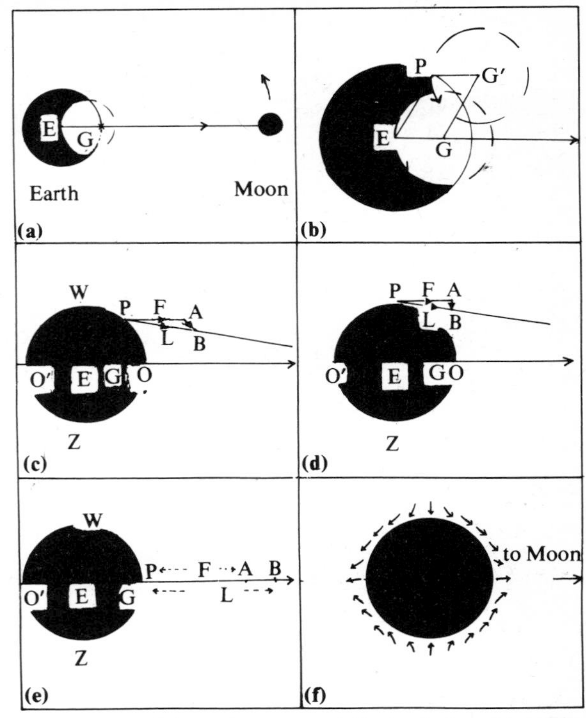

Figure 18.10 Schematic representation of the tide-raising forces. (a) The moon revolves about G, the common centre of gravity of the Earth and Moon, in an anticlockwise direction as viewed from the Pole Star. (b) The centre of the Earth, E, and any point P on its surface, describe circles of constant radius PG′, EG as the centre rotates about G. (c–e) PA is drawn parallel to EG and represents the centripetal force required to make the particle P move in its circle. PA is constant over the Earth's surface. The line PB represents the force of the Moon's attraction and by Newton's Law of Gravitation must decrease from equator to pole. The force PB is equivalent to the forces PA and AB. The force AB is the tide-raising force. The direction of AB varies from equator to pole as shown. (f) Summary of the directions of the tide-raising forces over the Earth's surface; the horizontal component gives rise to the equilibrium tide. (All after Tricker 1964.)

Thus far we have largely followed in the footsteps of Newton in our discussion of equilibrium tidal theory. In nature important modifying factors arise from the irregular distribution of land and sea, from the irregular shape and variable depth of the sea, from the Coriolis force produced by the Earth's rotation, from the inertia of the water mass, and from frictional effects.

The nature of the tidal oscillation depends critically on the natural periods of oscillation of the particular ocean basin. These coincide with the 12-hour tide-forming forces in the Atlantic Ocean giving **semi-diurnal tides**. The Gulf of Mexico oscillates in sympathy with the 24-hour forces giving **diurnal tides** whilst mixed tides result in the Pacific Ocean which does not oscillate so regularly.

To an observer fixed with respect to the Earth the tidal wave should seem to advance progressively from east to west. Locally, however, in straits and channels on continental shelves, the tidal wave may advance in any direction. Further, the tide often occurs as a standing wave whose tidal currents are zero in the centre of the oscillating water and reach a maximum at the margins (nodes). They are zero everywhere at high and low water and reach maximum values at half-water stages. The amplitude of the oscillation is greatest when the periods of the local sea and the oceanic tides coincide at about twelve hours.

Resonant effects greatly increase the oceanic tidal range (~0.5 m). The lengths of a strait or gulf which will resonate with a period of about twelve hours depends upon the depth, the critical resonant length of a gulf increasing with water depth. The Bay of Fundy has a standing resonant oscillation with a node at its entrance which causes the tidal range to increase from 3 m to 15.6 m along its length.

The semi-enclosed nature of many ocean basins and large shelves, combined with the Coriolis force, causes the tidal wave to become rotary, advancing anticlockwise in the northern hemisphere about a **nodal** (amphidromic) point of no displacement with maximum tidal displacements around the margins (Fig. 18.11). The crest of the tidal wave is a radius of circular basins and is also a **co-tidal** line along which tidal minima and maxima coincide. Concentric circles drawn about the node are **co-range** lines of equal tidal displacement. Tidal range is thus increased outwards from the node by this rotary action. Some semi-enclosed seas such as the North Sea generate a number of nodal points (Fig. 18.12). A broad division of shelf tidal ranges into **macro**- (range >4 m) **meso**- (range 2–4 m) and **microtidal** (range <2 m) is possible.

Considering a semi-enclosed basin with a rotary tidal wave, it is apparent that at any one place the velocity of the tidal current will vary in both magnitude and direction. This vector variation may be usefully summarised by means of a tidal-current ellipse whose ellipticity will be a direct function of tidal vector asymmetry (Fig. 18.13).

Important sedimentary consequences arise from the nature of the tidal ellipse. Marked periodicity in the strength of the current leads to alternating periods of sediment movement and no sediment movement after fall out of sediment from suspension occurs. Tidal laminations of distinctive character result from this cycle of events. Reversal of current direction will similarly produce telltale opposed cross laminations which, when combined

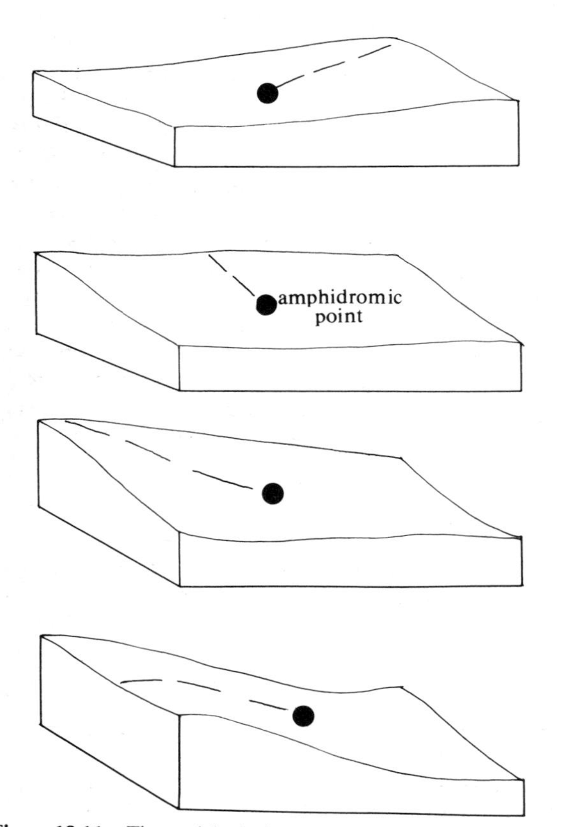

Figure 18.11 The anticlockwise motion of a tidal wave about an amphidromic point (after Komar 1976).

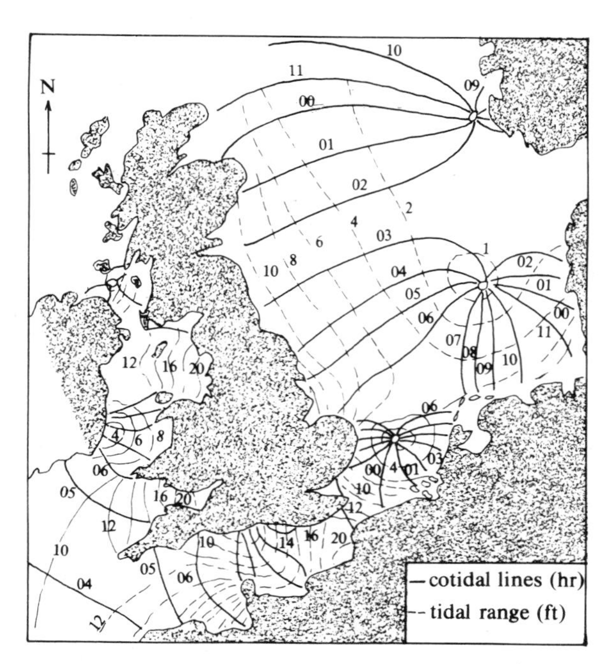

Figure 18.12 Amphidromic points, co-tidal lines and tidal ranges, as found on the NW European continental shelf (after Komar 1976).

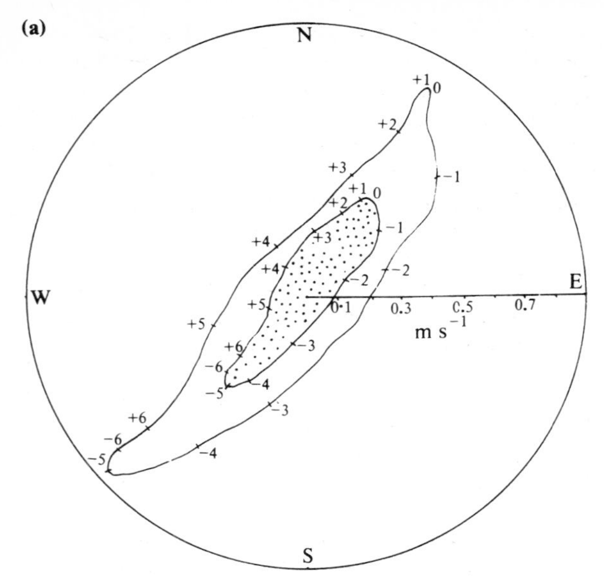

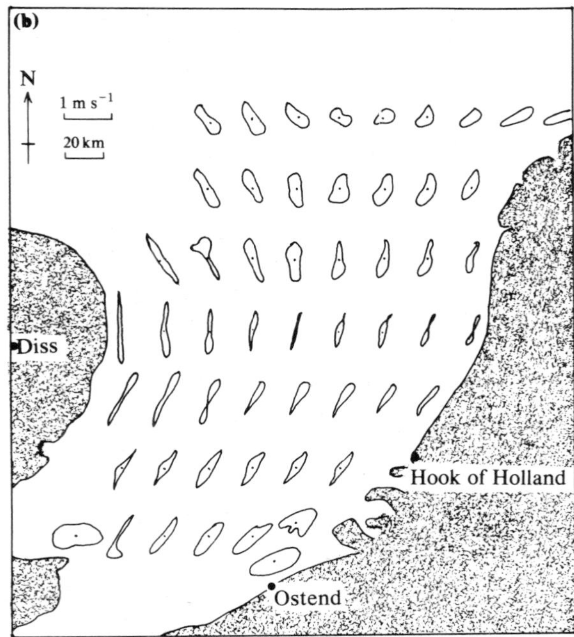

Figure 18.13 (a) A tidal current ellipse from the southern North Sea. Outer ellipse for near-surface flow; inner ellipse for near-bottom flow. The ellipse is constructed by drawing a velocity vector for a given constant height above the bed at various stages (0, +1, +2 etc.) in the tidal cycle. If the complete tidal cycle lasts 12 hours, then each vector is drawn at 360/12 degree intervals. Tidal currents thus vary in *both* magnitude and direction with time. Residual currents may also arise (see Ch. 21). (After McCave 1979.) (b) Arrays of tidal current ellipses (1 m above bed) for the southern North Sea (after McCave 1971).

with other evidence, may be highly suggestive of tidal forces acting in ancient sedimentary deposits.

A final point concerns the increased tides that result when very strong onshore winds from cyclones or hurricanes combine with the low barometric pressure to pile up water close to shore, raising tidal limits by up to 1000%. A return to better weather conditions causes an enormous seaward surge of water (ebb storm surge) which, as we shall see in a later chapter, may transport shallow water and intertidal detritus and fauna far out to sea.

18d Summary

Coastal and shelf processes are dominated by currents caused by wind-formed and tidally formed waves. Wind-formed waves cause purely oscillatory currents in deep water, but in shallow water a significant landward velocity vector for near-bed fluid is superimposed on this oscillatory flow. Deep-water waves react to the bottom as they pass onto the shelf. Wave speed and length decrease and height increases until oversteepening at the shoreline causes wave breakage. In addition to the to-and-fro motion of normally incident waves and the longshore component of motion of obliquely incident waves, rip-current cells may arise because of variations in breaker height caused by wave refraction and edge waves. Rip currents vigorously transport seawards a certain proportion of the sediment transported landwards by the residual currents of shoaling waves. The tidal wave arises due to the action of net residual forces on the oceans given by the difference between gravitational attraction and centripetal acceleration as the Moon revolves around the Earth and the two bodies revolve around the Sun. The advance of the tidal wave over irregular coastlines and shelves is accompanied by rotary deformation and amplification so that shallow-water tidal currents are extremely complex in both magnitude and direction.

Further reading

Very full and helpful accounts of physical coastal processes are given by Komar (1976) and in Stanley and Swift (1976). As noted in Appendix 18.1, Tricker (1964) gives a delightful elementary account based on general physics.

Appendix 18.1 Deep-water wave theory

Let us begin by assuming that our wave is of small amplitude A relative to water depth and that it is of simple harmonic type, say a sine or cosine curve. Then generally we have for the displacement y of a stationary curve (Fig. 18.14a)

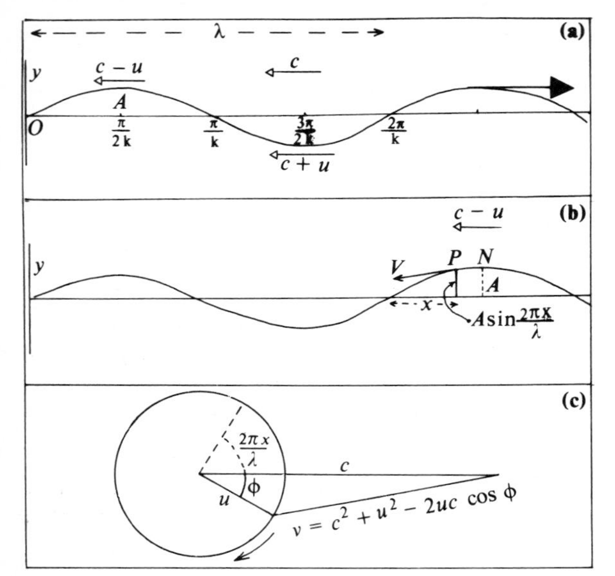

Figure 18.14 (a–c) Definition diagrams for deep-water wave theory (after Tricker 1964).

$$y = A \sin kx \qquad (18.17)$$

and since the wavelength λ is $2\pi/k$ we have

$$y = A \sin \frac{2\pi}{\lambda} x \qquad (18.18)$$

Now let us reduce our moving wave form to a stationary curve by the artifice of setting up an opposing current of equal velocity (c) to the wave velocity. The waves are thus brought to rest. If it were not for the opposing current, water particles would move horizontally forward with the waves at their crests and horizontally backwards in the troughs as part of their circular orbital motion. Assuming that the motion is symmetrical, we have these forward and backward motions as $+u$ and $-u$ respectively. Adding these to c we see that particles in the surface of the wave must move at velocities of $(c + u)$ and $(c - u)$ (Fig. 18.14a).

Let us now consider the energy relations of the wave form. The increase of potential energy as a particle moves from trough to crest, a distance $2A$, must be balanced by an equivalent loss of kinetic energy. In equation form

$$2mgA = \tfrac{1}{2}m[(c + u)^2 - (c - u)^2] \qquad (18.19)$$

or simplified

$$2mgA = 2mcu$$

so that

$$gA = cu \qquad (18.20)$$

Now consider water particles crossing the x axis. The horizontal component of velocity here will be c but there will also be a vertical velocity v, so that the resultant velocity will be $\sqrt{c^2 + v^2}$ and the kinetic energy of these particles will be $\tfrac{1}{2}m(c^2 + v^2)$. The energy relations here involve the equality of kinetic energy gained from wave crest to mid-point, and potential energy lost. Thus

$$mgA = \tfrac{1}{2}m(c^2 + v^2) - \tfrac{1}{2}m(c - u)^2 \qquad (18.21)$$

giving

$$gA = \frac{v^2 - u^2}{2} + uc \qquad (18.22)$$

but since $gA = cu$ from Equation 18.20 above

$$v^2 = u^2 \qquad (18.23)$$

Now it can be shown that near the x origin, close to P, the gradient y/x of the sine curve is given by $2\pi A/\lambda$. Thus

$$\frac{u}{c} = \frac{2\pi A}{\lambda} \qquad (18.24)$$

or

$$u = \frac{2\pi Ac}{\lambda} \qquad (18.25)$$

But we know from Equation 18.20 that

$$u = \frac{gA}{c}$$

so that

$$\frac{gA}{c} = \frac{2\pi Ac}{\lambda} \qquad (18.26)$$

or

$$c^2 = \frac{\lambda g}{2\pi}$$

or

$$c = \sqrt{\frac{\lambda g}{2\pi}} \qquad (18.27)$$

This is the equation for the velocity of waves in deep water. We see that c is directly proportional to the wavelength and independent of depth. Therefore long waves will travel faster than short waves.

In order to consider the motion of water particles in the surface, consider point P in Figure 18.14b. Where as before our wave form is kept stationary by an opposite current of velocity

c, the velocity at the crest is $c-u$ and the height of P is $A \sin 2\pi x/\lambda$. The velocity v at P is given by the abbreviated form of the energy relation

$$v^2-(c-u)^2=2gA\left(1-\sin\frac{2\pi x}{\lambda}\right) \tag{18.28}$$

but $gA=uc$, so

$$v^2-c^2-2uc-u^2=2uc\left(1-\sin\frac{2\pi x}{\lambda}\right) \tag{18.29}$$

or

$$v^2=c^2+u^2-2uc\sin\frac{2\pi x}{\lambda} \tag{18.30}$$

If we write (Fig. 18.14c)

$$\phi=\frac{2\pi x}{\lambda}-\frac{\pi}{2} \tag{18.31}$$

so that

$$\sin\frac{2\pi x}{\lambda}=\cos\phi \tag{18.32}$$

we have

$$v^2=c^2+u^2-2uc\cos\phi \tag{18.33}$$

This equation is shown geometrically in Figure 18.14c where it can be seen that v comprises the constant velocity c of the adverse current and a constant velocity u. As a particle passes over the wave form, $2\pi x/\lambda$ decreases. The velocity u thus rotates around the circle in Figure 18.14c in a clockwise direction. Removing the artifice c, we find that to a first approximation the paths of water particles on deep-water waves of small amplitude are circles.

The above derivations are taken from Tricker (1964), to which the reader is referred for a delightful and elegant introduction to wave and tidal theory from the viewpoint of general physics.

19 Deltas

19a Introduction

A good case could be made for regarding deltas as the single most important clastic sedimentary environment *solely* on the basis of the enormous reserves of coal, oil and natural gas that are located within ancient deltaic sediments. A similar case could have been made thousands of years ago since deltas were amongst the most prized areas of agricultural land in ancient times, witness the great civilisations that grew up based upon deltas such as the Nile and Tigris–Euphrates.

A study of over thirty modern deltas (Coleman & Wright 1975) reveals considerable variability in delta morphology which may be readily understood in terms of the extent of river, wave and tide dominance of the coastal area (Table 19.1, Fig. 19.1). It can be seen that the 'classical' Herodotus delta of the Nile is dominated by wave processes whereas the peculiar modern Mississippi birdsfoot delta is dominated by river processes. Thus we see at the outset that delta morphology is a result of the conflict between river input of water and sediment and wave/tidal coastal reworking.

Although delta morphology and processes are varied, we may distinguish two broad environments on all active deltas: the **delta front** and the **delta plain**. The delta plain comprises mainly the sub-aerial portion of the delta with its distributary channels, marshes, backswamps and lakes. The delta front comprises the distributary mouth areas of effluent discharge, lagoons, tidal channels, barrier beaches and interdistributary bays.

Table 19.1 Factors affecting delta regime, morphology and facies (after Elliot 1978a)

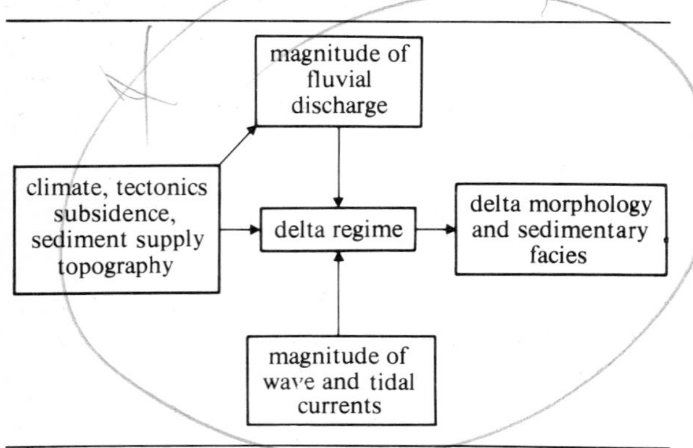

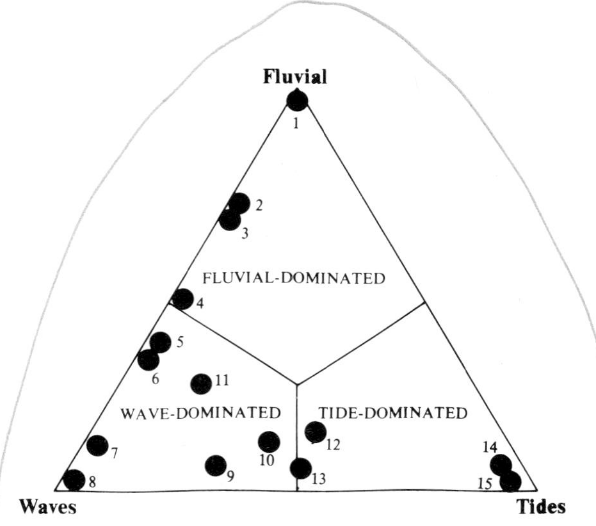

Figure 19.1 Ternary diagram to show how delta morphology may be qualitatively related to the nature of the predominant process(-es) at the delta front. Deltas: 1, Mississippi; 2, Po; 3, Danube; 4, Ebro; 5, Nile; 6, Rhône; 7, Sao Francisco; 8, Senegal; 9, Burdekin; 10, Niger; 11, Orinoco; 12, Mekong; 13, Copper; 14, Ganges–Brahmaputra; 15, Gulf of Papua. (After Galloway 1975, Elliott 1978b.)

19b Physical processes

Let us first consider the nature of the discharge of sediment and fresh water from the mouth of a major delta distributary. The degree to which this discharge is modified by wave and tide will control the gross morphology of the delta and hence its facies patterns. Initially we shall study a river-dominated delta front such as might develop in a lake, a microtidal sea of small extent, or in regions bordered by low-gradient delta front slopes that can attenuate incoming wave power (*q.v.*). Three factors may influence the nature of the sediment-laden freshwater jet: (a) inertia and turbulent diffusion, (b) turbulent bed friction and (c) buoyancy (Wright 1977).

Jets dominated by turbulent diffusion are **homopycnal** (negligible density difference between jet and ambient fluid) and show outlet Reynolds numbers greater than 3000, indicating turbulent dominance. Turbulent jet processes are summarised in Figure 19.2, together with the idealised depositional pattern that approximates to 'Gilbert-type' deltas characteristic of steep gradient streams entering deep lake bodies (Chs 14 & 16). Delta fronts of this type are relatively rare in marine environments.

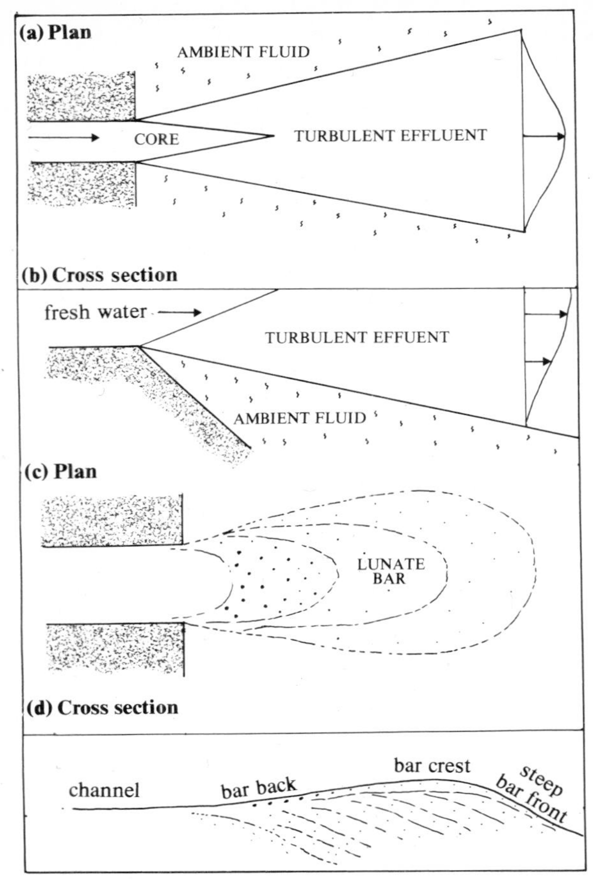

Figure 19.2 Plan and section views of inertia-dominated jets and their 'Gilbert-type' distributary mouth bars (after Wright 1977).

Since most marine basins shelve gently from shoreline regions, frictional effects arising from bottom drag on turbulent, homopycnal jet effluent are very important. Such jets experience rapid seaward spreading and deceleration (Fig. 19.3). Such **friction-dominated** jets quickly deposit a 'mid-ground' distributary mouth bar bordered on its margins by a Y-shape channel bifurcation. Bars of this type are particularly characteristic of subdelta growth in the Mississippi interdistributary bays (*q.v.*).

The extent of buoyancy influences on jet behaviour is expressed by the densimetric Froude number,

$$F' = \frac{\bar{u}}{\sqrt{gh'\gamma}}$$

where $\bar{u}$ is the mean effluent velocity, h' is the depth of the density interface from the surface of the jet and γ is the density ratio $1 - (\rho_f/\rho_s)$ where ρ_f and ρ_s are respectively the densities of fresh and sea water. Low values of F ($\leqslant 1$) suggest dominance by buoyant forces whereby the outflow spreads as a narrow expanding plume above a **salt wedge** that may extend for a considerable distance up the distributary channel. Such jets are termed **hypopycnal**. As we shall discuss further in the context of estuary behaviour in Chapter 20, salt wedges are best developed in deep channels with low tidal ranges. Internal waves that are generated at the salt-wedge/effluent boundary (Fig. 19.4) cause vertical mixing, rapid deceleration, and deposition of coarse sediment as a simple mouth bar. During periods of high river flow the salt wedge is expelled from the distributary channel to a position just seaward of the bar crest where bedload deposition occurs as the effluent separates from the salt water (Fig. 19.5). Continuing deposition of successively finer sediment occurs on the seaward bar slope. Good examples of such seaward-fining distributary mouth bars occur at the front of the major Mississippi outfalls.

Let us now turn to the modifying effects that waves and tides have on these simple jet models of delta front dynamics (Wright & Coleman 1973).

Wave power (Ch. 18) is substantially reduced as waves pass from offshore areas over very gently sloping offshore–nearshore zones; indeed some extremely gentle offshore slopes may cause complete dissipation of wave energy. In coastal areas of high wave power relative to river discharge, the effluent jets may be completely disrupted by wave reworking. On the other hand, in areas of low wave power relative to river discharge the effluent jets will remain dominant (Table 19.2). The effects of wave reworking include shoreward transport of sand to form swash bars around a broad crescentic mouth bar. Oblique wave incidence will cause beach–spit systems to prograde parallel to the coast (Fig. 19.6). Extensive beach–barrier systems are generated away from the channel mouths, fed by longshore transport of sands.

On macrotidal coastlines tidal currents will tend to disrupt and destroy the salt wedge pattern we noted above (see also Ch. 20) and will superimpose a significant to-and-fro motion upon the essentially unidirectional river input. The distributary mouths tend to be funnel shaped and to pass seawards into a zone of linear tidal current shoals which form from the disrupted mouth bar sands of fluvial origin (Fig. 19.7).

These discussions of physical processes enable us to look again at Figure 16.1 and to explain the gross morphology of the various delta types in terms of river, wave and tide processes.

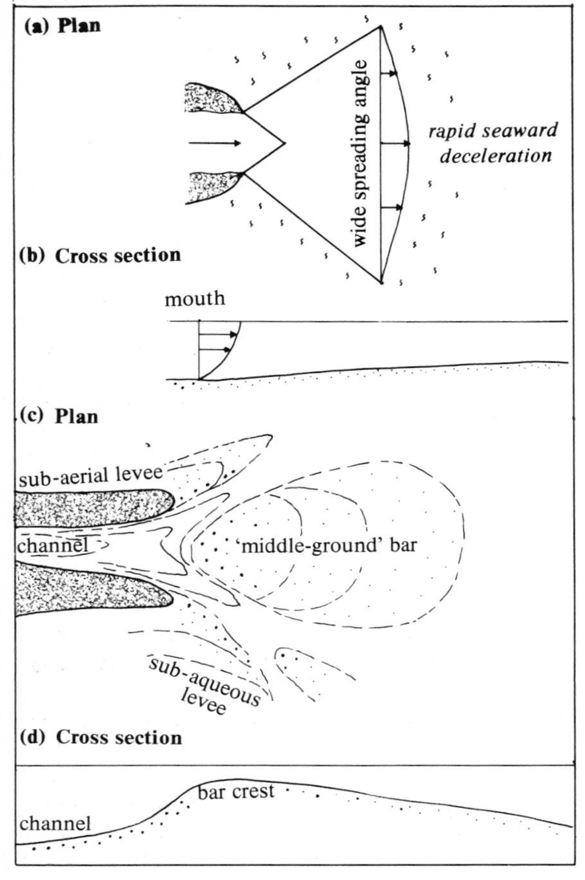

Figure 19.3 Plan and section views of friction-dominated jets and their 'middle-ground' distributary mouth bars (after Wright 1977).

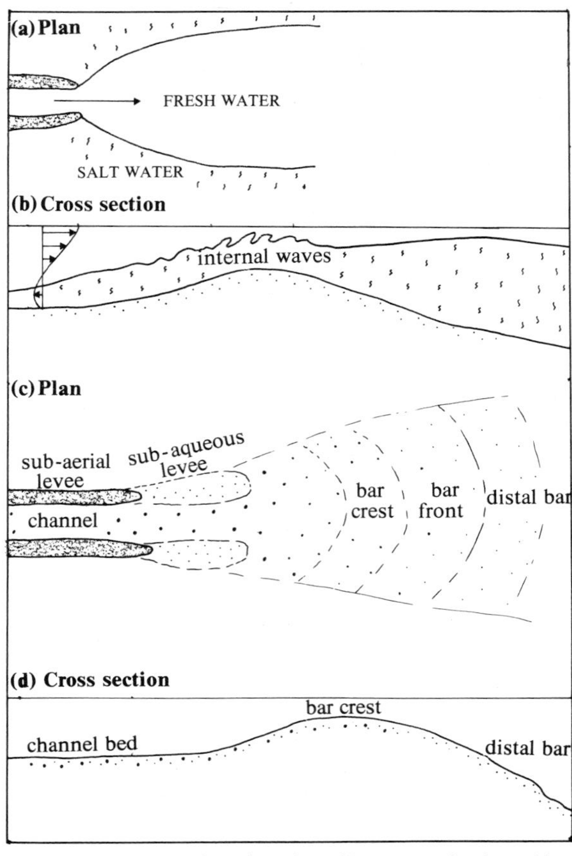

Figure 19.4 Plan and section view of buoyancy-dominated jets and their distributary mouth bars (after Wright 1977).

Table 19.2 To show river discharge:wave-power ratios for seven major deltas (after Wright & Coleman 1973). The **mean annual discharge effectiveness index** is obtained by dividing the discharge per foot of river mouth width (the total discharge in cubic feet per second divided by the total width of all outlet channels) by the wave power per foot of wave crest at the coast. The index, although dimensionally incorrect, thus gives an idea of the 'strength' of the river relative to wave power. The **mean annual attenuation ratio** is given by the ratio of deep-water wave power to nearshore wave power times the nearshore refraction coefficient (an index of wave power concentration or dissipation due to refraction). A value of 1.0 for the ratio indicates no reduction in wave power due to bottom friction. A value of 100 indicates that 1% of deep-water power is preserved into nearshore areas.

Delta	*Mean annual deepwater wave power (ft-lb/sec)*	*Mean annual nearshore wave power (ft-lb/sec)*	*Mean discharge (cusecs × 10^3)*	*Mean annual discharge effectiveness index*	*Mean annual attenuation ratio*
Mississippi	237.4	0.03	624.6	5477.0	7913.3
Danube	51.7	0.03	222.0	1171.0	2585.0
Ebro	168.8	0.11	19.5	267.8	1299.5
Niger	152.1	1.48	384.8	4.4	202.8
Nile	306.2	7.49	52.2	3.2	42.5
San Francisco	834.6	22.40	110.2	1.3	37.2
Senegal	351.9	84.60	27.2	0.3	4.2

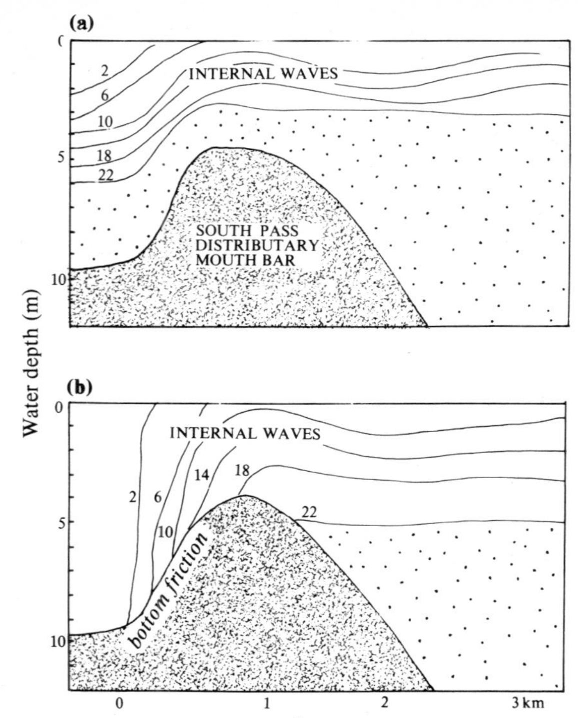

Figure 19.5 Longitudinal section through the South Pass distributary mouth (a) at low river stage and (b) at high river stage. Note the pronounced intrusive salt wedge present during low river stages. Sediment deposited below this wedge is mostly flushed out during high river discharges. Numbers refer to water salinity values (‰). (After Wright & Coleman 1973.)

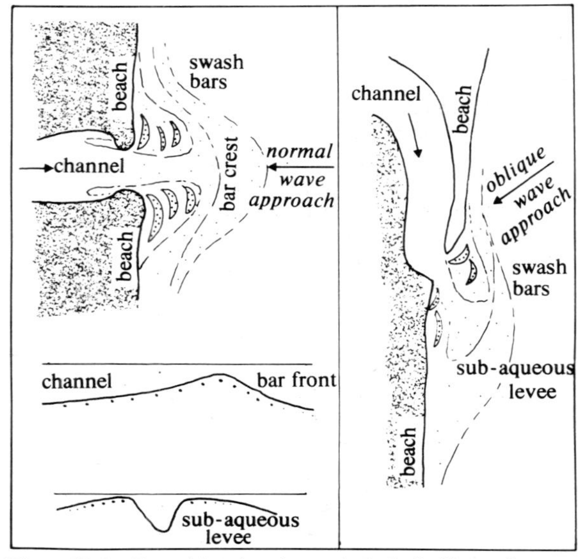

Figure 19.6 Depositional and morphological effects of strong wave activity at distributary mouths (after Wright 1977).

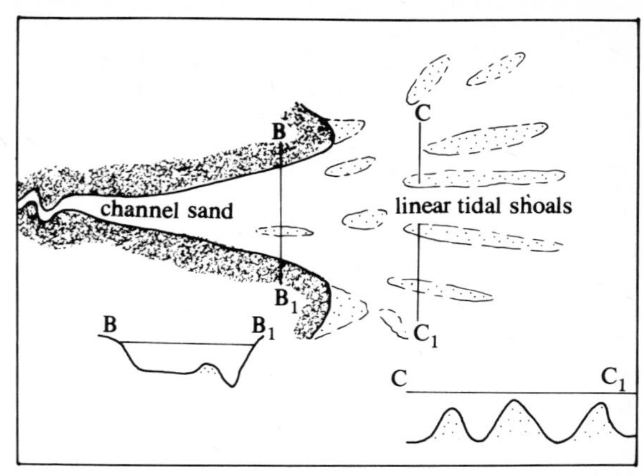

Figure 19.7 Depositional and morphological effects of strong tidal current activity at distributary mouths (after Wright 1977).

19c Modern deltaic facies

River-dominated deltaic environments, with well developed buoyant forces dominant during jet discharge, are best examined in the Mississippi delta where more than fifty years of detailed study provides a very large amount of both surface and subsurface data (Fisk 1944 & Fisk *et al.* 1954, Coleman & Gagliano 1964, Coleman *et al.* 1964). The 'exposed' delta plain is dominated by a small number of large distributaries and a host of minor ones (Fig. 19.8). Because of the very low slopes involved, these channels tend to be straight (Ch. 15b). Frequent avulsion occurs as the delta progrades, the channels periodically seeking new routes to the sea with gradient advantages. The 'birdsfoot' morphology of the delta results from these avulsions, with the 'claws' marked by channels and the 'webbed' connections marked by interdistributary bays. The bays are shallow, brackish to marine, and are gradually infilled by minor crevasse deltas and by overbank flooding until they became part of the sub-aerial marsh of the delta plain (Fig. 19.9). Cores through the prograding marshes and minor channels of the interdistributary bays reveal a variety of coarsening-upwards successions, capped by vegetation colonisation surfaces and sharp-based fining-upwards successions that resulted from deposition in the minor channels (Elliott 1974). At the delta front the distributaries flare slightly and pass into well developed lunate distributary mouth bars. Shoaling from the distributary to the mouth bar crest means that most of the latter will not become preserved and that continued delta progradation will cause the thick (50–150 m) coarsening-upwards mud-to-sand succession of the mouth bar to be cut by an erosive-based, fining-upwards distributary channel sand. Deposition of thick successions of muds in the delta front area

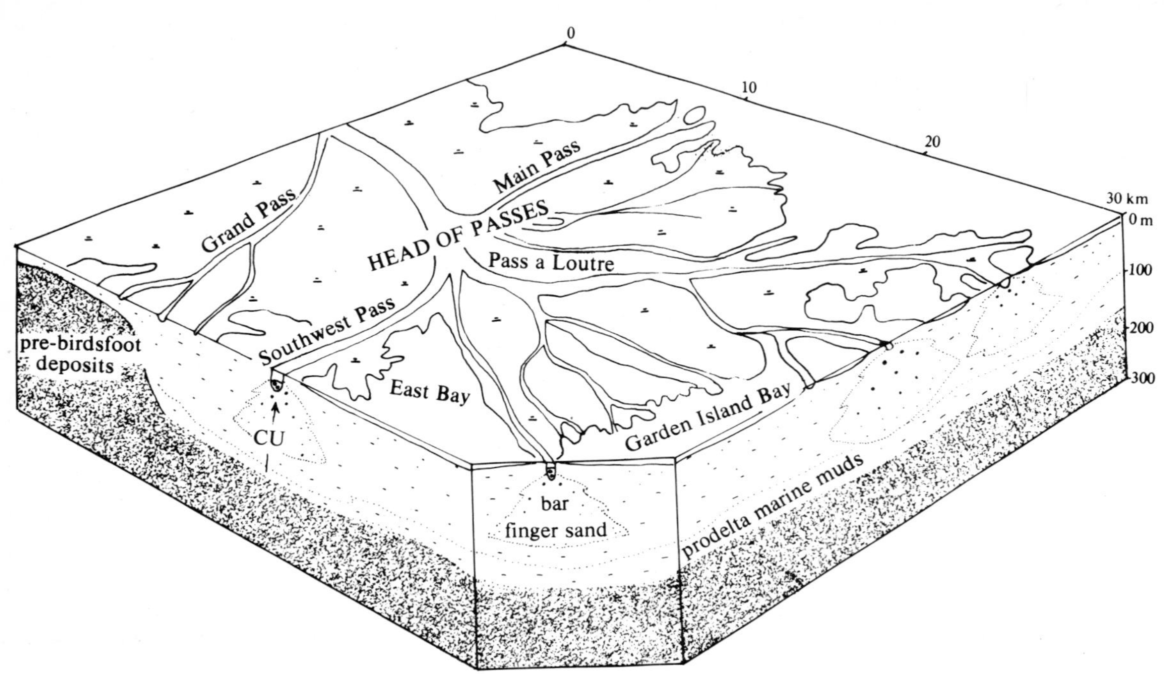

Figure 19.8 Block diagram to show the morphology and major facies present around the front of the modern Mississippi birdsfoot delta. CU – coarsening upwards. Bar finger sands form after continued progradation of distributary mouth bar sands. (After Fisk *et al.* 1954.)

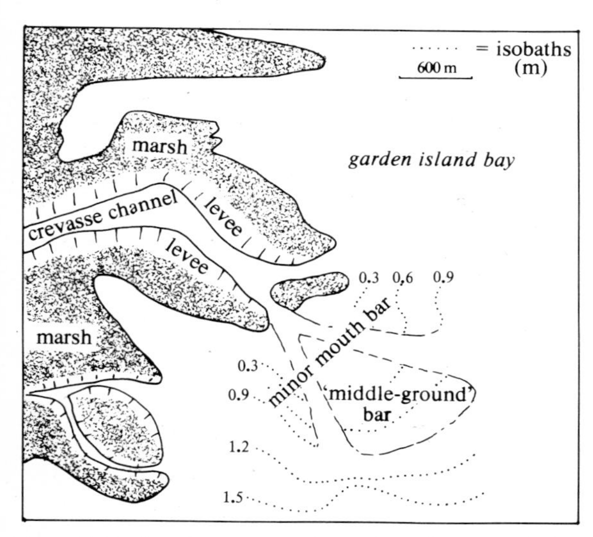

Figure 19.9 Sketch map to show the minor mouth bar (friction-dominated regime *cf.* Fig. 19.3) present at the front of a minor crevasse channel issuing into Garden Island interdistributary bay (after Coleman *et al.* 1964).

has encouraged the development of a wide range of soft sediment deformation, slump and growth fault features (Fig. 19.10).

Over the past few thousand years the active parts of the Mississippi delta have undergone periodic shifts (avulsions, see Ch. 15) along the coast of Louisiana as successive channels searched for gradient advantages over their precursors (Fig. 19.11). Transfer of the river to a new location (see Rouse *et al.* 1978 for a modern development) causes abandonment of the delta constructional system (Frazier 1967). Compaction of the pro-delta muds causes subsidence and further encourages wave reworking of the abandoned deltaic sediments. This process thereby leads to the production of a characteristic abandonment facies at the top of a delta-lobe facies association (Fig. 19.12). A mechanism thus exists within a switching delta for producing deltaic 'cycles', so long as the coastal plain is located in an area of net tectonic subsidence (as is indeed the case in the Gulf of Mexico).

By way of contrast to the Mississippi delta we now examine the mixed regime (tide/wave dominated) Niger delta (Allen 1965b, Oomkens 1974). The lower deltaic plain and coastal environments of the Niger delta are dominated by the occurrence of major coastal barrier islands separated by tidal inlets (Fig. 19.13). The great trunk stream of the braided Niger river (the seventh

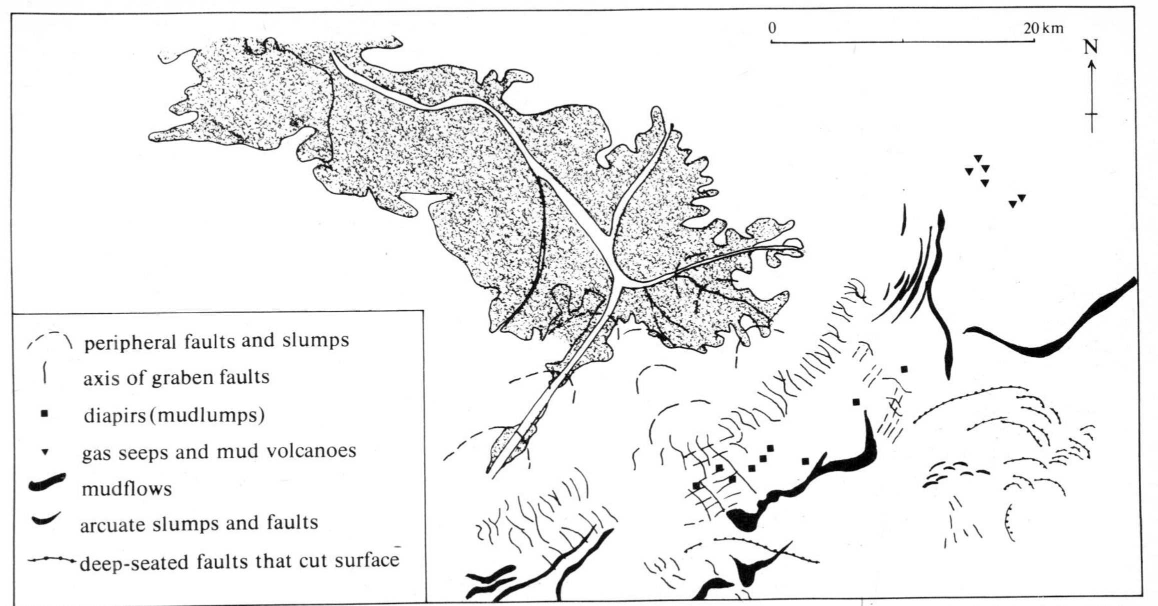

Figure 19.10 Distribution of sub-aqueous mass-movement features and sedimentary faults off the Mississippi delta (after Coleman 1976).

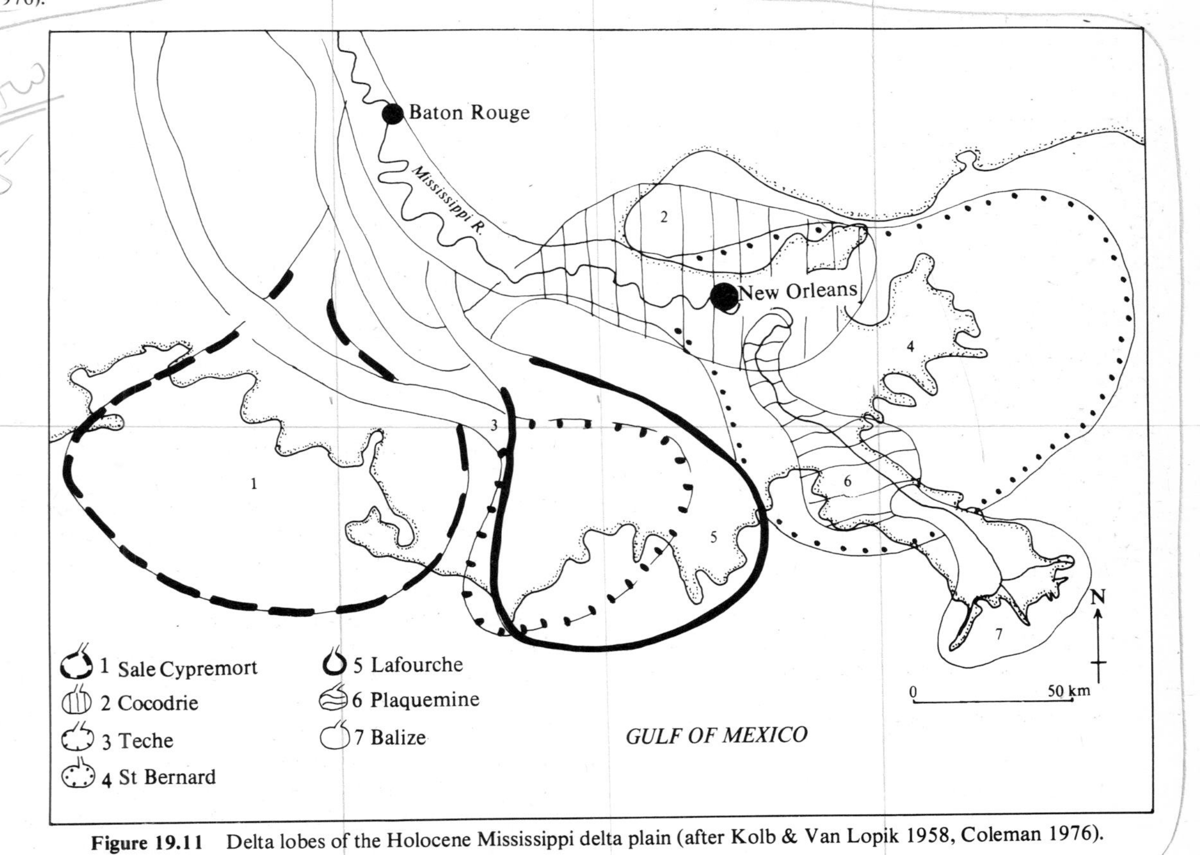

Figure 19.11 Delta lobes of the Holocene Mississippi delta plain (after Kolb & Van Lopik 1958, Coleman 1976).

largest in the world) thus breaks up into a host of smaller channels which are each tide-dominated. Sandy bedload deposited in ebb tidal deltas is redistributed along shore by severe wave action. Facies fence diagrams thus show a delta front dominated by tidal inlet channel fills and coastal barrier sands (Fig. 19.13). Continued delta progradation might result in some reworking and destruction of these facies at the expense of the upper delta plain facies.

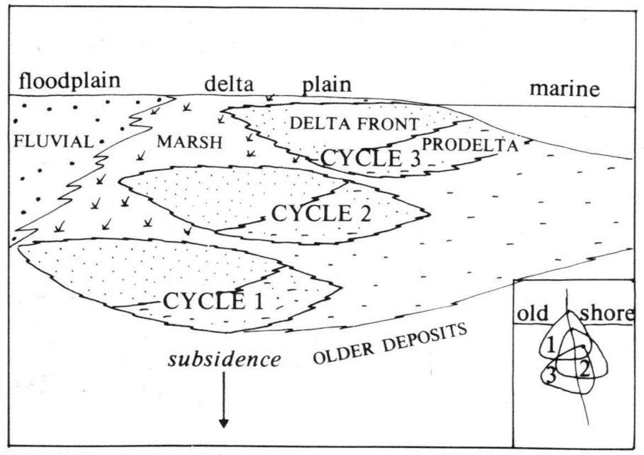

Figure 19.12 To show how successive deltaic cycles may be produced by delta lobe switching in a subsiding and prograding coastal plain (after Coleman & Gagliano 1964).

Deltas on macrotidal coasts where wave action is limited, such as the Ganges–Brahmaputra (Coleman 1969) and Gulf of Papua (Fisher *et al.* 1969), show a very distinctive delta front facies dominated by a dense network of tidal channels and islands which pass offshore into a network of coast-normal linear tidal current ridges (see Figs 19.7 & 21.3). Delta progradation causes a gradual emergence of the tidal current ridges so that they become coated with a sequence of intertidal and supratidal fine-grained sediments. The areas between the tidal current ridges eventually become tidal channels and ultimately fluvial channels as progradation continues.

19d Ancient deltaic facies

Many modern deltas are prograding into subsiding continental margins, particularly those passive margins of the Atlantic type. Subsurface exploration in the search for oil has revealed that many such deltas are simply the latest surface expression of a long-lived drainage basin outlet so that great thicknesses (2–8 km) of deltaic facies exist in the subsurface. Superb examples occur in the subsurface of both the Gulf of Mexico (Fig. 19.14) and the Nigerian Atlantic coast. Very detailed electric logging, facies analysis and micropalaeontology have facilitated

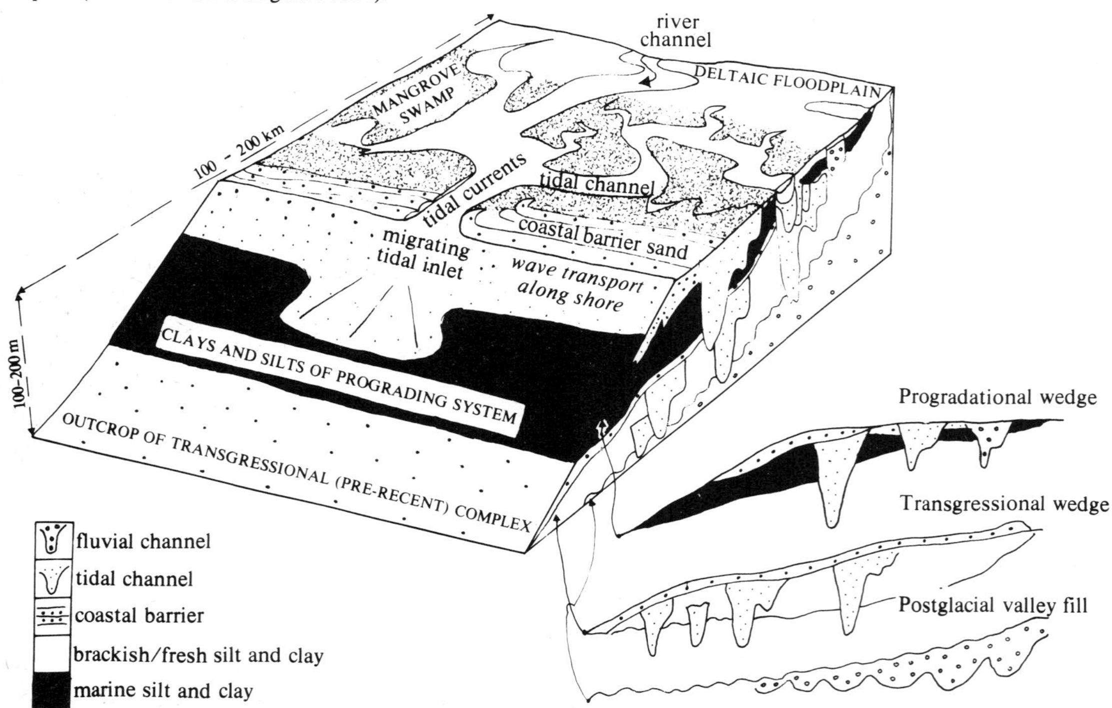

Figure 19.13 Lithofacies relations of the Late Quaternary Niger Delta (after Oomkens 1974). Discussion in text.

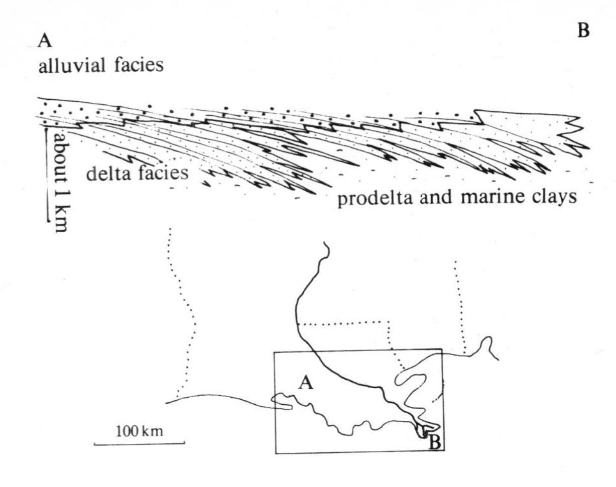

Figure 19.14 Schematic composite dip section through southern Louisiana to show successive Miocene progradation of proto-Mississippi delta lobes and periodic transgressive events. Note the diachronism of the three main facies. (After Curtis 1970.)

identification of contemporaneous delta systems and of successive delta lobe advancements. As we shall discuss later (Ch. 31f), such large-scale delta progradations form near-perfect conditions for oil entrapment. Electric log data for the Eocene Wilcox Group (Ch. 31) has enabled several authors to distinguish between fluvial and wave-dominated delta lobes (Fisher 1969, Fisher & McGowen 1969).

Studies of cores and mine sections through Carboniferous coal-bearing sediments in the mid-continent and Appalachian area of the USA have enabled impressive facies maps (Fig. 19.15) to be prepared showing the broad sweep of deltaic lithofacies over enormous areas (Wanless *et al.* 1970). The details of lateral changes across the Pocahontas and Dunkard basins in West Virginia and Pennsylvania (Ferm & Cavaroc 1969, Ferm 1974) reveal a wave-dominated delta front with extensive well-washed mature sandstones interpreted as barrier–beach facies (Fig. 19.16).

Two particularly detailed studies of ancient deltaic facies, both in northern England, are those of Elliott (1975) on a Yoredale cycle of Namurian age and of Allen (1960), Reading (1964), Walker (1966), Collinson (1969)

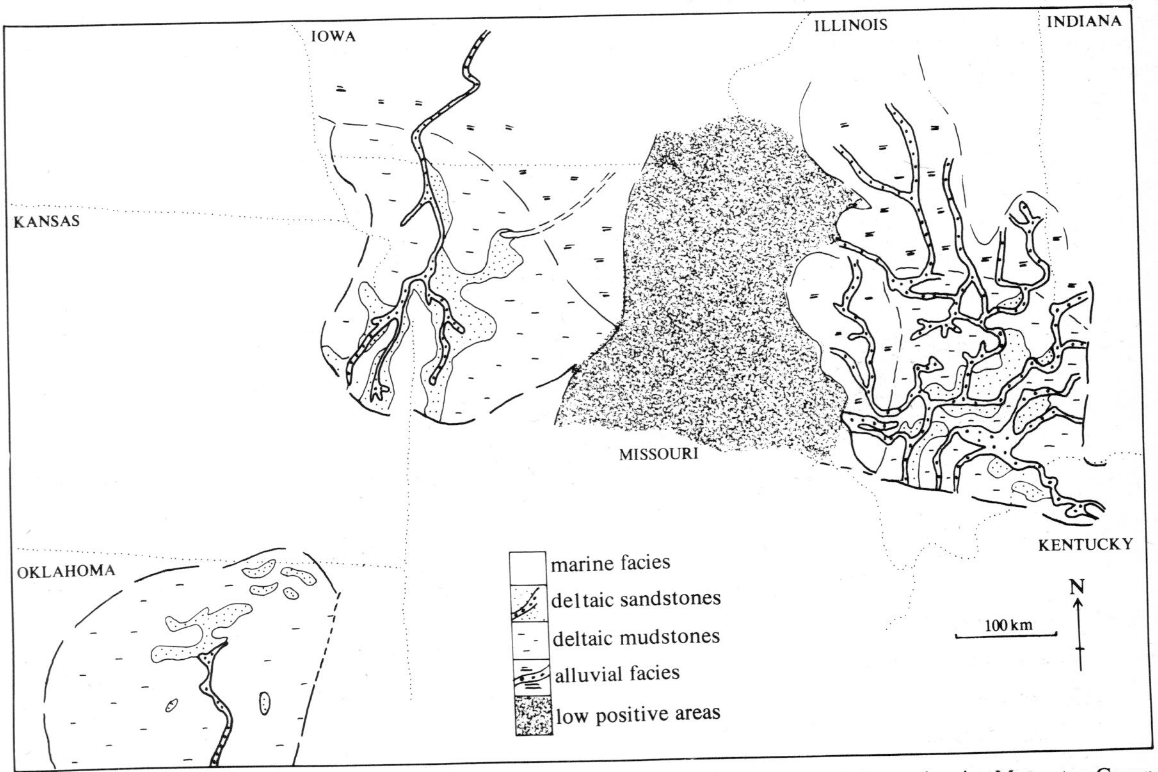

Figure 19.15 The distribution of facies associated with three huge delta complexes in the Middle Pennsylvanian Marmaton Group of northeastern USA. (After Wanless *et al.* 1970.) Note the scale.

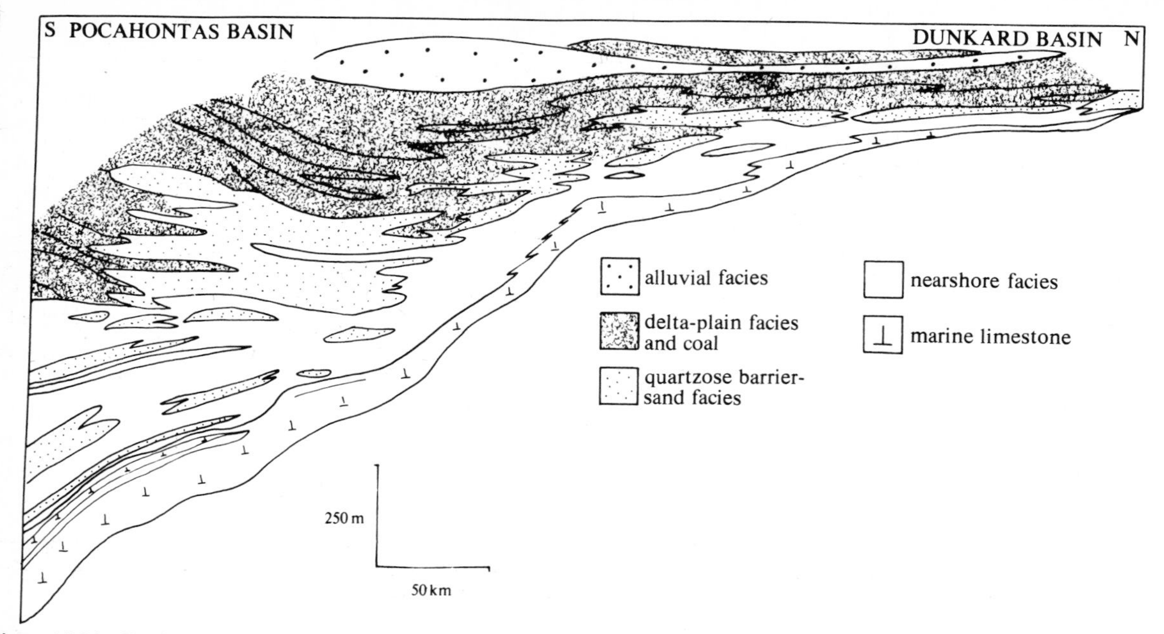

Figure 19.16 Environmental stratigraphic cross section across the Pocohontas and Dunkard basins in West Virginia and Pennsylvania to show deltaic facies with coastal barrier development at the delta front (after Ferm & Cavaroc 1969).

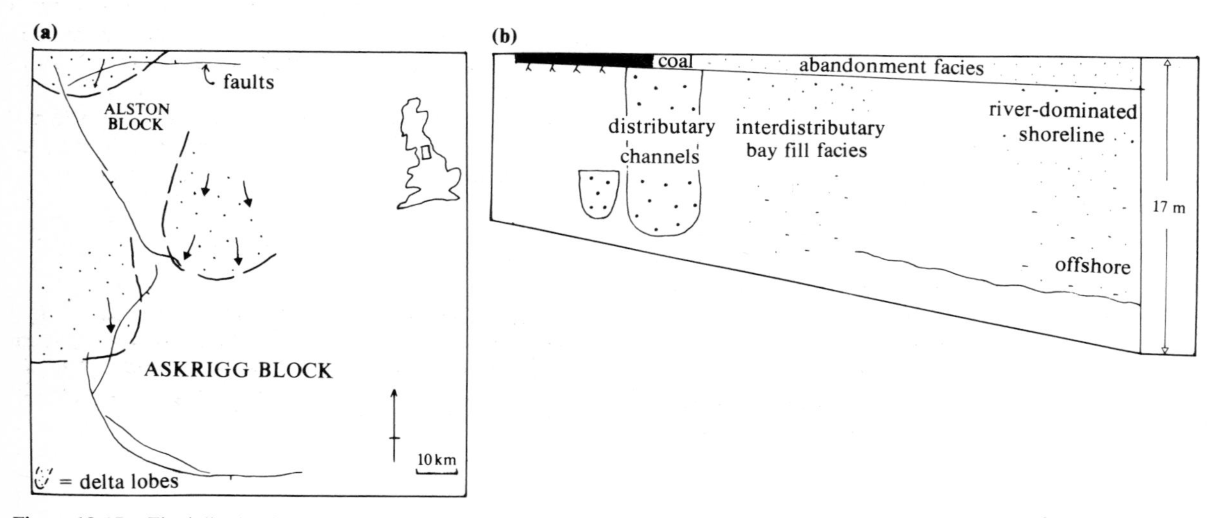

Figure 19.17 Fluvially dominated delta lobes from a Carboniferous Yoredale cyclothem in northern England. (a) Map to show three separate delta lobes and (b) longitudinal (N–S) section through the central lobe (after Elliott 1975).

and McCabe (1977) on the Namurian Millstone Grit deltas. The Yoredale cycle contains three fluvially dominated delta lobes whose recognition was possible by using lobe abandonment facies such as coals and beach/barrier sandstones as markers (analogous to the Chandeleur Islands of the Mississippi embayment) (Fig. 19.17).

The 'Millstone Grit' delta complex comprises two offlapping sequences which each pass upwards from basinal mudstones via turbiditic sandstones to delta plain sandstones deposited by very large low-sinuosity rivers. The delta slope was cut by deep submarine channels which fed submarine fans (Fig. 19.18). Little evidence for wave or tidal reworking of the delta front facies is seen.

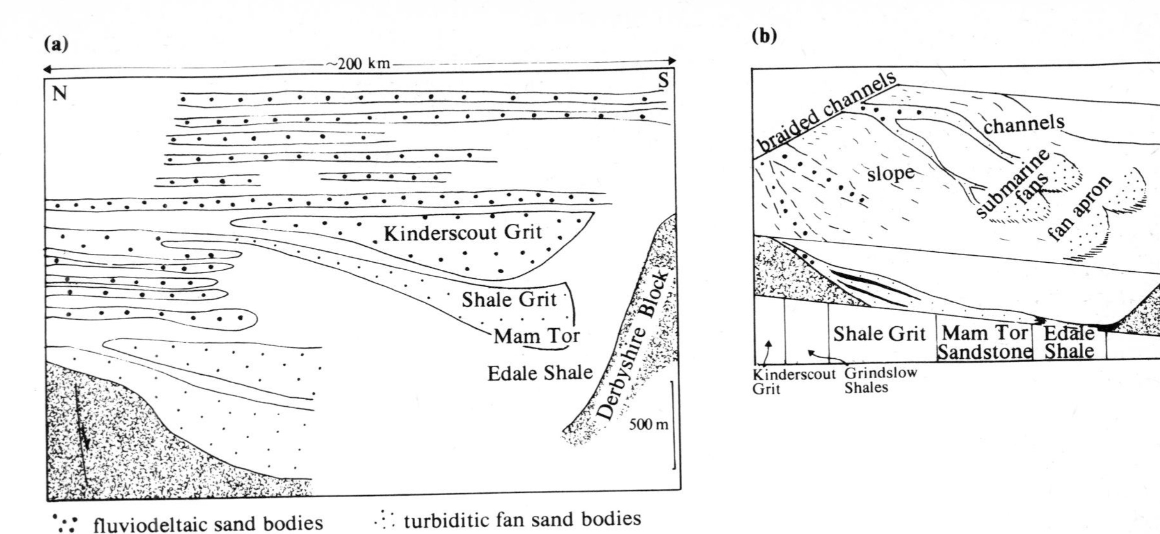

Figure 19.18 (a) N–S section through the Central Pennine Namurian basin of northern England to show the major lithostratigraphic units and their broad interpretation. (b) A palaeogeographic reconstruction of the Kinderscout delta system (after work by Allen 1960, Reading 1964, Walker 1966, Collinson 1969 and McCabe 1977, as summarised by Elliott 1978b).

Similar facies of Namurian age in western Ireland (Rider 1978) show spectacular developments of growth faults (see Ch. 11).

19e Summary

A rational approach to deltaic sedimentation involves the recognition that the jet of sediment-rich fresh water issuing from a distributary mouth is acted upon by (a) forces arising from deceleration due to diffusion and turbulence, (b) bed friction, (c) buoyancy forces, (d) wave currents and (e) tidal currents. The interaction between these various factors is responsible for the variety of deltaic morphology observed at the present day. Deltas may be more crudely divided into river, wave and tide dominated, and the sequences of sediment deposited vary accordingly. The coarsening-upwards cycles produced at the delta front are cut by both fining-upwards distributary channel sands of the delta plain or by estuarine tidal channel deposits of the delta front. The topmost part of a progradational deltaic wedge comprises the fine-grained sediment and peats deposited in the on-delta backswamps. The relatively coarse-grained deposits of the delta front and associated channels provide fine potential oil and gas reservoirs whilst most coal deposits owe their origin to peat growth in sediment-starved backswamps.

Further reading

The best updating of Bates' (1953) original application of the jet concept to deltas is given by Wright (1977). Modern and ancient deltaic sediments are well summarised by Coleman (1976) and Elliott (1978b). A host of interesting papers occur in the volumes edited by Morgan (1970) and Broussard (1975).

20 Estuaries

20a Introduction

An estuary may be defined as 'a semi-enclosed coastal body of water which has a *free* connection with the open sea and within which sea water is measurably diluted with fresh water of river origin' (Pritchard 1967). Note that this definition restricts the term to the dynamic interface between river and sea water so that tidal reaches of rivers are not necessarily estuarine, neither are large funnel-shaped marine basins such as the Fundy or Severn coastal inlets. It should also be noted that many delta distributary outlets show estuarine characters.

Drowned river valleys are by far the most important type of modern estuary and were formed in response to the early Holocene postglacial sea-level rise. Such estuaries are most common along the low-lying coastal plains of Atlantic-type (passive) continental margins. Frequently these estuaries are fronted on the marine side by barrier islands or spits, with marine interchange between offshore, bay and estuary through tidal inlets. Fine examples occur in New England and along the southern USA Atlantic coastal plain, whilst many funnel-shaped estuaries occur around the northwestern European coastline. The prominence of estuaries along such modern coasts may be a misleading clue as to their importance in the geological past, since many modern estuaries are being rapidly infilled because of high sedimentation rates. Although estuaries undoubtedly existed in the geological past, it is likely that they were much less obvious than those major drowned river valleys that dominate the coastal scene today.

20b Estuarine dynamics

The mixing of fresh and salt water causes estuarine circulation in response to density gradients. Sedimentary particles may be of both marine or river origin with both flocculation (see Kranck 1975, 1981) *and* resuspension of faecal pellet material as important control upon particle size (Schubel 1971b).

Water and sediment dynamics in estuaries are closely dependent upon the relative magnitude of tidal, river and wave processes. The most fundamental way of considering estuarine dynamics is through the salt-balance principle (Pritchard 1955) which states in mathematical form that the time rate of change of salinity at a fixed point is caused by two contrasting processes: **diffusion** and **advection**. Diffusion is restricted to the flux of salt by turbulent mixing whilst advection is the mass flux of both water and salt associated with circulation and internal breaking waves. Viewed in this way, sediment and water dynamics in estuaries may be conveniently represented by four end members which intergrade one with another (Pritchard & Carter 1971, Schubel 1971).

Type A estuaries (Fig. 20.1a) are those river-dominated estuaries where tidal and wave mixing processes are at a minimum. The system is dominated by an upstream tapering salt wedge over which the fresh river water flows as a buoyant plume. The picture is exactly that discussed previously with respect to river-dominated delta distributaries such as the modern Mississippi delta (Ch. 19). Internal waves form at the sharp salt-wedge/river-water interface and these cause limited upward mixing of salt water with fresh water (advection), but not vice versa. A prominent zone of shoaling at the tip of the salt wedge arises when sediment deposition from bedload occurs in both fresh and sea water. This zone of deposition shifts upstream and downstream in response to changes in river discharge and, to a much lesser extent, to tidal oscillation. Thus, deposited fine bedload sediment and flocculated suspended load is periodically flushed out of the system by turbulent shear during high river stage.

Type B estuaries (Fig. 20.1b) are termed 'partially mixed' because of the effects of tidal turbulence which destroy the upper salt wedge interface and produce a more gradual salinity gradient from bed to surface water by both advectional and diffusional mechanisms. Down-estuary changes in the salinity gradient at the mixing zone occur so that the zone moves upwards towards higher salinities. Earth rotational effects cause the mixing surface to be slightly tilted so that in the northern hemisphere the tidal flow up the estuary is nearer the surface and strongest to the left. Sediment dynamics will be strongly influenced by the upstream and downstream movement of the salt water over the various phases of the tidal cycle (Allen *et al.* 1976). Sediment particles of river origin, some flocculated, will undergo various transport paths, usually of a 'closed loop' kind, in response to settling into the salt layer and subsequent transport by the net up-stream tidal flow. This is the origin of the **turbidity maximum** found in the lower levels of estuarine water columns (Fig. 20.2).

It is important to note that many estuaries show mixed

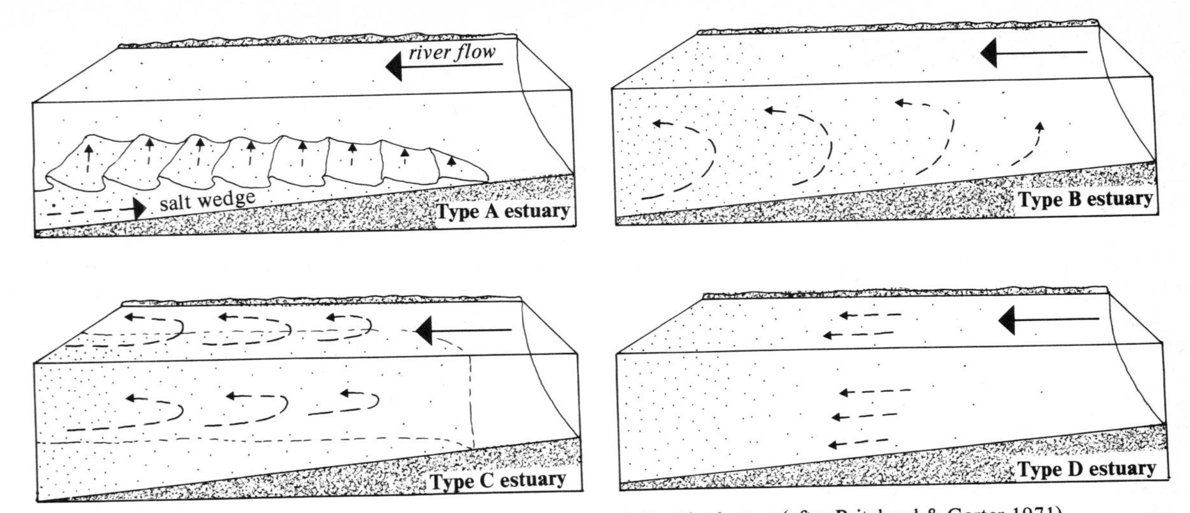

Figure 20.1 Schematic diagrams to show the four types of estuaries defined in the text (after Pritchard & Carter 1971).

A/B or B/C characteristics depending upon the relative magnitude of river and tidal flow at particular times of the year.

Type C estuaries (Fig. 20.1c) are termed vertically homogeneous because strong tidal currents completely destroy the salt-wedge/fresh-water interface over the entire estuarine cross section. Longitudinal and lateral advection and lateral diffusion processes dominate. Vertical salinity gradients no longer exist but there does exist a steady downstream increase in overall salinity. In addition, the rotational effect of the Earth still causes a pronounced lateral salinity gradient, as in Type B estuaries. Sediment dynamics are dominated by strong tidal flow, with estuarine circulation gyres produced by the lateral salinity gradient. Extremely high suspended sediment concentrations (caused by flocculation?) may occur close to the bed in the inner reaches of some tidally-dominated estuaries (e.g. Thames, Prentice *et al.* 1968). Generally, however, sediment traps caused by turbidity maxima and salt wedge effects should be absent because of the highly effective tidal mixing process.

Type D estuaries (Fig. 20.1d) are theoretical end members of the estuarine continuum and they show both lateral and vertical homogeneity of salinity. Under equilibrium conditions salt is diffused upstream to replace the salt lost by advective mixing. Sediment movement should be dominated entirely by tidal motions, again with no internal sediment trap.

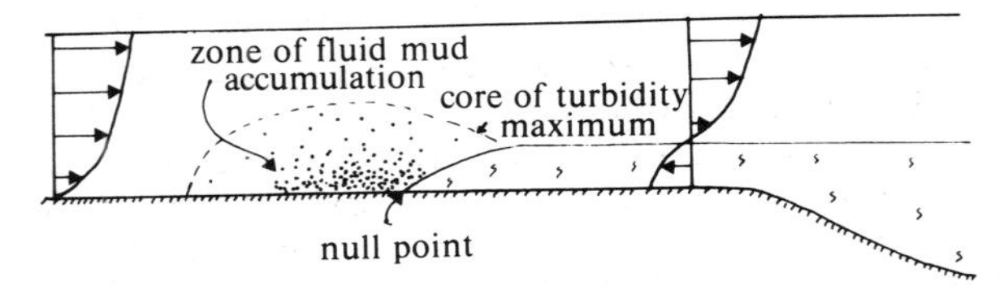

Figure 20.2 Schematic longitudinal section to show the development of a turbidity maximum and the zone of fluid mud accumulation around a null point in a partially-mixed estuary (after G. P. Allen 1971).

Our brief review of estuarine dynamics has stressed the interplay between river and tide. Most estuaries are those with Type-B characters and hence tend to act as trappers of sediment, particularly of fine grades. Whilst recognising the efficiency of such estuaries as sediment traps, it should also be pointed out that advective plumes of seaward-directed fine sediment may be driven by both residual tidal and wave currents far out on to the shelf (see Ch. 22).

20c Modern estuarine facies

Type B to D estuaries – those with appreciable tidal ranges – are bordered by coastal tidal flats with salt marsh (mangrove swamps in tropical estuaries) and minor drainage channels. Seaward progradation of funnel-shaped estuarine complexes on a coastal plain will thus cause the deeper-channel estuarine environments to become progressively overlain by fining-upwards tidal flat facies, as discussed in Chapter 21. Studies of the main estuarine channels in Type C estuaries (Hayes 1971) such as the Parker estuary, New England, show well developed ebb and flood tidal channels with extensive bordering dunefields. Sediment particles are dominantly derived from offshore. Thus in the Thames estuary, England, fully marine ostracods are found up to 20 km inland from their life habitats and show a decreasing mean grain size upstream (Prentice *et al.* 1968). Vertical sequences through such sediments would reveal alternating packets of ebb- and flood-orientated cross beds with clay interbeds resting on channel-floor erosion planes.

Important upstream changes in facies occur in many estuaries ending in tidally influenced point bars and lateral bars with much fine-grained material, including flaser, lenticular and wavy interlaminations (Howard *et al.* 1975). Both Type C and D estuaries merge seawards into tidally swept shelfs which may show tidal current ridges (Ch. 22).

Since estuaries are dominated by the mixing of fresh and salt water, the faunal assemblages present will be particularly adapted to salinity changes in both time and space. Ostracods are particularly useful in delineating salinity zones in estuarine facies and they also provide valuable indications of net transport, as dead transported valves are mixed with living assemblages (e.g. Prentice *et al.* 1968).

20d Ancient estuarine facies

In a prograding non-deltaic coastal plain truly estuarine facies should overlie and partly cut out a variety of marine nearshore facies. They, in their turn, will be overlain and partly cut out by fluvial facies. Within these broad limits the positive identification of truly estuarine facies will depend upon the recognition of tidal depositional processes and evidence for salinities intermediate between fresh and fully marine. Tidal processes are relatively easy to recognise in rock products. Recognition of a salinity spectrum will, of necessity, depend entirely upon biological inferences (e.g. Hudson 1963). It may therefore be impossible to recognise pre-Phanerozoic estuarine facies.

Bosence (1973) recognises the nature of the 'estuarine' problem in his study of the Eocene of the London Basin. Facies 1, composed of heterolithic silts with flaser and lenticular laminations and bimodal cross laminations, was ascribed to lateral accretion in a tidal channel environment. Facies 2, composed of mudflake conglomerates and planar cross stratification, was ascribed to vertical infilling of erosive scours. A simple marine tidal assignation of the beds was impossible because of the lack of body fossils and the rarity of burrows. An estuarine environment is possible for the sediments but this cannot be proven with available evidence (see further discussion by Goldring *et al.* 1978).

A rather different approach enabled Campbell and Oaks (1973) to postulate an estuarine origin for the Lower Cretaceous Fall River Formation of Wyoming. Here, supposed estuarine sandstones form seaward-imbricated complexes in which each younger body infills scours cut into older ones. The scour fills are enclosed by tidal flat or marine facies and pass up-dip into cross-stratified fluvial sandstones. Upper estuarine facies comprise ebb-orientated large-scale planar cross stratification whilst finer-grained lower estuarine facies comprise wave-rippled sands and flaser interlaminations. The location of the supposed estuarine facies in shallow erosive scours is attributed to periodic, storm-induced avulsions of major estuarine channels. Little palaeontological evidence was put forward to support this estuarine model. A fairly close modern analogue seems to be the Nith tidal estuary in the Solway Firth, Scotland, where periodic river/estuary channel avulsion (J. B. Wilson 1967) should lead to an interbedding of tidal-flat facies and estuarine/river channel facies dominated by tidal bedforms (de Mowbray 1980).

20e Summary

Estuaries are semi-enclosed coastal bodies of water which have a free connection with the sea. The mixing of fresh and salt water causes estuarine circulation in response to density gradients. Four types of estuaries may be defined according to the relative importance of freshwater and tidal mixing. Estuarine facies include channel muds, silts and sands moulded by seasonal river processes and tidal rhythms. These facies pass seawards into 'pure' tidal deposits and landwards into 'pure' river deposits. Estuaries are usually fringed by fine-grained tidal flats with salt marshes. Ancient estuarine facies must be identified with the aid of salinity-sensitive fauna and flora and by regional facies trends.

Further reading

The short course notes edited by Schubel (1971a) and the physical introduction by Dyer (1973) provide a good introduction to estuarine dynamics. A more advanced treatment is found in the two volumes edited by Wiley (1976) where papers on a variety of pure and applied estuarine problems may be found.

21 'Linear' clastic shorelines

21a Introduction

The major environments of coastal deposition comprise (a) **beach** and **intertidal flat**, (b) **barrier**, **spit** and **lagoon** complexes, (c) **chenier plain**, and (d) **shallow nearshore**. Complex mixtures of these environments may occur and all may coexist within major deltas and estuaries.

The distinction between beach and tidal flat has never been satisfactorily defined, but generally beaches occur as narrow intertidal to supratidal features dominated by wave action in which the sediment coarsens from offshore to onshore. In direct contrast, tidal flats are wide areas dominated by to-and-fro tidal motions in which the sediment fines from offshore to onshore. Tidal flats occur on open, macrotidal coasts or as part of back-barrier complexes on mesotidal coasts.

Coastal barriers do not occur on macrotidal coasts and they require some steady riverine or longshore supply of sand. Those barriers located on microtidal coasts are long and linear with sedimentary processes in the back-barrier lagoons dominated by **storm washover** effects. Barriers on mesotidal coasts are broken by frequent tidal inlets where large **flood tidal deltas** form in lagoons on the back-barrier side. True tidal flats occur in the back-barrier environment, sheltered by the barrier from most wave action.

Chenier plains are coastal plain environments of marsh and mud, broken by very extensive lateral shell ridges up to 50 km long and 3 m high. The classic cheniers occur in Louisiana, to the west of the modern lobe of the Mississippi delta. Periods of mudflat outbuilding coincide with abundant longshore mud supply from the Mississippi. Periods of mudflat erosion occur when mud supply is sparse and waves rework the nearshore mudflats and concentrate shell material as chenier storm ridges bordering the mudflats.

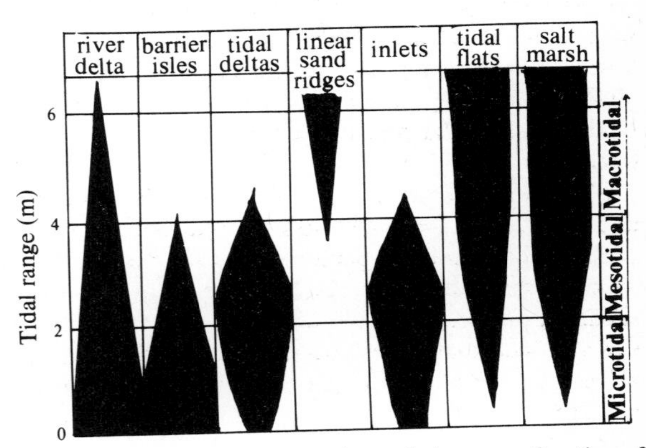

Figure 21.1 Variation of coastal morphology as a function of tidal range (after Hayes 1975).

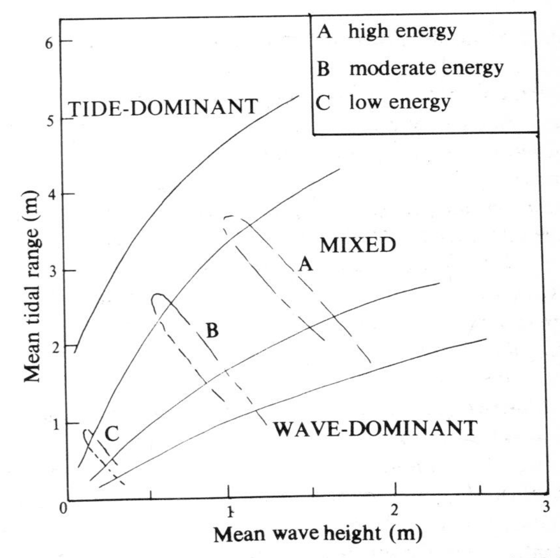

Figure 21.2 To show that shoreline embayments may be fully characterised by plotting local wave height against tidal range. High-energy arc A approximates to the German Bight, moderate energy arc B to the Georgia Bight, and low energy arc C to the west Florida Bight (after M. O. Hayes 1979).

21b Physical processes

We have previously discussed (Ch. 18) some aspects of wave and tide behaviour in general terms. Before we go on to examine some further details of these processes let us first look at the major controls upon coastal morphology. M. O. Hayes (1975, 1979) has shown that the variation of coastal morphology is a function of both tidal range (Fig. 21.1) *and* of wave energy (Fig. 21.2). Coastal environments with high tidal ranges show strong tidal currents. Assuming that sediment of all sizes is available, then a low-gradient coarsening-offshore tidal flat prism builds out by coastal progradation. Now not only is the incoming swell wave power greatly reduced by this low-gradient flat but also waves cannot break on any one part of the tidal flat for any length of time. The effectiveness of waves on such macrotidal coasts is thus

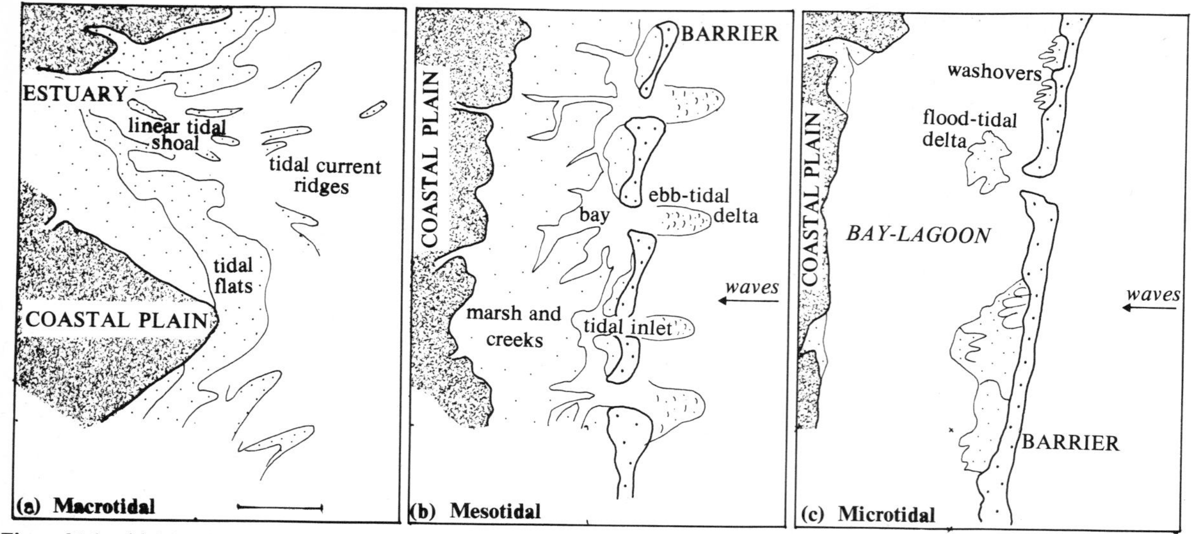

Figure 21.3 (a) Morphological summary of an idealised macrotidal coast with moderate wave energy. Note the absence of barrier islands and presence of tidal current ridges. (b) Idealised mesotidal coast with moderate wave energy. Note the barriers, tidal inlets and ebb-tidal deltas. (c) Idealised microtidal coast with moderate wave energy. Note the abundant washovers and the occasional flood-tidal deltas that lie inshore from the rare tidal inlets. (After M. O. Hayes 1979.)

greatly reduced. The opposite conclusions apply to microtidal coasts dominated by high wave power. Figure 21.3 summarises coastal morphology in terms of this model.

Turning to a more detailed consideration of beaches, let us briefly consider the nature of the beach profile (Fig. 21.4). Comparison of summer and winter beach profiles reveals major changes (Fig. 21.5). In summer, swell waves with low steepness values transport sediment onshore forming beach **berms**. In winter, storm waves with high steepness values transport sediment offshore forming **offshore bars**. Rip-current cells may occur in winter on steeper beaches. The beach face slope is governed by the asymmetry of offshore and onshore transport vectors. The seaward backwash is generally weaker than the shoreward movement of water from collapsing waves because of percolation and frictional drag on the swash. Sediment is thus moved up the beach slope until an equilibrium is established. Maximum percolation occurs on the most permeable gravel beaches and these usually have the highest slopes. Assuming that gravity opposes the net shoreward drift of sediment, Inman and Bagnold (1963) obtain the following relationship for local beach slope $\tan \beta$,

$$\tan \beta = \tan \phi \left(\frac{1 - c}{1 + c} \right) \qquad (21.1)$$

where ϕ is the coefficient of internal friction, c is the asym-

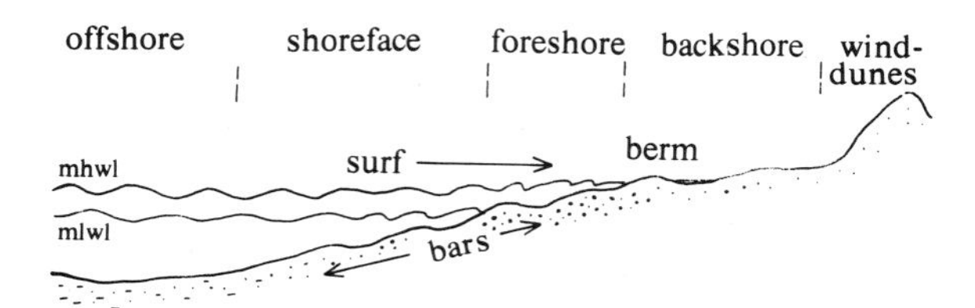

Figure 21.4 Morphological terms used in beach studies.

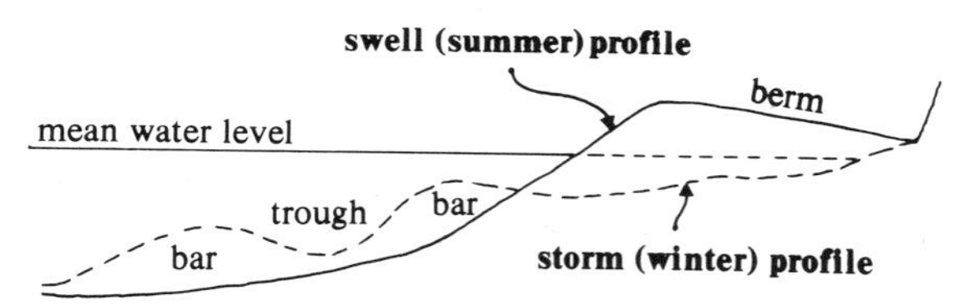

Figure 21.5 To show the contrast between summer and winter beach profiles (after Komar 1976).

metry term for offshore versus onshore wave energy. When $c = 1$ no asymmetry occurs and $\tan \beta = 0$. If the asymmetry is large then $c \rightarrow 0$ and $\tan \beta \rightarrow \tan \phi$, the beach slope approaching the angle of repose.

The coarsening-onshore trend found in almost all beach and nearshore systems is explained by the fact that the forward orbital motion under shallow-water wave crests is shortlived but powerful compared to the seaward return flow (Bagnold 1940). Thus coarse particles may be preferentially transported onshore towards the beach.

The origin and dynamics of shoreward-fining tidal flats

depend upon a steady supply of sediment, particularly of silt- and mud-grade, and a low degree of wave action. Fining-onshore is controlled by two related processes, termed scour lag and settling lag (Fig. 21.6) which encourage silt- and mud-grade sediment to accumulate on the upper tidal flats. Limited percolation of tidal waters because of sediment cohesion encourages surface runoff and hence meandering **tidal channel** networks are established. These channels then act as arteries and veins, funnelling the rising tide onto the tidal flat and transferring the residue back during the ebb. Seaward transport of rain water that falls onto the tidal flat during ebb or low tide also occurs (Bridges & Leeder 1976).

The origin of barrier systems is still controversial, despite over a hundred years of research. An early theory suggested that they resulted from the upbuilding of submerged offshore bars into emergent islands. The absence of offshore facies beneath modern lagoons usually discredits the theory, but some barriers clearly do have such an origin, witness the ring of barriers comprising the Chandeleur Islands that mark the reworked rim of the abandoned St Bernard subdelta of the Mississippi delta (see Ch. 19). Periodically some of these islands are destroyed by hurricanes but they reform again by emergence of the remnant submerged sand bars (Otvos 1979). A second theory maintains that barriers result from the partial drowning of normal coast-attached beaches during periods of transgression (Hoyt 1967). A third theory, originally proposed by Gilbert (1885), suggests that barriers result from the breaching and isolation of former spits.

A new and partly composite theory of barrier location has recently been proposed on the basis of extensive subsurface mapping along the Delmarva Peninsula of the mid-Atlantic Bight of Delaware, USA (Halsey 1979). This theory, although not tackling barrier *origin*, recognises the role played by inherited coastal topography generated during the low sea-level stand of the last glacial period. The regressive phase left a palaeochannel network separated by higher interfluve areas. As the Flandrian transgression began, beaches formed against the interfluves and estuaries formed along the river channel outlets. Depending upon the height of the interfluves and the density of palaeochannels, a variety of barrier, spit and lagoon–tidal flat environments evolved as transgression continued. This **inheritance** model may be found to be widely applicable in the study of barrier evolution (see also Oertel 1979).

It is evident that barrier origins are intimately connected with transgressional episodes but it is less evident how barrier–lagoon systems maintain an equilibrium morphology. Recent studies (Rampino & Sanders 1981) indicate that, in regions subject to transgression, rapid

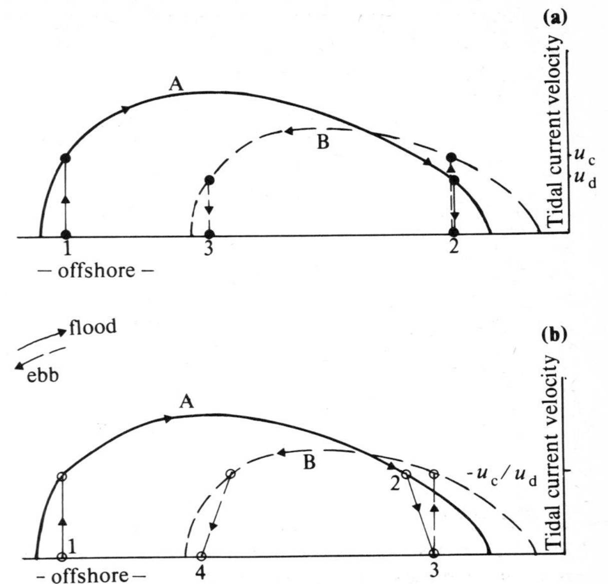

Figure 21.6 (a) *Scour lag.* A sediment grain at position 1 on the intertidal flat is entrained on the flood tide by water mass A, which has a velocity u_c at that point. At position 2 the deceleration of A to a velocity u_d renders it unable to support the grain which therefore comes to rest. When water mass A returns over position 2 during ebb it has not yet reached the critical velocity, u_c, required to set the grain in motion again. This velocity is attained only by water mass B, which does not return so far seawards as A. A net inshore movement of the grain from position 1 to position 3 thus occurs. (b) *Settling lag.* A sediment particle at position 1 on the intertidal flat is suspended on the flood tide by water mass A. At position 2 the decelerating water mass becomes unable to support the particle which therefore starts to settle out of suspension. While settling, the particle is still carried forward by the flow and eventually reaches the bed at position 3, shoreward of 2. Hence, water mass A is unable to resuspend the particle during ebb; it is in fact picked up by water mass B and eventually deposited at position 4, a net shoreward movement from position 1. (Both figures after de Mowbray 1980, based on the work of Van Straaten & Keunen 1957.)

sea-level rise and low sand supply favour the stepwise retreat of barriers, with the production of stranded defunct barriers and their lagoonal facies in the nearshore or shelf zone and the genesis of a new coastal barrier complex along the former lagoon inner margin. In areas subject to slow transgression and high sand influx (perhaps along high-relief hinterlands on destructive plate margins; see Bourgeois 1980) continuous shoreface retreat or even coastal progradation may occur.

Tidal inlets along barriers in mesotidal areas clearly play an important role, providing sediment to infill the back-barrier lagoons and channels and tidal currents to maintain an equilibrium. Barriers along microtidal coasts have no such dynamic equilibrium, their landward lagoons and bays tending to become infilled by storm washover sediments.

21c Recent facies of linear clastic shorelines

The major sub-environments of linear clastic shorelines are indicated on Figure 21.3. As noted previously there is a general offshore to onshore increase in sediment grain size on beaches. Shoreface and foreshore topography reflects the presence of various 'bars', ridges and troughs as well as wave-formed current ripples and dunes (Figs 21.7 & 8).

Offshore bars occur on all but the steepest high-energy beaches and are controlled in a complex way by the position of breaking waves. Successive coast-parallel or crescentic bars with wavelengths of tens to hundreds of metres tend to increase in height (up to 1.5 m) away from the shore, perhaps individually reflecting the average breaking position of waves of a certain height. The bars (Fig. 21.8) show variably dipping internal sets of tabular cross stratification directed landwards and the troughs show small-scale cross laminations produced by landward-migrating wave-current ripples (Davidson-Arnott & Greenwood 1974, 1976). Similar structures, though of smaller amplitude, occur in the foreshore zone of broad sandy tidal flats and beaches and are termed **ridge and runnel** topography. The very characteristic rip-current cells that occur on many beaches produce channels that may dissect offshore bars (Fig. 21.9). The fan-like terminations to rip-current channels deposit seaward-dipping cross sets.

Should a beach system prograde then a vertical upward-coarsening sequence will develop (Fig. 21.10). The detailed structures preserved in the shoreface environment will depend upon the complex interactions of fair and foul weather processes. The former might be expected to dominate since we have already discussed how such periods encourage net sediment accretion on the beach shoreface and foreshore.

The facies profile preserved by a seaward-prograding barrier island should bear many similarities to that deduced above for normal beach environments. Two major additions occur because of the presence of **back-barrier lagoons** and the **tidal inlets** of mesotidal barriers (Hubbard *et al.* 1979). Back-barrier lagoons vary tremendously in facies, depending upon climate, degree of tidal flushing by inlets, river inflow and extent of storm washover events. Lagoons in microtidal areas are dominated by storm washovers so that the bioturbated lagoonal silts and muds are intercalated with sheets up to 1.5 m thick of parallel-laminated sands derived from storm breaching of the exposed barrier profile. **Washover fans** produced in this way may show delta-like landward

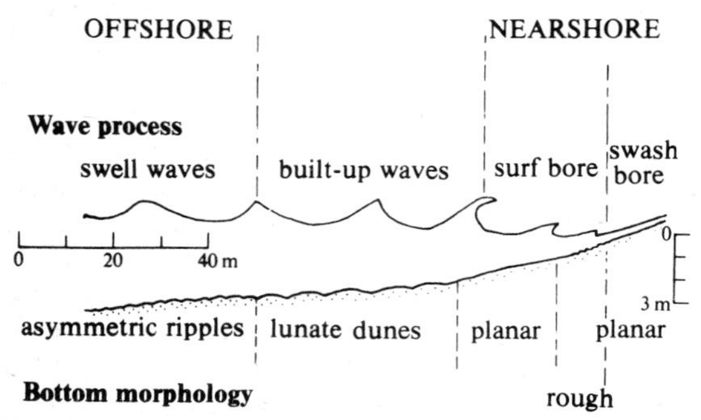

Figure 21.7 Relationship of bedforms to wave forms on the high-energy beaches of the Oregon coast (after Clifton *et al.* 1971).

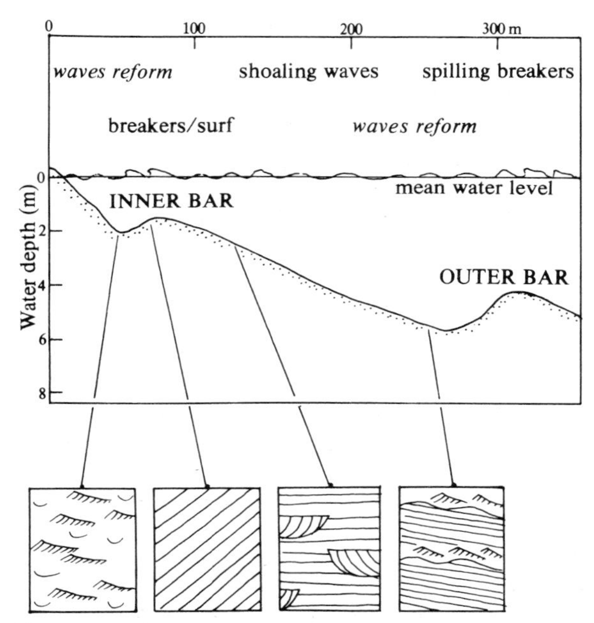

Figure 21.8 Sedimentary structures in relation to shallow offshore morphology and wave processes, New Brunswick (after Davidson-Arnott & Greenwood 1974).

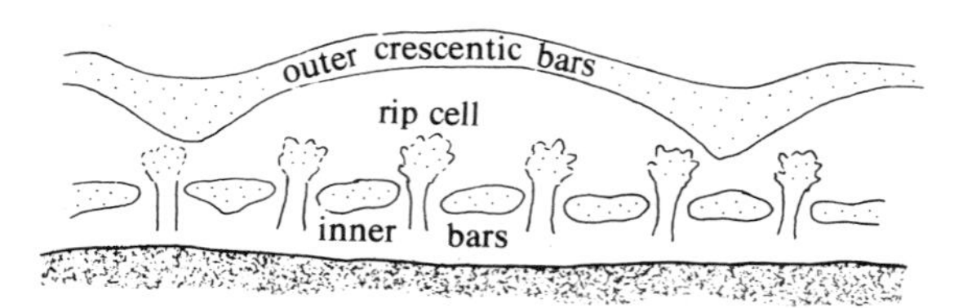

Figure 21.9 The combination of offshore bars and rip cells found on many coastlines (after Komar 1976).

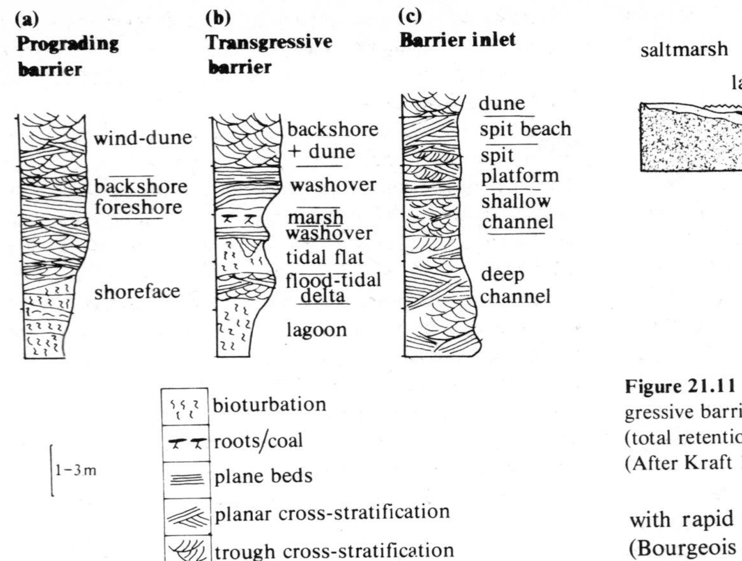

Figure 21.10 The three 'end-members' of successions produced by barrier island processes (after Reinson 1978).

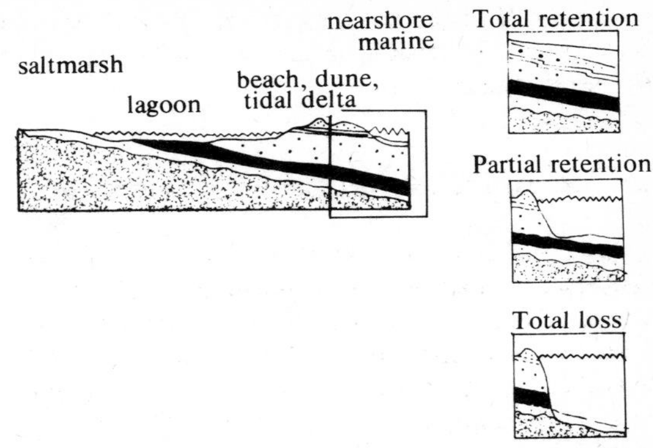

Figure 21.11 Diagram to show the varying retention of transgressive barrier facies according to whether sea-level rise is rapid (total retention), moderate (partial retention), or low (total loss). (After Kraft 1971.)

terminations with internal sets of landward-dipping planar cross stratification. (Schwartz 1975). Microtidal lagoons in semi-arid climates such as the Texan Laguna Madre show evaporite growth of sabkha types, carbonate precipitation as oöliths, and growth of algal mats (Fisk 1959, Rusnak 1960). Flood-tidal deltas in mesotidal lagoons occur on the inner sides of tidal inlets in response to flow expansion and deceleration (M. O. Hayes 1979). The subaqueous delta surface is covered by landward-directed dune bedforms. The remainder of the lagoon in such cases approximates to the physiography of a tidal flat as discussed below.

In summary, as barriers migrate seawards under conditions of net sediment supply, an upward-coarsening sequence is produced that may be broken by fining-upwards tidal-inlet channel facies (Fig. 21.10). The barrier may eventually be overlain by lagoonal or tidal-flat facies which reverse the coarsening-upwards trend. The behaviour of regressive barrier systems should be contrasted with that of transgressing barriers (see Kraft & John 1979). In the latter case little of the barrier facies themselves may be preserved if transgression is slow and sediment input low, erosional action at the shoreface zone producing a thin reworked offshore sand body that may eventually overlie the lagoon facies directly as transgression continues (Fig. 21.10 & 11, Kraft 1971). Lagoonal facies themselves may also be removed by the action of migrating tidal inlets on mesotidal barriers (Kumar & Sanders 1974). High rates of sediment supply combined with rapid transgression may cause barrier preservation (Bourgeois 1980).

The facies of tidal flats (e.g. Reineck 1967, Evans 1965) are dominated by the nearshore-to-offshore coarsening trend noted previously (Fig. 21.12). The supratidal salt-marsh zone with **halophytic** (salt-loving) plants passes gradationally outwards at a very low slope into a mudflat with a rich infauna. Seaward coarsening gives rise to a mixed sand-/mudflat with a variety of laminations including flasers (Ch. 8). Again, bioturbation by the abundant infauna is intense. The sand flats down towards mean low water mark show a great variety of wave- and current-formed ripple bedforms with complex interference forms caused by gravity runoff effects. Local dunes may result if tidal flows are strong enough. As noted previously many tidal flats, particularly the relatively impermeable upper mud- and mixed-flats, show a dense network of meandering **tidal creeks** (Fig. 21.13; Reineck 1958, Bridges & Leeder 1976). These rework much of the tidal-flat deposits and give rise to inclined **lateral accretion deposits** of interlaminated silts and muds (Fig. 21.14). Rapid deposition on the point bars discourages infaunas and hence these deposits are relatively free of bioturbation. In some areas, particularly off the Dutch and German macrotidal coasts and in the Bay of Fundy, the intertidal channels pass offshore into a subtidal zone of deep channels with major dune bedforms whose migration and accretion are dominated by the periodic ebb and flow of the tidal wave (Reineck 1967, 1972). Frequently ebb and flood channels are separate so that the resulting cross-stratified sand deposits tend to show either ebb or flood dominance, but rarely a mixture of the two. It can easily be appreciated that a seaward-prograding tidal flat system will produce an upward-fining sequence, broken by intertidal and subtidal channels and

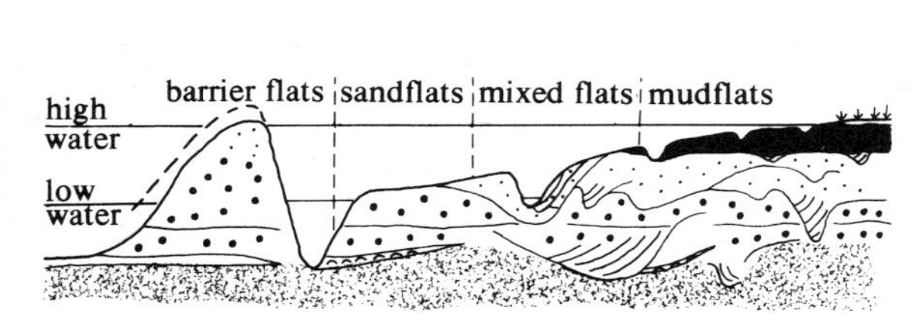

Figure 21.12 Section through a tidal flat complex to show the various subdivisions and the sort of sequence produced after tidal flat progradation (after Reineck & Singh 1973).

Figure 21.13 A meandering tidal channel with its cut bank and point bar from the intertidal mudflats of the Solway Firth, Scotland.

capped by a rootlet bed or peat accumulation of the salt marsh (Fig. 21.12).

Chenier sand or shell facies are dominated by storm washover effects that produce landward-dipping, low angle to planar cross sets on the landward (washover) side of the biconvex ridge. The nature of the base of the chenier succession varies from a sharp contact with marsh facies on the landward side of the ridge to a gradational contact with shallow-water or mudflat facies on the seaward side.

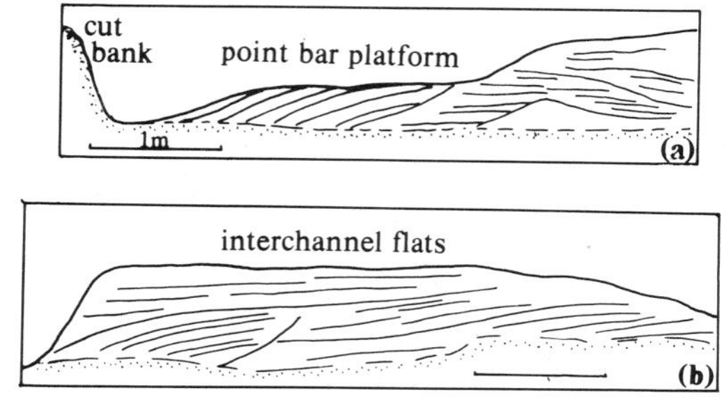

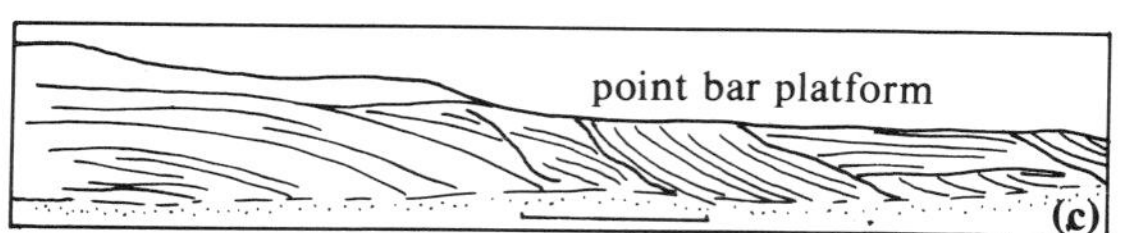

Figure 21.14 Sketches of excavations through silt- to mud-grade intertidal point bars to show the lateral accretion bedding (after Bridges & Leeder 1976). Note the complex internal erosion surfaces in (c).

21d Ancient clastic shoreline facies

Identification of true shoreline facies is of the utmost importance in palaeogeographic reconstructions, since firm limits may then be put on the extent of sea during a particular time interval. Additionally, identification of shoreline facies enables deductions to be made according to the magnitude of the tides, the relative importance of waves *versus* tides and the absolute bathymetry of the shoreline facies for palaeoecological studies. A few examples will illustrate these points.

Clifton *et al.* (1971) examined a thin sequence of Quaternary sands and gravels and were able to make close comparisons with modern high-energy shoreline facies forming today on the Oregon coast (Fig. 21.15). By way of contrast the Jurassic sediments of the topmost Lower Coal Series of Bornholm, Denmark (Sellwood 1972) are interpreted as forming by seaward progradation of sand, mixed sand/mud and salt marsh environments of a tide-dominated shoreline. The fining-upwards sequence so produced contains wavy-, lenticular-, and flaser-laminated units, bipolar (herringbone) cross laminations, preserved dunes with clay draped reactivation surfaces, and laterally accreted units of tidal creek point bar origins. As Sellwood points out, the Early Jurassic sea that covered northwest Europe is usually considered as epeiric and, according to Shaw (1964), such seas should be tideless. This idea is clearly refuted by the Bornholm facies. Following up an approach developed by Klein (1971) we may estimate the minimum tidal range for the Bornholm sequence by measuring the vertical sequence between salt marsh (= MHWM) and subtidal channels (= MLWM); a figure of around 6–8 m is obtained which clearly indicates a macrotidal regime.

Ancient barrier sandbodies form important oil reservoirs because of their high original porosity and permeability and of their persistence along strike. A good example is the Lower Cretaceous Muddy Sandstone of the Bell Creek oilfield in Montana, USA (Davies *et al.* 1971). Elliott (1975) describes a nice example of a prograding barrier–lagoon system which developed after a delta lobe abandonment in the Carboniferous of northern England. The time-transgressive and

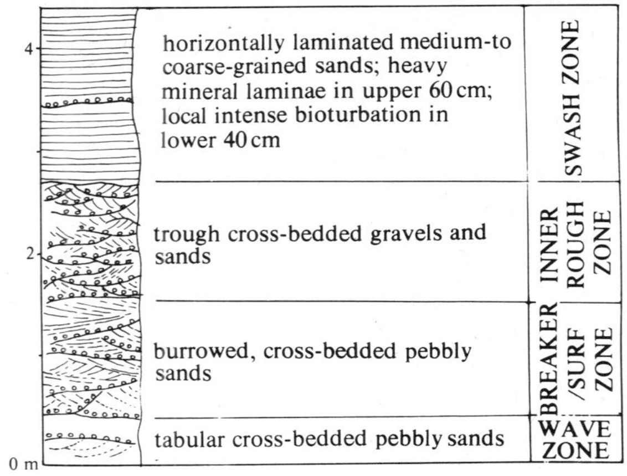

4
2
0 m
horizontally laminated medium-to coarse-grained sands; heavy mineral laminae in upper 60 cm; local intense bioturbation in lower 40 cm
SWASH ZONE
trough cross-bedded gravels and sands
INNER ROUGH ZONE
burrowed, cross-bedded pebbly sands
BREAKER /SURF ZONE
tabular cross-bedded pebbly sands
WAVE ZONE

Figure 21.15 Log of a stratigraphic sequence from the Quaternary of California that is interpreted to be of high-wave-energy beach origin. See also Figure 21.7. (After Clifton *et al.* 1971.)

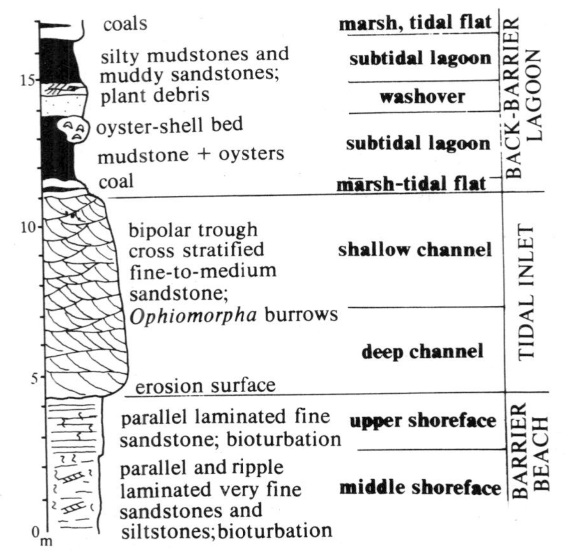

Figure 21.16 Log of a stratigraphic sequence from the Cretaceous of Alberta that is interpreted to be due to seaward progradation of a barrier–inlet–lagoon complex on a mesotidal coastline (after Young & Reinson 1975).

diachronous nature of barrier–beach facies is well illustrated by the Mesaverde Group (U. Cretaceous) of Utah and Colorado (Masters 1967) and by the Cretaceous Gallup Sandstone of New Mexico (Campbell 1971). The detailed interpretation of a barrier–inlet facies association is well illustrated by the Blood Reserve/St Mary River Formations of southern Alberta (Fig. 21.16, Young & Reinson 1975).

21e Summary

The morphology of linear clastic shorelines depends upon the relative importance of tide and wave currents and upon current magnitude. Open coastlines dominated by wave action feature beaches which show offshore-fining trends with a variety of offshore bars present. Sheltered or macrotidal coastlines are fringed by wide tidal flats where the phenomena of scour and settling lag cause onshore-fining trends from sand to mudflats with supratidal salt marshes. A variety of factors cause the formation of linear barrier coastlines with their lagoons and tidal inlets. On microtidal coasts barriers show few tidal inlets, and much coarse sediment is transferred into the lagoons during storm washover events. Barriers on mesotidal coasts show frequent tidal inlets whose migration causes much of the subsurface sequence to be dominated by inlet and tidal delta facies.

Further reading

Useful accounts of the physical processes which affect clastic shoreline sedimentation appear in the text by Komar (1976) and in the volumes edited by Hails and Carr (1975), R. A. Davis (1978) and K. S. Davis and Ethington (1976). The sedimentary facies of modern and ancient coastlines are well reviewed by Elliott (1978a) and by Reinson (1978). Well illustrated accounts of facies formed by modern coastlines are given by Reineck and Singh (1973) and in Ginsburg (1975). Barrier coastal facies, with particular relevance to the eastern USA, are described in a number of stimulating papers in the volume edited by Leatherman (1979).

22 Clastic shelves

22a Introduction

Perhaps more than any other single environment, the continental shelf typifies the dynamic 'input/output' aspect of Earth surface processes. Clastic sediment introduced onto the shelf must bypass the various nearshore sediment 'traps' such as estuaries, bays, lagoons, deltas and tidal flats. Once on the shelf a complex mixture of tidal, wave, oceanic and density currents disperse the sediment, allowing some proportion to 'escape' over the shelf edge into the deep ocean basins. Lest this picture sound too simplified it should also be pointed out that some 11 000 years ago sea level stood roughly at the shelf edge (Fig. 22.1). The ensuing Flandrian transgression advanced over an incised coastal plain with local glacial deposits. Thus some modern shelves are, to a greater or lesser extent, **relict** (Fig. 22.2) in the sense that pre-Holocene sediment is exposed and is being reworked by modern dispersal systems. Abundant evidence of relict morphology and progressive shoreline retreat is found on most modern shelves; indeed, shoreline retreat features enable a continuum to be traced on the shelf from intra-Flandrian coastline features to those of the present day (we have already discussed examples on barrier-fronted coasts in Ch. 21).

Shelves extend from the shallow offshore – say an arbitrary 10 m water depth – out to a prominent shelf edge break, at the top of the continental slope. The depth of the shelf edge (20–550 m) and the shelf width (2–1500 km) are tremendously variable, depending largely upon tectonic setting. Shelves on Atlantic-type ('passive') continental margins tend to be much wider than those on Andean-type or Pacific margins. The relatively smooth, gentle offshore slope to most shelves is basically a constructional feature moulded by shelf currents, deposited sediment and the accommodating effects of shelf subsidence. This latter factor is an especially important one since geophysical exploration has revealed quite clearly that most shelves are underlain by extremely thick sedimentary successions that lie in fault-bounded linear basins or in broader downsags. Most of these shelf–basin sedimentary rocks were deposited in quite shallow water and imply that shelves are prone to continued, gentle subsidence.

Although simple shelves with an oceanward-dipping prism of sediments are the most common type, important examples of shelf sediment 'damming' occur. Here the shelf sediment has built up behind a positive relief feature formed by block faulting (Pacific coast of Americas), reef growth (Red Sea, northeast Australia) or diapir intrusion (Gulf of Mexico).

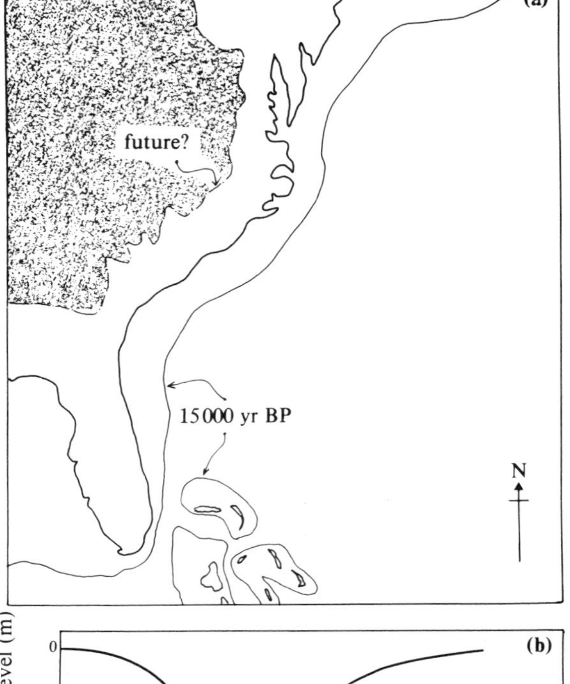

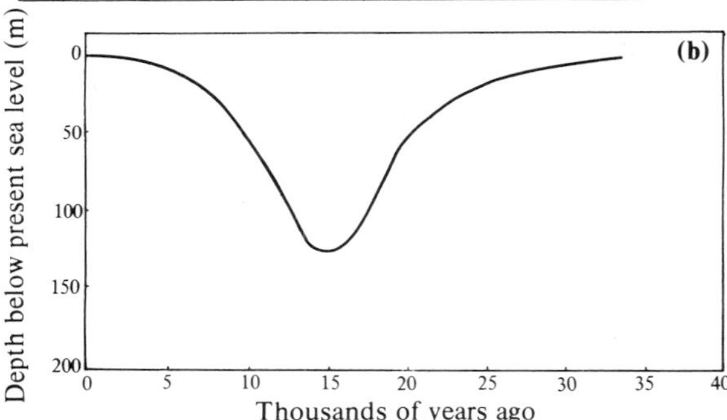

Figure 22.1 (a) Comparison of the Atlantic USA shoreline of 15 000 years BP, the present shoreline, and the probable shoreline if all polar ice were to melt. (b) Generalised sea-level:time curve based upon world wide data (after Emery 1969).

22b Shelf dynamics

As noted briefly above the dynamics of water and sediment movement on shelves is complex. The components of the shelf velocity field are summarised in Table 22.1.

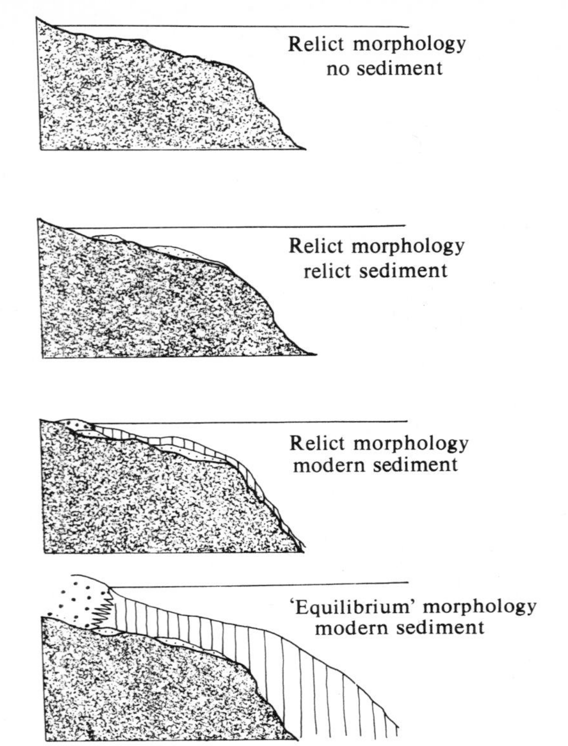

Figure 22.2 The change from a relict shelf to an equilibrium shelf following transgression and establishment of a physical equilibrium between wave, tide and introduced sediment (after Curray 1965).

The most important components are those of tide and wave, the latter including net mass transport of water by direct wind shear and by storm surges. Although most shelves are affected by both components, one or the other is usually dominant. This gives rise to a classification of shelves into **tide-dominated** and **weather-dominated** (the latter term being preferable to wave-dominated, Swift 1972), although most shelves will show a mixture of processes.

We have already seen (Ch. 18) that the rotary nature of a tidal wave about its amphidromic point rarely produces equal current vectors about the compass. Tidal current measurements at a point usually define a tidal current ellipse with inequality between tidal ebb and flood producing residual tidal currents of up to 0.5 m s^{-1}. Since sediment transport is a cube function of current velocity (Ch. 6) it can be appreciated that quite small residual tidal currents can cause appreciable net sediment transport in the direction of the residual current (Belderson *et al.* 1978). A further important consideration arises from the fact that turbulence intensities are higher during decelerating tidal flow than during accelerating tidal flow (McCave 1979). This arises from the greater intensity of the burst/sweep process (Ch. 5) in unfavourable pressure gradients. Increased bed shear stress during deceleration will thus cause increased sediment transport compared to that during acceleration, so that the net transport direction of sediment will lie at an angle to the long axis of the tidal ellipse (McCave 1979).

Detailed studies on the northwest European tide-dominated shelf (Stride 1963, Kenyon & Stride 1970, Belderson *et al.* 1978) have defined **tidal current transport paths** (Fig. 22.3) along which sediment particles move.

Table 22.1 Components of the shelf current velocity field (after Swift 1972).

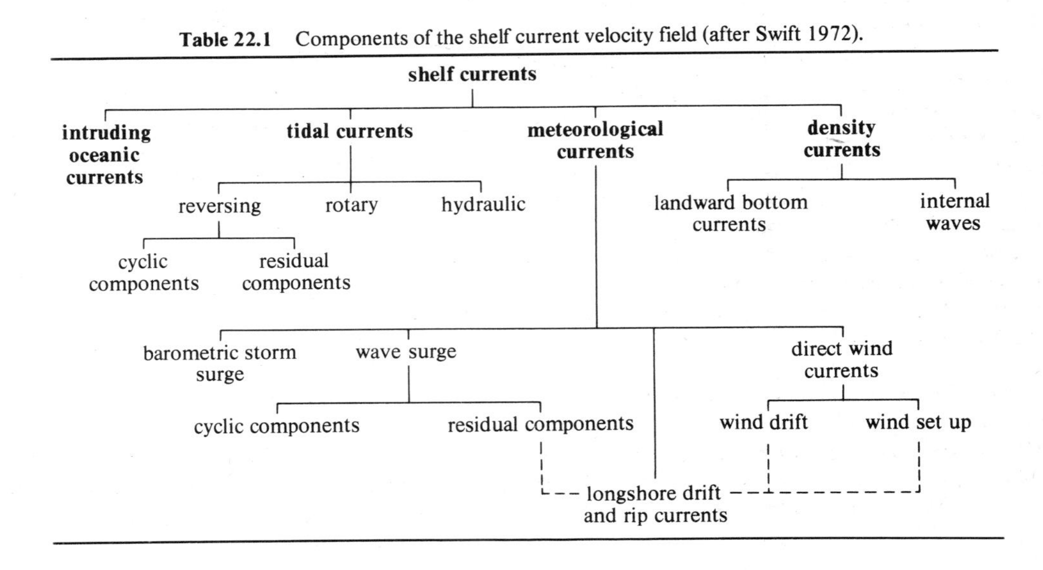

- **shelf currents**
  - **intruding oceanic currents**
  - **tidal currents**
    - reversing
      - cyclic components
      - residual components
    - rotary
    - hydraulic
  - **meteorological currents**
    - barometric storm surge
    - wave surge
      - cyclic components
      - residual components → longshore drift and rip currents
    - longshore drift and rip currents
    - direct wind currents
      - wind drift → longshore drift and rip currents
      - wind set up → longshore drift and rip currents
  - **density currents**
    - landward bottom currents
    - internal waves

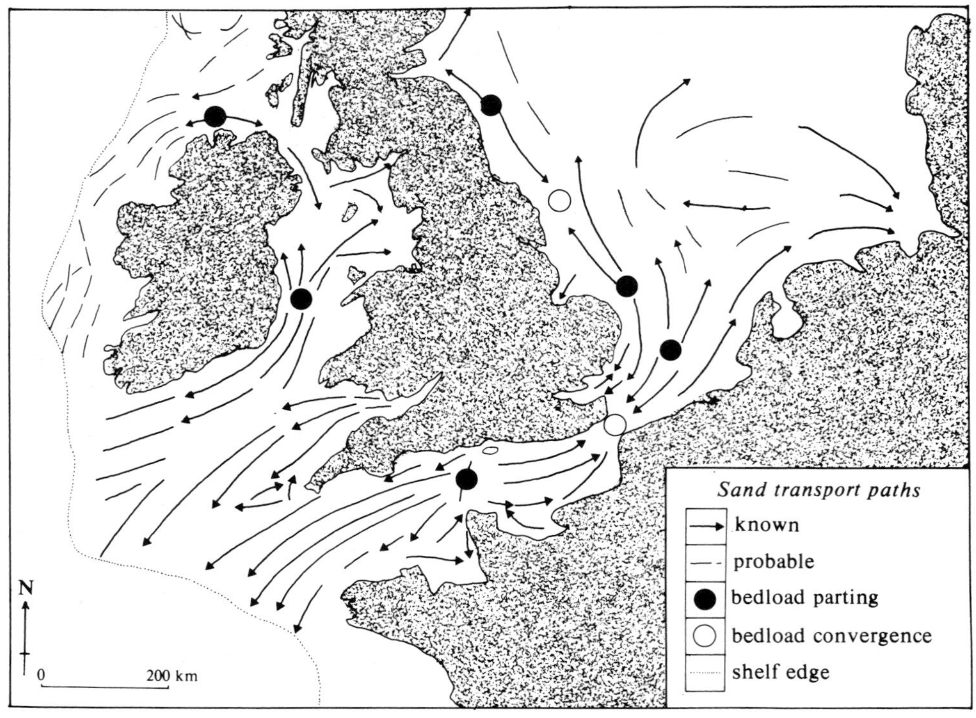

Figure 22.3 Tidal current transport paths on part of the complex northwest European shelf (after Stride 1963, Kenyon & Stride 1970).

These have been mapped out by a combination of observations on (a) surface tidal velocities, (b) elongation and asymmetry of tidal current ellipses, (c) facing direction of sandwaves, (d) trends of sand ribbons (*q.v.*), and grain size trends. (The reader should compare the methodology of this approach with the principles used to construct sand flow paths in the great Saharan desert ergs, as discussed in Ch. 13.)

Weather-dominated shelves usually show low tidal ranges (<3 m) and weak tidal currents (<0.3 m s^{-1}). Winter wind systems assume an overriding dominance, causing net residual currents arising from wind drift, wind set-up and wave surge. The oscillatory effects of surface waves upon the bottom sediments are simply to suspend fine sediment into the water column. This suspension may then be subjected to net transport by the residual currents noted above. On the Oregon continental shelf (Komar *et al.* 1972) long-period storm waves from the southwest stir the bottom to water depths of 200 m. The resuspension is then transported as plumes of sediment (Fig. 22.4) which bypass the shelf in surface and mid-water to deposit sediment along the continental slope. Net southward transport on the South Texas shelf (Shideler 1978) is attributed mainly to advection by residual drift currents which reflect a winter-dominated hydraulic regime. Frequent winter storms characterised by the relatively strong northerly winds that accompany the passage of cold fronts appear to be the dominant regional dispersal agents (Fig. 22.5).

It has been pointed out that since most fine-grained inorganic deep-sea sediment is derived from the continents, it must have crossed the shelves to reach the oceans (Schubel & Okabo 1972), yet the routes and mechanisms of this escape remain obscure (McCave 1972). Undoubtedly much fine sediment is lost by advective processes in turbid plumes which issue from coastal tidal inlets, estuaries and deltas (Figs 22.5 & 22.6). It should be noted however that very extreme runoff of oxidised terrigenous sediment from the Santa Barbara Mountains, California, during floods was of insufficient concentration to allow turbid layer flows to transport material over and out of the shelf. Most of this characteristic sediment was deposited on the shelf itself (Drake *et al.* 1972). Diffusional processes due to turbulence are also important, witness the broadly exponential decrease in suspended sediment concentration away from coasts measured by many workers (see summary in McCave 1972).

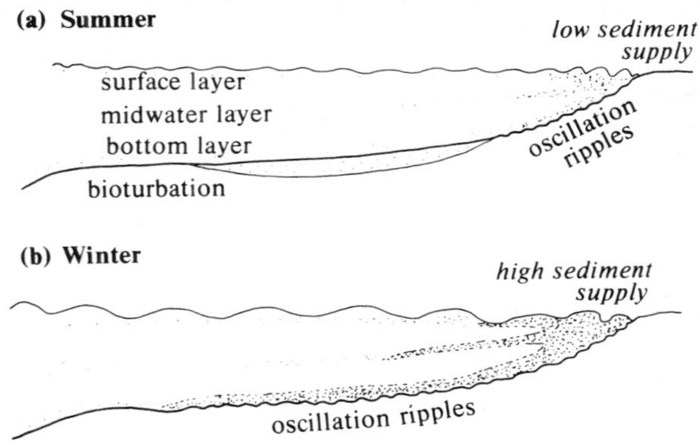

Figure 22.4 Fair and foul season sedimentation patterns on the Oregon shelf (after Kulm *et al.* 1975).

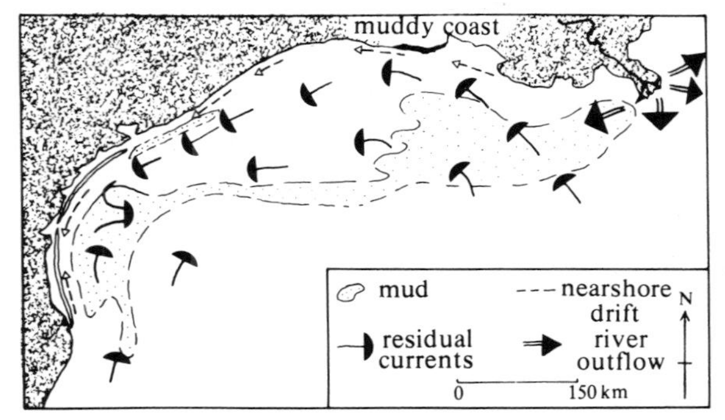

Figure 22.5 Mud deposits and residual currents in the northwest Gulf of Mexico: note the influence of Mississippi supply in the east (*cf.* cover photo; after McCave 1972 from data in Curray 1960 and van Andel & Curray 1960).

22c Recent shelf facies

The distribution of grain sizes on a tide-dominated shelf is complex, being dependent upon position with respect to local tidal current paths. There is always a general trend towards decreasing grain size down the tidal current paths, perhaps from coarse sand to mud. This trend is due to decreasing net current strength. The upstream parts of current paths, with velocities in excess of 1 m s^{-1}, may show **sand ribbon** bedforms up to 20 km long, 0.2 km wide and about 0.1 m thick (Fig. 22.7). These features occur in water depths of 20–100 m on gravel substrates that have a sparse cover of coarse sands. Simple parallel sand ribbons probably owe their existence to pairs of counter-rotating helical vortices (secondary flows, see Ch. 5). Another characteristic bedform along the higher-energy parts of tidal transport paths are large dune-like **sandwaves** 3–15 m high with wavelengths of up to 0.6 km. Given a sufficient supply of sand these bedforms will develop as asymmetric forms in areas of marked tidal

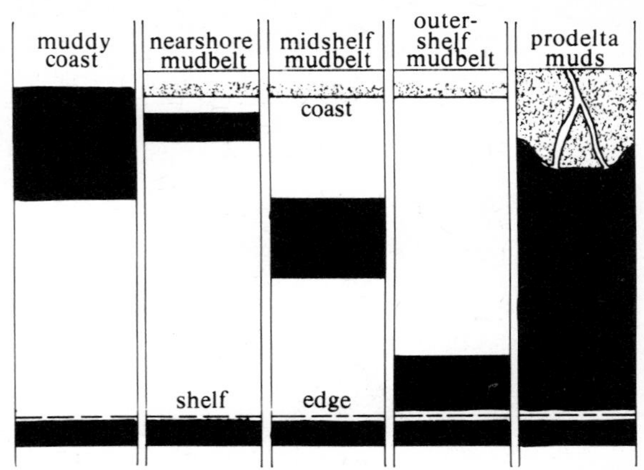

Figure 22.6 The various sites where shelf muds may accumulate (after McCave 1972).

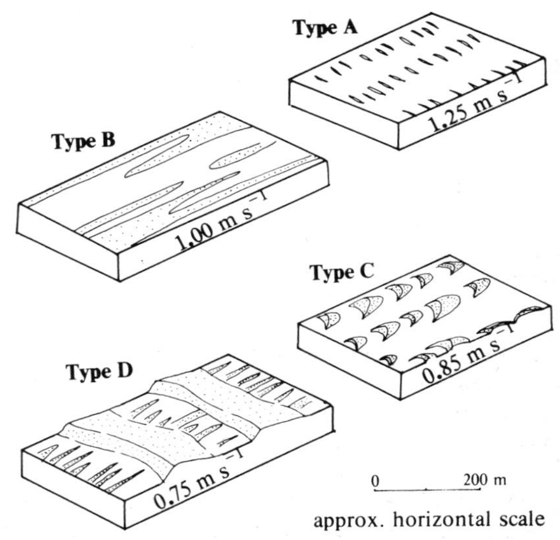

Figure 22.7 Types of sand ribbons seen on tidal-dominated shelves (after Kenyon 1970). See text for discussion.

ellipse asymmetry (Fig. 22.8) and as symmetric forms at bedload partings where ellipse asymmetry is absent (McCave 1971). Sandwaves are absent in nearshore areas where wave activity is high (Fig. 22.8). Little is known concerning the internal structures of these sandwaves but it may be inferred that they comprise dominantly unimodal large-scale cross stratification with perhaps smaller-scale sets with opposed orientations. Sandwaves with low-angle lee slopes may be expected to show numerous internal sets of cross stratification separated by downcurrent dipping set boundaries (Ch. 8; Reineck 1963, J. R. L. Allen 1968, Banks 1973). The distal ends of tidal transport paths comprise isolated sand patches and small sandwaves with numerous ripple bedforms and bioturbation features, the paths finally ending in areas of

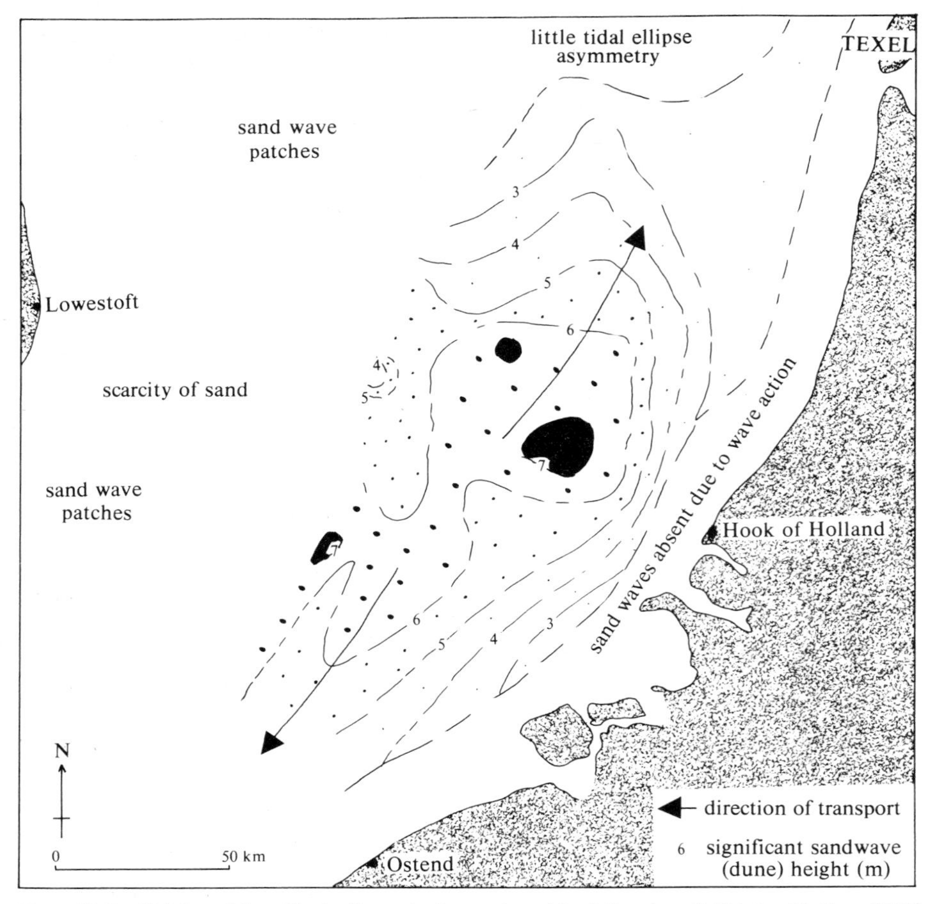

Figure 22.8 Heights of dune-like bedforms in the southern North Sea dune field (after McCave 1971).

mud deposition. The bioturbated mud deposits with rich infauna can occur only in relatively deep areas of low wave activity, the high deposition rates (3–5 mm a^{-1}) indicating continuous mud fall-out from suspension with important storm-produced suspensions contributing significantly (Reineck 1963, Gadow & Reineck 1969). Studies of the mud belt developed in the Heligoland Bight of the North Sea (Reineck *et al.* 1967, Reineck & Singh 1980) reveal frequent, thin, graded sand and shell layers attributed to storm surge density currents which transport intertidal sands and fauna up to 40 km out into the deep offshore (see Fig. 10.1).

A very prominent feature of the tide-dominated southern North Sea are the numerous parallel, large-scale linear **tidal ridges** orientated parallel to the direction of the residual tidal currents (Figs 22.9 & 10). These ridges are made up of shelly, well sorted medium sands and are up to 40 m high, 2 km wide, 60 km long and have spacings of between 5 km and 12 km. The ridges are asymmetric with the steep face inclined at a *maximum* of about 6°. Internal structure has been revealed by sparker surveys which show inclined low-angle foresets parallel to this steep face, indicating ridge migration in this direction (Fig. 22.10). Although the ridge systems are in equilibrium with the present tidal regime, they are thought to have formed periodically as nearshore linear bars separating ebb and flood tidal transport paths during the Flandrian transgression. Periodic detachment of the shoreface ridges created a **shoal retreat** complex that extends well out (200 km) into the North Sea (Robinson 1966, Houbolt 1968, Caston 1972, Swift 1974).

Weather-dominated shelves tend to show a general offshore decrease in grain size in response to attenuating wave power. This trend is well shown by the Bering, Oregon and southwestern Gulf of Mexico shelves (Sharma *et al.* 1972, Kulm *et al.* 1975, Shideler 1978). Mud-grade sediments tend to settle out close to the shelf edge break by processes of advection (caused by storm

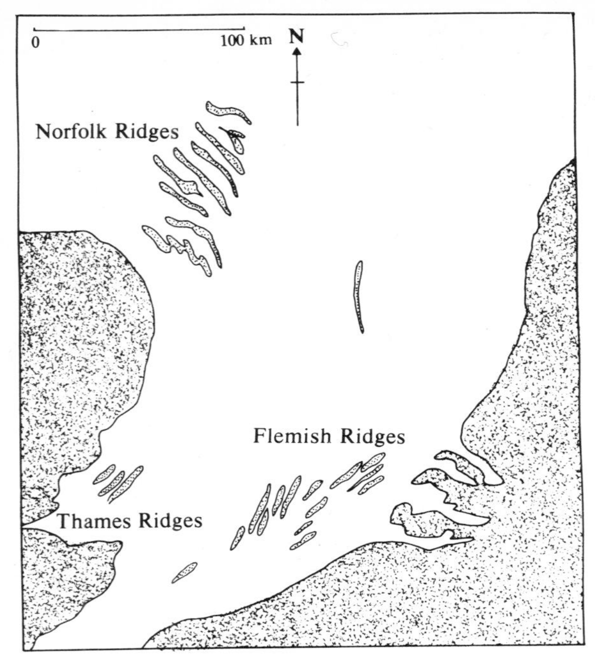

Figure 22.9 Areas of tidal sand ridges in the southern North Sea (after Houbolt 1968).

generated water-drift residual currents acting upon plumes of river and tidal inlet fines) and by diffusion. These fine sediments are then intermixed with partly reworked transgressive relict sands on the outer shelf. As already noted, wave-formed ripples can occur at depths up to 200 m on the Oregon shelf, indicating that rippled sand laminae may be expected to be common in many offshore areas. Fair weather reworking by burrowing organisms may destroy these storm laminae.

The Atlantic shelf off the eastern USA (Fig. 22.11) stands as a particularly well investigated weather-dominated shelf of complex morphology. The shelf shows incised pre-Recent river and estuarine channels and arcuate shoal retreat massifs formed at cape headlands during transgression (see Ch. 21b). The shelf is dominated by fields of linear, northeast-trending, shoal ridges up to 10 m high with slopes of a few degrees. Clusters of shoals merge with the modern shoreface in water only 3 m deep. The ridges make small angles (<35°) with the modern coastline and seismic profiles through the ridges reveal low-angle surfaces that dip to the southeast, the direction of broad ridge asymmetry. The active shoreface shoals are forming at the present time in response to storm-generated currents running approximately parallel to the shoal crests. Shoal detachment from the shoreface is thought to have occurred periodically *during the Flandrian transgression*, the detached shoals continuing to evolve at the present day in response to the storm wave surge and water drift currents (Swift *et al.* 1973, Field 1980). Little is known about the internal structure of these weather-produced shelf ridges, but it is likely that the inclined internal surfaces represent storm erosion planes. Successive internal planes may be separated by fine sediment showing small-scale cross laminations in the wave troughs produced by wave oscillations in normal weather conditions. The reader should note that these wave-dominated linear ridges are quite similar to the tide-dominated ridges noted previously. It may be difficult to distinguish between the two in terms of internal structures alone (Johnson 1977, 1978).

22d Ancient clastic shelf facies

Our first example of ancient shelf facies comes from the unfossiliferous late Precambrian Dalradian Super Group of Scotland – the Jura Quartzite and its correlatives (Anderton 1976). As shown in Figures 22.12 and 13 there is a downcurrent trend in this thick (up to 5 km) sheet-like deposit from cross-stratified quartzites with some

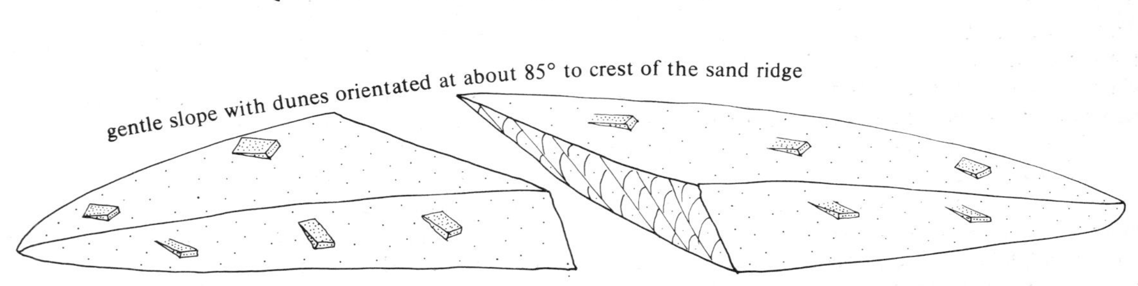

Figure 22.10 Morphology, currents and likely internal structure of a tidal sand ridge (modified from Houbolt's account of Wells Bank, 1968).

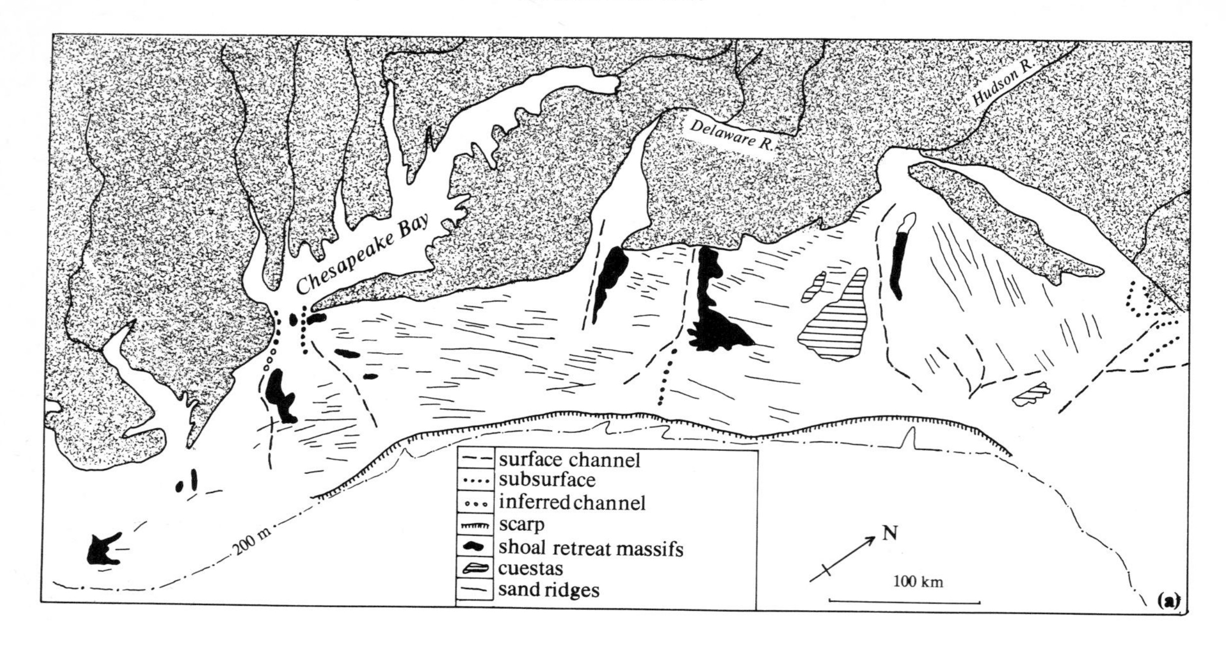

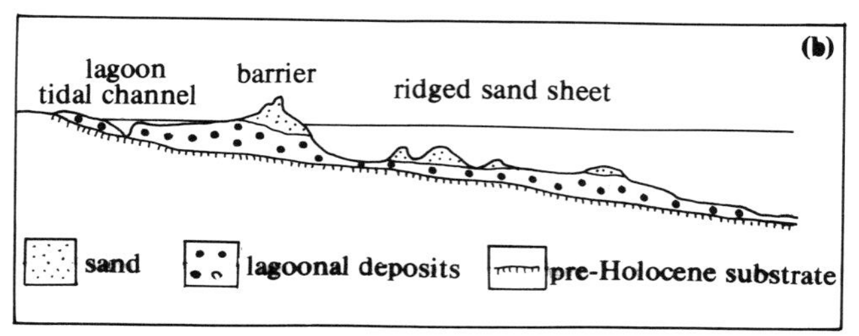

Figure 22.11 (a) Morphology of the Middle Atlantic Bight of the eastern USA shelf to show features of relict and transgressive origins. (b) Section across shelf to show transgressive barrier and progressively abandoned shelf sand ridges. (After Swift *et al.* 1973, Swift 1974.)

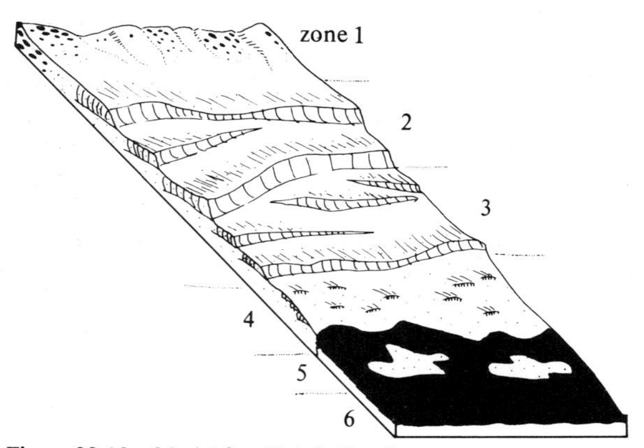

Figure 22.12 Model for tidal shelf sedimentation applied to the Jura Quartzite (after Anderton 1976). (See also Fig. 22.13.)

laminated sand and siltstone horizons to interbedded parallel and cross-laminated sandstones and mudstones. The whole sequence is interpreted as having accumulated along a persistent tidal-current transport path analogous to those present on a tide-dominated shelf such as the modern North Sea. Thus the coarse facies with tabular cross sets up to 4.5 m thick are attributed to deposition by migrating dunes and sandwaves on the high-energy part of a transport path. Periodic shallow scours and laterally persistent erosion surfaces are thought to be caused by enhancement of tidal current velocities by storm surges. The fine facies, with abundant evidence of decelerating currents in the thin (0.01–0.5 m) sharp-based sandstones, are attributed to decelerating storm currents at the downcurrent end of a transport path. Mud interbeds in the fine facies record more normal, post-storm mud deposition from suspension.

Vertical sections through individual sandwave complexes up to 20 m high in the marine Eocene Roda Sandstone Formation of the Spanish Pyrenees (Nio 1976) show a characteristic internal arrangement of facies

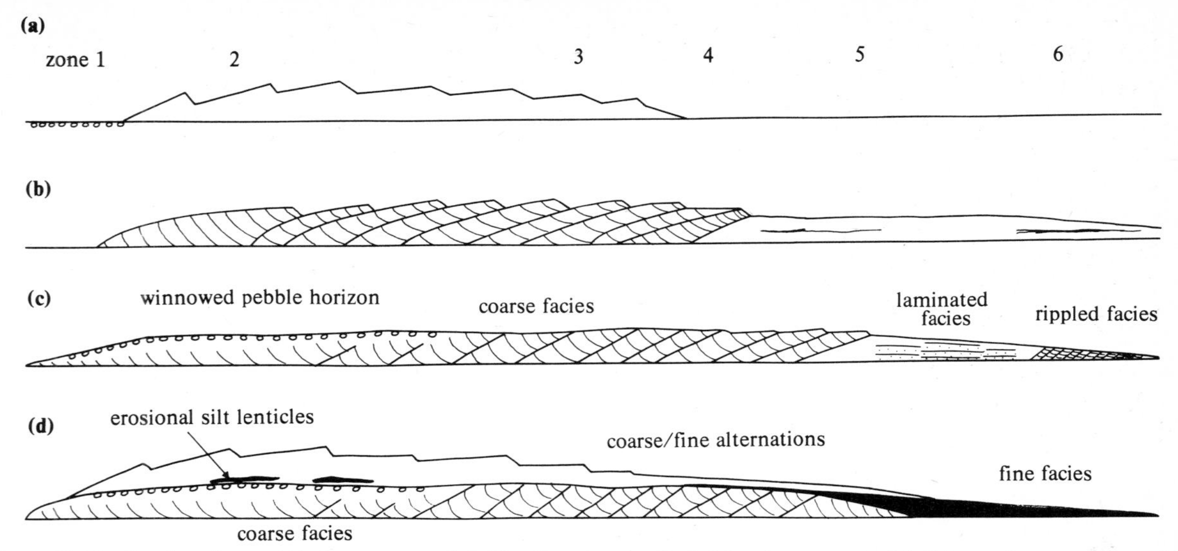

Figure 22.13 To show how vertical sequences of shelf sediments are affected by storm events. (a) Fair weather conditions; (b) moderate storms partly erode and transport sand downcurrent; (c) severe storms winnow zones 2 and 3 and radically shift dunes downcurrent; (d) re-establishment of fair weather system. (After Anderton 1976.)

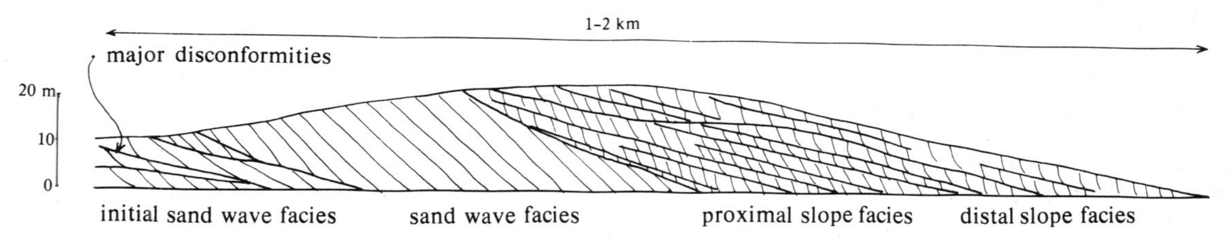

Figure 22.14 Schematic section through an idealised Roda Sandstone bedform (after Nio 1976, Johnson 1978). See text for discussion.

recording the initiation, growth and decay of a tidally produced bedform (Fig. 22.14). Subsurface recognition of (? tidal) current-produced shelf sandbodies are well documented from the Lower Cretaceous Shannon and Sussex Sandstones of Wyoming (Berg 1975, Spearing 1976). These show coarsening-upwards sequences formed by the downcurrent and lateral migration of linear sand ridges and comprising cross-bedded, well sorted and mature glauconitic sands with palaeocurrent modes parallel to the elongation of the sandbodies (up to 30 m thick, 60 km wide and 160 km long).

Interesting comparisons between the Flandrian transgression and the great Llandovery transgression in Wales are made by Bridges (1975). The Silurian example shows transgressive shelf sedimentation on an irregular hard rock shelf with storms generating many of the laminated sandstones and coquinas interbedded with muds (see also Goldring & Bridges 1973). At its graded climax condition the shelf showed a nearshore sand zone in the east, grading westwards into muds with thin sands.

Examples of purely wave-dominated facies are provided by de Raaf *et al.* (1977) from the Lower Carboniferous of County Cork, Ireland. Here a full range of wave-produced laminations (see Ch. 8) occurs in coarsening-upwards sequences which form streaked mudstones, lenticular laminations, cross laminations to parallel and low-angle laminated sandstones. The sequences are interpreted to have formed by lateral migration of shallow-water offshore bars dominated by oscillatory flow of wave origin.

22e Summary

Clastic sediment on the shelf is moulded by tidal and wave currents. The final pattern of modern shelf facies depends not only on the magnitude and direction of these currents but also upon the availability of sediment provided from the coast by various escape mechanisms and from relict continental sediments stranded on the shelf by the Holocene transgression. Tide-dominated shelves show facies patterns adjusted to residual tidal current vectors. Major downcurrent changes in both grain size and bedforms occur along tidal transport paths. On weather-dominated shelves, winter wind systems assume an overriding dominance, causing net residual currents to arise from wind drift, set-up and surge. Stirred-up bottom sediment is subjected to net transport by the currents down to depths approaching 200 m.

Further reading

Many papers on the physical aspects of shelf processes and sediments are to be found in the volumes edited by Swift *et al.* (1972) and by Stanley and Swift (1976). Johnson (1978) gives a thorough review of modern and ancient shelf facies.

23 Carbonate–evaporite shorelines, shelves and basins

23a Introduction

Although most shallow-water (non-pelagic) carbonate environments are physically closely comparable with siliciclastic environments, there are some important differences and contrasts of emphasis. These arise mostly because of the local biogenic origins of many carbonate grains (Table 23.1).

As sketched in Figures 23.1 and 23.2, the greatest source for carbonate grains is the warm shallow photic zone of the subtidal environment, whence storms and mass flows transfer detritus to the supratidal and deep basinal environments respectively. The high organic productivity of subtropical and tropical shallow shelves leads to the production of **carbonate platforms** with abrupt shelf margins marked by growth of organic buildups (Fig. 23.3). Such rimmed shelves characterise carbonate sedimentation in many Recent tropical areas. Major **offshore banks** completely isolated from terriginous clastic input occur as fragmented continental crustal 'microcontinents' bordered by abyssal plains and deep channels. Such an example is seen in the Bahamas Banks, one of the largest and most studied modern carbonate platforms.

J. L. Wilson (1975) has attempted to rationalise the study of carbonate sediments and rocks by erecting a generalised sequence of 'standard' facies belts (Fig. 23.4) whose pattern reflects slope, geological age, water-energy and climate. He points out that no single modern or ancient carbonate province will necessarily show all the facies belts.

Ancient carbonate facies cannot be understood without consideration of (a) palaeoecology of fauna and flora and (b) early diagenesis of carbonate sediments (Ch. 29). The generalised Wilson facies model for carbonate sediments (Fig. 23.4) serves as a general guide for ancient

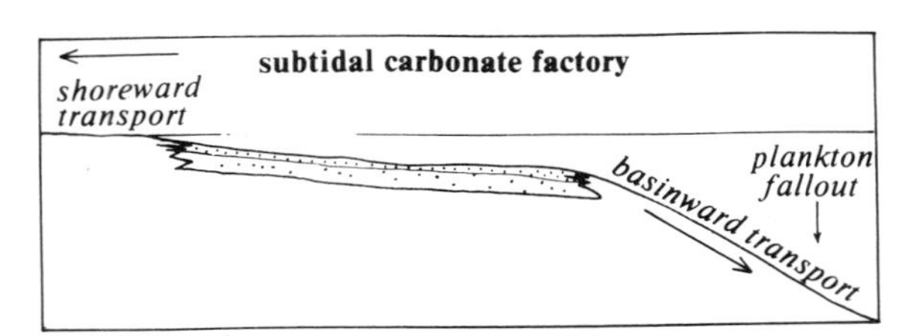

Figure 23.1 Sketch to illustrate the main locus of biogenic carbonate production (after James 1978a).

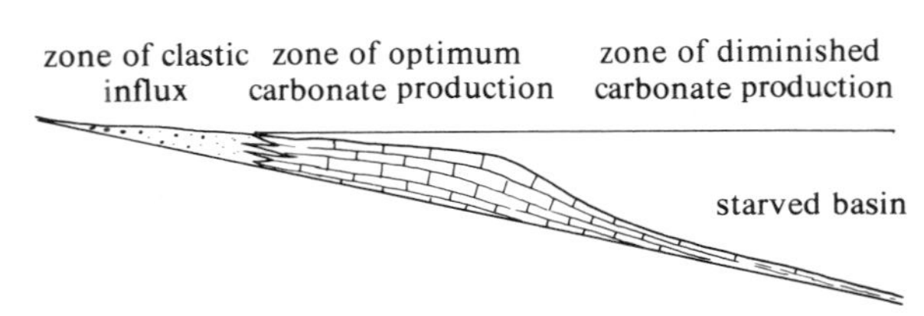

Figure 23.2 Creation of a shelf-to-basin topography caused by the high rate of shallow subtidal carbonate production (see Fig. 23.1) (after Meissner 1972, Wilson 1975).

Table 23.1 Some differences between carbonate and siliciclastic sediments (largely after James 1978a).

Carbonate sediments	*Siliclastic sediments*
The majority of sediments occur in shallow tropical environments	Sediments occur worldwide and at all depths
The majority of sediments are marine	Sediments are both terrestrial and marine
The grain size of sediments generally reflects the original size of calcified hard parts	The grain size of sediments reflects the hydraulic energy in the environment
The presence of lime mud often indicates the prolific growth of algae whose calcified portions are mud size crystal aggregates	The presence of mud indicates settling out from suspension
Sediment type has changed through time in response to evolution	Sediment grains unchanged through geological time
Shallow water lime sand bodies result primarily from localised physicochemical or biological fixation of carbonate	Shallow water sand bodies result from the interaction of currents and waves
Localised buildups of sediments without accompanying change in hydraulic regimen alter the character of surrounding sedimentary environments	Changes in the sedimentary environments are generally brought about by widespread changes in the hydraulic regimen
Sediments are commonly cemented on the sea floor.	Sediments remain unconsolidated in the environment of deposition and on the sea floor
Periodic exposure of sediments during deposition results in intensive diagenesis, especially cementation and recrystallisation	Periodic exposure of sediments during deposition leaves deposits relatively unaffected (duricrusts excepted)

(a)
highest wave energy
shelf
(b)
carbonate platform
shelf margin
buildup
(c)
shelf margins
shelf lagoon on platform
major offshore bank
starved basin
faults

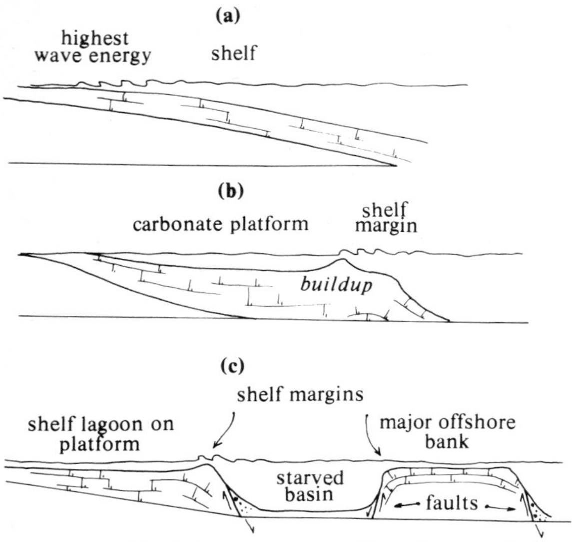

Figure 23.3 (a) Carbonate ramps; (b) carbonate platform protected by shelf margin build-up; (c) starved basins and offshore banks (after Wilson 1975).

limestone facies studies, but important local and regional details will hinge upon palaeogeographical, ecological and early diagenetic evidence. In the following account we shall therefore concentrate upon recent carbonate–evaporite facies.

23b Arid tidal flats and sabkhas

Arid tidal flats and sabkhas have been best studied along the southern shores of the Persian (Arabian) Gulf (e.g. Kinsman 1966, Evans *et al.* 1969) and around the deeply indented inlet margins of the Shark Bay area (Logan *et al.* 1970, 1974) in Western Australia (Ch. 23d). (We shall discuss the chemical diagenesis of supratidal sabkha environments in Chs 29 & 30.)

Both areas are dominated by evaporation due to a combination of extreme aridity and high annual temperatures. Thus rainfall in Shark Bay is a variable 230 mm a^{-1} with evaporation at 2200 mm a^{-1}. Rainfall in the Gulf is a sporadic 40–60 mm a^{-1} with evaporation 1500 mm a^{-1}. The major effect of such aridity on the supratidal and high intertidal sediments is greatly increased sediment pore-water salinity which leads to evaporite precipitation and dolomitisation.

In both areas, intertidal sedimentation is dominated by the growth of stromatolitic algal mats (see Ch. 10), which show well defined lateral zonation of growth forms due to variations in exposure. Around the Trucial coast the intertidal algal mat zone is up to 2 km wide, is broken up by an irregular network of channels, and is covered by discontinuous shallow ponds. Storm processes drive subtidal lagoonal sediments onto the intertidal flats and provide a major proportion of the pelletal sediment bound by the algal mats. Buried algal mat sections reached through pits in the prograding sabkha reveal that few of the detailed surface mat forms survive. This low preservation potential is caused by a combination of gypsum precipitation within the buried mat and to compaction and bacterial destruction of the organic-rich algal laminae (Park 1976, 1977).

Low-energy environments in Shark Bay (e.g. Nilemah Embayment, P. J. Woods & Brown 1975) are dominated by continuous algal mats. Well laminated sediments with fine laminoid **fenestrae** (Ch. 23c) occur beneath areas of smooth mat in the lower intertidal zone whilst poorly laminated sediments with irregular fenestrae occur beneath areas of pustular mat in the middle to upper intertidal zone. Dominant grain types in the tidal flats are

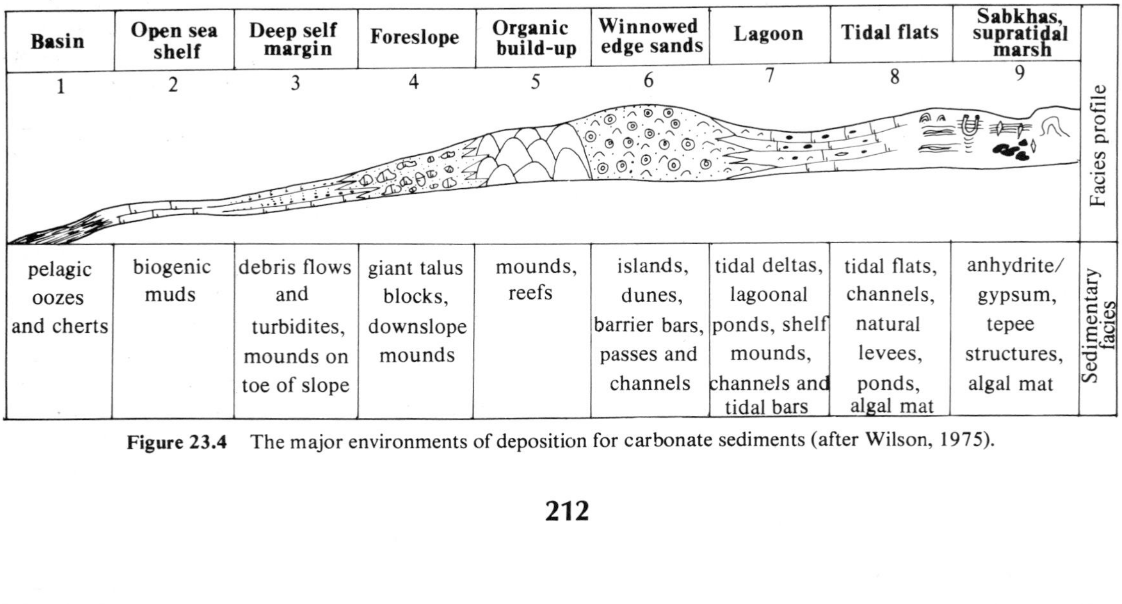

Basin
Open sea shelf
Deep self margin
Foreslope
Organic build-up
Winnowed edge sands
Lagoon
Tidal flats
Sabkhas, supratidal marsh
1
2
3
4
5
6
7
8
9
Facies profile
pelagic oozes and cherts
biogenic muds
debris flows and turbidites, mounds on toe of slope
giant talus blocks, downslope mounds
mounds, reefs
islands, dunes, barrier bars, passes and channels
tidal deltas, lagoonal ponds, shelf mounds, channels and tidal bars
tidal flats, channels, natural levees, ponds, algal mat
anhydrite/ gypsum, tepee structures, algal mat
Sedimentary facies

Figure 23.4 The major environments of deposition for carbonate sediments (after Wilson, 1975).

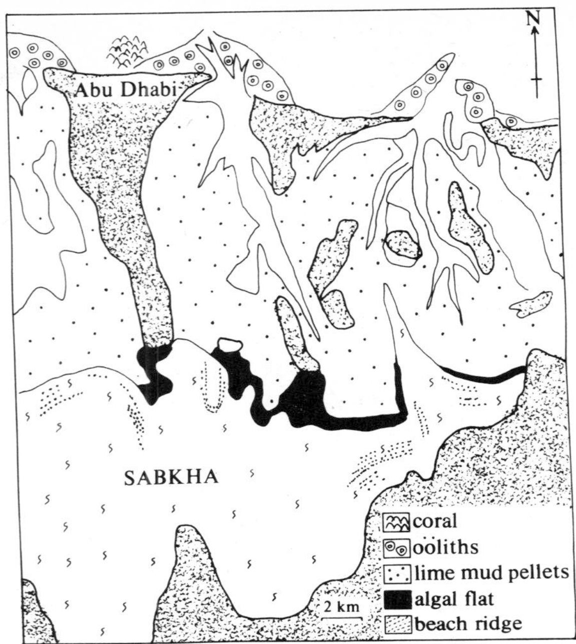

Figure 23.5 Map to show the distribution of lithofacies in the Abu Dhabi area of the Arabian Gulf. See also Fig. 23.8 (after Butler 1970).

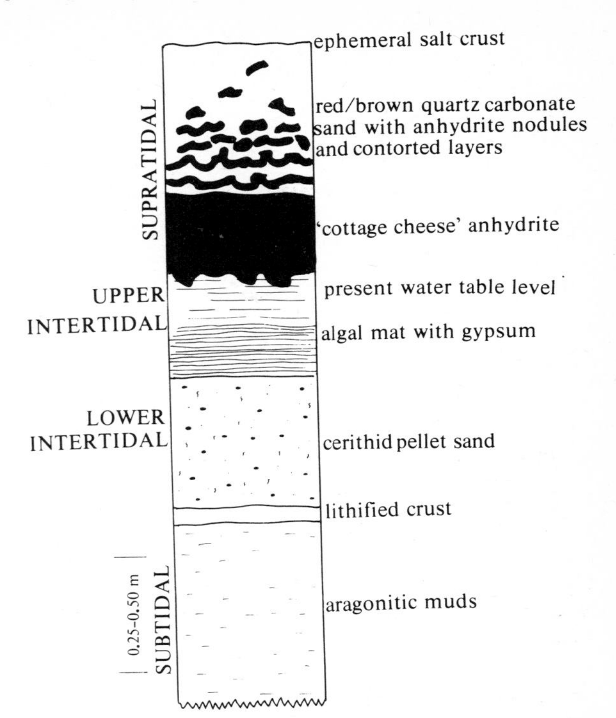

Figure 23.7 Sketch log to show regressive sabkha sequence developed in Figure 23.6 (after Till 1978).

pellets, altered skeletal grains and intraclasts, the last named being derived as storm rip-up clasts from areas of partially lithified sediment below algal mats in the high intertidal zone. Higher-energy environments in Shark Bay are typified by the northwestern margin of the Hutchison Embayment (Hagan & Logan 1975). Here lithified algal columns (Figs 10.1 & 2) and ridges form a stromatolitic reef that thickens seawards. The 'reef' is associated with a beach-ridge barrier comprising large-scale, cross-stratified molluscan coquinas.

The extensive supratidal **sabkhas** found in the Trucial Coast region of the Arabian Gulf (Figs 23.5 & 8) slope gently seawards at about 0.4 m km^{-1} and may be up to 16 km wide. As discussed in Chapters 29 & 30, dolomite and a well defined suite of evaporitic minerals occur in the shallow sabkha subsurface, the most distinctive feature being anhydrite and gypsum with chicken-mesh textures. The Quaternary sediments below the sabkha surface (Figs 23.6 & 7) reveal that a major transgressive event modified a sandy coastal desert zone at about 7000 years BP (Evans *et al.* 1969). The open coastal embayment so formed was changed into a lagoon/tidal flat complex when a small (? 1 m) sea-level fall caused the emergence of barrier islands and restricted circulation at about 3750 years BP. Subsequent tidal flat progradation caused sabkha flats to develop with their distinctive evaporitic suites. These are now subject to storm and storm-tidal processes which renew the interstitial pore waters

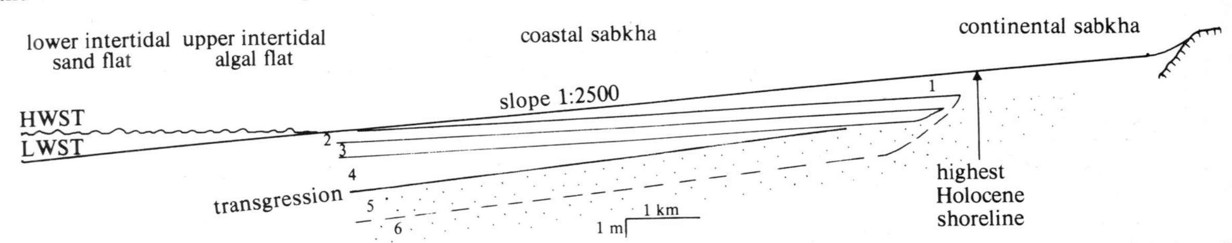

Figure 23.6 Composite section across Abu Dhabi sabkha. (1) Supratidal facies with evaporites; (2) upper intertidal facies with stromatolitic algal mats; (3) lower intertidal facies with muddy pellet-gastropod sands; (4) subtidal foram–bivalve muddy sands (**transgressive facies**); (5 & 6) Pleistocene wind-blown sands. Note superb regressive sequence over Holocene transgression plane. (After Evans *et al.* 1969.)

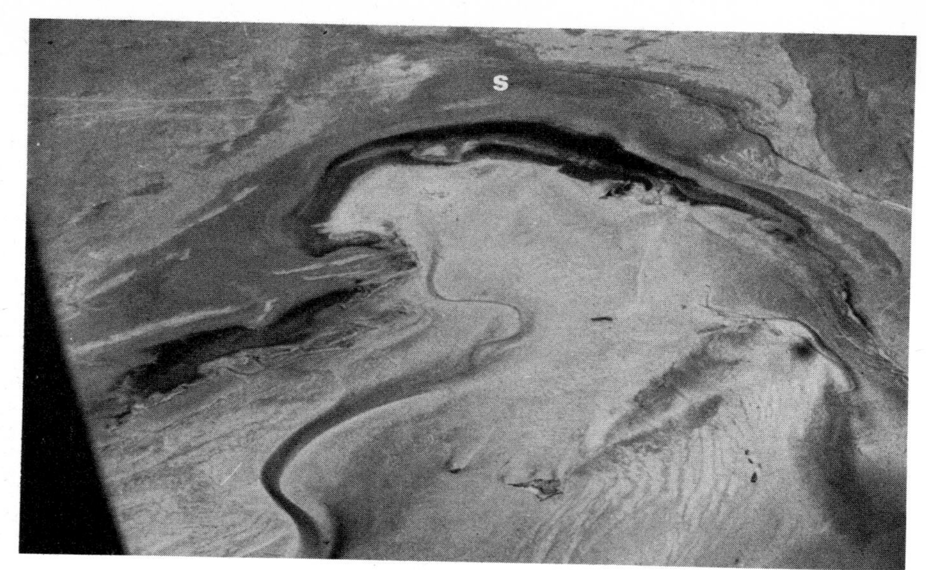

Figure 23.8 Aerial view of lagoon sands, tidal channel, intertidal algal belt (black) and sabkha(s), Abu Dhabi (photo R. Till).

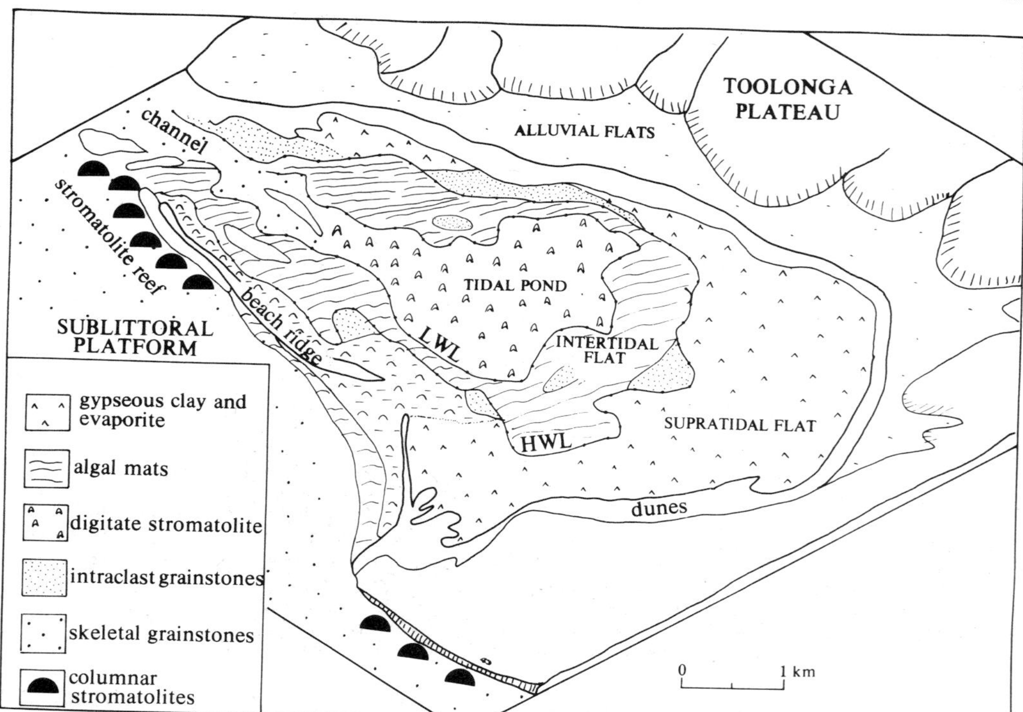

Figure 23.9 Map of Hutchison embayment, Shark Bay, W. Australia to show intertidal flat, lagoon and stromatolite facies (after Hagan & Logan 1975).

periodically. A generalised vertical facies sequence of the Quaternary sabkha facies is shown in Figure 23.7.

The Shark Bay tidal flats record a similar history of initial transgression (4000–5000 years BP), sea-level fall and coastal progradation. As the supratidal surface expanded, pore-water concentrations reached aragonite- and gypsum-precipitation levels. Gypsum is the major component in the upper intertidal and supratidal zones. A profile of the Hutchison Embayment tidal flat area is shown in Figure 23.9.

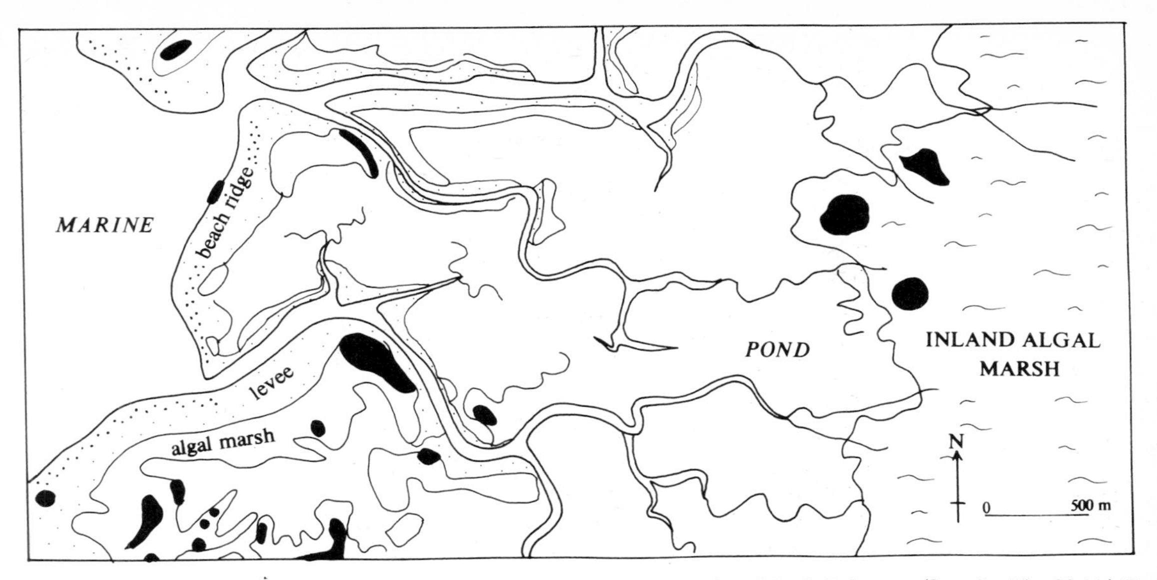

Figure 23.10 The environments of deposition present in the Three Creeks, Andros Island, Bahamas. (See also Fig. 23.11.) Dark patches indicate occurrence of cemented surface crusts. (After Hardie & Garrett 1977.)

Figure 23.11 Aerial photograph to show channelled tidal flats and marshes, Andros Island, Bahamas. Lagoon to top right (photo R. Till).

Ancient sabkha facies have been described from many areas and as far back in time as 2500 Ma. Shearman's classic pioneer study (1966) is a good starting point. Well illustrated accounts are given by Holliday and Shepard-Thorne (1974), West (1975), Wood and Wolfe (1969), Shearman and Fuller (1969) and Fuller and Porter (1969). The problems of ancient 'giant' sabkhas are discussed by Smith (1973), Leeder and Zeidan (1977), and Kendall (1978).

23c Humid tidal flats and marshes

The extensive tidal flats and supratidal marshes on the west, leeward, side of Andros Island, Bahamas (Hardie 1977) serve as the type examples of non-evaporitic flats (Figs 23.10–13). Similar examples occur around the Florida Coast. A tropical maritime climate prevails in the area with mean annual rainfall of about 130 cm a^{-1} (range 65–230 cm a^{-1}). This abundant rain water

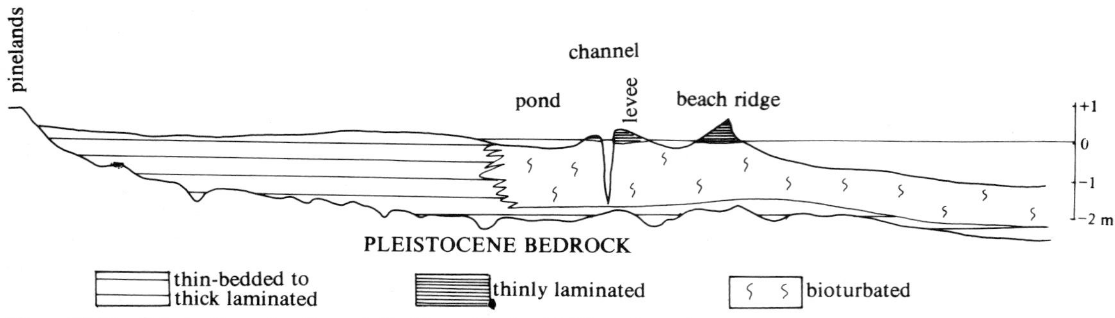

Figure 23.12 Schematic section through the Three Creeks tidal flats shown in Figure 23.10 (after Hardie & Ginsburg 1977).

	CORE	LAYER TYPE AND FEATURES	ENVIRONMENT	
laminite cap		smooth flat lamination with sandy lenses	washover crest	washover plain
		disrupted flat lamination with tiny mudcracks and intraclast lenses	washover backslope	
		crinkled fenestral lamination with lithified crust and tufa	high algal marsh	
tufa interval		algal tufa–peloidal mud interbeds with wide shallow mudcracks and intraclast pockets	low algal marsh	
burrowed unlayered base		thick bioturbated peloidal lime mudwith deep prism cracks, burrows, gastropod and foram. shells (very low faunal diversity)	intertidal pond and channel-fill	
		bioturbated peloidal lime mud with polychaete, worm and crustacean burrows and mollusc and echinoderm remains (moderate faunal diversity)	subtidal offshore lagoon or open bank	

Figure 23.13 Log to show the type of vertical succession that might result from continued seaward progradation of an Andros-type tidal flat. (After Hardie & Ginsburg 1977.)

freshens the supratidal marsh during summer months and prevents development of sabkha-type evaporites. Salinites of the tidal waters usually fall in the range 39–42‰ but may fall as low as 5‰ after heavy rainfall. This periodic freshwater 'flushing' creates a 'high stress' environment and is responsible for a restricted biota on the flats. The semi-diurnal tides have a mean maximum range of 0.5 m, but the tidal range is much affected by periodic storm surges. Wave action is not usually important because of the sheltered nature of Great Bahama Bank lagoon. The tidal flat sediments are dominantly pelleted carbonate muds, with $\lesssim$10% of skeletal material, dominantly foraminifera, and extensive algal mats. Three major sub-environments may be defined (Fig. 23.10) as (a) nearshore marine belt, (b) tidal flat complex of channels with levees and tidal pounds, and (c) inland supratidal algal marsh (Shinn *et al.* 1969, Hardie 1977).

The nearshore marine belt comprises thoroughly bioturbated, muddy pelletal sands loosely bound by a surface 'scum' of algae. Callianassid (crustacean) burrows are particularly common. The exposed shorelines between channel openings are beach ridges with terraces, clifflets and washover fans. The latter comprise intraclast gravels and rippled sands showing well developed internal laminations. Similar facies, with lithified intertidal beach-rocks (Ch. 29), make up much of the eastern, windward, coastline of Andros Island.

The intertidal flats, partly protected by the beach

ridges, are cut by a dense tidal channel network (Fig. 23.11). The channels are 1–100 m wide and 0.2–3 m deep. They meander but show little evidence of lateral migration, in contrast to channels on temperate siliciclastic tidal flats (Ch. 21). The channels contain lag gravels of skeletal debris, intraclasts, and fragments of Pleistocene bedrock. The channel banks and stationary point bars are heavily bioturbated by crabs and overgrown by mangroves and are covered by complex hemispherical stromatolite heads. Sections through these heads reveal well preserved domal laminae with abundant uncalcified filaments of the sediment-binding alga *Schizothrix calcicola*. The channel levees are only rarely covered by tidal waters and are coated by a thin algal mat. Sections reveal a fine millimetre-scale lamination without disruptions. The levee toes show small mudcracks which do disrupt the fine laminations.

In between adjacent channels lie extensive tidal ponds bounded by algal marshes. The ponds are frequently covered by tidal waters and the muddy sediment has a thin surface algal mat which is grazed by cerithid gastropods and polychaete worms. Sections reveal unlayered, bioturbated pelletal muds cut by deep (up to 30 cm) desiccation cracks that form during winter and spring low-water periods. The fringing algal marshes comprise a high marsh with continuous algal mats of the freshwater genus *Scytonema* and a low marsh with 'pincushion' growths of *Scytonema*. Patchy cementation by high-Mg calcite and aragonite occurs, sections through the mats revealing a well developed crinkly lamination with superb fenestrae (*q.v.*). The laminations seen in tidal flat sediments are thought to be due to sporadic onshore storms (generated by southward-pushing cold fronts) which suspend and transport lagoonal sediments onshore.

The inland algal marsh lies about 20 cm above the mean high-water level of the channelled tidal flats. A similar zonation of low 'pincushion' marsh to high 'carpet' marsh occurs as noted above from around the tidal ponds. No invertebrates live in this marsh, which may be up to 8 km wide. The *Scytonema* algal mats are frequently lithified by high-Mg calcite forming a discontinuous **algal tufa**. Desiccated mats give rise to characteristic polygon heads as the algae attempt to heal over the upturned polygon rims. Sections through the inland marsh reveal up to 1.7 m of laminated sediment with abundant fenestrae (see below). These marsh laminae (1–10 mm thick) are thought to be the result of periodic hurricane-driven sheet floods which carry lagoonal sediments over the entire marsh. The sediment is then bound by renewed *Scytonema* growth and the laminae preserved. The fenestrae are predominantly horizontal sheet cracks with subordinate vertical 'palisade' cracks. They form as primary voids from air pockets and as secondary voids from bacterial breakdown of algal filaments in vertical and horizontal layers and clusters. Lithification of the mats obviously enhances the preservation potential of the fenestrae. Indeed, spar-filled fenestrae ('birds-eye' fabric) in ancient carbonate sediments provide good evidence of high intertidal to supratidal origin.

Lateral and vertical sections through the entire Bahamian tidal flat complex are shown in Figures 23.12 and 23.13. Note that the supratidal zone contrasts markedly with arid tropical sabkhas. It is evaporite-free and it contains lithified algal tufa of freshwater-dominated marsh origin. The calcified *Scytonema* filaments and fenestrae in the latter serve to distinguish the stromatolites from the unlithified algal peats found in the intertidal zone of the Arabian Gulf.

Ancient analogues of Bahamian-type tidal flat and marsh deposits are to be found in the Triassic Lofer Cycles of Austria (Fischer 1964, 1975, Hardie 1977), the Devonian Manlius facies of the central Appalachians (Laporte 1971) and the Precambrian of South Africa (Eriksson 1977).

23d Lagoons and bays

Carbonate lagoons occur as relatively quiet-water environments separated from open marine environments by offshore islands of lithified Pleistocene limestones (Arabian Gulf), reefs (Honduras, Great Barrier), or a combination of the two (Florida, Bahamas). These fringing 'rims' protect the lagoons from onshore winds and hence the effects of waves. Tidal currents are forced to enter the lagoons via narrow inlets between islands or reefs and hence high current velocities occur. Efficient tidal exchange may keep lagoons close to oceanic salinity but in arid tropical areas high evaporation of the shallow water bodies may cause salinity to rise as high as 67‰ (Abu Dhabi). In humid tropical areas lagoonal water may be considerably freshened by freshwater runoff from the tidal flats and hinterland (West Coast of Andros Island; Florida Bay). Thus for the most part shallow coastal lagoons tend to be 'high stress' environments and a restricted biota occurs.

Lagoonal sediments usually comprise pelleted lime muds with decreasing amounts of mud as wave action increases in importance. The wave-stirred Trucial Coast lagoons, for example, are floored by pellet sands. The current-scoured outer Florida lagoon (Fig. 23.14) is floored by a winnowed lag of skeletal debris. Pellets are excreted by crabs, cerithid gastropods and polychaete worms. Aragonite mud is predominantly of algal origin in the inner Florida and Honduras lagoons, coccoliths being an important contributor in the latter area (Matthews

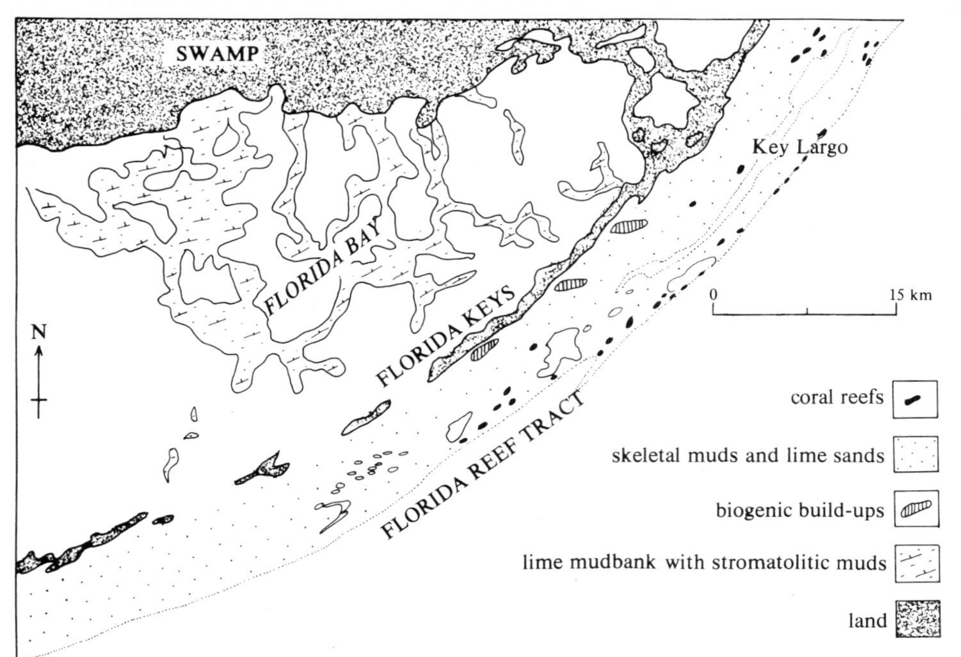

Figure 23.14 Recent sedimentary facies of South Florida (after Griffith *et al.* 1969).

1966). Controversy still rages as to the origin of the Bahamian aragonite muds (Ch. 2) but most is probably algal in origin. Minor skeletal debris usually comprises foraminifera and molluscs. Most lagoons support a thriving infauna, particularly crabs, which effectively destroy any primary laminations. *Thalassia* (sea grass) stands occur in the Florida, Bahamas and Arabian lagoons. In the inner part of Florida lagoon the *Thalassia* banks act as sediment baffles and have built up numerous winding mud mounds (Fig. 23.14). Patch reefs occur in many lagoons and are surrounded by a halo of coarse, reef-derived bioclastic grains. Variations in Holocene lagoonal sediment thickness are caused by differential topography of the lithified Pleistocene bedrock surfaces that underlie the lagoons. Some of this topography reflects the relief of buried karst surfaces (Purdy 1974, see Fig. 23.24).

Carbonate sedimentation in the bays of unrimmed, indented coastlines is best illustrated by reference to Shark Bay, Western Australia (Fig. 23.15). Shark Bay comprises a complex of embayments which are partially separated from the Indian Ocean by shallow sills and banks at their entrances. There is a general absence of freshwater input and the imperfect tidal flushing causes a general landward increase of salinity up to 70‰ (Fig. 23.16). The hypersaline portions of the bay are dominated by a monotypic coquina of the small bivalve *Fragum hamelini*, which is salinity tolerant. These subtidal coquinas fringe the arid tidal flats already discussed above (see Fig. 23.9). The metahaline and oceanic parts

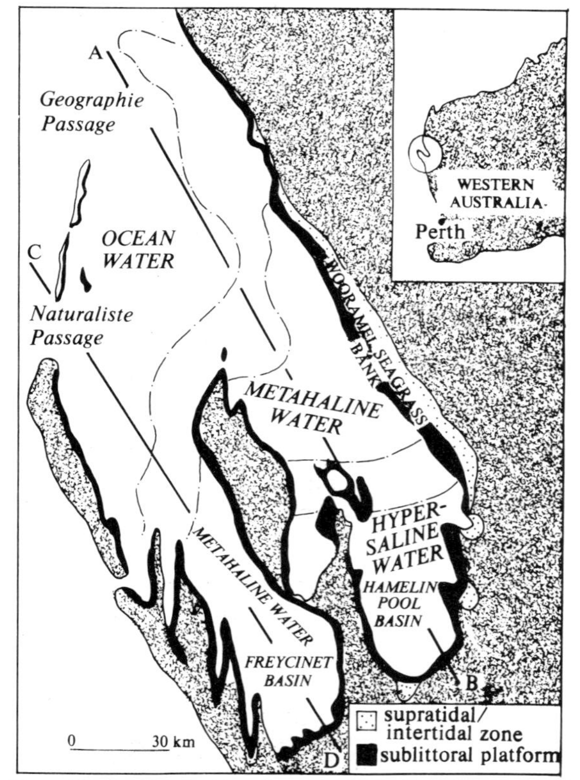

Figure 23.15 Shark Bay, Western Australia (after Logan & Cebulski 1972).

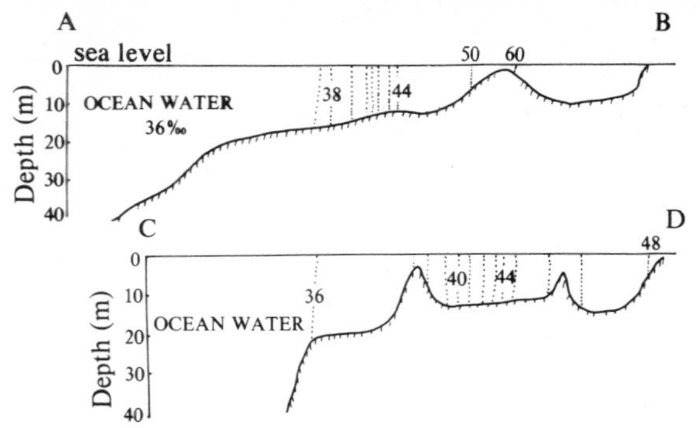

Figure 23.16 Schematic sections along the lines AB, CD in Figure 23.15 to show the increase in salinity observed between the Pacific Ocean and inner Shark Bay (after Logan & Cebulski 1972).

of the bay are dominated by spectacular seagrass-bound carbonate buildups (Fig. 23.17). These have topographic relief on the sea floor, lack an internal skeletal frame, and are composed of *in situ* and locally derived skeletal carbonate from the grass epibiota and sheltered benthos (Davies 1970, Hagan & Logan 1975). Skeletal breakdown causes much silt- and mud-grade material to be admixed with the coarser debris of molluscan, algal and forminiferal origins. The buildups occur as fringing, patch and barrier types. There is an upward trend within the buildups from matrix-rich skeletal packstones and wackstones to well washed skeletal grainstones. This upward trend arises as the seagrass meadows accrete upwards towards mean tide level. The buildups provide partial modern analogues (but see p. 225) for ancient carbonate mounds such as the Waulsortian 'reefs' of the Upper Palaeozoic. The baffling organisms in these ancient examples were probably crinoid meadows and bryozoan fronds since seagrasses evolved only during late-Mesozoic times.

Ancient lagoonal facies overlain by sabkha evaporites and/or tidal flat facies are described by Laporte (1971) and by D. B. Smith (1974).

23e Tidal delta and spillover oölite sands

There is a very close correlation between strong tidal currents and oölite formation. Tidal currents are amplified as they pass onto a rimmed shelf or into a lagoon through constrictions in reefal barriers and islands. Active oölite shoals result, with a variety of bedforms.

On the Bahamas Bank active shoals form at many localities around the perimeter (Illing 1954, Purdy 1963, Ball 1967, Harris 1979). The shoals take the form of **spillover lobes** directed towards the Bank interior and indicate dominance of the flood tidal currents and onshore storms (Fig. 23.18). The lobes have an axial channel and may be up to 1 km long and 0.5 km wide. The lobes terminate at steeply dipping noses and show internal large-scale foresets up to 1.75 m high. Smaller lobes superimposed on the larger forms show variable ebb

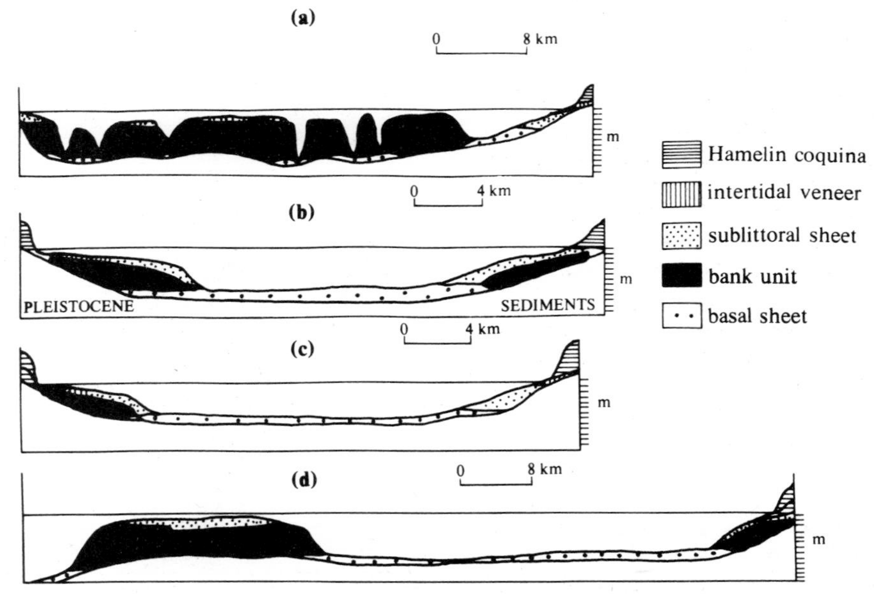

Figure 23.17 Cross sections across Hamelin Basin, Shark Bay, to show the development of carbonate banks (after Hagan & Logan 1974). For explanation see text.

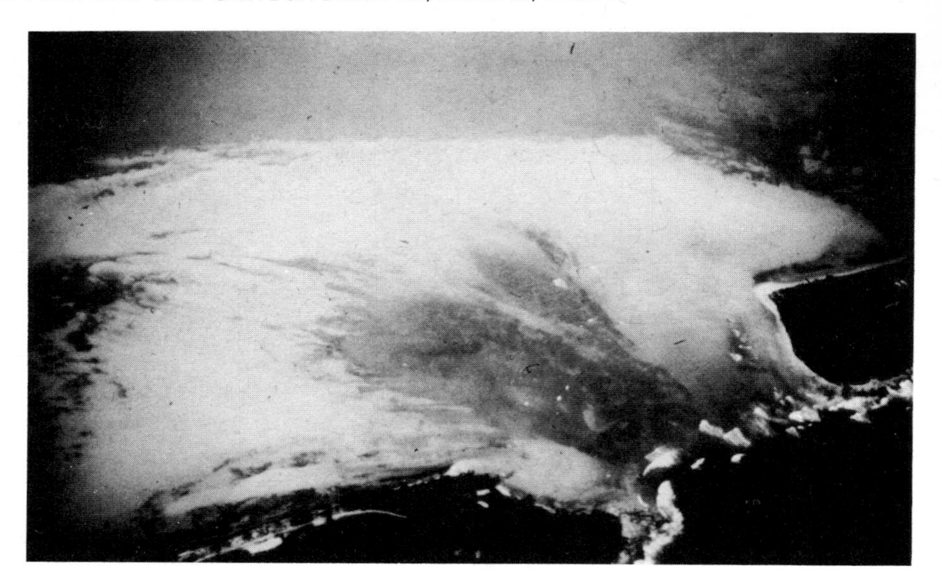

Figure 23.18 Aerial photograph to show oölitic spillover lobe developed between two islands as a flood tidal delta, Bimini, Bahamas. Lagoon to top left. (Photo: R. Till).

and flood orientations. Lobes are covered by current-ripple and dune bedforms which travel parallel to the lobe long axes. The larger spillover lobes may be active only when onshore storms or hurricanes assist the normal flood tidal currents. The active oölite shoals die out gradually towards the Bank interior where they are replaced by a stable oölitic and grapestone lithofacies covered with a thin subtidal algal mat and *Thalassia* stands that effectively stop any bedload movement.

These lateral facies changes from bank edge to lagoon are well illustrated by the Joulters Cay Shoal (Harris 1979, Fig. 23.19). Here the site of active oöid sands is located as a windward fringe 4 m thick with an extensive bankward spread of altered oöids mixed with skeletal grains and aragonitic muds. The muds are stabilised by grasses and algal films, being extensively bioturbated and up to 10 m thick. There is an upward trend towards less mud within the inactive interior shoal. Numerous horizons within the interior shoal show penecontemporaneous cementation (Ch. 29) with rim cements of acicular aragonite, and aragonitic or high-Mg calcite micrite. Cementation occurs in stabilised bottom areas covered by algal scum.

A further example of oölite shoals occurs in the Schooners Cay area at the north end of Exuma Sound (Ball 1967). Here the shoals take the form of **linear tidal ridges** (see Ch. 22) whose long axes are parallel to the dominant flood tidal currents. Individual ridges are up to 8 km long and 750 m wide with amplitudes of about 5 m. Spillover lobes occur with their long axes orientated sub-parallel to the ridge long axes. They indicate a component of on-bank flow that is reflected in the asymmetry of the

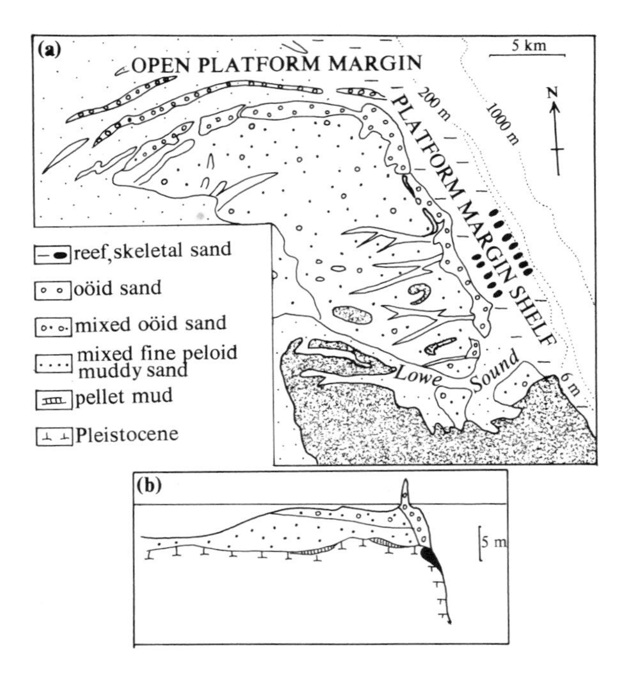

Figure 23.19 (a) Environments around Joulters Oölite Shoal, Bahamas. The shoal is a shallow sand flat (fine stipple), cut by tidal channels and fringed on the ocean side by mobile sands (coarse stipple). (b) Section to show the vertical and lateral distributions of facies; note the oceanward and vertical coarsening-upwards facies pattern. (Both after Harris 1979.)

Figure 23.20 Aerial photograph of ebb tidal oölite delta complex, Abu Dhabi, Arabian Gulf (photo R. Till).

ridges, whose steeper sides are directed bankwards. Ripples and dunes superimposed upon the ridges are also orientated bankwards. Flow in the channels separating ridges is dominantly parallel to the ridge long axes. In summary these linear ridges are very similar to those described from clastic tidally influenced shelves (Ch. 22) and are expected to show the same internal structures, i.e. cross sets dipping obliquely to perpendicularly with respect to the ridge long axis. The ridges overlie burrowed muddy pelletal sands. Penecontemporaneous cementation may occur in the channel floors between the active oölite ridges.

In the Trucial coast area, oölites are concentrated within spectacular ebb-tidal deltas at the mouths of barrier island inlets (Fig. 23.20). The purest oölite sands occur along the edges of the delta channels on levees where constant movement occurs in response to both ebb-tidal currents and onshore waves.

Ancient oölitic complexes are well described from the Pleistocene of Florida (Halley *et al.* 1977) and the Middle Jurassic of northwest Europe (e.g. Sellwood & McKerrow 1973, Purser 1979).

23f Open carbonate shelves

As noted in the introduction the majority of Recent carbonate sediments are being formed on rimmed platforms. Few studies of truly open carbonate shelves (Ginsburg & James 1974) with tidal flushing and wave action extending across the shelf have been undertaken, work being restricted to low productivity, partly relict examples (Florida, East Gulf of Mexico, Yucatan). The Yucatan shelf has an inner zone 130–90 km wide extending down to depths of 60 m where a zone of relict Quaternary buildups occurs along the shelf break. The modern sedimentary cover comprises a thin layer of molluscan debris, everywhere less than 1 m thick. At the shelf break the buildups are associated with relict Quaternary lime sands with oöids, peloids and lithoclasts. At greater depths these non-skeletal sands are increasingly diluted with the tests of winnowed pelagic foraminifera. This pattern of relict outer-shelf facies and contemporary inner-shelf molluscan debris occurs on most tropical or subtropical clastic-free shelves (Ginsburg & James 1974). This complex lateral lithofacies distribution is most useful as an indication of trends over ancient regressive/transgressive shelves. It also emphasises the extreme importance of the shallow subtidal carbonate 'factory' (James 1978a) as a sediment producer. Decreased shelf productivity out from the coastline encourages rimmed platform production, as was noted at the beginning of this chapter.

The Arabian Gulf provides perhaps the most studied modern examples of non-relict offshore-shelf carbonate sedimentation (Houbolt 1957, Pilkey & Noble 1967). Over most of its area the shelf waters show salinities between normal seawater values and 42‰. In shallow coastal areas (5–30 m deep) skeletal grainstones comprising well rounded and well sorted molluscan, foraminifera, algal and (localised) coral debris are accumulating. In deeper offshore areas (>30 m) sorting becomes poorer and skeletal fragments are more angular, their sharp fracture surfaces perhaps being caused by *in situ* mechanical breakdown. Increasing admixtures of silt and mud-grade low-Mg calcite occur in deeper areas, giving rise to packstone and wackstone fabrics and, ultimately, marls. The fines are thought to be derived from wind-blown dusts that originate in the Mesozoic and Cainozoic limestone mountains and deserts of central Arabia and Iran. Although 'whitings' of precipitated aragonite mud (Ch. 2) occur periodically in the Gulf, no trace of this aragonite has been recorded in the offshore sediments. As noted in Chapter 29 much of the Gulf shelf is covered by a thin, lithified subtidal hardground that supports a specialised epifauna adapted to hard substrate life. This cemented horizon again stresses the relatively low productivity, and hence sedimentation rates, that may occur in offshore carbonate shelves. Low sedimentation rates and current winnowing of fines actively encourages cementation and further discourages a 'normal' carbonate-producing fauna.

Many examples of ancient open carbonate shelf facies have been described, including the studies of Townson (1975) on the Portlandian of southern England, Wilson's summary of work on the Smackover Formation of Texas (1975), and Talbot (1973) on the Corallian carbonate cycles of southern England.

23g Platform margin reefs and buildups

Ancient and modern carbonate platforms are very frequently rimmed by carbonate buildups which control the resultant distribution of carbonate facies on the platform itself. Wilson (1975) defines the following terms which are used, somewhat variably, by authorities on 'reefs':

carbonate buildup – mostly organic bodies of locally formed and laterally restricted carbonate sediment which possesses topographic relief
mound – equidimensional or ellipsoidal buildup
pinnacle – conical or steep-sided upward tapering mound
patch reef – isolated circular buildup in shallow water
knoll reef – *ditto* in deeper water
atoll – ring-like organic accumulation surrounding a lagoon
barrier reef – curvilinear belt of organic accumulation situated somewhat offshore and separated from the coast by a lagoon
fringe reef – belt of organic accumulation built out directly from the shoreline.

Any account of carbonate buildups through geological time must take into account the evolution of organic 'reefal' communities (Fig. 23.21). Such a vast approach is beyond the scope of the present text; instead we will

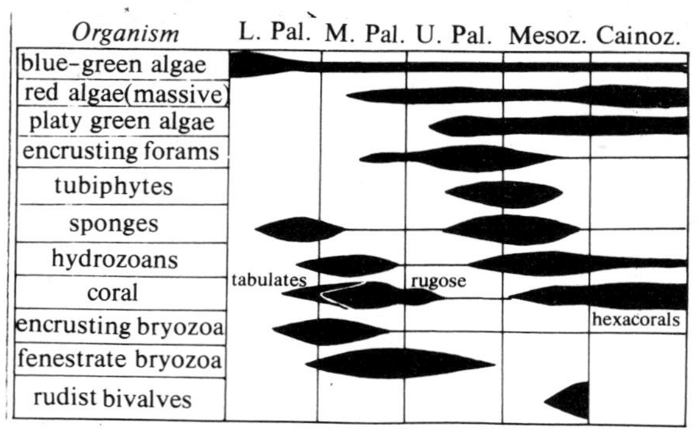

Figure 23.21 The predominance of various organic groups within carbonate buildups through geological time (after Heckel 1974).

follow J. L. Wilson (1975) in defining three types of carbonate platform margin, each with its own organic signature.

Type 1 margins (Fig. 23.22) comprise **downslope buildups** of carbonate mud and organic detritus which is trapped by baffling organisms. The buildups may take the form of mounds, linear fringes or barriers and need not be confined to the photic zone. The *Thalassia*-bound buildups of Shark Bay and Florida Bay (p. 218) represent modern examples in 'lagoonal' environments, but no whole platform rim is made of these buildups at the present day. The deepwater buildups recently discovered on the flanks of the Bahamas platform (Neumann *et al.* 1977, Mullins & Neumann 1979; see Ch. 23h) may provide important modern analogues to some ancient lime mud mounds, including perhaps the Carboniferous Waulsortian buildups of North America and Europe. Figure 23.23 shows a section through an idealised mud mound of this latter sort. The trapping and baffling organisms may be sponges and algae (Cambro–Ordovician), bryozoans (Ordovician–Permian), platy algae (U. Carboniferous), crinoids (Silurian–Carboniferous), rudist bivalves (Cretaceous), and marine grasses (Tertiary–Recent). Proof of topographic relief on ancient mud mounds revolves around recognition of contemporary, non-tectonic dips using internal sediment and sparry calcite **spirit levels** and talus spreads tonguing out from the buildup flanks.

Type 2 margins (Fig. 23.22) comprise linear belts of **framebuilt knoll reefs** located on the gentle slopes of some shelf margins. The association is adjusted to fairly gentle currents and it shows branching and fasciculate rudist, coral, sponge and stromatoporoid colonies which may pass upwards into more massive encrusting forms. Much interreef debris occurs. Examples of knoll reef ramps include the great Cretaceous rudist reefs of the circum-Gulf of Mexico, (Wilson 1975).

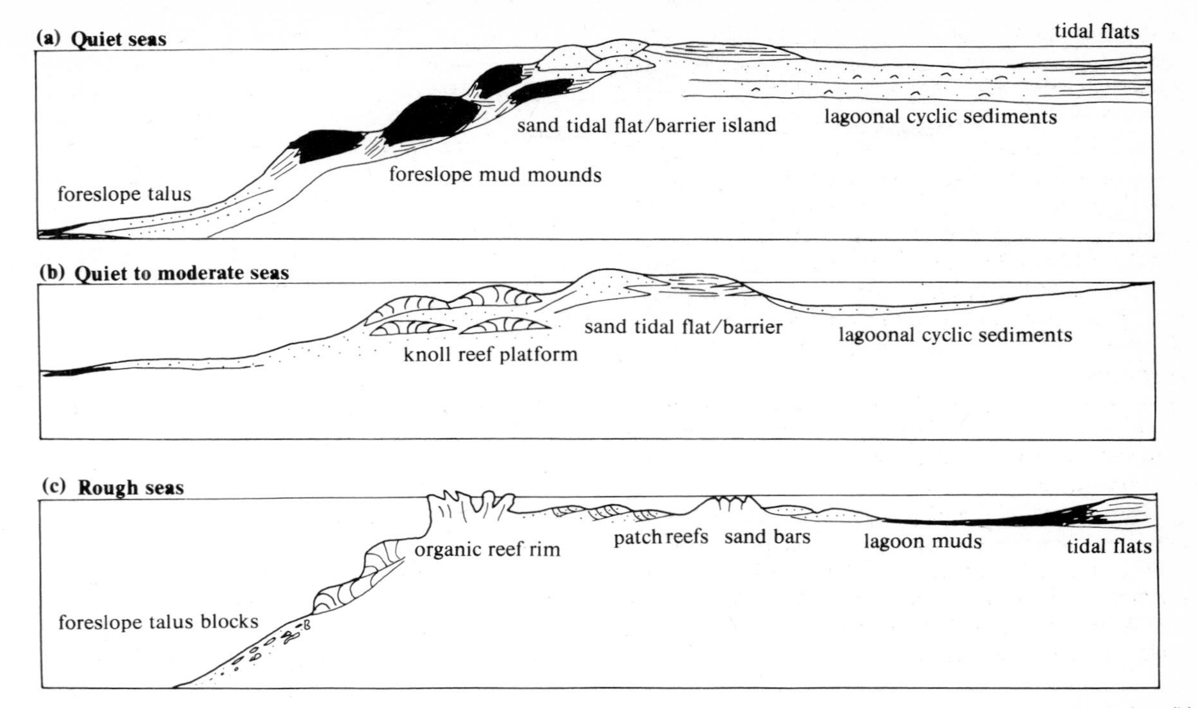

Figure 23.22 The three types of carbonate platform/shelf margins defined by Wilson (1975). (a) Downslope mud accumulation; (b) knoll reef ramp or platform; (c) organic reef rim.

Type 3 margins (Fig. 23.22) comprise **framebuilt reef rims** which grow up to or close to mean sea level and hence into the zone of greatest wave energy. They form barrier or fringing reefs and are zoned ecologically in parallel belts, with their hexacoral growth forms precisely reflecting light intensity, sediment concentration, exposure level and energy levels (see summary in James & Ginsburg 1979 and Chappell's (1980) recent model). The reefs usually show steep seaward slopes with abundant reef talus. Such reef rims predominate today and reflect the major influence of glacio-eustatic sea level changes upon reef rim erosion and accretion (see Ch. 23h). On the Bahamas and Florida carbonate platforms the framebuilt reefs support not only a highly diverse coral community but also a flourishing reefal epifauna of molluscs, echinoids, coralline algae and foraminifera. Sand fringes in the back-reef and fore-reef areas are dominated by calcareous algal fragments, the coral not being a good sand former. In the Floridan reef tract the extensive back-reef environment, with its patch reefs and sublittoral coral/algal sand spreads, grades into the back-reef lagoon.

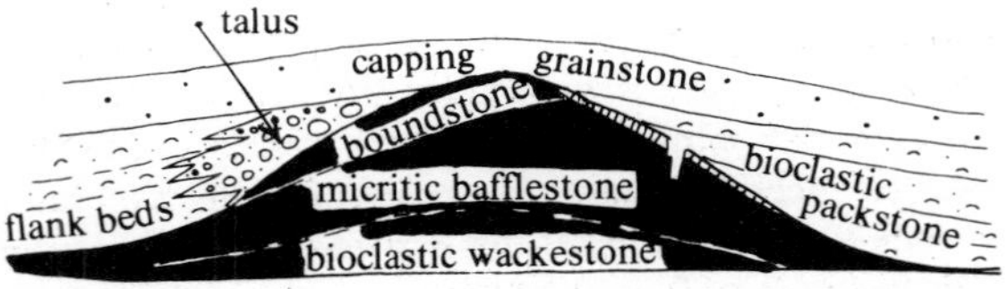

Figure 23.23 Idealised section through a typical carbonate mound buildup (after Wilson 1975).

In a stimulating paper, Purdy (1974) has drawn attention to the effects that Quaternary sea level oscillations have had upon reef development. He amplifies MacNeil's (1954) hypothesis of karst-induced effects upon atoll and barrier reef morphology. His **antecedent karst theory** envisages sub-aerial exposure of a limestone platform surrounded by a relatively steep structural or depositional slope during a sea level low. $CaCO_3$ dissolution is concentrated in the middle of atoll-like offshore banks and on the landward flanks of barrier-like ramps. Karstic rims and tower karsts acted as nucleii for coral growth as sea level rose to its present position, producing the present day atoll rims, barrier reefs and lagoonal pinnacle reefs (Fig. 23.24).

Ancient reef facies are well summarised by J. L. Wilson (1975) and James (1978b), both sources containing full bibliographies.

23h Platform margin slopes and basins

Relatively little is known about modern platform margin slopes and their associated basins, although much has

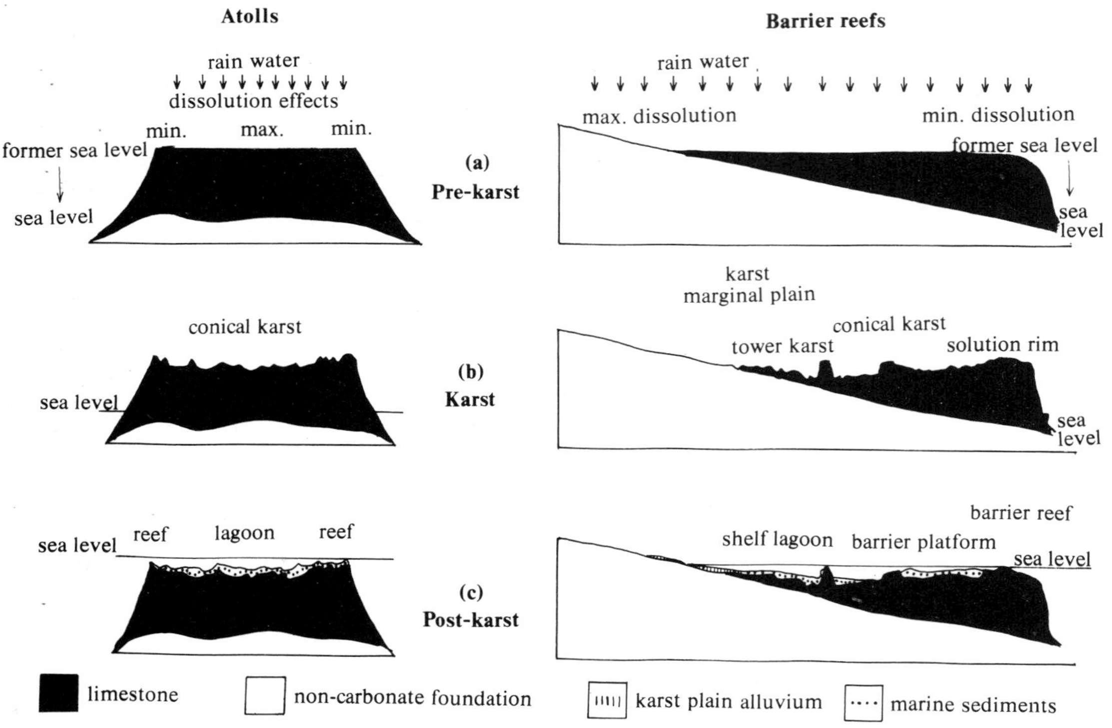

Figure 23.24 Diagrammatic evolution of atolls and barrier reefs according to antecedent karst theory. Both sequences begin with sub-aerial exposure of carbonate banks or platforms after sea-level fall and end with renewed transgression and reef growth (after Purdy 1974).

been inferred from ancient carbonate platform complexes (see review by James 1978b).

In their outstanding monograph on the seaward margins to the Belize barrier and atoll reefs James and Ginsburg (1979) recognise four facies belts passing seawards from the reef front. The **reef front** down to 70 m depth comprises coarse coral and *Halimeda* sands and conglomerates with grainstone fabrics. The **reef wall** (65–120 m) is made up of well cemented coral-rich limestones which yield ages in the range 8000–15 000 years BP. The **fore-reef** talus fans comprise muddy Halimeda sands showing packstone and wackstone fabrics. The **deep basin** sediments are made up of pelagic carbonate muds. No reef-derived grains are found further than 4 km out from the reef wall. Cements in the reef wall include common high-Mg calcite and subordinate aragonite. Isotopic and trace-element analyses prove the marine origins of these cements. James and Ginsburg propose an accretionary model for the seaward margin to the Belize platform, with erosion having occurred during periods of low sea level, and rapid coral growth 'plastering' the reef wall during periods of rising sea level. Reef wall growth is thus envisaged as a discontinuous lateral accretion process, with submarine cementation occurring after each period of accretion.

Recent exploration of the margins and basins around the Bahamas Bank has identified a number of interesting features. In their study of the slopes and basins bordering the Little Bahama Bank and the northern Great Bahama Bank (Figs 23.25 & 26) Mullins and Neumann (1979) recognise the following controls upon sedimentation at bank margins: (a) presence or absence of bank margin faults, (b) direction and magnitude of off-bank sediment transport, (c) amount of mass flows and pelagic deposits, (d) nature of the oceanographic circulation, (e) degree of submarine cementation, and (f) presence of deep water organic buildups.

Basement faulting may control the initial area of major offshore carbonate banks during oceanic rifting. The shallow-water platform acts as the major source of carbonate sediment for the deep basins and their margins. Along shallow windward margins sediment is transported on to the adjacent platform; along leeward margins it may be transported offshore. Pelagic and gravity-flow processes dominate bank margin sedimentation. Pelagic carbonates are important only when not winnowed by

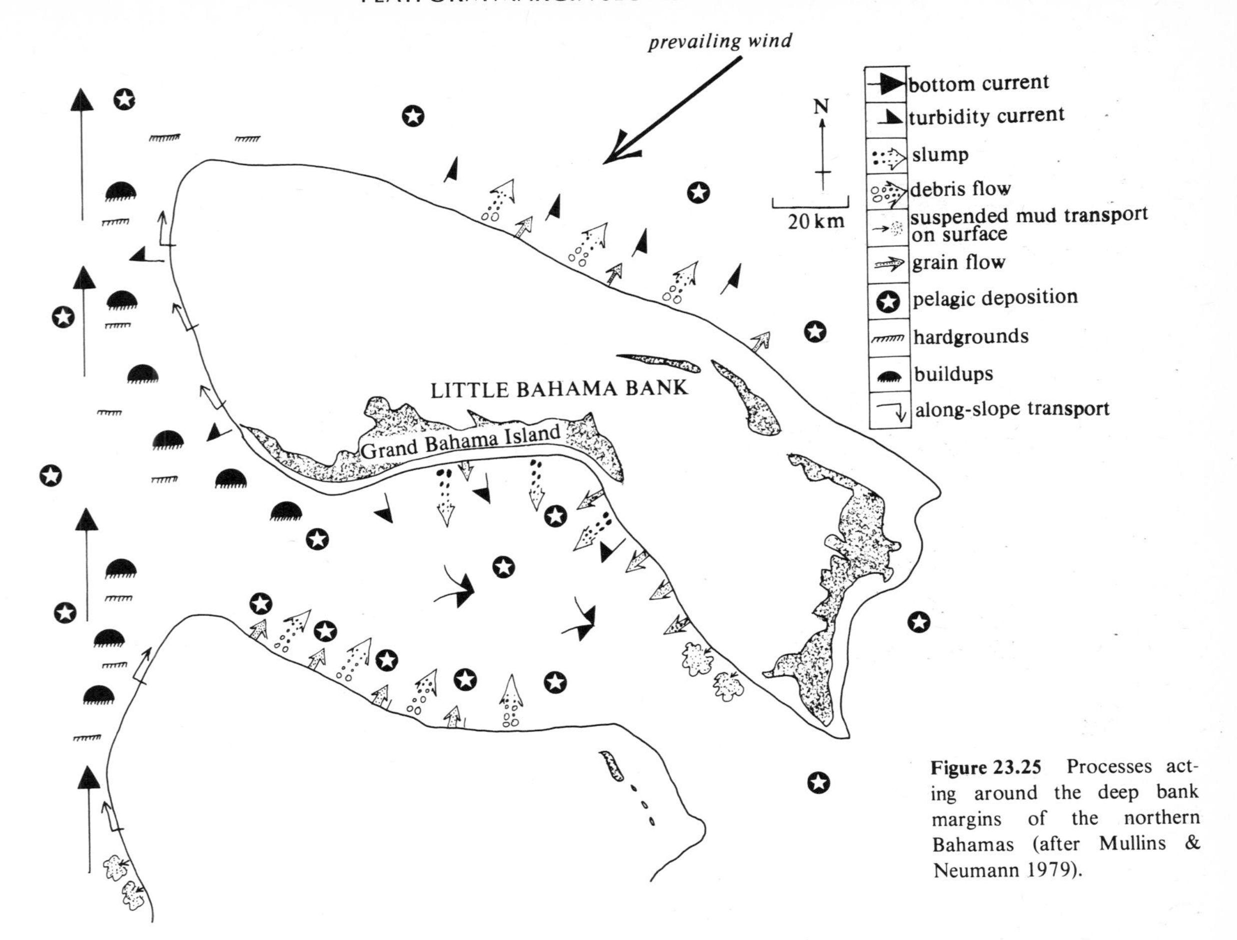

Figure 23.25 Processes acting around the deep bank margins of the northern Bahamas (after Mullins & Neumann 1979).

bottom currents, diluted by gravity flows or dissolved below the carbonate compensation depth. Gravity flows are important where slopes are steep. Thick (~0.5 m) graded carbonate turbidites occur on the lower slopes around the Little Bahamas Bank and as thinner beds (~0.17 m) in the basins where they are interbedded with pelagic oozes. Slope breccias of debris flow origin occur on the gentle muddy slopes whilst grain flows occur at the base of very steep (~18°) slopes around the marginal escarpment. Very extensive slump, debris flow and turbidity flow deposits have recently been discovered in Exuma Sound basin, Bahamas (Crevello & Schlager 1981). Ancient analogues to these gravity flows are described and discussed by Mountjoy *et al.* (1972).

Very extensive areas of submarine cementation occur west of the northern Bahamas down to depths of 500 m or so (Fig. 23.26). These lithified slopes are very stable and the cementation undoubtedly helps to maintain the steep gradients observed. In general the degree of cementation decreases down slope from well lithified hardgrounds at depths <375 m, to lithified nodules in a softy muddy matrix from depths between 375 m and 500 m, and soft oozes at depths greater than 500 m (Mullins *et al.* 1980). The nodules are multigeneration intramicrites to intramicrudites cemented by high-Mg calcite in layers up to 1.5 m thick. It is thought to be no coincidence that the cementation occurs along slopes where the Florida Current flows. This strong bottom current enhances cementation decreases down slope from well lithified hardgrounds at depths <375 m, to lithified nodules in a softy (c) providing unlimited sea water to source the cement ions (Mullins *et al.* 1980).

Cementation and bottom currents also play a role in localising the spectacular 'ribbon' of deep-sea buildups (Fig. 23.25) which extends over 200 km from the Blake Plateau along the western margin of the Little Bahama Bank to Bimini (Neumann *et al.* 1977, Mullins & Neumann 1979). These lithified buildups occupy a zone some 15 km wide in water depths of 600–700 m. They show up to 50 m relief and may be hundreds of metres

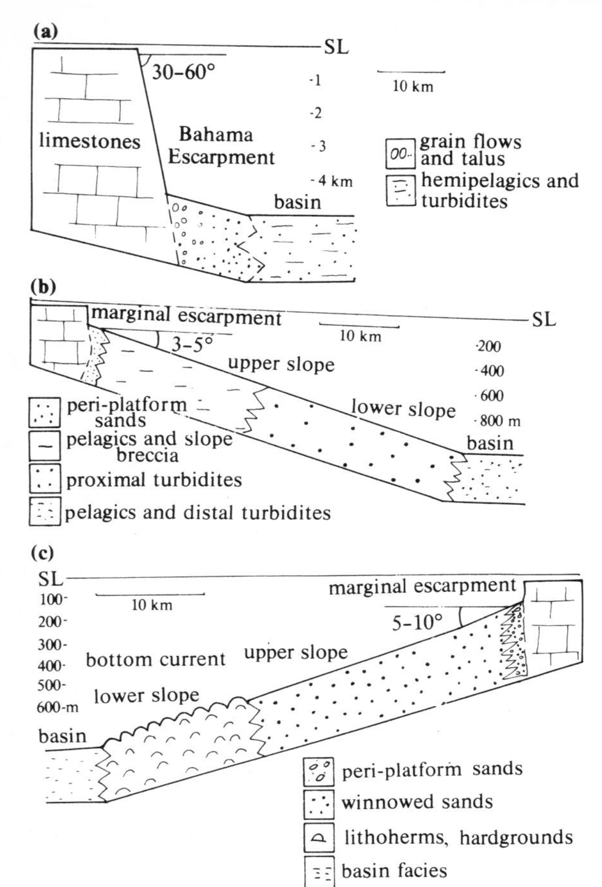

Figure 23.26 Sketch sections across deep bank margins in the northern Bahamas. (a & b) Margins on windward oceanic coasts; (c) margins on leeward seaway margins with contour currents. (After Mullins & Neumann 1979.)

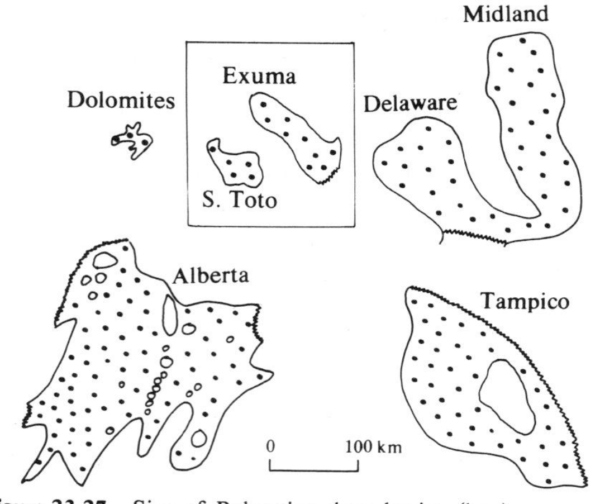

Figure 23.27 Size of Bahamian deep basins (box) compared with some ancient carbonate basins (after Schlager & Chermak 1979).

long, significantly orientated parallel to the northerly deep current flow. Observations from submersibles confirm their contemporary origin and reveal a dense and diverse benthic community of crinoids, ahermatypic corals and sponges which must baffle and trap sediment provided by the bottom currents. The buildups are constructed *in situ* by lithification of successive layers of trapped sediment by micritic high-Mg calcite.

This summary of Bahamian platform margins and basins stresses the importance of sediment supply, slope magnitude and contour currents in shaping the sedimentary framework (see also Maldonado & Stanley (1979) on the Menorca Fan, western Mediterranean). In their analysis of ancient carbonate slopes, McIlreath and James (1978) define two main groups of platform margins, both of which may be recognised in the Bahamas. **Depositional** margins are generally gentle and the slope decreases basinwards to merge with the basin floor. **Bypass margins** are astride a marginal scarp so that sediment is transported from shallow to deep water, bypassing much of the slope through channels and canyons. The latter is the commonest type on the margins of modern platforms, a particularly good example being the Tongue of the Ocean basin described by Schlager and Chermak (1979). The size of modern Bahamian basins is compared with some well documented ancient examples in Figure 23.27.

Ancient platform slope and basin plains are discussed by McIlreath and James (1978). Good case histories are given by Enos (1977) and Hopkins (1977) and in several other papers in SEPM Publication no. 25. Mountjoy *et al.* (1972) is a good review of the significance of gravity flow deposits in the elucidation of ancient platform margins. Modern platform/reef slopes and their deposits around the Honduras margin are well described by James and Ginsburg (1979).

23i Sub-aqueous evaporites

Until the sabkha evaporite model was outlined in the mid-1960s it was almost universally assumed that evaporitic salts in the geological record were 'straight' chemical precipitates from bodies of standing brines. As outlined in Chapter 3 the classical model of sub-aqueous evaporite precipitation involved a barred shallow basin in which excess of evaporation over freshwater influx causes 'topping-up' of the basin by one-way exchange over the barrier with the open ocean. Salinities thus build up within the basinal brines until the solubility product of $CaSO_4$ is exceeded and gypsum is precipitated. Increased brine concentration then gives rise to halite precipitation and so on until a residual bittern precipitates potash salts. In this way characteristic evaporite cycles of variable thickness and composition are produced.

The crux of the modern dilemma regarding sub-aqueous evaporites hinges around (a) the extremely large area over which some ancient evaporites may be traced and (b) the scarcity of modern examples of such large evaporite basins. The problem is compounded by the lack of knowledge concerning the evolution of sabkha plains with time, for it has been postulated that continued sabkha progradation over periods of up to 10^6 years produces standing evaporitic brine bodies that might cause sub-aqueous evaporite facies to overlie true sabkha anhydrite facies (Leeder & Zeidan 1977).

Some light is shed on the matter of scale by reference to the early evaporite phase that many opening and closing oceans seem to pass through. As we shall discuss in Chapter 26, marine transgression into an incipient oceanic rift often seems to have been followed by a period of massive evaporite precipitation from shallow to deep brine bodies. Continued seafloor spreading ultimately encourages better exchange with the parent oceanic mass, evaporite precipitation ceases and the thick halite-dominated evaporite succession is overlain by 'normal' oceanic sediments. Similarly, closing oceans are prone to salinity buildups should they become isolated from the world ocean. The famous Messinian evaporites of the western Mediterranean are thought to have originated in this way (Ch. 26).

Leaving aside these examples of oceanic evaporite sequences, it is apparent that recognition of ancient sub-aqueous evaporites must be made on as broad a basis as possible (Figs 23.28 & 29). Marine-derived brines may form in a multitude of settings, including on-sabkha depressions, barred lagoons, intra-platform basins and intra-buildup basins. Consideration of local and regional facies should serve to delimit such settings. Lateral facies changes may allow basinal sub-aqueous evaporitic units to be traced into contemporary nearshore facies with evidence of shallow-water deposition. In such cases a *prima facie* case for basin topography may emerge. Sub-aqueous non-evaporitic facies that overlie and underlie an evaporite unit may also be helpful in this respect, although these will not necessarily prove that the evaporite unit is sub-aqueously deposited. In the last resort it is the features of the evaporites themselves that may prove decisive. Let us briefly examine features that might indicate sub-aqueous evaporite deposition:

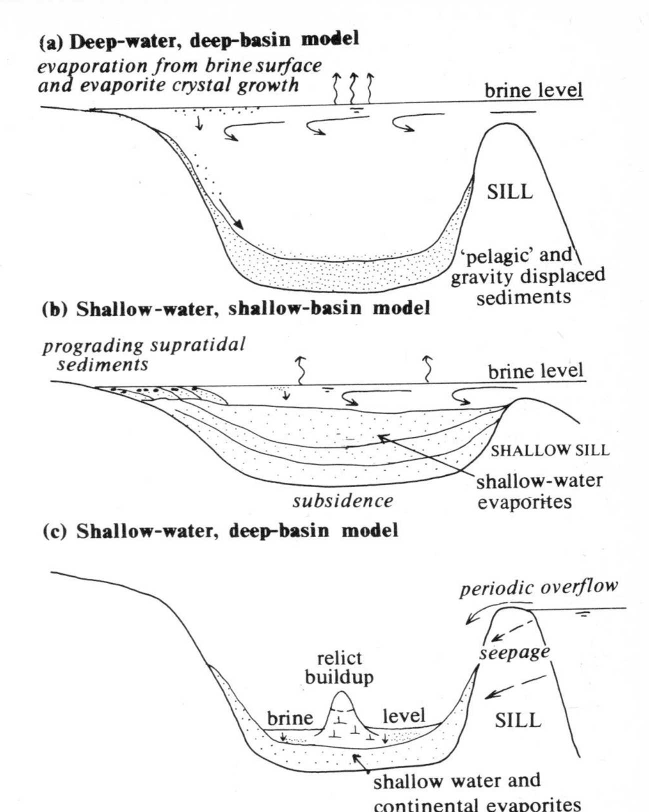

Figure 23.28 Depositional models for basin-central evaporites (after Kendall 1978).

(a) Shallow brine bodies should be affected by wave-driven and, perhaps, tidal currents. Primary bedforms and sedimentary structures should therefore result. Clastic textures within the evaporite will also be formed, but these, together with some of the current-produced laminations, may be completely destroyed during burial diagenesis, particularly if the gypsum → anhydrite change is involved.

(b) Should sub-aqueous evaporite be subjected to penecontemporaneous differential applied stresses then secondary structures will be formed, ranging from sedimentary deformation (load casts) to slump-like mass flows. Should rapidly moving mass flows be subject to mixing with the ambient brine, then redeposited evaporite beds with internal turbidite characteristics will be produced (Schreiber *et al.* 1976). Some redeposited evaporites may have been derived from pre-existing evaporites that have been uplifted and eroded.

(c) There is no doubt that sub-aqueous evaporite deposition will favour the production of widely traceable (10s to 100s of km) varve-like laminations (Ch. 3).

(d) Sub-aqueous evaporites should show evidence of crystal growth either at the brine/air interface or at the brine/sediment interface. Delicate rafts of hopper halite crystals form at the brine/air interface whilst upward-growing **chevron** halite growths form at the

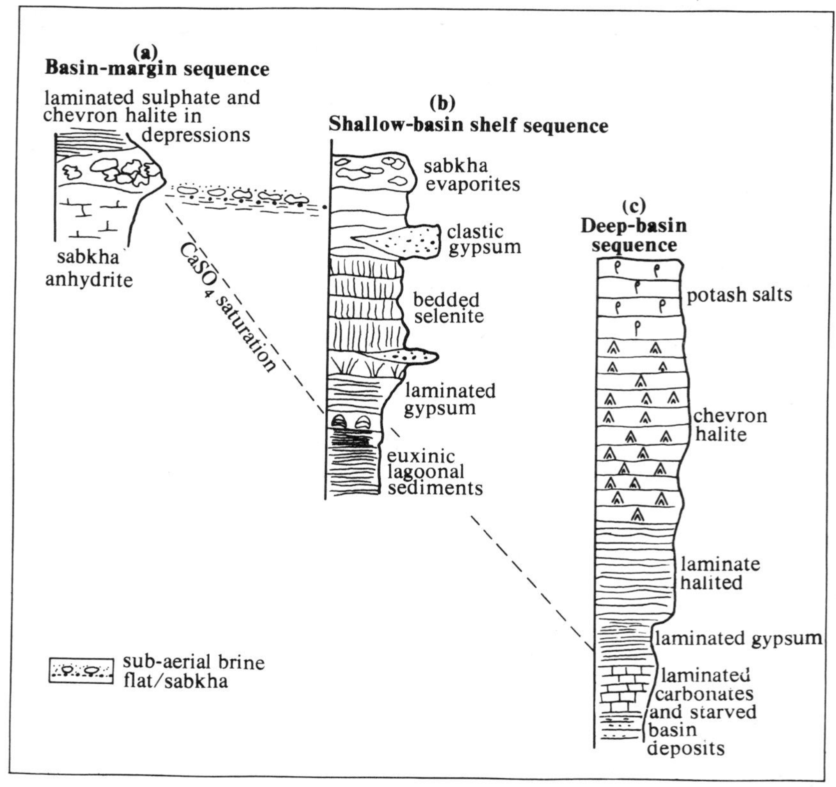

Figure 23.29 Hypothetical evaporite successions (not to scale) in different parts of evaporite basins (after Kendall 1978).

brine/sediment interface from capsized hopper rafts (Shearman 1970, Arthurton 1973). Layers of large, vertically standing, elongate gypsum crystals with internal horizontal inclusion trails and dissolution surfaces are also thought to indicate growth upwards from the sediment surface into brine (Schreiber *et al.* 1976). All such growth fabrics may be subject to recrystallisation during evaporite burial.

The three main (intergradational) models for subaqueous evaporite deposition are summarised in Figures 23.28 & 29). Careful all-round geological and sedimentological reasoning is required to distinguish them in the sedimentary record. Possible evaporite successions that might result from basin margins, shallow basins and deep basins are summarised in Figure 23.29.

23j Summary

Calcium carbonate grains of biological origin are produced most abundantly in the photic zones of warm tropical to subtropical seas. Continued production leads to carbonate platform development with a relatively steep seaward platform margin. Carbonate–evaporite facies belts across a platform closely reflect wave and tidal power as well as the composition of marine waters and sediment pore waters. Sabkhas prograde seawards in arid climates. Pore-water ionic concentrations build up during evaporation, and distinctive evaporite suites result. Lagoonal facies have been deposited behind platform edge reefs or islands of Pleistocene limestones. Strong tidal currents at platform edges encourage oöid shoals to develop. The platform edge-to-ocean slope in the Bahama Banks is marked by sediment gravity flows, bioherm growth and differential cementation.

Further reading

As noted previously, Bathurst (1975) is the starting point for all serious students of carbonate sedimentology. It is particularly good on recent environments and carbonate diagenesis. Ancient limestone facies and facies models are more thoroughly discussed by J. L. Wilson (1975) and are critically reviewed by James (in Walker 1978b). Both texts contain very full bibliographies, to which the reader is referred. A particularly detailed and well illustrated account of a rimmed shelf margin is given by James and Ginsburg (1979).

PART SEVEN OCEANIC ENVIRONMENTS AND FACIES ANALYSIS

Like as the waves make towards the pebbled shore,
So do our minutes hasten to their end;
Each changing place with that which goes before
In sequent toil all forwards do contend.

From Sonnet LX *(Shakespeare)*

Plate 7 Turbiditic sandstones (section about 75 m thick) deposited on the mid-fan to fan apron of submarine fans advancing into deep water intracratonic basin of Namurian age in central England (Mam Tor, Derbyshire, England).

Theme

The oceans are systems of great complexity. They act as a sink for ions liberated during continental weathering and transported seawards in river water. A vast global circulation of oceanic water occurs in response to the planetary wind system and to variations in water temperature from the Equator polewards. The most important non-clastic oceanic sediments originate in areas of high organic productivity where nutrient-rich deep ocean water is brought to the surface by upwelling or by oceanic divergence. It is now well established that the oceanic crust has a magnetic 'memory' because of the production of new ocean crust by sea-floor spreading. The ability to reconstruct the size, shape and depth of the Mesozoic and Cainozoic ocean basins means that oceanic sediments may be correctly placed in their exact physical and chemical palaeo-environments. Another important aspect of oceanic environments is the tectonic setting of their continental margins. The distribution, geometry and character of oceanic clastic facies is largely controlled by tectonic influences. In contrast to pelagic oceanic facies, clastic facies form thick successions around the ocean basin margins where they were deposited by turbidity and thermohaline current systems.

24 Oceanic processes

24a Introduction

Major advances in our understanding of ocean dynamics have occurred since 1945, replacing the notion that the oceans and deep seas beyond the edge of the continental shelf are stagnant, moving only at the surface in response to wind stresses at the ocean/atmosphere boundary. The oceans are now seen as dynamic systems of great complexity in which surface currents coexist with deep currents of some strength. A section through any ocean reveals distinct water masses which exhibit sharp discontinuities in temperature, salinity and velocity vectors. Examination of hundreds of Deep Sea Drilling Project cores, particularly by marine micropalaeontologists and stable-isotope geochemists, has led to an appreciation of the effects on ocean currents and sediments of both climatic change and changing oceanic configurations.

The chemical nature and preservation of pelagic deep-sea sediments is now known to be closely controlled by variations in the depth of the oxygen minimum layer and the carbonate compensation depth, whilst areas of maximum accumulation of planktonic biogenic material correspond closely to areas of high organic productivity in the warm surface layer of the ocean where mixing of distinct water masses occurs. Since the 1950s it has been postulated that the fringe of thick clastic sediments around the continental slopes and inner abyssal plains were deposited by turbidity current flows. More recent recognition of the depositional and erosional importance of deep ocean currents, and of debris flows and sediment slumps, has blurred the issue a little. Sedimentation around the ocean margins is now known to be closely related to the type of plate-tectonic boundary present, in addition to the regional nature of the sediment dispersal system of coastal plain and continental shelf.

24b Physical processes

As briefly noted above the oceans are not just deep stationary water masses. Solar heating drives a complex current system caused by direct wind drag and by density differences produced during heating, cooling and evaporation. Before we consider the broad pattern of oceanic circulation it is necessary to review the nature of the forces acting upon the ocean reservoirs.

External forces are caused by direct *wind shear* on the surface oceanic layers and by the **Coriolis force.** The latter arises only when there is velocity relative to the Earth's surface. The magnitude of the horizontal component of the Coriolis force is given by $2\omega \sin Uv$, where ω is the angular velocity of the Earth, U is latitude, and v is the horizontal velocity of an object. There is no horizontal component of the Coriolis force at the Equator since here $\sin U = 0$. The Coriolis force is directed normal to the object velocity, to the right in the northern hemisphere and to the left in the southern hemisphere. The Coriolis force thus deflects every physical object that moves horizontally over the Earth's surface. Slowly-moving air and water masses are particularly subject to deflection.

Internal forces are caused by horizontal pressure gradients and by frictional boundary effects, the former arising from variations in the height of the oceanic surface relative to some datum level.

The main motivators for surface ocean currents are thus wind stress giving rise to **drift currents,** and horizontal pressure gradients giving rise to **gradient currents**. The Coriolis force and internal friction act as a result of movements set up by wind and pressure. They cannot cause the water masses to move, but they have a strong effect on the resultant motion.

The movement of surface waters by direct wind stress is more complicated than appears at first sight. Comparisons between the velocity paths of icebergs and the prevailing wind in Arctic seas showed an angular difference of 20–40° to the right of the wind for iceberg (and hence surface-water) travel. The physicist Ekman established a quantitative solution to this problem by a consideration of viscous accelerations and Coriolis acceleration. His solution requires that the surface water velocity is directed at 45° to the right of the surface wind stress, the angular difference increasing with depth until a small reverse flow occurs. The magnitude of the drift current falls off exponentially with depth so that below a certain depth the effects of surface winds upon water motion ceases. This depth is latitude dependent as well as velocity dependent. As shown in Figure 24.1 the velocity vector of a wind drift current is in the form of a spiral. The effective depth limit to this **Ekman spiral** is around 100 m in moderate latitudes.

It is important to note that Ekman's drift-current theory assumes no horizontal pressure gradients to be present in the ocean. Should such gradients be present,

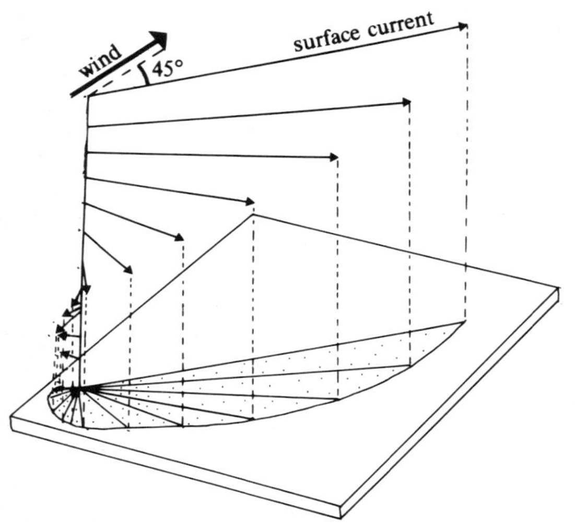

Figure 24.1 The Ekman spiral produced by wind action upon a deep water mass.

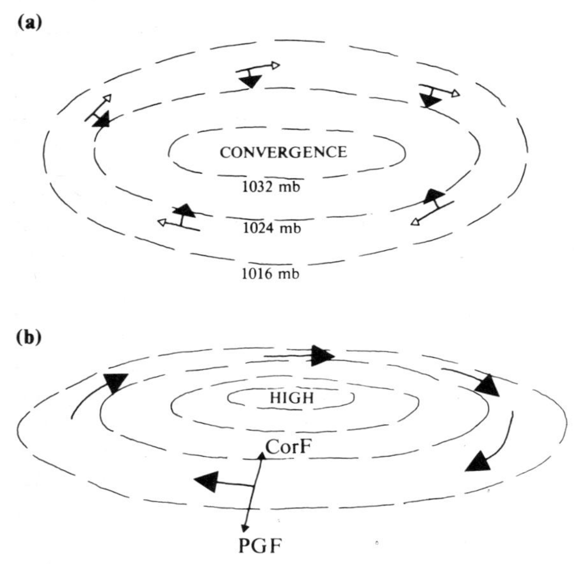

Figure 24.2 Water movements associated with anticyclonic winds in the northern hemisphere. (a) Pressure, winds and Ekman transport to the right; (b) topography of sea surface and associated gradient currents below zone of frictional influence. CorF – Coriolis force; PGF – pressure gradient force. (All after Harvey 1976.)

because of variations in density or to driving up by Ekman transport (drift currents causing gradients), then the further class of gradient currents can occur. As shown in Figure 24.2, the gradient current runs not from areas of high to low pressure but parallel to the isobaric lines, as in air flow. This is due to compensation of the pressure gradient force by the Coriolis force.

24c Chemical and biochemical processes

Our purpose here is to focus on two important 'fences' that exist in ocean water, the junctions between carbonate saturation and undersaturation and between oxygen enrichment and depletion.

As noted previously (Ch. 2) the existence of carbonate compensation depths (CCD) for aragonite and calcite is caused by an increase of $CaCO_3$ dissolution with depth which occurs in response to increasing water pressure and decreasing water temperature. The actual depth of the CCD is dependent upon mass balance of supply and withdrawal, bathymetric change and nature of water mass chemistry. The CCD is the local depth where the rate of supply of $CaCO_3$ to the sea floor is balanced by the rate of dissolution so that there is no net accumulation (Bramlette 1961). In practice the CCD is mapped out at the level at which % $CaCO_3 \rightarrow 0$. The concept of a **lysocline** (Berger 1971) was introduced to denote a well defined junction between well preserved and poorly preserved foraminifera in deep-sea surface sediment and it is regarded as the level of maximum change in composition of calcareous fossils due to differential solution. Regarding the factors that ultimately control the CCD (Berger 1974), it is clear that a balance must exist between upper oceanic productivity in the warm photic layer and deep oceanic dissolution. Cooling, compression and CO_2 uptake produce bottom waters that are sufficiently undersaturated to redissolve the excess supply of biogenic $CaCO_3$ from the warm surface layers. Thus in the long run a dynamic steady state is achieved. This is beautifully illustrated by events at the Eocene/Oligocene boundary as revealed in deep sea cores (Kennett & Shackleton 1976). Oxygen isotope measurements on deep benthic forams revealed a rapid temperature fall caused by the onset of the deep Antarctic bottom current at this time. Onset of deep oceanic circulation (see below) caused increased oceanic turnover which caused increased calcareous biogenic production in the Central Pacific. This was closely followed by a major and apparently rapid deepening of the carbonate compensation depth (Fig. 24.3).

The regions of the ocean that show particularly high productivity, and hence high sedimentation rates, are along the eastern ocean margins where cold nutrient-rich waters upwell (see below) and at the convergence of the major oceanic gyres in equatorial and polar regions.

The second important chemical 'fence' marks the junction between the upper oxygenated levels of the ocean and the lower oxygen-deficient levels (Fig. 24.4). Oxygen in

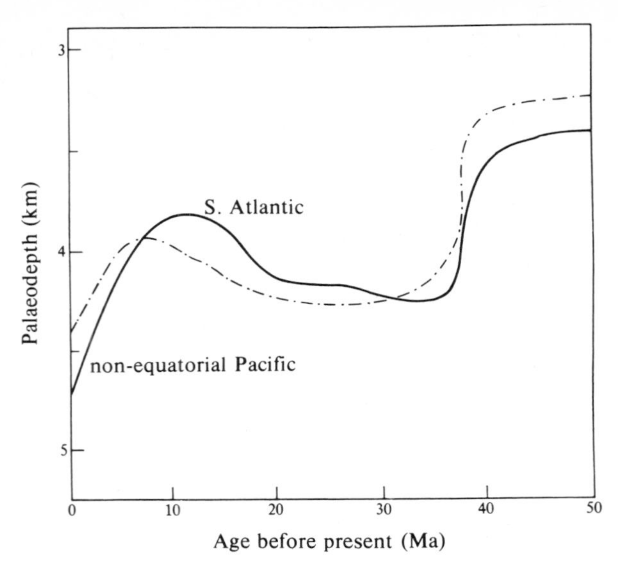

Figure 24.3 Variation of the CCD with time in the South Atlantic and equatorial Pacific oceans (after van Andel *et al.* 1977).

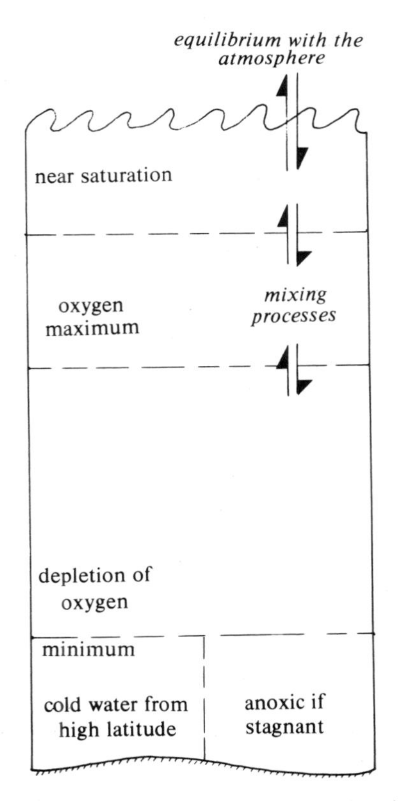

Figure 24.4 The patterns of O_2 concentration in oceanic waters (after Horne 1969).

sea water is derived from the atmosphere and as a byproduct of photosynthesis. Saturation is usually achieved in near-surface levels, but oxygen contents generally fall with increasing depth (50–100 m). Below the photosynthetic zone, processes such as the oxidation of organic material reduce the oxygen content. In deeper water the oxygen content may rise again if cold, oxygen-rich thermohaline currents are present. A major development in the past twenty years has been the location of vast areas of open ocean that show marked oxygen deficiency or absence at depths of between 2 m and 1000 m (reviews in Deuser 1975, Demaison & Moore 1980). It is no coincidence that all such areas currently identified are close to regions of upwelling and therefore of high organic productivity. The absence of anoxic waters in the open Atlantic ocean is attributed to more efficient flushing by ocean currents and to the general paucity of phosphate which limits primary production levels. The large area of anoxic ocean water found in the Arabian Sea is noteworthy because it is accompanied by abundant hydrogen sulphide in the bottom sediments.

Anoxic water masses also exist in basins that have a narrow connection with the open ocean across a shallow sill. Intermittent anoxic conditions occur in many fjords and small bays whilst permanent anoxic conditions occur in the Black Sea. In all cases anoxism is caused by the development of mid-water stratification produced by salinity or temperature gradients, often a result of saline density currents intruding into the basin over the entrance sill. Oceanographic research has identified periodic anoxic events in many oceans; examples will be discussed further in Chapter 26.

Sediment below anoxic water is usually rich in organic material and it provides a broad modern analogue to some **black shales** which are a frequent occurrence in the geological record. Such organic-rich sediments are important potential oil source rocks. The high organic content is not thought to be caused by lower rates of organic decomposition but is a consequence of the enhanced organic productivity which produced the anoxic conditions in the first place. Significant occurrences of light hydrocarbon produced by bacteria occur in sediment below anoxic water.

24d Surface currents and circulation

The surface currents of the ocean are closely related to the planetary wind system, the major features of which are shown diagrammatically in Figure 24.5. Wind movement around the great anticyclonic 'highs' causes Ekman transport of surface waters in towards their centres. This transport causes the ocean surface to slope outwards from the centre of the high, setting up gradient currents that run parallel with the wind (Fig. 24.2). Thus the two great subtropical gyres of water transport are formed which dominate circulation in both Atlantic and Pacific Oceans (Fig. 24.6). Along the intertropical zone of convergent light winds, a return eastward-flowing equatorial counter current occurs. Complex interactions in this zone cause powerful divergences and marked ocean water mixing. Another very important current boundary occurs around Antarctica between 50° and 60°S which is known as the Antarctic convergence where cold dense water sinks below the surface to form Antarctic intermediate water.

One of the most striking features of the general oceanic wind-driven circulation described above is the intense crowding of streamlines near the western borders of the oceans. These are manifest as the great western boundary currents such as the Gulf Stream, Kuro Shio and Agulhas currents. This asymmetry of the subtropical gyres is explicable in terms of vortex theory (Stommel 1948; see Harvey 1976 for an introduction). The effect adds to the existing vorticity (equal to the Coriolis force) of the symmetrical wind-driven motions on the western sides of oceans and opposes it on the eastern sides. To establish a steady state the total vorticity cannot rise without some

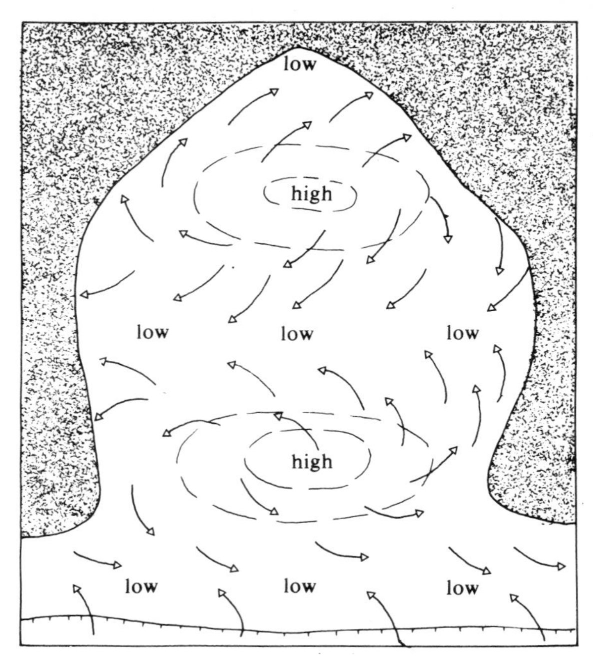

Figure 24.5 Idealised pressure distribution and winds over an idealised ocean (after Harvey 1976).

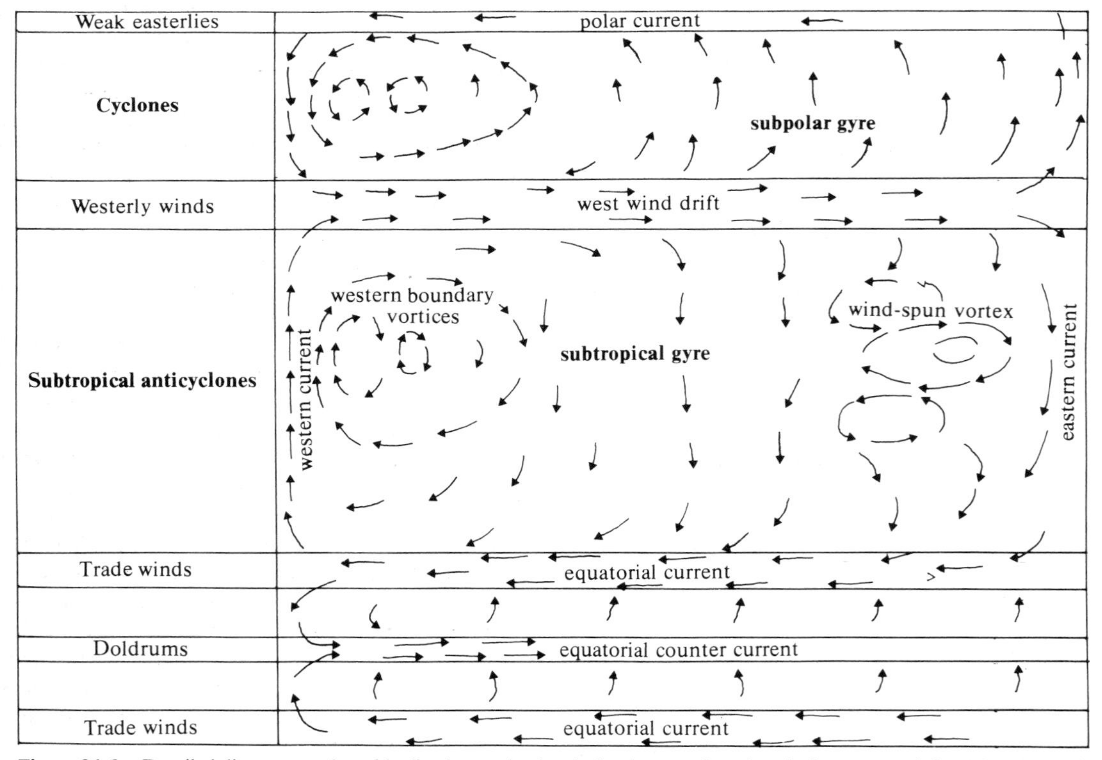

Figure 24.6 Detailed diagram to show idealised oceanic circulation in a northern hemisphere ocean (after Munk 1950a).

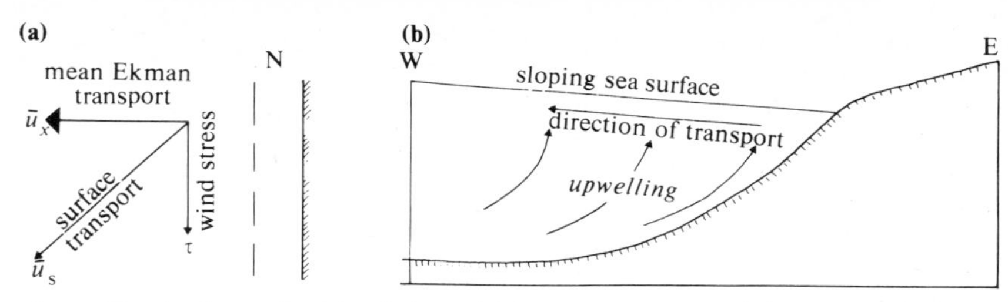

Figure 24.7 Coastal upwelling in the northern hemisphere. (a) Ekman transport and surface transport caused by coast parallel winds; (b) section to show the resultant sloping sea surface and upwelling deep waters that replace the offshore-moving surface waters.

limiting effect. The breaking action is provided by frictional effects ($\propto$ velocity^2) which cause the western current to be extremely strong, up to ten times that of the eastern currents. In recent years it has become clear that major western currents such as the Gulf Stream affect the ocean floor; thus major erosive events on the Blake Plateau have been directly attributed to Gulf Stream flow during glacial epochs when the current was at its strongest (Kaneps 1979).

Important effects arise when the dominant winds blow parallel to the coastline. Consider a north wind blowing parallel to the west side of a continent, as shown in Figure 24.7. The wind generates a surface Ekman current to the right so that the surface waters are driven away from the coast, causing a depression of the ocean surface towards the land. Compensation for this surface flow comes in the form of deeper, cold water from below. This cold upwelling water may come from depths of up to 300 m and it brings with it nutrients which cause a greatly increased plankton biomass to be supported in the surface waters. Distinctive sedimentological 'signatures' of upwelling include well preserved diatom and radiolarian assemblages, increased amount of fish debris and the occurrence of phosphorite grains (see Ch. 4). The rate of upwelling is only 1–2 m per day so that current-produced structures will be absent. Upwelling is particularly important off the coasts of Peru, north and southwest Africa, and California.

24e Structure, deep currents and circulation

The surface oceanic circulation described above is merely the most obvious visible sign of a deeper, more thoroughgoing circulation system that affects the whole ocean mass (Fig. 24.8). Detailed temperature, density and isotopic studies have revealed a complex network of deep, cold, relatively saline waters that dominate ocean floor processes. Far from being inactive below the level of surface wind effects, the world's oceans are in constant flux at *all* levels. The sedimentological consequences of this deep circulation are profound, since steady current velocities of up to 0.5 m s^{-1} have been recorded in some areas where the normally slow ($\sim$0.05 m s^{-1}) thermohaline currents are accelerated on the western sides of oceans and in topographic 'constrictions'.

The basic reasons for the existence of active deep oceanic currents are to be found in the extreme temperature variations that exist today from the Equator polewards. The vast continental ice cap of Antarctica and, to a lesser extent, the Greenland and Arctic ice masses, provide cold saline water masses that sink to become the Antarctic bottom water (ABW) and North Atlantic deep water (NADW) respectively. Because of topographic damming effects in the North Atlantic, the ABW dominates the world's oceans as the deep cold current source. The ABW moves northwards along the western boundaries of the Atlantic and also moves around Antarctica to the east to provide western boundary undercurrents around the Indian and Pacific oceans. On the eastern sides of the oceans, the cold water wells up into the warm surface waters where it is driven by wind currents back to the poles so that the whole system is in dynamic equilibrium.

The NADW is important on the western margin of the North Atlantic ocean where the current system and its sedimentary products have been particularly well studied in recent years. The current is termed the Western Boundary Undercurrent in these areas and it flows parallel to and over the continental rise at velocities of up to 0.5 m s^{-1} (Fig. 24.9). Seismic studies and coring in the area dominated by the deep current reveals a thick (Km) sediment 'drift' comprising alternations of thin, very fine sands, silts and muds. The sands and silts are thinly bedded, ungraded, well sorted, and may contain heavy mineral placers in small-scale cross laminations (Heezen & Hollister 1963, Hollister & Heezen 1972, Bouma & Hollister 1973). These deposits have been termed **contourites**, that is deposits laid down by (thermohaline) contour-following undercurrents. Their good sorting and lack of grading distinguishes them from thin distal turbidites, but recent discussions stress the difficulties of telling the two deposits apart in the ancient geological record, since muddy contourite 'drifts' have also been

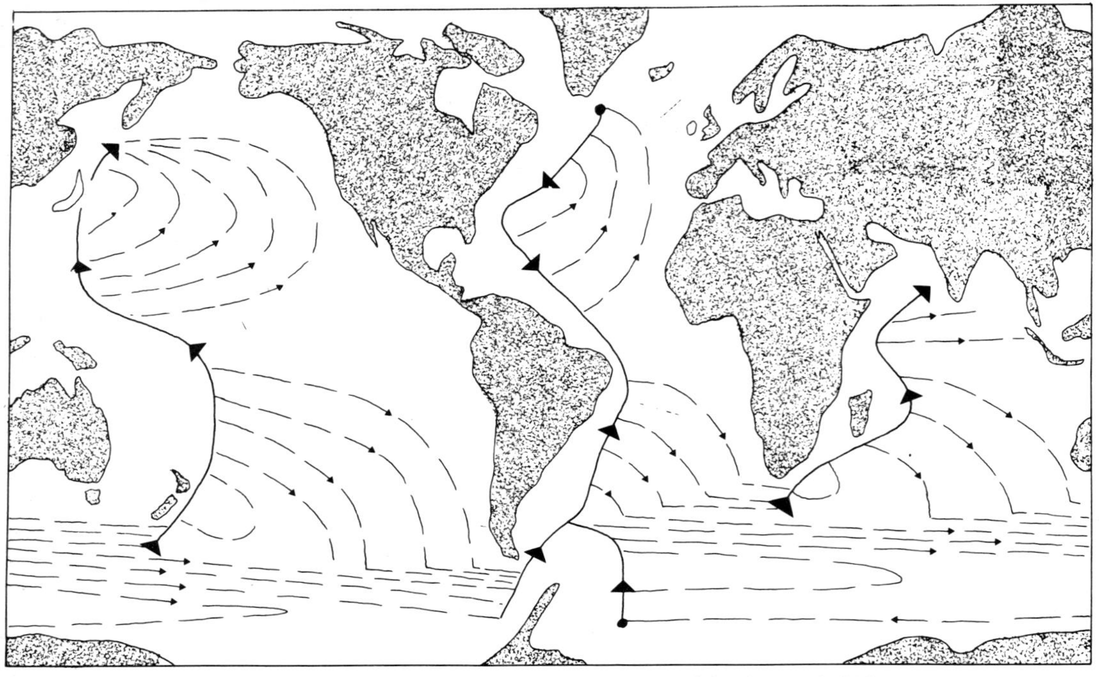

Figure 24.8 The deep circulation of the world ocean (after Stommel 1957).

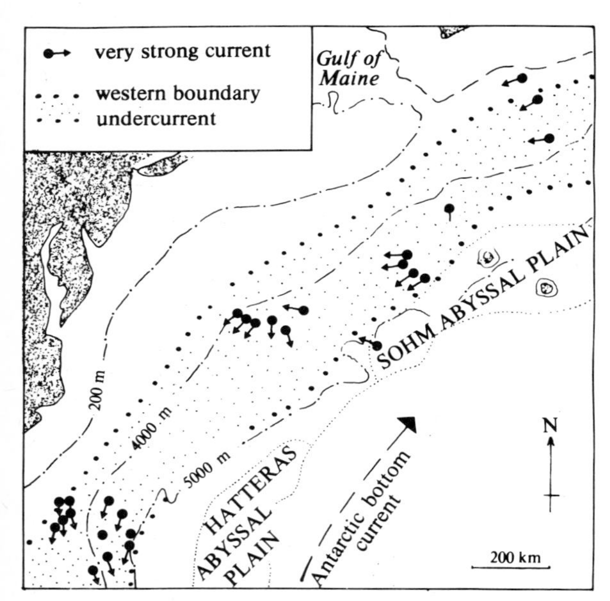

Figure 24.9 The Western Boundary Current as defined along the continental rise of North America (after Heezen & Hollister 1971).

identified (discussion in Stow & Lovell 1979 and McCave *et al.* 1981). The erosional effects of cold undercurrents are extremely important, many stratigraphic gaps in deep sea sediment cores from Oligocene times onwards being attributed to the onset of contour-current erosion. Recent recognition of erosive effects of the Gulf Stream (see above) may considerably complicate the interpretation of deep-sea erosional and depositional features that were formerly attributed to the effects of thermohaline currents alone.

A final feature of deep ocean waters is attributed in part to the erosive action of thermohaline currents. This is the phenomenon of increased suspended material in near-bottom water, revealed by light-scattering techniques. These are termed **nepheloid layers** (Fig. 24.10). The most intense near-bottom nepheloid layers may be caused by the frictional effects of thermohaline currents on the ocean floor muds, although turbiditic muds from dilute distal turbidity-current flows probably have a significant role. Some nepheloid layers may be up to 2 km thick, although 1–300 m is a more usual figure (Eittreim *et al.* 1975). Nepheloid layers may keep grains up to 12 μm suspended with concentrations of 0.3–0.01 mg l^{-1} (McCave & Swift 1976, Pierce 1976, Swift 1976).

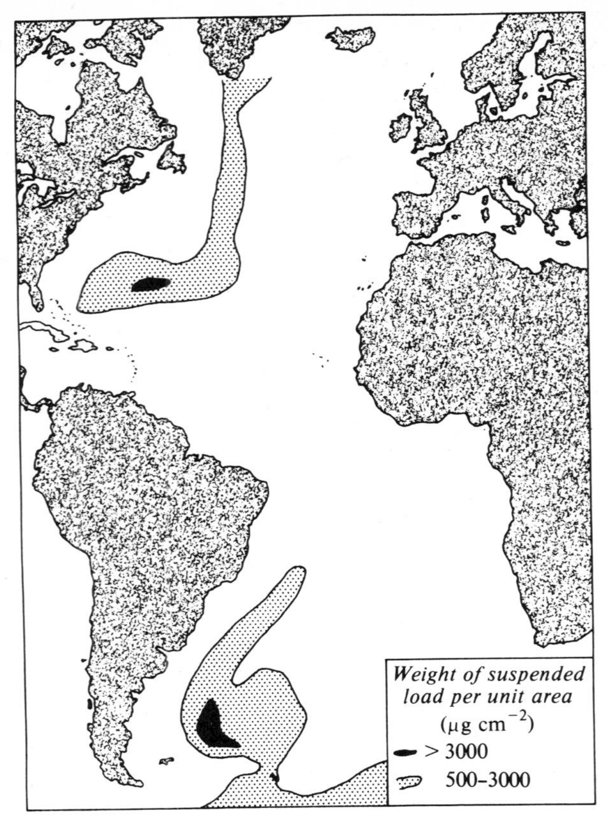

Figure 24.10 The high suspended load content of bottom waters defines the nepheloid layer (due to thermohaline currents) of the deep Atlantic ocean (after Biscaye & Eittreim 1977).

24f Slumps, debris flows and turbidity currents

We now turn our attention briefly to the major mechanisms whereby appreciable volumes of terrigenous clastic materials may reach the continental rise and inner abyssal plains. As noted previously (Ch. 7) there may be a continuum between slumps, debris flows and turbidity currents, the series passing from proximal to distal in space. A fundamental point concerns the degree of instability of outer shelf and continental slope sediment, especially sediment with an admixture of mud-grade material (see Fig. 25.3). Studies of the continental slope off the northeastern USA (Keller *et al.* 1979) revealed deposits dominated by silty clays in an area with gradients of between 3° and 10° heavily dissected by submarine canyons. The sediments are thought to be gravitationally stable but the generally high water content encourages periodic mass movement by shocks and abnormal pressures produced by infrequent earthquakes and internal waves. Lower values of shear strength in the vicinity of submarine canyons are related to a combination of increased concentrations of organic matter and fine-grained sediment. Some idea of the effects of passing surface waves on bottom sediment may be gained from the calculations of Watkins and Kraft (1978) that large storm waves, particularly during hurricanes, may induce pressure anomalies with wavelengths of 300 m and amplitudes of 70 kN m^{-2} in water depths of 60 m. Although the amplitude of these wave-induced pressure anomalies should decrease in deeper water, the effect is considered important on shelf-edge sediment liable to failure.

Detailed exploration of the ocean margins reveals that debris flows are probably a much more important depositional process on the sea floor than was previously suspected (see Embley 1976, Flood *et al.* 1979). For example Embley (1976) identified debris flow deposits of enormous extent that were generated by large sediment slides off the Spanish Sahara on the northwest African continental margin (Fig. 24.11). The debris flow travelled on a slope as low as 0.1° for a distance of several hundred kilometres. The deposits cover an area of about 30 000 km^2 and originated from a massive slump of volume 600 km^3 on the upper continental rise where a prominent slide scar now exists. Recognition of the debris flow deposits is based on a characteristic geometry, a distinctive acoustic character, a pebbly mudstone fabric and sharp angular contacts in cores, and an undulating surface morphology revealed by bottom photographs.

Three decades of ocean floor exploration and coring have confirmed Keunen's initial prognosis that the deposits of turbidity currents dominate terriginous clastic

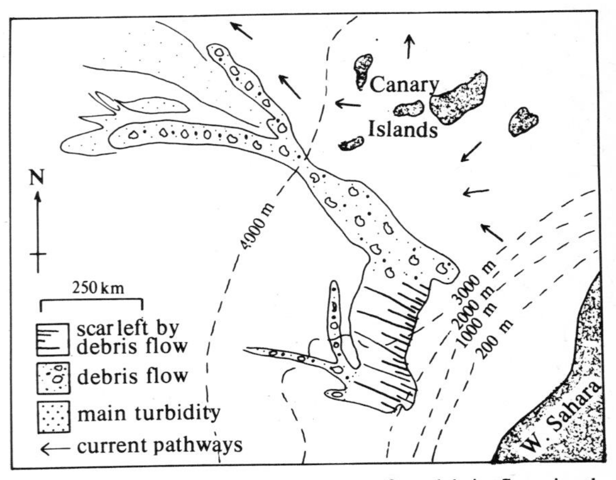

Figure 24.11 The enormous extent of a debris flow in the Atlantic Ocean off northwest Africa (after Embley 1976).

successions along many continental margins. The slump and turbidity current generated by the 1929 Grand Banks earthquake still stands as the classic example (Figs 24.12 & 13) although many others originated from both shocks and river influx have since been documented (Heezen & Hollister 1971). Direct measurement of down-canyon surges have been interpreted as low-speed turbidity currents whose velocities may reach 1 m s^{-1} (Shepard 1979). These may occur at intervals of as little as a few days wherever rivers are discharging considerable loads of sediment into the ocean in close proximity to submarine canyon heads. Major problems remain concerning the distinction between fine-grained distal turbidites and contourites, particularly those of mud and silt grade (see Piper 1978). Stow and Lovell (1979) present some possible distinguishing features, and the interested reader is referred to this review.

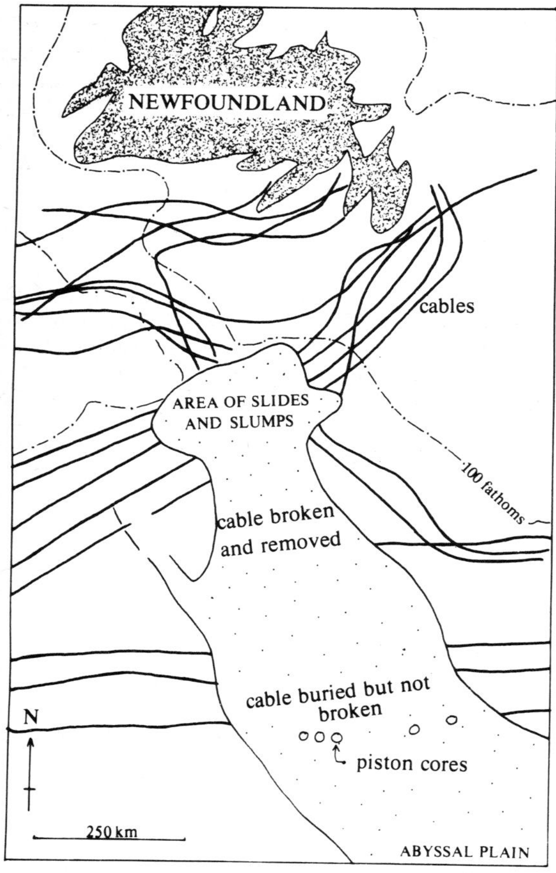

Figure 24.12 The extent of the Grand Banks turbidity current of 1929 (after Heezen & Hollister 1971).

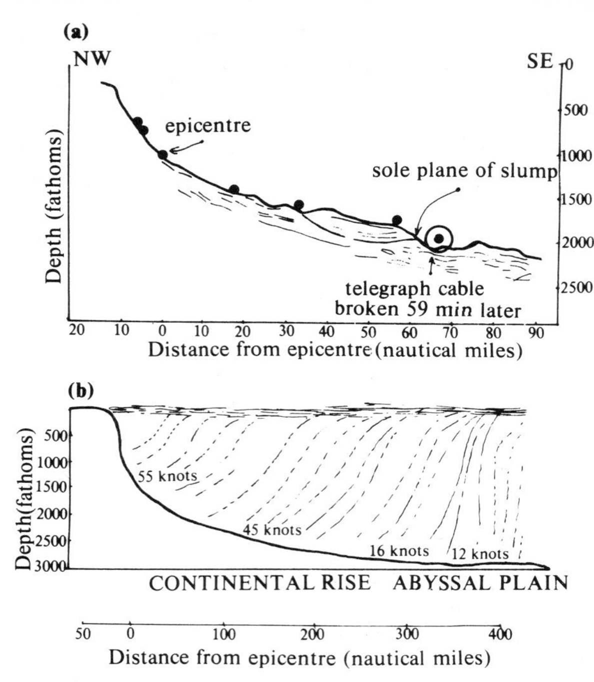

Figure 24.13 (a) Seismic reflection profile to show the slump induced by the Grand Banks earthquake and the positions of the cables (circles) broken by the slump and the turbidity flows. (b) Profile to show the estimated speed of turbidity flow as determined from the intervals between cable breaks. (Both after Duxbury 1971.)

24g Palaeo-oceanography

Oceanic sediments and their contained fauna provide a superb memory bank which can be used to trace the physical and chemical evolution of oceanic water masses. We shall examine particular facies models for opening oceans in Chapter 26 and will now briefly outline some of the principles used in palaeo-oceanographic reconstructions.

A prerequisite for reconstruction is some knowledge of the evolution of oceanic shape, size and depth. Linear ocean-crust magnetic anomalies enable shape and size to be reconstructed (e.g. A. G. Smith & Briden 1977) whilst models for ocean crust cooling and subsidence enable palaeodepths to be estimated (Sclater *et al.* 1971, 1977). Estimates of the carbonate compensation depth (CCD) in ancient oceans depend upon plots of $CaCO_3$ accumulation rates against palaeodepth. The position of the CCD can then be computed from a regression equation, being the intercept of depth as the accumulation rate tends to zero (see van Andel *et al.* 1977). Curves relating CCD to time show strong similarities between all oceans, implying control by circulation of deep oceanic waters (Fig. 24.3). Evidence for ocean anoxic events comes from widespread black shale horizons whilst upwelling is recorded by cherts, phosphorite and fish debris. Very extensive cherts record high equatorial productivity of radiolarian species. Contour-current activity is indicated by thick continental-rise 'contourites' whilst submarine fans of turbiditic origin witness the growth of terrestrial drainage systems on continental areas adjacent to the opening ocean.

Oxygen isotope compositions of planktonic organisms (chiefly foraminifera) provide the most valuable evidence for palaeo-oceanic temperatures once corrections have been made to account for ocean water stored as ice (see review by Hudson 1978b). The onset of deep cold currents of Antarctic origin in the Cainozoic (Kennett & Shackleton 1976, Kennett *et al.* 1974) is documented in the isotopic composition of deep-dwelling benthonic forams which give evidence for a 5°C decline in bottom-water temperature over about 100 000 years at the Eocene/Oligocene boundary. Kennett (1977) has pointed out that the present world oceanic circulation depends upon the particular topographic evolution of the ocean basins, the position of the continents and the world climate. The current circulation developed because of thermohaline deep currents following glaciation *and* the development of a circum-Antarctic seaway after seafloor spreading. The former process provided the necessary cold saline water whilst the latter allowed deep dispersal of this water into the world's oceans.

Major effects arise through the reconnection with the world ocean of temporarily isolated ocean basins (Thierstein & Berger 1978) or through reconnection of adjacent semi-isolated basins (Thunell *et al.* 1977). Isolated basins may supply either hypersaline or fresh-brackish waters into the main oceans, causing major facies changes and extinctions amongst planktonic and benthonic foraminifera.

24h Summary

The oceans are dynamic systems in which surface currents coexist with deep currents of some strength. Surface currents are caused by direct wind stress and by horizontal pressure gradients, the resultant motion being strongly affected by the Coriolis force and internal friction. The complex network of deep oceanic currents arises from the sinking of cold dense waters at the poles. Their Coriolis-induced tendency to flow along western oceanic boundaries at velocities up to 0.5 m s^{-1} causes much sediment transport, erosion and deposition along the continental rises, particularly in the Atlantic ocean. Important chemical 'fences' in the oceans are the junctions between $CaCO_3$ saturation and undersaturation and between O_2 enrichment and depletion. Clastic sediment is supplied to the oceans by slumps, debris flows and turbidity flows issuing from the continental margin, by turbid plumes advancing from the shelf edge, and by wind-blown dusts. Although ocean dynamics are controlled by the surface and deep currents noted above, these variables depend ultimately upon the age, size, morphology, 'interconnectedness' and position of opening and closing oceans. Analysis of deep sea cores has shed much light upon oceanic water evolution in response to seafloor spreading and climatic change in the past 100 Ma or so.

Further reading

The physical aspects of ocean currents are introduced by von Arx (1962), Gröen (1967) and Harvey (1976). More advanced treatments are given by McLellan (1965), Pond and Pickard (1978) and, especially recommended, Neumann and Pierson (1966). The techniques and approaches of palaeo-oceanography are given in a stimulating, though disjointed, fashion by Schopf (1980). Oceanic chemistry and ocean modelling are well outlined by Broecker (1974).

25 Clastic oceanic environments

25a Introduction

To a very large degree the distribution, geometry and nature of clastic facies in the ocean are controlled by the type of plate tectonic setting.

Passive- or '**Atlantic**'-type continental margins show a relatively simple morphology of continental slope, continental rise and broad abyssal plain (see Fig. 25.1). Sediment input occurs from the shelf as suspended plumes; from the very numerous submarine canyons and valleys as slumps and turbidity currents; from massive slope failure as slumps, debris flows and turbidity currents; from large-scale abyssal cones which front major drainage systems; and from thermohaline current systems on the western sides of oceans. Mature opening oceans such as the Atlantic show broad continental shelves whose strong wave and tidal-current systems rework detritus derived from continental 'old lands'. Material reaching the continental rise and extensive abyssal plains (Fig. 25.2) thus tends to be relatively mature (see Ch. 1).

Active continental margins are dominated by subduction processes and show some or all of the following features: back-arc basin, volcanic arc (island chain or continental margin cordillera), fore-arc basin, trench and narrow abyssal plain (see Fig. 25.11). Sediment input is dominated by material of volcaniclastic and/or cordilleran origin in the fore-arc basin and trench, and is

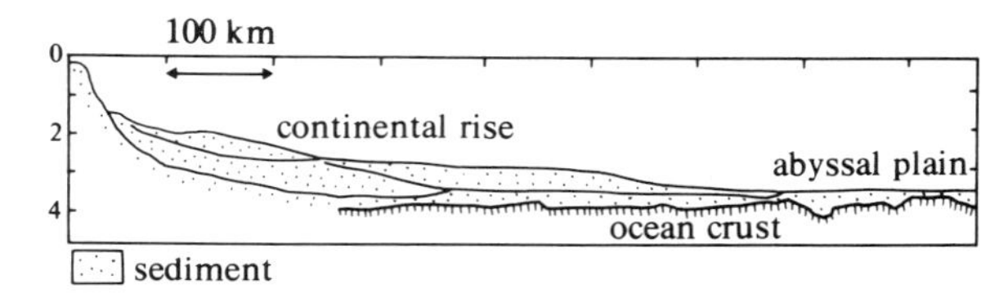

Figure 25.1 Diagrammatic section of the Atlantic continental rise off New York. Note the oceanward migrating progression of continental rise sediments stacked against the base of the continental slope. The sediment is of turbiditic and pelagic origin, modified by the deep thermohaline Western Boundary Undercurrent (see Fig. 25.4). (After Heezen & Hollister 1972.)

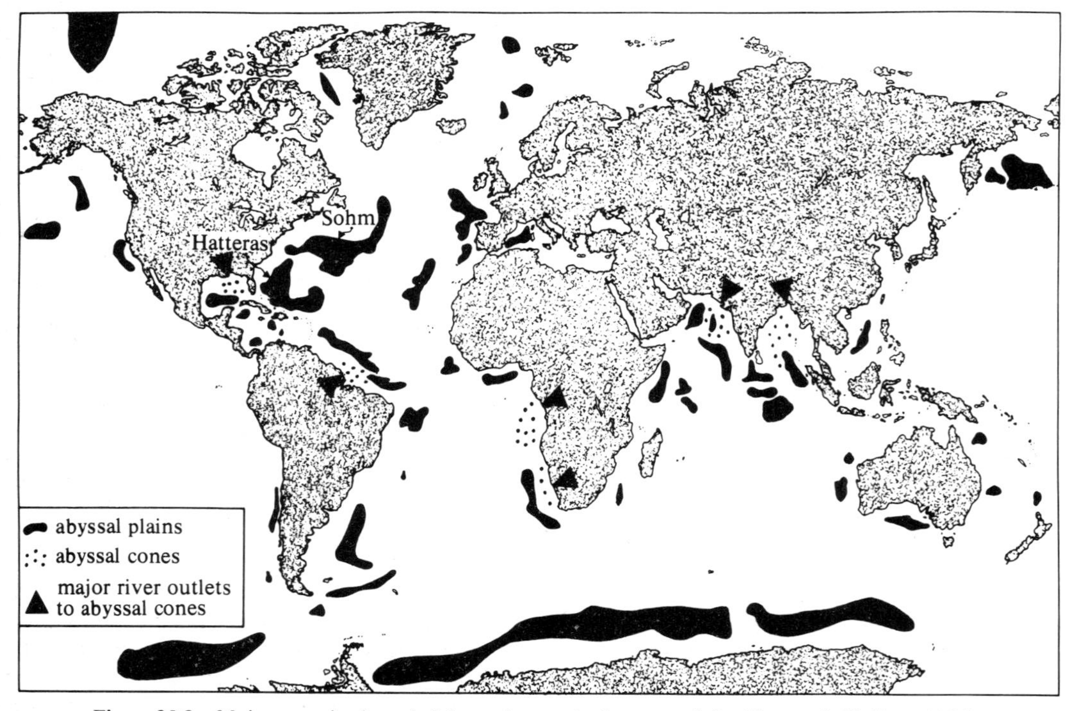

Figure 25.2 Major oceanic abyssal plains and cones in the oceans (after Heezen & Hollister 1971).

usually immature (see Ch. 1). Some trenches show very thin sediment accumulations, particularly those that lie adjacent to desert coastlines where sediment input from the continental area is low (e.g. Peruvian–Chilean Andes). Back-arc basins resemble passive margins on their continental sides but receive much volcaniclastic detritus on their island-arc sides.

Active strike-slip continental margins are typified by the Californian example where splays of the San Andreas and related faults have created a block-and-basin topography in the continental crust. Nearshore basins receive their sediment input directly from the nearshore zone where longshore currents and rip currents transport material directly into canyon heads. Small submarine fans are filling in these basins while more offshore basins are relatively sediment starved.

In the discussion that follows we shall examine modern and ancient sedimentation in a variety of these tectonic settings.

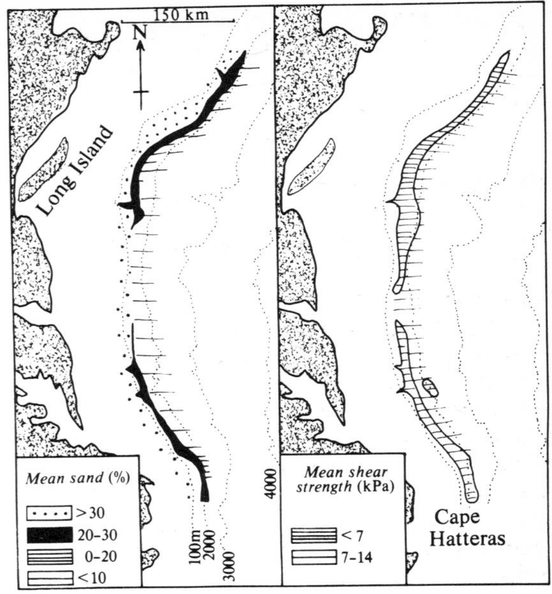

Figure 25.3 Sand percentage and shear stress distributions for the eastern Atlantic continental margin slope (after Keller *et al.* 1979).

25b Continental slopes and rises of passive margins

These are typified by the much-studied continental margin off the eastern USA. The continental slope in this region has gradients of 3–10° and is heavily dissected by submarine canyons and valleys of Pleistocene age. The slope proper is clearly defined by the 120–140 m and 2000 m isobaths. The overall sedimentary framework of the slope comprises a thick prograded Tertiary succession which has been deposited conformably with the present slope (Fig. 25.1). Away from the active canyons off New England (Veatch, Hudson, Hydrographer), where coarser sediments occur, there is a predominant downslope fining trend from the wave-affected sands of the outer shelf to the silty clays of the lower slope (Fig. 25.3). Large areas of the slope are affected by slumping (Keller *et al.* 1979), emphasising the role of the slope as both a receiver of fine sediment from the shelf and a provider of sediment through mass flow to the continental rise and abyssal plain. Strong bottom currents (up to 0.7 ms^{-1}) occur along part of the slope, but the origin and effects of these currents are not well established.

The broad continental rise of the eastern seaboard of the USA is dominated at the present day by the effects of the thermohaline Western Boundary Undercurrent that sweeps around the rise from Newfoundland to the Blake–Bahama Outer Ridge and thence to the Antilles off northeastern South America (Fig. 25.4). Current measurements show velocities between 0.05 and 0.4 $m\ s^{-1}$, the higher velocities being quite capable of transporting mud, silt and very fine sands. Confirmation of the

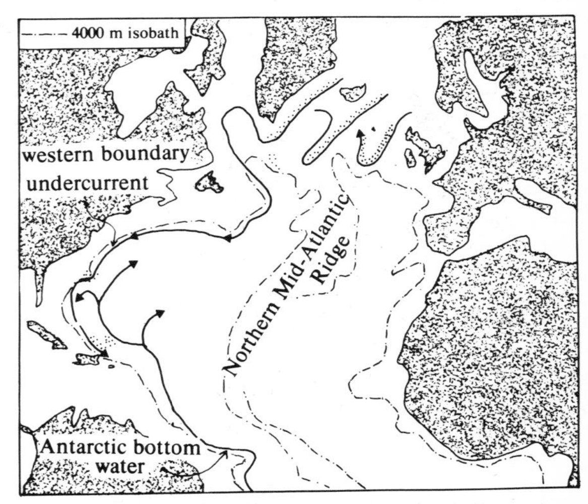

Figure 25.4 Thermohaline bottom currents in relation to deep-sea sediment ridges (stippled) in the North Atlantic. Circles indicate recovery of thick 'contourite' deposits. Contour drawn for 4000 m depth. (After Stow & Lovell 1979.)

erosive and transport capacities of the current come from underwater photographs of erosive grooves and lineations. (Hollister *et al.* 1974). Contour currents are thought to have deposited much of the sediment mantle (up to 2 km thick) that underlies the continental rise, particularly those long sediment ridges found on the Blake–Bahama

Ridge and Antilles Ridge (Fig. 25.4). Further north the rises off eastern Greenland and around the Rockall and Hatton Troughs are also thought to be of contour-current origin (McCave *et al.* 1981). Giant mud waves have been recorded on these ridges (e.g. Hollister *et al.* 1974, Roberts & Kidd 1979), with heights of 10–40 m and wavelengths 1–5 km, although their origins are not well known.

Mineralogical studies give clear evidence that these continental-rise sediment accumulations have a northerly provenance. Particularly informative work on the Greater Antilles Outer Ridge (Tucholke 1975) reveals sediment enriched in chlorite and illite which was derived from the northeastern continental margin of the USA and transported south by the Western Boundary Undercurrent. The sediment itself is fine grained ($<2\mu m$), homogeneous and brown coloured, with cyclic variations in carbonate content. Bioturbation and micromanganese nodules are common. The sediment is too fine for current-produced structures to have developed and it is thought to have been deposited by suspension fallout in the slow-moving (<0.1 m s^{-3}) bottom currents. Deposition rates for the sediment are in the range 0.06–0.3 mm a^{-1}. Thick sediment ridges composed of muddy contourites in other areas may have formed by the interaction of a thermohaline bottom current with relatively static adjacent water masses or even with an adjacent current moving in the opposite direction (Bryan 1970, Markle *et al.* 1970). In all cases, contour-current deposition has been active only since the mid-Tertiary when Arctic cold waters began to penetrate into the western North Atlantic.

Continental rises with no vigorous bottom circulation are typified by the eastern Atlantic Ocean off Africa where, away from the abyssal cone of the Niger, sedimentation may be dominated by suspension fallout onto the slope and by slumping onto the rise. As noted previously, major slide scars and debris flows have recently been located in the area (Embley 1976).

25c Submarine fans and cones

Continental margins of all structural types may show the delta-shaped accumulations of submarine fans and cones. These accumulations issue from a point source such as a submarine canyon or valley. They may be fed directly by river input, as is the case in many great abyssal cones such as the Ganges–Brahmaputra, Niger and Mississippi (Fig. 25.2), or by shelf or coastal input as off the eastern USA and California respectively. The Laurentian cone (see Fig. 25.10) is an example of a cone fed by Pleistocene meltwaters (from the St Lawrence river). Feeder canyons cut the continental margins of many continents and have diverse origins including direct river- or estuary-channel erosion during low glacial sea levels, turbidity current erosion, slumping, and erosion of faulted gullies. They serve as arteries, transporting coarse sediment oceanwards by slumping, debris flow and turbidity flow. Thus coarse sediment effectively *bypasses* the continental slopes whose deposits are dominantly fine grained.

The submarine fan (Fig. 25.5) is directly analogous to sub-aerial alluvial fans (Ch. 14). The dispersal of coarse-grained sediment across the fan is controlled by a migrating distributary channel system separated by levees and interchannel areas (Normark & Piper 1972, Normark 1970, 1974, 1978). The *upper fan* contains the main feeder channel bordered by major levees. The *middle fan* is a prominent depositional bulge appearing as a convex-upwards segment on radial profiles (Normark 1970). The main channel splits into numerous distributaries which may meander or braid between interchannel areas bounded by levees. As with sub-aerial fans, the fan shows active and inactive areas with depositional lobes issuing from the mid-lower fan and inactive lobes covered by a bioturbated veneer of hemipelagic muds.

The overall depositional pattern of a fan involves a down-fan fining from thick coarse sand or gravel-grade turbidites and debris flows in the upper fan channels to thin, very fine sand or silt-grade turbidites on the lower fan apron. Levees comprise thin turbidites whilst interchannel flats comprise bioturbated muds and thin silts. Sections normal to the fan axis on the mid-fan show channelised turbidites separated by fine interchannel areas. Fan progradation is postulated to cause a gross

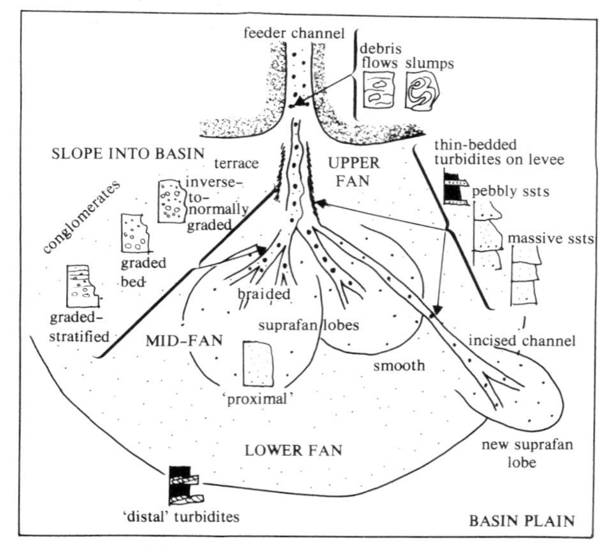

Figure 25.5 Schematic environmental model of a submarine fan (after Walker & Mutti 1973).

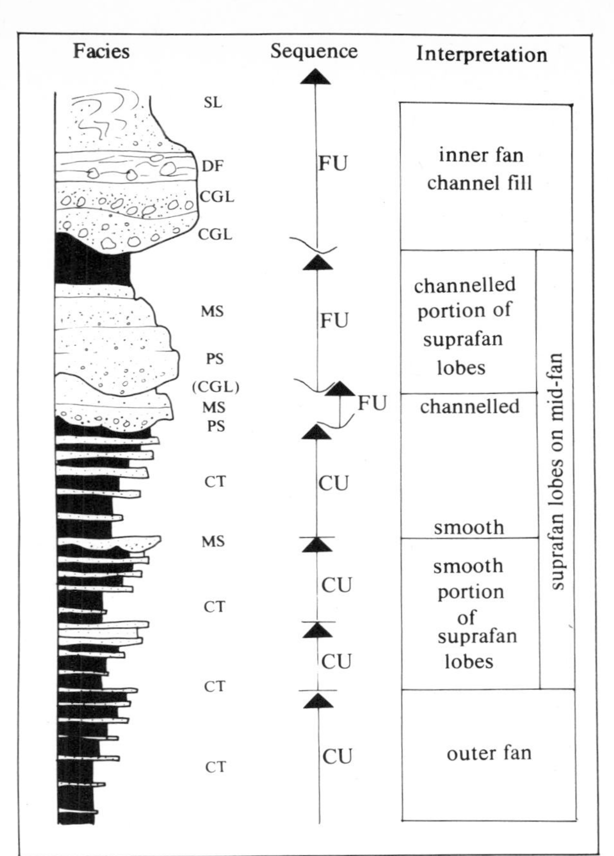

coarsening-upwards succession (R. G. Walker & Mutti 1973) as progressively more proximal fan facies overlie the fan apron (Fig. 25.6).

A particularly fine ancient fan sequence has been documented by Rupke (1977) from the Upper Carboniferous of the Cantabrian Mountains, northern Spain. The fan comprises at least eight mappable depositional lobes which fine basinwards and show downslope changes in depositional processes (Fig. 25.7).

Submarine fans abut against the continental slope and may extend across the continental rise into the abyssal plain. Many fans have become inactive since the Holocene sea-level rise as their canyon heads became far removed from direct riverine input of sediment. The

Figure 25.6 Hypothetical submarine fan stratigraphic sequence produced by fan progradation. CT – 'classical' turbidite with Bouma divisions; MS – massive sandstone; PS – pebbly sandstone; DF – debris flow, CU – coarsening-upwards; FU – fining-upwards. (After Walker and Mutti 1973.)

Figure 25.7 The Upper Carboniferous Pesaguero submarine fan from northern Spain. The sequence is vertical and youngs towards the south-west. Note down-strike fining, succession of lobes (numbers 1–8) and radial currents. Inset shows a 3D model of the fan facies: a facies triplet (mudstone blanket, sandstone lobe, conglomerate tongue) represents a complete cycle of progradation of a major fan lobe. A new lobe then forms by lateral avulsion. (After Rupke 1977.)

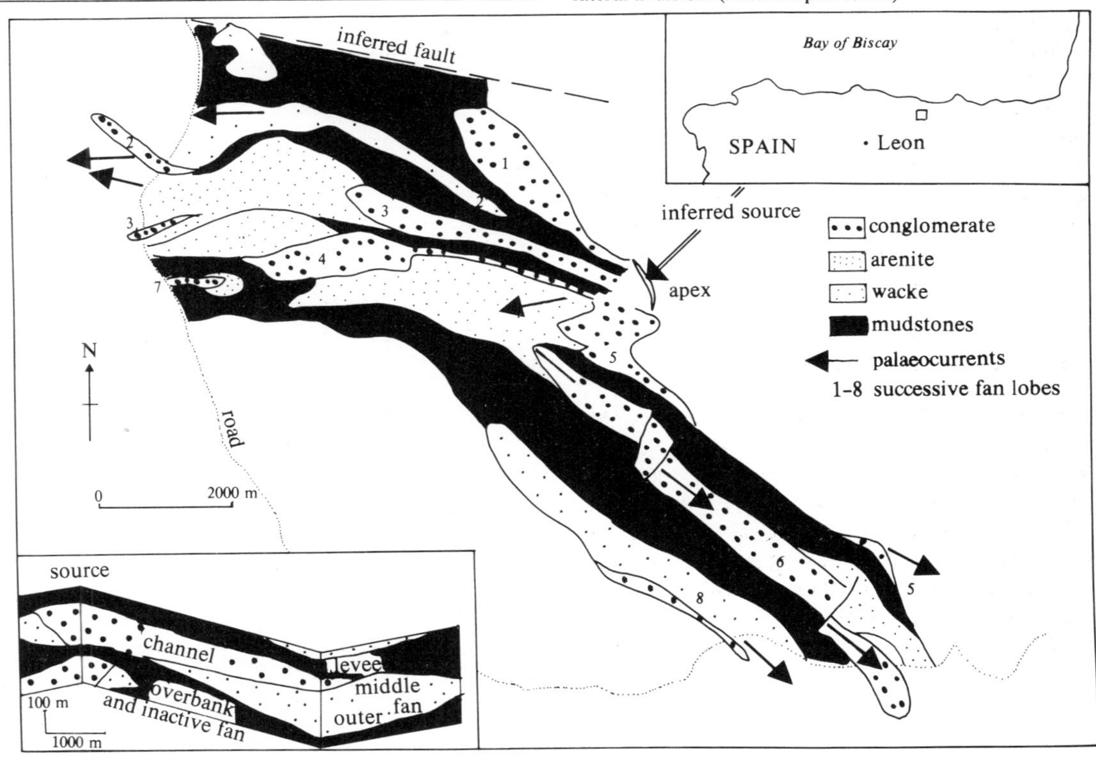

Table 25.1 Some statistics for the Amazon, Bengal and Mississippi submarine fans (after Moore *et al.* 1978).

Fan	*Submarine canyon or trough* Head depth (m)	Mouth depth (m)	*Fan length (km)*	*Fan width (km)*	*Depth fan base (m)*	*Surface area* (10^3 km^2)	Volume (10^3 km^2)	*Sediment thickness (km)* Apex	Margin	*Approx. age* (10^6 a)
Amazon	50	1500	520	600	4800	215	710	14	1	8.15
Bengal	45	1600	3000	1000	5000	3000	10 000	12.5	1	50
Mississippi	75	1200	350	600	3300 and 3500	170	85	?	12	3.5–6

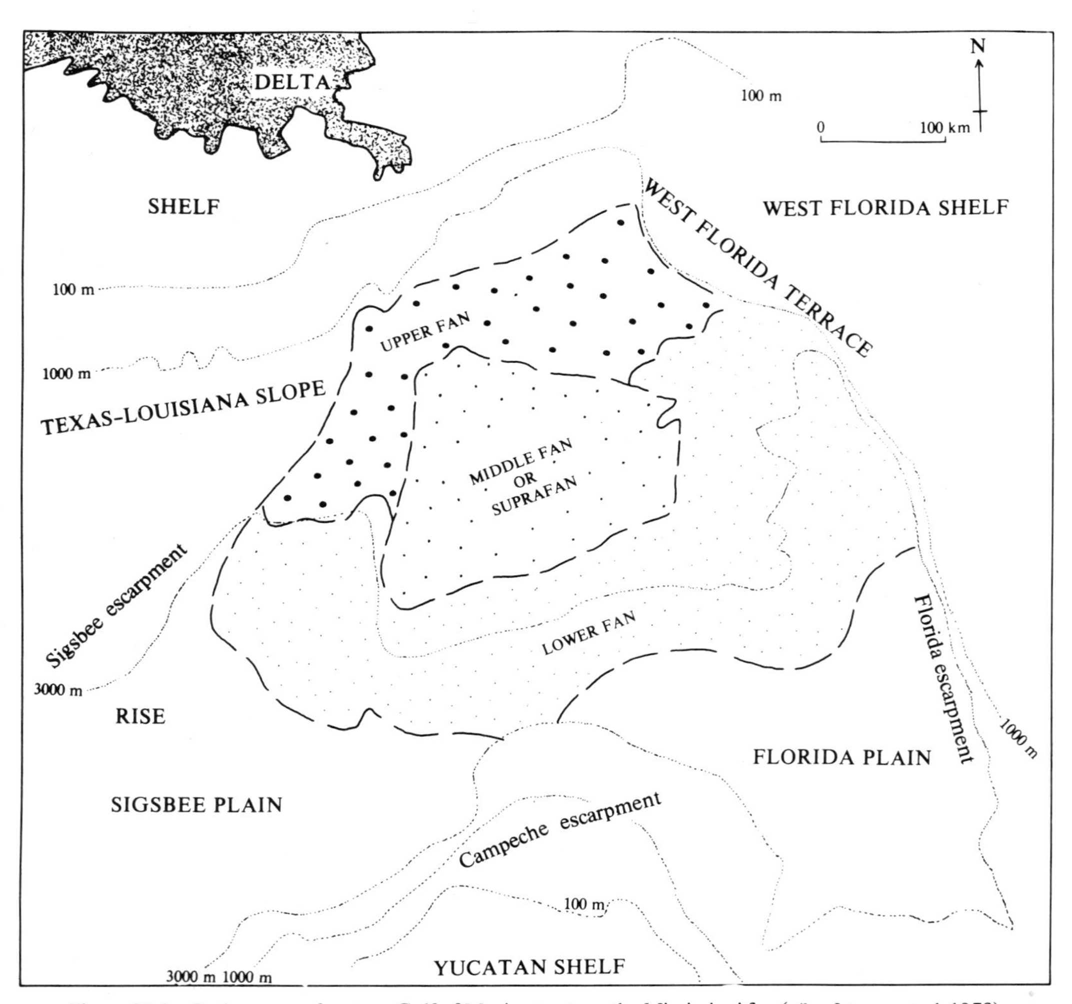

Figure 25.8 Bathymetry of eastern Gulf of Mexico to show the Mississippi fan (after Moore *et al.* 1978).

world's largest and most impressive fans are those low-gradient accumulations termed abyssal cones. These have been active throughout the history of most ocean basins and they show staggering dimensions and sediment thicknesses (Table 25.1, Fig. 25.8). They grade gently up on to the continental slope and thus substantially modify the normal pattern of continental margin isobaths (Fig. 25.8).

25d Abyssal plains

As shown in Figure 25.1, abyssal plains are most common along passive-type ('Atlantic') continental margins although they also occur along the inner flanks of back-arc basins as in the Bering, Japan and South China seas. Abyssal plains represent the most oceanward sediment trap for continent-derived sediment. Their level floors, with gradients less than 1:1000, have been built up layer by layer by successive turbidity-current deposits and hemipelagic muds (Heezen & Laughton 1963, Horn *et al.* 1971, Rupke 1975). Seismic studies reveal variable thickness of sediment (100–2000 m) with successions thinning and fining oceanwards against rugged sea-floor topography with only a thin cover of pelagic sediment.

Abyssal plain sedimentation is best illustrated by the Hatteras and Sohm plains of the eastern USA which are sourced by major submarine canyons.

The Hatteras abyssal plain (Fig. 25.9) lies at a depth of about 6 km and is up to 200 km wide and 1000 km long. Patterns of sediment size and bed thickness indicate turbidity current influxes from the Hatteras and Hudson canyons. Fan-like accumulations at the toes of these canyons contain ungraded medium sand and pass oceanwards into the abyssal plain. It is evident that the turbidity currents turned parallel to the continental margin along a southerly slope after leaving the canyon toe (Fig. 25.9). In the proximal area of the abyssal plain, graded fine sands occur in beds that are generally 0.5–2 m thick, though some may reach 6 m. There is a vague downslope thinning trend apparent with a better-developed downslope fining trend towards graded silts. Contrary to expectation it has proved impossible to correlate individual turbidites from cores over the abyssal plain, which suggests that individual turbidity currents are localised within the plain, perhaps by very shallow depressions and mounds created by successive turbidite deposition.

The Sohm abyssal plain (Fig. 25.10) is a T-shape plain lying south of Nova Scotia and the Grand Banks. The plain lies south of the gigantic Laurentian cone whose turbidity current activity at the present day has already been

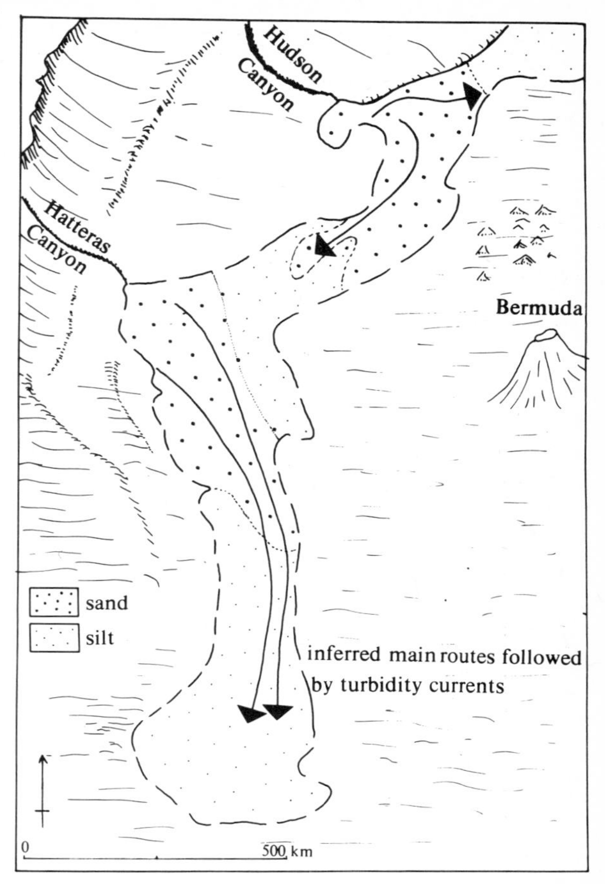

Figure 25.9 The Hatteras abyssal plain and the pattern of near-surface grain size (after Horn *et al.* 1971).

noted (Ch. 24). Many thick graded and ungraded sand beds lie close to the surface of the plain, but trends in sediment size and bed thickness are not readily apparent, probably because of multiple sourcing. Turbidity flows originate from the Gulf of Maine to the north-west and from the Mid-Atlantic Channel to the north east, as well as from the Laurentian Cone. Recent studies of the mid-Atlantic Channel (Chough & Hesse 1976) reveal its great importance as a source of turbidity currents. It is also likely that individual turbidity flows were diverted around numerous obstacles such as projecting sea mounts.

The above discussion of the Sohm and Hatteras abyssal plains shows that sedimentation of coarser units can be complicated by the multiple sourcing of turbidites. The mud units which often form the majority of abyssal plain successions are of turbidity-current, nepheloid-layer *and* pelagic origin. In the Balearic abyssal plain, pelagic

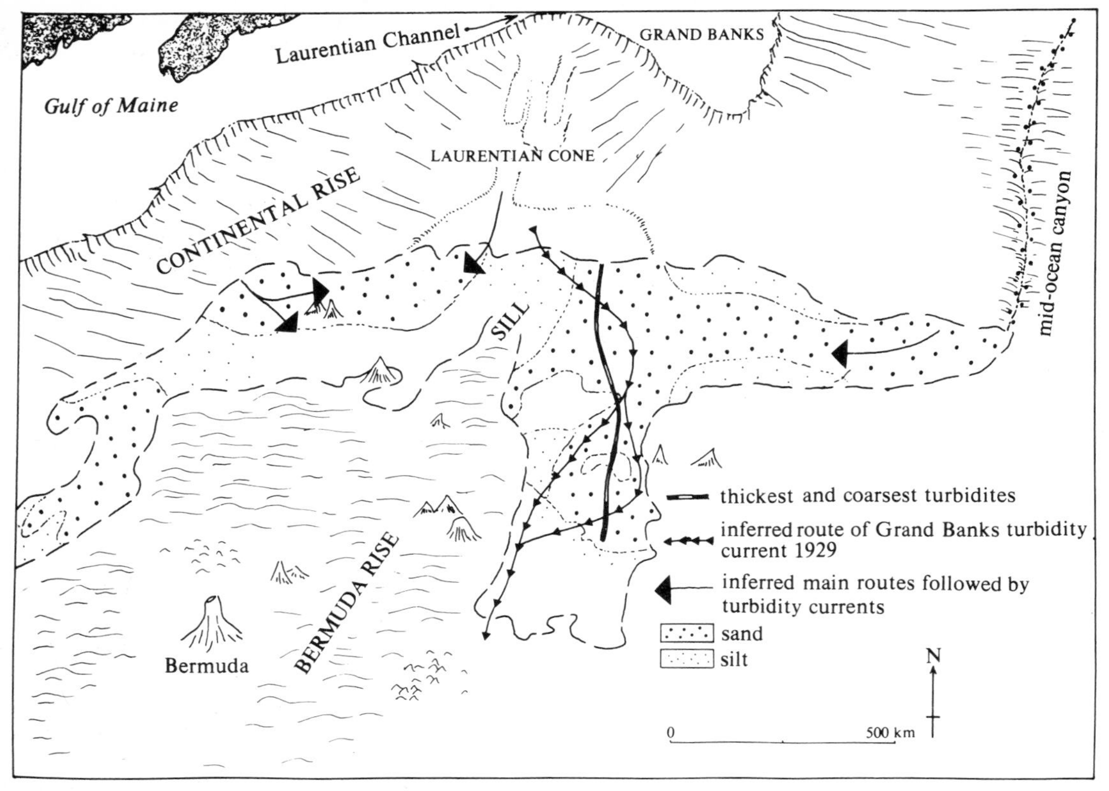

Figure 25.10 The Sohm abyssal plain and the pattern of near-surface grain size (after Horn *et al.* 1971).

muds overlie the turbiditic mud units which cap silt or sand turbidites. They show generally higher carbonate contents and contain sand-grade pelagic foram and pteropod tests (Rupke 1975).

Recognition of ancient abyssal plain successions depends heavily upon the occurrence of laterally extensive, relatively thin turbidite sands and silts interbedded with appreciable thicknesses of turbiditic and pelagic muds (Walker & Mutti 1973, Mutti 1977). It is doubtful, however, whether all basin-plain turbidite sands are as laterally extensive as some authors believe. We have already seen above how difficult it is to correlate turbidites on modern abyssal plains. However, Ricci-Lucci and Valmori (1980) have recently correlated ancient turbidites of presumed basin-plain origin over distances up to 300 km down current and 110 km normal to current. Palaeocurrents may be both longitudinal, as in the Hatteras abyssal plain, or radial, as in the Sohm abyssal plain. Abyssal plain facies should grade up current into lower submarine fan slopes, the distinction between the two being difficult.

25e Trenches and fore-arc basins of active margins

Figure 25.11 shows an idealised section through an active margin. Morphologically the system is dominated by the trench and the volcanic arc. The subduction zone of active tectonism with its **accretionary wedges** of trench and oceanic pelagic sediments occupies an uplifted belt on the arc side of the trench axis. Between the accretionary subduction complex and the volcanic arc may be a broad fore-arc basin formed in response to subsidence of oceanic-remnant crust trapped by the subduction process (Dickinson & Seely 1979). Let us examine sedimentation trends in this complex environment by means of a transect from ocean to arc along Figure 25.11.

The **outer rise** is a broad upwarp formed in response to lithospheric flexuring as the subduction zone is approached. The rise carries a pelagic succession whose facies will vary according to the pelagic depositional history of the oceanic plate (see Ch. 26b). The shallow ($<5°$) **trench outer slope** passes down into the trench proper whose floor may be a narrow **trench abyssal** plain

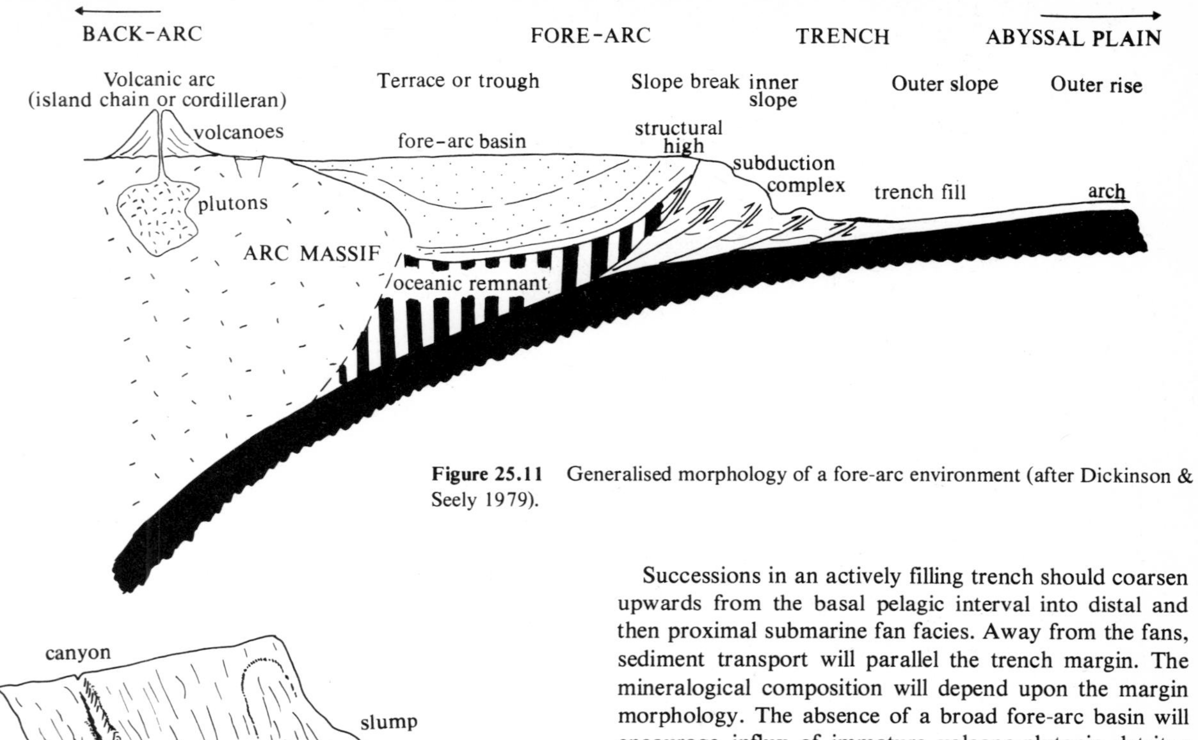

Figure 25.11 Generalised morphology of a fore-arc environment (after Dickinson & Seely 1979).

Figure 25.12 Sedimentation in the Aleutians trench (after Piper *et al.* 1973).

filled with a relatively thin (0–700 m) succession of predominantly clastic sediments. The trench may be filled by turbiditic sands and silts derived from submarine fans (e.g. Aleutian trench, Piper *et al.* 1973) draining down the trench inner slope or by slumps and debris flows from the subduction complex (Fig. 25.12). However, many trenches show almost no fill and here the trench is simply a sharp feature line with no flat floor. The amount of infill depends critically upon the influx of detritus from the adjacent cordillera or subduction complex and upon the local plate convergence rate (Schweller & Kulm 1978). Oblique spreading margins and high influx rates combine to give thick trench sequences in areas such as the northwestern United States.

Successions in an actively filling trench should coarsen upwards from the basal pelagic interval into distal and then proximal submarine fan facies. Away from the fans, sediment transport will parallel the trench margin. The mineralogical composition will depend upon the margin morphology. The absence of a broad fore-arc basin will encourage influx of immature volcano-plutonic detritus from the volcanic arc or cordillera. The presence of a fore-arc basin and uplifted subduction complex will result in sediment being reworked from previous trench offscrapings. The trench inner slope may show marked topographic irregularities caused by faulting, with small 'perched' basins which may trap slumps, debris flows and minor turbidity currents (Fig. 25.13). Abundant 'perched' basins may cause sediment starvation of the actual trench (Moore & Karig 1976).

Fore-arc basins, where present, show a thick fill dominated by clastic deposits derived from a volcanic arc or cordillera. Although not yet well documented they might be expected to show an upward trend from oceanic pelagic facies overlying the trapped ocean crust to deep-water, arc-derived montmorillonitic muds and pyroclastic falls. The latter may grade up into turbidite fan facies dominated by volcaniclastic and plutonic debris. On their inner edges the basin sediments would interfinger with lavas, lahars and pyroclastic flows (Dickinson & Seely 1979).

Recognition of ancient, subduction-dominated continental margins is possible only by a combination of regional–structural and sedimentological synthesis. Figure 25.14 shows the interpretation by Dickinson and Seely (1979) of the evolution of the Californian margin since Mesozoic times. Note the important Great Valley

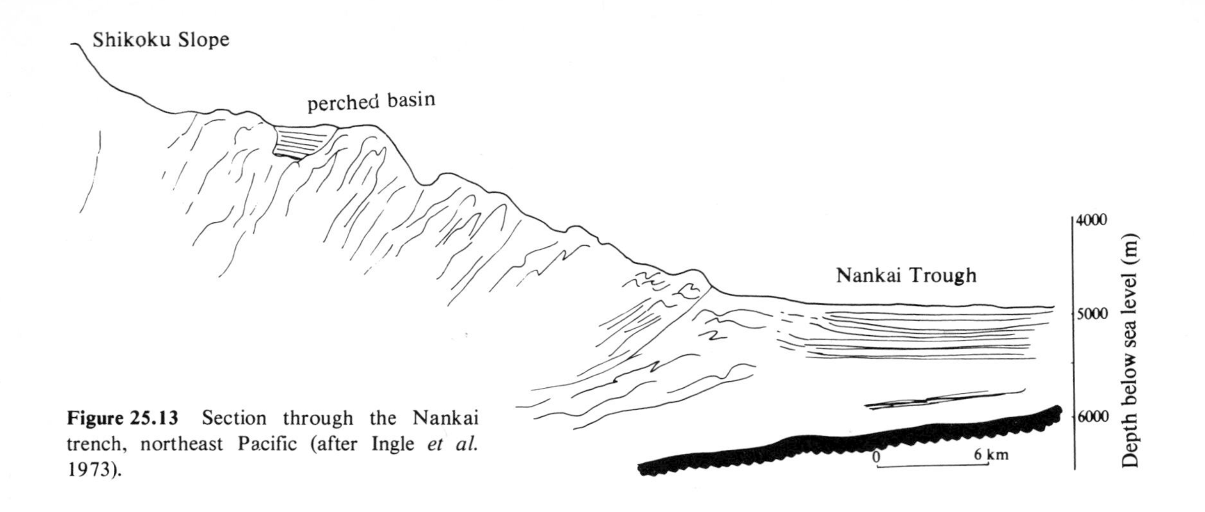

Figure 25.13 Section through the Nankai trench, northeast Pacific (after Ingle *et al.* 1973).

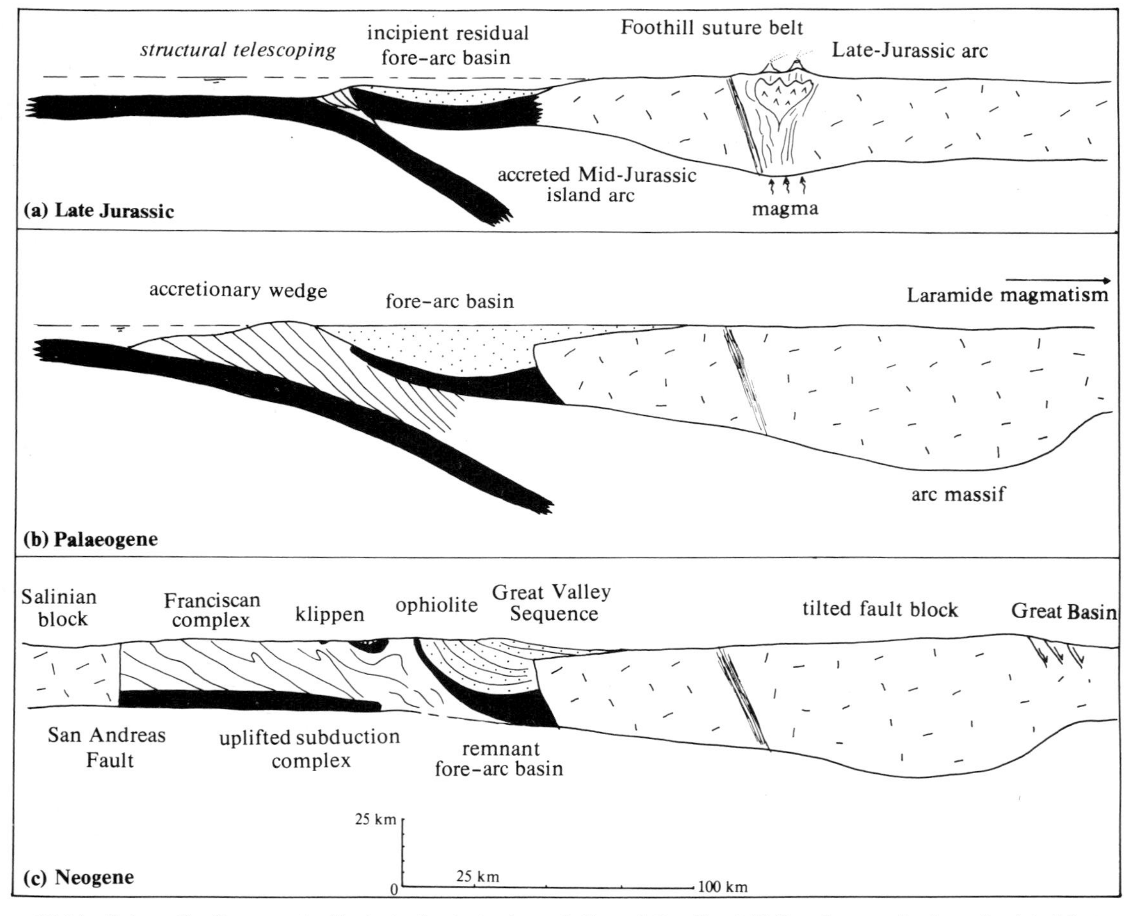

Figure 25.14 Schematic diagrams to illustrate the tectonic evolution of the Great Valley fore-arc basin and related features in northern California (after Dickinson & Seely 1979).

fore-arc basin with its deep fill (and important gasfields). The identification of subduction complexes hinges critically upon the recognition of successive thrust sheets of trench and submarine fan facies arranged in oceanward-younging slices, the sediments within each slice younging landwards. A very elegant model for the Southern Uplands of Scotland is based upon this model (McKerrow *et al.* 1977).

25f Summary

The distribution, geometry and nature of oceanic clastic facies is largely controlled by the type of tectonic setting of the ocean and its continental hinterland. The continental slopes and rises of passive oceanic margins receive fine sediment from turbid coastal/shelf plumes and are subjected to mass movements leading to turbidity flows. Major depositional and erosional effects along continental rises are caused by thermohaline (contour) currents. Continental margins of all types show submarine fans and cones which mark the oceanward spread of sediment introduced from point sources at the bases of submarine canyons as turbidity currents and debris flows. Fan progradation is postulated to cause coarsening-upwards successions. Abyssal plains are most common along passive margins, their horizontal floors being built up layer by layer by successive turbidites and pelagic muds. The trenches and fore-arc basins of destructive margins are fed by fans which tap cordillera or subduction complexes, sediment from the former giving ample evidence of derivation from calc-alkaline volcanic and plutonic terrains. Identification of subduction complexes hinges critically upon the recognition of successive thrust sheets of trench and submarine fan facies arranged in oceanward-younging slices, the sediments within each slice younging landwards.

Further reading

The interrelationship between clastic sedimentation and tectonics in ocean environments is discussed in an important paper by Mitchell and Reading (1969) and in a later version, with historical perspectives, by the same authors (Mitchell & Reading 1978). Dickinson and Seely (1979) give the most useful review of sedimentation and tectonics at active margins. The most valuable general review of modern and ancient oceanic clastic sediments is given by Rupke (1978). Vertical sequence analysis of submarine fan and abyssal plain environments is well summarised by Walker and Mutti (1973) and Mutti (1977). Problems and perspectives in the 'contourite' saga are given in the review by Stow and Lovell (1979). Many papers of interest are to be found in the volumes edited by Stanley and Kelling (1978), Bouma *et al.* (1978) and Watkins *et al.* (1979), the latter being especially recommended. A valuable discussion of submarine fan processes, including the terms proximal and distal as applied to turbidites, is given by Nilsen, Walker and Normark (1980).

26 Pelagic oceanic sediments

26a Sediment types

In oceanic environments below the carbonate compensation depth (CCD) **red clays** (actually chocolate to red-brown silty clays) accumulate at very slow rates of between 0.0001 and 0.001 mm^{-1}. These deposits are predominantly of clay minerals (see Ch. 1) whose compositions reflect continental climatic regimes (illite, chlorite, kaolinite) or intra-oceanic basic igneous source rocks (montmorillonite) (see Ch. 1). Slow-growing manganese nodules (Ch. 30) are common in certain areas of red clay deposition. Adjacent to trade wind deserts such as the Sahara, appreciable amounts of wind-blown silt (Ch. 13) occur in red clay facies, much of it wüstenquartz (with a characteristic iron oxide coating). Studies of aeolian dust in North Atlantic cores has proved of great help in elucidating desert expansion and contraction during Tertiary times.

Above the CCD, biogenic **calcareous oozes** predominate, with the main culprits being the coccolithophores, foraminifera and pteropods which fall through the ocean column in faecal pellet aggregates. As noted previously (Ch. 24) mapping out of the distribution of calcareous oozes in subsurface oceanic sediments provides critical evidence for the chemical dynamics of the ocean with time (see also below).

As noted in Chapter 3 the components of **siliceous oozes** are the opaline skeletons of diatoms, silicoflagellates and radiolarians. Diatom oozes are typical high-latitude deposits, with radiolarians more common in low latitudes. The deposits typically occur in high-fertility areas of the ocean marked by either surface water divergence (as in equatorial regions) or coastal upwelling. In both areas high phosphate contents arise from thermocline breakdown and deep-water mixing processes. Preservation of opal is largely independent of

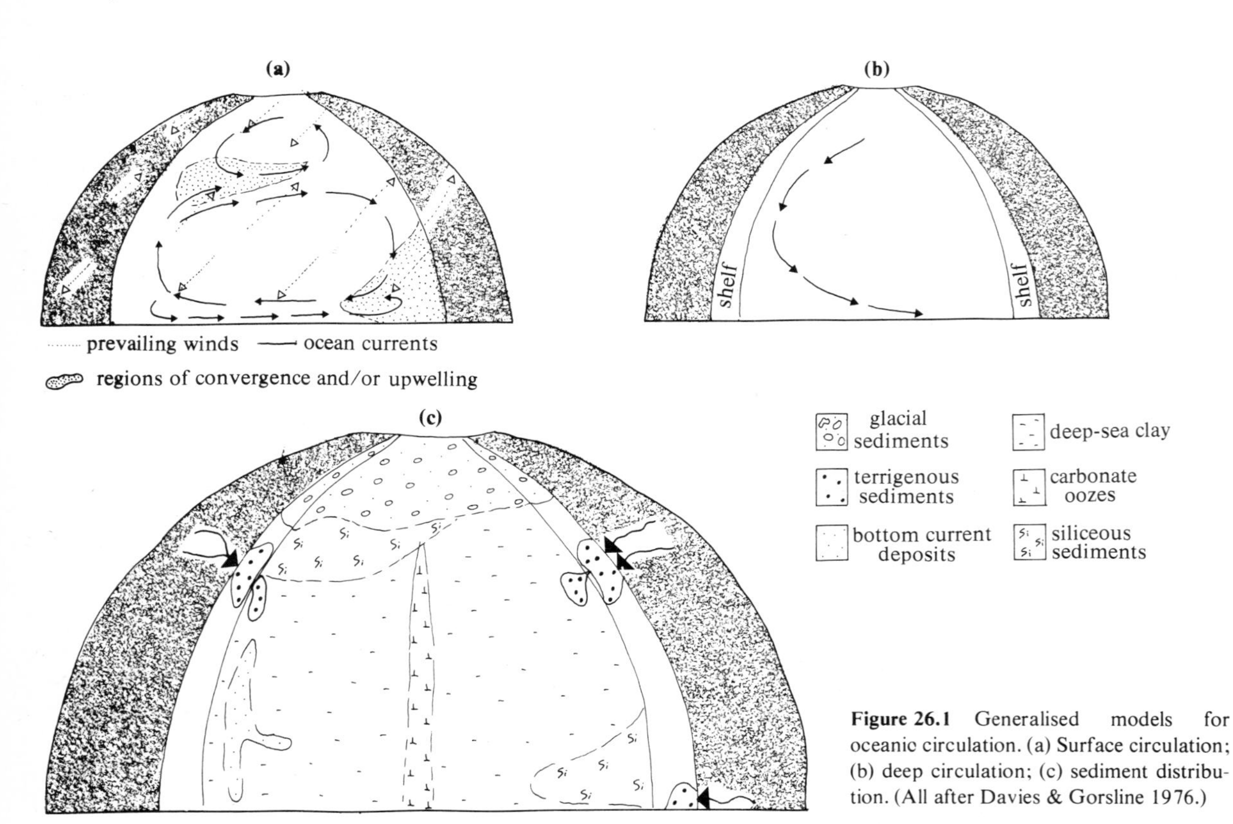

Figure 26.1 Generalised models for oceanic circulation. (a) Surface circulation; (b) deep circulation; (c) sediment distribution. (All after Davies & Gorsline 1976.)

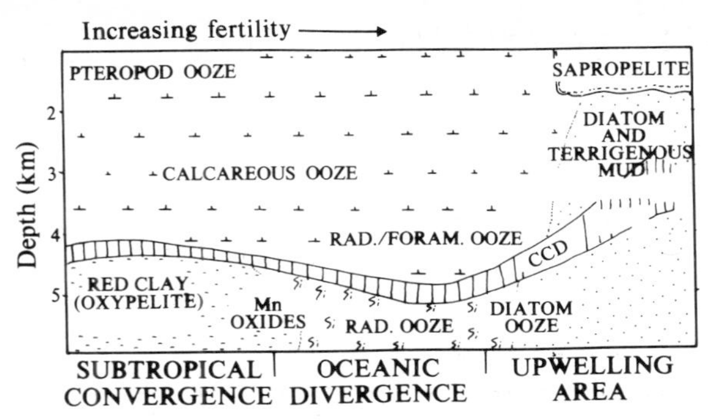

Figure 26.2 Major pelagic oceanic facies in a depth–fertility grid. Based upon sediment patterns in the eastern central Pacific (after Berger 1974).

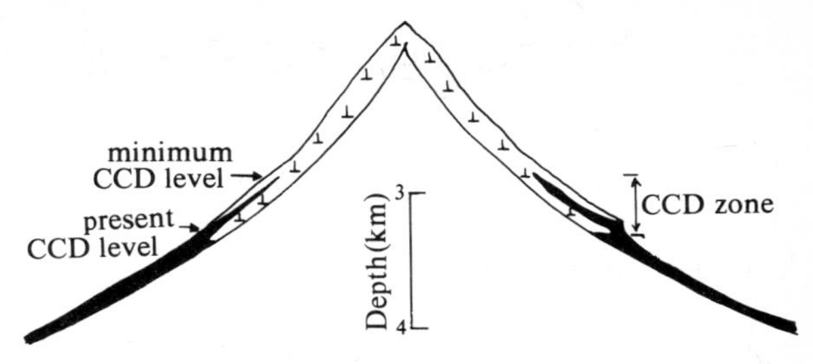

Figure 26.3 Accumulation of carbonate on a non-spreading ridge with a fluctuating CCD (after Berger & Winterer 1974).

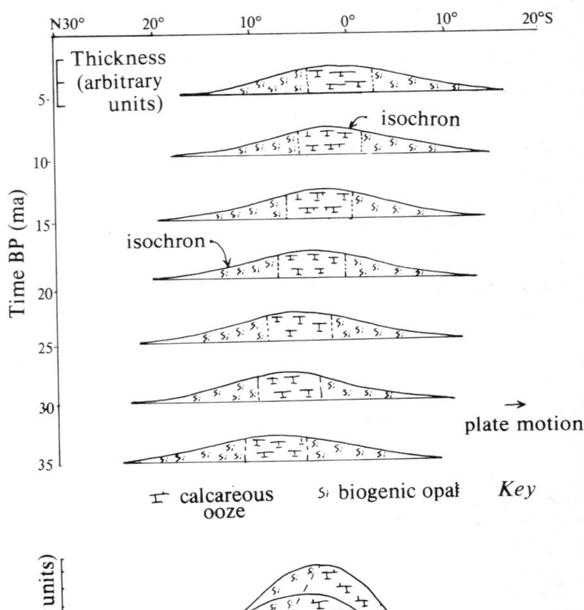

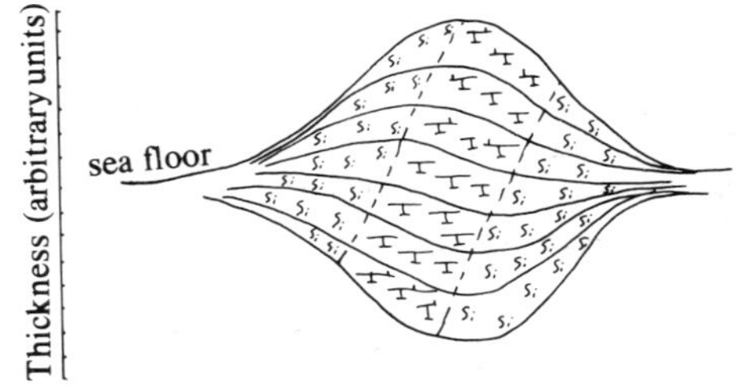

Figure 26.4 Schematic model for generating isochrons, facies patterns and sediment thickness within the fertile Pacific equatorial region. Orthoequatorial facies: carbonate/silica cycles. Para-equatorial facies: siliceous oozes. (After Berger & Winterer 1974.)

water depth, i.e. there is no silica compensation depth, hence siliceous biogenic sediments are a good indicator of ocean surface productivity. Studies of biogenic silica in Pacific cores, for example, reveal that the maximum accumulation rates have occurred at the Equator for the past 50 Ma, indicating the persistence of equatorial upwelling caused by divergence (Leinen 1979). However, once deposited the opaline skeletons of plankton are vulnerable to early diagenetic dissolution by pore waters, with much silica diffusion back into the ocean water (Ch. 30).

Although the majority of calcareous and siliceous oozes represent 'passive' fall outs from the oceanic water column, local redeposition of oozes does occur as small turbidity currents formed in response to slumping in the heavily faulted ridge-and-basin topography of slow spreading mid-ocean ridges such as the Mid-Atlantic Ridge.

The idealised distribution of sediment types in an ocean is shown in Figure 26.1, with a depth–fertility grid for oceanic facies shown in Figure 26.2.

26b Oceanic facies successions

As noted in Chapter 24 the succession of pelagic oceanic facies found at any one locality will depend upon a number of factors, including changing ocean morphology, water depth and circulation patterns. For example, consider a simple ridge system with the fluctuating CCD (Fig. 26.3). Over a period of time a complex intercalation of calcareous and non-calcareous facies will occur. Similarly, if a continental plate moves under an area of high productivity, then a maximum in sedimentation rate can be traced out in the accumulated sediment pile (Fig. 26.4). Large time gaps at erosion planes may occur in the ocean basins and on elevated plateaux. Thus Neogene oozes around Antarctica (Kennett *et al.* 1974) and in the South Atlantic ocean (van Andel *et al.* 1977) show many unconformities caused by the onset of the deep cold Antarctic bottom current system (see Ch. 24).

In order to appreciate the development of sedimentary successions below the ocean floor it is necessary to

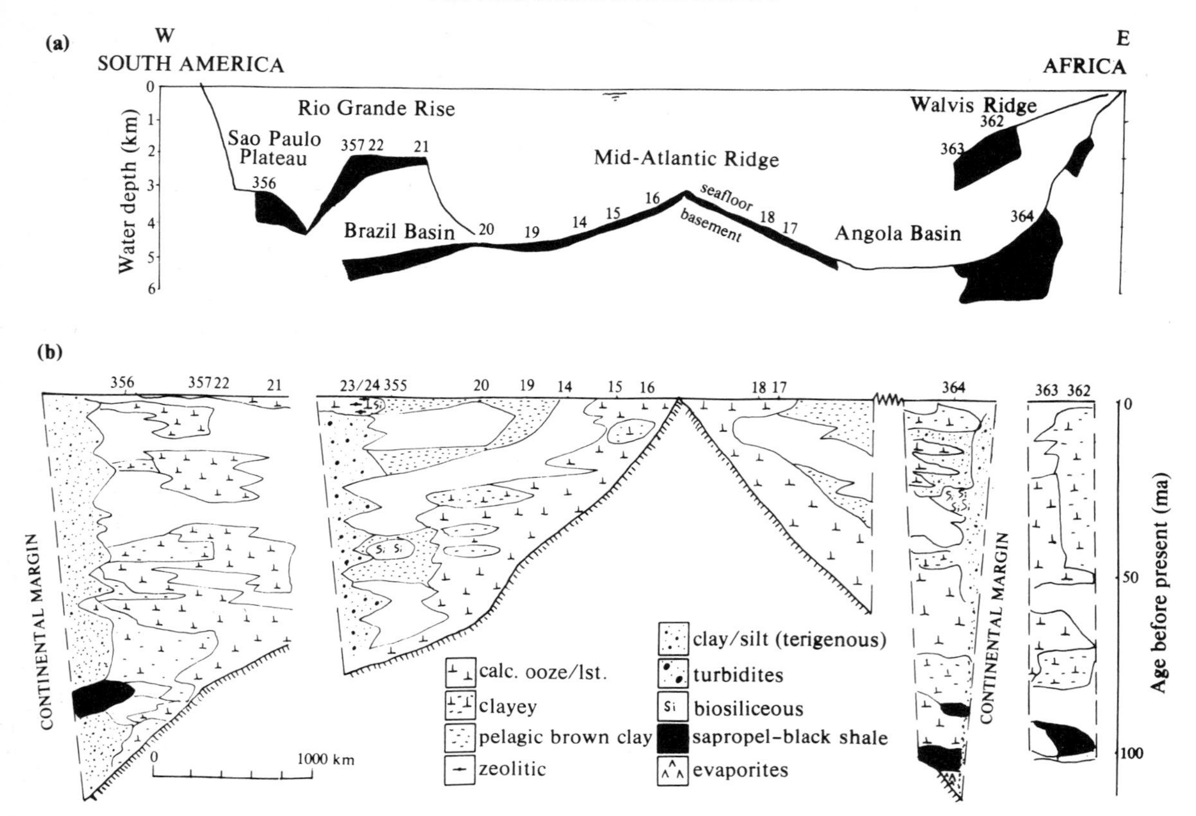

Figure 26.5 (a) Section across the South Atlantic to show bathymetry and sediment thicknesses (*black* areas). (b) Stratigraphic section across the South Atlantic to show oceanic facies. Blank areas are hiatuses. (All after van Andel *et al.* 1977.)

undertake a palaeo-oceanographic reconstruction, using some of the principles discussed in Chapter 24. In their superb study of the history of sedimentation in the South Atlantic Ocean, van Andel *et al.* (1977) used DSDP data to reconstruct the history of variables such as the spatial and temporal distribution of lithofacies, CCD, surface fertility and erosional events. Figure 26.5 shows the distribution of lithofacies in a stratigraphic section. Early rifting gave rise to a narrow northern basin with Aptian evaporites, separated by a ridge from a more open southern basin with normal pelagic sediments. Free circulation of surface water between the two basins and the North Atlantic occurred late in the Mesozoic and in the early Cainozoic. During the early and middle Mesozoic the South Atlantic had its own oceanographic character with terrigenous sedimentation and two anoxic black mudstone phases in the Albian and Santonian. The latter are attributed to a strong oxygen minimum in mid-water. In Cainozoic times the distribution of pelagic facies had been controlled by increasing width and water depth of the ocean and by fluctuations in the CCD. For example, the rapid fall in the CCD from about 3.75 km to 4.75 km in the past 10 Ma (Fig. 24.3) is paralleled by a marked expansion of calcareous oozes over the flanks of the mid-ocean ridge system (Fig. 26.5). At the Eocene/Oligocene boundary, 50 Ma ago, the onset of thermohaline circulation of cold Antarctic bottom water is clearly marked by erosional events, a sharp drop in the level of the CCD (Fig. 24.3) and the production of siliceous oozes in the Argentine basin. Evidence for coastal upwelling is apparent in the Upper Miocene deposits off southwest Africa with well preserved diatom and radiolarian assemblages, low values for the ratio of planktonic to benthonic forminifera, increased amounts of fish debris and presence of phosphorite grains (Dieter-Haas & Schrader 1979).

26c Anoxic oceans and oceanic events

As noted previously (Ch. 24) oceans may experience anoxic events caused by the development of oxygen

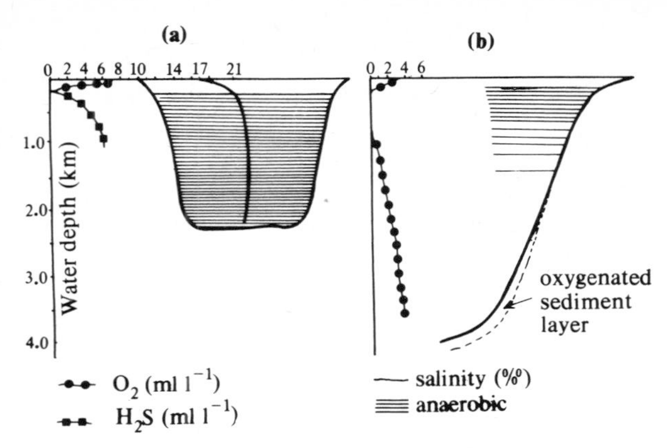

Figure 26.6 Schematic sections of aerobic/anaeorobic water masses in the Black Sea (a) and Indian Ocean (b) (after Thiede & van Andel 1977).

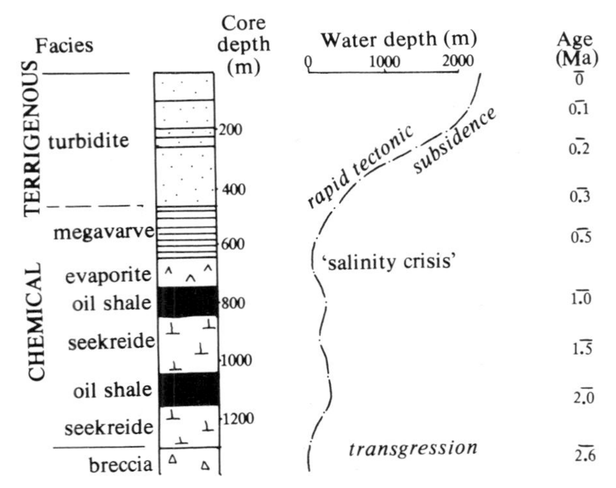

Figure 26.7 Schematic core section from the central Black Sea with suggested water depths and ages (after Degens & Stoffers 1980).

minimum layers. Anoxic sediments (black organic-rich shales) then develop either at the intersection point of the O_2 minimum layer with the ocean margin (Fig. 26.6, Thiede & van Andel 1977) or with oceanic plateaux or continental shelves (Schlanger & Jenkyns 1977).

By way of contrast to the above examples the Black Sea (Degens & Ross 1974) is the only major modern example of a silled basin that has turned anoxic at depth (Fig. 26.6). The Black Sea is up to 2200 m deep, with the O_2/H_2S interface at a mean depth of about 200 m. Surface salinities in the O_2 layer are 17.5–19‰ whilst the remainder is at about 22‰. Rapid tectonic subsidence of the Black Sea about 300 000 years BP changed the old shallow-water sea/lake into its present deep inland oceanic form. (The subsidence rate reached an astonishing maximum of 5 mm a^{-1} during this time.) The early Quaternary shallow-water deposits (Fig. 26.7) comprise 'megavarves' (10–100 mm thick), evaporites, chalky oozes (seekreide) and oil shales formed in a stratified water body fluctuating between fresh and saline, perhaps analogous to the Green River Formation environment discussed in Chapter 16 (Degens & Stoffers 1980). The late Quaternary to Recent deposits (Fig. 26.7) are dominantly terrigenous turbidites deposited in oxic environments, with five sapropel (black shale) and chalky ooze intercalations that mark anoxic conditions. The sapropels are a few centimetres to decimetres thick with ~10% organic matter. Well developed varves are present, with dark microlaminae originating from seasonal mass-mortality of planktonic bacteria. The sapropels formed over short time intervals (~5000 years) during times of warm climate when successive saline spills from the Mediterranean ocean formed a rising front that gradually moved through the water mass. Termination of sapropel deposition occurred when the O_2/H_2S interface was constant so that the permanent density stratification produced a planktonic community adjusted to the new stable habitat (Degens & Stoffers 1980). Thus the present stable conditions in the Black Sea have lasted for about a thousand years, the youngest sapropel being overlain by annually varved coccolith oozes that continue to form today.

Periodic Quaternary anoxic events in the eastern Mediterranean (Thunell *et al.* 1977) provide an interesting link with those described from the Black Sea above. Periodic sapropel layers here are marked by the complete absence of benthonic microfossils and by the presence of an abnormal planktonic forminifera containing a high proportion of a particular salinity-sensitive form. The deposition of sapropels was synchronous in the eastern Mediterranean and the onset of anoxic conditions is thought to have been caused by flushing of fresh water from the Black Sea into the area during interglacial periods when enormous surface runoff from Eastern Europe occurred. The low-salinity overspill formed a surface layer that prevented oxygenation of the basin at depth. Whether the freshwater overspill *out* of the Black Sea was coincident and compensation for the saltwater underspill *into* the Black Sea from the Mediterranean noted above is an intriguing but unanswered question. Nevertheless the development of anoxic conditions in marginally connected basins is clearly of great interest.

26d Hypersaline oceans

After cratonic rifting young ocean basins are invaded by marine waters from pre-existing oceans. Poor circulation

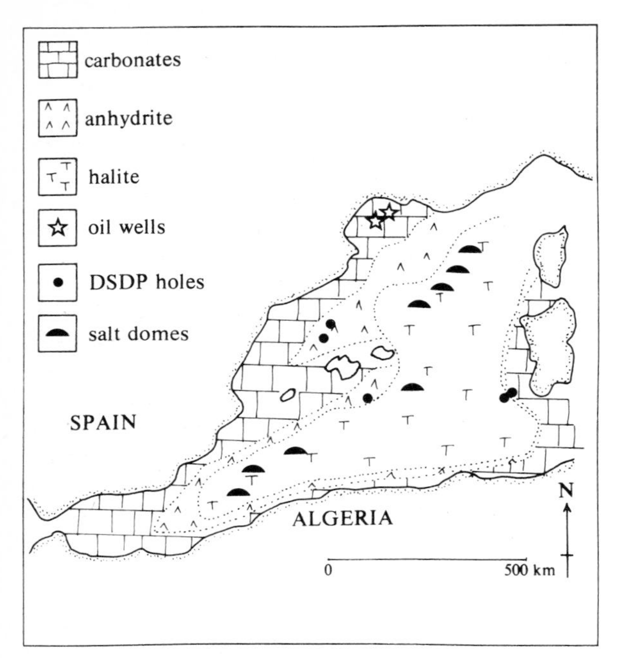

Figure 26.8 The distribution of Messinian evaporites in the western Mediterranean (after Hsü 1972).

in these proto-oceans will encourage evaporite formation if the climatic regime is suitable (Kinsman 1975a & b). Tremendous thicknesses of evaporites, particularly halite, may form. For example in the margins of the Red Sea and in the adjacent Danikil Depression of Ethiopia up to 3 km of Miocene halite are recorded in seismic and borehole sections. Thoroughgoing oceanic circulation then followed in Pliocene times, giving rise to a succession of more normal pelagic deposits on top of the evaporites. As noted previously, evaporites also developed during the early history of the South Atlantic ocean.

Perhaps the most spectacular development of evaporites in a major ocean basin occurred in the western Mediterranean during the Miocene (Hsü *et al.* 1972 & 1977). Deep Sea Drilling Project cores revealed deep pelagic and hemipelagic oceanic sediments overlain by anhydrite and halite evaporites (Fig. 26.8) up to 1500 m thick with clear evidence of shallow-water playa to sub-aerial sabkha precipitation (desiccation cracks, stromatolites, chicken-mesh texture). The implication was clear to Hsü that the Mediterranean had completely desiccated during Miocene times following tectonic blockage at the western outlet to the Atlantic which today is the Straits of Gibraltar. A gigantic deep basin over a thousand metres below sea level was formed, with streams and rivers deeply incising their valleys into the basin margins. The evaporites are overlain by deep-water pelagic and hemipelagic sediments which give evidence of an early Pliocene reconnection of the ocean to the Atlantic. It has been estimated (Thierstein & Berger 1978) that the Miocene evaporites of the Mediterranean contained about 6% of the total world oceanic salts during Miocene times and that this massive salt extraction may have affected the diversity of contemporary planktonic fauna.

26e Continental outcrops of ancient facies

Identification of pelagic facies in continental outcrops as truly oceanic depends upon correct interpretation of associated basic and ultrabasic igneous rocks as oceanic crustal remnants. Thus ancient pelagic oceanic facies usually occur as parts of obducted or subducted ophiolite complexes. One of the most thoroughly documented ophiolites is the Troodos Massif in Cyprus, a part of the former Tethyan ocean floor. Overlying the ophiolite proper is a pelagic facies of variable thickness (up to 800 m) (Robertson & Hudson 1974, Robertson 1975, 1977). Fe-, Mn- and heavy-metal-rich mudstones known as **umbers** rest directly on an irregular surface of pillow lavas. The umbers are thought to be derived from the

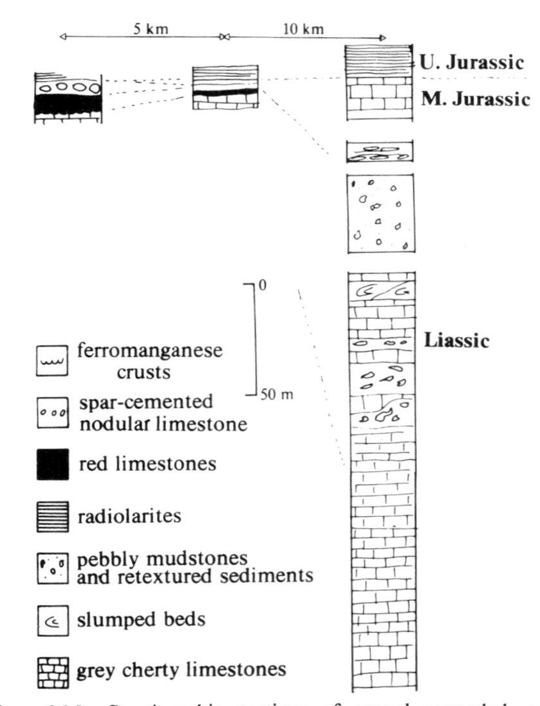

Figure 26.9 Stratigraphic sections of coeval expanded and condensed Jurassic pelagic successions in the eastern Alps of Austria (after Bernoulli & Jenkyns 1970).

pillow lavas and basalts by leaching from convection cells of recirculating sea water and juvenile fluids at or close to active spreading ridges. Modern analogues occur in the East Pacific Rise, Carlsberg Ridge and Mid-Atlantic Ridge. The umbers are overlain by radiolarian cherts and mudstones which pass up into a variable succession of chalks with and without bedded cherts. These pelagic facies represent deposition generally above the CCD and they contain many redeposited chalk units (derived from ridge ponding?) which acted as preferential sites for chertification.

Good examples of turbiditic radiolarian chert beds of suspected ridge origin occur in the Othris Mountains of central Greece (Nisbet & Price 1974) whilst the bedded cherts of the Ordovician Ballantrae ophiolite in southern Scotland show fine soft-sediment deformation folds (Bluck 1978), related perhaps to early slumping and melange formation.

Recognition of condensed pelagic carbonates, cherts and marls overlying shallow water Bahamian-type carbonate platforms in the late Triassic–Jurassic of the Tethyan ocean in Europe (now exposed in Alpine terrain) enabled Bernoulli and Jenkyns (1974) to postulate an important regional tensional regime related to Atlantic opening and subsidence of the European Alpine shelves. Block faulting and differential subsidence affected many carbonate platforms (Fig. 26.10) and gave rise to a seamount-and-basin topography on which coeval condensed and expanded sequences (Fig. 26.9) were deposited. The condensed facies show deposition rates of 0.0005–0.007 mm a^{-1} (cf. red clay rates discussed above) and comprise, in part, red biomicritic limestones rich in ammonites and Fe–Mn crusts. Thicker expanded successions include radiolarites and grey-to-white pelagic limestones. Slumped and turbiditic beds occur at the basin margins with much evidence for fissuring and fissure-

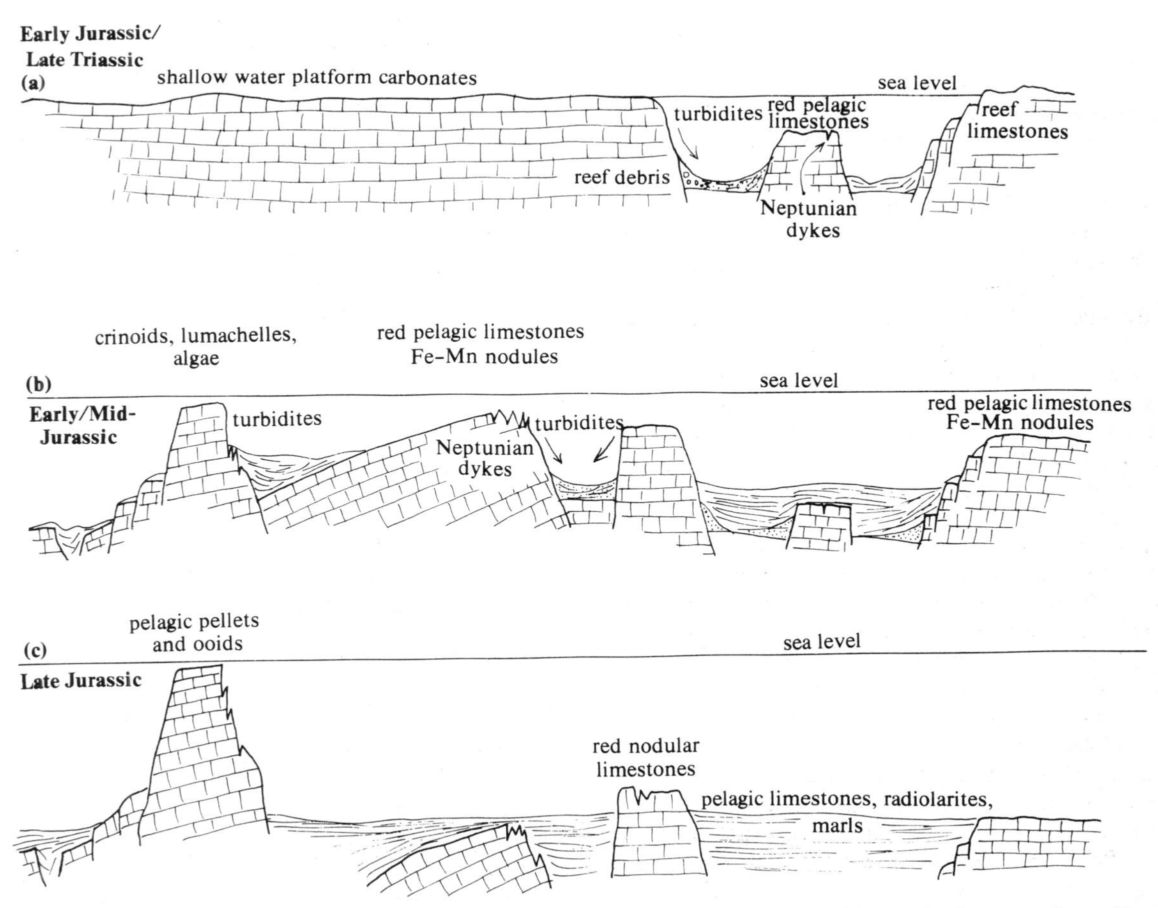

Figure 26.10 The evolution of the Tethyan continental margin during the early Mesozoic. Note increasing fragmentation and break-up of the shallow-water carbonate platform by block faulting and differential subsidence (After Bernoulli & Jenkyns 1974.)

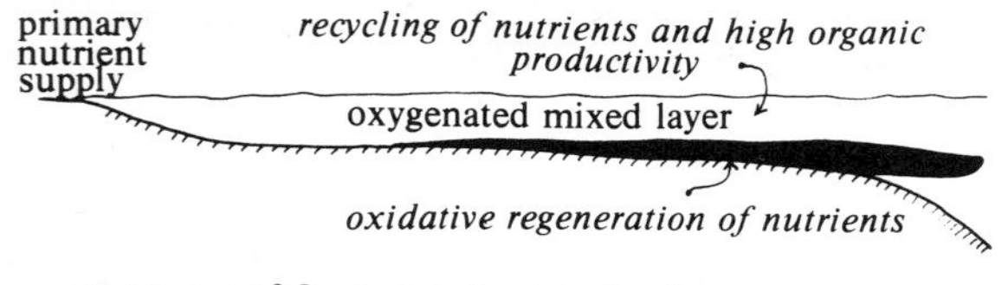

Figure 26.11 How an anoxic water mass may arise after shelf transgression because of oxidation of abundant organic matter stranded on the shelf below the level of turbulent mixing. Black shales may result which overlie 'normal' shelf facies. (After Jenkyns 1980.)

filling in both the pelagic facies and the underlying platform carbonates. Modern oceanic hardgrounds of erosional origin with Fe/Mn crusts and fissures analogous to those noted above have recently been discovered on Carnegie Ridge in the eastern Pacific (Malfait & van Andel 1980). Here the erosional surfaces are due to deep ocean currents and dissolution.

Almost identical facies to the Mesozoic seamount-and-basin example discussed above occur in the upper Palaeozoic Rheno-Hercynian zone of southwest England and Germany (Tucker 1973, 1974).

We must now mention the occurrence of pelagic facies deposited over continental shelves, the prime example being the Cretaceous chalks of north America, northwest Europe and the Middle East. A major worldwide transgression in the Cenomanian flooded vast areas of the world's continental shelves and coastal plains. Cut off from terrigenous input the Chalk sea became highly productive with a flourishing coccolith population in the photic zones which gave rise to deposition rates of the order of 0.01–0.05 mm a^{-1}. These rates are comparable with those seen today under the fertile surface waters of the equatorial Pacific where foram oozes are being formed. The source of nutrients to keep up such high rates of deposition from the planktonic biomass are unknown, but the scale of production implies a more dynamic oceanographic regime in the Chalk sea than is usually envisaged (Funnell 1978). Major hiatuses are known from almost all chalk sequences and they comprise penecontemporaneously lithified surfaces of hardground type (see Ch. 29). These have their own ecological groups of borers and cementing organisms and may contain glauconite and phosphate nodules.

Transgression may also give rise to anoxic conditions on shelves (Fig. 26.11).

26f Summary

Pelagic oceanic sediments comprise red clays (with their Mn nodules (Ch. 30)) and calcareous and siliceous oozes. Oceanic facies successions depend upon a number of factors, including changing oceanic morphology, water depth and circulation patterns. These successions record for posterity the history of any ocean basin when combined with data on spreading rates and palaeobathymetry. 'Abnormal' events in oceans include development of hypersalinity, leading to the production of oceanwide suites of evaporites (saline giants), and of O_2 deficiency leading to widespread deposition of organic-rich 'black shales'.

Further reading

A useful overview of ancient and modern pelagic oceanic sediments is given by Jenkyns (1978), and many interesting papers may be found in the volume edited by Hsü and Jenkyns (1974). The use of DSDP cores in reconstructing oceanic history is best illustrated by the work of van Andel *et al.* (1977). Hsü (1972) gives a lively account of the desiccating Miocene Mediterranean.

PART EIGHT DIAGENESIS: SEDIMENT INTO ROCK

Full fathom five thy father lies;
Of his bones are coral made;
Those are pearls that were his eyes;
Nothing of him that doth fade
But doth suffer a sea-change
Into something rich and strange.

From Ariel's Song in The Tempest *(Shakespeare)*

Plate 8 View and close-up looking seawards from present beach at Kpone, E. Region, Ghana to show Holocene beachrock overlying a cemented Pleistocene aeolianite (coastal dune). The aeolianite (4000–4500 years BP) contains sparry calcite cements typical of meteoric vadose and phreatic environments, cementation having occurred in a freshwater lens within the dune. The aeolianite was then exposed, pitted, abraded and encrusted by marine organisms as a result of coastal transgression. The surface (see close-up, below) was then blanketed by beach sands and gravels during a phase of coastal progradation within the past 1200 years. The beach sediments during this time have lithified to form beach-rock with micritic and acicular aragonite rim cements which also overlie earlier freshwater calcite cements in some aeolianite pores. (Photos and description by M. R. Talbot.)

Theme

Diagenesis used to be thought of as a rather staid subdiscipline of sedimentology, with its woolly generalisations gleaned from microscopic and X-ray work. The growth of a modern diagenetic approach in the past 25 years has firmly placed the study back into the centre of sedimentological attention. *All* aspects of sedimentology combine in diagenetic studies and it is fitting that this book should end on the subject. Once deposited, sediment grains undergo changes during burial as pore waters are progressively removed from their depositional realm. Increased temperature and pressure cause the subsurface movement of formation water down fluid-potential gradients. Grains dissolve and new minerals precipitate as cement; muds compact and undergo mineralogical change with burial. Organic matter undergoes chemical changes in its conversion from tissue to hydrocarbon. Gradually sediment changes into rock. Examination of rock collected at outcrop or from coring enables these diagenetic changes to be deduced. Just as we have facies analysis, so we have diagenetic facies analysis.

An all round approach to sedimentology thus enables the sedimentologist to trace the journeyings of a sand grain or shell fragment from source to depositional environment and hence on its journey underground during burial and subsidence and back up again during uplift. The story is an interrelated one which leaves no room for extensive specialisation.

27 Diagenesis: general considerations

27a Definitions

The transformation of a freshly deposited sediment into a rock is a complex process. **Diagenesis** is the name given to the many chemical and physical processes which act upon sediment grains in the subsurface. **Halmyrolysis** refers to the more restricted aspect of chemical changes operative at the sediment/water interface. Diagenetic changes give way eventually to metamorphic recrystallisation processes at higher temperature and pressure but there is no clear-cut boundary between the two realms (Fig. 27.1).

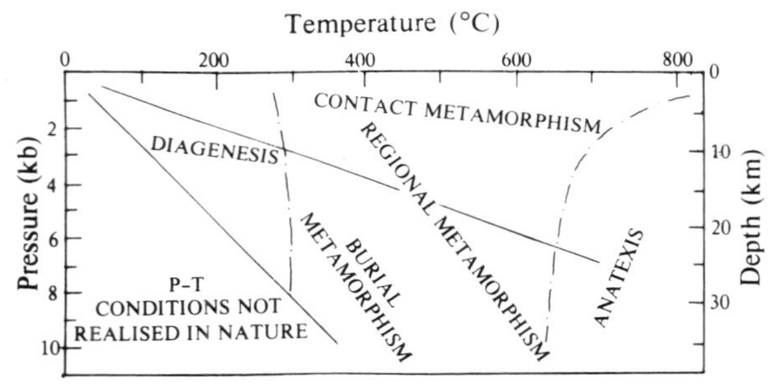

Figure 27.1 The diagenetic realm in relation to the metamorphic and anatexic realms (after Pettijohn *et al.* 1972).

Diagenetic routes are many and varied, depending upon initial sediment composition and grain size, depositional environment, temperature and pressure conditions during progressive burial and depth of burial. Taken to its limits, diagenesis changes water-saturated, porous aggregates of grains into tightly fitting aggregates of the most stable grains bonded together by newly formed diagenetic minerals and grain-to-grain welding: the sediment is lithified into rock. Diagenesis is usually accompanied by massive throughflow of water transferred from compacting muds into more porous horizons. These solutions transfer ions in solution through rock pores. Should local mineral precipitation occur, an indication is then provided of the pore-water composition at that instant in time. Diagenetic mineral reactions thus take place predominantly under open system conditions.

27b Subsurface pressure and temperature

The raised pressures and temperatures below the Earth's surface are responsible for driving many diagenetic reactions involving both mineral solution and precipitation.

When discussing pressures caused by overburden in sedimentary deposits, it is important to distinguish between **bulk pressure** and **pore pressure.**

The mean bulk pressure, P_b, exerted by overburden on the grain-to-grain contacts over unit area of sediment is given by

$$P_b = \sigma_{bw} g h \qquad (27.1)$$

where σ_{bw} is the bulk wet density of sediment, h is the height of the sedimentary column above the unit area in question, and g is the gravitational acceleration. The bulk wet density of sediment is the sum of the partial densities of pure water ρ and pure sediment σ given by

$$\sigma_{bw} = p\rho + (1-p)\sigma \qquad (27.2)$$

where p = porosity. Although porosity, and hence bulk wet density, vary with depth because of compaction, it is usually assumed that the P_b increases linearly with depth by use of a mean value for the bulk wet density. Exact calculation of the pressure at depth h_2 from the surface (h_0) will require evaluation of the integral

$$P = g \int_{h_2}^{h_0} \sigma_{bw} \, dh \qquad (27.3)$$

The pore (or interstitial) pressure P_p may be regarded as the pressure exerted within the fluid-filled pores of unit bed area by overburden. When the pores are not connected to their neighbours, as is commonly the case in compacted mudstones or shales, then P_p is simply equal to the bulk overburden pressure P_b, and the distribution of pressure with depth follows a mean **geostatic** pressure gradient (Fig. 27.2). However, when the pores are interconnected to the surface (regardless of the tortuosity of such connections) then P_p in any pore is said to be **hydrostatic** (Fig. 27.2) and is due simply to the weight of the overlying water column, given by

$$P_p = \rho g h \qquad (27.4)$$

The distribution of hydrostatic pore pressure with depth is linear (neglecting the effects of temperature, pressure and dissolved salts upon water density) and it defines the mean hydrostatic pressure gradient.

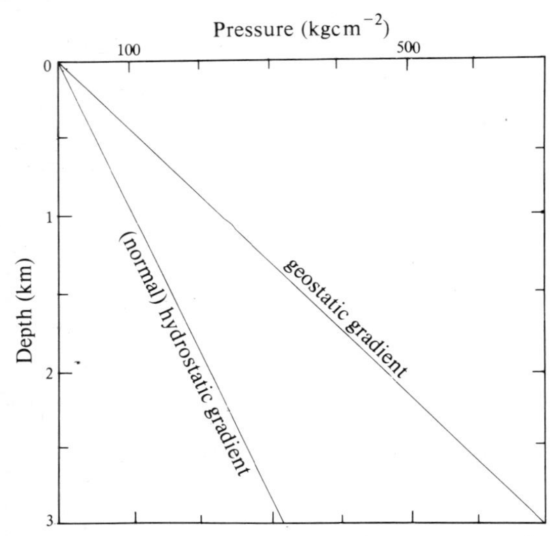

Figure 27.2 Hydrostatic and geostatic pressure gradients (after Chapman 1976).

It can be seen from Figure 27.2 that P_p and P_b, defining the hydrostatic and geostatic gradients respectively, are limiting values of subsurface pressures. The magnitude of the two sorts of pressure is an important control upon rates of solution caused by pressure transfer (Ch. 27g) and the route of pore-fluid migrations (Chs 27f & 13).

The transfer of radioactive heat from the convecting Earth's asthenosphere into the upper levels of the lithosphere is achieved mainly by *conduction*. Additional high level *convection* may occur in porous and permeable water-saturated rocks in areas of high heat flow close to magma sources such as volcanic magma 'chambers' and mid-ocean ridges. Measurements of temperature in deep boreholes through impermeable rocks below near-surface levels ($h \gtrsim 300$ m) enable temperature gradients to be defined. Typical gradients for continental areas lie in the region of 20–30°C km^{-1}.

27c Petrography in diagenetic studies

A major part of diagenetic studies involves the description and identification of textures and mineral precipitates produced by chemical and physical processes. Perhaps the single most important aspect is the recognition of true **cement** fabrics, where the term cement includes all passively precipitated, space-filling crystals which grow attached to a grain surface surrounding a pore space (Bathurst 1975). Cement minerals are thus newly formed from pore-fluid ions and are said to be **authigenic**. There is a host of cement minerals but the most important are *calcite*, *dolomite*, *quartz* and *clay minerals*. When present as true cement in the pores of sediment or rock, these minerals might show some or all of the following features which distinguish cement fabrics from primary depositional or recrystallisation fabrics (see Bathurst 1975):

(a) the crystals occur in pore space between deposited solid sediment grains or within hollow fossil grains;
(b) the crystals grow out normally from the grain surfaces;
(c) if present as an incomplete or rim cement around a pore, the crystals show well formed growth habits;
(d) when completely filling a pore space the cement crystals commonly increase in size away from the pore wall;
(e) because of the competitive nature of cement growth, intercrystalline boundaries are of a compromise nature and show plane interfaces;
(f) many of the intercrystalline triple boundaries are of **enfacial** type, with one of the three angles at 180°;
(g) there may be two or more generations of cements lining a pore space, so that the order of precipitation may be established by the principle of superposition, i.e. the cement generation closest to the pore wall is the oldest.

One of the most difficult tasks for the sedimentary petrographer is the correct identification of recrystallisation fabrics and their distinction from cement (Bathurst 1958, 1971, Folk 1965). The problem is most acute in limestones and hence discussion of this topic is postponed to Chapter 29.

Both transmitted light and scanning electron microscopes may be used to examine diagenetic fabrics. The latter are of great importance in studies of clay mineral cementation in sandstones since the tiny clay crystals are often difficult to resolve using the traditional light microscope and may be destroyed during the slide making process. Modern analytical techniques using the electron microprobe enable simultaneous morphological and compositional measurements to be carried out.

27d Stable isotopes in diagenetic studies

Isotopes of a given element have an identical nuclear charge, (i.e. the same atomic number) but differ in atomic mass because of a difference in the number of neutrons in the nucleus. The most familiar isotopes are those whose nuclei undergo spontaneous disintegration, changing to more stable nuclei and giving rise to emissions of particles and rays. There are, however, many stable isotopes in nature whose relative abundance in chemical and biochemical systems varies slightly according to the general nature and conditions of the chemical reactions involved. Oxygen, for example, has three isotopes ^{16}O, ^{17}O and ^{18}O which exist in relative percentage abundance 99.759:0.0374:0.2039 in the atmosphere (Epstein 1959). The ^{18}O:^{16}O ratio in nature actually varies by about 10%. This is because of slight differences in chemical properties of compounds which contain ^{18}O and ^{16}O, e.g. $H_2{}^{18}O$, $H_2{}^{16}O$. The thermodynamic properties are, to a small but important extent, influenced by the masses of the atoms involved.

Stable-isotope analysis of minerals and rocks is a very valuable tool for interpreting conditions of formation and subsequent alteration. The two most common isotopic ratios used in sedimentary studies are ^{13}C:^{12}C and ^{18}O:^{16}O. In both cases we are not so much interested in the absolute abundances of the isotopes or in the actual ratios but in the magnitude of the ratios compared to the ratios of some standard sample. This comparison is calculated in the delta (δ) terminology, e.g. for stable carbon isotopes

$$\delta^{13}C‰ = \frac{^{13}C/^{12}C(\text{sample}) - {}^{13}C/^{12}C(\text{standard})}{^{13}C/^{12}C(\text{standard})} \times 10^3$$

and similarly for oxygen. Note that a positive delta value indicates enrichment in the heavy isotope, vice versa for a negative delta value. 1% enrichment of ^{13}C is equivalent to a $\delta^{13}C$ of +10‰. For carbonate rocks and minerals the PDB standard is normally used for both carbon and oxygen isotopes. This standard was prepared from the marine Cretaceous Peedee belemnite so that its isotopic composition is $\delta^{13}C = 0$ and $\delta^{18}O = 0$ by definition. For water and for other minerals containing oxygen, standard mean ocean water (SMOW) is used. SMOW has a $\delta^{18}O$ of −30.8 on the PDB scale, the conversion factor between the two standards being

$$\delta^{18}O\ \text{SMOW} = 1.031\ \delta^{18}O\ \text{PDB} + 30.8 \quad (27.5)$$

Let us now briefly consider the principles of interpretation of oxygen and carbon stable-isotope ratios in sedimentary rocks and minerals with especial reference to calcium carbonate. Some of the points discussed below are summarised in Figures 27.3 and 27.4.

The oxygen isotope composition of calcium carbonate precipitated from natural waters depends upon the composition of the aqueous phase and upon temperature. The behaviour of the isotopes in natural waters depends in large part upon the fact that $H_2{}^{16}O$ is more volatile than the heavier $H_2{}^{18}O$. Present-day sea water has a rather narrow range of isotopic composition, but when

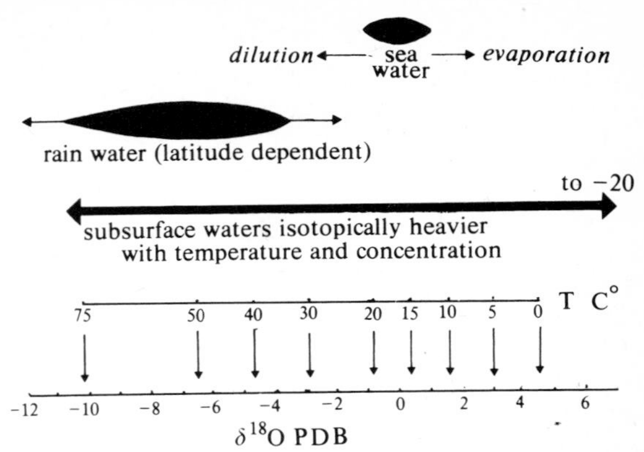

Figure 27.3 Factors that control oxygen isotope values in natural waters (after Hudson 1977a). The temperature scale is valid for precipitation of carbonates of $\delta^{18}O$PDB shown on the lower scale from water of $\delta^{18}O$SMOW = 0 (average sea water).

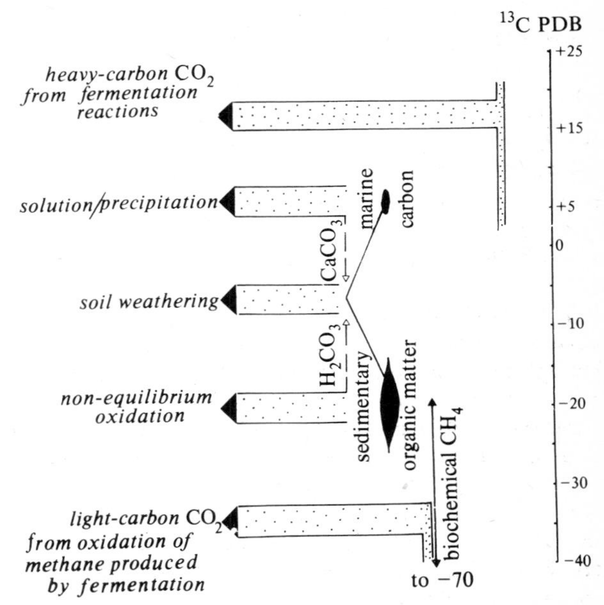

Figure 27.4 Factors that control carbon isotope values in $CaCO_3$ precipitated from natural waters (after Hudson 1977a with Hudson's personal corrections).

extrapolating this composition back in time to non-glacial epochs a small correction (~1‰) must be made to account for the volume of isotopically light water stored as glacial ice. Rain water derived by evaporation of sea water is depleted in ^{18}O, the more so at high latitudes or altitudes, and hence shows $-\delta^{18}O$, a composition shared by near-surface ground waters that have been directly derived from rain water. Conversely, brines that are the residues of evaporation of normal sea water are enriched in $\delta^{18}O$ and hence show $+\delta^{18}O$. Deep formation waters

(Ch. 27h) have undergone the most complex changes since their entombment as sea water within deposited sediments and may show quite variable isotopic ranges.

$CaCO_3$ concentrates $\delta^{18}O$ relative to the water it is precipitated from and this process is negatively temperature dependent. This provides a palaeotemperature scale for marine carbonates and also means that temperature increase caused by burial tends to decrease the $\delta^{18}O$ of late-diagenetic carbonates (Ch. 29).

Biochemical fractionation processes, including photosynthesis, play the dominant role in controlling carbon isotopic ratios and the divergence of $\delta^{13}C$ values from the zero value of the PDB marine calcium carbonate standard. Atmospheric CO_2 has a $\delta^{13}C$ of −7 but the ^{13}C:^{12}C ratio in plants is some 2% lower than in the atmosphere because of more frequent collisions between the more energetic $^{12}CO_2$ molecules and the photosynthesising systems compared with $^{13}CO_2$. Thus organic carbon is isotopically light, averaging about $\delta^{13}C = -24$ for land plants. Reactions between such light carbon present in organically derived acids and marine carbonates will produce light calcium carbonate precipitates, as in soil limestones and marine limestones undergoing diagenesis in the freshwater zone (Fig. 27.5). Thus

$$\underset{\text{marine}}{CaCO_3} + \underset{\text{organically derived } CO_2}{H_2CO_3} \rightarrow Ca(HCO_3)_2 \qquad (27.6)$$

$$\delta^{13}C \approx 0 \qquad \delta^{13}C \approx -24 \qquad \delta^{13}C \approx -12$$

Further reaction of the Ca $(HCO_3)_2$ with soil-derived CO_2 will cause $\delta^{13}C \rightarrow -24$ or with atmospheric-derived CO_2 causing $\delta^{13}C \rightarrow -9$.

Very light carbon down to $\delta^{13}C = -80$ is produced in methane (CH_4) derived by anaerobic *bacterial fermentation* of organic materials in marsh environments or in organic-rich muds of the shallow subsurface in marine environments. Oxidation of this methane liberates light carbon for uptake into calcium carbonate lattices. By way of contrast the heavy bicarbonate residue produced by fermentation reactions causes heavy carbon to be made available during carbonate precipitation (Curtis 1977).

The various possibilities summarised above and in Figures 27.3 to 27.5 provide most valuable criteria for the determination of the site of diagenetic precipitation processes (further discussion in Chs 28 & 29).

Problems in stable-isotope studies include (a) the extent of post-crystallisation ionic diffusion of particular isotopic species (usually assumed negligible), (b) interpretation of individual mineral reactions from 'whole rock' analyses (overcome in recent studies by extracting single mineral

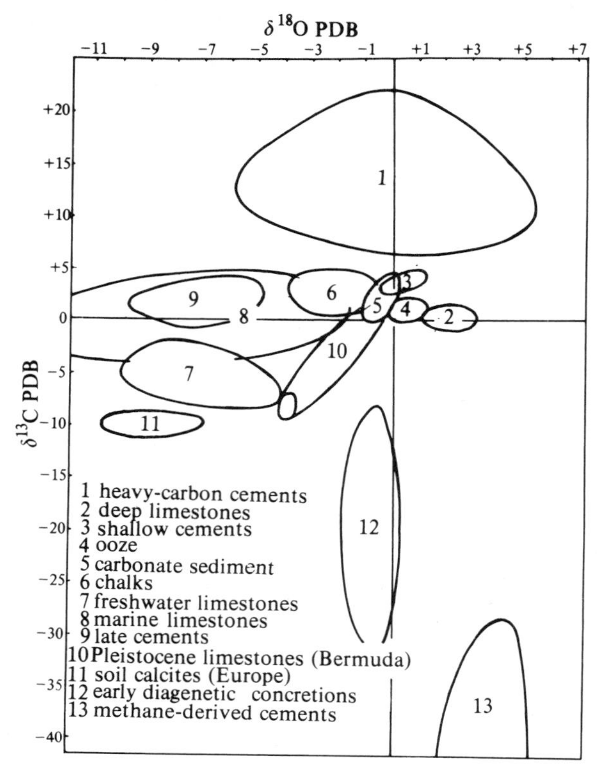

Figure 27.5 ^{13}C–^{18}O isotope grid to show the compositional ranges of $CaCO_3$ sediment, rock and cements (after Hudson 1977a).

samples by dissection; see Dickson & Coleman 1980), and (c) uncertainty as to the pre-Mesozoic isotopic composition of the oceans (unresolvable).

27e Eh–pH phase diagrams in diagenetic studies

We have already briefly introduced the concepts of Eh–pH diagrams in Chapter 1. When the stability fields for various mineral and ionic solution phases are transferred onto such diagrams, valuable information is provided for understanding diagenetic and depositional mineral stabilities (Garrels & Christ 1965). For every oxidation–reduction system we can write the following reaction

$$\text{reduced state} = \text{oxidised state} + n \text{ electrons}$$

The Eh of the reaction may be calculated using the Nernst equation. For example, if we consider the stability of water at 25°C and 1 atmosphere,

$$2H_2O = 2H_2 + O_2 \tag{27.7}$$

We can write this equation in terms of hydrogen ions and/or electrons so that

$$2H_2O = O_2 + 4H^+_{aq.} + 4e \tag{27.8}$$

which gives the relationship

$$Eh = 1.23 - 0.059\ pH \tag{27.9}$$

from the Nernst equation for the equilibrium between water and oxygen at a partial pressure of 1 atmosphere. The equilibrium is represented as a straight line on an Eh–pH plot, with a slope of –0.059 volts per pH unit and an intercept of Eh = +1.23 (Fig. 27.6a).

As further examples of the construction of Eh–pH phase diagrams let us consider the stability of iron, magnetite and haematite in the presence of water at 25 °C and 1 atm total pressure.

For native iron

$$\underset{\text{iron}}{3Fe} + 4H_2O = \underset{\text{magnetite}}{Fe_3O_4} + 8H^+ + 8e \tag{27.10}$$

$$Eh = -0.084 - 0.059\ pH \tag{27.11}$$

and for magnetite

$$\underset{\text{magnetite}}{2Fe_3O_4} + H_2O = \underset{\text{haematite}}{3Fe_2O_3} + 2H^+ + 2e \tag{27.12}$$

$$Eh = 0.221 - 0.059\ pH \tag{27.13}$$

The two lines defined by Equations 27.11 and 27.13 are plotted on Figure 27.6a. We have now defined the stability fields of water, haematite + water and magnetite + water as functions of pH and Eh. Note that the line for native-iron–magnetite stability falls below the water stability line for pH_2 of 1 atmosphere and therefore the reaction cannot take place stably in the presence of water, i.e. the stability field of iron cannot be reached in the presence of water, if equilibrium is maintained.

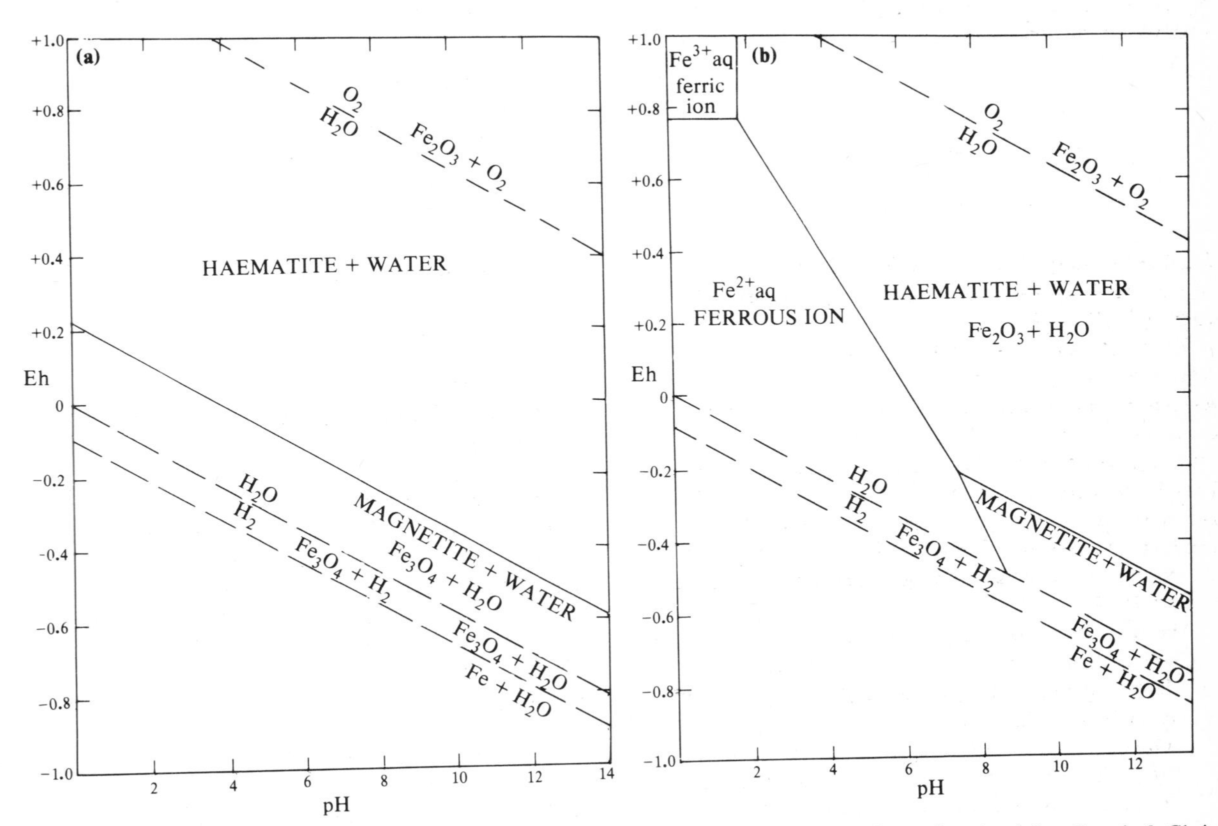

Figure 27.6 (a & b) Eh–pH diagrams constructed for the system Fe^{2+}, Fe^{3+}, H_2O; see text for explanation (after Garrels & Christ 1965).

Now, it is very important to realise that Figure 27.6a is of little further use unless we consider the extent to which dissolved iron ionic species may be present in equilibrium with the various solid mineral phases such as magnetite and haematite. If the activity of a dissolved species in equilibrium with a given solid is less than about 10^{-6}, then the solid will behave as an immobile constituent in its environment (Garrels & Christ 1965). This value is valid for depositional waters at or above the sediment/water interface, but since cationic activities are maintained at much higher levels in interstitial waters below the interface, a value of 10^{-3} will be more appropriate in diagenetic systems (Curtis & Spears 1968). Let us briefly examine the equilibrium of the ferric ion with haematite

$$\underset{\text{haematite}}{Fe_2O_3} + 6H^+_{aq.} = \underset{\text{ferric acid}}{2Fe^{3+}} + 3H_2O \qquad (27.14)$$

The equilibrium constant for this reaction may be found from thermodynamic data which eventually yields the relation

$$\log [Fe^{3+}] = -0.72 - 3\,pH \qquad (27.15)$$

The log of ferric ion activity in equilibrium with haematite is thus a linear function of pH alone and contours of Fe^{3+} will lie parallel to the ordinate of Figure 27.6b.

Similar exercises for the ferrous ion with haematite and magnetite (in which Eh is also a variable in Equations such as 27.15) yield Eh–pH phase diagrams such as that in Figure 27.6b in which fields of dissolved ions and solid phases are defined. Although the details do not concern us here (see Ch. 30) note the small field of stability for Fe^{3+} and the large fields for Fe^{2+} and haematite. Even now, however, we cannot be satisfied that the phase boundaries and mineral stability fields are complete, since we have completely ignored the presence of two abundant and important anions, HS^- and HCO_3^-. The activities of both these species will exert further controls upon the types of diagenetic mineral products expected to be present (Curtis & Spears 1968). To be precise these anions may encourage pyrite and siderite stability respectively (Ch. 30).

Regarding typical Eh–pH values for depositional and interstitial waters we may generalise by stating that marine depositional waters are usually characterised by positive Eh and are slightly alkaline, whereas interstitial waters are characterised by negative Eh and approximate neutrality. The contrast in Eh values is especially important in understanding the nature of diagenetic reactions since ions and solids liberated from oxygenated weathering zones or newly formed in sea water or at the sediment/water interface must almost inevitably be subjected to reduction as deposition continues.

In conclusion we may say that the stability of natural diagenetic minerals depends not simply upon Eh and pH as implied by the basic form of the phase diagram, but also upon the activities of dissolved HS^-, HCO_3^- and metallic cations present in the diagenetic environments.

27f Compaction and fluid migration

The compaction of deposited sediment is predominantly a pressure-controlled mechanical process which causes the centres of overlying and underlying grains to be brought closer together with time. The sediment is, in effect, subjected to *pure strain* by loading stresses. Compaction causes a reduction in porosity, permeability and electrical conductivity, and an increase in bulk density and seismic velocity. Fine-grained sediments comprising organic matter and clay minerals compact most, coarse-grained sediments least. Compaction is always accompanied by fluid expulsion, the expelled fluids playing a vital role in transferring ions and organic molecules from 'donor' mudstone or lime-mudstone beds into 'acceptor' sandstone or lime-sandbeds.

Very many measurements have been made in coreholes of the change of porosity with depth in fine-grained clastic sediments (Fig. 27.7). Although there is an appreciable

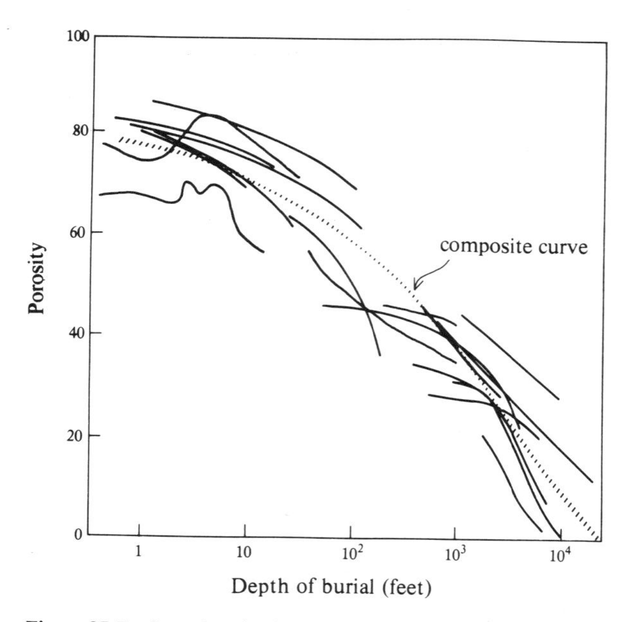

Figure 27.7 Porosity–depth relationships determined from a variety of sedimentary sequences, with Baldwin's composite curve superimposed (after Baldwin 1971).

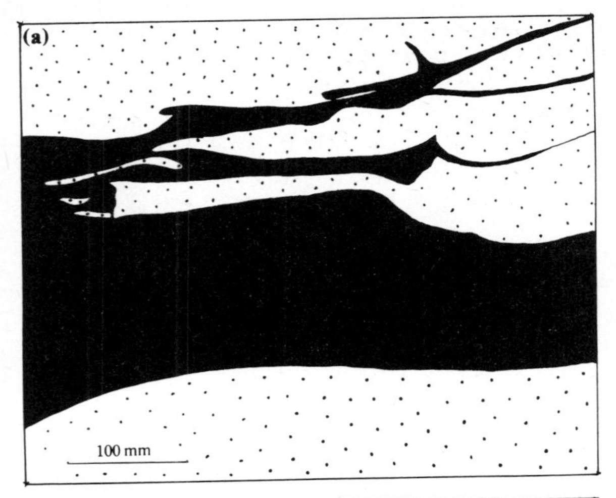

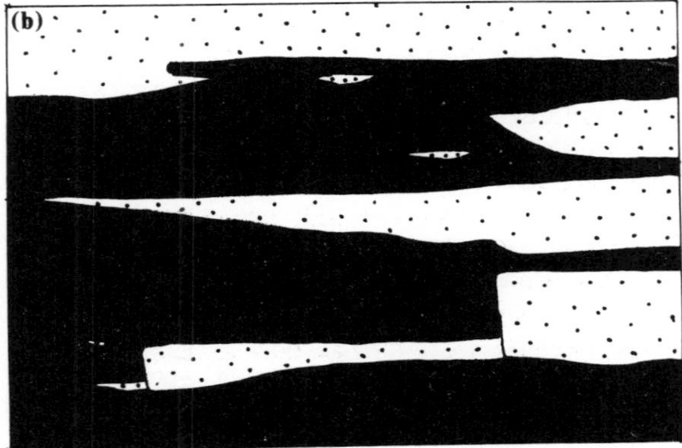

Figure 27.8 The influence of compaction upon laminae shape and thickness: (a) present geometry; (b) precompaction geometry. (After Baldwin 1971.)

scatter in the curves of various investigators a first-order decay rate of the form

$$P = P_0 e^{-cy} \tag{27.16}$$

fits the data trend, where P = porosity at depth y, P_0 = porosity at surface ($y = 0$) and c = constant. The mean porosity/depth curve of Figure 27.7 may be fitted by the polynomial (Bridge & Leeder 1979)

$$P = 0.78 - 0.043 \ln (y + 1) - 0.0054 [\ln (y + 1)]^2 \tag{27.17}$$

This relationship probably also applies to the compaction of fine-grained carbonate sediment since laboratory experiments show compaction rates similar to those of clastic clays. Complications may arise in the natural case, however, since early lithification of carbonate muds (Ch. 29) will tend to inhibit compaction.

The geologist should remember that differential compaction must always have occurred in the successions of interbedded mudstones and sandstones that he observes in the stratigraphic record. The severe effects that result when channel-shaped sandstones are compacted are illustrated in Figure 27.8. It is useful to apply decompaction procedures in such cases and also in examples where information on the changing rate of sediment thickness with time is required (Perrier & Quiblier 1974).

In many areas of the world, zones of **undercompacted** sediments are known to exist at depth. Such zones are abnormally pressured and are particularly characteristic of rapidly deposited deltaic sediments preserved in deep sedimentary basins. Under normal conditions fluid pressures in rock pores are hydrostatic (Ch. 27b), i.e. in equilibrium with the weight of a salt water column extending to the surface. If the fluid in adjacent rock pores is isolated, compaction is prevented and the pressures are geostatic. Horizons of rapidly deposited muds sealed by impermeable horizons or by differential compaction and/or cementation will therefore show overpressuring and exhibit higher porosity, lower density, lower seismic velocity, increased rate of drilling penetration, increased temperature, increased electrical conductivity and lower salinity than overlying or underlying normally pressured rocks (Figs 27.9 & 10). The overpressured horizons present major problems to oil drilling (blowouts, well caving) and cause the subsurface development of mud diapirs because of the anomalously low density of the overpressured units.

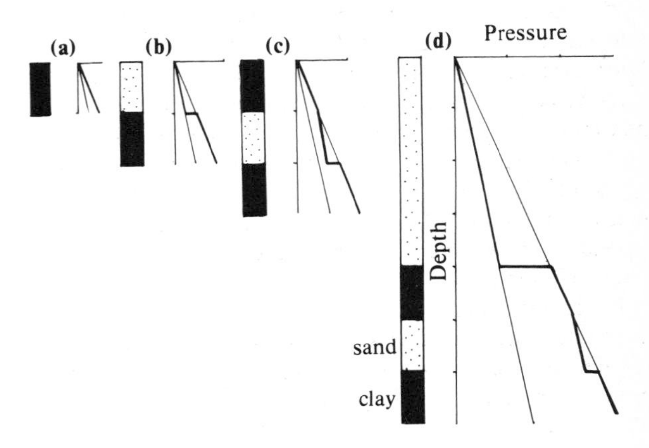

Figure 27.9 Schematic graphs (nominal scales) to show the development of overpressuring in alternating porous sand and mud formations during progressive (a–d) burial (after Chapman 1976).

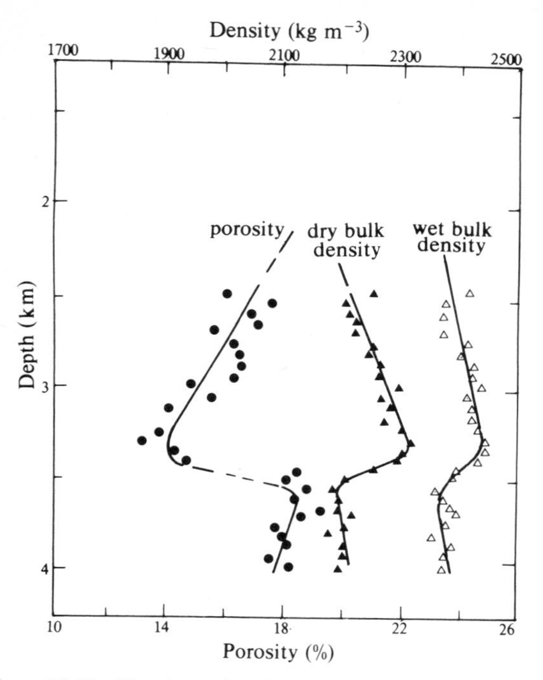

Figure 27.10 To show the sharp increase of porosity and decrease in shale density at the top of an overpressured horizon in the Manchester Field, Calcasiev Parish, Louisiana, USA. (After Schmidt 1973.)

In addition to the compactional explanation for overpressuring, three other factors must be taken into account. The first is **aquathermal pressuring** (Barker 1972) caused by the pressure increase produced by the expansion of water in a closed pore space by heating. The second is **montomorillonite dehydration** (Powers 1967), applicable only to clastic mudstones, in which water is given off to pores during the burial reaction montmorillonite → illite (Ch. 28). The third is **methane generation** (Hedberg 1974), caused by the subsurface addition of methane gas to pore volumes generated by biochemical and thermochemical processes (Ch. 31). Overpressured muds are frequently very rich in natural gas pockets.

Before we finally consider the timing and routes of pore-fluid migration it is necessary to discuss briefly the origin of the fluid phase. Formation water is a useful nongenetic term for any sort of aqueous phase present in sediment or rocks (White 1965). **Meteoric** water, i.e. water recently involved in atmospheric circulation, may be present in relatively near-surface levels. **Connate** or fossil water has been out of contact with the atmosphere for at least an appreciable part of a geological period and it consists of fossil interstitial water plus water driven from adjacent sediments. Original ocean water evolves to

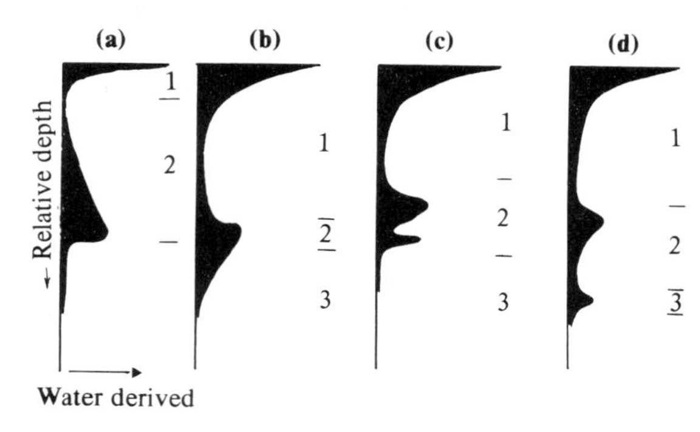

Figure 27.11 Schematic patterns of water escape from compacting muds, as postulated by (a) Powers (1967), (b) Burst (1969), (c) Perry & Hower (1972), high geothermal gradient, (d) ibid., low geothermal gradient. Zone 1 – mechanical compaction; zone 2, 3 – smectite lattice de-watering (stepped process?).

connate water in response to complex chemical changes as the fluid migrates from compacting fine-grained sediments and travels along fluid-potential gradients. Connate water is usually very much enriched in salts compared with sea water (Table 27.1) and may be as dense as 1100 kg m^{-3}. Further discussion of the process whereby this enrichment occurs may be found in Chapter 28.

In the beginning of this section it was implied that compaction, and therefore water expulsion, was a continuous process. It is probable, however, that water expulsion arises not only from interstitial fluid but also from the thermally activated explusion of interlayer montmorillonite water at deeper burial depths when most of the interstitial water has been expelled (Fig. 27.11). This second aqueous phase will initially consist of almost pure water, in contrast to the brines produced during progressive filtration of original sea water through the clay particles of compacting mudstone beds.

Compaction of an homogeneous clastic or carbonate mud succession or of a mud succession with perfectly horizontal clastic or carbonate sand interbeds will cause a predominant upward flow of expelled fluid with local downward flow along pressure gradients if overpressuring occurs. Fluid migration will always occur normal to surfaces of equal pore pressure, down the local pore-pressure gradient (from high to low pressure). In mud successions overlain by carbonate or clastic sands, or with sands dipping up towards the basin margin, flow will occur from the muds into and then along the permeable sand beds up towards the basin margin (Fig. 27.12, Magara 1976). In both cases the expelled water and brine from the fine-grained sediment will carry ions and oil and gas molecules into permeable and porous horizons where precipitation of diagenetic minerals or entrapment of oil may occur (Chs 28, 29, 31).

Table 27.1 Major constituents of chemical analyses of some saline 'connate' formation waters, with oceanic water for comparison (after White 1965).

	1	*2*	*3*	*4*	*5*	*6*	*7*	*8*	*9*	*10*
Brine type	*Ocean*	*'Type connate'*	*Connate Na–Ca*	*Connate Na–Ca*	*Connate Na–Cl(?)*	*Connate NaCl(?)*	*Connate Na–Ca*	*Connate Na–Ca*	*Connate Na–Ca*	*Connate Na–Cl(?)*
SiO_2	7.0	<10	6.9	22	52	47	14	80	63	20
Al	1.9	5.4	1.5	4.1	0.9	0.4	0.2	30	1.2	0.6
Fe	0.02	1.0	5.6	0.0	6.4	1.2	61	2.7	0.1	15.0
Mn	0.01	2.5	0.2	2	0.3	0.08	2	0.2	7.0	0.05
Ca	400	62 900	10 100	5750	325	373	3040	3400	12 200	57
Mg	1272	179	1920	1070	123	115	49	43	275	23
Sr	13.3	320	279	456	21	82	66	40	320	9
Ba	0.05	4	<2	3	7.2	7.1	8.7	1.4	110	13
Na	10 560	11 900	42 000	31 500	6150	5820	6710	4310	13 600	6300
K	380	38	323	585	136	132	113	80	404	11
NH_4	0.07	<10	42	140	45	51	51	23	134	11
HCO_3	140	24	72	140	666	535	287	795	80	1010
SO_4	2649	88	990	180	4.1	1.6	31	1030	16	1.1
Cl	18 980	128 000	90 300	60 400	9940	9840	15 300	11 100	44 000	9690
Br	65	997	347	257	35	30	46		238	128
I	0.05	3.2	17	12	20	23	29		56	105
total as reported	34 475	204 000	146 440	100 579	17 700	17 100	25 900	21 045	71 576	17 407
specific conductance	–	–	147 000	109 000	25 200	25 800	31 800	28 500	–	27 200
pH (lab. determination)	8.1	6.5	6.2	6.8	7.1	7.5	7.2	6.48	5.70	7.6
temperature producing zone, °C	(~4.0)	–	–	49	49	81	81.4	89	104	–
density, 20°C, kg l^{-1}	(~1021)	1174	1112	1073	1009	1009	1016	1016	1054	1009
evaporated residue 180°C, ppm	–	225 000	157 000	106 000	17 800	18 600	28 500	26 800	79 100	17 000

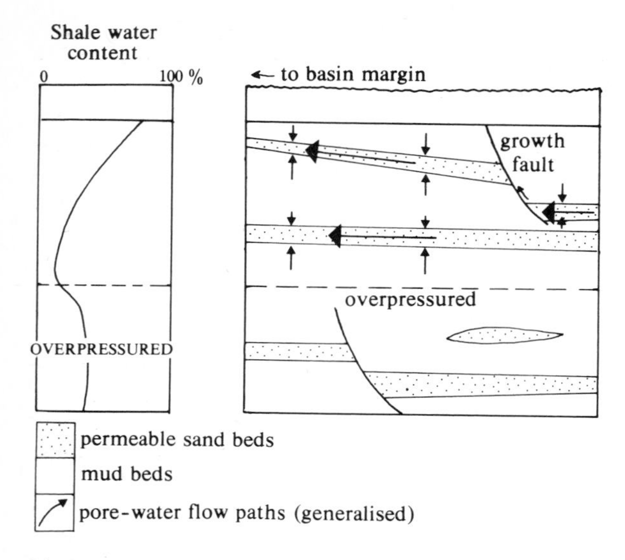

Figure 27.12 Schematic sections to show mud water content versus depth towards overpressured shales and the flow paths of expelled water during compaction (after J. B. Hayes 1979).

27g Pressure solution

Thin-section examination of many clastic and carbonate rocks reveals frequent occurrences of grain-to-grain **sutured** contacts which indicate that significant volumes of solid material have somehow been removed. Spectacular examples on a larger scale are provided by pitted pebbles in conglomerates and by the laterally extensive **stylolites** seen in limestones. These features are not restricted to deeply buried sediments but may occur in tectonically stressed rocks where they are associated with metamorphic mineral reactions and production of spaced-cleavage zones.

The above fabrics are caused by the phenomenon known as **pressure solution** or solution transfer. Pressure solution arises because parts of a solid under stress show a higher solubility than those free of stress. Therefore around the perimeter of grains under a load stress there will be high stresses at grain contacts and normal (hydrostatic) stresses around the fluid-filled perimeter of the pore space. The stress gradient so produced gives rise to a gradient in chemical potential, and diffusive transfer of material from the areas of high stress to those of low stress dissipates this gradient (Durney 1972, 1976, de Boer 1977). Mineral material at grain-contact points thus dissolves and grain material in contact with pore solutions grows, i.e. the pore space is gradually infilled, leading to a loss of porosity and an increase in strength as the pressure-solution process decays.

There are two theories for the mechanism of the dissolution/precipitation process. Bathurst's theory states that dissolution can occur only in places where the grains are in direct contact with the pore fluid. The rims of the grain-to-grain contacts are then preferentially leached because these locations are subjected to high shear stresses. The rim then dissolves and the leached grain-to-grain contact collapses, the small grain fragments dissolve and the cycle starts again. Weyl's theory states that normal compressive stresses, not shear stresses, cause dissolution. Since compressive stresses are only effective inside the grain-to-grain boundaries and since dissolved material must be able to reach the pore solutions, then an adsorbed water layer must be present over the entire grain/grain contact area. This layer must be strong enough to withstand pressure differences in the diagenetic realm. The dissolved ions diffuse through the adsorbed water layer into the pore space under hydrostatic pressure where mineral material can precipitate.

Theoretical and experimental evidence comes out strongly in favour of Weyl's theory (de Boer 1977). The experiments simulating pressure-solution processes in quartz sands show (a) water is a prerequisite for pressure solution, (b) pressure-solution effects are increased by an increase in temperature, (c) pressure solution is not affected by pore-water composition, and (d) pressure solution causes supersaturation in free pore spaces high enough to give precipitation. In terriginous clastic sediments the diffusion of dissolved silica from the pressurised locations inside the grain-to-grain contacts to the pore solution emerges as the rate-determining step in the pressure-solution process. There is still some uncertainty, however, as to the exact route of 'escape' for dissolved ions. Many 'pitted' pebbles show a marginal zone around the 'pit', much altered by chemical change, implying ion diffusion through the altered zone, perhaps encouraged by microfractures (McEwan 1978).

27h Diagenetic realms

Four main diagenetic realms may be distinguished based upon the nature of the pore-filling aqueous phase (see Folk 1973).

The **vadose** subdivision of the **meteoric-derived** realm lies above the continental water table, defined as the level of pore saturation with water. Pore spaces are thus in contact with atmospheric gases and the partially water-filled pores show positive Eh conditions. Near-surface diagenetic reactions in deposited sediments may sometimes be strongly affected by soil-forming processes involving biogenic contributions.

The **phreatic** subdivision of the meteoric-derived realm

lies below the continental water table. There is also a phreatic subdivision of the near-surface marine realm. Pore spaces are permanently filled with pore water, showing negative Eh conditions. Pressures and temperatures do not substantially depart from those at the Earth's surface. Movement of pore water occurs in response to fluid-potential gradients, and a cycle is set up involving rainfall, seepage, subsurface flow within reservoirs, outflow at the surface, and evaporation. Meteoric water in the phreatic zone is characterised by a particular, though variable, residence interval within the subsurface, and during this time it is not substantially altered in composition from rain water unless subjected to evaporation, as in playas or on sabkha surfaces.

The **marine-derived** realm shows a zone of mixing where it is in contact with the meteoric realm. Apart from a narrow intertidal zone, pore space is entirely filled by waters of oceanic composition. Pressures and temperatures do not substantially differ from those at the Earth's surface. A very thin upper zone in contact with ocean water shows positive Eh conditions, elsewhere the pore waters are usually reducing. Pore-fluid chemistry is very strongly influenced by bacterial oxidation and reduction reactions. Stagnant marine basins may show euxinic characters with a thick water mass dominated by reducing conditions (Ch. 24). Coasts adjacent to areas of upwelling, such as Peru or Namibia, are dominated by phosphate production. Deep oceanic areas will include a contribution from primary 'hydrothermal' fluids of exhalative type, particularly in areas close to active spreading centres.

The **subsurface (burial)** realm is the deepest and most extensive (and least known) diagenetic environment. It has a gradational upper contact with marine- and meteoric-derived waters. As noted previously the composition of the waters present in this diagenetic realm differs substantially from that of fresh or sea water since compactional strain has forced the original fossil sea water through a sort of filter press which has changed the composition irreversibly. Subsurface fluids take no part in the meteoric cycle and are subjected to temperatures and pressures that may substantially exceed those of the Earth's surface. Eventually (conventional wisdom tells us) there comes a burial depth when extensive recrystallisation of sedimentary minerals begins to occur and the process of metamorphism takes over from diagenesis. It is a moot point, however, exactly where and why any distinction between the two processes should be made.

27i Summary

Diagenesis refers to the sum of chemical and physical processes which act upon deposited sediment grains. Once buried, the sediment is acted upon by increased temperatures and pressures. Cementing minerals are precipated within pore space by migrating pore fluids. Stable-isotope analysis of cements enables interpretation of the physical and chemical conditions operating during progressive precipitation to be deduced. The stability of diagenetic minerals depends not simply upon Eh and pH but also upon the activities of dissolved HS^-, HCO_3^- and metallic cations present in the pore waters. Massive amounts of water are given off during compaction of muds and as a result of the montmorillonite → illite change during deep burial. These waters act as carriers for inorganic and organic ions and compounds formed during diagenesis.

Further reading

Several sections on the general chemistry of diagenetic matters are to be found in Krauskopf (1979). More advanced accounts are given by Berner (1971, 1980). Hudson (1977a) provides the best general introduction to stable-isotope analysis in diagenetic studies whilst Faure (1977) is the most rigorous elementary account of chemical principles of stable isotope behaviour. Garrels and Christ (1965) is a classic and essential reference on many aspects of chemical diagenesis, including the uses and misuses of Eh–pH diagrams.

28 Terrigenous clastic sediments

28a Introduction

Until very recently there was a tendency to apply diagenetic studies to carbonate and evaporite sediments, with only a passing nod of acknowledgement towards clay mineralogists. Remarkable progress in understanding terriginous clastic diagenesis in the past ten years has come about chiefly from the impetus provided by studies on reservoir properties inspired by the oil industry and from the application of scanning electron microscopy, electron-microprobe and stable-isotope techniques. Although progress has been great on the observation and interpretation side, there is still a significant lack of unified diagenetic theories for clastic sediments, a state of affairs that not only results from the great complexity of clastic diagenesis but also from the continued tendency of workers to study numerous case histories on an '*ad hoc*' basis. A further drawback to many studies is the too-rigid separation of sandstone and mudstone diagenetic processes. As already discussed (Ch. 27), such separation is unfortunate since compacting mud sequences provide the fluids and ions in solution that must eventually be forced through porous sandstone horizons, causing pore-filling cementation to occur. A final point concerns the great importance of diagenetic studies in shedding light on rational sandstone classifications. The composition of lithified sandstones is clearly the product of a source area, a transport mechanism, an environment of deposition *and* a diagenetic history. The scientist aware of Blake's exhortation 'To see a world in a grain of sand' should be reminded that the pore-filling matrix and cement surrounding the grains of sand are just as revealing!

28b Marine mud diagenesis

Information concerning the early stages of mud diagenesis comes from studies of near-surface deposits, whereas the later stages, as the mud becomes an anhydrous mudstone, must be studied in the geological record. Needless to say, the early diagenesis of marine muds is known in much greater detail. There is particular ignorance concerning the timing and mechanism of diagenesis of organic material and the production of hydrocarbons. These processes are intimately interwoven with mud diagenesis and fluid expulsion; we shall discuss them further in Chapter 31. Similarly, the origin of diagenetic iron and manganese minerals in mudstones will be discussed in more detail in Chapter 30.

Figure 28.1 and Table 28.1 present a general summary model (mostly after Curtis 1977) for marine mud diagenesis based upon a number of depth-defined zones.

Zone 1 is a very thin zone where the pore waters of the freshly deposited muds are kept oxygenated by downward diffusion of dissolved oxygen from the overlying depositional waters. Oxidation of organic material by bacteria produces isotopically light CO_2, but carbonate supersaturation is unlikely to occur because of bicarbonate diffusion into the depositional waters. Porosity may be up to 80% in this zone, with the undisturbed mud structure dominated by flocculated and pelleted fabrics. Bioturbation in zone 1 and in the top few centimetres of zone 2 plays an important role in controlling levels of sulphate activity and also may physically change the flocculated clay fabric into aggregates of faecal pellets.

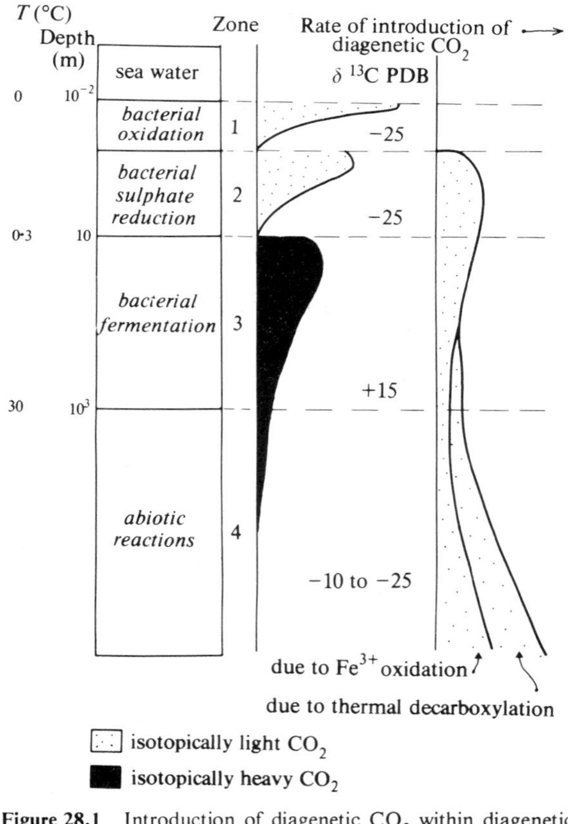

Figure 28.1 Introduction of diagenetic CO_2 within diagenetic zones 1–4 in mudstone sequences (after Irwin *et al.* 1977). The isotopic signatures of each zone are preserved in carbonate concretions (see text for discussion).

Table 28.1 Diagenetic zones for marine mud successions. ΔT°C – increase in temperature with depth below sediment/water interface due to a gradient of 27.5°C km^{-1} (after Curtis 1977).

Depth (km)	*ΔT (°C)*	*Porosity (%)*	*Diagenetic zones (minerals formed)*
			1 **oxidation**
0.0005			
			2 **sulphate reduction** pyrite calcite dolomite (low Fe carbonates ^{12}C enriched) kaolinite? phosphates?
0.01	0.2	80	
			3 **fermentation** high Fe-carbonates calcite dolomite ankerite siderite ^{13}C enriched
1.00	28	31	
			4 **decarboxylation** siderite
2.50	69	21	
			5 **hydrocarbon formation** (a) wet – oil (b) dry – methane montmorillonite → illite (a) disordered, (b) ordered
7.00	192	9	
			6 **metamorphism** (a) 200°C chlorite (b) 300°C mica, feldspar, epidote?

Clay minerals resting on the sea floor or within zone 1 may have time to equilibrate with sea water; degraded illites, for example, may take up K^+ from sea water into their lattices so as to restore their stoichiometric compositions.

Zone 2 is dominated by the bacterial reduction of sulphate anions (Goldhaber & Kaplan 1974) present in pore waters diffused from the overlying depositional waters. The bacteria is the genus *Desulphovibrio*. The zone of sulphate reduction may extend down to a depth of about 10 m but is best developed in the top 0.5 m or so. The reduction of sea water sulphate may be represented in a simple form by

$$\underset{\text{organic matter}}{2CH_2O} + \underset{\text{sulphate}}{SO_4^{2-}} \xrightarrow[\text{catalysts}]{\text{bacterial}} \underset{\text{light } \delta^{14}C}{2CO_2} + \underset{\text{sulphur}}{S^{2-}} + 2H_2O \qquad (28.1)$$

giving rise to isotopically light bicarbonate and hydrogen sulphide. The pore-water sulphate is gradually exhausted with depth and the reaction ceases. Biological sulphate reduction is an anoxic process which may occur in stagnant depositional water bodies, but it is most common in the diagenetic environment.

The hydrogen sulphide released by bacterial reduction is toxic to all respiratory organisms and, together with the other products of bacterial metabolism (CO_2, NH_3, PO_4), is chemically reactive so that diagenetic processes such as metal sulphide precipitation, carbonate precipitation, pH modification and methane generation are all results of bacterial activity. (*NB*. The transfer of dissolved sulphate from the oceans to precipitated sulphides in sediments is a major component of the balancing system which keeps the sulphur content of ocean water roughly constant and in balance with that supplied by erosion of the continents.) As we shall discuss further in Chapter 30, ferric ions are reduced to ferrous, with pyrite formation eventually resulting from the additional bacterially produced sulphide. Studies of the inner zones of Jurassic calcite concretions thought to have formed in zone 2 reveal relict pelletal textures with pyrite enrichment in the pellets reflecting the abundance of organic material for bacterial reduction (Hudson 1978). The oxygen isotope data shows clearly that the concretions grew within contact of sea water in the sediment pores. The light carbon isotopes indicate a source of bicarbonate from oxidation of organic material involved in the sulphate reduction process. The non-ferrous nature of some of the concretionary calcite is attributed to the low Fe^{2+} activity encouraged by precipitation of insoluble iron sulphides, although the exact reason for the non-partition of Fe^{2+} between carbonate and sulphide remains unclear. Other concretions may be ferroan and some may have originally been of high-Mg calcite composition because of their significantly high Mg contents (>2 wt %).

In *Zone 3* sulphate is exhausted and the precipitation of isotopically light carbonate as concretions and of pyrite ceases. Organic fermentation reactions now begin as sulphate reduction ceases, possibly because the bacteria involved cannot tolerate dissolved sulphides. Fermentation reactions may be approximated by

$$\underset{\text{organic material}}{CH_2O} \rightarrow \underset{\text{methane (light carbon)}}{CH_4} + \underset{\text{carbon dioxide (heavy carbon)}}{CO_2} \qquad (28.2)$$

Fermentation is characterised by marked carbon isotope fractionation leading to very light methane (down to −75‰ PDB) and heavy carbon dioxide (~ +15‰). Precipitation of heavy carbonate in the presence of ferrous ions produced by further iron reduction produces the mineral phases ferroan calcite, ferroan dolomite, ankerite and siderite in concretionary bodies. Studies of such concretions from progressively deeper levels in the Jurassic Kimmeridge Clay of England (Irwin *et al.* 1977) shows a gradual lightening of the $\delta^{13}C$ values from above +8 to −6. This trend is due to the decreasing importance of fermentation reactions as zone 3 is descended and the increasing importance of abiotic reactions that dominate zone 4 and provide light carbon again. Bacterial fermentation is believed to extend down to a depth of about 1 km when it is prevented by either increasing temperatures or exhaustion of suitable organic materials.

In successions containing basic or intermediate submarine volcanics the low-temperature shallow surface alteration of the volcanics, especially any glassy shards, leads to the formation of smectite clay minerals, causing Mg-depletion and Ca-enrichment of the pore waters. Volcanic ash layers are thus progressively altered to smectite-rich bentonites with time.

Zone 3 is also the depth range within which much of the interstitial pore water in mud sediments is expelled. This is an exponential process tailing off at depths of about 1.5–2 km. At this stage the mud density will be increased from about 1320 to 1960 kg m^{-3} and the porosity reduced from 70–80% to about 10–20%. It must be emphasised however that up to 30% of water by volume still remains in the compacting muds, the majority as clay interlayer water (Fig. 28.2). Pore-water expulsion continues into *Zone 4* which is dominated by inorganic processes breaking down the remaining organic material. One such process is decarboxylation which is represented by

$$R.CO_2H \rightarrow RH + CO_2 \qquad (28.3)$$

giving rise to light carbon dioxide again (Curtis 1977). Most models of clay dehydration propose that at these depths (1–2.5 km) much of the pore water has already been driven off.

Zone 5 is an exciting but little known interval where major mineralogical changes begin to take place amongst clay mineral species and when hydrocarbon generation begins in earnest. The zone extends from about 2.5 km to 5 km or more in depth (temperature range of 70–190°C). Studies of deep borehole cores through the mud sediments of the Gulf Coast of the USA (e.g. Perry & Hower 1970, 1972, Hower *et al.* 1976) and in many other areas show a number of changes including the important conversion of mixed-layer montmorillonite–illite clays (with up to 80% montmorillonite) to illite and illite-rich mixed layers. The process is accompanied by a final phase of water loss from interlayer positions which probably provides the main carrier for dissolved hydrocarbon species in primary migration (Ch. 31). The K^+ also required for this change

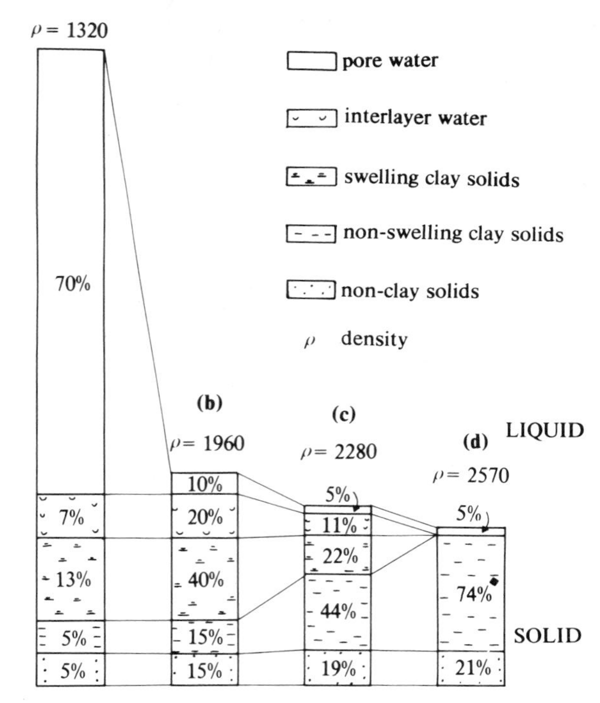

Figure 28.2 The changing aspect of marine mud bulk composition during burial and compaction. (a) Recent muds, (b) after first dehydration, (c) after second dehydration, (d) after third dehydration. (After Burst 1969.) See also Figure 27.11b.

probably comes from the dissolution of tiny K-feldspar particles. It is useful to write the notional equation (Hower *et al.* 1976)

$$\text{smectite} + \text{K-feldspar} \rightarrow \text{illite} + Si^{4+} + Mg^{2+} + Fe^{2+} \qquad (28.4)$$

for this reaction. It is important to realise, after the previous discussion, that the alteration of smectitic to illitic clays, involving the exchange of interlayer water for K^+, cannot be accomplished by pressure alone. Temperature is the main controlling factor since the heat of reaction for clay dehydration is negative and must be supplied by geothermal sources. The temperatures necessary for the initiation of the alteration lie between 70°C and 90°C. There is also a progressive change from disordered to ordered illite within zone 5, the degree of illite crystallinity obtained from X-ray diffraction traces being a good estimator of burial depth (de Segonzac 1970). Many cores show complete dissolution of particulate $CaCO_3$ in zone 5 (Hower *et al.* 1976). It is not known why this should occur since most early diagenetic calcite concretions seem to remain stable in this zone.

In summary we see therefore that zone 5 processes provide H_2O, SiO_2, Mg^{2+}, Fe^{2+} and hydrocarbons as mobile phases. These may be carried upwards by compaction into zone 4 or transferred into interbedded sand or limestone formations as potential sources of cement material (Chs 28e & 29).

Zone 6 represents the incoming of truly metamorphic reactions with illite recrystallised to sericite and hence to muscovite by 300°C and kaolinite recrystallising to dickite or nacrite or combining with Mg^{2+} and Fe^{2+} to form chlorite above 200°C. It is most important to take the depth limits of zone 6 with a 'pinch of salt' since our diagenetic scheme has assumed that pressure is entirely load-produced. Directed pressures resulting from rock deformation may cause metamorphic-type reactions in cleavage zones at relatively shallow depths.

In summary we can say that although the above zonal diagenetic scheme provides a useful framework in which to consider clay diagenesis, the boundaries of temperature, pressure and reaction mechanisms to each zone should be regarded as gradational and approximate – many local conditions may serve to modify them (see below). It is perhaps most important to consider the effects of the rate of sedimentation upon the characters of the resulting clay sediment (Curtis 1977) since this will control the length of 'residence interval' that a particular volume of sediment spends within a particular diagenetic zone (Table 28.2).

A final point in this section concerns the role of clays in changing the composition of the formation waters that are gradually squeezed out during compaction (White 1965). Analyses of formation waters show that clays are not equally permeable to all constituents, some ions have greater 'mobility' than others. The filtering effect is dependent upon the different charges and ionic and 'molecular' radii of the hydrated ions. The salinity of the formation brines (density 1100 kg m^{-3}) increases with age and the $Ca^{2+}:Cl^{2-}$ ratio increases with salinity. Oxygen isotopes prove that the salinity change is not the result of evaporation. The filtration process must inevitably lead to the production of (a) escaping, low-salinity filtered waters and (b) retained, high-salinity concentrated waters. Both groups may take part in diagenetic mineral reactions.

28c Non-marine mud diagenesis

Although many of the diagenetic changes noted in the previous section also occur in non-marine mud successions, there are a number of important differences.

Since fresh waters contain two orders of magnitude less SO_4^{2-} than sea water (Table 2.1), there is correspondingly less potential sulphide to be made available by sulphate-reducing bacteria in diagenetic zone 2 of fresh water subaqueous mud sequences. This lessening of sulphide activity encourages siderite precipitation rather than 'pyrite' precipitation to occur in the near-surface diagenetic zones, provided that carbonate activities remain high (see Ch. 30). The progressive growth of siderite nodules is recorded in boreholes through fresh water marsh and lake muds of the Mississippi delta (Ho & Coleman 1969). Particularly relevant data on siderite concretion growth has come from studies of Pennsylvanian mudstones from Yorkshire, England (Oertel & Curtis 1972, Curtis *et al.* 1972). Combined insoluble residue, fabric analysis and carbon isotope studies of small samples taken across a nodule (Fig. 28.3) reveal initial siderite precipitation as a pore-filling phase within flocculated clay mineral grains. This initial siderite, now seen in the nodule centre, is enriched in heavy carbon which must result from bacterial fermentation reactions, as noted previously (p. 271). It thus seems that fermentation reactions involving organic material occur at shallower depths in fresh water muds than in marine mud, probably because of the lack of bacterial poisoning by sulphide production. Heavy carbon (up to +7.64 $^{13}C_{PDB}$) also occurs in the Holocene siderite nodules from the Mississippi fresh water sequences noted previously. The outer portion of siderite in the Carboniferous nodule includes more detrital material and is composed of progressively lighter carbonate, due to lessening of bacterial activity with depth.

A further feature of non-marine mud facies are the opportunities open for alteration of deposited clay mineral species and other grains by leaching and sub-aerial weathering. For example, acidic weathering in poorly drained, organic-rich marsh soils will rapidly degrade illitic, smectitic or mixed-layer clays to kaolinites. Similar conditions will alter thin volcanic ash layers to **tonsteins** rich in kaolinite (Spears & Kanaris-Sotirious 1979). It seems likely that tonsteins, which are present abundantly over very wide areas in the non-marine coal swamp facies of the European Pennsylvanian, are the sub-aerial equivalents of bentonites. The occurrence of alkaline pore waters on the other hand may encourage the shallow subsurface precipitation of smectitic clays.

28d Classification of mudrocks

Once compacted and lithified, pure clay-grade clastic sediment becomes known as **claystone**. Since it is more usual for fine-grained sediments to contain admixtures of very fine sand-, silt- and clay-sized grains the more general term **mudrock** has come into general use in recent years. It should be noted however that usage of terminology for fine-grained sediments varies widely, mudstone being a common synonym for mudrock. Fissile mudrocks are often referred to as shales, but here again usage varies widely since some authors use the term as yet another synonym for mudrock. Neither claystones nor mudrocks are composed entirely of clay minerals, the amount of quartz and chert grains present averaging about 30% by weight. The average mudrock has a size distribution appropriate to a poorly sorted, medium-to-fine siltstone containing approximately $\frac{1}{8}$th sand-, $\frac{6}{8}$ths silt- and $\frac{1}{8}$th clay-grade grains. Since mudrocks form nearly 75% of all clastic sedimentary rocks it seems clear that the majority by weight of the Earth's detrital silica occurs as silt fractions. In addition to detrital silica, chert and clay-mineral grains, mudrocks may contain significant proportions of fine-grained organic material, in the form of **kerogen** (Ch. 31) and of $CaCO_3$. The various intergradations between the pure end members of these compositions are shown in Table 28.3. Mudrocks contain 95% of the organic material present in clastic sedimentary rocks, the amount varying from 0–40% by weight in particular samples with a grand mean of about 1%. Organic-rich mudrocks are usually dark in colour.

Table 28.2 Links between sediment mineralogy and burial rate in marine muds (after Curtis 1977).

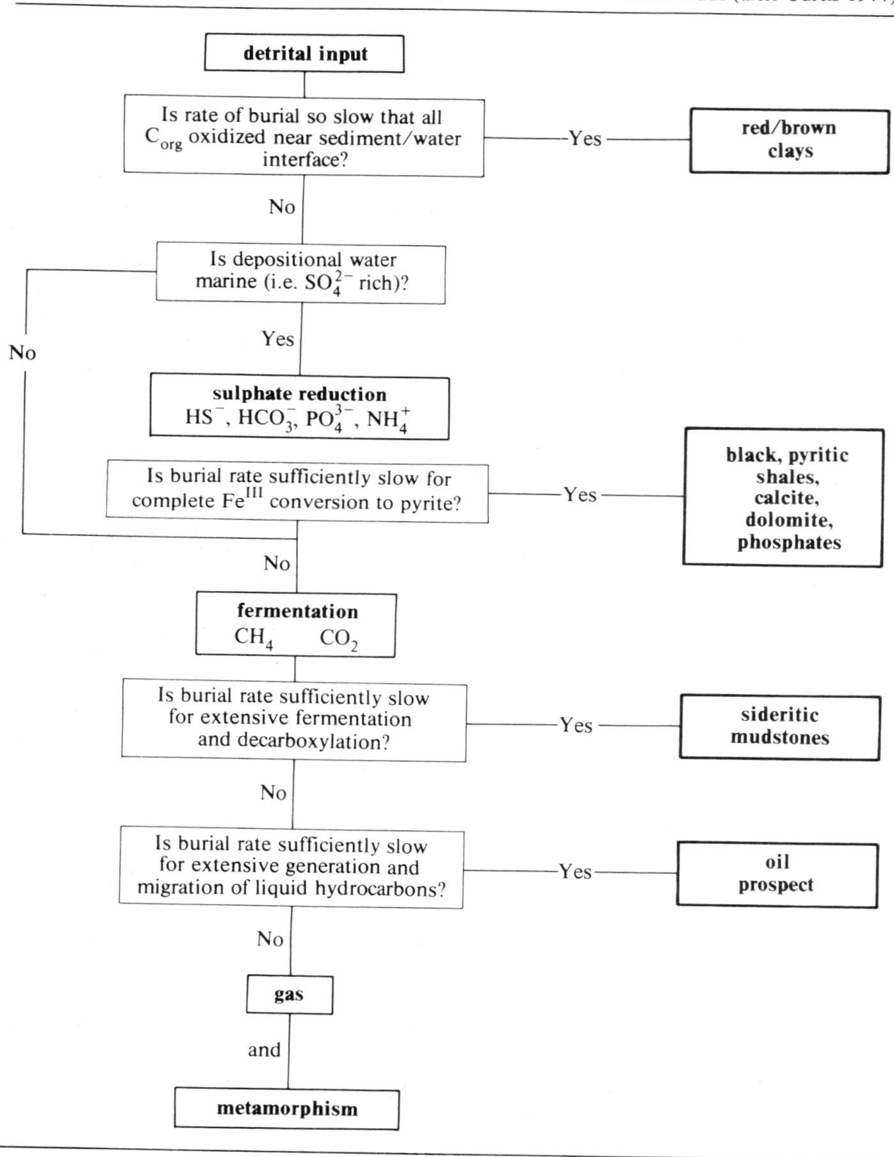

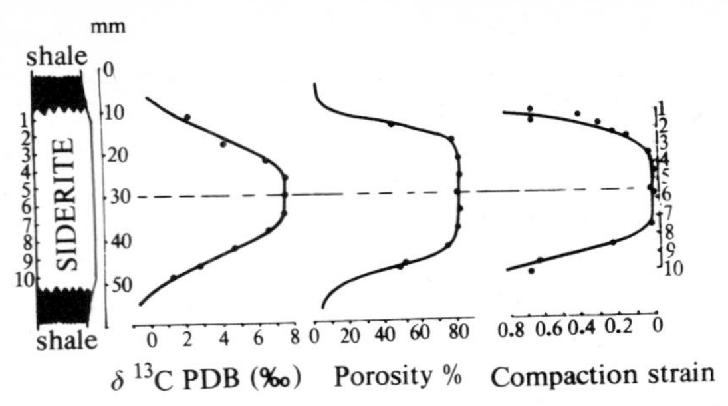

Figure 28.3 Distribution of stable carbon isotopes, porosity and compactional strain in a sideritic concretion in Pennsylvanian (Westphalian-A) mudstones from Yorkshire, England (after Curtis *et al.* 1972, Oertel & Curtis 1972). Discussion in text.

Abundant organic-rich material in mudrocks is favoured by high or rapid sedimentation rates so that near-surface oxidation in diagenetic zone 1 is prevented (Table 28.2).

The origin of **fissility** in mudrocks is still controversial (see review by Moon 1972). There is little correlation between fissility and burial depth although all compaction leads to general orientation of clay mineral flakes parallel to bedding (Curtis *et al.* 1980). It is obvious that the detailed microstructure of the flake aggregates must play a vital role. Even after compaction, flocculated fabrics will tend to discourage fissility. Many non-fissile kaolinite mudrocks are of this type. Dispersed fabrics will tend to encourage fissility, particularly when abundant organic ions are present which neutralise clay-flake surface charges and thereby discourage edge-to-face type bonding. Some fissile mudrocks show very fine light/dark laminations of a varve-like character when seen in unweathered samples (Spears 1976). Upon weathering, these laminations cause marked fissility. Such laminations probably reflect varying organic contributions with time and certainly imply fairly rapid sedimentation rates and the absence of bioturbation. It is likely that bioturbation is responsible for the lack of fissility in many originally laminated mudrocks.

Table 28.3 Schematic ternary diagram to illustrate the nomenclature and composition of mudrocks (mostly after Selley 1976).

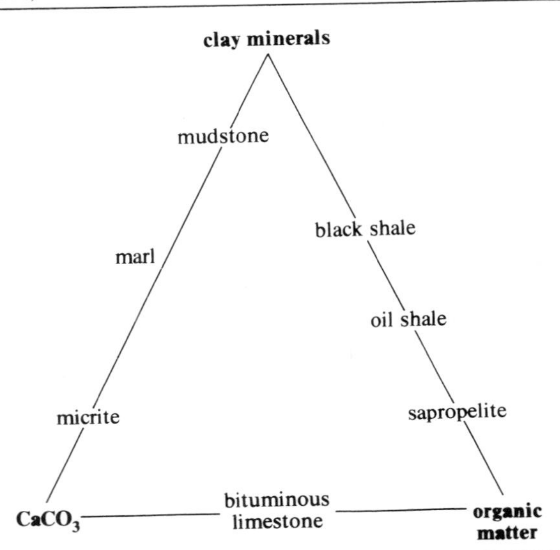

28e Near-surface sand diagenesis

Sediments deposited in semi-arid climates (deserts, alluvial fans and ephemeral river flood courses) frequently remain within the oxygenated vadose zone for long periods until subsidence takes them below the continental water table. During their residence interval within the vadose zone the coarser grained sediments, which are most abundant in these environments, are subjected to clay infiltration, intrastratal detrital mineral solution, authigenic mineral precipitation and reddening by ferric pigments (Fig. 28.4; Walker *et al.* 1978, Walker 1976). A significant mineralogical factor that largely controls the above reactions is the immaturity of many deposited sands and gravels in semi-arid areas, particularly those **first-cycle sediments** derived directly from igneous and/or metamorphic hinterlands.

Clay infiltration occurs when floodwaters charged with suspended washload detrital clays drain through the porous and permeable framework of sands or gravels on alluvial fans. Clay minerals are deposited on detrital grains in the shallow subsurface in such a way that the clay platelets are aligned parallel to the grain surfaces. Mechanically infiltrated clays of this sort are most abundant above relatively impermeable horizons. The clays change the original texture of the alluvial deposits so as to mimic the textures of matrix-rich deposits of debris flows (Ch. 7).

Intrastratal detrital mineral solution occurs preferentially on minerals low in the Goldich stability series (Ch. 1) such as pyroxene, amphibole (Fig. 28.5d) and plagioclase feldspar. Electron microscope studies show intense pitting caused by solution etching and the formation of dissolution voids around and within grains. Dissolution of whole grains leaves characteristic voids surrounded by a thin layer of infiltrated clays (Fig. 28.5a) which prove more chemically stable to the rigours of the diagenetic pore solutions than do the mineral grains themselves. In addition to dissolution, mineral grains may be replaced *in situ* by clay minerals along planes of

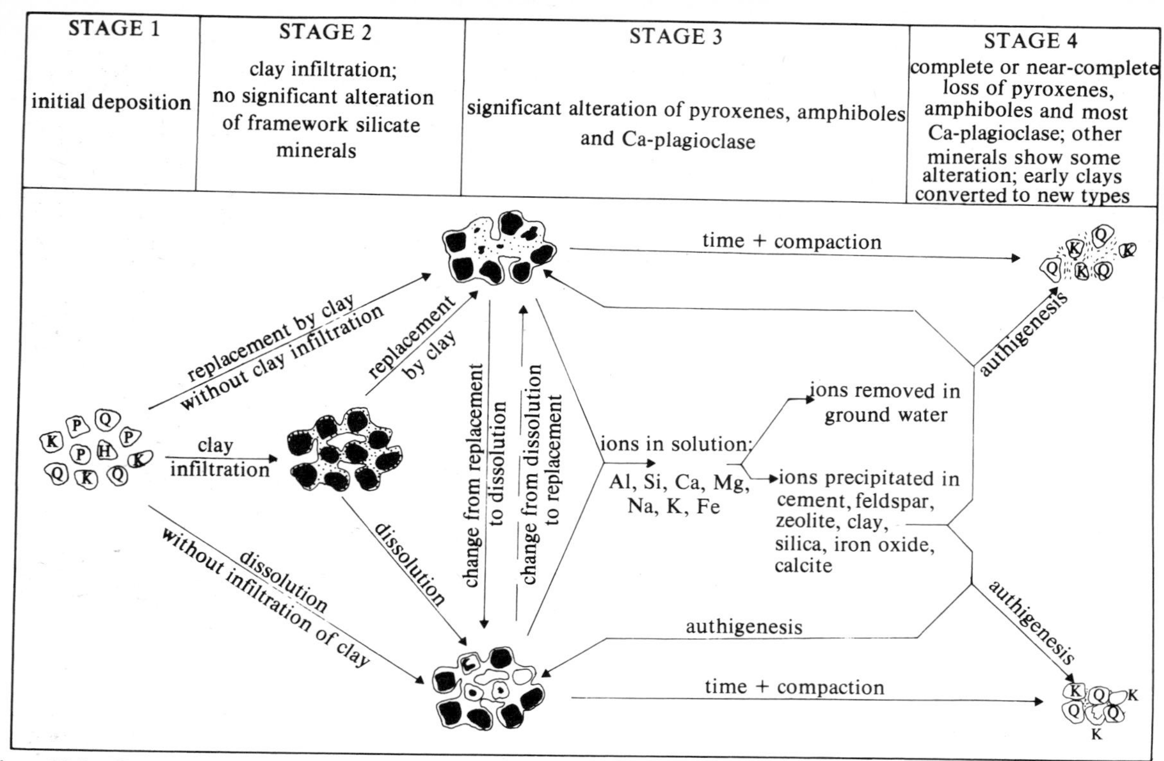

Figure 28.4 Observed and inferred diagenetic alterations in first-cycle desert alluvium (after T. R. Walker *et al.* 1978). K – feldspar, P – plagioclase, H – hornblende, Q – quartz.

weakness in the grains (Fig. 28.5b). The clays involved are randomly interstratified mixed-layer montmorillonite–illite with 80–95% expandable layers.

The exact reaction mechanisms involved in the mineral solutions noted above are not known, but may be approximated by some of those noted in the section on chemical weathering (Ch. 1d, e). Cations liberated by the solution process include Ca^{2+}, Mg^{2+}, Fe^{2+}, Si^{4+}, K^{+}. These cations may be entirely removed from the system in solution or may recombine within the rock pores to precipitate as newly formed or **authigenic mineral species**. The authigenic minerals are best seen using SEM techniques with an attached energy-dispersive X-ray analyser for the determination of the elemental composition and the standard X-ray diffractometer for mineralogical identification of authigenic mineral concentrate. The commonest authigenic minerals are K-feldspar (Fig. 28.5c), zeolite, mixed-layer illite–montmorillonite, quartz, haematite and calcite. All these mineral phases occur as delicate crystalline cement infills to cavities and dissolution voids with the K-feldspar. Quartz also forms overgrowths on pre-existing detrital silica grains.

The liberation of oxidised Fe^{3+} from ferrous silicates such as biotite, amphibole, pyroxene and olivine causes the red staining of mineral grains identified in the description **red bed**. It is now known that many red beds form during early diagenesis in this manner (see review in Turner 1980). Time is needed, however, for the continued precipitation of Fe^{3+} as limonite and the transformation of the oxide to haematite. In areas of the southwestern USA for example first-cycle arkoses increase in 'redness' with age from the Recent through the Pleistocene to the Pliocene–Miocene (Walker 1976).

The early vadose diagenetic changes noted above are summarised in Figure 28.4. It is not known what will happen to the diagenetic fabric during compaction and burial but it is likely that all pore space will be gradually eliminated and that the authigenic and infiltrated clays will be merged into a 'clay matrix' of mixed pedigree. Unless protected by overgrowths of quartz or feldspar, haematite rims may become reduced and the diagenetic reddening eliminated. Compaction will change the clay composition of illite and eventually, if burial proceeds deeply enough or if acid flushing occurs, to kaolinite.

Pure aeolian sands preserved in dunes may undergo diagenetic changes similar to those deduced above for

Figure 28.5 Photomicrographs to illustrate arid-climate continental vadose zone diagenesis in first-cycle alluvium. (a) Thin-section photomicrograph to show peripherally dissolved plagioclase grain (Pl) with dissolution voids (dv) and a clay skin (cs) marking the original grain outline. Scale bar = 250 μm. Pliocene fanglomerate, Baja California. (b) Thin-section photomicrograph to show plagioclase grain (P) that has been irregularly replaced by clay (rc). Scale bar = 250 μm. Location as (a). (c) SEM photomicrograph of bright-red interstitial matrix consisting of a mixture of mechanically infiltrated clay, authigenic clay and authigenic K-feldspar (AF). Scale bar = 5 μm. Location as (a). (d) SEM photomicrograph to show partially dissolved hornblende grain (Ho) with well developed dissolution 'needles' and dissolution voids (dv). Scale bar = 50 μm. Location as (a). (Photographs by T. R. Walker; see Walker *et al.* 1978).

alluvial fans. Development of a reddened clay pellicle surrounding well rounded quartz grains may be the result of infiltration of wind-blown dust after desert storms. The pellicle may be followed by the development of spectacular silica overgrowths on the quartz grains (Fig. 28.6; Waugh 1970a & b). The development of these optically continuous (**syntaxial**) overgrowths is governed by the atomic structure and crystal orientation of the detrital grains. Initial growth starts with the appearance of numerous orientated rhombohedral and prismatic projections on the grain surfaces. Merging and overlap of the projections results in the formation of large crystal faces whose form is dependent upon the initial location of the projections with respect to the internal crystallographic axes, growth being particularly rapid along the *c*-axis. The source of this secondary quartz is thought to be siliceous dusts produced during aeolian abrasion. Alkaline desert ground waters are believed to dissolve the dust, and the silica is then precipitated out upon evaporation. Needless to say such a process is most effective today when acting upon stabilised dunes formerly related to pre-Holocene wind systems.

The surface or shallow subsurface crusts described above are a variety of **silcrete** (see also Ch. 30). If the desert sands are deflated flush to the surface of locally high water tables, then halite and gypsum may precipitate as local cement phases as part of the process of development of interdune sabkhas (Ch. 17).

Both siliceous and evaporitic cementation in aeolian sandstones may severely affect the otherwise excellent oil or gas reservoir properties of these sand bodies. On the other side of the coin, however, the development of an early diagenetic siliceous rim cement may preserve much porosity in the face of compaction.

It should finally be noted that additional cementation and dissolution phases may occur in continental sand and gravel deposits in response to deep burial in sedimentary basins. These processes are considered further in Section 28f below.

28f Subsurface sand diagenesis

Subsurface sand diagenesis is exceedingly complex since the end product is dependent upon initial composition, environment of deposition, nature of interbedded sediment, composition of formation waters and depth of burial – to name but a few of the controlling variables. Two processes may be said to dominate: (a) the alteration, dissolution and mechanical compaction of framework grains and (b) the growth (and sometimes the dissolution) of precipitated cement minerals in pore space (see Fig. 28.7). An important point is that sand horizons provide the pathway for the fluids expelled from associated compacting mud sequences (see Fig. 27.11), hence the source of the ions involved in process (b) (Fig. 28.8) may be located far distant from the actual pore space where cementation occurs.

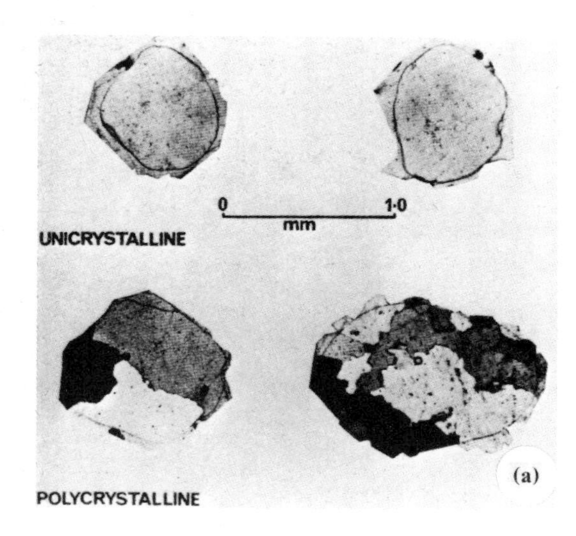

Figure 28.6 Quartz overgrowths. (a) Thin sections of unicrystalline and polycrystalline quartz grains with their corresponding overgrowths. Note the optical continuity of the overgrowths with the corresponding subcrystals in the polycrystalline grains (view under cross-nicols). (b) Completed unicrystalline overgrowths, illustrating the double pyramidal terminations to the prism faces. Note the rounded outline of the detrital quartz grain in D. (All after Waugh 1970a.)

Figure 28.7 SEM photographs of authigenic clays precipitated in sandstone pore spaces. (a) Stacked kaolinite plates showing face-to-face arrangement and pseudo-hexagonal outlines of individual plates; Eocene Frio Sand, Texas. (b) Authigenic illite with unusually long lath-like projections; Permian Rotliegendes Sandstone, North Sea. (c) Honeycomb growth forms of chlorite as a coating to sand grains. Dark spots on grains represent points of contact between adjacent grains, some of which were removed during sample preparation. Jurassic Norphlet Sand, Florida. (d) Mixed-layer smectite–illite showing a crystal habit very similar to pure illite with its short lath-like projections; Cretaceous Mesaverde Group, Colorado. (All after Wilson & Pittman 1977.)

During burial, feldspar grains and volcanic lithic fragments are particularly prone to alteration and dissolution by acidic formation waters (Surdam & Boles 1979). Feldspar breaks down to illite and hence to kaolinite with the liberation of Si^{4+} and K^{+}. In wells through Gulf coast sands, K-feldspar rarely occurs below depths of about 2.5 km (Fig. 28.9). Basic volcanic lithic fragments break down to smectite clays and volcanic glass devitrifies as burial proceeds. In the latter case zeolites may crystallise from pore-water solutions. These have been used to define a series of depth-related burial metamorphic zones (Hay 1966, Packham & Crook 1960). Sands initially rich in feldspar or lithic clasts will thus have a significant portion of their rigid framework broken down to soft clay-mineral aggregates which will tend to 'smear' and flatten during compaction to form a clay matrix which severely reduces sand permeability. A similar fate awaits initial soft mudstone intraclasts and detrital mica flakes which are

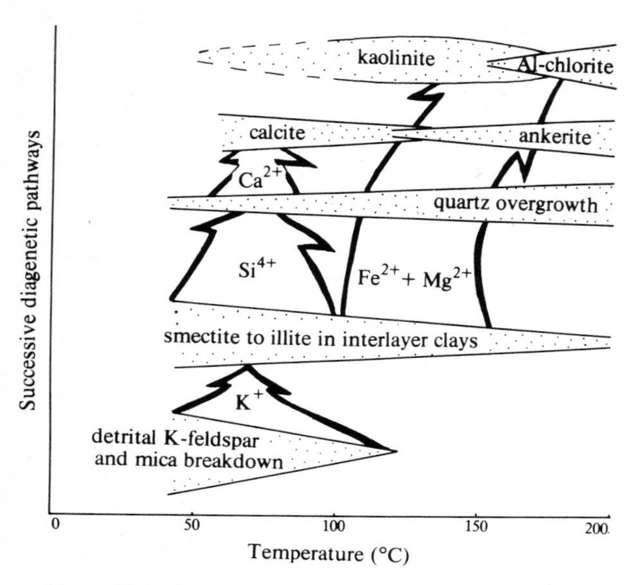

Figure 28.8 Schematic diagram to show the ions (large arrows) made available for sandstone cementation by reactions amongst interlayered detrital clay minerals and detrital feldspars during diagenesis (after Boles & Franks 1979).

bent around detrital quartz grains in a characteristic manner. The production of diagenetic matrix has great implications regarding permeability trends, sandstone classification (Ch. 28h) and textural interpretation (Nagtegaal 1978).

The provision of silica in solution from K-feldspar breakdown and from the smectite→illite change in associated compacting muds (Fig. 28.8) may lead to the growth of silica rim cements and overgrowths over pre-existing quartz grains (Land & Dutton 1978). Such overgrowths may appear identical to those produced in near-surface vadose diagenesis in semi-arid and arid climates, but in many cases overgrowth formation has been inhibited by early authigenic clay rims, in contrast to the vadose examples. Early production of quartz overgrowths greatly reduces the likelihood of later pressure-solution processes reducing porosity at burial depths down to 4–8 km.

Over 90% of all sandstones contain authigenic clay minerals (Fig. 28.7) precipitated in pores during diagenesis (Wilson & Pittman 1977). Montmorillonite, mixed-layer clays and chlorite may precipitate in pore space very early in diagenesis, close to the seawater/sediment interface. This is particularly so when

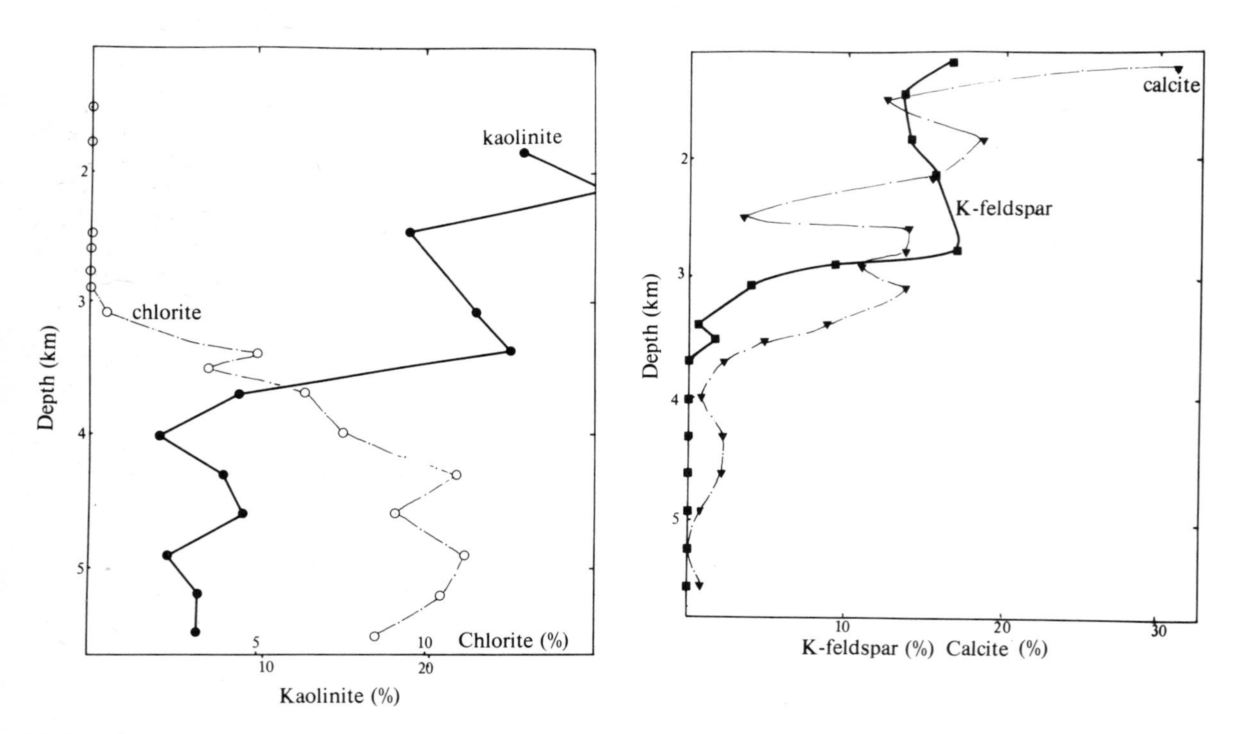

Figure 28.9 (a & b) Changes of abundance in primary.detrital (calcite, K-feldspar, kaolinite) and diagenetic (chlorite) phases observed within a 5+ km borehole section through marine muds and sandstones of Oligocene–Miocene age of the USA Gulf Coast (after Hower *et al.* 1976).

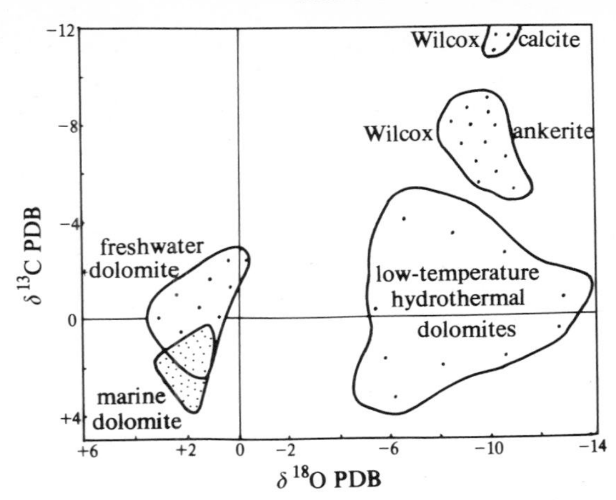

Figure 28.10 Stable isotope compositions of carbonate cements in marine-deltaic sandstones of the Wilcox Formation in the subsurface Eocene of southwest Texas (after Boles 1978).

the sands are rich in basic, volcaniclastic fragments that undergo alteration almost immediately below the interface. At burial depths in excess of 2.5 km the mixed-layer clays change to illites with the K^+ supplied from the breakdown of K-feldspar as already described. If acidic formation waters are present, or if the sands are encroached by meteoric waters during uplift, then kaolinite will form the clay-mineral cement phase. Subsequent burial and flushing by alkaline pore waters may lead to illitisation. In many oil reservoirs in the British North Sea, oil migration was earlier than subsequent illitisation so that oil-filled pores contain authigenic kaolinite whilst others below the oil zone contain authigenic illite (Sommer 1978, Hancock & Taylor 1978). It is clear that the oil has 'protected' the kaolinite from the influence of K^+-rich formation waters. Studies in the USA Gulf Coast (Boles and Franks 1979) show that kaolinite becomes progressively less abundant with increasing burial depth (between 3 and 4.5 km) whilst over roughly the same interval chlorite becomes more abundant (Fig. 28.9). This distribution pattern suggests a reaction relationship between the two minerals with the excess Fe^{2+} and Mg^{2+} required coming from the illitisation of smectite and interlayer clays (Eq. 28.4).

Carbonate cemented sandstones are relatively common and it is often the case that carbonate is the major cement material that cuts down porosity and permeability in many economic and near-economic sandstone hydrocarbon reservoirs (but see Section 28g below). Calcium, iron and magnesium are all supplied to migrating pore fluids from the change smectite→illite in associated compacting mudrock successions. Calcite, ferroan calcite, ferroan dolomite, ankerite and siderite are the common carbonate cements and in the majority of cases they are fairly late phases. In the Wilcox sandstones of the USA Gulf Coast (Boles 1978) calcite is replaced by ankerite at burial depths in excess of 2.5–3 km, corresponding to temperatures of about 120°C. Isotopic analysis (Fig. 28.10) shows that these ankerites are light in both ^{18}O and ^{13}C, the latter presumably because the source of CO_2 is thermal decarboxylation of organic material present in associated mudrock successions. The light oxygen values indicate similarities with the field of hydrothermal dolomites (Fig. 28.10).

The importance of muds as 'donors' of ions for cementation in porous sand formations is emphasised in the common enrichment of carbonates close to mudrock/sandstone contacts and of their dominance in most thin sandstones interbedded within mudrock sequences.

Pressure solution converts some grain boundaries into concavo-convex and sutured contacts. Suturing seems to be encouraged by the presence of thin illitic clay coats on the framework grains, these presumably speeding up ion diffusion which is the rate-determining step in the pressure-solution process.

28g Secondary porosity and sandstone diagenesis

In addition to the primary porosity, that results from the failure of cement minerals to totally infill available pore space, a variety of mechanisms may produce secondary porosity (reviewed by Schmidt & MacDonald 1979) during the course of sandstone diagenesis (Fig. 28.11). This secondary porosity may develop at any stage during diagenesis but is thought to develop most effectively during deep-burial diagenesis following dissolution of the carbonate-cementing minerals calcite, dolomite and siderite (Figs 28.12 & 13). Such **decarbonatisation** results from decarboxylation reactions involving organic material, the CO_2 so produced combining with formation waters to form acidic solutions. The large volumes of dissolved Mg^{2+}, Ca^{2+}, HCO_3^- and CO_3^{2-} ions so produced may migrate upwards during compaction and be precipitated again as carbonate cements at higher levels in the deposited sedimentary pile. It is thought that primary hydrocarbon migration (Ch. 31) follows closely after the decarbonatisation reactions because the main phase of hydrocarbon generation follows the culmination of decarboxylation. Thus it can be seen that the close association of source rock and reservoir favours hydrocarbon accumulation in secondary pore space. This pore space is created at depths far below the generally accepted limits to effective primary porosity.

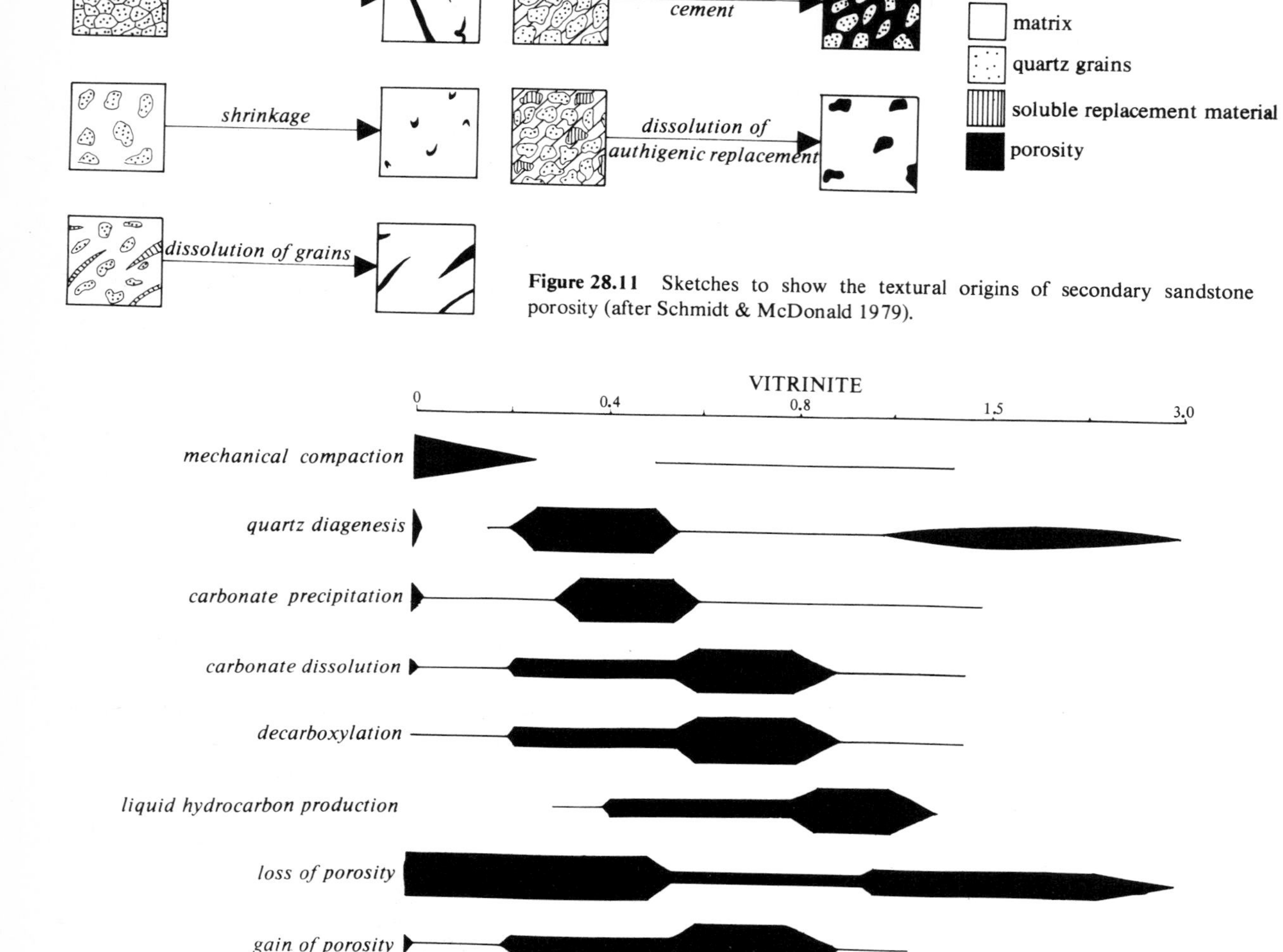

Figure 28.11 Sketches to show the textural origins of secondary sandstone porosity (after Schmidt & McDonald 1979).

Figure 28.12 Diagram to show the various mechanisms responsible for burial diagenesis of quartz arenites. Values of vitrinite reflectance indicate progressive intensification of diagenesis (see Ch. 31). (After Schmidt & McDonald 1979.)

28h Classification

The most useful sandstone classification is a dual one, based upon both modal composition of grains and texture (Fig. 28.14). Compositions of sandstone cements are therefore not relevant in this approach but clearly may be described by an adjectival prefix, e.g. calcareous sandstone for calcite-cemented sandstone. Composition is described on a triangular graph by the three main components: quartz, feldspar and rock fragments (Dott 1964, Pettijohn *et al.* 1972) **Arenites** are the major clan of sandstones having less than 15% fine-grained matrix material, i.e. arenites are well sorted. Arkosic, lithic and quartz arenites are rich in feldspar, rock fragments and quartz respectively. Quartz arenites (synonymous with orthoquartzites) are the most mature sandstone group and are frequently multicycle deposits (Ch. 1). It is possible to subdivide the lithic arenites further depending upon the composition of the rock fragments involved.

The second major clan of sandstones are the **wackes**, which have greater than 15% matrix. Greywacke is an outdated general term used to describe wackes as a whole. Wackes are divided into feldspar, lithic and quartz types, which are, respectively, rich in feldspar, rock fragments

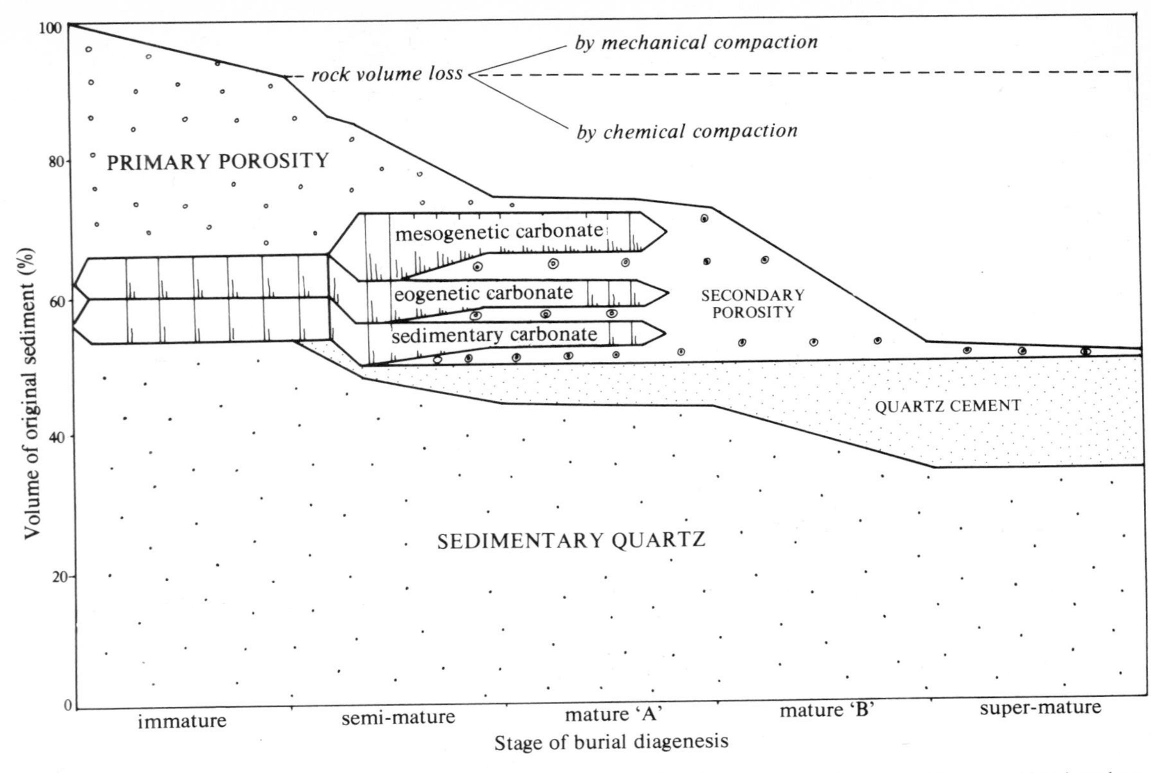

Figure 28.13 Diagram to show the burial diagenesis of a quartz arenite in relation to secondary porosity created by decarbonatisation of primary carbonate and secondary carbonate cements (after Schmidt & McDonald 1979).

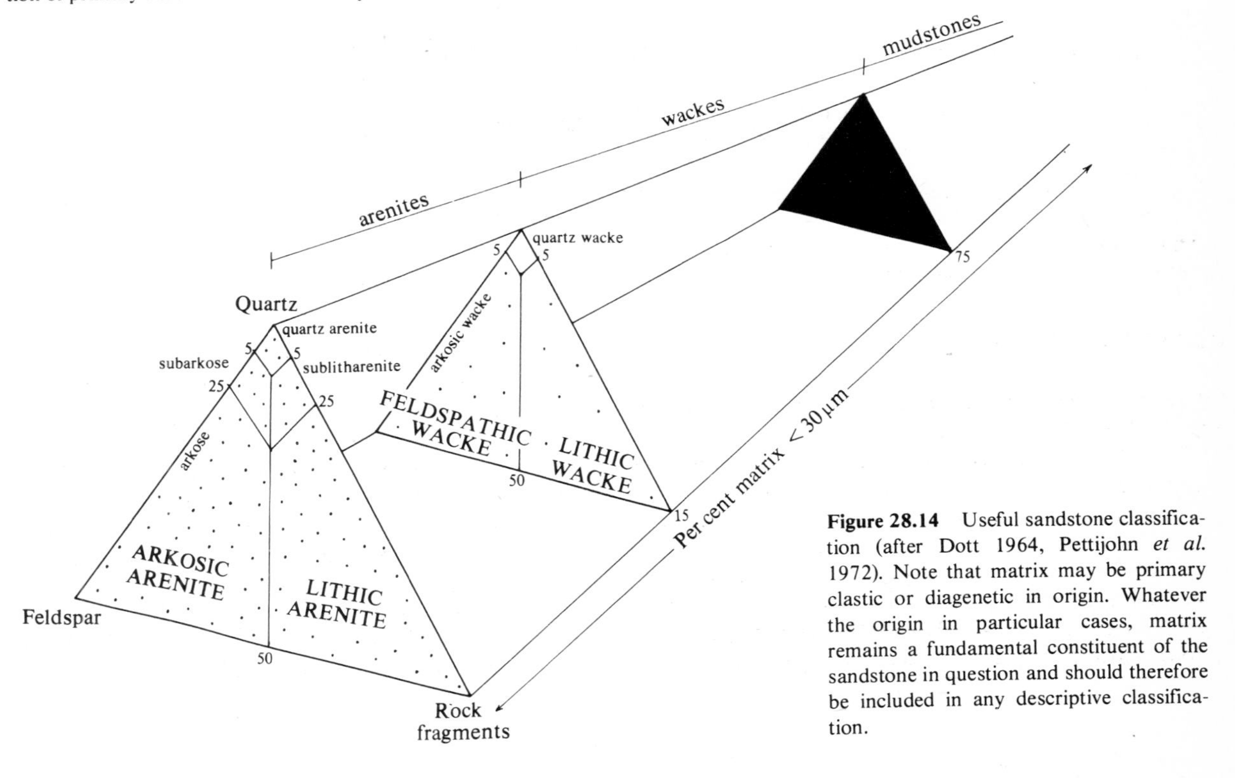

Figure 28.14 Useful sandstone classification (after Dott 1964, Pettijohn *et al.* 1972). Note that matrix may be primary clastic or diagenetic in origin. Whatever the origin in particular cases, matrix remains a fundamental constituent of the sandstone in question and should therefore be included in any descriptive classification.

and quartz. Further subdivision of the lithic wackes is possible depending upon the composition of the rock fragments involved. With increasing content of fine-grained matrix the wackes grade into sandy mudrocks (>50% matrix) and hence into mudrocks (>75% matrix).

Thus far we have avoided any discussion of the term **matrix**. Let us first consider the grain-size limits to matrix. Some authors would restrict the term to truly clay-size grains, i.e. <4μm, which are essentially unresolvable as individual grains under the ordinary light microscope. Others set the upper size limit to matrix higher, mostly around 20–30 μm, i.e. in the medium silt range. The latter size range is probably better since, as we have noted above, the mean grain size of mudrocks is about medium silt on the Wentworth scale, and the classification outlined above allows for intergradation between wackes and mudrocks.

Controversy as to the origin of matrix in wackes (Cummins 1962) has led to the so-called *greywacke problem*. Following the advent of turbidity current theory, many geologists identified ancient greywackes as muddy sandstones deposited by turbidity currents. Although the majority of Palaeozoic turbidites were indeed of wacke composition it was later shown that certain Recent and Tertiary turbiditic sand deposits were arenites singularly lacking in matrix. Recent studies of Pacific and Atlantic turbidites do show, however, that many turbidites contain a significant amount of primary mud and silt matrix. It can also be shown that many non-turbiditic sandstones are wackes.

The problem is resolvable if one postulates that some matrix in wackes is of diagenetic origin as briefly discussed above (Ch. 28f). Breakdown and compaction of mineral and rock fragments and clay mineral precipitation both contribute towards the production of diagenetic matrix. The distinction between primary and secondary matrix in ancient sandstones is extremely difficult (Dickinson 1970).

When discussing the composition and texture of ancient sandstones, particularly those which have been deeply buried, *the reader should take the utmost care in drawing conclusions as to provenance and depositional mechanism. In many cases the present character of sandstones will tell more about the diagenetic history than either the original provenance or depositional mechanism.* Studies aimed at inferring provenance in ancient rocks should therefore concentrate on identifying 'key' remnant grains of tell-tale provenance rather than an absolute or relative abundance of feldspars or rock fragments.

Further reading

Many helpful papers on various aspects of clastic diagenesis are to be found in Scholle & Schluger (1979) and in *J. Geol Soc.* London, Volume 135 (1978).

29 Carbonate sediments

29a Introduction

At first sight the diagenesis of carbonate sediments would seem to be a simple affair as compared to terriginous clastic diagenesis. The original sediment grains are $CaCO_3$ of marine origin in contrast to multi-ionic and multimineralic terriginous clastic grains, and they will be most commonly bathed in a solution of seawater origin. Progressive burial of such a diagenetic system might be expected to lead to predictable depth-related changes based upon progressive pore-water fractionation.

Inevitably perhaps the above assumptions hide processes of greater complexity since: (a) calcium carbonate, unlike quartz, is easily precipitated and dissolved on a large scale in a variety of Earth surface and near-surface conditions; (b) the polymorphs of $CaCO_3$, aragonite and calcite (the latter with variable Mg-content) show contrasting stability fields in the various diagenetic realms; (c) organic processes such as photosynthesis or decay frequently involve usage or production of CO_2 so that carbonate equilibria are often dependent upon the degree of organic influence, particularly in the microenvironment of the pore space or cavity that characterises the diagenetic environment below the sediment/water interface; (d) the precipitation of dolomite is governed by kinetic factors rather than by sensible considerations of ionic concentrations in surface and near-surface waters, and (e) many carbonate sequences are underlain by, interbedded with or overlain by clastic mudstones, so that an entirely separate source of ionic species may be supplied to the carbonate pore spaces at various times during burial diagenesis.

Despite the above complications it may be said that carbonate diagenetic studies are much further advanced than clastic studies, the historical reasons for which need not concern us in detail. Undoubtedly the attraction of thin-section studies of limestones, the staining technique for mineral identification, the use of stable isotopes, the location of many early diagenetic carbonate products in sunny subtropical climates, and the obvious role of carbonate cements in reducing carbonate reservoir porosity and permeability have all played their part.

Before we examine the various environments of carbonate diagenesis in detail, let us briefly list the nature of the carbonate grains awaiting diagenesis in a subtropical carbonate province. Biogenic grains will include low-Mg calcite, high-Mg calcite and aragonite. Certain shells will exhibit micrite envelopes or gradations to completely micritised amorphous lumps. Aragonitic pellets will show some degree of interstitial hardening following aragonite precipitation in micropores. Aragonitic oöliths will again show varying degrees of internal micritisation leading to partial destruction of their primary internal radial or tangential fabrics. Near-surface carbonate muds in quiet-water environments will comprise aragonite needles of varying pedigree, some derived by algal breakdown, some probably chemical precipitates from sea water. Isotopically the bulk compositions of the carbonate sediment will be around $\delta^{18}O = -2$ and $\delta^{13}C = +2$ (Fig. 27.5). Variations will arise because of organic influences on isotope fractionation, e.g. micrite envelopes will show higher $\delta^{13}C$ than will shell aragonite.

29b Early meteoric diagenesis

Marine carbonate sediments may become stranded in the meteoric realm after sea level fall, tectonic uplift or coastline progradation (see Ch. 23). Many areas of Pleistocene sediments and partly cemented limestones exist today which are subject to meteoric diagenesis because of the first two factors. As already noted the meteoric realm may be divided into the *vadose* and *phreatic* zones.

To a large degree **vadose meteoric diagenesis** involves the solution and reprecipitation of the unstable carbonate minerals (high-Mg calcite and aragonite) by rain water that has passed through a soil zone and which has approached equilibrium with the ambient CO_2 pressure in the soil, pCO_2 in soils being higher than that of the atmosphere (Thorstenson *et al.* 1972). The acidic waters dissolve $CaCO_3$ to attain equilibrium. As the solution moves downwards into an environment of reduced pCO_2, the more stable and less soluble $CaCO_3$ (of low-Mg composition) is then precipitated out of solution as the stable cement phase of low-Mg calcite. The ^{12}C-enriched soil CO_2, derived from organic sources, gives this precipitated calcite cement distinctive negative $\delta^{13}C$ values. Negative $\delta^{18}O$ values are caused by equilibrium with the local meteoric waters. Although $\delta^{13}C$ values between −4 and −10 are common in published analyses, recent detailed sampling on vadose-zone sediments in Barbados, West Indies (Allan & Mathews, 1977), shows that values less than −6‰ PDB occur only in the surface zones of outcrops, most borehole samples away from the surface zone falling in the range −1 to −5‰. The depth/$\delta^{13}C$ plot thus shows a steady trend towards less negative $\delta^{13}C$ (Fig. 29.1).

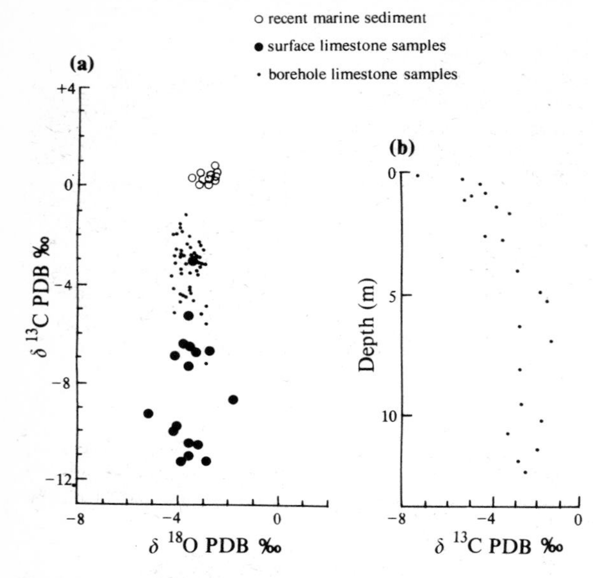

Figure 29.1 (a) Stable isotope composition of Barbados Pleistocene diagenetic limestones (vadose zone) and Holocene marine sediments. (b) ^{13}C versus depth for borehole limestone samples. See text for discussion. (After Allan & Mathews 1977.)

Detailed studies of vadose meteoric diagenesis (work of Land 1966 as summarised in Bathurst 1975) in Bermudan Pleistocene limestones defines five broad stages (Fig. 29.2) of alteration of primary carbonate sediment (*stage 1*) comprising mixtures of aragonite (molluscs, corals, *Halimeda*) and high-Mg calcite (foraminifera, coralline algae, echinoderms).

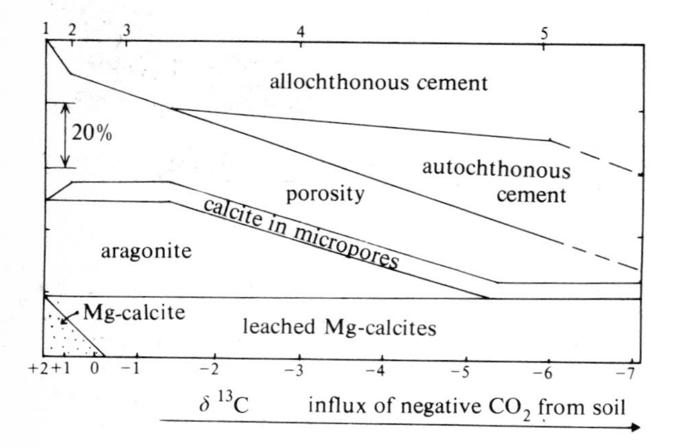

Figure 29.2 Meteoric diagenetic zones (1–5) and their characteristic compositions; based upon Bermudan Pleistocene limestones (after Land 1966, with isotopic results of Friedman 1964 as summarised by Bathurst 1975).

Stage 2 involves the formation of a partially lithified and still-friable limestone in which low-Mg calcite occurs as an intergranular and intragranular rim cement showing mildly negative $\delta^{13}O$ and $\delta^{18}O$ values as noted above. Since the pore spaces of the vadose zone are largely air filled, carbonate precipitation may take place by CO_2-degassing from capillary-held water at grain-to-grain contacts. A distinctive meniscus-type rim cement results in the pore throats and, more rarely, a pendant-type cement on the bases of larger grains (see Fig. 29.4). The cementation process is further developed in fine-grained sediments that have the greatest capillarity (Land 1970). The source of the $CaCO_3$ for stage 2 cements is thought to be dissolved carbonate at the near-surface zone of dissolution, i.e. the cement $CaCO_3$ is allochthonous.

Stage 3 sees the change of biogenic particles comprising high-Mg calcite to low-Mg calcite. Such a change is accompanied by little loss of structural detail within the shells and tests involved, but the process must have proceeded by complete repopulation of the lattice by all ions (not just an exchange of Ca^{2+} for Mg^{2+}) since the oxygen and carbon isotopes show that the low-Mg calcite is in complete isotopic equilibrium with the vadose diagenetic waters.

Stage 4 involves *local* aragonite dissolution and reprecipitation as low-Mg calcite cement and also calcitisation of aragonite shells by migrating solution fronts without an intermediate cavity phase (Pingitore 1976). In the vadose zone the latter process takes place across micron or narrower water films, aragonite being dissolved on one side and calcite precipitated on the other (Fig. 29.3). Chemical exchange of such trace elements as Sr^{2+} with pore waters is limited so that concentrations may build up in the calcite. The calcite fabric of corals may mimic fairly closely that of the original aragonite fabric. Scanning electron microscopy of the replaced aragonite reveals tiny, very abundant aragonite inclusions in the neomorphic (newly formed) calcite. During stage 4, micrite envelopes (Ch. 2) remain as stable frameworks resisting dissolution whilst their enclosed aragonitic skeletons may become replaced by calcite or dissolved away entirely (Bathurst 1966). In the latter case the micrite envelopes remain as framework moulds ready to receive a cement fill at some later diagenetic stage or to be collapsed by increasing overburden load during burial.

Stage 5 culminates in the production of a well lithified rock with *c.* 20% porosity and the particles and cements composed largely of low-Mg calcite (Fig. 29.5). Vadose diagenesis of this sort is usually envisaged as being very rapid (thousands of years for completion) but in many areas high-Mg calcite and aragonite have survived the rigours of vadose processes for upwards of 100 000 years, especially where the passage of ground water is

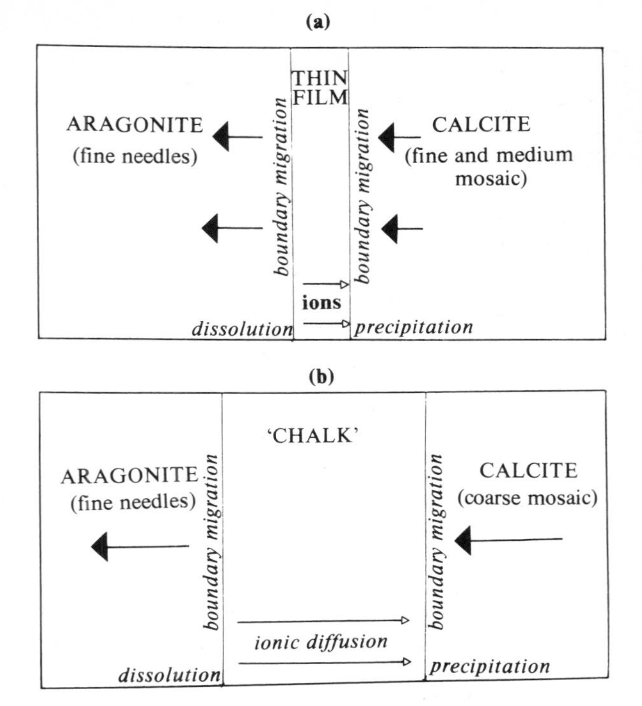

Figure 29.3 (a) Diagrammatic representation of thin film transformation of aragonite to calcite leading to relatively good preservation of coral skeletal fabric; (b) ditto for 'chalk' transformation of aragonite to calcite leading to relatively poor preservation of coral skeletal fabric (Pingitore 1976).

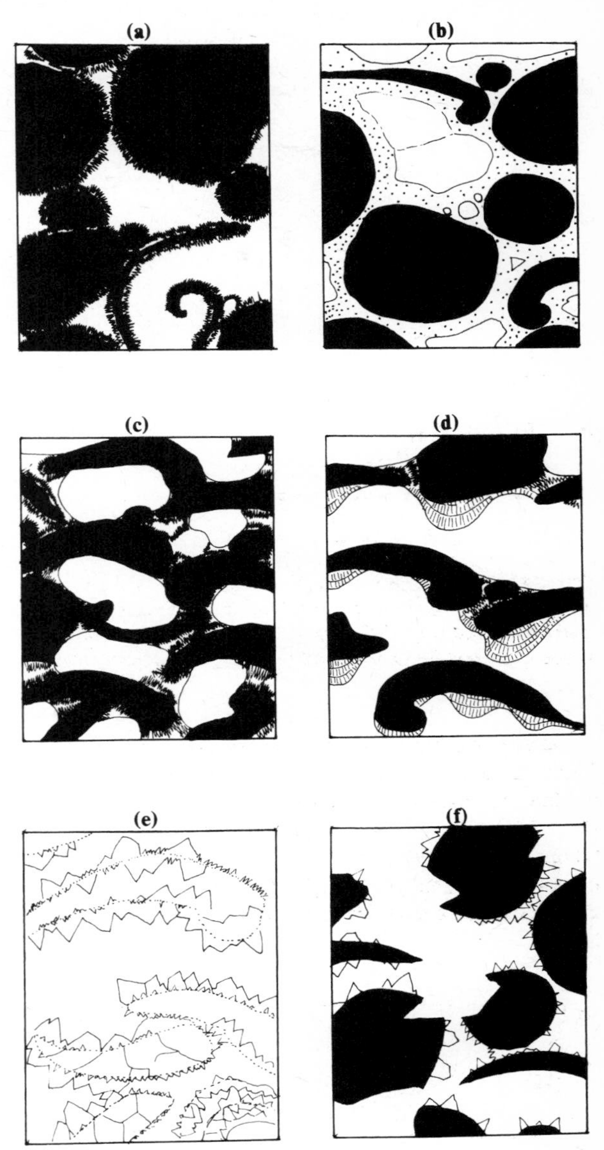

Figure 29.4 Sketches from thin sections to illustrate cement fabrics in Recent carbonates. (a) Marine isopachous fibrous aragonite; (b) isopachous micritic aragonite or magnesian calcite; (c) inter-supratidal fibrous grain-contact cement; (d) inter-supratidal fibrous microstalactitic cement; (e) meteoric blocky calcite and dissolution fabric; (f) meteoric dissolution of dedolomite giving flat-sided secondary pore-spaces. (All after Purser 1978.)

held up by clastic or carbonate mudstone interbeds. As we shall see below, meteoric phreatic diagenesis is a more rapid process than vadose diagenesis.

Before we leave vadose diagenesis it is necessary to consider briefly the processes enacted in the topmost part of the vadose zone where calcification is characteristically developed. Caliches develop upon carbonate rock or sediment substrates causing characteristic profiles to develop which deepen and become more intense with age. At the surface in impermeable outcrops or at the rock/soil interface a characteristic thin laminated crust of micrite develops up to 50 mm thick (Harrison & Steinen 1978). Stringers and lenses of micrite also extend downwards from the surface along cracks and joints in porous substrates. Tepee-like fold structures are common (see Assereto and Kendall 1977) with a subsurface brecciated zone most characteristic of older profiles where subangular-to-rounded fragments of host carbonate are embedded in a matrix of pelleted micrite. Deposition of micritic calcite around these clasts gives rise to **vadose pisoliths** up to 10 cm diameter (Fig. 29.5e) which may frequently show dripstone textures consisting of micrite

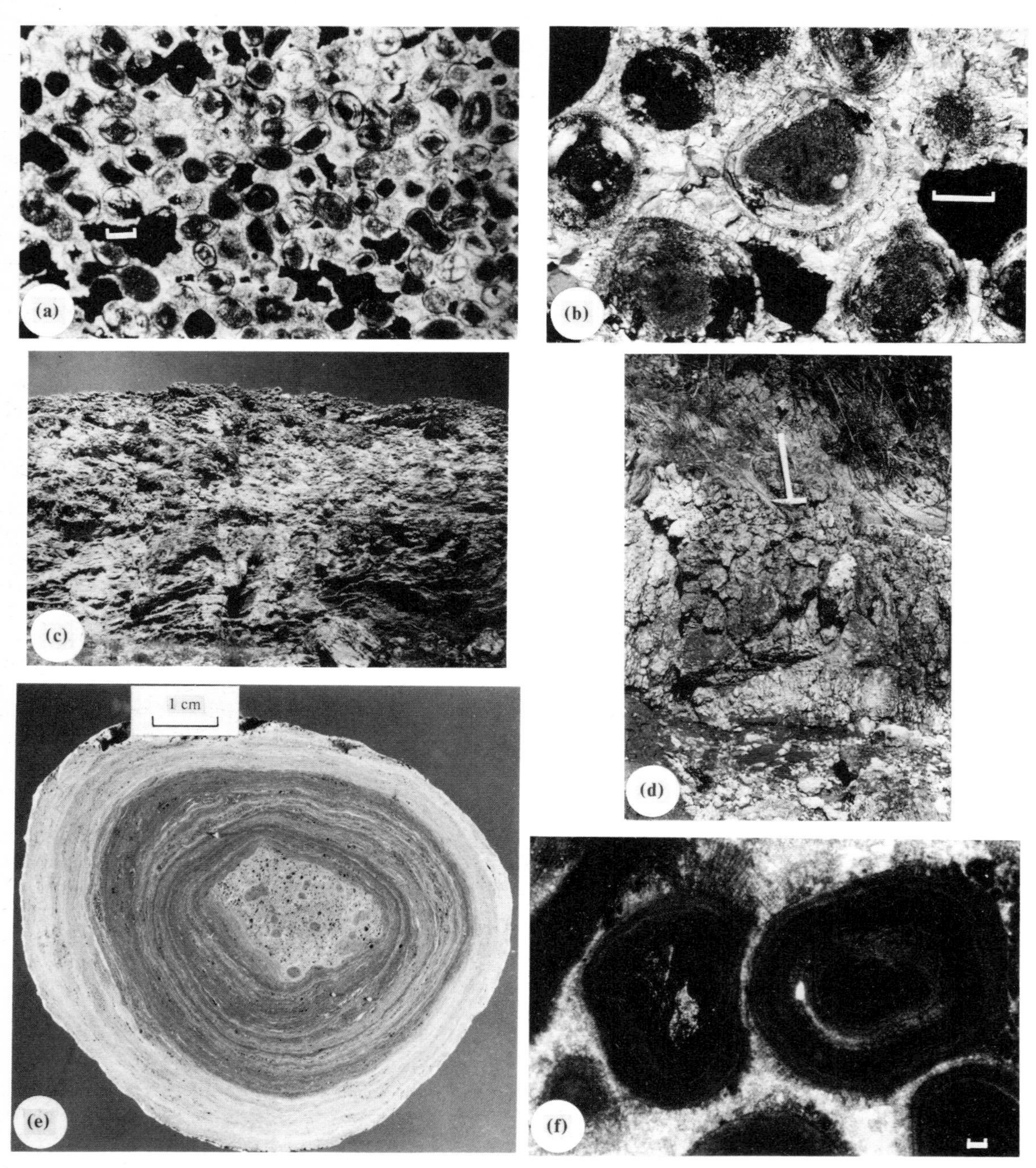

Figure 29.5 (a) Thin section under crossed-nicols to show a Pleistocene oölitic limestone (aeolianite) with partial dissolution of aragonite and co-precipitation of low-Mg calcite as a blocky cement in the pore spaces. Note remaining pore space (black). Meteoric zone, Bimini, Bahamas. Scale bar = 100 μm. (Coll. R. Till.) (b) Close-up of (a) to show dissolution voids in the aragonitic oölite cortexes and the blocky low-Mg cement linings to pores. (c) Extremely thick (~12 m) composite calcrete profile overlying the Dwyaka Tillite, Bogogobe, Botswana (photo L. Watts). (d) Mature calcrete with cylindroids and nodules; Lower Old Red Sandstone, Lydney, Gloucestershire, England. (e) Extraordinarily large vadose pisolith from Quaternary calcrete, Shark Bay, Western Australia (coll. E. J. W. Van der Graaf). (f) Thin-section view of vadose pisoliths associated with Capitan 'backreef' facies, Walnut Canyon, New Mexico; scale = 500 μm (coll. D. B. Smith).

laminae hanging from the bases of pisoliths (Fig. 29.4). Percolation of water through the vadose zone of these caliche profiles may wash detrital crystals of **vadose silt** into pore space and cavities (Dunham 1969a & b).

Extensive areas of calcrete (Fig. 29.5d–f) are developed on *non-carbonate* substrates in semi-arid regions where they form extensive resistance caprocks of **duricrust** type (Goudie 1973). The Holocene calcretes occur most commonly where there is a reasonably distributed, but not excessively peaked, annual rainfall of 100–500 mm. The calcretes generally form by carbonate precipitation and replacement in the vadose zone in the C soil horizon, usually on non-depositional surfaces. The carbonate mineral is predominantly low-Mg calcite of micritic to microsparitic grade (but see Watts 1980 on high-Mg calcite occurrences) and gives abundant evidence in thin section of its replacement origins after pre-existing silicate mineral grains. True cement fabrics and displacive carbonate also occur (Watts c. 1980). The carbonate profiles thicken with time, as scattered filaments and small nodules are joined by larger, coalescing nodules. Ultimately a plugged profile develops with a finely laminated caprock and associated vadose pisoliths, brecciation and expansion features and chalcedony lenses. Radiocarbon dating suggests that thick plugged calcretes are at least 10^4 years old (Gile & Hawley 1969). Many thick calcrete profiles (e.g. Fig. 29.5c) are probably remnants from Pleistocene times; some may record episodic carbonate accumulation in response to Plio-cene–Pleistocene climatic fluctuations.

In contrast to the meteoric vadose zone, the **meteoric phreatic zone** consists of sub-water-table environments where pore spaces are filled with fresh water. The water in the freshwater phreatic zone should not be imagined as being static and stagnant, however, since large-scale circulations occur, involving percolating meteoric water, flow of phreatic water in response to regional hydrologic gradients from recharge areas to coasts, and mixing of phreatic fresh water with marine-derived sea water. As noted previously, phreatic diagenesis is considered to be rapid compared to vadose diagenesis because of constant water movement in very large quantities (Land 1970). It gives rise to a coarser-grained sparite cement of low-Mg calcite (crystals up to 250 μm). Evidently the dilute phreatic waters encourage the slow uninterrupted precipitation of large calcite crystals as CO_2 degassing takes place. By contrast, the episodic pore-water occurrence in the vadose zone leads to rapid growth of tiny crystals at many nucleation sites. Further, in the phreatic zone the cements tend to be (Figs 29.4 & 29.6 b & d) of equal thickness around the grains (**isopachous**). Where the phreatic cements completely fill pore space it has been noted that the final cement mosaic does not display good enfacial junctions (Land 1970), as it should do according to theory and much petrographic evidence. No reason has been adduced for this fact.

Skeletal aragonite alteration in the meteoric phreatic zone proceeds by a different route than in the vadose zone (Pingitore 1976). A zone of 'chalky' aragonite occurs between the aragonite and replacing calcite where there is extensive temporary development of secondary porosity and where Ca^{2+} and CO_3^{2-} ions diffuse from dissolution sites in the chalk area to precipitate on the calcite mosaic (Fig. 29.3b). Should the ions diffuse out or be transported out of the system in ground water then large dissolution vugs will develop as seen in many Pleistocene, Tertiary and Mesozoic coral colonies. The 'chalkification' process of aragonite replacement leads to a uniformly coarse calcite spar mosaic with poor preservation of internal structures. Sr^{2+} contents in the phreatically replaced skeleton and in the cement are low and uniform in contrast to the vadose zone.

Studies of fabrics developed in aragonite muds subjected to marine–phreatic diagenesis reveal patches of coarser-grained microspar (defined as crystals of 4–50 μm) set in a micritic matrix (Steinen 1978). These microspar crystals are revealed by electron microscopy to be true cement crystals, filling in tiny dissolution voids, rather than recrystallisation (neomorphic) fabrics. Thus partial early cementation of lime muds can occur in the phreatic zone and it will help to prevent compaction of the muds as burial proceeds. This is an important point to consider when comparing the burial behaviour of clastic and carbonate muds and their porosity/depth functions in relation to compaction (see Ch. 27).

We have seen in this section how mineralogical, chemical and textural alterations occur in the meteoric zone. The zone is 'open' chemically and allows output of Mg^{2+} and Sr^{2+} ions plus other trace elements from high-Mg calcite and aragonite rejected by low-Mg calcite that precipitates in equilibrium with fresh waters.

29c Early marine diagenesis

In many tropical, subtropical and, to a lesser extent, temperate intertidal zones irregular patches of lithified sand- and gravel-grade carbonate sediments exist. The lithified sediment is known as **beach-rock** (Fig. 29.6a) and occurs in layers just below the sediment surface or as partially eroded masses at the sediment surface showing erosional features such as solution pits, enlarged cracks and mechanically eroded potholes. The irregularly cemented beach-rock is usually a few decimetres or less in thickness and is proven to be of Recent origin by the inclusion of dateable human-derived flotsam and jetsam

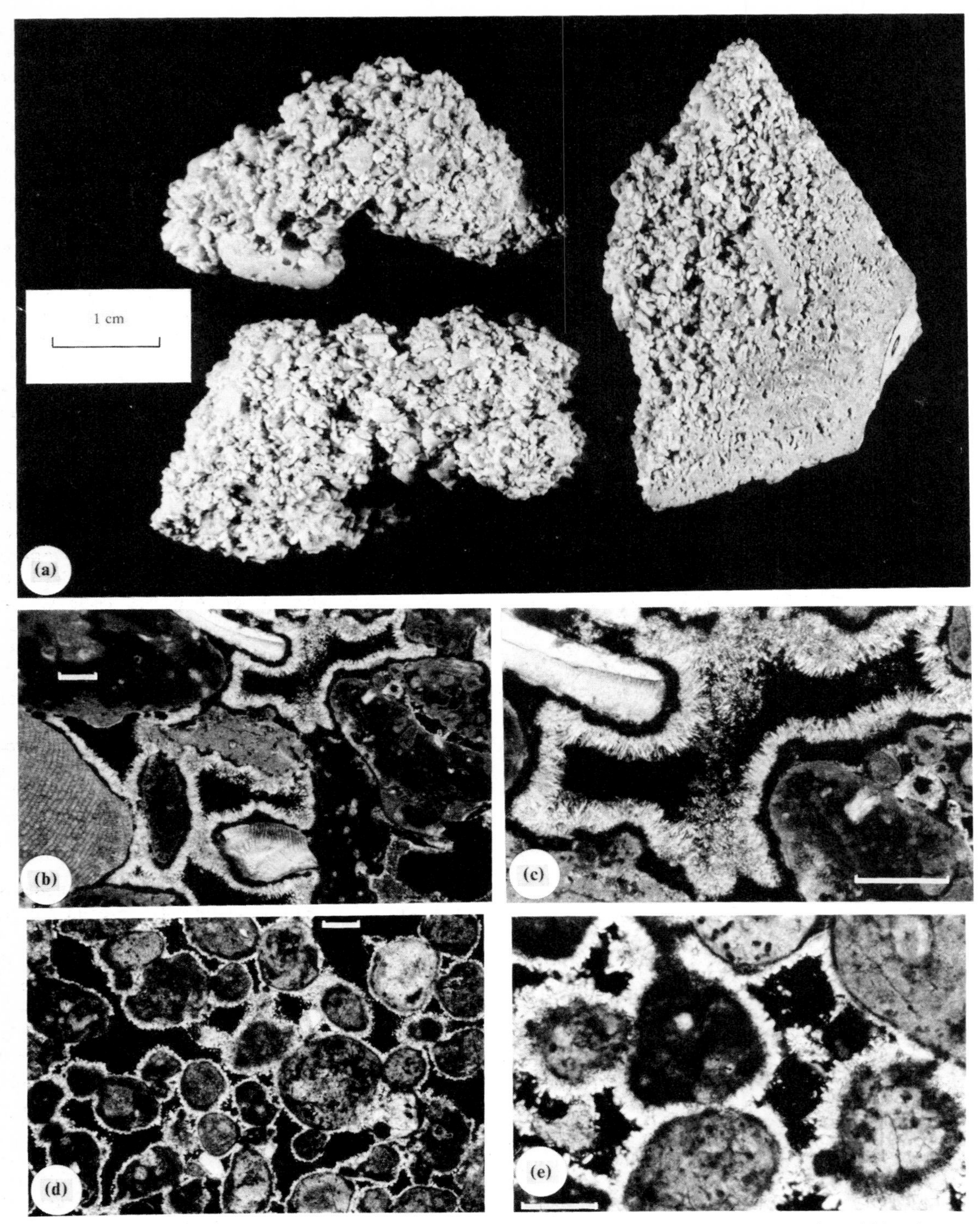

Figure 29.6 (a) Fragments of lithified Recent beachrock, Bimini, Bahamas (coll. R. Till). (b) Thin-section view of (a) under crossed-nicols to show algal bioclasts cemented by isopachous micritic and needle-fringes of aragonite rim cement. Scale bar 100 μm. (c) Close-up of (b) to show isopachous rim cements. (d) Thin-section view under crossed-nicols of a subtidal 'hardground' to show pellets and degraded skeletal fragments cemented by a fringe of high-Mg calcite rim cement; scale bar = 100 μm; Arabian Gulf (coll. R. Till). (e) Close-up of (d). Note abundant remaining pore-space (black).

and by ^{14}C dating of the carbonate cements themselves. The latter are predominantly of acicular aragonite (Fig. 29.6 b & c) or cryptocrystalline high-Mg calcite. Cement fabrics include grain-contact meniscus and micro-stalactitic types (Fig. 29.4).

The simplest model to account for beach-rock cementation notes that between the frequently agitated surface sediment and the deep sediment with stagnant pore waters there is a zone in which the mechanically stable grains are bathed in sea water which may partially evaporate during falling and low tides causing carbonate precipitation (Bathurst 1975, Milliman 1974). This model ignores the obvious complications of water flow in the beach zone (Fig. 29.7) where meteoric-derived phreatic waters also exist close to the beach surface at the intersection of the continental water table with the land surface (Hannor 1978). There have been suggestions that beach-rock cementation occurs as a response to the mixing of meteoric/phreatic and marine pore waters, but this is not supported by either mixing theory or experiment. An alternative hypothesis (Hannor 1978) postulates that the cementation is caused by degassing of CO_2-rich carbonate-saturated ground waters as they pass into shore areas having collected sufficient CO_2 from the organic decay of soil material. The CO_2-degassing from the concentrated ground water to the atmosphere is brought about by tidal oscillation and pumping effects. Although low-Mg calcite has been precipitated from such ground water, it has yet to be proved that this mechanism is capable of aragonite- or high-Mg-calcite precipitation.

Additional physical structures found in intertidal and supratidal sediments are **fenestrae** or 'birds-eyes'. These (see also Ch. 23) are small, millimetre-sized irregular cavities variously attributed to the decay of algal films, gas accumulation, desiccation shrinkage and lithification. Provided sufficient early diagenetic cementation occurs, then these cavities may persist during burial compaction and may be filled by later cement generations.

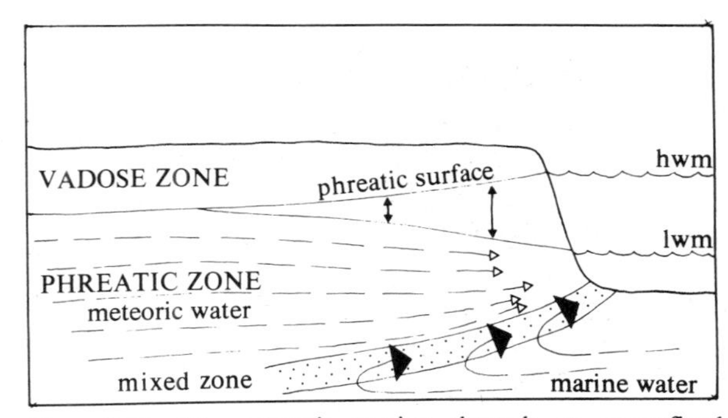

Figure 29.7 Diagrammatic section through an unconfined coastal aquifer to show seaward flux of meteoric water and the marine mixing zone (after Hanor 1978).

In addition to the many known examples of intertidal cementation there is increasing evidence for substantial areas of penecontemporaneous shallow subtidal cementation in tropical and subtropical areas. The subtidal **hardgrounds** so produced are perhaps best documented from an area of some 70 000 km^2 in the 1–60 m deep waters of the Persian Gulf where 0.05–10 m thick crusts are cemented by isopachous rim cements of high-Mg calcite and minor aragonite (Fig. 29.6d & e, Shinn 1969). Where exposed the hardground upper surfaces are smooth due to abrasion and may be bored or encrusted by hard substrate organisms (Fig. 29.8 a & b). The lower surfaces are usually irregular with abundant burrows and borings penetrating into the sediment below. The cemented layers probably accrete by the coalescence of lithified patches. Extensive polygonal fracture systems give evidence for surface expansion caused by inter-particle cementation.

The most widely quoted model for hardground cementation postulates that extensive interparticle or intraparticle cement precipitation may occur in the shallow subsurface of warm subtropical or tropical seas as long as sediment agitation is kept to a minimum and deposition rates are low. A further prerequisite is the presence of suitable carbonate nuclei. There is some evidence that organic control such as photosynthesis is important because of $\delta^{13}C$ values that range from +3 to +4‰. Pore-water circulation is also needed so that organic pumping of marine water amongst burrows and borings can occur. Many hardgrounds (up to 50 cm thick) found in ancient limestones are located at horizons of extensive hiatus. They have associated coatings and impregnations of glauconite and phosphate, and they record the change from a burrowed soft sediment to a bored lithified rock via a phase of lithified nodule development (for deep sea examples see Ch. 23h).

The above 'hiatus' or 'slow sedimentation' model for hardground development has recently received a severe setback through the discovery of hardgrounds over a whole spectrum of high-energy environments in the Eleuthera Bank area of the Bahamas platform (Dravis 1979). Here, centimetre to decimetre crusts occur on the crests and flanks of actively moving bedforms. There is a decrease in lithification from the surface downwards. Oöliths within the crusts are dull, corroded and much micritised. The cements are composed of 10–100 μm thick isopachous aragonite needles with subordinate microcrystalline high-Mg calcite. Important algal influences upon cementation are provided by the presence of algal filaments which both bind and cement grains by filament calcification and subsequent micrite precipitation. This apparent anomaly of lithified crust development in a high-energy environment is not yet understood, but it may involve very rapid colonisation of irregular scoured areas in sub-aqueous bedforms by scum algal mats which protect the surface grains from traction (see Scoffin 1970) and encourage carbonate precipitation. Further developments undoubtedly hinge upon the study of the hydrodynamics and traction characteristics of these tidal bedforms.

A further occurrence of shallow submarine cementation is within the pores and cavities of scleractinian coral reefs (Schroeder 1972, MacIntyre 1971, James *et al.* 1976). Upward growth of the coral causes abandonment of tissue-filled cavities which become the site of infiltration by very fine sediment to form the so-called **geopetal cavity fills** (or 'spirit-levels') which always show their upper surfaces parallel to the Earth's surface. Many types of cement fabric also occur, including radial fibrous

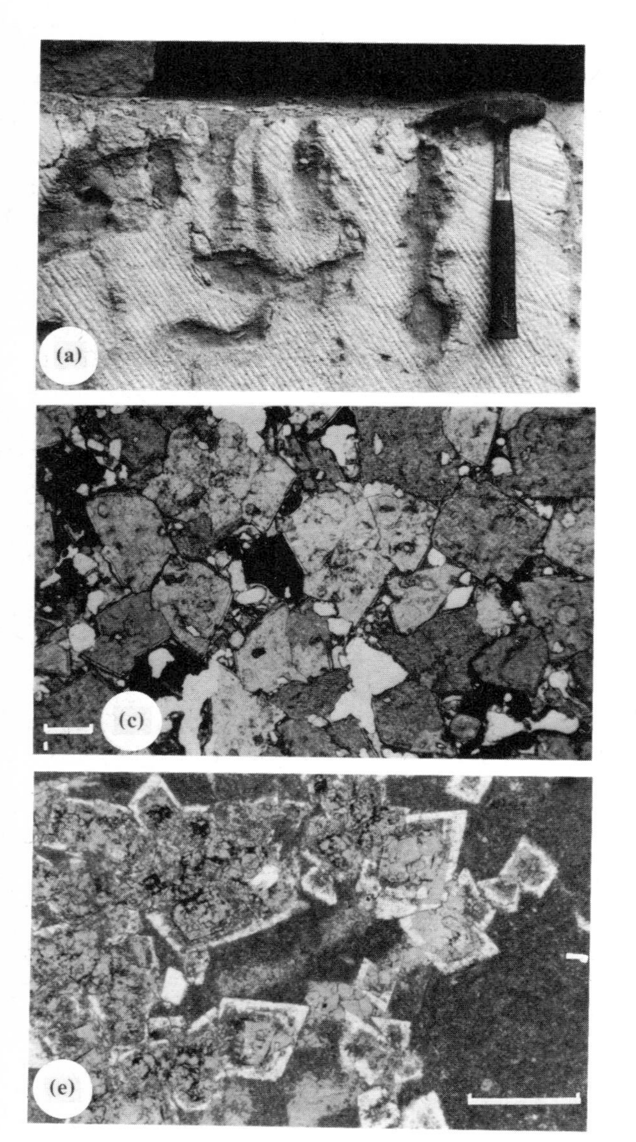

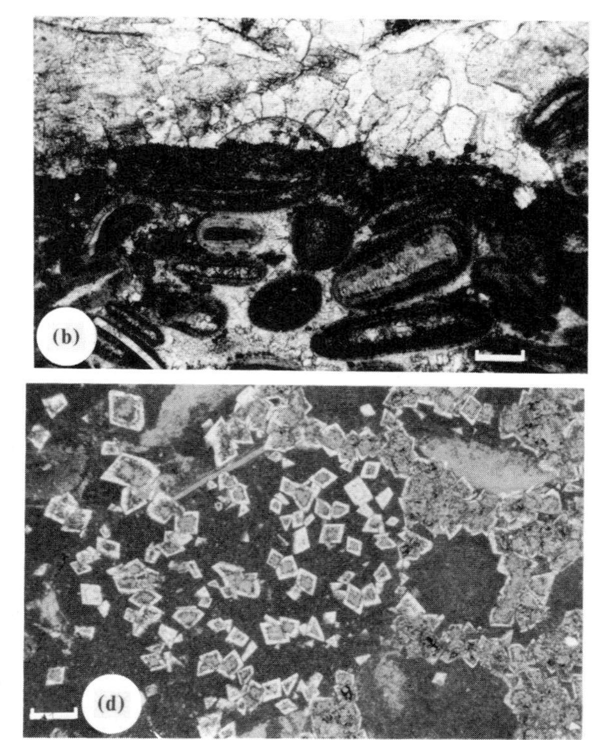

Figure 29.8 (a) Vertical section through a Jurassic hardground. The bedding plane under the hammer head shows cemented oysters. The prominent unfilled burrows were presumably abandoned soon after cementation began. Doulting Stone, Upper Inferior Oölite, Nr Bristol, England. (b) Thin section through thin micritic crust developed upon a corraded hardground surface of ?subtidal origin. Note later nucleation of calcite spar crystals from hardground surface. Scale bar = 100 μm. Dinantian, Bewcastle, Cumberland, England. (c) Thin section under crossed-nicols of a coarse dolsparite with considerable intergranular porosity. Note clear outer rims to crystals. Scale bar = 100 μm. Magnesian Limestone, Weatherby, England. (d & e) Micritic limestones with well formed dolomite rhombs showing internal granular calcite spar due to partial dedolomitisation. Note coalescence of calcite defining large areas of spar in certain areas. Jubaila Formation, Upper Jurassic of Central Saudi Arabia. Scale bars 100 μm. (Coll. R. Zeidan.)

acicular and spherulitic aragonite and microcrystalline high-Mg calcite. Cementation in these reefs is undoubtedly influenced by organic tissue decay causing increasing alkalinity, but probably more important is a 'pumping' mechanism which can circulate marine water through the porous reef frameworks. Such mechanisms are most likely caused by tidal or wave forces and in this context it is most significant that maximum lithification tends to occur within the seaward reef framework where throughflow and porewater replenishment are most effective (see also Ch. 23h).

In addition to numerous examples of shallow marine cementation there is now much evidence for contemporaneous lithification in deep oceanic pelagic carbonate sediments (see also Ch. 23). This is particularly true of partly enclosed marine basins such as the Red Sea and Mediterranean Sea where opportunities exist for development of supersalinity and warm bottom waters, sometimes caused by proximity to spreading centres (Milliman 1974). Aragonitic pteropod sediments are cemented by fibrous and cryptocrystalline aragonite whilst calcitic planktonic foraminifera sediments are cemented by high-Mg calcite. However, aragonitic cements are in general rather rare in deep-sea sediments because of the inhibiting effects of low temperatures and relatively low pH upon aragonite precipitation. Interesting examples of lithified high-Mg calcite nodule growth in Mediterranean pelagic carbonates are thought to have been triggered by high salinities, high temperatures and low sedimentation rates (Müller & Fabricius 1974). These horizons are often encrusted or bored and they provide interesting similarities to the supposed nodular growth and spread of shallow subtidal hardgrounds (see above) and to ancient nodular limestones (e.g. Knollenkalk, Ammonitico Rosso).

Thus in conclusion we see that there are numerous examples of marine penecontemporaneous cementation of carbonate sediments, from the beach zone down to the abyssal plains. It should be stressed at the same time, however, that this early lithification is the exception rather than the rule. Evidently the inhibition effects upon $CaCO_3$ precipitation in marine waters (noted in Ch. 2) still exist in the very shallow subsurface zones of diagenesis, albeit to a lesser extent.

29d Subsurface diagenesis by formation waters

'Subsurface' in this case is taken to mean areas affected by formation waters that have remained out of contact with the surface hydrological cycle for some time (see Ch. 27). Such waters, in equilibrium with elevated temperatures and pressures, are neither pure fresh water nor pure sea water and they may have originated at points distant from their eventual active role in subsurface carbonate diagenesis. Burial-zone processes must be largely inferred from the textural relationships observed in ancient limestone successions. This inevitably leads to problems and a multitude of hypotheses. It would be fair at the outset to say that subsurface diagenetic processes are still not well understood.

A number of important diagenetic changes affect aragonite skeletons and early cement phases in the subsurface realm. Most aragonite that has escaped alteration in the meteoric realm will be increasingly prone to wet recrystallisation to calcite by the solution film model noted previously (pp. 286, 289) although the aragonite may persist for several million years. Finely structured aragonite skeletons and their proteinaceous matrices are thus gradually changed (rate unknown) to aggregates of coarse-grained neomorphic sparry calcite which may preserve inclusion trails to indicate the original structure and which may show a pseudopleochroism caused by inclusions of residual organic material (Hudson 1962). The coarse neomorphic spar also shows characteristic curved to wavy intercrystalline boundaries with rare enfacial junctions (Bathurst 1975). These various properties help to distinguish neomorphic fabrics from cement fabrics (see also Ch. 27). Rare preservation of skeletal aragonite in ancient ($\leqslant$300 Ma) rocks is usually facilitated by the presence of impermeable organic-rich muds in which polar amino-acid molecules are thought to coat the aragonite with a protective water-repelling monolayer which prevents access of water to the crystals (Kennedy & Hall 1967).

Aragonite rim cements are replaced by calcite rim cements, usually before compaction breakage (Bathurst 1964) because of evidence from broken shells where only a later, second generation, cement occurs nucleated onto the broken fracture surface.

It is now thought that the distinctive radiaxial fibrous cement mosaics result from the calcite replacement of pre-existing bundles of acicular aragonite or high-Mg calcite cements (Kendall & Tucker 1973). The radiaxial fibrous fabric (Fig. 29.9) is composed of calcite mosaics whose crystals possess curved twin lamellae and glide planes, optic axes that converge from cavity walls and subcrystals that diverge from cavity walls. A replacement origin for the mosaic is indicated by evidence from truncated cavities in early lithified crusts (Fig. 29.10) by the presence of 'ghost' crystals picked out by inclusion patterns (Fig. 29.9), and by the non-planar nature of the intercrystalline boundaries. Replacement of pre-existing acicular cement bundles proceeded from the solid wall outwards into the cement by means of a thin-film solution

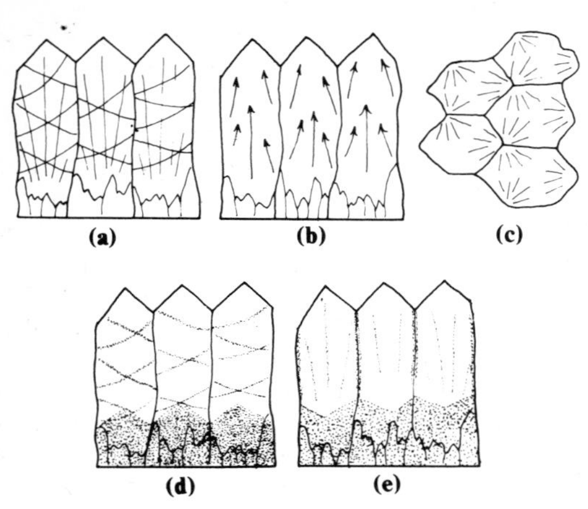

Figure 29.9 Radiaxial fibrous calcite. (a) Subcrystals divergent away from substrate and twin lamellae convex towards substrate; (b) optic axes convergent away from substrate; (c) transverse section showing radiating optic axes shared by adjacent crystals; (d) inclusion patterns along twin lamellae and along the substrate boundary; (e) inclusions along intercrystalline and inter-subcrystal boundaries. Note the inclusion-defined terminations in (d) and (e) do not always correspond with present fibrous crystals. (After Kendall & Tucker 1973).

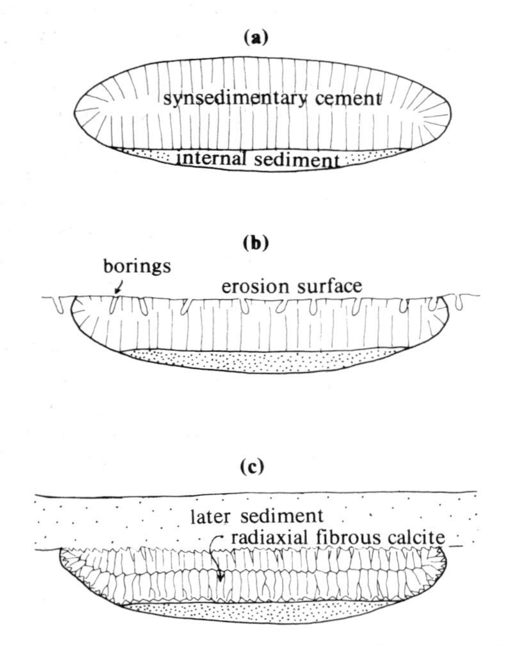

Figure 29.10 Diagrams illustrating the formation of an idealised truncated cavity and its later fill by radiaxial fibrous calcite: (a) cavity infilled with internal sediment and fibrous aragonite cement; (b) erosion and boring of cavity fill; (c) resumed sedimentation followed by replacement of aragonite cement to radiaxial fibrous calcite. (After Kendall & Tucker 1971.)

front. A variety of other calcite replacement fabrics of marine acicular aragonite are reviewed by Mazzullo (1980).

High-Mg calcite skeletal debris may persist unaltered from the marine diagenetic realm into the subsurface where a process of pseudomorphism by low-Mg calcite faithfully reproduces the details of the skeletal structures, with the Mg^{2+} ions precipitated locally as microdolomite inclusions (Lohmann & Myers 1977). It has been found that skeletal debris of high-Mg calcite origins tends to be replaced by ferroan calcite (Richter & Fuchtbauer 1978). This is further evidence that wholesale pseudomorphism is involved in the low- to high-Mg calcite change and that the process does not involve simple Mg-diffusion out of the crystal lattices. A further important deduction regarding the timing of this change in the subsurface realm also follows. Ferrous ions are not readily available in marine pore fluids during early diagenesis since sulphate reduction leads to their incorporation as ferrous sulphide phases. It follows that ferroan calcite replacement of high-Mg calcite skeletal debris must occur during burial diagenesis when such processes no longer restrict the availability of Fe^{2+}. It has been suggested that the $\delta^{18}O$ values of −5.4‰ in ferroan calcites indicate a meteoric origin for the replacement process (Richter & Fuchtbauer 1978), but such values are also likely at higher temperatures in the subsurface realm.

Let us now turn to the increasingly vexed question of **oöid diagenesis**. Two basic hypotheses involve contrasting interpretations of the simple fact that almost all modern oöids are aragonitic whilst almost all ancient oöids are calcitic.

The first hypothesis states that all oöids start off as aragonite but are changed during diagenesis to calcite (Shearman *et al.* 1970). Now, most ancient calcitic oöids show a rather finely preserved microfabric of radial calcite fibrous crystals, so that exponents of the replacement hypothesis have to explain this fact. The template hypothesis of Shearman *et al.* (1970) attempts to do this by saying that the replacement process takes place between the internal organic-rich layers in the oölith and involves dissolution of aragonite enmeshed in organic material, closely followed by precipitation of calcite which grows radially out from the concentric mucilage rings (Fig. 29.11). A severe disadvantage of this hypothesis is that the postulated mode of replacement is quite contrary to that deduced for all other aragonite allochems that are replaced by calcite, viz. large calcite crystal size, inclusion of organic material as trails proving the inability of organic layers to control crystal growth directions, and lack of crystal alignment.

A number of further petrographic observations support

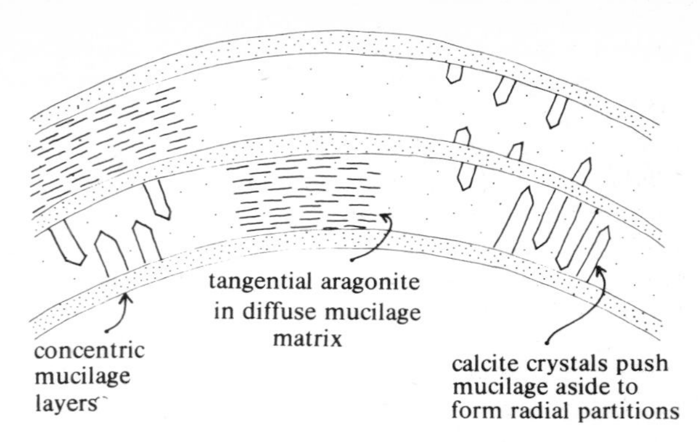

Figure 29.11 Diagram to illustrate the 'template' hypothesis for oöid diagenesis (after Shearman *et al.* 1970).

the non-uniformitarian postulate (Sorby 1879) that most ancient oöliths were calcitic with a radial microstructure (Sandberg 1975, Wilkinson & Landing 1978). Evidence from compacted oöliths indicates that the aragonitic nuclei were removed by solution in early diagenesis to give hollow elliptical shells which were crushed during burial (Fig. 29.12). The following textural evidence (Fig. 29.12) suggests that the present composition and fabric of

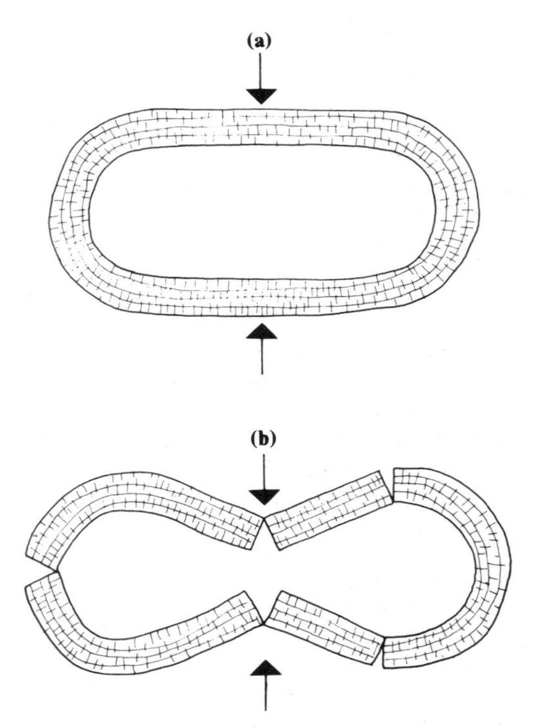

Figure 29.12 Eggshell diagenesis of oöids to show how (a) a primary radial calcitic fabric responds to (b) loading during burial (after Wilkinson & Landing 1978).

calcite is primary: (a) the calcite void-filling to the dissolved nuclei began only after compaction and the cortex was rigid at this time; (b) if the cortex was aragonite originally, why wasn't it dissolved away with the nucleus?; (c) the fractures caused by compaction are now perfectly parallel to the calcite radial laths suggesting that the fabric at this time was also radial; and (4) the fractures sometimes cut obliquely across the calcite laths, suggesting their presence before fracture. It is thus possible that such ancient oöids were of high-Mg calcite. Rare modern calcitic oöids (Laguna Madre, Barrier Reef) are also of high-Mg calcite composition.

As noted previously (Ch. 2) the predominance of primary calcitic oöids in the ancient (pre-Cainozoic) past implies an earlier lower oceanic Mg:Ca ratio (~2:1). This ratio must have progressively changed to the modern value of 5:1 which now encourages aragonite rather than calcite precipitation in the world's oceans (Sandberg 1975; see Ch. 2). This change may be the result of wholesale removal of Ca^{2+} from the oceans in planktonic foraminifera and coccoliths which have undergone an explosive evolution since the late Mesozoic.

The next point in our discussion concerns the origin of the so-called **second-generation cement phase**. The early diagenetic phases in the meteoric and marine realms have been stated to leave about 20% porosity. Since many ancient limestones show minimal ($\lesssim$5%) porosity there must be a very large-scale additional source of $CaCO_3$ to provide the remaining cement ions. The identity of this source is not clear, but before further discussion it should be noted that the above requirement for ~15% by volume of extra $CaCO_3$ is probably excessive since it was mooted before the recognition of rapid phreato-meteoric cementation, which process leads to considerable porosity decreases early on in diagenesis.

The source of the extra second-generation cement is often assumed to be the result of pressure-solution processes (Ch. 27) which provide a continuing dilute source of ions for eventual precipitation as large equant, second-generation calcite spar crystals in the remaining pore space (see Hudson 1975, Purser 1978). The large spar crystals are evidently not subjected to the lateral growth restrictions found in marine environments where the high Mg^{2+} content seems to inhibit calcite crystallisation in favour of aragonite (Folk & Land 1975). As noted previously (Ch. 27) formation waters are usually much poorer in Mg (Mg/Ca < 1) and hence the inhibition mechanism to large spar-crystal growth is absent. The relatively late age for this cement is proven by its relation to broken shell fragments already coated by a (first-generation) rim cement. The fact that much of this late cement is ferroan calcite in sequences containing argillaceous limestones (Oldershaw & Scoffin 1967) is a further indication of the importance of the burial–pressure-solution process. The Fe^{2+} ions are probably obtained by reduction and solution of adsorbed Fe^{3+} on clay mineral platelets, the latter actively encouraging the pressure-solution process.

In addition to the pressure-solution source it is likely that large-scale migration of dissolved Ca^{2+}, Fe^{2+}, Mg^{2+} and CO_3^{2-} ions from deeply buried compacting mudstone sequences is important in cementation. We have already seen (Ch. 28) how the amount of particulate $CaCO_3$ ($\delta^{13}C \approx 0$) in shales (up to 20%) may be reduced to zero below about 3.5 km depth in Gulf Coast successions. Carbonate formations frequently occur interbedded with such shales or are located up-dip at basin edges. Therefore, precipitation of the spar cement can occur in the remaining limestone pores, probably as a ferroan calcite phase at shallow depths, with $\delta^{13}C$ around zero and additional Fe^{2+} provided by the smectite→illite change noted previously (Ch. 28). There is some evidence from carbonate cements in sandstones that ferroan calcites formed in this way are progressively changed to ferroan dolomites and ankerites below depths of about 2.5 km, the extra Fe^{2+} and Mg^{2+} being provided again by the smectite→illite change induced by burial (Boles 1978).

Our penultimate topic in subsurface diagenesis of carbonates concerns the fate of aragonite muds (Ch. 2) and their conversion to calcitic micrite limestones. As noted previously, patches of calcitic microspar (3–8 μm) seem to cement finer aragonitic muds in the phreatic environment, thus providing enough 'rigidity' so that compaction effects are minimised. The widespread absence of compaction effects (squashed fossils, etc.) in ancient calcitic micrites (Bathurst 1975) lends credence to claims that such widespread phreatic cementation may be the rule rather than the exception for fine-grained carbonate rocks. A persistent problem in the interpretation of ancient micrites is the origin of patches of coarser microspar calcite crystals which are set in a groundmass of true micritic crystals or grains. This structure is termed **structure grumeleuse** when the micrite aggregates are completely surrounded by microspar, giving a peloidal appearance to the micrite. These aggregates or pools of coarser crystals are usually interpreted as having resulted from a recrystallisation process in which larger crystals grew at the expense of the smaller ones (Bathurst 1975). Such a process is thermodynamically sound since larger grains have less surface energy per unit mass and hence can enlarge at the expense of small neighbouring grains. Such a process of **aggrading neomorphism** must be a wet change in the diagenetic context, but the details, kinetics and energetics of the change are unknown. Indeed it is at present uncertain just how many of the apparently neomorphic microspar fabrics in micrites are in fact the result of phreato-marine grain dissolution and calcite reprecipitation as true cement.

In contrast to the originally aragonitic lime muds discussed above, the pelagic Cretaceous chalks are a group of originally calcitic lime muds formed by the accumulation of vast numbers of tiny pelagic coccoliths and pelagic foraminifera skeletons in fairly shallow waters ($\gtrsim$ 250 m). The absence of cementation fabrics in many chalks has been ascribed to the lack of unstable aragonite or high-Mg calcite in the original sediments. Chalks buried below about 2000 m are usually well cemented because of pressure solution and reprecipitation. The lack of such cementation in some deeply buried chalks has been attributed to the presence of sufficient Mg^{2+} ions in the pore solutions to delay $CaCO_3$ cementation onto low-Mg calcite detrital nuclei (Neugebauer 1974). This is a further example of our old-friend the 'Mg-inhibition mechanism' at work again (Ch. 2). Where observed, chalk cements are syntaxial, that is the calcite cements overgrow the detrital nuclei in optical continuity. Certain very well lithified chalks such as those of Northern Ireland give $\delta^{18}O$ values of −6‰ PDB and provide evidence for cementation under meteoric influences under a very thick cover of Tertiary basalts (Scholle 1974).

An exception to the general statement above concerning the under-lithified nature of most chalks is the occurrence of thin hardgrounds throughout many chalk successions. These horizons are analogous to those hardgrounds discussed above forming in the shallow marine subtidal realm of diagenesis. They contain ample faunal evidence for early cementation and concomitant water shallowing, the two processes being in some (unknown) way connected.

29e Summary of limestone diagenesis

Ancient limestones will have passed through more than one diagenetic realm on their journey from the Earth's surface to the subsurface. This progression of diagenetic conditions will cause successive generations of cement to occur which may lend themselves to stratigraphic investigation when traced out over an area. The cement generations will be recognisable on the basis of form, fabric and composition. Although bulk isotopic analysis of the final limestone will tend to obscure the internal complexities of the diagenetic process, it is evident that most ancient limestones have a fairly narrow range of carbon isotope compositions around $\delta^{13}C \approx 0$ (± 2‰). Without knowing the detailed compositions of individual cement phases (but see results of Dickson & Coleman

1979, Marshall & Ashton 1980) and their exact numerical proportions of the whole rock it is rather difficult to conclude that most marine limestones have been necessarily wholly cemented by marine-derived CO_3^{2-} ions, but it is equally clear that very large-scale additions of light or heavy carbon have not occurred (Hudson 1975).

In concluding these remarks on limestone diagenetic processes the evocative story of the evolution of an 'ideal' (English) Jurassic limestone as told by Hudson (1977a) is apposite:

> It was deposited in the Jurassic as a carbonate sand composed of a mixture of oöids, aragonitic bivalve and gastropod shells, and calcitic oyster shells. It formed an offshore 'sand'-wave at a depth of 10 m and a temperature of 20°C. Salinity was normal oceanic and $\delta^{18}O$ of the seawater was −1 SMOW, there being no ice around in those days. The sediment had a bulk $\delta^{18}O$ of −2 and $\delta^{13}C$ of +2.
>
> While on the sea-floor, the margins of many of the shell fragments were micritized producing micrite envelopes with higher $\delta^{13}C$ than the bulk sediment.
>
> Next, sea-level fell, and the sediment was exposed for a few thousand years to freshwater emanating from low limestone islands, although for most of the time it remained in the phreatic zone. Most of the aragonite in the sediment dissolved, and a little non-ferroan, low-Mg calcite was precipitated around the allochems which thus became very lightly lithified. The new cement had a $\delta^{13}C$ of −5 and $\delta^{18}O$ of −4, reflecting its freshwater origin and some input of soil CO_2.
>
> Subsidence resumed, and the limestone was gradually buried in what became a shallow sedimentary basin. Some compaction occurred, rupturing some of the micritic envelopes and their cement fringes.
>
> The stratigraphical column beneath the limestone contained fossiliferous shales. Much water was compacted out of these, and it carried bicarbonate in solution, due partly to solution of aragonite and to pressure-solution of shells where they were brought into contact along developing microstylolites. The water also contained a little ferrous iron in solution, because conditions were by now reducing. The water passing slowly through my limestone, eventually many thousands of pore-volumes of it, was not all simply sea-water; a lot of it was meteoric rain-water that had slowly, driven by a hydrologic head on-shore, been pushed out underneath the still subsiding sedimentary basin.
>
> Gradually, very gradually, the pores of the limestone filled up with ferrous sparry calcite with $\delta^{13}C$ = +1, $\delta^{18}O$ = −7; the temperature was by now, at 500 m burial, 35°C, and the $\delta^{18}O$ of the water −3.25. Lithification was complete. Isotopic evolution was finished. Or was it? – with one tectonic leap my limestone was free: or at least at outcrop. A crack appeared in it, and was filled up with calcite $\delta^{13}C$ = −10, $\delta^{18}O$ = −7, in England, almost yesterday. The limestone now has a $\delta^{13}C$ of +1.5, $\delta^{18}O$ = −6. The oyster shell still has $\delta^{13}C$ = +2, $\delta^{18}O$ = +1.2, just as it always had. One day I, or my reader, will collect this limestone.

29f Models for dolomitisation

Dolomite rock and partially dolomitised limestones are very common in the geological record, yet instances of primary dolomite precipitation at the present day number, perhaps, only two. In one of these cases, at Deep Spring Lake, USA, the shallow evaporative lake contains sediments with tiny (<1 μm) euhedral dolomite crystals which may be proved by ^{14}C dating to have grown at the astonishingly slow rate of 0.09 μm Ka^{-3}. In most Earth surface environments, crystals with such a slow growth rate would be overwhelmed by other sediment grains and precipitation so as to form an insignificant proportion of the accumulating sediment.

The other case of suspected primary precipitation is at Coorong Lagoon, Australia. In this system of isolated, periodically hypersaline lakes dolomite is precipitated from the alkaline lake waters during dry periods. The lake water itself is derived from resurgence of saline groundwater. The dolomite is deposited as a 'gelatinous, brine-saturated, yogurt-like slurry' (Von der Borch & Lock, 1979; Muir *et al.* 1980), but the exact mechanism and time of precipitation is still unclear. The fine-grained dolomite is associated with algal mats, chert, indurated crusts and desiccation polygons, but in contrast to sabkha dolomite is *not* associated with evaporite deposits (p. 298). Muir *et al.* have used a Coorong model for dolomitic facies to explain many features of Proterozoic dolomite formation in Northern Territory, Australia.

The extreme rarity of primary precipitated dolomite today might lead us to conclude that modern sea water is very undersaturated with respect to dolomite. Such a conclusion is in error. Consider the following reversible reaction.

$$Ca^{2+} + Mg^{2+} + 2CO_3^{2-} \rightleftharpoons \underset{\text{dolomite}}{CaMg(CO_3)_2}$$

Although subject to some uncertainty, the equilibrium

constant, $K_{dol.}$, and the ion activity product in sea water IAP_{dol}, have the following values

$$K_{dol.} \approx 10^{-16.7} \approx [Ca^{2+}]\,[Mg^{2+}]\,[CO_3^{2-}]^2$$

$$IAP_{dol.} \approx 10^{-15.0}$$

indicating that sea water is supersaturated with respect to dolomite and therefore dolomite should be precipitating widely (Hsü 1966). Further, Hsü shows that dolomite is more stable than either calcite or aragonite in sea water by considering the following reaction

$$Ca^{2+} + \underset{\text{dolomite}}{CaMg(CO_3)_2} \rightleftharpoons Mg^{2+} + \underset{\text{calcite}}{2CaCO_3}$$

whereby $K = 0.67 = [Mg^{2+}]/[Ca^{2+}]$. In sea-water $[Mg^{2+}]/[Ca^{2+}] = 5.7$, indicating that the reverse reaction above should proceed spontaneously to equilibrium, i.e. carbonate sediments should be dolomitised by contact with ordinary sea water.

The incompatibility of the observed and theoretical behaviour of dolomite in sea water is the crux of what has been called 'the dolomite problem' (for the most up-to-date review, see Zenger & Dunham 1980). The inability of dolomite to precipitate from sea water must be caused by some crystallisation difficulty, i.e. a **kinetic factor**. In the case of dolomite this difficulty is thought to arise from the extreme regularity of the crystal lattice (Fig. 29.13), comprising alternate layers of Ca^{2+} ions, CO_3^{2-} ions, Mg^{2+} ions, CO_3^{2-} ions, and so on. The Ca^{2+} and Mg^{2+} ions have such similar properties and sizes (Ca = 1.08 Å; Mg = 0.80 Å) that they are in close competition for lattice sites during precipitation. At normal Earth surface temperatures a magnesian calcite results by rapid spiral growth (Fig. 29.13a, Deelman 1975). Special conditions are required for the face-by-face growth of dolomite (Fig. 29.13b). These are thought to include slow growth and dilute ionic solutions (see below).

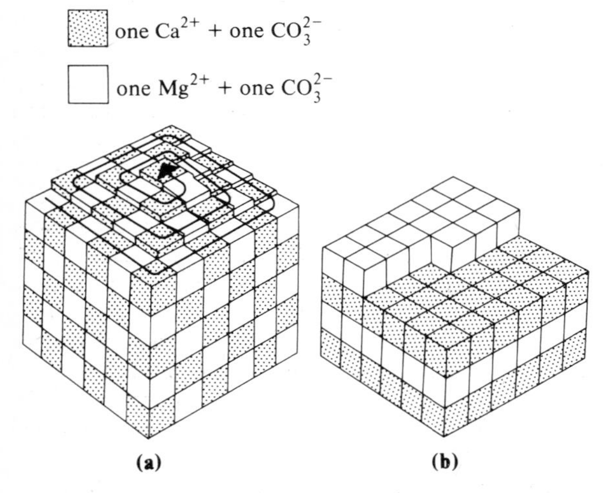

Figure 29.13 (a) Spiral growth of high-Mg calcite in bicarbonate solutions containing both Ca^{2+} and Mg^{3+}; (b) layer-by-layer growth required for dolomite formation from the same solution. (Both after Deelman 1978.)

Attempts to precipitate dolomites under laboratory conditions at Earth surface temperatures lead to the formation of magnesian calcites which lack the ordering of the true dolomite lattice. Under certain conditions, however, a metastable form of dolomite may be produced by primary precipitation or by the alteration of pre-existing aragonite or calcite. The **protodolomites** thus formed are best defined formally as 'metastable single-phase rhombohedral carbonates which deviate from the composition of the dolomite that is stable in a given environment, or are imperfectly ordered or both, but which possess a high degree of cation order as witnessed by the unambiguous presence of order reflections in X-ray diffraction patterns' (Gaines 1977). Protodolomite appears as a precursor to dolomite in replacement reactions. Protodolomites are in effect calcian dolomites with an excess of Ca^{2+} in their lattices $(Ca_{1.05}\,Mg_{0.95}\,(CO_3)_2$. High temperatures (~200°C) are needed to precipitate pure dolomite in the laboratory.

Having discussed the various chemical difficulties involved in direct dolomite precipitation from sea water, let us turn to discuss the three major secondary dolomitisation mechanisms which have been proposed in recent years. These are conveniently referred to as the evaporite brine-residue model, the fresh phreatic/marine groundwater mixing model, and the formation water model.

The **evaporite brine-residue model** is based upon the observed chemical and mineralogical changes taking place just under the surface of the broad supratidal plains around the Arabian Gulf known as **sabkhas** (see also Chs 23 & 30). Very widespread dolomitisation has taken place in these sediments, involving the replacement of aragonitic sediment by a very fine-grained protodolomite mudrock (Illing *et al.* 1965, Kinsman 1966). Analysis of the pore waters (Fig. 29.14) at various points over the surface of the sabkha (from the lagoon inland) reveals systematic chemical changes indicative of large-scale removal of Ca^{2+} and SO_4^{2-} ions as gypsum, followed by massive dolomitisation of aragonitic sediment grains of the former lagoon sediments. The gypsum is precipitated as nodules within the dolomitised carbonates (Ch. 30). The chemical changes resulting in dolomitisation are caused by progressive concentration of sea water by evaporation ($\delta^{18}O = +3$ to $+7$) from pores at the sabkha surface (Kinsman 1966). Pore-water replenishment

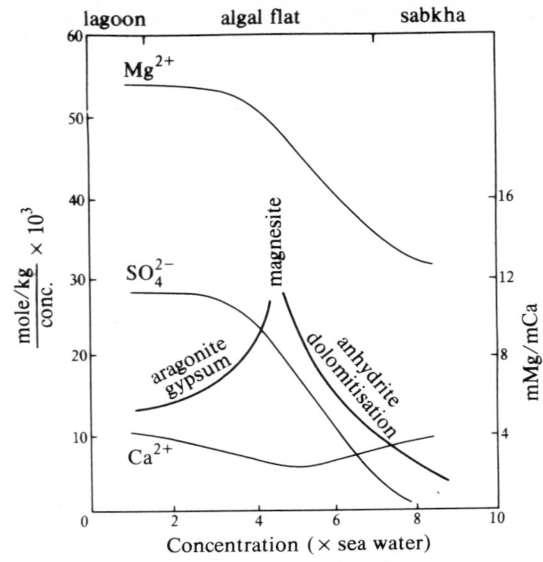

Figure 29.14 Relations between molar concentration of ions, the molar ratio Mg^{2+}:Ca^{2+} and brine concentration in the Abu Dhabi sabkha (after Kinsman 1966).

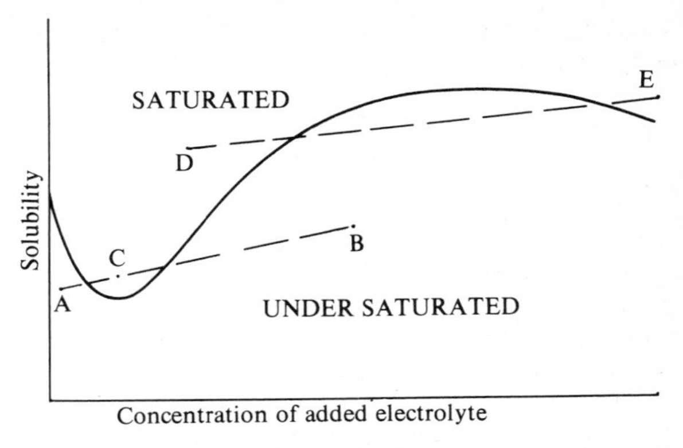

Figure 29.15 Hypothetical solubility curve showing how mixing of waters undersaturated with mineral phase *x* (A–B) may result in a supersaturated water (C), and how mixing of supersaturated waters (D & E) may result in undersaturation (after Runnells 1969).

occurs by periodic storm flood events, areas furthest from the lagoon being replenished less frequently and hence showing greatest concentrations of Mg^{2+}.

From the above evidence it seems that the kinetic constraints upon dolomite precipitation are overcome in systems that contain a high-order ratio of Mg:Ca, thus facilitating replacement dolomitisation of a $CaCO_3$ precursor. The Mg:Ca ratio reaches a value of over 10 on some portions of the sabkha. Increased Mg:Ca ratios of this magnitude are only possible by massive removal of Ca^{2+} as gypsum and anhydrite evaporites. It is likely that SO_4^{2-} removal is essential in any case since dolomite is rapidly dissolved in the face of SO_4-rich ground waters (see below).

Examples of sabkha brine dolomitisation are widespread in the geological record but inevitably after the introduction of the model in the early 1960s it was applied willy-nilly to ancient dolomite occurrences that hardly fitted the intimate constraints demanded by the model (see Zenger 1972). A particular ruse was to appeal to the subsurface sinking or reflux of high Mg:Ca ratio brines which could then dolomitise vast areas of the vadose and parts of the phreatic zones. Such models were applied even in the face of the relatively feeble presence of a few evaporitic bands in very thick, regionally extensive dolomites.

The **groundwater mixing model** for dolomitisation has revolutionised the field, providing a ready and rational explanation for the numerous examples of ancient dolomites of regional extent that seemed to defy explanation by the sabkha model. The model also accounts for the very low Mg:Ca ratios found in formation waters (1:2 to 1:4 typically). The model is based in principle upon the non-linearity of solubility curves when contrasting solutions are mixed (Fig. 29.15; Runnels 1969).

Experimental data on the solubility of calcium carbonate as a function of added salts indicates that mixing of solutions which differ only in their content of dissolved electrolytes may cause either precipitation or dissolution. For dolomitisation (Hanshaw *et al.* 1971, Land 1973), calculations show that mixing meteoric ground waters (essentially fresh water) with up to 30% sea water causes undersaturation with respect to calcite, whereas dolomite saturation increases continuously (Fig. 29.16). Therefore in the range 5–30% sea water the mixed solution can cause calcite to be replaced by dolomite or, indeed, can cause primary dolomite precipitation (Badiozamani 1973). The ability of the mixed solution to cause dolomite formation must be related to the ionic dilution produced by the mixing effect (Fig. 29.17; Folk & Land 1975). Note for example that the dilution of one part of a sabkha brine with nine parts of typical river water would only reduce the Mg:Ca ratio from 7:1 to 6:1. Such severe ionic dilution is postulated to cause slow precipitation of dolomite, causing the perfect ordering necessary for dolomite lattice formation to take place on a layer-by-layer basis (see Deelman 1975).

The great advantage of the mixing model is that very large-scale dolomitisation may result at the junction of the fresh phreatic and marine groundwater realms. Modern examples of dolomitisation at these junctions are known

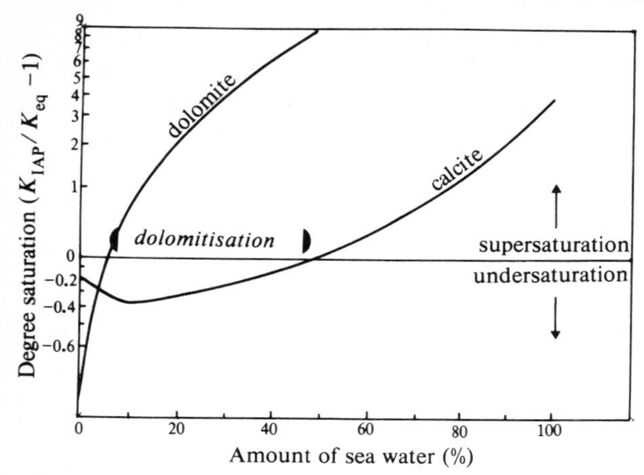

Figure 29.16 The mixing of meteoric fresh water with sea water may cause dolomitisation in a zone (5–30% sea water) which is undersaturated with respect to calcite and supersaturated with respect to dolomite (after Badiozamani 1973).

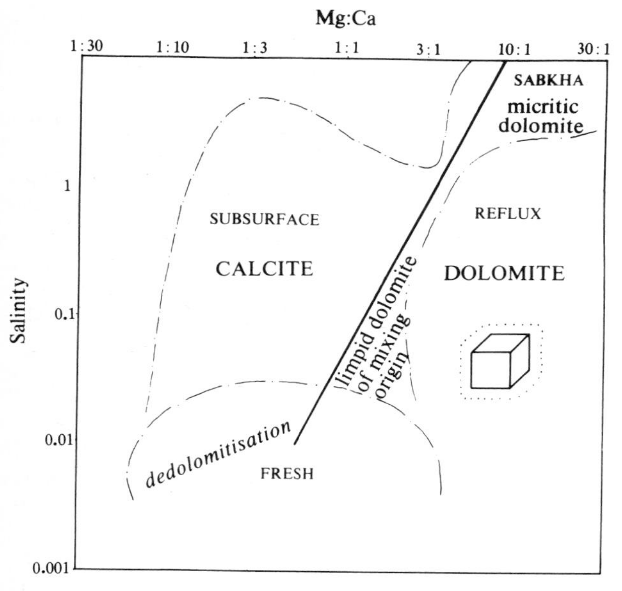

Figure 29.17 Folk & Land's hypothesis concerning the influence of salinity and Mg:Ca ratios upon dolomite crystal size and stability (after Folk & Land 1975).

in the Jamaican and Floridan aquifers (Land 1973, Hanshaw *et al.* 1971, Randazzo & Hickey 1978). The ancient examples are likely to be legion, particularly at horizons associated with sedimentary regression and/or sea level fall. The dolomite spar and microspar crystals that are thought to result from the mixing process are perfectly formed, clear and euhedral if precipitated in cavities. They show perfect stoichiometry because of slow and careful precipitation. They are light in $\delta^{18}O$ and dissolve much less readily in dilute acid than do sabkha-type dolomites which were formed by rapid, imperfect crystal growth via a protodolomite precursor. They have been appropriately termed **limpid dolomites** by Folk and Land (1975). Large-scale removal of Mg^{2+} from ground waters to form these dolomites may partly explain the low Mg^{2+} contents found in formation waters (Ch. 26), although it is likely that other mechanisms also apply.

The final dolomitisation model, the **formation water model**, is not yet fully worked out but is likely to be of some importance where deeply buried limestone beds are receiving pore-water volumes from compacting mudstones. Liberation of Mg^{2+} and Fe^{2+} from smectite clays in their change to illite (Ch. 28) may cause dolomitisation or ankeritisation of pre-existing calcite minerals or precipitation of dolomite or ankerite cements if a source of Ca^{2+} and CO_3^{2-} ions is available, e.g. from oxidation of organic materials. The dolomite produced is always likely to be ferroan and may show strongly negative $\delta^{13}C$ because of incorporation of ^{13}C-poor methane or thermally induced decarboxylation reactions (Ch. 28). Because it is formed by a deep burial reaction, the dolomite should show fabric evidence for late precipitation or replacement.

An interesting but relatively small-scale example of dolomitisation is provided by the occurrence of finely interlaminated dolomite and calcite in many ancient stromatolites, the dolomite concentrating in the dark, algal-rich layers of the structures (Gebelein & Hoffman 1972). The algal filaments are thought to accumulate Mg^{2+} preferentially and to encourage high-Mg calcite precipitation, perhaps aided by bacterial action. In the reducing environment within the dead part of the stromatolite, high pH causes the CO_2 produced by photosynthesis and algal decay to go into CO_3^{2-} rather than HCO_3^-. Dolomite precipitation is then encouraged by the high alkalinity. The importance of high alkalinity for dolomite precipitation is illustrated by the fact that at both the Coorong Lagoon, Australia, and in the Bonaire lagoons of the Dutch Antilles, dolomite formation occurs in brines with high CO_3^{2-} contents whereas in normal sea water CO_3^{2-} is much outnumbered by HCO_3^- and hence is not available for incorporation in the dolomite lattice (Fig. 2.2).

Let us end this section with some brief discussion of a number of further points concerning dolomitisation. Faced with a dolomite rock thin section the investigator should look carefully for evidence of replacement such as 'ghosts' of replaced allochems. If the dolomite has formed in a partially closed system (without additional CO_3^{2-}) by replacement of limestone and migration of excess Ca^{2+} ions, then a volume decrease should have occurred because of the increase in dolomite density compared to calcite (2860 kg m^{-3} compared to 2720 kg m^{-3}). Characteristic intergranular porosity of about 10% should be present. Evidence for multi-stage dolomite growth is provided by internal growth halt lines within dolomite crystals (Fig. 29.7). If the early dolomite is ferroan, then a line of oxidised Fe^{3+} 'scalings' at a growth halt may indicate temporary exposure to an oxidising environment. Dolomite dissolution and replacement by calcite – the so-called **dedolomitisation** (Fig. 29.7d & e) – is common in areas where dolomite rocks have been washed through by vadose meteoric waters rich in SO_4^{2-} ions. Such a situation happens frequently below a dissolved-out evaporite sequence, spectacular examples being known from the Jurassic of central Arabia (Zeidan unpubl.) and the Zechstein of Europe. Evidently the sulphate ions encourage dolomite to go into solution and calcite to precipitate. This is an expected reversal of the trend noted above where we saw the SO_4^{2-} precipitation as gypsum was a prerequisite for dolomitisation.

29g Classification

Two useful and slightly different modern classifications have been proposed for limestones.

Folk's classification (1962) divides limestones into two major clans – micrites and sparites – depending upon the predominance of each sort of 'matrix' in the rock. The clans are then subdivided according to the predominant allochems (Table 29.1) present and may be given further descriptive tags regarding spatial density of allochems and of allochem sorting. The classification as it stands is not genetic and does not imply that all the micrites tend to be fairly quiet-water sediments, since micrite is sometimes of cement origin. Another point is that rocks such as oömicrites will tend to be met rather rarely since the evidence for agitation provided by the oölite allochems obviously clashes with the apparently quiet-water origin of the micrite. A number of solutions may present themselves, e.g. the oöliths are washed into lagoonal muds by storms (Ch. 23); the micrite is resedimented from above or is a cement phase. Perhaps the most confusing sort of rock to fit into the classification is one with a largely neomorphic texture in which the spar or microspar is not a cement but which may have formed from an originally micritic primary matrix. Careful observation of thin sections is needed to spot such 'demons' and to classify the rock according to its pre-neomorphic state, if required.

A disadvantage of the Folk classification is that it rather ignores recent carbonate sediments and partially lithified limestones with abundant pore space.

Dunham's (1962) simple alternative classification (Table 29.2) overcomes these difficulties. Note that a grainstone may be 10% lithified or 100% lithified, there being no term used for cement content. The classification is thus entirely a depositional one and it does not depend upon (sometimes rather difficult) decisions as to the primary or secondary nature of the spar matrix and what effect this should have upon the status of the rock name. Problems still arise, however, when the *depositional* textures of ancient limestones have to be deduced.

The point about both classifications is that they require active effort with hand lens and acid bottle in the field. The very act of classification releases a torrent of

Table 29.1 Classification of carbonate rocks according to depositional texture (after Dunham 1962).

DEPOSITIONAL TEXTURE RECOGNISABLE					*DEPOSITIONAL TEXTURE NOT RECOGNISABLE*
original components not bound together during deposition				original components were bound together during deposition . . . as shown by intergrown skeletal matter, lamination contrary to gravity, or sediment-floored cavities that are roofed over by organic or questionably organic matter and are too large to be interstices	*crystalline carbonate* (subdivide according to classification designed to bear on physical texture or diagenesis)
contains mud (particles of clay and fine silt size)			lacks mud and is grain-supported		
mud-supported		grain-supported			
less than 10% grains	more than 10% grains				
mudstone	*wackestone*	*packstone*	*grainstone*	*boundstone*	

Table 29.2 Classification of carbonate rocks according to the nature and proportions of lime mud matrix and pore-filling spar cement. Rock terms are illustrated by biogenic grains; for other dominant grains substitute the suffixes oö-, pel-, intra-, e.g. oömicrite, pelsparite, etc. Also shown is Folk's earlier terminology (after Folk 1962).

	OVER 2/3 LIME MUD MATRIX				SUB-EQUAL SPAR & LIME MUD	OVER 2/3 SPAR CEMENT		
Percent allochems	0–1%	1–10%	10–50%	OVER 50%		SORTING POOR	SORTING GOOD	ROUNDED & ABRADED
Representative rock terms	MICRITE & DISMICRITE	FOSSILI-FEROUS MICRITE	SPARSE BIOMICRITE	PACKED BIOMICRITE	POORLY WASHED BIOSPARITE	UNSORTED BIOSPARITE	SORTED BIOSPARITE	ROUNDED BIOSPARITE
1959 terminology	micrite & dismicrite	fossiliferous micrite	biomicrite		biosparite			
Terrigenous clastic analogues	claystone		sandy claystone	clayey or immature sandstone		submature sandstone	mature sandstone	supermature sandstone

potentially useful information about both the depositional and diagenetic history of the rock. More detailed deductions concerning diagenetic fabrics must await thin-section study.

29h Summary

Diagenetic modifications to carbonate sediments take place rapidly in the meteoric and phreatic zones. Meteoric diagenesis involves the solution of high-Mg calcite and aragonite by rain water that has approached equilibrium with the ambient pCO_2 in the soil. $CaCO_3$ is then precipitated locally as stable low-Mg calcite cement. Phreatic cementation produces a coarse-grained sparite cement of low-Mg calcite, the dilute phreatic waters encouraging slow uninterrupted precipitation of large calcite crystals as CO_2 degassing takes place. Early cementation by aragonite or high-Mg calcite occurs at or near the sediment/water interface in the intertidal zone and in shallow subtidal and deep oceanic environments where low sediment deposition rates occur. During burial diagenesis a number of changes take place, including: aragonite dissolution and reprecipitation as low-Mg calcite; Mg-diffusion from high-Mg calcite; massive precipitation of low-Mg calcite (often ferroan) as a second cement phase; and neomorphic growth of calcite microspar. Dolomite may form as a secondary phase after $CaCO_3$ due to the action of sabkha brine, groundwater mixing or formation water.

Further reading

Again, Bathurst (1975) is obligatory reading whilst Longman's (1980) recent review contains an interesting synthesis of near-surface carbonate diagenesis. Hudson (1977a) is the best modern review on the application of stable isotopes to limestone diagenesis. Scholle (1978) illustrates many diagenetic textures most beautifully. The most up-to-date volume on dolomite formation is edited by Zenger *et al.* (1980). Especially recommended in this volume is the thoughtful trace element and isotopic review by Land and a number of case history studies of ancient dolomite occurrences.

30 Evaporites, silica, iron and manganese

30a Evaporites

As discussed in Chapter 3, the most common precipitated salts are halite, gypsum and the complex group of potash minerals. To those we must add a group of salts precipitated within deposited clastic and carbonate sediments as early diagenetic phases. These include anhydrite, magnesite and celestite, as well as gypsum and minor halite. Both groups of salts may undergo changes during burial, including dehydration, recrystallisation and plastic flow. Uplift of deeply buried evaporites induces yet more changes, including rehydration, calcitisation and solution. Thus a hand specimen of evaporite collected at outcrop may have had an exceedingly complex diagenetic history.

A major breakthrough in evaporite studies occurred in the mid-1960s when it was discovered that evaporite minerals were forming as diagenetic growths in the shallow subsurface of arid supratidal plains around the southern shores of the Arabian Gulf (Shearman 1966, Kinsman 1966). These arid plains are known as **sabkhas**, from the Arabic word for salt flat. As discussed in Chapter 23, these sabkhas border lagoonal water bodies and record the progradation of sediment across the lagoons since the great Flandrian sea level rise. The featureless sabkha flats are subject to great evaporation so that their interstitial pore waters become progressively concentrated with salts. Lagoonal waters are periodically driven across the seaward sabkha fringe as a shallow sheet flood during storms (Fig. 30.1). The sea water percolates into the subsurface to join the phreatic pore water at depths of up to 1 m. It is also possible that water is drawn up from the water table under the sabkha surface by a process termed **evaporative pumping** (Hsü & Siegenthaler 1969; McKenzie *et al.* 1980). With increasing distance from the lagoon the pore waters show progressive changes in composition because of progressive evaporative concentration and diagenetic mineral precipitation and dissolution (Fig. 30.2). Gypsum and minor aragonite precipitation (Fig. 29.14) begins in the shallow subsurface of the upper intertidal zone. Continued gypsum precipitation in the lower sabkha surface causes the Mg:Ca ratio of the pore-water brines to increase until, at values of about 10, the primary aragonitic sediments in the shallow subsurface become dolomitised (see Ch. 29f) and ultimately magnesite forms.

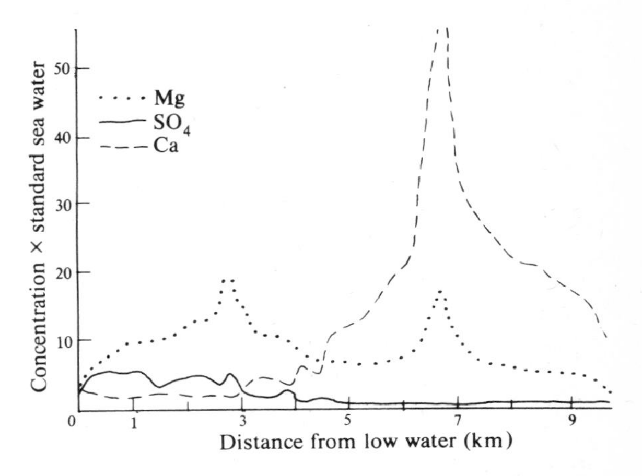

Figure 30.1 Changes in the ionic composition of sabkha brines with distance from low water mark (after Bush 1973).

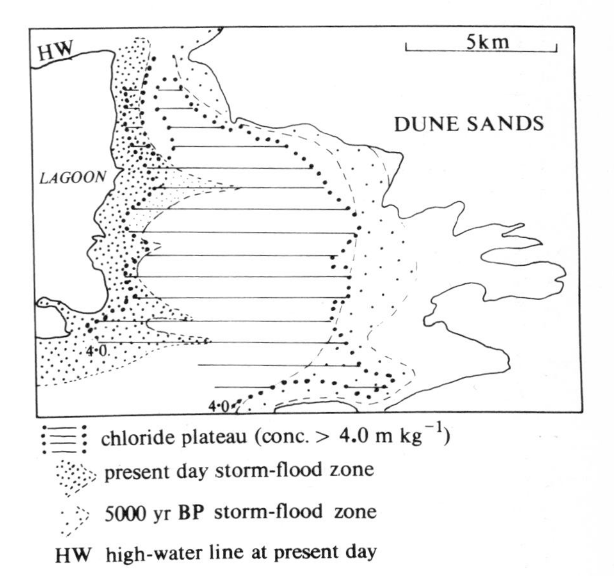

Figure 30.2 Sketch map of Abu Dhabi sabkha to show the chloride 'plateau' (>4 moles Cl^- kg^{-1} solution) together with the present-day zone of marine storm flooding and the probable extent of flooding 5000 years BP (after Patterson & Kinsman 1976).

Some celestine ($SrSO_4$) is formed from the Sr^{2+} released during the dolomitisation of aragonite. Anhydrite begins to dominate the diagenetic evaporites with increasing distance from the normal high-water level (Fig. 30.3). It replaces gypsum as pseudomorphs and grows as a primary precipitate. The anhydrite occurs as distinctive coalescing nodular growths, the **macrocel** or 'chicken-mesh' texture (Fig. 30.5), and as folded **enterolithic** layers, both growth habits testifying to a largely displacive growth mechanism. As the landward margin of the sabkha is approached, the anhydrite becomes hydrated back to gypsum under the influence of brines of continental origin (*q.v.*). Halite occurs only as ephemeral crusts on the sabkha surface after marine sheet-flood events and potash salts are entirely absent. Kinsman (1976) has postulated that there is an important relative humidity control upon the precipitation of halite and K-salts so that only evaporite basins with high continentality (and therefore lower mean relative humidity) are likely to precipitate such salts.

Evaporite diagenesis in sabkhas is a delicate function of brine composition (Patterson & Kinsman 1977). The seaward sabkha margin is fed by marine brines with $Cl^-:Br^-$ ratios of less than 1000 whilst the landward sabkha margin is fed by continental brines with $Cl^-:Br^-$ ratios of greater than 5000 (Fig. 30.1). The mixed brine zone over much of the sabkha shows Cl^- concentrations of $\geqslant 4$ moles kg^{-1} and it is these brines that are in equilibrium with the characteristic subsurface nodular anhydrite noted above. The continental brines show much decreased Cl^- concentrations, causing anhydrite to hydrate to gypsum. Long-term preservation of most of the anhydrite in a prograding sabkha is possible only if the rate of progradation is greater than that of the rate of movement of the boundary between the continental brine and mixed brine. Studies of the Abu Dhabi sabkhas show that this is so (Patterson & Kinsman 1977).

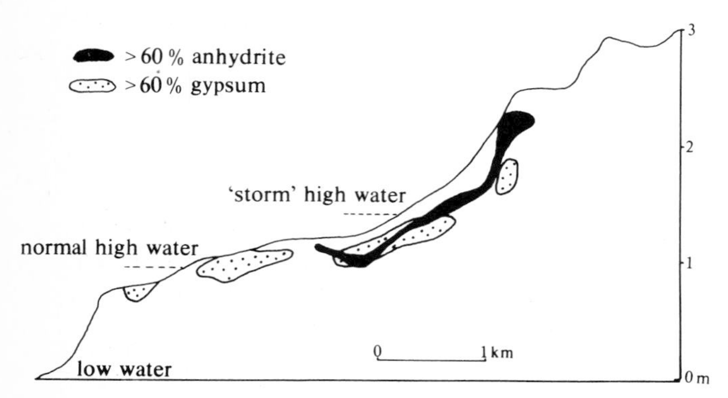

Figure 30.3 Distribution of anhydrite and gypsum in sabkha sediments of Abu Dhabi (after Bush 1973).

Textures (Fig. 30.4) in modern and ancient sabkha evaporites are complex (Holliday 1968, 1973). Primary interstitial gypsum occurs as replacive and displacive lozenge-shaped crystals (often in algal mats) ranging from hundreds of millimetres to centimetres in length. These may be replaced as pseudomorphs by fine-grained anhydrite laths and granules defining an **aphanitic** texture. Aphanitic anhydrite nodules are often a delicate glacier ice-blue colour and they show variable replacement by white, coarser-grained granular or felted-lath fabrics. Crystals in the latter group are up to 1.5 mm long showing delicate fibroradiate or 'wheatsheaf' aggregates. Interstitial anhydrite laths may show displacive relations to earlier fabrics and are interpreted as primary anhydrite precipitates. Laths may also show recrystallisation to granular or coarser lath textures.

Diagenetic textures of the utmost complexity occur in other evaporite minerals, particularly the K-salts, because of the action of bittern brines on pre-existing halite and gypsum, as well as their own precipitates. Thus polyhalite in the British Zechstein evaporites (Steward 1949) occurs as fine-grained (1–200 μm) aggregates of irregular grains which may show elongate grains or fibrous aggregates. The mineral replaces anhydrite and halite during the reactions of K- and Mg-rich brines. Primary carnallite is replaced by sylvite, halite and anhydrite. Numerous other replacement reactions effectively destroy any idealised pattern of primary bittern precipitates that might be predicted from physicochemical theory (see Ch. 3).

Major changes affect evaporite successions that have been buried down to depths in excess of 1 km or so. Gypsum unaffected by early diagenetic dehydration to anhydrite completely recrystallises to anhydrite with up to 38% volume loss. This change may obscure any original depositional textures within the gypsum rock. Increasing pressure and temperature cause most evaporite minerals, particularly halite and K-salts, to deform plastically, causing growth of diapiric salt domes and pillars. Flowage causes a foliated fabric to form in the salts and may very severely disrupt evaporite beds, causing rapid and unpredictable fluctuations in thickness and composition. Such effects make potash mining a particularly hazardous and expensive operation (see Woods 1979, Smith & Crosby 1979).

Evaporite sequences brought back to (or near to) the Earth's surface by uplift and erosion undergo a further series of diagenetic changes. Highly soluble salts, such as halite and the K-salts, may be dissolved rapidly in the meteoric zone. Slower dissolution of anhydrite and gypsum also occurs, the resulting SO_4^{--} rich ground waters reacting with dolomite to form dedolomitisation features (Ch. 29f). Massive regional or local collapse breccias result from dissolution. Abundant SO_4^{--}-reducing

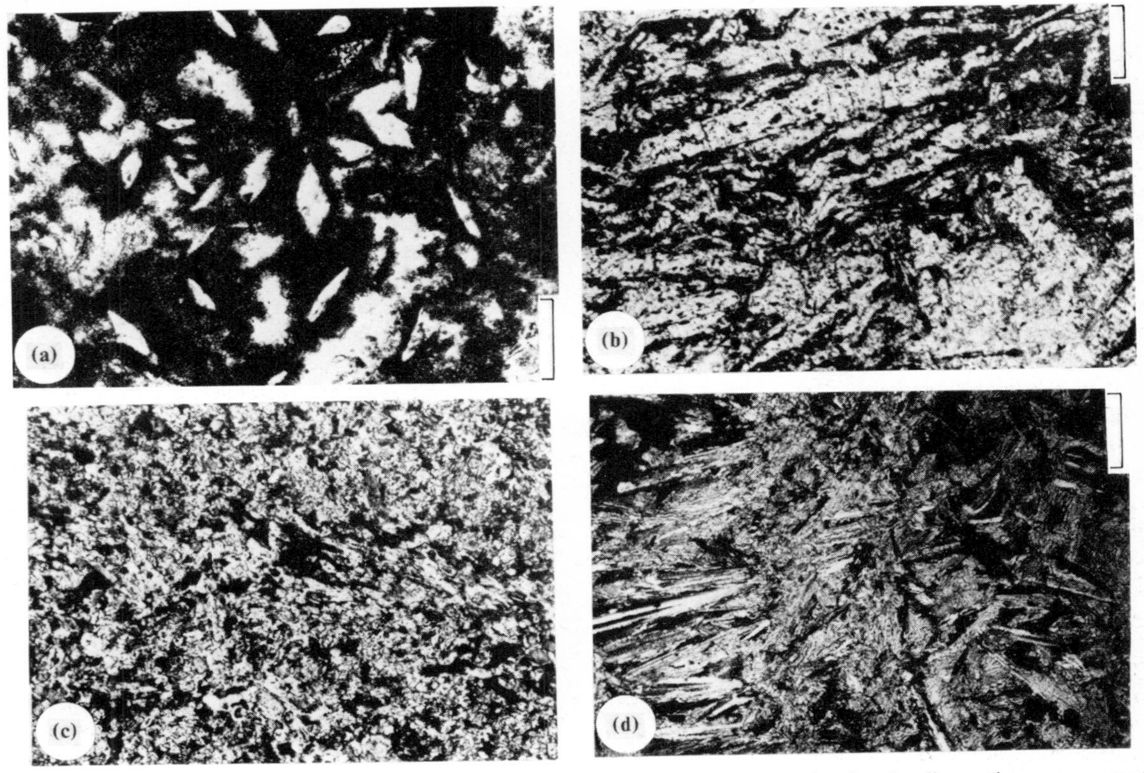

Figure 30.4 Thin-section photomicrographs of evaporite textures. (a) Interstitial growth of early diagenetic gypsum crystals: pseudomorphs of aphanitic anhydrite after gypsum in micrite matrix. Scale bar = 500 μm. Basal Purbeck evaporites (Upper Jurassic), Brightling mine, Sussex, England. (b) Interstitial growth of primary lath anhydrite, with no *in situ* gypsum precursor, in fine-grained opaque carbonate matrix. Scale bar = 250 μm. Basal Purbeck evaporites (Upper Jurassic) Fairlight borehole, Sussex. (c) Aphanitic anhydrite under crossed-nicols with some recrystallised laths. Scale bar = 100 μm. Locality as (a). (d) Aphanitic anhydrite under crossed-nicols, largely recrystallised to ragged laths on right and to 'wheatsheaf' fabric to left. Scale bar = 500 μm. Billingham Main Anhydrite (Upper Permian), Billingham mine, Co. Durham, England. (All after Holliday, 1973.)

bacteria may cause calcitisation of anhydrite and production of elemental *sulphur*. The great sulphur deposits of the USA Gulf Coast occur in the contact zone between anhydrite and limestone caps to shallow salt domes.

Secondary gypsum rocks produced by near-surface hydration of anhydrite may be divided into two types (Holliday 1970). **Porphyroblastic secondary gypsum** shows abundant relict corroded grains of anhydrite surrounded by single uniformly extinguishing gypsum crystals. The porphyroblasts may be euhedral or anhedral and of various shapes and sizes. **Alabastrine secondary gypsum** (Fig. 30.5a) comprises the bulk of secondary gypsum rock, being fine aggregates of granular crystals which may show undulose to irregular extinction. There is little fabric evidence of volume increase during gypsification of anhydrite, and the process is thought to occur by a solution–precipitation replacement process with excess sulphate removed in solution. Minor amounts of displacive gypsum occur in veins with a distinctive fibrous 'satin-spar' habit, the fibres being orientated with their long axes normal to the vein margins (Fig. 30.5d).

Silicification of evaporites is discussed in the next section.

30b Silica diagenesis

As noted previously (Ch. 4), siliceous oozes of pelagic marine origin comprise the vast majority of recent siliceous sediments. In the geological record, however, large amounts of nodular cherts also occur in essentially shallow-water carbonate facies. These nodular cherts are clearly of diagenetic origin and are mostly replacive. Partially-to-wholly lithified oceanic cherts and porcellanites of late Mesozoic to Cainozoic age have now been recovered from sites in all the major ocean basins by

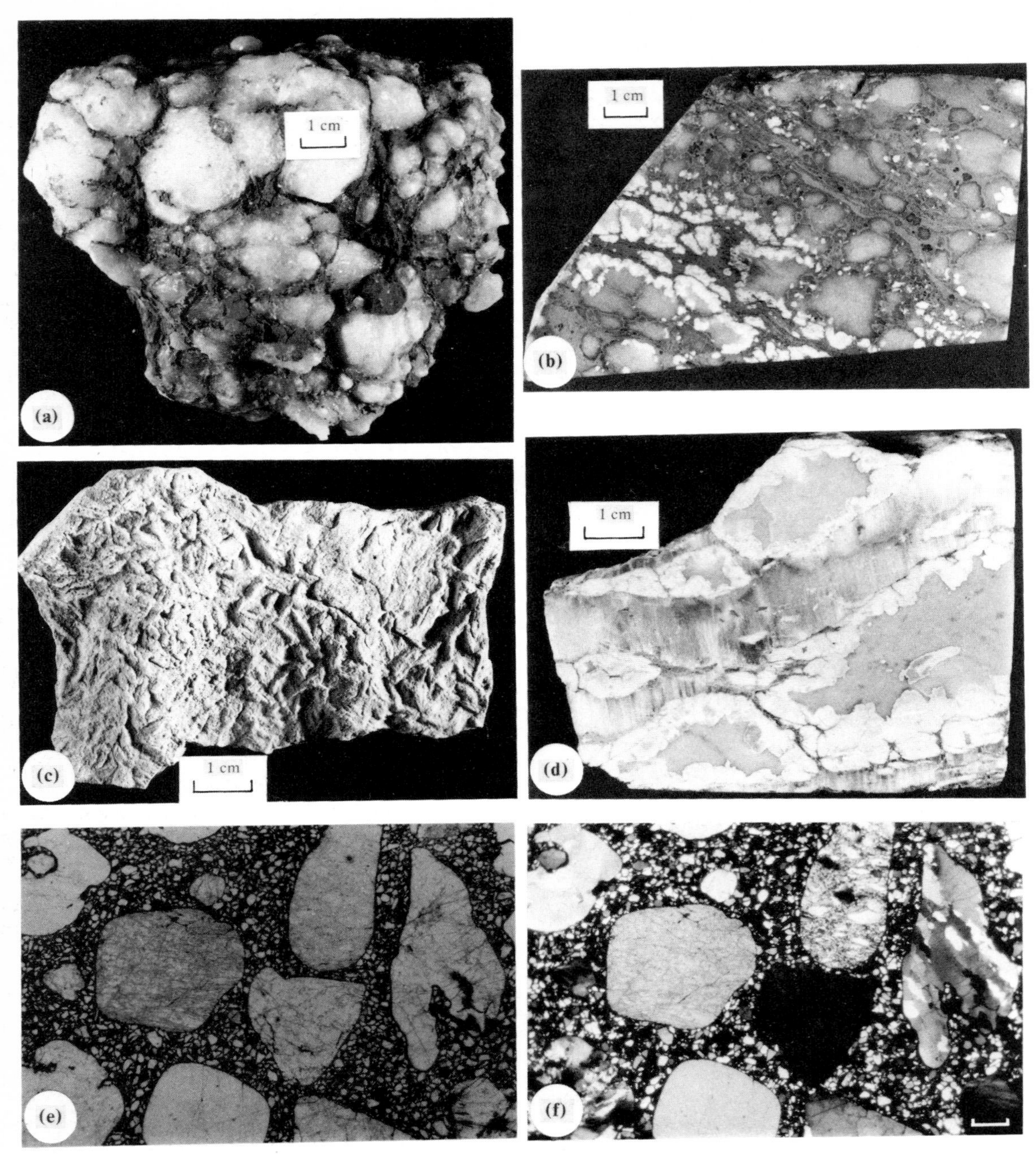

Figure 30.5 (a) Chicken-mesh texture of alabastrine gypsum (after anhydrite) in red mudstone matrix. Rhaetic, Severn Bridge, Avon, England. (b) Chicken-mesh texture of aphanitic anhydrite in dolomite. Note the secondary white selenitic gypsum crystals forming rims around certain nodules. Hith Anhydrite, Dahl Hith, Nr Riyadh, Saudi Arabia. (c) Pseudomorphs of dolomite after wedge-shaped gypsum crystals in algar-laminated facies. Saharonim Formation (Jurassic), Ramon, Israel. (d) Chicken-mesh anhydrite with rims of secondary selenitic gypsum and prominent veins of 'satin-spar' gypsum. Locality as (b). (e & f) Terrazo-type silcrete under plane polarised light and under crossed-nicols. Tertiary of Western Australia (coll. E. J. W. Van der Graaf). Scale bar = 100 μm.

the DSDP cruises (see Calvert 1977, Riech and von Rad 1979). Quartz and opal-CT are the principal silica phases, the latter occurring as the spherical microcrystalline aggregates of bladed crystals known as **lepispheres**. Opal-CT is thought to form as a metastable intermediate stage in the conversion of amorphous biogenous opal-A to microcrystalline or chalcedonic quartz. The opal-A→opal-CT change occurs by a dissolution–reprecipitation mechanism which obscures the organic origin of the ooze by partly destroying the actual radiolarian tests. Silica is thus highly mobile during early diagenesis. In fact the concentration of dissolved silica in the pore waters of marine sediments is *higher* than in the associated bottom water. The concentration gradient in dissolved silica across the sediment/water interface implies a flux of silica from the sediments into the overlying water by both diffusion and advection. The magnitude of the flux is calculated to be very large and is just as important a source of dissolved silica to the oceans as continental runoff (Ch. 3; review by Calvert 1974). The opal-CT–quartz change is now also thought to be a solution–redeposition process rather than a solid state change.

All pre-Jurassic cherts are composed entirely of quartz, but the concept of a gradual maturation process with time (and hence depth in a given succession) from a soft biogenous ooze (opal-A) through porcellanite (opal-CT) to quartzitic chert is complicated by the preferential occurrence of porcellanites in clay-rich sediments and of quartzitic cherts in purer carbonate sediments. Experiments show that the transformation rate of opal-A to opal-CT is much higher in carbonate than in clay-rich sediments and that opal-CT lepisphere formation is aided by the precipitation of nuclei with magnesium hydroxide as an important component (Kastner *et al.* 1977). The more disordered opal-CT associated with clays is thought to be less prone to quartzification than that formed in pure carbonates. The importance of Mg^{2+} ions in the process indicates that dolomitic carbonates should be preferentially silicified. As well as many observations in the ocean basins, support for the above experimental approach also comes from outcrops of the English Chalk of Upper Cretaceous age where chalks high in montmorillonitic clay residues contain few chert nodules (flint) but are richer in opal-CT lepispheres. Abundant nodules occur in the purer chalks which are more common as the chalk succession is ascended. Similarly, in the Mesozoic pelagic sediments of Cyprus quartzitic cherts occur in pure limestone turbidite beds whereas opal-CT lepispheres and well preserved radiolaria occur abundantly in the fine-grained clay-rich sediments (Robertson 1977). The release of Mg^{2+} during the change opal-CT→quartz also explains the occurrence of small amounts of the Mg-rich clay palygorskite and of dolomite in younger and older cherts respectively.

The second major occurrence of cherts in the rock record is as nodular cherts in marine shallow-water limestones. Field and petrographic data show clearly that such cherts are replacive. In order to explain such replacements it is necessary to find an environment where diagenetic waters are simultaneously supersaturated with respect to crystalline silica and undersaturated with calcite. The explanation should also set out to account for the surprising fact that stable hydrogen and oxygen isotopes of such cherts indicate a major meteoric-water component in their formation (Knauth & Epstein 1976, Kolodny *et al.* 1980). One attractive model appeals to a meteoric/marine mixing hypothesis (Knauth 1979; Fig. 30.6 & 7) much as that discussed previously for dolomitisation (Ch. 29).

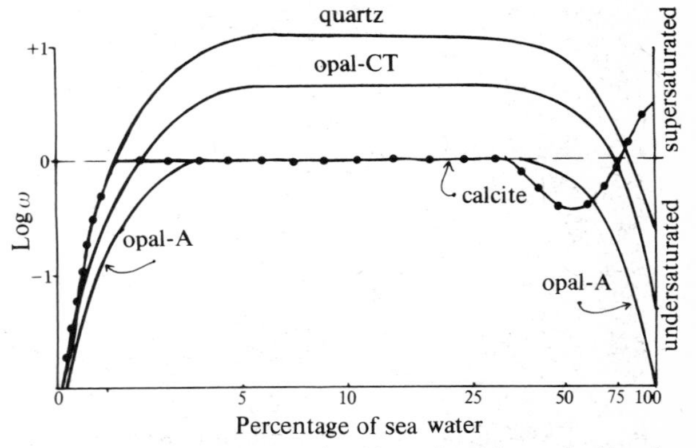

Figure 30.6 Solubility relationships of calcite and silica in mixed meteoric-marine ground waters closed with respect to CO_2. ω is the ratio of IAP to K, negative values indicating undersaturation. In the hypothetical case shown, mixing of meteoric and marine waters has produced water undersaturated with respect to calcite and supersaturated with respect to silica (opal-cT/quartz). (After Knauth 1979.)

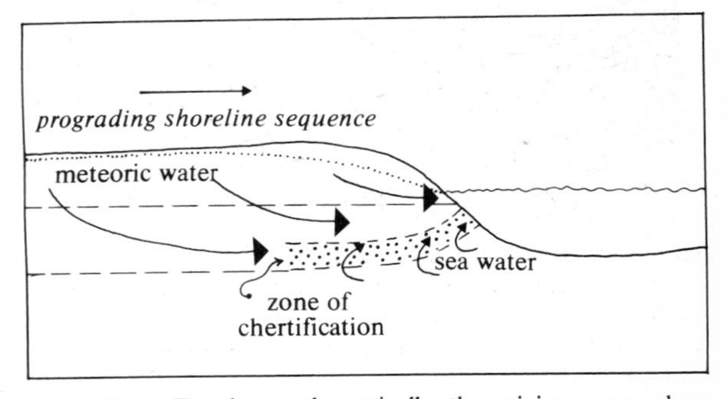

Figure 30.7 To show schematically the mixing zone where chertification may proceed in a seaward-prograding carbonate shoreline (after Knauth 1979).

Rather special sorts of cherts may replace anhydrite or gypsum nodules (West 1964, Folk & Pittman 1971). These contain abundant length-slow chalcedony of the lutecite and quartzine varieties. Normally chalcedony is length-fast, possibly because at neutral or low pH, or when SO_4^{2-} ions are lacking, the silica tetrahedra are polymerised into spiral chains which lie down flat on the surface of accumulation so that the *c*-axes are tangential to the growing surface (Folk & Pittman 1971). At high pH or when polluted by SO_4^{2-} ions the silica tetrahedra are solitary and thus precipitate one by one as chalcedony to form the normal quartz orientation (length-slow). Quartz nodules that have replaced evaporite nodules frequently show many minute inclusions of evaporite minerals or the quartz may show pseudomorphs after the crystal habits of anhydrite or gypsum. These last features, together with observations concerning the gross morphology of the quartz replacements, should always be used as necessary supporting evidence for the evaporite-replacement nature of the length-slow chalcedony since such chalcedony may occur in unequivocally non-evaporitic cherts (Chowns & Atkins 1974).

Major accumulations of silica occur in Australia and South Africa as **silcrete** duricrusts (Langford-Smith 1978). These very resistant horizons up to 5 m thick form prominent plateaux and mesas, and they preserve conspicuous remnants of Tertiary to Quaternary planation surfaces. There is general agreement that silcrete formed in a warm and more humid climate than that existing at present over much of interior Australia and that the silica was derived from weathering of silicate minerals in many different kinds of host rocks. Some authors envisage a major source of Si^{4+} from **lateritisation**. Sub-basaltic silcretes in New South Wales show geochemical evidence for silica leaching from the basalts through percolating ground waters (Taylor & Smith 1975). Ollier (1978) envisages silica released from weathering travelling in both surface and ground waters. Silcretes form locally if quartz sands act as nuclei, or regionally if the silica-charged waters reach zones of internal drainage where accumulation and evaporation may occur. Hutton *et al.* (1978) stress that silcrete can only form as a result of prolonged pedogenic activity under stable environmental conditions. They develop quite slowly and the evolution of thick profiles was possible only with pedological, geological, geomorphological and climatic stability such as occurred over much of Australia in the early Cainozoic. Smale (1978) has divided silcretes into a number of types including **terrazo** (quartz or lithic grains in chert cement, Fig. 30.5e & f) and **quartzitic** (authigenic quartz overgrowths on quartz grains).

30c Iron minerals

Five variables control the nature of chemical reactions involving Fe^{2+} and Fe^{3+}. These are Eh, pH and activities of dissolved HS^-, HCO_3^- and Fe^{2+}. Eh–pH diagrams (Fig. 30.8) show clearly that haematite and other ferric species are the only iron minerals that can exist in equilibrium with depositional waters above the sediment/water interface.

The major source of detrital iron in sediments is goethite derived from weathering and soil reactions, including **lateritisation**. Clay minerals may contain Fe^{2+} in their lattices or as adsorbed species on the clay mineral surfaces (Carroll 1958). Once deposited, **goethite** must reach an equilibrium with **haematite** by the dehydration reaction (Berner 1969)

$$\underset{\text{goethite}}{2HFeO_2} \rightarrow \underset{\text{haematite}}{Fe_2O_3} + \underset{\text{liquid}}{H_2O}$$

ΔG^0 (see Ch. 1) for this reaction is always negative, the exact value being dependent upon goethite crystallinity. Thus limonitic goethite is always unstable relative to

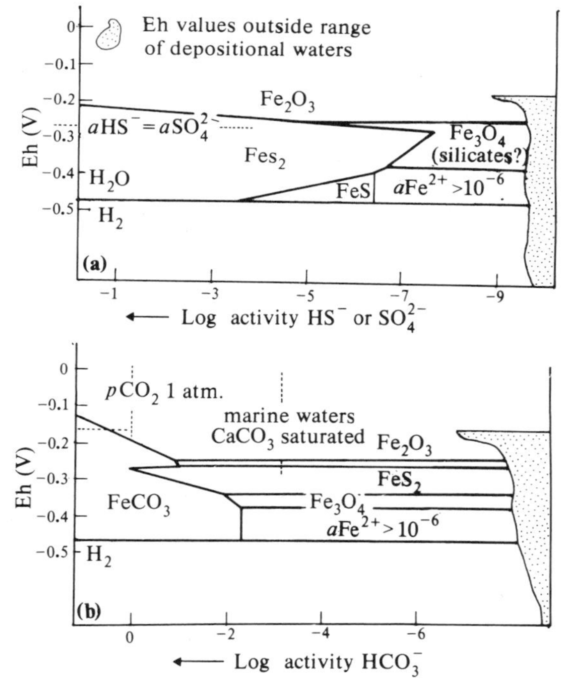

Figure 30.8 (a) Iron mineral stability fields as a function of Eh, aHS^-. (b) Iron mineral stability fields as a function of Eh and $aHCO_3^-$. (Both after Curtis & Spears 1968.)

haematite + water under diagenetic conditions. This explains the complete absence of yellow to brown limonite/goethite sediments in the subsurface. We have already seen (Ch. 28) that arid-zone vadose diagenesis encourages red haematitic pigment production with time in near-surface environments. The haematite may remain stable as long as organic matter remains absent. The presence of organic material will encourage reduction of Fe^{3+} to Fe^{2+} and the red pigment will disappear.

Pyrite is a common diagenetic iron mineral in many marine sediments. As noted in Chapter 28, below the thin zone of oxygenated pore waters, anaerobic bacterial reduction of sulphate occurs. The H_2S which forms reacts with iron to form the iron monosulphide FeS. Continued H_2S production encourages sulphur-oxidising bacteria to form elemental sulphur. This sulphur reacts with the FeS over a period of years (Berner 1970) to form microscopic (0.5–100 μ) aggregates of pyrite crystals termed **framboids**. The most important factor limiting pyrite formation is the availability of organic matter that can be metabolised by bacteria. Both iron and sulphate species are usually present in abundance in marine diagenetic pore waters. Thus the greater the amount of organic material, the greater the amount of pyrite produced in zone 2 of the diagenetic scheme established above (Ch. 28). Marine black shales rich in organic material are frequently pyritous in the geological record. By way of contrast, non-marine diagenetic pore waters are generally very low in dissolved SO_4^{2-} and hence pyrite is usually absent. Rapid marine deposition will tend to inhibit pyrite formation since there will be a limited time available for SO_4^{2-} diffusion from the overlying marine reservoir. The SO_4^{2-} available is restricted to that contained within the buried pore waters.

Siderite will form as a diagenetic mineral only where a very low dissolved sulphide concentration is coupled with high dissolved carbonate, high Fe^{2+}/Ca^{2+}, low Eh and near-neutral pH. As discussed in Chapter 28, these conditions usually restrict siderite formation to non-marine diagenetic environments (low SO_4^{2-}) where abundant Fe^{2+} is present (tropical-zone weathering). The mineral is particularly common in deltaic swamp facies where concretions give evidence for continued growth during progressive burial (Ch. 28). For siderite to be stable relative to calcite, the iron concentration must be greater than 5% that of calcium. In sea water it is less than 0.1%. The occurrence of sideritic beds in certain marine mudstones points to somewhat unusual conditions. Associated burrows and fauna often indicate minor depositional gaps (Sellwood 1968). The siderite may have formed much later in diagenesis from an oxide precursor when marine SO_4^{2-} was exhausted and when the Ca^{2+} ions were used up.

Chamosite occurs most characteristically as oöids and as mud with siderite in **minette-type** iron ores. Associated fauna indicate fully marine conditions with agitation to encourage oöid growth. However, the environmental requirements for ferrous silicate stability resemble those of siderite modified by low carbonate activity and saturation with some active silica form. They suggest reducing conditions below the sediment/water interface. It is therefore likely that the chamosite oöids were originally formed of some early mineral (or minerals) that was subsequently converted to chamosite during diagenesis. Sorby (1856) postulated that the oöliths were originally calcitic (see also Kimberley 1979), but detailed fabric studies do not support this conclusion (see Bradshaw *et al.* 1980). It is possible that the oöids were formed as an Fe–Al-rich gel within wave-agitated lagoons and that the gel was subsequently converted to chamosite during shallow burial (see Curtis & Spears 1968, Talbot 1973). However, details of the process remain obscure.

Glauconite formation is poorly understood. It forms as a marine phase in areas of much-reduced sedimentation where it fills in shell cavities and replaces faecal pellets. Berner (1971) notes that it forms slowly at the sediment/water interface where it is associated with organic matter and generally positive but fluctuating Eh conditions. Glauconites vary tremendously in composition from K-poor smectites to K-rich glauconitic micas with a general trend towards increasing K with time.

A summary of iron mineral diagenetic occurrences is given in Figure 30.9.

30d Manganese

As is well known, deep sea exploration has located deposits of manganese nodules in many of the world's

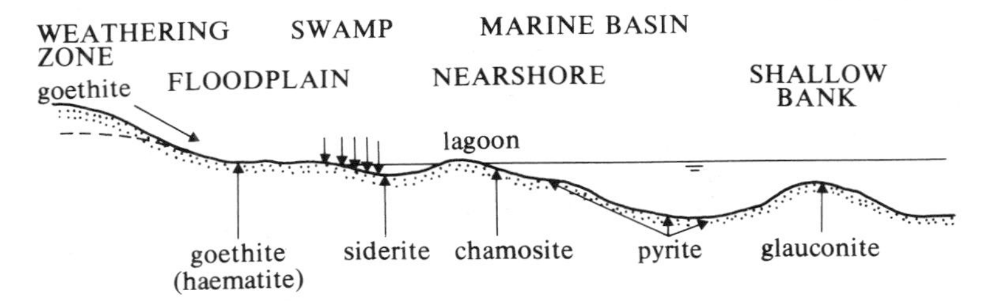

Figure 30.9 Schematic section indicating the various environments where diagenetic iron minerals may form (after Berner 1970). Note haematite forms in the weathering zone of semi-arid climates (Ch. 28).

oceans. The nodules range in size from a few millimetres up to a decimetre or more and have grown by the addition of successive concentric rings around a nucleus such as a volcanic rock fragment. They grow extremely slowly, perhaps as little as 3 mm per 10^6 years, and their occurrence as surface phenomena in areas of red clay deposition up to a thousand times this accretion value implies that weak current systems must periodically erode the red clay. In addition to manganese (in the form MnO_2) the nodules contain substantial amounts of iron and are much enriched in the trace elements Ni, Co and Cu.

Precipitation of manganese requires that the dispersed insoluble Mn^{4+} in the sediments be reduced by organic matter to the soluble Mn^{2+} ion and that this then be oxidised back to Mn^{4+}. Manganese ions are derived from both depositional and diagenetic waters, the ultimate sources being Mn from streams, rivers and the oceanic volcanogenic flux associated with basalt–seawater interactions (review in Elderfield 1976). Manganese migration occurs by processes of diffusion, advection and reaction in oceanic waters and pore waters. Note the shapes of the schematic dissolved Mn profiles in Figure 30.10, indicating reduction of Mn^{4+} and remobilisation of soluble Mn^{2+} in lower levels and Mn^{2+} escape by advection and diffusion to be precipitated as insoluble Mn^{4+} in the upper levels.

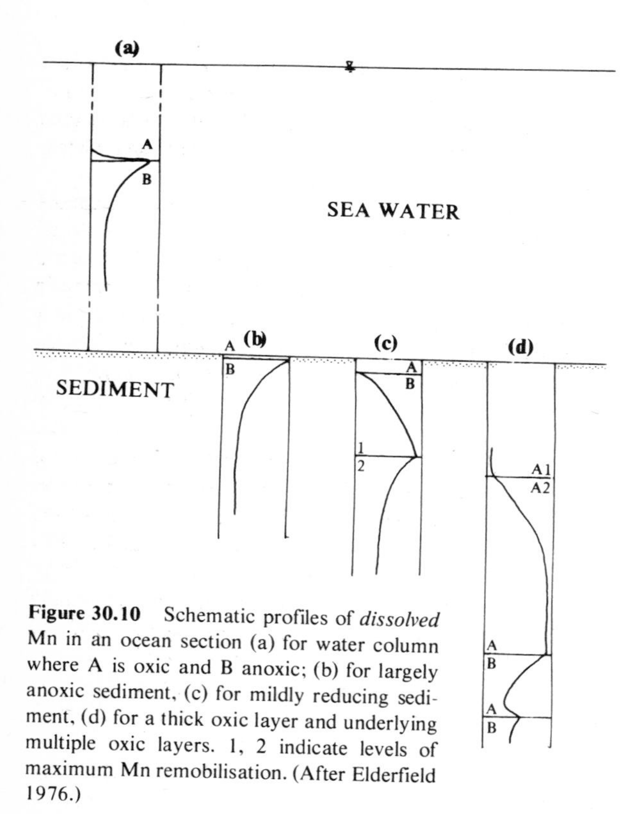

Figure 30.10 Schematic profiles of *dissolved* Mn in an ocean section (a) for water column where A is oxic and B anoxic; (b) for largely anoxic sediment, (c) for mildly reducing sediment, (d) for a thick oxic layer and underlying multiple oxic layers. 1, 2 indicate levels of maximum Mn remobilisation. (After Elderfield 1976.)

30e Summary

Diagenetic growth of gypsum and anhydrite occurs in the shallow subsurface of supratidal plains (sabkhas) in arid climatic regimes. Anhydrite forms as an alteration product from gypsum and also as a primary phase. Complex diagenetic changes occur in the precipitated K-salts because of later bittern percolation, solution and reprecipitation. During burial evaporites undergo plastic flow, causing cataclastic fabrics to form; hydrated species such as gypsum lose their molecular water. Rehydration occurs, as anhydrite is brought back to near-surface levels by uplift and erosion; secondary gypsum rocks then result.

During burial siliceous oozes undergo mineralogical transformations from opal-A of biogenic origin to microcrystalline or chalcedonic quartz via the metastable intermediary, opal-CT. The transformation proceeds by a dissolution–reprecipitation mechanism and is aided by the presence of Mg^{2+} ions in carbonate-rich oozes. Replacement cherts in shallow-water limestones may be caused by the mixing of meteoric with marine waters in the phreatic zone, enabling the pore waters to become supersaturated with respect to crystalline silica whilst undersaturated with respect to calcite. Length-slow chalcedonic silica may frequently replace evaporite nodules, although not all such chalcedony is of this origin. Silcrete duricrusts develop in continental areas subjected to leaching over long time intervals.

The major source of detrital iron is goethite derived from weathering and soil reactions, but the goethite is always unstable relative to haematite + water under diagenetic conditions. Pyrite is produced in low-Eh marine pore water from byproducts of SO_4^{2-}-reducing and H_2S-oxidising bacteria. Siderite forms where very low dissolved sulphide concentrations are coupled with high dissolved CO_3^{2-}, high Fe^{2+}:Ca^{2+} ratio, low Eh and near-neutral pH. Chamosite probably forms in low-Eh pore waters from buried oölitic Fe-rich gels.

Mn^{2+} ions are provided to the sediment/water interface by processes of diffusion and advection. Here the Mn oxidises and slowly precipitates as insoluble Mn^{4+} around nuclei to form manganese nodules.

Further reading

Berner (1971) is essential reading on all aspects of diagenesis discussed in this chapter. Silica diagenesis is well summarised by Calvert (1974) and by Riech and Von Rad (1979).

31 Hydrocarbons

31a Introduction

There can be no doubt that economic applications of sedimentology are most important in the coal and hydrocarbon industries, with a large number of specialists being employed, particularly by the oil companies. The laboratories of these companies have produced much important research and in some countries, chiefly the USA and Canada, there is a healthy cross-fertilisation between company and university. The concentration of organic matter to form coal, oil or natural gas is a process which involves almost all of the subdisciplines of sedimentology, with overlaps into many other chemical and biological fields.

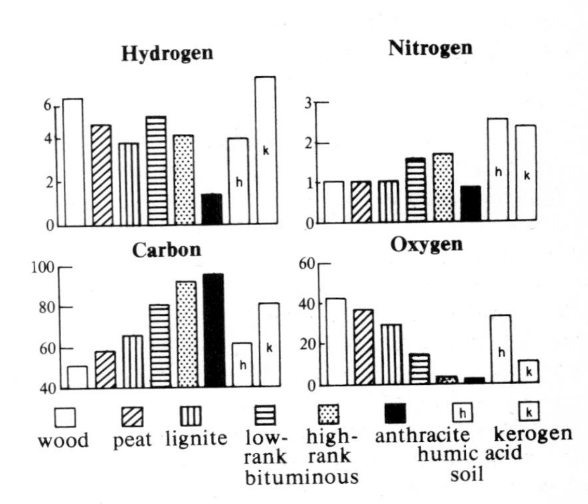

Figure 31.1 Carbon, oxygen, nitrogen and hydrogen content in various coals, humic acids and kerogens (after data collected by Degens 1965).

31b Coal composition and rank

Coal is basically macroscopic plant debris that has undergone progressive physical and chemical alteration through geological time. A variety of coal types exist, the most important of which may be listed in terms of increasing carbon content into a *rank* series. Figure 31.1 shows the main elemental composition of various coal types – the higher the rank the more distinct the coal has become from the initial starting material (Table 31.1). The main coal series are termed **humic coals**. These laminated coals have passed through a peat stage where woody materials have been affected by the humification process whereby micro-organisms convert original plant **lignin** into humic acids and their residues. By way of contrast the dull, unstratified **sapropel coals** originate as subaqueous organic muds rich in algal remains (boghead coal) or in plant spores (cannel coal). Increasing temperature aided by compaction is the major factor in the coalification process after the initial microbiological changes effected in the near-surface zones. Coals of all ranks and ages show similar carbon isotopic compositions of around –25‰ PDB. This is similar to the composition of modern wood and indicates little fractionation during the coal-forming process.

Coal contains a variety of plant tissue in different states of preservation. The tissues are the petrographic constituents of coal and are termed **macerals**. Three main groups of macerals may be defined. The **huminite–vitrinite** group are those macerals of woody and humic origin. Huminite occurs in low-rank coals and vitrinite occurs in medium- to high- rank coals. **Liptinite** macerals are the remains of lipid-rich plant relics. **Inertinite** macerals are hard, brittle carbon-rich relics and they include charcoal debris and liptinite relics subjected to high-rank alterations.

A petrographic technique known as **reflectance measurement** enables the rank of coal or of dispersed organic matter (e.g. spores) to be estimated. Reflectance is the amount of incident standard light (monochromatic green) reflected from a maceral surface compared to that reflected from a standard surface of known reflectance. Reflectance is correlated to rank by volatile, chemical or moisture measurements (Table 31.1). The huminite to vitrinite components are usually selected for reflectance measurement, a mean reflectance value being obtained from a number of individual measurements. Liptinite macerals have a relatively low reflectance whilst the inertinite macerals show a high reflectance. The huminite macerals in brown coals show intermediate reflectance which increases as the cellular tissues are gradually impregnated and infilled by humic substances. At about 0.4% reflectance the huminite component is termed vitrinite. Use of reflectance indices of organic diagenesis and metamorphism is also widespread in estimating the maturity of potential oilfield source rocks and may be used as a comprehensive diagenetic indicator (e.g. Fig. 28.12).

Table 31.1 Rank stages and important petrographic characteristics of coals (after Tissot & Welte 1978).

Rank stages	% reflectance of vitrinite	Important microscopic characteristics	% C in vitrinite
Peat		large pores	50
		details of initial plant material still recognisable	
		free cellulose	
Brown coal		no free cellulose	
soft brown coal		plant structures still recognisable (cell cavities frequently empty)	60
dull brown coal	c. 0.3	marked gelification and compaction takes place	
bright brown coal		plant structures still partly recognisable (cell cavities filled with collinite)	
	c. 0.5		75
Hard coal		exinite becomes markedly lighter in colour	
bituminous hard coal		('coalification jump')	
	c. 2.2	exinite no longer distinguishable	90
anthracite		from vitrinite in reflected light	
graphite	11.0	reflectance anisotropy	100

← rank increasing

The use of rank and reflectance studies to estimate subsurface palaeotemperatures is well illustrated by the study by Deroo *et al.* (1977) of the western Canadian basin in Alberta. Here there is a gradual rank increase from lignites in the east to high-volatile bituminous coals in the west (Fig. 31.2). The true relationship between depth and rank can be obtained only when the *maximum* burial depth is known. A general correlation graph between moisture content and burial depth then enables a relationship to be drawn concerning rank and maximum burial depth (Fig. 31.3). A coalification diagram is then used to estimate maximum temperature obtained during coal diagenesis (Fig. 31.4). Studies such as these also shed great light upon the maximum temperature of oil formation since the mudrock interbeds may be potential source rocks. In this particular case the temperature limits to oil formation were 60–120°C.

Methane generation is an important process during coalification. It begins to become most important in the rank range of medium volatile bituminous coal (1.3–1.4% reflectance). Major natural gas deposits result if the methane can be trapped in a suitable reservoir (e.g. the Rotliegendes aeolian sands that overlie the Coal Measures in the southern North Sea area; Fig. 13.6).

31c Coal-forming environments

Potential coal-bearing sites are restricted to those environments of high plant productivity, low sediment influx, limited plant-tissue oxidation and bioregeneration, and high preservation potential. Thick modern freshwater peats indicate a massive plant productivity with limited surface oxidation and re-use of plant litter debris by higher and lower members of the local food chain. A prerequisite for thick peat development appears to be a water-saturated swampland far removed from sediment influxes. Limited oxidation may occur in many peat beds, but further down the profiles microbiological activity dominates, with production of humic breakdown substances that progressively infiltrate the compacting plant tissues. With increasing depths peat changes from brown to black in colour and assumes a gel-like consistency. Percolating acids derived from the peat-forming process

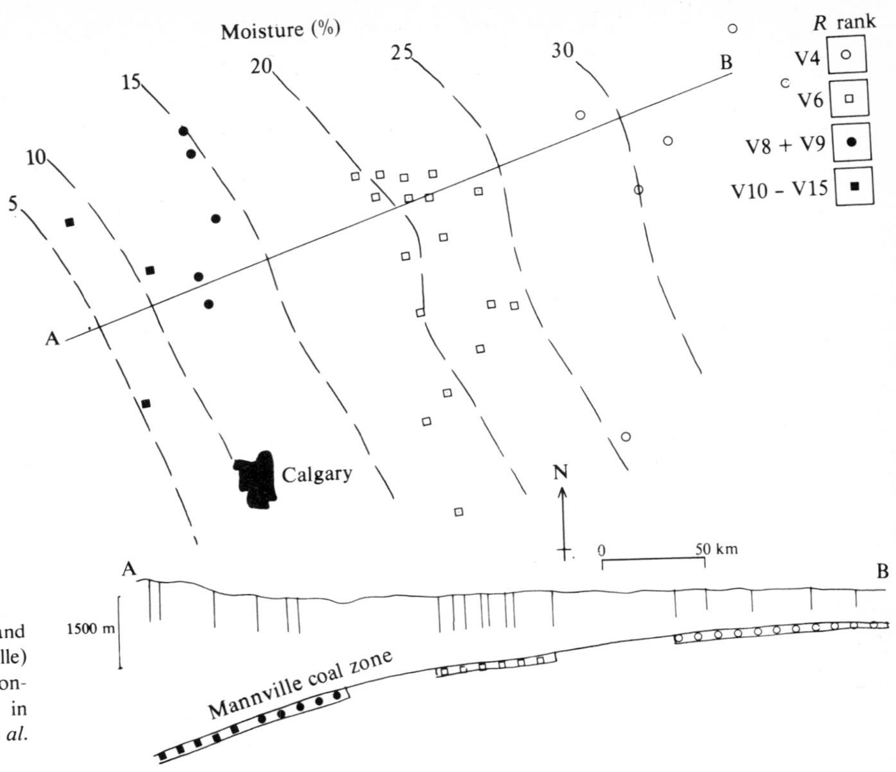

Figure 31.2 Location, depth and rank of Cretaceous (Mannville) coal samples and isomoisture contours of near-surface coals in western Canada (after Deroo *et al.* 1977).

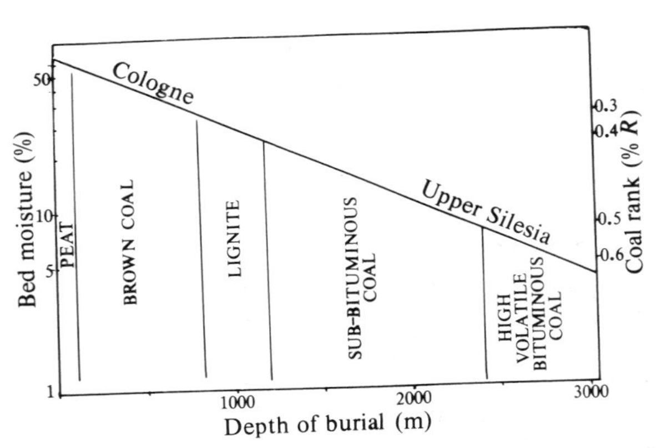

Figure 31.3 Relationship of moisture content of coal, depth of burial and reflectance (R) (from data assembled by Deroo *et al.* 1977).

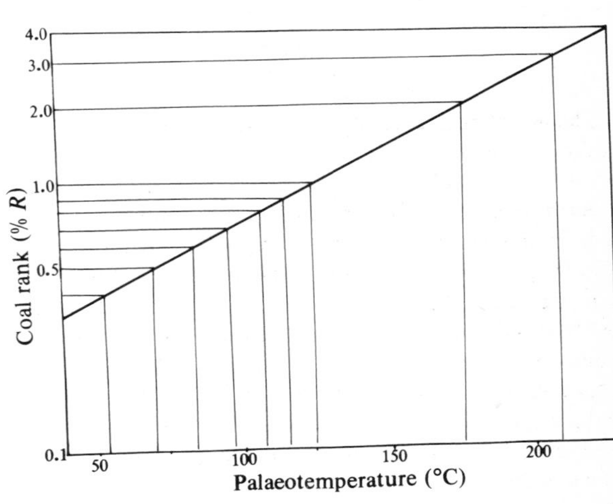

Figure 31.4 Relationship of rank and palaeotemperatures of Mannville coals (after Deroo *et al.* 1977). See Figure 31.2.

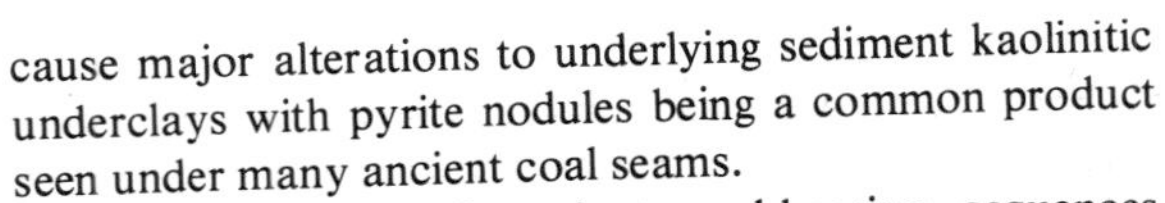

cause major alterations to underlying sediment kaolinitic underclays with pyrite nodules being a common product seen under many ancient coal seams.

Facies analysis of ancient coal-bearing sequences reveals predominant peat formation in freshwater swamps, bogs and marshes on wet tropical coastal plains. On-delta backswamps, low-lying alluvial plains and back-barrier swamps seem to be the most important sub-environments of coal formation. The Florida Swamps serve as a modern partial analogue to many ancient peat swamps (see review in McPherson *et al.* 1976, Cohen & Spackman 1977). Here fresh, brackish and marine peats are accumulating in a very complex coastal plain.

Another interesting modern analogue occurs in the Snuggedy Swamps of coastal Carolina (Staub & Cohen 1979). Here (Fig. 31.5) up to 4.5 m of Holocene peats have accumulated in the lagoonal depressions between abandoned Pleistocene barrier islands. The peats spread across salt marsh as coalescing peat islands and form a regressive sequence above fine-grained lagoonal sediments showing coarsening-upwards trends (Fig. 31.5). The peats are often 'split' by fine-grained crevasse splays from adjacent tidal channels. Some splays originate after extensive peat fires cause local depressions in which sediments may then accumulate. These splay deposits may be distinguished in vertical section by means of a thin basal layer of charcoal debris.

Many detailed studies of sedimentary facies in coal-bearing sediments have been completed in recent years, notably in the Pennsylvanian of the USA (e.g. Wanless *et al.* 1970, Ferm 1970, 1974; see Ch. 19) and the Westphalian of Britain (Scott 1979, Haszeldine & Anderton 1980). The work of Scott provides a particularly important link between sedimentary processes and reconstructions of floral communities from palaeoecological studies. Interesting floral studies without a sedimentary input are provided by the work of M. and R. Teichmuller (1968) on the German brown coal deposits.

There is little doubt that coalfield exploration and exploitation is much helped by a thorough sedimentological approach. Thus prediction of channel sandstone 'washouts', coarsening-upwards lake infill sequences and other seam-splitting sediments are of much value in efficient extraction of both underground and opencast coals. These features may often be ignored in areas of thick persistent seams, but in more marginal fields early recognition of 'abnormal' sediments at or about the seam level has important repercussions in terms of both mining practice and coal extraction.

31d Oil and gas – organic matter, source rocks and diagenesis

Oil and gas owe their origins to the biological and low-temperature thermal breakdown of disseminated organic

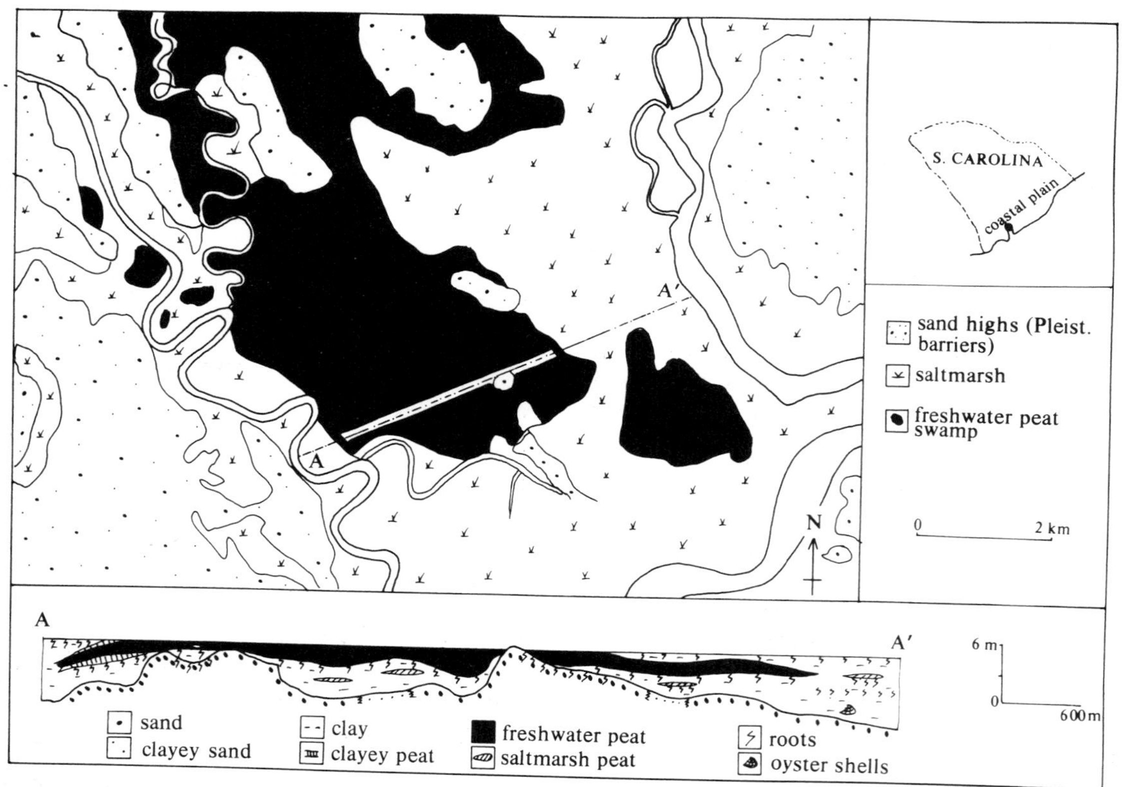

Figure 31.5 The Snuggedy Swamp of South Carolina, USA – facies map and section (after Staub & Cohen 1979). See text for discussion.

materials present in fine-grained sedimentary rocks. Organic matter is present as 2.1% of shales, 0.29% of carbonates and 0.05% of sandstones (mean values). The total disseminated organic matter present in all the world's sediments has been calculated to total 3.8 x 10^{15} tonnes, of which 3.6 x 10^{15} tonnes is present in shales (Degens 1965). By way of comparison the total tonnage of coal is estimated to be 6 x 10^{12} tonnes and that of oil as 2 x 10^{11} tonnes.

The disseminated organic material present in shales and muds comes from a huge variety of sources, but plankton of various sorts probably supply the most important proportion. Modern plants and animals are composed of varying proportions of protein, carbohydrates, fatty lipids, pigments and lignins (only in higher plants). During diagenesis these biopolymers present in primary organic matter are changed to geopolymers collectively called **kerogen**. Early on in diagenesis micro-organisms act upon the primary starting materials to form amino acids and sugars. The residues not used by the micro-organisms for their own life processes recombine by polycondensation and polymerisation to form brown compounds similar to the poorly known fulvic and humic acids. During burial it is thought that thermal maturation acts upon these residues and produces (Fig. 31.6) the inert kerogen and crude oil and natural gas as byproducts (Fig. 31.7). The maturation process, from organic starters with no hydrocarbons to crude oil with a huge range of hydrocarbons, is obviously a process of great complexity which is still poorly understood. The interested reader is referred to the summary discussions of Degens (1965) and Tissot and Welte (1978) for further details.

Much evidence gathered from the geological content of oil occurrences and from rank studies of buried organic matter indicates that liquid hydrocarbon genesis begins at about 65°C and ends at between 135°C and 150°C. This so-called 'liquid window' corresponds to mean depths of around 3 km in areas of normal thermal gradients, with natural gas formation occurring at generally higher temperatures of between 120°C and 200°C. Referring back to our discussions of diagenesis in clayey sediments (Ch. 28) we thus see that oil formation occurs in diagenetic zones 4 and 5. It is evident from many studies that oil source rocks need not be particularly rich in organic matter to begin with. The main problem is to prevent very early diagenetic oxidation. This may be achieved by rapid sedimentation rates (but dilution of organics results) or by a deficit of O_2 in the depositional or diagenetic waters. Notwithstanding the above comments on organic content, it is evident that the most prolific source rocks for petroleum genesis will be those thick mudstone sequences that were deposited in areas of high organic productivity. As noted previously (Ch. 24) high organic productivity will then cause O_2-deficiency and encourage organic preservation; the resulting black shales are thus prime hydrocarbon source rocks (see Demaison & Moore 1980 for the most up-to-date review of anoxic environments and oil-source bed genesis).

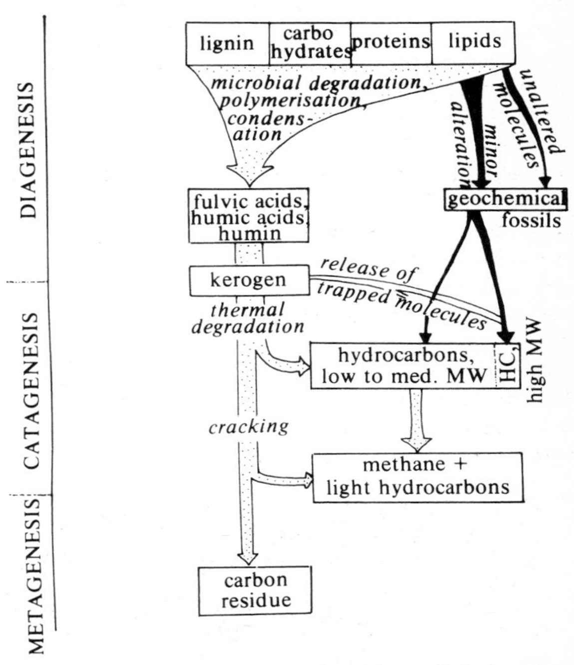

Figure 31.6 Sources of hydrocarbons in geological environments (after Tissot & Welte 1978).

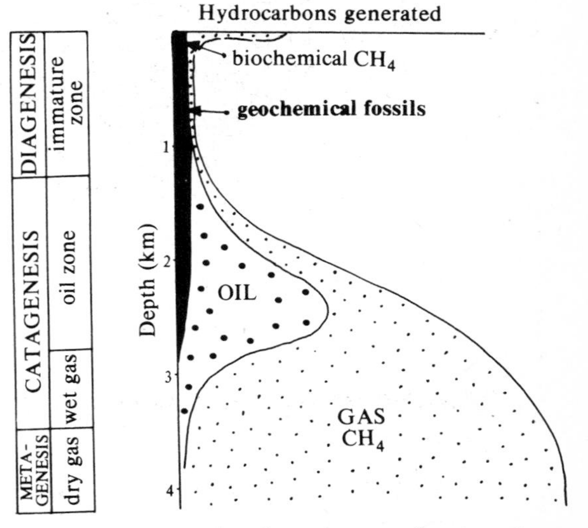

Figure 31.7 Hydrocarbon formation as a function of source rock burial; depths only approximate (after Tissot & Welte 1978).

31e Oil and gas migration

Given that an organic-rich mudrock has been buried to sufficient depth to encourage hydrocarbon generation, the next step is migration of the hydrocarbons from source to porous reservoir rock (*primary* migration) and then through the porous rock into a suitable hydrocarbon trap (*secondary* migration) where the hydrocarbon is effectively sealed in. It has been variously assumed that during primary migration the hydrocarbon moves as either discrete globules, as colloidal suspensions (micelles), or in solution. It is likely that all three modes may contribute under different circumstances. The hydrocarbon moves within an aqueous phase in response to the differential stresses set up during compaction. It has been suggested (Burst 1969) that redistribution of hydrocarbon along fluid-potential gradients reaches a maximum during the release of interlayer water from mixed-layer clays (see Chs 27 & 28). Any porous interbed within a compacting claystone sequence will act as a conduit for the hydrocarbons present in moving connate fluids.

As summarised by Tissot and Welte (1978) three factors influence secondary migration. These are (a) the tendency for buoyant rise of oil and gas in the water-saturated rock pores, (b) the capillary pressures that determine multiphase flow, and (c) the hydrodynamics of pore-fluid flow. Oil globules or gas bubbles larger than any given pore throat diameter have to be distorted before they can squeeze through the narrow passage. The interfacial tension between oil–gas and water must overcome the capillary pressure. Petroleum trapped in a porous rock thus represents an equilibrium between buoyant/hydrodynamic driving forces and capillary resisting forces. A trap (*q.v.*) must be sealed by sediments whose pore diameters are sufficiently small to exert a capillary pressure in excess of the driving forces.

The direction of hydrocarbon movement is determined largely by the direction of fluid movement along local or regional pressure gradients; upward, lateral and downward migrations are all possible depending upon local conditions. The hydrocarbons will continue to migrate as long as driving forces predominate. Some studies indicate hydrocarbon migrations of a hundred kilometres or more. Liquid migration in the subsurface is much aided by the low viscosities of hot crude and by the effects of gases carried in solution.

31f Oil and gas traps and reservoir studies

A variety of hydrocarbon traps occur in nature, the most important of which are summarised in Figure 31.8. The

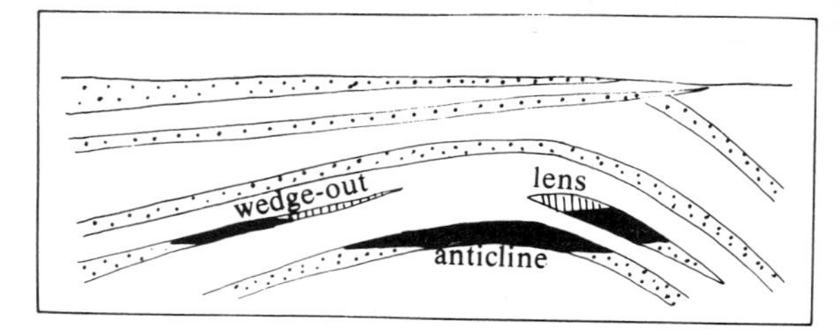

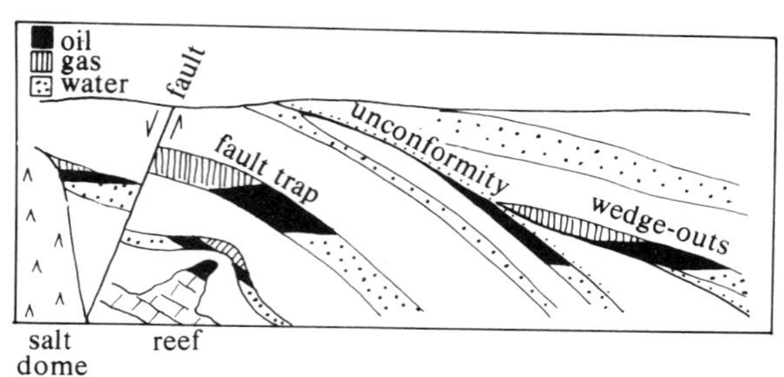

Figure 31.8 Various traps for oil and gas (largely after Hobson & Tiratsoo 1975).

stratigraphic traps are of most direct sedimentological interest since they depend upon lateral facies changes for their existence. However, studies of 198 giant oilfields (ultimate recovery $>500 \times 10^6$ barrels each) reveals a great preponderance of structural anticline traps (Moody 1975). At the same time it should be noted that such fields are relatively simple to detect in the subsurface and that increasing attention is being paid to the more subtle stratigraphic traps, particularly in such intensively explored areas as the USA (see, for example, White 1980). Traps are usually sealed by compacted mudrocks or evaporites, although leakage of oil and gas may often occur during the early stages of formation of an oil or gas pool when compaction has not proceeded to its most efficient limit. Leakage may also occur along some fault planes (Fig. 31.9). Relatively late movements of this kind are termed **tertiary migrations**.

Regarding reservoir studies, it is here that most direct sedimentological interest lies. The extent of reservoir rocks will always be governed by facies changes as modified by diagenetic processes of pore plugging by cementation. The best reservoir rocks will comprise extensive, well sorted sand- to gravel-grade sediments such as braided river-channel sheet sandstones, desert sands of erg origin, spreads of littoral to sublittoral carbonate and clastic sands, delta front sands, shelf sands, reefs and reef talus, and proximal deposits of submarine fans. Each facies has its own environmental signature (see Chs 12–26) that may be recognised and interpreted from drill-hole cores and well-log techniques. Indeed, most modern oilfield discoveries are subjected to a detailed facies analysis in order that maximum understanding of the

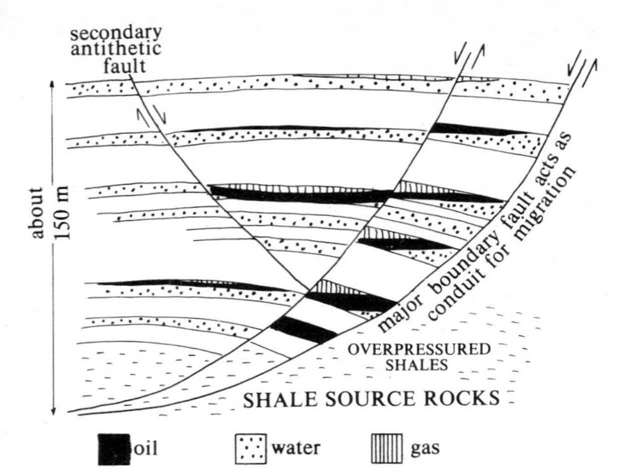

Figure 31.9 Schematic section through a growth fault in the Niger delta showing the position of hydrocarbon pools and possible migration routes (after Weber & Daukoru 1975).

reservoir is reached for both production and future exploration purposes. Even though most oilfields are structurally trapped, the oil occurs only where the anticlinal areas coincide with the local development of a suitable reservoir. Fine examples are given by Harms (1966), and over a larger area (Fig. 31.10) by Fisher and McGowen (1969). Examples of growth fault traps in deltaic facies are shown in Figure 31.9 based upon work by the Shell Group. Stratigraphic traps in non-reefal carbonate–evaporite facies are shown in Figure 31.11, based upon the work of Illing *et al.* (1967). Facies studies are becoming even more important in secondary recovery

Figure 31.10 Location of oil and gas fields with respect to depositional facies in the lower Wilcox Group of Texas, USA. (After Fisher & McGowen 1969).

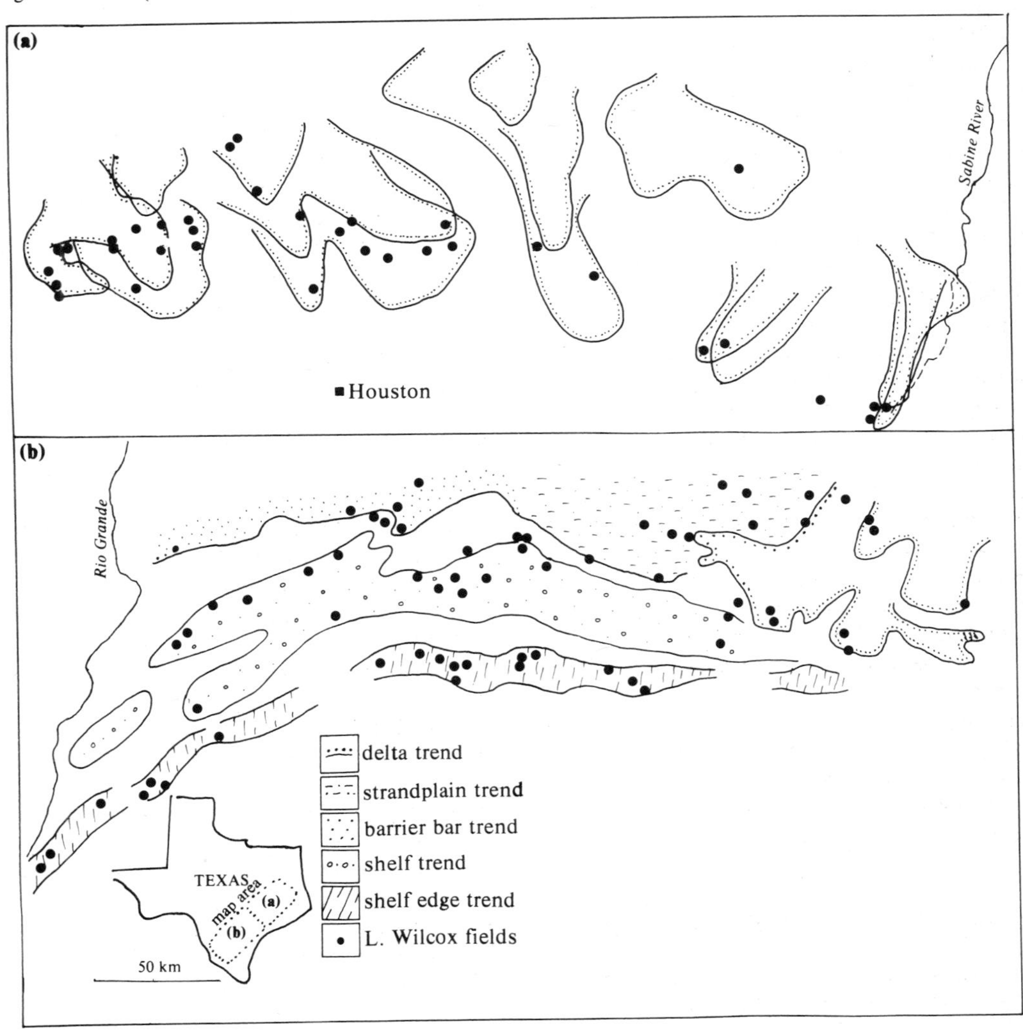

programmes where water injection processes depend upon a detailed interpretation of the reservoir. Thus the problem of interconnectedness of river-channel sandstones (Bridge & Leeder 1979) mentioned in Chapter 15 assumes great importance in such studies (e.g. Van Veen 1977).

A final aspect of reservoir studies concerns diagenetic controls upon permeability and porosity. Cement minerals tend to restrict primary pore space, particularly pore throats, and lead to a decrease in effective porosity and permeability. Thus, cemented rocks in potential traps cannot receive hydrocarbon, and partly cemented reservoirs in traps cannot efficiently release their contained oil. The most common cements are diagenetic clay minerals (Ch. 28) in sandstone reservoirs and carbonate cements in limestone reservoirs. As we have seen previously, all porous sediments tend to undergo cementation as they are buried, so it is important that oil migration should occur as early as possible before this process carries on too far. Evidence from many fields suggests that, once inside reservoir pore space, hydrocarbons actually prohibit further cementation. Good examples occur in the porous lime sands of the Arab-zone reservoirs of Saudi Arabia, Qatar and Abu Dhabi. Here only a thin rim cement is present around the oil-filled pores, whereas in adjacent outcrop and non-oilfield subcrops the pores are almost closed by second-generation calcite cements.

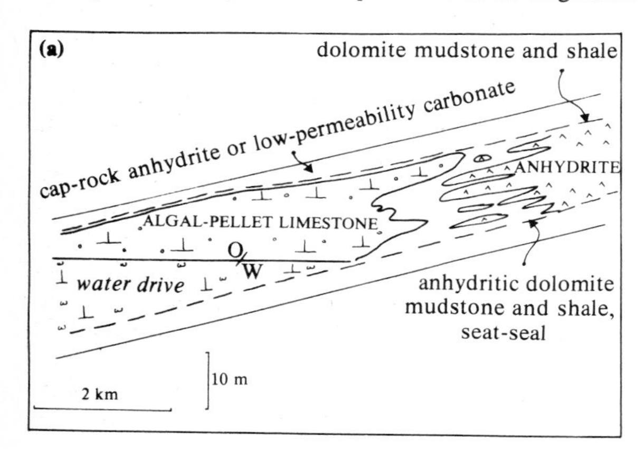

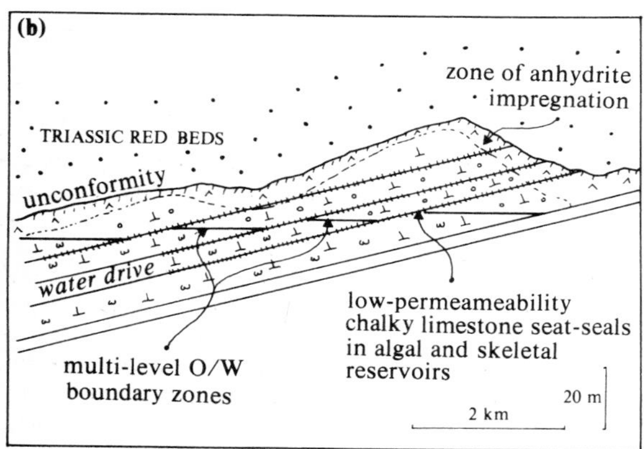

Figure 31.11 Stratigraphic traps caused by facies changes in (a) sabkha–lagoon facies, (b) carbonate reservoirs under an evaporite-impregnated unconformity. These are typical examples from the Mississippian of Saskatchewan and North Dakota, but many similar examples occur in the Middle East. (After Illing *et al.* 1967.)

A process known as diagenetic sealing (Wilson 1977) may occur as a result of cementation below the oil/water contact in both carbonate and clastic stratigraphic traps. This has important repercussions in both secondarily tilted traps and in secondary recovery programmes.

Some reservoirs owe their porosity to secondary dissolution or to recrystallisation processes. We have already discussed the example of secondary pore generation in sandstones (Ch. 28g). Other examples include the dissolution of dolomite or evaporite minerals in pores and vugs, and the dissolution of aragonite, both processes occurring when carbonate–evaporite rocks are invaded by meteoric waters. Major porosity in the Canadian Devonian reefs resulted from this latter process (*q.v.*). Another classic method of producing secondary porosity is by thoroughgoing dolomitisation proceeding in a closed system in limestones. The process leads to a decrease in rock volume and hence to an increase in porosity of up to a maximum of about 10%.

Another kind of porosity is caused by relatively late-stage rock fracturing on a minute scale. The fracturing may be due to gentle folding of the already fully cemented reservoir or the fractures may provide more efficient connections between otherwise isolated pores. Many Middle Eastern carbonate reservoirs show this kind of fracture porosity (e.g. Iran, Iraq). Interesting examples also occur in the Cretaceous chalks of the North Sea. These rocks are usually fairly porous but show extremely small permeabilities. In the Ekofisk field microfracturing of the chalk above a salt intrusion has created a widespread reservoir of high permeability.

31g Tar sands

The term **tar sand** is applied to those deposits of heavy crude oil that cannot be recovered by conventional production methods. The crude is dense, highly viscous and sour, and it simply sticks in the pores of the reservoir rocks. Truly enormous reserves of heavy crude lie 'locked-up' in these deposits. It has been estimated that only sixteen tar deposits contain as much oil as that presently known in all conventional fields (Demaison 1977).

Tar sands thus represent an important reserve of petroleum for future generations when methods of extraction become more advanced. The Athabaska tar sands in western Canada have reserves estimated at more than four times those present in the world's largest producing oilfield, the Ghawar field of Saudi Arabia.

Tar sands originate when low-temperature ($\lesssim 90°C$), oxygen-rich meteoric waters rich in bacteria come into contact with fluid, medium-gravity crude oils. Water-washing removes the more soluble, light hydrocarbon fractions, especially the aromatic groups. Biodegradation by bacteria removes the normal paraffins. The resulting heavy tar begins as a tar 'mat' at the oil/water contact zone and progressively invades the whole reservoir. CO_2 produced during the biochemical reactions is light and is easily detected by isotopic analysis. Tar sands tend to occur at basin margins or in near-surface traps where faults enable meteoric waters to pass through caprock sequences.

31h Oil shales

It has been estimated that oil shales contain 600 times the current known reserves of liquid petroleum. This vast resource still awaits large-scale production, although significant tonnages of petroleum are produced yearly by China and the Soviet Union (Estonia). The problem lies in the cost-effectiveness and environmental effects of large-scale shale retorting – the oil can be liberated only by heating to above 500°C.

Oil shales are kerogen-rich mudrocks whose high organic content is provided by algal remains. The algae represent periodic planktonic blooms in relatively quiet-water lacustrine or shallow-shelf environments. The algal slime so produced is saved from oxidation by the development of anoxic conditions within the depositional waters or at the sediment/water interface or by brine-induced stratification below an upper, photic oxic layer. As discussed in Chapter 16, the world's largest oil shale reserve occurs in the Green River Formation of the western USA. Here the oil shales outcrop and subcrop over vast areas. They originated as facies in a playa–lacustrine environment that periodically expanded and contracted (Fig. 16.7). These and most other oil shales show delicate internal varve-like laminations of algal-rich and algal-poor sediments. Varves of coccolith origin make up marine oil shales such as the Kimmeridge Clay of northwestern Europe. Deep burial of oil shales causes large-scale liberation of fluid hydrocarbons, much of the northern North Sea reserves being derived in this way from the Kimmeridge source rocks.

31i Summary

Solid hydrocarbons such as the coal series form by the progressive thermal diagenesis of organic peats. Coal rank is assessed by reflectance measurements and is largely determined by depth of burial and heat flow. Coals occur in a variety of coastal and alluvial lithofacies but are most common within the backswamp environments of ancient coastal plains and deltas.

Liquid and gaseous hydrocarbons result from the biogenic and low-temperature diagenesis of disseminated organic materials (biopolymers) to the geopolymer, kerogen. Reflectance studies and geological evidence indicate that oil genesis takes place at temperatures of 65–150°C. Primary migration from source to porous sedimentary rock occurs with aqueous phases expelled during compactional de-watering of both connate and molecular water. The fluids move along potential gradients until they are halted by a structural or stratigraphic feature termed a trap. Reservoir rocks in traps are many and varied, including alluvial channel, desert dune, delta front, barrier, shelf, submarine fan, reef, and platform-margin carbonate. Hydrocarbon yields and flow rates also depend upon reservoir permeability and porosity. Diagenetic precipitation of cement minerals and compaction reduces the value of these parameters. Early hydrocarbon migration often occurs, and dissolution of cements and grains during burial (secondary porosity) may increase porosity and permeability to acceptable levels.

Further reading

The most up-to-date account of hydrocarbon geology, though with most emphasis on organic metamorphism and fluid migration, is by Tissot and Welte (1978). Degens (1965) has a good section on organic diagenesis. Accounts of oil geology are also given by Chapman (1976) and by Hobson & Tiratsoo (1975). Any number of case histories documenting sedimentological approaches to oil- and gas-field development are to be found in the pages of the American Association of Petroleum Geologists Bulletin for the past twenty years or so. An interesting series of papers on several aspects of petroleum geology, including a good review of fluid migration by Magara and an account of organic geochemical trends during diagenesis by Tissot, appeared in Hobson (1977).

References

Abbott, J. E. and J. R. D. Francis 1977. Saltation and suspension trajectories of solid grains in a water stream. *Phil Trans R. Soc. Lond.* (A) **284**, 225–54.
Ager, D. V. 1973. *The nature of the stratigraphical record.* London: Macmillan.
Alexandersson, E. T. 1976. Actual and anticipated petrographic effects of carbonate undersaturation in shallow seawater. *Nature* **262**, 653–57.
Allan, J. R. and R. K. Mathews 1977. Carbon and oxygen isotopes as diagenetic and stratigraphic tools: surface and subsurface data, Barbados, W. Indies. *Geology* **5**, 16–20.
Allen, G. P. 1971. Déplacement aisonniers de la lentille de 'Crême de Vase' dans l'estuarie de la Gironde. *Comptes Rend. de l'Acad. Sci. Paris* **273**, 2429–31.
Allen, G. P., G. Sauzay and P. Castaing 1976. Transport and deposition of suspended sediment in the Gironde estuary, France. In Wiley (1976), 63–81.
Allen, J. R. L. 1960. The Mam Tor sandstones: a 'turbidite' facies of the Namurian deltas of Derbyshire, England. *J. Sed. Petrol.* **30**, 193–208.
Allen, J. R. L. 1964. Primary current lineation in the Lower Old Red Sandstone (Devonian), Anglo-Welsh Basin. *Sedimentology* **3**, 89–108.
Allen, J. R. L. 1965a. A review of the origin and characteristics of recent alluvial sediments. *Sedimentology* **5**, 89–191.
Allen, J. R. L. 1965b. Late Quaternary Niger delta and adjacent areas: sedimentary environments and lithofacies. *Bull. AAPG* **49**, 547–600.
Allen, J. R. L. 1966. On bedforms and palaeocurrents. *Sedimentology* **6**, 153–90.
Allen, J. R. L. 1968. *Current ripples.* Amsterdam: North-Holland.
Allen, J. R. L. 1969a. Some recent advances in the physics of sedimentation. *Proc. Geol. Ass.* **80**, 1–42.
Allen, J. R. L. 1969b. Erosional current marks of weakly cohesive mud beds. *J. Sed. Petrol.* **39**, 607–23.
Allen, J. R. L. 1970a. The avalanching of granular solids on dune and similar slopes. *J. Geol.* **78**, 326–51.
Allen, J. R. L. 1970b. *Physical processes of sedimentation.* London: George Allen & Unwin.
Allen, J. R. L. 1970c. Studies in fluviatile sedimentation: a comparison of fining-upwards cyclothems, with special reference to coarse-member composition and interpretation. *J. Sed. Petrol.* **40**, 298–323.
Allen, J. R. L. 1971a. Mixing at turbidity current heads, and its geological implications. *J. Sed. Petrol.* **41**, 97–113.
Allen, J. R. L. 1971b. Transverse erosional marks of mud and rock: their physical basis and geologic significance. *Sed. Geol.* **5**, 167–385.
Allen, J. R. L. 1972. A theoretical and experimental study of climbing-ripple cross-lamination, with a field application to the Uppsala esker. *Geog. Annlr* **53A**, 157–87.
Allen, J. R. L. 1973. Phase differences between bed configuration and flow in natural environments, and their geological relevance. *Sedimentology* **20**, 323–9.
Allen, J. R. L. 1974. Studies in fluviatile sedimentation: implications of pedogenic carbonate units, Lower Old Red Sandstone, Anglo-Welsh outcrop. *Geol J.* **9**, 181–208.
Allen, J. R. L. and N. L. Banks 1972. An interpretation and analysis of recumbent-folded deformed cross-bedding. *Sedimentology* **19**, 257–83.
Allen, J. R. L. and J. D. Collinson 1974. The superimposition and classification of dunes formed by unidirectional aqueous flows. *Sed. Geol.* **12**, 169–78.
Allen, J. R. L. and M. R. Leeder 1980. Criteria for the instability of upper-stage plane beds. *Sedimentology* **27**, 209–17.
Allen, P. 1967. Origin of the Hastings facies in north-western Europe. *Proc. Geol. Ass.* **78**, 27–105.
Allen, P. 1972. Wealden detrital tourmaline: implications for north-western Europe. *J. Geol. Soc. Lond.* **128**, 273–94.
Allen, T. 1968. *Particle size measurements.* London: Chapman & Hall.
Andel, T. H. van and J. R. Curray 1960. Regional aspects of modern sedimentation in northern Gulf of Mexico and similar basins, and paleogeographic significance. In *Recent sediments: N.W. Gulf of Mexico*, F. P. Shepard, F. B. Phleger & T. H. van Andel (eds), 345–64. Tulsa, Okla.: AAPG.
Andel, T. H. van, J. Thiede, J. G. Sclater and W. W. Hay 1977. Depositional history of the S. Atlantic Ocean during the last 125 million years. *J. Geol.* **85**, 651–98.
Anderton, R. 1976. Tidal shelf sedimentation: an example from the Scottish Dalradian. *Sedimentology* **23**, 429–58.
Arthurton, R. S. 1973. Experimentally produced halite compared with Triassic layered halite-rock from Cheshire, England. *Sedimentology* **20**, 145–60.
Arx, W. S. Von 1962. *An introduction to physical oceanography.* Reading, Mass.: Addison-Wesley.
Ashley, G. M. 1975. Rhythmic sedimentation in glacial Lake Hitchcock, Massachusetts–Connecticut. In Jopling & McDonald (1975), 304–20.
Assereto, R. L. A. M. and C. G. St C. Kendall 1977. Nature, origin and classification of peritidal tepee structures and related breccias. *Sedimentology* **24**, 153–210.

Badiozamani, K. 1973. The Dorag dolomitisation model – application to the Middle Ordovician of Wisconsin. *J. Sed. Petrol.* **43**, 465–84.
Bagnold, R. A. 1935. *Libyan sands.* London.
Bagnold, R. A. 1940. Beach formation by waves: some model experiments in a wave tank. *J. Inst. Civ. Engrs* **15**, 27–52.
Bagnold, R. A. 1946. Motion of waves in shallow water: interactions between waves and shallow bottoms. *Proc. R. Soc. Lond.* (A) **187**, 1–18.
Bagnold, R. A. 1954a. Experiments on a gravity-free dispersion of large solid spheres in a Newtonian fluid under shear. *Proc. R. Soc. Lond.* (A) **225**, 49–63.
Bagnold, R. A. 1954b. *The physics of blown sand and desert dunes*, 2nd edn. London: Chapman & Hall.
Bagnold, R. A. 1956. The flow of cohesionless grains in fluids. *Phil Trans R. Soc. Lond.* (A) **249**, 335–97.
Bagnold, R. A. 1962. Auto-suspension of transported sediment: turbidity currents. *Proc. R. Soc. Lond.* (A) **265**, 315–19.
Bagnold, R. A. 1963. Mechanics of marine sedimentation. In *The sea*, M. N. Hill (ed.), 507–23. New York: Wiley.
Bagnold, R. A. 1966a. The shearing and dilation of dry sand and the 'singing' mechanism. *Proc. R. Soc. Lond.* (A) **295**, 219–32.
Bagnold, R. A. 1966b. *An approach to the sediment transport problem from general physics.* USGS Prof. Pap., no. 422-I.

Bagnold, R. A. 1968. Deposition in the process of hydraulic transport. *Sedimentology* **10**, 45–56.

Bagnold, R. A. 1973. The nature of saltation and of 'bed-load' transport in water. *Proc. R. Soc. Lond.* (A) **332**, 473–504.

Bagnold, R. A. 1977. Bedload transport by natural rivers. *Water Resources Research* **13**, 303–12.

Baker, V. R. 1973. *Paleohydrology and sedimentology of Lake Missoula flooding in eastern Washington*. Geol. Soc. Am. Spec. Pap., no. 144.

Baker, V. R. 1974. Paleohydraulic interpretation of Quaternary alluvium near Golden, Colorado. *Quat. Res.* **4**, 94–112.

Baldwin, B. 1971. Ways of deciphering compacted sediments. *J. Sed. Petrol.* **41**, 293–301.

Ball, M. M. 1967. Carbonate sand bodies of Florida and the Bahamas. *J. Sed. Petrol.* **37**, 556–91.

Ballance, P. F. and H. G. Reading (eds) 1980. *Sedimentation in oblique-slip mobile zones*. Spec. Publ. Int. Ass. Sed., no. 4.

Banerjee, I. and B. C. McDonald 1975. Nature of esker sedimentation. In Jopling & McDonald (1975), 132–54.

Banks, N. L. 1973. The origin and significance of some downcurrent-dipping cross-stratified sets. *J. Sed. Petrol.* **43**, 423–7.

Barker, C. 1972. Aquathermal pressuring – role of temperature in development of abnormal-pressure zones. *Bull. AAPG* **56**, 2068–71.

Basu, A. S., W. Young, L. J. Suttner, W. C. James and G. H. Mack 1975. Re-evaluation of the use of undulatory extinction and polycrystallinity in detrital quartz for provenance interpretation. *J. Sed. Petrol.* **45**, 873–82.

Bates, C. C. 1953. Rational theory of delta formation. *Bull. AAPG* **37**, 2119–61.

Bathurst, R. G. C. 1958. Diagenetic fabrics in some British Dinantian limestones. *Geol J.* **2**, 11–36.

Bathurst, R. G. C. 1964. The replacement of aragonite by calcite in the molluscan shell wall. In *Approaches to paleoecology*, J. Imbrie & N. D. Newell (eds), 357–76. New York: Wiley.

Bathurst, R. G. C. 1966. Boring algae, micrite envelopes and lithification of molluscan biosparites. *Geol J.* **5**, 15–32.

Bathurst, R. G. C. 1968. Precipitation of oöids and other aragonitic fabrics in warm seas. In *Recent developments in carbonate sedimentology in Central Europe*, G. Muller & G. M. Friedman (eds), 1–10. Berlin: Springer.

Bathurst, R. G. C. 1975. *Carbonate sediments and their diagenesis*, 2nd edn. Amsterdam: Elsevier.

Beard, D. C. and P. K. Weyl 1973. Influence of texture on porosity and permeability of unconsolidated sand. *Bull. AAPG* **51**, 349–69.

Beaty, C. B. 1963. Origin of alluvial fans, White Mountains, California and Nevada. *Ann. Ass. Am. Geogs* **53**, 516–35.

Belderson, R. H., M. A. Johnson and A. H. Stride 1978. Bedload partings and convergences at the entrance to the White Sea, USSR and between Cape Cod and Georges Bank, USA. *Mar. Geol.* **28**, 65–75.

Berg, R. R. 1975. Depositional environment of Upper Cretaceous Sussex Sandstone House Creek Field, Wyoming. *Bull. AAPG* **59**, 2099–110.

Berger, W. H. 1971. Sedimentation of planktonic foraminifera. *Mar. Geol.* **11**, 325–58.

Berger, W. H. 1974. Deep-sea sedimentation. In *The geology of continental margins*, C. A. Burk & C. L. Drake (eds), 213–41. New York: Springer.

Berger, W. H. and E. L. Winterer 1974. Plate stratigraphy and the fluctuating carbonate line. In Hsü & Jenkyns (1974), 11–48.

Berner, R. A. 1969. Goethite stability and the origin of red beds. *Geochim. Cosmochim. Acta* **33**, 267–73.

Berner, R. A. 1970. Sedimentary pyrite formation. *Am. J. Sci.* **208**, 1–23.

Berner, R. A. 1971. *Principles of chemical sedimentology*. New York: McGraw-Hill.

Berner, R. A. 1975. The role of magnesium in the crystal growth of calcite and aragonite from sea water. *Geochim. Cosmochim. Acta* **39**, 489–504.

Berner, R. A. 1976. The solubility of calcite and aragonite in seawater at atmospheric pressure and 34.5‰ salinity. *Am. J. Sci.* **276**, 713–30.

Berner, R. A. 1980. *Early diagenesis: a theoretical approach*. Princeton, NJ: Princeton Univ. Press.

Berner, R. A., J. T. Westrich, R. Graber, J. Smith and C. S. Martens 1978. Inhibition of aragonite precipitation from supersaturated seawater. A laboratory and field study. *Am. J. Sci.* **278**, 816–37.

Bernoulli, D. and H. C. Jenkyns 1970. A Jurassic basin: the Glasenbach Gorge, Salzburg, Austria. *Verh. Geol. Bundesanst. Wien* **1970**, 504–31.

Bernoulli, D. and H. C. Jenkyns 1974. Alpine, Mediterranean and Central Atlantic Mesozoic facies in relation to the early evolution of the Tethys. In *Modern and ancient geosynclinal sedimentation*, R. H. Dott & R. H. Shaver (eds) 129–60. SEPM Spec. Pubn no. 19.

Beuf, S., B. Biju-Duval, O. de Charpal, P. Rognon, O. Oariel and A. Bennacef 1971. *Les gres du Palaeozoique Inferieur au Sahara*. Paris: Ed. Technip.

Bigarella, J. J. 1972. Eolian environments: their characteristics, recognition and importance. In *Recognition of ancient sedimentary environments*, J. K. Rigby & W. K. Hamblin (eds), 12–62. SEPM Spec. Pubn no. 16.

Bigarella, J. J. 1973. Paleocurrents and the problem of continental drift. *Geol. Runds.* **62**, 447–77.

Biscaye, P. E. and S. L. Eittreim 1977. Suspended particulate loads and transports in the nepheloid layer of the abyssal Atlantic Ocean. *Mar. Geol.* **23**, 155–72.

Blatt, H., G. V. Middleton and R. Murray 1980. *Origin of sedimentary rocks*, 2nd edn. Englewood Cliffs, NJ: Prentice-Hall.

Bluck, B. J. 1964. Sedimentation of an alluvial fan in southern Nevada. *J. Sed. Petrol.* **34**, 395–400.

Bluck, B. J. 1965. The sedimentary history of some Triassic conglomerates in the Vale of Glamorgan, South Wales. *Sedimentology* **4**, 225–45.

Bluck, B. J. 1971. Sedimentation in the meandering River Endrick. *Scott. J. Geol.* **7**, 93–138.

Bluck, B. J. 1978. Geology of a continental margin: the Ballantrae Complex. In *Crustal evolution in NW Britain and adjacent regions*, D. R. Bowes & B. E. Leake (eds), 151–62. *Geol. J.* Spec. Issue, no. 10.

Bluck, B. J. 1979. Structure of coarse-grained braided stream alluvium. *Trans R. Soc. Edinb.* **70**, 181–221.

de Boer, A. B. 1977. On the thermodynamics of pressure solution – interaction between chemical and mechanical forces. *Geochim. Cosmochim. Acta* **41**, 249–56.

Boersma, J. R. 1967. Remarkable types of mega cross-stratification in the fluviatile sequence of a subRecent distributary of the Rhine, Amerongen, the Netherlands. *Geol. Mijn.* **46**, 217–35.

Boles, J. R. 1978. Active ankerite cementation in the subsurface Eocene of Southwest Texas. *Contrib. Mineral. Petrol.* **68**, 13–22.

Boles, J. R. and S. G. Franks 1979. Clay diagenesis in Wilcox

Sandstones of SW Texas: implications of smectite diagenesis on sandstone cementation. *J. Sed. Petrol.* **49**, 55–70.

Boothroyd, J. C. and G. M. Ashley 1975. Processes, bar morphology and sedimentary structures on braided outwash fans, northeastern Gulf of Alaska. In Jopling & McDonald (1975), 193–222.

Borch, C. von der and D. Lock 1979. Geological significance of Coorong dolomites. *Sedimentology* **26**, 813–24.

Borchert, H. and R. O. Muir 1964. *Salt deposits*. London: Van Nostrand Reinhold.

Bosence, D. W. J. 1973. Facies relationships in a tidally-influenced environment: a study from the Eocene of the London Basin. *Geol. Mijn.* **52**, 63–7.

Bott, M. H. P. 1976. Formation of sedimentary basins of graben type by extension of the continental crust. *Tectonophysics* **36**, 77–86.

Boulton, G. S. 1968. Flow tills and related deposits on some West Spitsbergen glaciers. *J. Glaciol.* **7**, 391–412.

Boulton, G. S. 1972a. The role of thermal regime in glacial sedimentation. *Spec. Pubn Inst. Brit. Geogs* **4**, 1–19.

Boulton, G. S. 1972b. Modern Arctic glaciers as depositional models for former ice sheets. *Q. J. Geol Soc. Lond.* **128**, 361–93.

Boulton, G. S. and N. Eyles 1979. Sedimentation by valley glaciers; a model and genetic classification. In *Moraines and varves*, C. Schluchter (ed.), 11–24. Rotterdam: Balkema.

Bouma, A. H. 1969. *Methods for the study of sedimentary structures*. New York: Wiley (reprinted by Krieger, NY, in 1979).

Bouma, A. H. and C. D. Hollister 1973. Deep ocean basin sedimentation. In *Turbidites and deep water sedimentation*, 79–118. SEPM. Short course Anaheim.

Bouma, A. H., G. T. Moore and J. M. Coleman (eds) 1978. *Framework, facies and oil-trapping characteristics of the Upper Continental Margin*. Tulsa, Okla.: AAPG (Studies in Geology, no. 7).

Bourgeois, J. 1980. A transgressive shelf sequence exhibiting hummocky stratification: the Cape Sebastion Sandstone (U. Cretaceous), SW Oregon. *J. Sed. Petrol.* **50**, 681–702.

Bowen, A. J. 1969. Rip currents, 1: theoretical investigations. *J. Geophys. Res.* **74**, 5467–78.

Bowen, A. J. and D. L. Inman 1969. Rip currents, 2: laboratory and field observations. *J. Geophys. Res.* **74**, 5479–90.

Bowen, A. J., D. L. Inman and V. P. Simmon 1968. Wave 'set down' and 'set up'. *J. Geophys. Res.* **73**, 2569–77.

Bowler, J. M. 1977. Aridity in Australia: age, origins and expression in aeolian landforms and sediments. *Earth Sci. Rev.* **12**, 279–310.

Bradshaw, M. J., S. J. James and P. Turner 1981. Origin of oölitic ironstones – discussion. *J. Sed. Petrol.* **50**, 295–9.

Braithwaite, C. J. R. 1968. Diagenesis of phosphatic carbonate rocks on Remire, Amirantes, Indian Ocean. *J. Sed. Petrol.* **38**, 1194–212.

Braithwaite, C. J. R. 1973. Settling behaviour related to sieve analysis of skeletal sands. *Sedimentology* **20**, 251–62.

Bramlette, M. N. 1961. Pelagic sediments. In *Oceanography*, M. Sears (ed.), 345–66. Pubn Am. Assoc. Adv. Sci., no. 67.

Bridge, J. S. 1976. Bed topography and grain size in open channel bends. *Sedimentology* **23**, 407–14.

Bridge, J. S. 1977. Flow, bed topography, grain size and sedimentary structures in open channel bends: a three dimensional model. *Earth Surf. Proc.* **2**, 401–16.

Bridge, J. S. 1978a. Palaeohydraulic interpretation using mathematical models of contemporary flow and sedimentation in meandering channels. In Miall (1978), 723–42.

Bridge, J. S. 1978b. Origin of horizontal lamination under turbulent boundary layers. *Sed. Geol.* **20**, 1–16.

Bridge, J. S. and J. Jarvis 1976. Flow and sedimentary processes in the meandering River South Esk, Glen Clova, Scotland. *Earth Surf. Proc.* **1**, 303–36.

Bridge, J. S. and J. Jarvis 1982. The anatomy of a river bend: a study in flow and sedimentary processes. *Sedimentology*, in press.

Bridge, J. S. and M. R. Leeder 1979. A simulation model of alluvial stratigraphy. *Sedimentology* **26**, 617–44.

Bridges, P. H. 1975. The transgression of a hard substrate shelf: the Llandovery (L. Silurian) of the Welsh Borderland. *J. Sed. Petrol.* **45**, 79–94.

Bridges, P. H. and M. R. Leeder 1976. Sedimentary model for intertidal mudflat channels with examples from the Solway Firth, Scotland. *Sedimentology* **23**, 533–52.

Broecker, W. S. 1974. *Chemical oceanography*. New York: Harcourt Brace Jovanovich.

Broecker, W. S. and T. Takahashi 1966. Calcium carbonate precipitation on the Bahama Banks. *J. Geophys. Res.* **71**, 1575–602.

Brookfield, M. 1970. Dune trends and wind regime in Central Australia. *Z. Geomorph.*, supp. no. 10, 121–53.

Brookfield, M. E. 1977. The origin of bounding surfaces in ancient aeolian sandstones. *Sedimentology* **24**, 303–30.

Broussard, M. L. (ed.) 1975. *Deltas: models for exploration*. Houston: Geol. Soc. Houston.

Bryan, G. M. 1970. Hydrographic model of the Blake Outer Ridge. *J. Geophys. Res.* **75**, 4530–45.

Bull, W. B. 1972. Recognition of alluvial fan deposits in the stratigraphic record. In *Recognition of ancient sedimentary environments*, J. K. Rigby & W. K. Hamblin (eds), 63–83. SEPM Spec. Pubn, no. 16.

Burst, J. F. 1969. Diagenesis of Gulf Coast clayey sediments and its possible relation to petroleum migration. *Bull. AAPG* **53**, 73–93.

Bush, P. 1973. Some aspects of the diagenetic history of the sabkha in Abu Dhabi, Persian Gulf. In *The Persian Gulf*, B. H. Purser (ed.), 395–407. Berlin: Springer.

Butler, G. P. 1970. Recent gypsum and anhydrite of the Abu Dhabi sabkha, Trucial Coast: an alternative explanation of origin. In *Third Salt Symposium*, J. L. Rau & L. F. Dellwig (eds), 120–152. Cleveland: Northern Ohio Geol. Soc.

Callander, R. A. 1978. River meandering. *Ann. Rev. Fluid. Mech.* **10**, 129–58.

Calvert, S. E. 1974. Deposition and diagenesis of silica in marine sediments. In Hsü & Jenkyns (1974), 273–300.

Calvert, S. E. 1977. Mineralogy of silica phases in deep-sea cherts and porcelanites. *Phil Trans R. Soc. Lond.* (A) **286**, 239–52.

Campbell, C. V. 1971. Depositional model – Upper Cretaceous Gallup beach shoreline, Ship Rock area, NW New Mexico. *J. Sed. Petrol.* **41**, 395–409.

Campbell, C. V. 1976. Reservoir geometry of a fluvial sheet sandstone. *Bull. AAPG* **60**, 1009–20.

Campbell, C. V. and R. Q. Oakes 1973. Estuarine sandstone filling tidal scours, Lower Cretaceous Fall River Formation, Wyoming. *J. Sed. Petrol.* **43**, 765–78.

Cant, D. J. and R. G. Walker 1978. Fluvial processes and facies sequences in the sandy braided South Saskatchewan River, Canada. *Sedimentology* **25**, 625–48.

Carlston, C. W. 1965. The relation of free meander geometry to stream discharge and its geomorphic implications. *Am. J. Sci.* **263**, 864–85.

Carroll, D. 1958. Role of clay minerals in the transportation of iron. *Geochim. Cosmochim. Acta* **14**, 1–27.
Carroll, D. 1970. *Rock weathering*. New York: Plenum.
Carson, M. A. 1971. *The mechanics of erosion*. London: Pion.
Carver, R. E. 1971. *Procedures in sedimentary petrology*. New York: Wiley.
Caston, V. N. D. 1972. Linear sand banks in the southern North Sea. *Sedimentology* **18**, 63–78.
Chapman, R. E. 1976. *Petroleum geology: a concise study*. Amsterdam: Elsevier.
Chappell, J. 1980. Coral morphology, diversity and reef growth. *Nature* **286**, 249–52.
Chave, K. E. and E. Suess 1970. Calcium carbonate saturation in seawater: effects of organic matter. *Limnol. & Oceanogr.* **15**, 633–7.
Chepil, W. S. 1961. The use of spheres to measure lift and drag on wind-eroded soils. *Proc. Soil Sci. Soc. Am.* **25**, 343–5.
Chough, S. and R. Hesse 1976. Submarine meandering thalweg and turbidity currents flowing for 4000 km in the NW Atlantic Mid-Ocean Channel, Labrador Sea. *Geology* **4**, 529–33.
Chowns, T. M. and J. E. Elkins 1974. The origin of quartz geodes and cauliflower cherts through the silicification of anhydrite nodules. *J. Sed. Petrol.* **44**, 885–903.
Clemmey, H. 1976. Discussion. In Donovan & Archer (1975).
Clemmey, H. 1978. A Proterozoic lacustrine interlude from the Zambian Copperbelt. In Matter & Tucker (1978), 259–78.
Clifton, H. E., R. E. Hunter and R. L. Phillips 1971. Depositional structures and processes in the non-barred, high energy nearshore. *J. Sed. Petrol.* **41**, 651–70.
Cloud, P. E. 1962. *Environment of calcium carbonate deposition west of Andros Island, Bahamas*. USGS Prof. Pap., no. 350.
Cohen, A. D. and. and W. Spackman 1977. Phytogenic organic sediments and sedimentary environments in the Everglades mangrove complex Part II. The origin, description and classification of the peats of S. Florida. *Palaeontographica* (B) **162**, 71–114.
Colbeck, S. C. (ed.) 1980. *Dynamics of snow and ice masses*. New York: Academic Press.
Coleman, J. M. 1969. Brahmaputra River: channel processes and sedimentation. *Sed. Geol.* **3**, 129–239.
Coleman, J. M. 1976. *Deltas: processes of deposition and models for exploration*. Champaign, Ill.: Continuing Education Publishing.
Coleman, J. M. and S. M. Gagliano 1964. Cyclic sedimentation in the Mississippi river delta plain. *Trans Gulf Coast Assoc. Geol Socs* **14**, 67–80.
Coleman, J. M. and L. D. Wright 1975. Modern river deltas: variability of processes and sand bodies. In *Deltas, models for exploration*, M. L. Broussard (ed.), 99–149. Houston: Houston Geol. Soc.
Coleman, J. M., S. M. Gagliano and J. E. Webb 1964. Minor sedimentary structures in a prograding distributary. *Mar. Geol.* **1**, 240–58.
Collins, J. I. 1976. Approaches to wave modelling. In Davis & Ethington (1976), 54–68.
Collinson, J. D. 1969. The sedimentology of the Grindslow Shales and the Kinderscout Grit: a deltaic complex in the Namurian of northern England. *J. Sed. Petrol.* **39**, 194–221.
Collinson, J. D. 1970. Bedforms of the Tana River, Norway. *Geog. Ann.* **52A**, 31–56.
Collinson, J. D. and D. B. Thompson 1982. *Sedimentary structures*. London: George Allen & Unwin.
Cooke, R. U. 1979. Laboratory simulation of salt weathering processes in arid environments. *Earth Surf. Proc.* **4**, 347–59.
Cooke, R. U. and A. Warren 1973. *Geomorphology in deserts*. London: Batsford.
Costello, W. R. 1974. *Development of bed configurations in coarse sands*. Cambridge, Mass.: Earth & Planet. Sci. Dept, MIT, Rept 74.1.
Crans, W., G. Mandl and J. Haremboure 1980. On the theory of growth faulting: a geomechanical delta model based on gravity sliding. *J. Petrol Geol.* **2**, 265–307.
Creager, J. S. and R. W. Sternberg 1972. Some specific problems in understanding bottom sediment distribution and dispersal on the continental shelf. In *Shelf sediment transport: process and pattern*, D. J. P. Swift, D. B. Duane & O. H. Pilkay (eds), 333–46. Stroudsburg, Pa: Dowden, Hutchinson & Ross.
Crevello, P. D. and W. Schlager 1981. Carbonate debris sheets and turbidites, Exuma Sound, Bahamas. *J. Sed. Petrol.* **50**, 1121–48.
Crimes, T. P. (ed.) 1970. *Trace fossils*. Liverpool: Seel House Press.
Crimes, T. P. and J. C. Harper (eds) 1977. *Trace fossils, 2*. Liverpool: Seel House Press.
Crowell, J. C. 1973. *Ridge Basin Southern California, sedimentary facies changes in Tertiary Rocks California Transverse and Southern Coast Ranges*. Soc. Econ. Pal. Mineral field trip guide 1–7.
Csanady, G. T. 1978. Water circulation and dispersal mechanisms. In Lerman (1978), 21–64.
Cummins, W. A. 1962. The greywacke problem. *Geol J.* **3**, 51–72.
Curray, J. R. 1960. Sediments and history of the Holocene transgression, continental shelf, Gulf of Mexico. In *Recent sediments, NW Gulf of Mexico*, F. P. Shepard, F. B. Phleger & T. H. van Andel (eds), 221–66. Tulsa, Okla.: AAPG.
Curray, J. R. 1964. Transgressions and regressions. In *Papers in marine Geology*, R. L. Mitter (ed.), 175–203. New York: Macmillan.
Curray, J. R. 1965. Late Quaternary history, continental shelves of the United States. In *The Quaternary of the United States*, H. E. Wright & D. G. Fry (eds) 723–35. Princeton, NJ: Princeton Univ. Press.
Curtis, C. D. 1976. Stability of minerals in surface weathering reactions: a general thermochemical approach. *Earth Surf. Proc.* **1**, 63–70.
Curtis, C. D. 1977. Sedimentary geochemistry: environments and processes dominated by involvement of an aqueous phase. *Phil Trans R. Soc. Lond.* (A) **286**, 353–72.
Curtis, C. D. 1978. Possible links between sandstone diagenesis and depth-related geochemical reactions occurring in enclosing mudstones. *J. Geol Soc. Lond.* **135**, 107–17.
Curtis, C. D. and D. A. Spears 1968. The formation of sedimentary iron minerals. *Econ Geol.* **63**, 257–70.
Curtis, C. D., C. Petrowski and G. Oertel 1972. Stable carbon isotope ratios within carbonate concretions: a clue to time and place of origin. *Nature* **235**, 98–100.
Curtis, C. D., S. R. Lipshie, G. Oertel and M. J. Pearson 1980. Clay orientation in some U. Carboniferous mudrocks, its relationship to quartz content and some inferences about fissility, porosity and compactional history. *Sedimentology* **27**, 333–40.
Curtis, D. M. 1970. Miocene deltaic sedimentation, Louisiana Gulf Coast. In Morgan (1970), 293–308.

Dalrymple, D. W. 1966. Calcium carbonate deposition associated with blue-green algae mats, Baffin Bay, Texas. *Publns Inst. Mar. Sci. Univ. Tex.* **10**, 187–200.

Davidson-Arnott, R. G. D. and B. Greenwood 1974. Bedforms and structures associated with bar topography in the shallow water wave environment, Kouchibougvac Bay, New Brunswick, Canada. *J. Sed. Petrol.* **44**, 698–704.

Davidson-Arnott, R. G. D. and B. Greenwood 1976. Facies relationships on a barred coast, Kouchibougvac Bay, New Brunswick, Canada. In Davis & Ethington (1976), 149–68.

Davies, D. K., F. G. Ethridge and R. R. Berg 1971. Recognition of barrier environments. *Bull. AAPG* **55**, 550–65.

Davies, G. R. 1970. Algal-laminated sediments, Gladstone embayment, Shark Bay, Western Australia. *Mem. AAPG* **13**, 169–205.

Davies, P. J., B. Bubela and J. Ferguson 1978. The formation of oöids. *Sedimentology* **25**, 703–30.

Davies, T. A. and D. S. Gorsline 1976. Oceanic sediments and sedimentary processes. In *Chemical oceanography*, J. P. Riley & R. Chester (eds), 2nd edn, **5**, 1–80. London: Academic Press.

Davis, K. S. and J. A. Day 1964. *Water: the mirror of science.* London: Heinemann.

Davis, R. A. and R. L. Ethington 1976. *Beach and nearshore sedimentation.* SEPM Spec. Pubn, no. 24. Tulsa.

Davis, R. A. (ed.) 1978. *Coastal sedimentary environments.* New York: Springer.

Deelman, J. C. 1975. Dolomite synthesis and crystal growth. *Geology* **3**, 471–2.

Deelman, J. C. 1978. Experimental oöids and grapestones: carbonate aggregates and their origin. *J. Sed. Petrol.* **48**, 503–12.

Degens, E. T. 1965. *Geochemistry of sediments.* Englewood Cliffs, NJ: Prentice-Hall.

Degens, E. T. and D. A. Ross (eds) 1974. *The Black Sea – geology, chemistry and biology.* Mem. AAPG, no. 20.

Degens, E. T. and P. Stoffers 1980. Environmental events recorded in Quaternary sediments of the Black Sea. *J. Geol. Soc. Lond.* **137**, 131–8.

Demaison, G. J. 1977. Tar sands and supergiant oil fields. *Bull. AAPG* **61**, 1950–61.

Demaison, G. J. and G. T. Moore 1980. Anoxic environments and oil source bed genesis. *Bull. AAPG* **64**, 1179–209.

Denny, C. S. 1967. Fans and pediments. *Am. J. Sci.* **265**, 81–105.

Deroo, G., T. G. Powell, B. Tissot and R. G. McCrossan 1977. The origin and migration of petroleum in the Western Canadian sedimentary basin, Alberta. *Bull. Geol Surv. Can.* 262.

Deuser, W. G. 1975. Reducing environments. In *Chemical oceanography*, J. P. Riley & G. Skirrow (eds), 1–60. London: Academic Press.

Dickinson, W. R. 1970. Interpreting detrital modes of graywacke and arkose. *J. Sed. Petrol.* **40**, 695–707.

Dickinson, W. R. and D. R. Seely 1979. Structure and stratigraphy of fore-arc regions. *Bull. AAPG* **63**, 2–31.

Dickson, J. A. D. and M. L. Coleman 1980. Changes in carbon and oxygen isotope composition during limestone diagenesis. *Sedimentology* **27**, 107–18.

Deeter-Haas, L. and H. J. Schrader 1979. Neogene coastal upwelling history off NW and SW Africa. *Mar. Geol.* **29**, 39–53.

Donovan, R. N. 1975. Devonian lacustrine limestones at the margin of the Orcadian Basin, Scotland. *J. Geol. Soc. Lond.* **131**, 489–510.

Donovan, R. N. and R. J. Foster 1972. Subaqueous shrinkage cracks from the Caithness flagstone series (Middle Devonian) of Northeast Scotland. *J. Sed. Petrol.* **42**, 309–17.

Donovan, R. N. and R. Archer 1975. Some sedimentological consequences of a fall in the level of Haweswater, Cumbria. *Proc. Yorks. Geol Soc.* **40**, 547–62.

Donovan, R. N., R. J. Foster and T. S. Westoll 1974. A stratigraphic revision of the Old Red Sandstone of northeastern Caithness. *Trans R. Soc. Edinb.* **69**, 167–201.

Dott, R. L. 1964. Wacke, greywacke and matrix – what approach to immature sandstone classification? *J. Sed. Petrol.* **34**, 625–32.

Drake, D. E., R. L. Kolpack and P. J. Fischer 1972. Sediment transport on the Santa Barbara – Oxnard shelf, Santa Barbara Channel, California. In *Shelf sediment transport*, D. J. P. Swift, D. B. Duane & O. H. Pilkey (eds), 307–31. Stroudsburg, PA: Dowden, Hutchinson & Ross.

Dravis, J. 1979. Rapid and widespread generation of recent oölitic hardgrounds on a high energy Bahamian Platform, Eleuthera Bank, Bahamas. *J. Sed. Petrol.* **49**, 195–208.

Dreimanis, A. 1979. The problems of waterlain tills. In *Moraines and varves*, C. Schluchter (ed.), 167–78. Rotterdam: Balkema.

Dunham, R. J. 1962. Classification of carbonate rocks according to depositional texture. In *Classification of carbonate rocks*, W. E. Ham (ed.), 108–21. Tulsa, Okla.: AAPG.

Dunham, R. J. 1969a. Early vadose silt in Townsend mound (reef) New Mexico. In *Depositional environments in sedimentary rocks*, G. Friedman (ed.), SEPM Spec. Pubn, no. 14, 139–81. Tulsa.

Dunham, R. J. 1969b. Vadose pisolite in the Capitan Reef (Permian), New Mexico and Texas. In *Depositional environments in sedimentary rocks*, G. Friedman (ed.) SEPM Spec. Pubn 14, 182–91.

Durney, D. W. 1972. Solution-transfer, an important geological deformation mechanism. *Nature* **235**, 315–17.

Durney, D. W. 1976. Pressure-solution and crystallisation deformation. *Phil Trans R. Soc. Lond.* **283**, 229–40.

Duxbury, A. C. 1971. *The Earth and its oceans.* Reading, Mass.: Addison-Wesley.

Dyer, K. R. 1972. *Estuaries: a physical introduction.* Chichester: Wiley.

Dzulynski, S. and E. K. Walton 1965. *Sedimentary features of flysch and greywackes.* Amsterdam: Elsevier.

Edwards, M. B. 1975. Glacial retreat sedimentation in the Smalfjord Formation, Late Precambrian, North Norway. *Sedimentology* **22**, 75–94.

Edwards, M. B. 1976. Growth faults in upper Triassic deltaic sediments, Svalbard. *Bull. AAPG* **60**, 341–55.

Einstein, H. A. and H. Li 1958. Secondary currents in straight channels. *Trans Am. Geophys. Union* **39**, 1085–94.

Eittreim, S., P. E. Biscaye and A. F. Amos 1975. Benthic nepheloid layers and the Ekman thermal pump. *J. Geophys. Res.* **80**, 5061–7.

Elderfield, H. 1976. Manganese fluxes to the oceans. *Mar. Chem.* **4**, 103–32.

Elliott, T. 1974. Interdistributary bay sequences and their genesis. *Sedimentology* **21**, 611–22.

Elliott, T. 1975. The sedimentary history of a delta lobe from a Yoredale (Carboniferous) cyclothem. *Proc. Yorks. Geol Soc.* **40**, 505–36.

Elliott, T. 1978a. Clastic shorelines. In Reading (1978), 143–77.

Elliott, T. 1978b. Deltas. In Reading (1978), 97–142.

Embleton, C. 1980. Glacial processes. In *Process in geomorphology*, C. Embleton & J. Thornes (eds) 272–306. London: Edward Arnold.

Embley, R. W. 1976. New evidence for occurrence of debris flow deposits in the deep sea. *Geology* **4**, 371–4.

Emery, K. O. 1969. The continental shelves. *Scient. Am.* **221**, 106–22.

Emery, K. O. 1978. Grain size in laminae of beach sand. *J. Sed. Petrol.* **48**, 1203–12.

Enos, P. 1977. Tamabra limestone of the Poza Rica trend, Cretaceous, Mexico. In *Deep water carbonate environments*, H. E. Cook & P. Enos (eds), 273–314. SEPM Spec. Pubn, no. 25. Tulsa.

Epstein, S. 1959. The variations of the O^{18}/O^{16} ratio in nature and some geological implications. In *Researches in geochemistry*, P. H. Abelson (ed.) 217–40. New York: Wiley.

Eriksson, K. A. 1977. Tidal flat and subtidal sedimentation in the 2250 Ma Malmani Dolomite, Transvaal, South Africa. *Sed. Geol.* **18**, 223–44.

Ethridge, F. G. and S. A. Schumm 1978. Reconstructing paleochannel morphologic and flow characteristics: methodology, limitations and assessment. In Miall (1978), 703–21.

Eugster, H. P. and L. A. Hardie 1975. Sedimentation in an ancient playa-lake complex: the Wilkins Peak Member of the Green River Formation of Wyoming. *Bull. Geol Soc. Am.* **86**, 319–34.

Evans, G. 1965. Intertidal flat sediments and their environments of deposition in the Wash. *Q. J. Geol Soc. Lond.* **121**, 209–45.

Evans, G., V. Schmidt, P. Bush and H. Nelson 1969. Stratigraphy and geologic history of the sabkha, Abu Dhabi, Persian Gulf. *Sedimentology* **12**, 145–59.

Ewald, P. P., T. Püschl and L. Prandtl 1930. *The physics of solids and fluids*. London: Blackie.

Faure, G. 1977. *Principles of isotope geology*. New York: Wiley.

Ferguson, J., B. Bubela and P. J. Davies 1978. Synthesis and possible mechanism of formation of radial carbonate oöids. *Chem. Geol.* **22**, 285–308.

Ferm, J. C. 1974. Carboniferous environment models in eastern United States and their significance. In *Carboniferous of the SE United States*, G. Briggs (ed.), 79–96. Geol Soc. Am. Spec. Pap., no. 148.

Ferm, J. C. and V. V. Cavaroc 1969. *A field guide to Allegheny deltaic aspects in the upper Ohio valley, with a commentary on deltaic aspects of Carboniferous rocks in the northern Appalachian Plateau.* Pittsburgh and Ohio Geol. Socs, Guidebook for Annual Field Trip.

Field, M. E. 1980. Sand bodies on coastal plain shelves: Holocene record of the US Atlantic inner shelf off Maryland. *J. Sed. Petrol.* **50**, 505–28.

Fischer, A. G. 1964. The Lofer cyclothems of the Alpine Triassic. In *Symposium on cyclic sedimentation*, D. F. Merriam (ed.), 107–49. Bull. Geol. Surv. Kansas. no. 169.

Fischer, A. G. 1975. Tidal deposits, Dachstein Limestone of the North Alpine Triassic. In Ginsburg (1975), 235–42.

Fisher, W. L. 1969. Facies characteristics of Gulf Coast Basin delta systems with some Holocene analogues. *Trans Gulf Coast Ass. Geol Socs* **19**, 239–61.

Fisher, W. L. and J. H. McGowen 1969. Depositional systems in Wilcox Group (Eocene) of Texas and their relation to occurrence of oil and gas. *Bull. AAPG* **53**, 30–54.

Fisher, W. L., L. F. Brown, A. J. Scott and J. H. McGowen 1969. *Delta systems in the exploration for oil and gas.* Austin, Texas: Bureau Economic Geol.

Fisk, H. N. 1944. *Geological investigations of the alluvial valley of the Lower Mississippi River.* Vicksberg, Miss.: Miss. Riv. Comm.

Fisk, H. N. 1959. Padre Island and the Laguna Madre flats, coastal South Texas. *National Acad. Sci.-Nat. Res. Council, 2nd Coastal Geography Conf.*, 103–51.

Fisk, H. N., E. McFarlan, C. R. Kolband and L. J. Wilbert 1954. Sedimentary framework of the modern Mississippi delta. *J. Sed. Petrol.* **24**, 76–99.

Flood, R. D. and C. D. Hollister 1974. Current controlled topography on the continental margin off the eastern USA. In *The geology of continental margins*, C. S. Burk & C. L. Drake (eds), 197–205. New York: Springer.

Flood, R. D., C. D. Hollister and P. Lonsdale 1979. Disruption of the Feni sediment drift by debris flows from Rockall Bank. *Mar. Geol.* **32**, 311–34.

Folk, R. L. 1962. Spectral subdivision of limestone types. In *Classification of carbonate rocks*, W. E. Ham (ed.), 62–84. Tulsa, Okla.: AAPG.

Folk, R. L. 1965. Some aspects of recrystallisation in ancient limestones. In *Dolomitisation and limestone diagenesis.* L. C. Pray & R. C. Murray (eds), SEPM Spec. Pubn, no. 13, 14–48.

Folk, R. L. 1971. Longitudinal dunes of the northwestern edge of the Simpson Desert, Northern Territory, Australia. 1: Geomorphology and grain size relationships. *Sedimentology* **16**, 5–54.

Folk, R. L. 1973. Carbonate petrography in the post-Sorbian age. In *Evolving concepts in sedimentology*, R. N. Ginsburg (ed.) 118–58. Baltimore: Johns Hopkins Press.

Folk, R. L. 1974a. *Petrology of sedimentary rocks.* Austin, Tex.: Hemphills.

Folk, R. L. 1974b. The natural history of crystalline calcium carbonate: effect of magnesium content and salinity. *J. Sed. Petrol.* **44**, 40–53.

Folk, R. L. and J. S. Pittman 1971. Length-slow chalcedony: a new testament for vanished evaporites. *J. Sed. Petrol.* **41**, 1045–58.

Folk, R. L. and L. S. Land 1975. Mg:Ca ratio and salinity: two controls over crystallisation of dolomite. *Bull. AAPG* **59**, 60–8.

Fournier, F. 1960. *Climat et érosion: la relation entre l'erosion du sol par l'eau et les précipitations atmospheriques.* Paris.

Francis, J. R. D. 1969. *A textbook of fluid mechanics.* London: Edward Arnold.

Francis, J. R. D. 1973. Experiments on the motion of solitary grains along the bed of a water stream. *Proc. R. Soc. Lond.* (A) **332**, 443–71.

Frazier, D. E. 1967. Recent deltaic deposits of the Mississippi delta: their development and chronology. *Trans Gulf Coast Ass. Geol Socs* **17**, 287–315.

Frey, R. W. 1975. *The study of trace fossils.* Berlin: Springer.

Friedman, G. M. 1961. Distinction between dune, beach and river sands from their textural characteristics. *J. Sed. Petrol.* **31**, 514–29.

Friedman, G. M. 1964. Early diagenesis and lithification of carbonate sediments. *J. Sed. Petrol.* **34**, 777–813.

Friend, P. F. 1978. Distinctive features of some ancient river systems. In Miall (1978), 531–42.

Friend, P. F. and M. Moody-Stuart 1972. Sedimentation of the Wood Bay Formation (Devonian) of Spitsbergen: regional analysis of a late orogenic basin. *Norsk. Polar. Skr.* **157**, 1–77.

Fryberger, S. G., T. S. Ahlbrandt and S. Andrews 1979. Origin, sedimentary features and significance of low-angle eolian 'sand-sheet' deposits. Great Sand Dunes National Monument and vicinity, Colorado. *J. Sed. Petrol.* **49**, 733–46.

Fuller, A. D. 1979. Phosphate occurrences on the western and southern coastal areas and continental shelves of Southern Africa. *Econ. Geol.* **74**, 221–31.

Fuller, J. G. C. M. and J. W. Porter 1969. Evaporite formations with petroleum reservoirs in Devonian and Mississippian of Alberta, Saskatchewan and N. Dakota. *Bull. AAPG* **53**, 909–26.

Funnell, B. M. 1978. Productivity control of chalk sedimentation. In Friedman (ed.) 1, 228 *Abstracts, 10th Int. Congress of Sedimentology*, Jerusalem.

Gadow, S. and H. E. Reineck 1969. Ablandiger sand transport bei Sturmfluten. *Senckenberg. Marit.* **3**, 103–33.

Gaines, A. M. 1977. Protodolomite redefined. *J. Sed. Petrol.* **47**, 543–6.

Galloway, W. E. 1975. Process framework for describing the morphologic and stratigraphic evolution of the deltaic depositional systems. In *Deltas, models for exploration*, M. L. Broussard (ed.), 87–98. Houston: Houston Geol. Soc.

Galvin, C. J. 1968. Breaker type classification on three laboratory beaches. *J. Geophys. Res.* **73**, 3651–9.

Garrels, R. M. and M. E. Thompson 1962. A chemical model for sea water at 25°C and one atmosphere total pressure. *Am. J. Sci.* **260**, 57–66.

Garrels, R. M. and C. L. Christ 1965. *Solutions, minerals and equilibrium*. New York: Harper & Row.

Garrett, P. 1970. Phanerozoic stromatolites: noncompetitive ecologic restriction by grazing and burrowing animals. *Science* **169**, 171–3.

Gasiorek, J. M. and W. G. Carter 1967. *Mechanics of fluids for mechanical engineers*. London: Blackie.

Gebelein, C. D. and P. Hoffman 1973. Algal origin of dolomite laminations in stromatolitic limestone. *J. Sed. Petrol.* **43**, 603–13.

Gibbs, R. J., M. D. Mathews and D. A. Link 1971. The relationship between sphere size and settling velocity. *J. Sed. Petrol.* **41**, 7–18.

Gilbert, G. K. 1885. The topographic features of lake shores. *Ann. Rept USGS* **5**, 75–123.

Gile, L. H. and J. W. Hawley 1969. Age and comparative development of desert soils at the Gardner Spring radiocarbon site, New Mexico. *Proc. Soil Sci. Soc. Am.* **32**, 709–16.

Gill, W. D. and P. H. Keunen 1958. Sand volcanoes on slumps in the Carboniferous of County Clare, Ireland. *Q. J. Geol Soc. Lond.* **113**, 441–60.

Ginsburg, R. N. (ed.) 1975. *Tidal deposits*. Berlin: Springer.

Ginsburg, R. N. and N. P. James 1974. Holocene carbonate sediments of continental shelves. In *The geology of continental margins*, C. A. Burk and C. L. Drake (eds), 137–55. Berlin: Springer.

Glennie, K. W. 1970. *Desert sedimentary environments*. Amsterdam: Elsevier.

Glennie, K. W. 1972. Permian Rotliegendes of northwest Europe interpreted in light of modern desert sedimentation studies. *Bull. AAPG* **56**, 1048–71.

Goldhaber, M. B. and I. R. Kaplan 1974. The sulfur cycle. In *The sea* **5**, E. D. Goldberg (ed.), 569–655. New York: Wiley.

Goldich, S. S. 1938. A study in rock weathering. *J. Geol.* **46**, 17–58.

Goldring, R. 1964. Trace fossils and the sedimentary surface. In *Developments in sedimentology 1: deltaic and shallow marine deposits*, L. M. J. U. Van Straaten (ed.), 136–43. Amsterdam: Elsevier.

Goldring, R. and P. H. Bridges 1973. Sublittoral sheet sandstones. *J. Sed. Petrol.* **43**, 736–47.

Goldring, R., D. J. W. Bosence and T. Blake 1978. Estuarine sedimentation in the Eocene of southern England. *Sedimentology* **25**, 861–76.

Gole, C. V. and S. V. Chitale 1966. Inland delta building activity of Kosi River. *J. Hyd. Div. Am. Soc. Civ. Engrs* **92**, 111–26.

Goudie, A. 1973. *Duricrusts in tropical and subtropical landscapes*. Oxford: Oxford Univ. Press.

Goudie, A. S., R. U. Cooke and I. S. Evans 1970. Experimental investigation of rock weathering by salts. *Area* **4**, 42–8.

Grace, J. T., B. T. Grothaus and R. Ehrlich 1978. Size frequency distributions taken from within sand laminae. *J. Sed. Petrol.* **48**, 1193–202.

Grass, A. J. 1970. Initial instability of fine bed sand. *J. Hyd. Div. Am. Soc. Civ. Engrs* **96**, 619–32.

Grass, A. J. 1971. Structural features of turbulent flow over smooth and rough boundaries. *J. Fluid Mech.* **50**, 233–55.

Gray, W. A. 1968. *The packing of solid particles*. London: Chapman & Hall.

Green, P. 1967. *The waters of the sea*. New York: Van Nostrand Reinhold.

Gregory, K. J. (ed.) 1977. *River channel changes*. Chichester: Wiley.

Griffin, J. J., H. Windom and E. D. Goldberg 1968. The distribution of clay minerals in the world ocean. *Deep-sea Res.* **15**, 433–59.

Griffith, L. S., M. G. Pitcher and G. W. Rice 1969. Quantitative environmental analysis of a Lower Cretaceous Reef Complex. In *Depositional environments in carbonate rocks*, G. H. Friedman (ed.), 120–37, Tulsa, SEPM Spec. Pubn no. 14.

Grim, R. E. 1968. *Clay mineralogy*, 2nd edn. New York: McGraw-Hill.

Gulbrandsen, R. A. 1969. Physical and chemical factors in the formation of marine apatite. *Econ. Geol.* **69**, 365–82.

Gunatilaka, A. 1975. Some aspects of the biology and sedimentology of laminated algal mats from Mannar lagoon, Northwest Ceylon. *Sed. Geol.* **14**, 275–300.

Gunatilaka, A. 1976. Thallophyte boring and micritisation within skeletal sands from Connemara, W. Ireland. *J. Sed. Petrol.* **46**, 548–54.

Guy, H. P., D. B. Simons and E. V. Richardson 1966. *Summary of alluvial channel data from flume experiments, 1956–61*. USGS Prof. Pap., no. 462-I.

Hagan, G. M. and B. W. Logan 1974. Development of carbonate banks and hypersaline basins, Shark Bay, Western Australia. In Logan *et al.* (1974), 61–139.

Hagan, G. M. and B. W. Logan 1975. Prograding tidal-flat sequences Hutchison Embayment, Shark Bay, Western Australia. In Ginsburg (1975), 215–22.

Hails, J. and A. Carr (eds) 1975. *Nearshore sediment dynamics and sedimentation*. London: Wiley.

Hallam, A. 1969. Tectonism and eustasy in the Jurassic. *Earth Sci. Rev.* **5**, 45–68.

Halley, R. B. 1977. Oöid fabric and fracture in the Great Salt Lake and the geologic record. *J. Sed. Petrol.* **47**, 1099–120.

Halley, R. B., E. A. Shinn, J. H. Hudson and B. H. Lidz 1977. Pleistocene barrier bar seaward of oöid shoal complex near Miami, Florida. *Bull. AAPG* **61**, 519–26.

Halsey, S. D. 1979. Nexus: new model of barrier island development. In Leatherman (1979), 185–210.

Hamilton, W. and D. Krinsley 1967. Upper Paleozoic glacial

deposits of South Africa and Southern Australia. *Bull. Geol Soc. Am.* **78**, 783–800.

Hampton, M. A. 1972. The role of subaqueous debris flow in generating turbidity currents. *J. Sed. Petrol.* **42**, 775–93.

Hancock, N. J. and A. M. Taylor 1978. Clay mineral diagenesis and oil migration in the middle Jurassic Brent Sand Formation. *J. Geol Soc. Lond.* **135**, 69–72.

Hanor, J. S. 1978. Precipitation of beach rock cements: mixing of marine and meteoric waters *vs* CO_2-degassing. *J. Sed. Petrol.* **48**, 489–501.

Hanshaw, B. B., W. Back and R. G. Deike 1971. A geochemical hypothesis for dolomitisation by ground water. *Econ. Geol.* **66**, 710–24.

Harbaugh, J. W. and G. Bonham-Carter 1970. *Computer simulation in geology*. New York: Wiley.

Hardie, L. A. 1967. The gypsum–anhydrite equilibrium at one atmosphere pressure. *Am. Mineral.* **52**, 171–200.

Hardie, L. A. 1968. The origin of the Recent non-marine evaporite deposit of Saline Valley, Inyo County, California. *Geochim. Cosmochim. Acta* **32**, 1279–301.

Hardie, L. A. (ed.) 1977. *Sedimentation on the modern carbonate tidal flats of NW Andros Island, Bahamas*. Baltimore: Johns Hopkins Press.

Hardie, L. A. and P. Garrett 1977. General environmental setting. In Hardie (1977), 12–49.

Hardie, L. A. and R. N. Ginsburg 1977. Layering: the origin and environmental significance of lamination and thin bedding. In Hardie (1977), 50–123.

Hardie, L. A., J. P. Smoot and H. P. Eugster 1978. Saline lakes and their deposits: a sedimentological approach. In Matter & Tucker (1978), 7–42.

Harland, W. B., K. Herod and P. H. Krinsley 1966. The definition and identification of tills and tillites. *Earth Sci. Rev.* **3**, 225–56.

Harms, J. C. 1966. Stratigraphic traps in a valley fill, W. Nebraska. *Bull. AAPG* **50**, 2119–49.

Harrell, J. A. and K. A. Eriksson 1979. Empirical conversion equations for thin-section and sieve derived size distribution parameters. *J. Sed. Petrol.* **49**, 273–80.

Harris, P. M. 1979. Facies anatomy and diagenesis of a Bahamian oöid shoal. *Sedimenta* **7**. Comparative Sedimentology Laboratory, University of Miami, Fl.

Harrison, R. S. and R. P. Steinen 1978. Subaerial crusts, caliche profiles and breccia horizons: comparison of some Holocene and Mississippian exposure surfaces, Barbados and Kentucky. *Bull. Geol Soc. Am.* **89**, 385–96.

Harvey, J. G. 1976. *Atmosphere and ocean: our fluid environments*. Sussex: Artemis Press.

Haszeldine, R. S. and R. Anderton 1980. A braid plain facies model for the Westphalian-B Coal Measures of NE England. *Nature* **284**, 51–3.

Hay, R. L. 1966. *Zeolites and zeolite reactions in sedimentary rocks*. Geol Soc. Am. Spec. Pap., no. 85.

Hayes, J. B. 1979. Sandstone diagenesis – the hole truth. In Scholle & Schluger (1979).

Hayes, M. O. 1971. Geomorphology and sedimentation of some New England estuaries. In Schubel (1971), 1–71.

Hayes, M. O. 1975. Morphology of sand accumulations in estuaries. In *Estuarine research*, L. E. Cronin (ed.), 3–22. New York: Academic Press.

Hayes, M. O. 1979. Barrier island morphology as a function of tidal and wave regime. In Leatherman (1979), 1–27.

Hays, J. D. and W. C. Pitman 1973. Lithospheric plate motions, sea-level changes and climatic and ecological consequences. *Nature* **246**, 18–22.

Hays, J. D., J. Imbrie and N. J. Shackleton 1976. Variations in the Earth's orbit: pacemaker of the ice ages. *Science* **194**, 1121–32.

Heath, G. R. 1974. Dissolved silica and deep-sea sediments. In *Studies in paleo-oceanography*, W. W. Hay (ed.) 77–93. SEPM Spec. Pubn., no. 20.

Heckel, P. H. 1974. Carbonate build-ups in the geological record: a review. In *Reefs in time and space*, L. F. Laporte (ed.), 90–154. SEPM Spec. Pubn., no. 18.

Hedberg, H. D. 1974. Relation of methane generation to undercompacted shales, shale diapirs and mud volcanoes. *Bull. AAPG* **58**, 661–73.

Heezen, B. C. and C. D. Hollister 1963. Evidence of deep sea bottom currents from abyssal sediments. *Int. Union Geod. Geophys.* **6**, 111.

Heezen, B. C. and A. S. Laughton 1963. Abyssal plains. In *The sea*, M. N. Hill (ed.) **3**, 312–64. New York: Wiley.

Heezen, B. C. and C. D. Hollister 1971. *The face of the deep*. New York: Oxford Univ. Press.

Heward, A. P. 1978a. Alluvial fan and lacustrine sediments from the Stephanian A and B (La Magdalena, Cinera-Matallana and Sabero) coalfields, northern Spain. *Sedimentology* **25**, 451–88.

Heward, A. P. 1978b. Alluvial fan sequence and megasequence models: with examples from Westphalian D–Stephanian B coalfields, northern Spain. In Miall (1978), 669–702.

Hickin, E. J. 1974. The development of meanders in natural river channels. *Am. J. Sci.* **274**, 414–42.

Hinte, J. E. Van 1978. Geohistory analysis – application of micropaleontology in exploration geology. *Bull. AAPG* **62**, 201–22.

Ho, C. and J. M. Coleman 1969. Consolidation and cementation of recent sediments in the Atchafalaya Basin. *Bull. Geol Soc. Am.* **80**, 183–92.

Hobson, G. D. (ed.) 1977. *Developments in petroleum geology – 1*. London: Applied Science.

Hobson, G. D. and E. N. Tiratsoo 1975. *Introduction to petroleum geology*. Beaconsfield: Scientific Press.

Holliday, D. W. 1968. Early diagenesis in Middle Carboniferous nodular anhydrite of Spitsbergen. *Proc. Yorks. Geol Soc.* **36**, 277–92.

Holliday, D. W. 1970. The petrology of secondary gypsum rocks: a review. *J. Sed. Petrol.* **40**, 734–44.

Holliday, D. W. 1973. Early diagenesis in nodular anhydrite rocks. *Trans Inst. Min. Metall.* **82**, 81–4.

Holliday, D. W. and E. R. Shephard-Thorne 1974. *Basal Purbeck evaporites of the Fairlight Borehole, Sussex*. Rept Inst. Geol Sci., no. 74/4.

Hollister, C. D. and B. C. Heezen 1972. Geological effects of ocean bottom currents: western North Atlantic. In *Studies in physical oceanography*, A. L. Gordon (ed.), 37–66. New York: Gordon and Breach.

Hollister, C. D., R. D. Flood, D. A. Johnson, P. Lonsdale and J. B. Southard 1974. Abyssal furrows and hyperbolic echo traces on the Bahama Outer Ridge. *Geology* **2**, 395–400.

Honji, H., A. Kaneko and N. Matsunaga 1980. Flows above oscillatory ripples. *Sedimentology* **27**, 225–9.

Hooke, R. LeB. 1967. Processes on arid-region alluvial fans. *J. Geol.* **75**, 438–60.

Hooke, R. LeB. 1972. Geomorphic evidence for Late Wisconsin and Holocene tectonic deformation, Death Valley, California. *Bull. Geol. Soc. Am.* **83**, 2073–98.

Hopkins, J. C. 1977. Production of foreslope breccia by differential submarine cementation and downslope displacement of carbonate sands. In *Deep water carbonate environments*, H. E. Cook & P. Enos (eds), 155–70. SEPM Tulsa.

Horn, D., M. Ewing, B. M. Horn and M. N. Delach 1971. Turbidites of the Hatteras and Sohm Abyssal Plains, Western North Atlantic. *Mar. Geol.* **11**, 287–323.

Horne, R. A. 1969. *Marine chemistry*. New York: Wiley.

Horowitz, A. S. and P. E. Potter 1971. *Introductory petrography of fossils*. Berlin: Springer.

Houboult, J. J. H. C. 1957. *Surface sediments of the Persian Gulf near Qatar Peninsular*. Thesis, Univ. Utrecht.

Houboult, J. J. H. C. 1968. Recent sediments in the southern Bight of the North Sea. *Geol. Mijn.* **47**, 245–73.

Howard, J. D., C. A. Elders and J. F. Heinbotel 1975. Animal–sediment relationships in estuarine point bar deposits, Ogeechol River – Ossabaw Sound, Georgia. *Senckenberg. Marit.* **7**, 181–203.

Hower, J., E. V. Eslinger, M. E. Hower and E. A. Perry 1976. Mechanism of burial metamorphism of argillaceous sediment. 1: Mineralogic and chemical evidence. *Bull. Geol Soc. Am.* **87**, 725–37.

Hoyt, J. H. 1967. Barrier island formation. *Bull. Geol Soc. Am.* **78**, 1125–36.

Hsü, K. J. 1966. Origin of dolomite in sedimentary sequences: a critical analysis. *Mineral. Depos.* **2**, 133–8.

Hsü, K. J. 1967. Chemistry of dolomite formation. In *Carbonate rocks, physical and chemical aspects*, G. V. Chilingar, H. J. Bissell & R. W. Fairbridge (eds), 169–91. Amsterdam: Elsevier.

Hsü, K. J. 1972. Origin of saline giants: a critical review after the discovery of the Mediterranean evaporite. *Earth Sci. Rev.* **8**, 371–96.

Hsü, K. J. and C. Siegenthaler 1969. Preliminary experiments on hydrodynamic movement induced by evaporation and their bearing on the dolomite problem. *Sedimentology* **12**, 11–25.

Hsü, K. J. and H. C. Jenkyns (eds) 1974. *Pelagic sediments: on land and under the sea*. Int. Ass. Sed. Spec. Pubn, no. 1. Oxford: Blackwell Scientific.

Hsü, K. J., L. Montadert, D. Bernoulli, M. B. Cita, A. Erikson, R. E. Garrison, R. B. Kidd, X. Melieres, C. Muller and R. Wright 1977. History of the Mediterranean salinity crisis. *Nature* **267**, 399–403.

Hubbard, D. K., G. Oertel and D. Nummedal 1979. The role of waves and tidal currents in the development of tidal inlet sedimentary structures and sand body geometry: examples from N. Carolina, S. Carolina and Georgia. *J. Sed. Petrol.* **49**, 1073–92.

Hudson, J. D. 1962. Pseudo-pleochroic calcite in recrystallised shell-limestones. *Geol Mag.* **99**, 492–500.

Hudson, J. D. 1963. The recognition of salinity-controlled mollusc assemblages in the Great Estuarine Series (middle Jurassic) of the Inner Hebrides. *Palaeontology* **6**, 318–26.

Hudson, J. D. 1975. Carbon isotopes and limestone cement. *Geology* **3**, 19–22.

Hudson, J. D. 1977a. Stable isotopes and limestone lithification. *Q. J. Geol Soc. Lond.* **133**, 637–60.

Hudson, J. D. 1977b. Oxygen isotope studies on Cenozoic temperatures, oceans and ice accumulations. *Scott. J. Geol.* **13**, 313–26.

Hudson, J. D. 1978. Concretions, isotopes and the diagenetic history of the Oxford Clay (Jurassic) of central England. *Sedimentology* **25**, 339–70.

Hunter, R. E. 1977. Basic types of stratification in small eolian dunes. *Sedimentology* **24**, 361–87.

Huntly, D. A. and A. J. Bowen 1973. Field observations of edge waves. *Nature* **243**, 160–1.

Huntly, D. A. and A. J. Bowen 1975. Comparison of the hydrodynamics of steep and shallow beaches. In *Nearshore sediment dynamics and sedimentation*, J. Hails & A. Carr (eds). New York: Wiley.

Hutchinson, G. E. 1957. *A treatise on limnology. 1: geography, physics and chemistry*. New York: Wiley.

Hutton, J. T., C. R. Twidale and A. R. Milnes 1978. Characteristics and origin of some Australian silcretes. In *Silcrete in Australia*, T. Langford-Smith (ed.), 19–40. Univ. of New England.

Illing, L. V. 1954. Bahamian calcareous sands. *Bull. AAPG* **38**, 1–95.

Illing, L. V., A. J. Wells and J. C. M. Taylor 1965. Penecontemporaneous dolomite in the Persian Gulf. In *Dolomitisation and limestone diagenesis: a symposium*, L. C. Pray & R. C. Murray (eds), 89–111. SEPM Spec. Pubn, no. 13. Tulsa.

Illing, L. V., G. V. Wood and J. G. C. M. Fuller 1967. Reservoir rocks and stratigraphic traps in non-reef carbonates. *Proc. 7th World Petrolm Conf.* (Mexico) 487–99.

Ingle, J. C., D. E. Karig and A. H. Bouma 1973. Leg 31, Western Pacific floor. *Geotimes* **18**, 22–5.

Inman, D. L. and R. A. Bagnold 1963. Littoral processes. In *The sea*, M. N. Hill (ed.) **3**, 529–83. New York: Wiley.

Inman, D. L. and A. J. Bowen 1963. Flume experiments on sand transport by waves and currents. *Proc. 8th Conf. on Coast Engng* 137–50.

Irwin, H., M. Coleman and C. D. Curtis 1977. Isotope evidence for several sources of carbonate and distinctive diagenetic processes in organic-rich Kimmeridgian sediments. *Nature* **269**, 209–13.

Jackson, R. G. 1975. Velocity-bedform texture patterns of meander bends in the lower Wabash River of Illinois and Indiana. *Bull. Geol Soc. Am.* **86**, 1511–22.

Jackson, R. G. 1976a. Sedimentological and fluid-dynamic implications of the turbulent bursting phenomena in geophysical flows. *J. Fluid. Mech.* **77**, 531–60.

Jackson, R. G. 1976b. Depositional model of point bars in the lower Wabash River. *J. Sed. Petrol.* **46**, 579–94.

Jackson, R. G. 1978. Preliminary evaluation of lithofacies models for meandering alluvial streams. In Miall (1978), 543–76.

James, N. P. 1978a. Introduction to carbonate facies models. In Walker (1978b), 105–8.

James, N. P. 1978b. Reefs. In Walker (1978b), 121–32.

James, N. P., R. N. Ginsburg, D. S. Marszalek and P. W. Choquette 1976. Facies and fabric specificity of early subsea cements in shallow Belize (British Honduras) reef. *J. Sed. Petrol.* **46**, 523–44.

James, N. P. and R. N. Ginsburg 1979. *The seaward margin of Belize barrier and atoll reefs*. Int. Ass. Sed. Spec. Pubn, no. 3.

John, J. E. A. and N. L. Habermann 1980. *Introduction to fluid mechanics*. 2nd edn. Englewood Cliffs, NJ: Prentice-Hall.

Jenkyns, H. C. 1978. Pelagic environments. In Reading (1978), 314–71.

Jenkyns, H. C. 1980. Cretaceous anoxic events: from continents to oceans. *J. Geol Soc. Lond.* **137**, 171–88.

Johansson, C. E. 1976. Structural studies of frictional sediments. *Geog. Ann.* **58**, 201–300.

Johnson, A. M. 1970. *Physical processes in geology*. San Francisco: Freeman, Cooper.

Johnson, H. D. 1977. Shallow marine sand bar sequences: an example from the late Precambrian of N. Norway. *Sedimentology* **24**, 245–70.

Johnson, H. D. 1978. Shallow siliciclastic seas. In Reading (1978), 207–58.

Jones, C. M. and P. J. McCabe 1980. Erosion surfaces within giant fluvial cross-beds of the Carboniferous in N. England. *J. Sed. Petrol.* **50**, 613–20.

Jopling, A. V. and B. C. McDonald (eds) 1975. *Glacioflucial and glaciolacustrine sedimentation.* SEPM Spec Pubn, no. 23.

Kahle, C. F. 1974. Oöids from Great Salt Lake, Utah, as an analogue for the genesis and diagenesis of oöids in marine limestones. *J. Sed. Petrol.* **44**, 30–9.

Kaneps, A. G. 1979. Gulf Stream: velocity fluctuations during the late Cenozoic. *Science* **204**, 297–301.

Kastner, M., J. B. Keene and J. M. Gieskes 1977. Diagenesis of siliceous oozes. 1: Chemical controls on the rate of opal-A to opal-CT transformation – an experimental study. *Geochim. Cosmochim. Acta* **41**, 1041–54.

Keller, G. H., D. N. Lambert and R. H. Bennett 1979. *Geotechnical properties of continental slope deposits – Cape Hatteras to Hydrographer Canyon.* In *Geology of continental slopes*, L. J. Doyle & O. H. Pilkey (eds). SEPM Spec. Pubn, no. 27, 131–51.

Keller, W. D. 1954. Bonding energies of some silicate minerals. *Am. Mineral.* **39**, 783–93.

Kelts, K. and K. J. Hsü 1978. Freshwater carbonate sedimentation. In Lerman (1978) 295–321.

Kendall, A. C. 1978. Subaqueous evaporites. In Walker (1978b), 159–74.

Kendall, A. C. and M. E. Tucker 1971. Radiaxial fibrous calcite as a replacement after syn-sedimentary cement. *Nature Phys. Sci.* **232**, 62–3.

Kendall, A. C. and M. E. Tucker 1973. Radiaxial fibrous calcite: a replacement after acicular carbonate. *Sedimentology* **20**, 365–89.

Kennedy, J. F. 1963. The mechanics of dunes and antidunes on erodible-bed channels. *J. Fluid Mech.* **16**, 521–44.

Kennedy, W. J. and A. Hall 1967. The influence of organic matter on the preservation of aragonite in fossils. *Proc. Geol Soc. Lond.* **1643**, 253–5.

Kennett, J. P. 1977. Cenozoic evolution of Antarctic glaciation, the circum-Antarctic ocean, and their impact on global palaeo-oceanography. *J. Geophys. Res.* **82**, 3843–60.

Kennett, J. P., R. E. Houtz, P. B. Andrews, A. R. Edwards, V. A. Gostin, M. Hajos, M. A. Hampton, D. G. Jenkins, S. V. Margolis, A. T. Overshine and K. Perch-Nielson 1974. Development of the circum-Antarctic current. *Science* **186**, 144–7.

Kennett, J. P. and N. J. Shackleton 1976. Oxygen isotope evidence for the development of the psychrosphere 38 Ma ago. *Nature* **260**, 513–15.

Kenyon, N. H. 1970. Sand ribbons of European tidal seas. *Mar. Geol.* **9**, 25–39.

Kenyon, N. H. and A. H. Stride 1970. The tide-swept continental shelf sediments between the Shetland Isles and France. *Sedimentology* **14**, 159–73.

Kerr, P. F. 1959. *Optical mineralogy.* New York: McGraw-Hill.

Kersey, D. G. and K. J. Hsü 1976. Energy relations and density current flows: an experimental investigation. *Sedimentology* **23**, 761–90.

Keulegan, G. H. 1957. *Thirteenth progress report on model laws for density currents. An experimental study of the motion of saline water from locks into freshwater channels.* US Natl Bur. Stand. Rept 5168.

Keunen, P. H. 1964. Experimental abrasion of pebbles, 4: eolian action. *J. Geol.* **69**, 427–49.

Keunen, P. H. 1965. Value of experiments in geology. *Geol. Mijn.* **44**, 22–36.

Kimberley, M. M. 1979. Origin of oölitic iron formations. *J. Sed. Petrol.* **49**, 111–32.

Kinsman, D. J. J. 1966. Gypsum and anhydrite of Recent age, Trucial Coast, Persian Gulf. In *Second symposium on salt.* J. L. Rau (ed.), 302–26. Cleveland Northern Ohio Geol Soc.

Kinsman, D. J. J. and R. K. Park 1976. Algal belt and coastal sabkha evolution, Trucial Coast, Persian Gulf. In Walter (1976), 421–33.

Kinsman, D. J. J. 1975a. Salt floors to geosynclines. *Nature* **255**, 375–8.

Kinsman, D. J. J. 1975b. Rift valley basins and sedimentary history of trailing continental margins. In *Petroleum and global tectonics*, A. G. Fischer & S. Judson (eds), 83–126. Princeton NJ: Princeton Univ. Press.

Kinsman, D. J. J. 1976. Evaporites: relative humidity control of primary mineral facies. *J. Sed. Petrol.* **46**, 273–9.

Klein, G. de V. 1971. A sedimentary model for determining paleotidal range, *Bull. Geol Soc. Am.* **82**, 92.

Kline, S. J., W. C. Reynolds, F. A. Schraub and P. W. Runstadler 1967. The structure of turbulent boundary layers. *J. Fluid Mech.* **30**, 741–73.

Knauth, L. P. 1979. A model for the origin of chert in limestone. *Geology* **7**, 274–7.

Knauth, L. P. and S. Epstein 1976. Hydrogen and oxygen isotope ratios in nodular and bedded cherts. *Geochim. Cosmochim. Acta* **40**, 1095–108.

Kobluk, D. R. and M. J. Risk 1977. Calcification of exposed filaments of endolithic algae, micrite envelope formation and sediment production. *J. Sed. Petrol.* **47**, 517–28.

Kolb, C. R. and J. R. Van Lopik 1958. *Geology of the Mississippi River deltaic plain.* US Corps Engrs. Waterways Expt. Sta. Tech. Repts, 3–483, 3.484.

Kolodny, Y., A. Taraboulos and U. Frieslander 1980. Participation of fresh water in chert diagenesis: evidence from oxygen isotopes and boron α-track mapping. *Sedimentology* **27**, 305–16.

Komar, P. D. 1971. The mechanics of sand transport on beaches. *J. Geophys. Res.* **76**, 713–21.

Komar, P. D. 1972. Mechanical interactions of phenocrysts and the flow differentiation of igneous dykes and sills. *Bull. Geol Soc. Am.* **83**, 973–88.

Komar, P. D. 1975. Nearshore currents: generation by obliquely incident waves and longshore variations in breaker height. In *Nearshore sediment dynamics and sedimentation* J. Hails & A. Carr (eds) 17–46. New York: Wiley.

Komar, P. D. 1976. *Beach processes and sedimentation.* Englewood Cliffs, NJ: Prentice-Hall.

Komar, P. D. and D. L. Inman 1970. Longshore sand transport on beaches. *J. Geophys. Res.* **75**, 5914–27.

Komar, P. D., R. H. Neudeck and L. D. Kulm 1972. Observations and significance of deep water oscillatory ripple marks on the Oregon continental shelf. In *Shelf sediment transport: process and pattern*, P. J. P. Swift, D. B. Duane & O. H. Pilkey (eds), 601–19. Stroudsburg, Pa: Dowden, Hutchinson & Ross.

Kraft, J. C. 1971. Sedimentary facies patterns and geologic history of a Holocene marine transgression. *Bull. Geol Soc. Am.* **82**, 2131–58.

Kraft, J. C. and C. J. John 1979. Lateral and vertical facies relations of transgressive barrier. *Bull. AAPG* **63**, 2145–63.
Kranck, K. 1975. Sediment deposition from flocculated suspensions. *Sedimentology* **22**, 111–23.
Kranck, K. 1981. Particulate matter grain-size characteristics and flocculation in a partially mixed estuary. *Sedimentology* **28**, 107–14.
Krauskopf, K. B. 1979. *Introduction to geochemistry*, 2nd edn. New York: McGraw-Hill.
Krumbein, W. C. 1934. Size frequency distributions of sediments. *J. Sed. Petrol.* **4**, 65–77.
Kulm, L. D., R. C. Rousch, J. C. Harlett, R. H. Neudeck, D. M. Chambers and E. T. Runge 1975. Oregon continental shelf sedimentation: interrelationships of facies distribution and sedimentary processes. *J. Geol.* **83**, 145–76.
Kumar, N. and J. E. Sanders 1974. Inlet sequences: a vertical succession of sedimentary structures and textures created by the lateral migration of tidal inlets. *Sedimentology* **21**, 491–532.

Lambe, T. W. and R. V. Whitman 1969. *Soil mechanics*. New York: Wiley.
Land, L. S. 1966. *Diagenesis of metastable skeletal carbonates*. Thesis Lehigh Univ. Pa.
Land, L. S. 1970. Phreatic versus vadose meteoric diagenesis of limestones: evidence from a fossil water table. *Sedimentology* **14**, 175–85.
Land, L. S. 1973. Holocene meteoric dolomitisation of Pleistocene limestones, N. Jamaica. *Sedimentology* **20**, 411–24.
Land, L. S. 1980. The isotopic and trace element geochemistry of dolomite: the state of the art. In Zenger *et al.* (1980), 87–110.
Land, L. S. and S. Epstein 1970. Late Pleistocene diagenesis and dolomitisation, N. Jamaica. *Sedimentology* **14**, 187–200.
Land, L. S. and S. P. Dutton 1978. Cementation of Pennsylvanian deltaic sandstone: isotopic data. *J. Sed. Petrol.* **48**, 1167–76.
Langbein, W. B., and S. A. Schumm 1958. Yield of sediment in relation to mean annual precipitation. *Trans Am. Geophys. Union* **39**, 1076–84.
Langford-Smith, T. (ed.) 1978. *Silcrete in Australia*. Dept. Geography, Univ. of New England.
Laporte, L. F. 1971. Palaeozoic carbonate facies of the Central Appalachian Shelf. *J. Sed. Petrol.* **41**, 724–40.
Leatherman, S. P. (ed.) 1979. *Barrier islands*. New York: Academic Press.
Leeder, M. R. 1973. Fluviatile fining upward cycles and the magnitude of palaeochannels. *Geol Mag.* **110**, 265–76.
Leeder, M. R. 1974. Tournaisian fluvio-deltaic sedimentation and the palaeogeography of the Northumberland basin. *Proc. Yorks. Geol Soc.* **40**, 129–80.
Leeder, M. R. 1975a. Pedogenic carbonates and flood sediment accretion rates: a quantitive model for alluvial arid-zone lithofacies. *Geol Mag.* **112**, 257–70.
Leeder, M. R. 1975b. Lower Border Group (Tournaisian) stromatolites from the Northumberland basin. *Scott. J. Geol.* **3**, 207–26.
Leeder, M. R. 1977. Bedload stresses and Bagnold's bedform theory for water flows. *Earth Surf. Proc.* **2**, 3–12.
Leeder, M. R. 1979. 'Bedload' dynamics: grain–grain interactions in water flows. *Earth Surf. Proc.* **4**, 229–40.
Leeder, M. R. 1980. On the stability of lower stage plane beds and the absence of current ripples in coarse sands. *J. Geol Soc. London.* **137**, 423–30.
Leeder, M. R. and A. Zeidan 1977. Giant late Jurassic sabkhas of Arabian Tethys. *Nature* **268**, 42–4.
Leeder, M. R. and M. Nami 1979. Sedimentary models for the non-marine Scalby Formation (M Jurassic) and evidence for late Bajocian/Bathonian uplift of the Yorkshire Basin. *Proc. Yorks. Geol Soc.* **42**, 461–82.
Lees, A. 1975. Possible influences of salinity and temperature on modern shelf carbonate sedimentation. *Mar. Geol.* **19**, 159–98.
Leinen, M. 1979. Biogenic silica accumulation in the Central equatorial Pacific and its implications for Cenozoic palaeo-oceanography: Summary. *Bull. Geol Soc. Am.* **90**, 801–3.
Leliavsky, S. 1955. *An introduction to fluvial hydraulics*. London: Constable.
Leopold, L. B. and M. G. Wolman 1960. River meanders. *Bull. Geol Soc. Am.* **71**, 769–94.
Leopold, L. B., M. G. Wolman and J. P. Miller 1964. *Fluvial processes in geomorphology*. San Francisco: W. H. Freeman.
Lerman, A. (ed.) 1978. *Lakes: physics, chemistry and geology*. New York: Springer.
Levey, R. A. 1978. Bedform distribution and internal stratification of coarse-grained point bars Upper Congaree River, S. C. In Miall (1978), 105–27.
Lindholm, R. C. and R. B. Finkleman 1972. Calcite staining: semiquantitive determination of ferrous iron. *J. Sed. Petrol.* **42**, 239–42.
Lindsay, J. F. 1970. Depositional environment of Paleozoic glacial rocks in the central Transantarctic mountains. *Bull. Geol Soc. Am.* **81**, 1149–72.
Lippmann, F. 1973. *Sedimentary carbonate minerals*. New York: Springer.
Lisitzin, A. P. 1967. Basic relationships in distribution of modern siliceous sediments and their connection with climatic zonation. *Int. Geol Rev.* **9**, 631–52.
Livingstone, D. A. 1963. *Chemical composition of rivers and lakes*. USGS Prof. Pap., no. 440G.
Logan, B., R. Rezak and R. N. Ginsburg 1964. Classification and environmental significance of algal stromatolites. *J. Geol.* **72**, 68–83.
Logan, B. W., G. R. Davies, J. F. Read and D. E. Cebulski 1970. *Carbonate sedimentation and environments, Shark Bay, Western Australia*. Mem. AAPG no. 13.
Logan, B. W. and D. E. Cebulski 1970. Sedimentary environments of Shark Bay, W. Australia. In Logan *et al.*, (1970), 1–37.
Logan, B. W., P. Hoffman and C. F. Gebelein 1974a. Algal mats, cryptalgal fabrics and structures, Hamelin Pool, Western Australia. In Logan *et al.* Mem. AAPG, no. 22, 140–94.
Logan, B. W., J. F. Read, G. M. Hagan, P. Hoffman, R. G. Brown, P. J. Woods and C. D. Gebelein 1974b. *Evolution and diagenesis of Quaternary carbonate sequences, Shark Bay, W. Australia*. Mem. AAPG, no. 22.
Lohmann, K. C. and W. J. Myers 1977. Microdolomite inclusions in cloudy prismatic calcites: a proposed criterion for former high-Mg calcites. *J. Sed. Petrol.* **47**, 1078–88.
Longman, M. W. 1980. Carbonate diagenetic textures from nearsurface diagenetic environments. *Bull. AAPG* **64**, 461–87.
Longuet-Higgins, M. S. 1953. Mass transport in water waves. *Phil Trans R. Soc. Lond.* (A) **245**, 535–81.
Longuet-Higgins, M. S. 1970. Longhore currents generated by obliquely incident sea waves. *J. Geophys. Res.* **75**, 6778–801.

Longuet-Higgins, M. S. and R. W. Stewart 1964. Radiation stress in water waves; a physical discussion with applications. *Deep-sea Res.* **11**, 529–63.

Loreau, J.-P. and B. H. Purser 1973. Distribution and ultrastructure of Holocene oöids in the Persian Gulf. In *The Persian Gulf – Holocene carbonate sedimentation and diagenesis in a shallow epicontinental sea.* B. H. Purser (ed.), 279–328. Heidelberg: Springer.

Lowe, D. R. 1975. Water escape structures in coarse-grained sediments. *Sedimentology* **22**, 157–204.

Lowe, D. R. 1976. Grain flow and grain flow deposits. *J. Sed. Petrol.* **46**, 188–99.

Lowe, D. R. and R. D. Lopiccolo 1974. the characteristics and origins of dish and pillar structures. *J. Sed. Petrol.* **44**, 484–501.

Lowenstam, H. A. 1963. Biologic problems relating to the composition and diagenesis of sediments. In *The Earth sciences – problems and progress in current research,* T. W. Donnelly (ed.), 137–95. Chicago: Univ. Chicago Press.

MacIntyre, I. G. 1977. Distribution of submarine cements in a modern Caribbean fringing reef, Caleta Point, Panama, *J. Sed. Petrol.* **47**, 503–16.

MacNeil. F. S. 1954. Organic reefs and banks and associated detrital sediments. *Am. J. Sci.* **252**, 385–401.

Magara, K. 1976. Water expulsion from clastic sediments during compaction-directions and volumes. *Bull. AAPG* **60**, 543–53.

Mainguet, M. 1978. The influence of trade winds, local air-masses and topographic obstacles on the aeolian movement of sand particles and the origin and distribution of dunes and ergs in the Sahara and Australia. *Geoforum* **9**, 17–28.

Mainguet, M. and L. Canon 1976. Vents et paloevents du Sahara. Tentative d'approche paleoclimatique. *Rev. Geog. Phys. Geol. dyn.* **18**, 241–50.

Majewske, D. P. 1969. *Recognition of invertebrate fossil fragments in rocks and thin sections.* Leiden: Brill.

Maldonado, A. and D. J. Stanley 1979. Depositional processes and late Quaternary evolution of two Mediterranean submarine fans: a comparison. *Mar. Geol.* **31**, 215–50.

Malfait, B. T. and T. H. van Andel 1980. A modern oceanic hardground on the Carnegie Ridge in the eastern Equatorial Pacific. *Sedimentology* **27**, 467–96.

Mantz, P. A. 1978. Bedforms produced by fine, cohesionless, granular and flakey sediments under subcritical water flows. *Sedimentology* **25**, 83–104.

Mardia, K. V. 1972. *Statistics of directional data.* London: Academic Press.

Margolis, S. V. and D. H. Krinsley 1974. Processes of formation and environmental occurrence of microfeatures on detrital quartz grains. *Am. J. Sci.* **274**, 449–64.

Markle, R. G., G. M. Bryan and J. I. Ewing 1970. Structure of the Blake–Bahama Outer Ridge. *J. Geophys. Res.* **75**, 4539–55.

Marshall, J. D. and M. Ashton 1980. Isotopic and trace element evidence for submarine lithification of hardgrounds in the Jurassic of E. England. *Sedimentology* **27**, 271–90.

Masey, B. S. 1979. *Mechanics of fluids*, 4th edn. New York: Van Nostrand Reinhold.

Masters, C. D. 1967. Use of sedimentary structures in determination of depositional environments, Mesaverde formation, William Fork Mountains, Colorado. *Bull. AAPG* **51**, 2033–43.

Matter, W. A. and M. E. Tucker (eds) 1978. *Modern and ancient lake sediments.* Int. Ass. Sed. Spec. Pubn, no. 2.

Matthews, R. K. 1966. Genesis of Recent lime mud in British Honduras. *J. Sed. Petrol.* **36**, 428–54.

Matthews, R. K. 1974. *Dynamic stratigraphy.* Englewood Cliffs, NJ: Prentice-Hall.

Mazzullo, S. J. 1980. Calcite pseudospar replacive of marine acicular aragonite, and implications for aragonite cement diagenesis. *J. Sed. Petrol.* **50**, 409–22.

McCabe, P. J. 1977. Deep distributary channels and giant bedforms in the Upper Carboniferous of the Central Pennines, northern England. *Sedimentology* **24**, 271–90.

McCall, J. G. 1960. The flow characteristics of a cirque glacier and their effect on glacial structure and cirque formation. In *Norwegian cirque glaciers,* W. V. Lewis (ed.), 39–62. R. Geog. Soc. Res. Ser., no. 4.

McCave, I. N. 1971. Sand waves in the North Sea off the coast of Holland. *Mar. Geol.* **10**, 199–225.

McCave, I. N. 1972. Transport and escape of fine-grained sediment from shelf areas. In *Shelf sediment transport: process and pattern,* D. J. P. Swift, D. B. Doane & O. H. Pilkey (eds), 225–48. Stroudsburg, Pa: Hutchinson & Ross.

McCave, I. N. 1979. Tidal currents at the North Hinder lightship, southern North Sea: flow directions and turbulence in relation to maintenance of sand bars. *Mar. Geol.* **31**, 101–14.

McCave, I. N. and J. Jarvis 1973. Use of the Model-T Coulter Counter in size analysis. *Sedimentology* **20**, 305–16.

McCave, I. N. and S. A. Swift 1976. A physical model for the rate of deposition of fine-grained sediments in the deep sea. *Bull. Geol. Soc. Am.* **87**, 541–6.

McCave, I. N., P. F. Lonsdale, C. D. Hollister and W. D. Gardner 1981. Sediment transport over the Halton and Gardar contourite drifts. *J. Sed. Petrol.* **50**, 1049–62.

McEwen, T. J. 1978. Diffusional mass transfer processes in pitted pebble conglomerates. *Contr. Min. Petrol.* **67**, 405–15.

McIlreath, I. A. and N. P. James 1978. Carbonate slopes. In Walker (1978b), 133–44.

McKee, E. D. 1966. Structure of dunes at White Sands National Monument, New Mexico. *Sedimentology* **7**, 1–61.

McKee, E. D. (ed.) 1978. *A study of global sand seas.* USGS Prof. Pap., no. 1052.

McKee, E. D. and G. C. Tibbits 1964. Primary structures of a seif dune and associated deposits in Libya. *J. Sed. Petrol.* **34**, 5–17.

McKenzie, J. A., K. J. Hsü and J. F. Schneider 1980. Movement of subsurface waters under the sabkha, Abu Dhabi, UAE, and its relation to evaporative dolomite genesis. In Zenger *et al.* 1980, 11–30.

McKerrow, W. S., J. K. Leggett and M. H. Eales 1977. Imbricate thrust model of the Southern Uplands of Scotland. *Nature* **267**, 237–9.

McLellan, H. J. 1965. *Elements of physical oceanography.* Oxford: Pergamon.

McPherson, B. F., G. Y. Hendrix, H. Klein and H. M. Tyas 1976. *The environment of S. Florida, a summary report.* USGS Prof. Pap., no. 1011.

Meier, M. F. 1960. *Mode of flow of Saskatchewan Glacier, Alberta, Canada.* USGS Prof. Pap., no. 351.

Meissner, F. F. 1972. Cyclic sedimentation in Middle Permian strata of the Permian basin, West Texas and New Mexico. In *Cyclic sedimentation in the Permian basin.* J. C. Elam & S. Chuber (eds), 203–32. West Texas Geol. Soc., Texas.

Miall, A. D. 1973. Markov chain analysis applied to an ancient alluvial plain succession. *Sedimentology* **20**, 347–64.

Miall, A. D. 1974. Paleocurrent analysis of alluvial sediments – a discussion of directional variance and vector magnitude. *J. Sed. Petrol.* **44**, 1174–85.

Miall, A. D. 1977. A review of the braided river depositional environment. *Earth Sci. Rev.* **13**, 1–62.

Miall, A. D. (ed.) 1978. *Fluvial sedimentology*. Mem. Can. soc. Petrolm Geol., no. 5.

Middleton, G. V. 1965. Antidune cross-bedding in a large flume. *J. Sed. Petrol.* **35**, 922–7.

Middleton, G. V. 1966a. Experiments on density and turbidity currents. 1: Motion of the head. *Can. J. Earth. Sci.* **3**, 523–46.

Middleton, G. V. 1966b. Experiments on density and turbidity currents. 2: Uniform flow of density currents. *Can. J. Earth Sci.* **3**, 627–37.

Middleton, G. V. 1966c. Experiments on density and turbidity currents. 3: Deposition of sediment. *Can. J. Earth Sci.* **4**, 475–505.

Middleton, G. V. 1970. Experimental studies related to problems of flysch sedimentation. In *Flysch sedimentology in N. America*, J. Lajoie (ed.), 253–72. Geol. Assoc. Can. Spec. Pap., no. 7.

Middleton, G. V. 1976. Hydraulic interpretation of sand size distributions. *J. Geol.* **84**, 405–26.

Middleton, G. V. and M. A. Hampton 1973. Sediment gravity flows: mechanics of flow and deposition. In *Turbidites and deep water sedimentation*, 1–38, AGI-SEPM short course lecture notes.

Middleton, G. V. and J. B. Southard 1978. *Mechanics of sediment movement*. Tulsa, Okla: SEPM short course, no. 3.

Miller, M. C., I. N. McCave and P. D. Komar 1977. Threshold of sediment motion under undirectional currents. *Sedimentology* **24**, 507–28.

Miller, M. C. and P. D. Komar 1980. Oscillation sand ripples generated by laboratory apparatus. *J. Sed. Petrol.* **50**, 173–82.

Milliman, J. D. 1974. *Marine carbonates*. New York: Springer.

Mitchell, A. H. G. and H. G. Reading 1969. Continental margins, geosynclines and seafloor spreading. *J. Geol.* **77**, 629–46.

Mitchell, A. H. G. and H. G. Reading 1978. Sedimentation and tectonics. In Reading (1978), 439–76.

Monty, C. L. V. 1967. Distribution and structure of Recent stromatolitic algal mats, Eastern Andros Island, Bahamas. *Ann. Soc. Geol. Belg.* **90**, 55–100.

Moody, J. D. 1975. Distribution and geological characteristics of giant oil fields. In *Petroleum and global tectonics*, A. G. Fischer & S. Judson (eds), 307–20. Princeton, NJ: Princeton Univ. Press.

Moon, C. F. 1972. The microstructure of clay sediments. *Earth Sci. Rev.* **8**, 303–21.

Moore, G. T. 1979. Mississipi river delta – April 9, 1976 – from Landsat 2. *Bull. AAPG* **63**, 660–7.

Moore, G. T., G. W. Starke, L. C. Bonham and H. O. Woodbury 1978. Mississippi Fan, Gulf of Mexico – physiography, stratigraphy and sedimentational patterns. In *Framework, facies and oil-trapping characteristics of the upper continental margin*, G. T. Moore & J. M. Coleman (eds), 155–91. Studies in Geology, no. 7. Tulsa, Okla: AAPG.

Moore, J. C. and D. E. Karig 1976. Sedimentology, structural geology, and tectonics of the Shikiku subduction zone, southwestern Japan. *Bull. Geol Soc. Am.* **87**, 1259–68.

Morgan, J. P. (ed.) 1970. *Deltaic sedimentation modern and ancient*. SEPM Spec. Pubn, no. 15.

Morse, J. W. and R. A. Berner 1972. Dissolution kinetics of calcium carbonate in seawater. II: A kinetic origin for the lysocline. *Am. J. Sci.* 272, 840–51.

Moss, A. J., P. H. Walker and J. Hutka 1973. Fragmentation of granitic quartz in water. *Sedimentology* **20**, 489–512.

Mountjoy, E. W., H. E. Cook and L. C. Pray 1972. Allochtonous carbonate debris flows – worldwide indicators of reef complexes, banks or shelf margins. *Proc. 24th Int. Geol Cong.* **6**, 172–89.

Mowbray, T. de 1980. *Sedimentary processes of recent intertidal channels*. Unpubl. PhD thesis, Univ. Leeds.

Muir, M., D. Lock and C. Von der Borch 1980. The Coorong model for penecontemporaneous dolomite formation in the middle Proterozoic McArthur Group, Northern Territory, Australia. In Zenger *et al.* (eds), 1980, 51–67.

Muller, J. and F. Fabricius 1974. Magnesian-calcite nodules in the Ionian deep sea: an actualistic model for the formation of some nodular limestones. In Hsü & Jenkyns (1974), 235–48.

Mullins, H. T. and A. C. Neumann 1979. Deep carbonate bank margin structure and sedimentation in the northern Bahamas. In *Geology of continental slopes*, L. Doyle & D. H. Pilkey (eds), 165–92. SEPM Spec. Pubn, no. 27.

Mullins, H. T., A. C. Neumann, R. J. Wilber and M. R. Boardman 1980. Nodular carbonate sediment on Bahamian slopes: possible precursors to nodular limestones. *J. Sed. Petrol.* **50**, 117–31.

Munk, W. H. 1950a. On the wind-driven ocean circulation. *J. Meteorol.* **7**, 79–93.

Munk, W. H. 1950b. Origin and generation of waves. *Proc. 1st Conf. Coast Engng*, 1–4. Berkeley, Ca: Council on Wave Research.

Mutti, E. 1977. Distinctive thin-bedded turbidite facies and related depositional environments in the Eocene Hecko Group (south central Pyrenees, Spain). *Sedimentology* **24**, 107–32.

Nagtegaal, P. J. C. 1978. Sandstone-framework instability as a function of burial diagenesis. *J. Geol Soc. Lond.* **135**, 101–5.

Nami, M. 1976. An exhumed Jurassic meander belt from Yorkshire. *Geol Mag.* **113**, 47–52.

Nami, M. and M. R. Leeder 1978. Changing channel morphology and magnitude in the Scalby Formation (M. Jurassic) of Yorkshire, England. In Miall (1978), 431–40.

Nanson, G. C. 1980. Point bar and floodplain formation of the meandering Beatton River, northeastern British Columbia, Canada. *Sedimentology* **27**, 3–29.

Needham, R. S. 1978. Giant-scale hydroplastic deformation structures formed by the loading of basalt on to water-saturated sand, Middle Proterozoic, northern Territory, Australia. *Sedimentology* **25**, 285–96.

Neev, D. 1978. Messinian and Holocene gypsum deposits of relatively deep water. *Abs. 10th Int. Cong. Sed., Jerusalem* **2**, 459.

Neev, D. and K. O. Emery 1967. The Dead Sea: depositional processes and environments of evaporites. *Israel Geol Surv. Bull.* **41**, 1–147.

Neugebauer, J. 1974. Some aspects of cementation in chalk. In Hsü & Jenkyns (1974), 149–76.

Neumann, A. C. and L. S. Land 1975. Lime mud deposition and calcareous algae in the Bight of Abaco, Bahamas: a budget. *J. Sed. Petrol.* **45**, 763–86.

Neumann, A. C., J. W. Kofoed and G. H. Keller 1977. Lithoherms in the Straits of Florida. *Geology* **5**, 4–10.

Neumann, G. and W. J. Pierson 1966. *Principles of physical oceanography*. Englewood Cliffs, NJ: Prentice-Hall.

Nilsen, T. H., R. G. Walker and W. R. Normark 1980. Modern and ancient submarine fans: discussion and replies. *Bull. AAPG* **64**, 1094–1113.

Nio, S.-O. 1976. Marine transgressions as a factor in the formation of sand wave complexes. *Geol. Mijn.* **55**, 18–40.

Nisbet, E. G. and I. Price 1974. Siliceous turbidites: bedded cherts as redeposited ocean ridge-derived sediments. In Jenkyns & Hsü (1974), 351–66.

Normark, W. R. 1970. Growth patterns of deep-sea fans. *Bull. AAPG* **54**, 2170–95.

Normark, W. R. and D. J. W. Piper 1972. Sediments and growth pattern of Navy deep-sea fan, San Clemente Basin, California Borderland. *J. Geol.* **80**, 198–223.

Oertel, G. E. and C. D. Curtis 1972. Clay-Ironstone concretion preserving fabrics due to progressive compaction. *Bull. Geol Soc. Am.* **83**, 2597–606.

Oertel, G. F. 1979. Barrier island development during the Holocene recession, SE United States. In Leatherman (1971), 273–90.

Offen, G. R. and S. J. Kline 1975. A proposed model of the bursting process in turbulent boundary layers. *J. Fluid Mech.* **70**, 209–28.

Oldershaw, A. E. and T. P. Scoffin 1967. The source of ferroan and non-ferroan calcite cements in the Halkin and Wenlock Limestones. *Geol J.* **5**, 309–20.

Ollier, C. D. 1978. Silcrete and weathering. In *Silcrete in Australia*, T. Langford-Smith (ed.), 13–8. Univ. of New England.

Olphen, H. Van 1963. *An introduction to clay colloid chemistry*. New York: Wiley.

Oomkens, E. 1974. Lithofacies relations in the Late Quaternary Niger delta complex. *Sedimentology* **21**, 195–222.

Otvos, E. G. 1979. Barrier island evolution and history of migration, North Central Gulf Coast. In Leatherman (1979), 291–319. New York: Academic Press.

Packham, G. H. and K. A. W. Crook 1960. The principle of diagenetic facies and some of its implications. *J. Geol.* **68**, 392–407.

Pantin, H. M. 1979. Interaction between velocity and effective density in turbidity flow: phase plane analysis, with criteria for autosuspension. *Mar. Geol.* **31**, 59–99.

Park, R. K. 1976. A note on the significance of lamination in stromatolites. *Sedimentology* **23**, 379–93.

Park, R. K. 1977. The preservation potential of some recent stromatolites. *Sedimentology* **24**, 485–506.

Passega, R. 1964. Grain size representation by C.M. patterns as a geological tool. *J. Sed. Petrol.* **34**, 830–47.

Paterson, W. S. B. 1969. *The physics of glaciers*. Oxford: Pergamon.

Patterson, R. J. and D. J. J. Kinsman 1977. Marine and continental groundwater sources in a Persian Gulf coastal sabkha. 381–97. *Studies in Geology* 4, Tulsa, Okla: AAPG.

Perrier, R. and J. Quiblier 1974. Thickness changes in sedimentary layers during compaction history; methods for quantitive evaluation. *Bull. AAPG* **58**, 507–20.

Perry, E. and J. Hower 1970. Burial diagenesis in Gulf Coast pelitic sediments. *Clays and Clay Mineral.* **18**, 165–77.

Perry, E. A. and J. Hower 1972. Late-stage dehydration in deeply buried pelitic sediments. *Bull. AAPG* **56**, 2013–21.

Pettijohn, F. J. 1975. *Sedimentary rocks*, 3rd edn. New York: Harper & Row.

Pettijohn, F. J. and P. E. Potter 1964. *Atlas and glossary of primary sedimentary structures*. Berlin: Springer.

Pettijohn, F. J., P. E. Potter and R. Siever 1972. *Sand and sandstone*. New York: Springer.

Pierce, J. W. 1976. Suspended sediment transport at the shelf break and over the outer margin. In Stanley & Swift (1976), 437–58.

Pilkey, O. H. and D. Noble 1967. Carbonate and clay mineralogy of the Persian Gulf. *Deep-sea Res.* **13**, 1–16.

Pingitore, N. E. 1976. Vadose and phreatic diagenesis: processes, products and their recognition in corals. *J. Sed. Petrol.* **46**, 985–1006.

Piper, D. J. W. 1978. Turbidite muds and silts on deep sea fans and abyssal plains. In Stanley & Kelling (1978), 163–75.

Piper, D. J. W., R. Von Heune and J. R. Duncan 1973. Late Quaternary sedimentation in the active eastern Aleutian Trench. *Geology* **1**, 19–22.

Pittman, E. D. 1969. Destruction of plagioclase twins by stream transport. *J. Sed. Petrol.* **39**, 1432–7.

Plummer, L. N. and F. T. Mackenzie 1974. Predicting mineral solubility from rate data: application to the dissolution of magnesian calcites. *Am. J. Sci.* **274**, 61–83.

Pond, S. and G. L. Pickard 1978. *Introductory dynamic oceanography*. London: Pergamon.

Potter, P. E. and F. J. Pettijohn 1978. *Paleocurrents and basin analysis*. New York: Academic Press.

Powers, M. C. 1953. A new roundness scale for sedimentary particles. *J. Sed. Petrol.* **23**, 117–9.

Powers, M. C. 1967. Fluid-release mechanisms in compacting marine mudrocks and their importance in oil exploration. *Bull. AAPG* **51**, 1240–54.

Pratt, C. J. 1973. Bagnold approach and bed-form development. *J. Hyd. Div. A.S.C.E.* **99**, 121–37.

Prentice, J. E., I. R. Beg, C. Colleypriest, R. Kirby, P. J. C. Sutcliffe, M. R. Dobson, B. d'Olier, M. F. Elvines, T. I. Kilenyi, R. J. Maddrell and T. R. Phinn 1968. Sediment transport in estuarine areas. *Nature* **218**, 1207–10.

Pritchard, D. W. 1955. Estuarine circulation patterns. *Proc. Am. Soc. Civ. Engrs* **81**, 1–11.

Pritchard, D. W. 1967. What is an estuary: physical viewpoint. In *Estuaries*, G. H. Lauff (ed.) Am. Assoc. Adv. Sci.

Pritchard, D. W. and H. H. Carter 1971. Estuarine circulation patterns. In Schubel (1971), 1–17.

Puigdefabrigas, C. and A. Van Vleit 1978. Meandering stream deposits from the Tertiary of the Southern Pyrenees. In Miall (1978), 469–86.

Purdy, E. G. 1963. Recent calcium carbonate facies of the Great Bahama Bank. 2: sedimentary facies. *J. Geol.* **71**, 472–97.

Purdy, E. G. 1974. Reef configurations: cause and effect. In *Reefs in time and space*, L. F. Laporte (ed.), 9–76. SEPM Spec. Pubn, no. 18.

Purser, B. H. 1978. Early diagenesis and the preservation of porosity in Jurassic limestones. *J. Petrol. Geol.* **1**, 83–94.

Purser, B. H. 1979. Middle Jurassic sedimentation on the Burgundy Platform. *Symp. Sed. Jurass. W. Europe, A.S.F. Pubn. 1*, 75–84.

Raaf, J. F. M. de, J. R. Boersma and A. Van Gelder 1977. Wave-generated structures and sequences from a shallow marine succession. Lower Carboniferous, County Cork, Ireland. *Sedimentology* **24**, 451–83.

Rampino, M. R. and J. E. Sanders 1981. Evolution of the barrier islands of Southern Long Island, New York. *Sedimentology* **28**, 37–48.

Randazzo, A. F. and E. W. Hickey 1978. Dolomitisation in the Floridan aquifer. *Am. J. Sci.* **278**, 1177–84.

Roa, D. B. and T. S. Murty 1970. Calculation of the steady-state wind-driven circulation in Lake Ontario. *Arch. Meteor. Geophys. Bioklim.* **A19**, 195–210.

Raudkivi, A. J. 1976. *Loose boundary hydraulics*. Oxford: Pergamon.

Raudkivi, A. J. and D. L. Hutchinson 1974. Erosion of kaolinite clay by flowing water. *Proc. R. Soc. London.* (A) **337**, 537–54.

Raymond, C. F. 1971. Flow in a transverse section of Athabaska glacier, Alberta, Canada. *J. Glaciol.* **10**, 55–84.

Rayner, D. H. 1963. The Achanarras Limestone of the middle Old Red Sandstone, Caithness, Scotland. *Proc. Yorks. Geol Soc.* **34**, 117–38.

Reading, H. G. 1964. A review of the factors affecting the sedimentation of the Millstone Grit (Namurian) in the Central Pennines. In *Deltaic and shallow marine deposits*, L. M. J. U. Straaten (ed.), 26–34. Amsterdam: Elsevier.

Reading, H. G. 1978 (ed.). *Sedimentary environments and facies*. Oxford: Blackwell.

Reading, H. G. 1978. Facies. In Reading (1978), 4–14.

Reineck, H. E. 1958. Longitudinale schragschit im Watt. *Geol. Rdsch.* **47**, 73–82.

Reineck, H. E. 1963. Sedimentgefuge in Bereich der Sudlichen Nordsee. *Abh. Sebck. Naturforsch. Ges.* **505**, 138 pp.

Reineck, H. E. Layered sediments of tidal flats, beaches and shelf bottoms of the North Sea. In *Estuaries*, G. D. Louff (ed.), 191–206. Washington DC: Am. Ass. Adv. Sci.

Reineck, H. E. 1972. Tidal flats. In *Recognition of ancient sedimentary environments*, J. K. Rigby & W. K. Hamblin (eds), 146–59. SEPM Spec. Pubn, no. 16.

Reineck, H. E. and I. B. Singh 1973. Genesis of laminated sand and graded rhythmites in storm-sand layers of shelf mud. *Sedimentology* **18**, 123–8.

Reineck, H. E. and I. B. Singh 1980. *Depositional sedimentary environments*. 2nd edn. Berlin: Springer.

Reineck, H. E. and F. Wunderlich 1968a. Zur unter scheidung von asymmetnschen oszillationrippeln und Stromungsrippeln. *Senck. Leth.* **49**, 321–45.

Reineck, H. E., W. F. Gutman and G. Hertweck 1967. Das schlickgebiet südlich Helgoland als Beispiel rezenter Schelfablagerungen. *Senck. Leth.* **48**, 219–75.

Reineck, H. E., J. Dorjes, S. Gadow and G. Herweck 1968. Sedimentologie, Founenzonierung und Faziesabfolge vor der Ostkuste der inneren Deutschen Bucht. *Senck. Leth.* **49**, 261–309.

Reinson, G. E. 1978. Barrier island systems. In Walker (1978), 57–74.

Ricci-Lucci, F. and E. Valmori 1980. Basin-wide turbidites in a Miocene, oversupplied deep-sea plain: a geometrical analysis. *Sedimentology* **27**, 241–70.

Richardson, J. F. and W. N. Zaki 1958. Sedimentation and fluidisation. *Trans Inst. Chem. Engrs* **32**, 35–53.

Richter, D. K. and H. Fuchtbauer 1978. Ferroan calcite replacement indicates former magnesian calcite skeletons. *Sedimentology* **25**, 843–60.

Richter-Bernberg, G. 1955. Uber salinaire sedimentation 2. *Dtsch. Geol. Ges.* **105**, 593–6.

Rider, M. H. 1978. Growth faults in Carboniferous of Western Ireland. *Bull. AAPG* **62**, 2191–213.

Riech, V. and U. Von Rad 1979. Silica diagenesis in the Atlantic Ocean: diagenetic potential and transformations. In M. Ewing Series, 3, *Am. Geophys. Union* 315–40.

Riggs, S. R. 1979. Petrology of the Tertiary phosphorite system of Florida. *Econ. Geol.* **74**, 195–220.

Roberts, D. G. and R. B. Kidd 1979. Abyssal sediment wave fields on Feni Ridge Rockall Trough: long-range sonar studies. *Mar. Geol.* **33**, 175–91.

Robertson, A. H. F. 1975. Cyprus umbers: basalt–sediment relationships on a Mesozoic ocean ridge. *J. Geol Soc. Lond.* **131**, 511–31.

Robertson, A. H. F. 1977. The origin and diagenesis of cherts from Cyprus. *Sedimentology* **24**, 11–30.

Robertson, A. H. F. and J. D. Hudson 1974. Pelagic sediments in the Cretaceous and Tertiary history of the Troodos Massif, Cyprus. In Hsü & Jenkyns (1974), 403–36.

Robinson, A. H. W. 1966. Residual currents in relation to sandy shoreline evolution of the East Anglian coast. *Mar. Geol.* **4**, 57–84.

Rodine, J. D. and A. M. Johnson 1976. The ability of debris, heavily freighted with coarse clastic materials, to flow on gentle slopes. *Sedimentology* **23**, 213–34.

Rouse, L. J., H. H. Roberts and R. H. W. Cunningham 1978. Satellite observation of the subaerial growth of the Atchafalaya Delta, Louisiana. *Geology* **6**, 405–8.

Rowe, P. W. 1962. The stress-dilitancy relation for static equilibrium of an assembly of particles in contact. *Proc. R. Soc. Lond.*, **A269**, 500–27.

Rozovskii, I. L. 1963. *Flow in bends of open channels*. Jerusalem: Israel Programme for Scientific Translations.

Runnells, D. D. 1969. Diagenesis, chemical sediments and the mixing of natural waters. *J. Sed. Petrol.* **39**, 1188–201.

Rupke, N. A. 1975. Deposition of fine-grained sediments in the abyssal environment of the Algero-Balearic Basin, W. Mediterranean Sea. *Sedimentology* **22**, 95–109.

Rupke, N. A. 1977. Growth of an ancient deep-sea fan. *J. Geol.* **85**, 725–44.

Rupke, N. A. 1978. Deep clastic seas. In Reading (1978), 372–411.

Rusnak, G. A. 1960. Sediments of Laguna Madra, Texas. In *Recent sediments, NW Gulf of Mexico*, F. P. Shepard, F. B. Phleger and T. H. van Andel (eds), 153–96. Tulsa, Okla: AAPG.

Rust, B. R. 1975. Fabric and structure in glaciofluvial gravels. In Jopling & McDonald (1975), 238–48.

Sandberg, P. A. 1975. New interpretations of Great Salt Lake oöids and of ancient nonskeletal carbonate mineralogy. *Sedimentology* **22**, 497–537.

Sanderson, I. D. 1974. Sedimentary structures and their environmental significance in the Navajo Sandstone, San Rafael Swell, Utah. *Brigham Young Univ. Geol. Studies* **21**, 215–46.

Schäfer, W. 1972. *Ecology and palaeoecology of marine environments*. Edinburgh: Oliver & Boyd.

Schlager, W. and A. Chermak 1979. Sediment facies of platform-basin transition, Tongue of the Ocean, Bahamas. In *Geology of continental slopes*, L. Doyle & O. H. Pilkey (eds), 193–208. SEPM Spec. Pubn, no. 27.

Schlanger, S. O. and H. C. Jenkyns 1976. Cretaceous oceanic anoxic events: causes and consequences. *Geol. Mijn.* **55**, 179–84.

Schmidt, G. W. 1973. Interstitial water composition and geochemistry of deep Gulf Coast shales and sandstones. *Bull. AAPG* **57**, 321–31.

Schmidt, V. and D. A. MacDonald 1979. Texture and recognition of secondary porosity in sandstones. In Scholle & Schluger (1979), 209–25.

Schminke, H. V., R. V. Fisher and A. C. Waters 1975. Antidune and chute-and-pool structures in the base surge deposits of the Laacher See area, Germany. *Sedimentology* **20**, 553–74.

Scholle, P. A. 1974. Diagenesis of Upper Cretaceous chalks from England, N. Ireland and the North Sea. In Hsü & Jenkyns (1974), 177–210.

Scholle, P. A. 1978. *A color illustrated guide to carbonate rock constituents, textures, cements and porosites.* Mem. 27, Tulsa, Okla: AAPG.

Scholle, P. A. and P. R. Schulger (eds) 1979. *Aspects of diagenesis.* SEPM Spec. Pubn, no. 26.

Schopf, T. J. M. 1980. *Palaeoceanography.* Cambridge, Mass: Harvard Univ. Press.

Schreiber, B. C., G. M. Friedman, A. Decima and E. Schreiber 1976. Depositional environments of Upper Miocene (Messianian) evaporite deposits of the Sicilian basin. *Sedimentology* **23**, 729–60.

Schroeder, J. H. 1972. Fabrics and sequences of submarine carbonate cements in Holocene Bermuda Cup reef. *Geol. Rundsch.* **61**, 708–30.

Schubel, J. R. (ed.) 1971a. *The estuarine environment.* Washington DC: American Geol. Inst.

Schubel, J. R. 1971b. A few notes on the agglomeration of suspended sediment in estuaries. In Schubel (1971a), X.1–X.29.

Schubel, J. R. 1971c. Estuarine circulation and sedimentation. In Schubel (1971a) VI.1–17.

Schubel, J. R. and A. Okabo 1972. Comments on the dispersal of suspended sediment across the continental shelves. In *Shelf sediment transport: process and pattern,* D. J. P. Swift, D. B. Duane and O. H. Pilkey (eds), 333–46. Stroudsburg, Pa: Dowden, Hutchinson & Ross.

Schumm, S. A. 1960. The effect of sediment type on the shape and stratification of some modern river deposits. *Am. J. Sci.* **258**, 177–84.

Schumm, S. A. 1963a. Sinuousity of alluvial channels on the Great Plains. *Bull. Geol Soc. Am.* 74, 1089–100.

Schumm, S. A. 1963b. *Disparity between present rates of denudation and orogeny.* USGS Prof. Pap., no. 454.

Schumm, S. A. 1968a. Speculations concerning paleohydrologic controls of terrestrial sedimentation. *Bull. Geol Soc. Am.* 79, 1573–88.

Schumm, S. A. 1968b. *River adjustment to altered hydrologic regimen – Murrumbidgee River and paleochannels, Australia.* USGS Prof. Pap., no. 598.

Schumm, S. A. 1972. Fluvial paleochannels: In *Recognition of ancient sedimentary environments,* J. K. Rigby & W. K. Hamblin (eds), 98–107. SEPM Spec. Pubn, no. 16.

Schumm, S. A. 1973. Geomorphic thresholds and complex response of drainage systems. In *Fluvial geomorphology,* M. Morisawa (ed.), 299–310. London: George Allen & Unwin.

Schumm, S. A. 1977. *The fluvial system.* New York: Wiley.

Schumm, S. A. and H. R. Khan 1971. Experimental study of channel patterns. *Nature Phys. Sci.* **233**, 407–9.

Schumm, S. A. and M. A. Stevens 1973. Abrasion in place: a mechanism for rounding and size reduction of coarse sediments in rivers. *Geology* **1**, 37–40.

Schumm, S. A., H. R. Khan, B. R. Winkley and L. G. Robbins 1972. Variability of river patterns. *Nature Phys. Sci.* **237**, 75–6.

Schwarz, H-U, G. Einsele and D. Herm 1975. Quartz-sandy, grazing contoured stromatolites from coastal embayments of Mauritania, W. Africa. *Sedimentology* **22**, 534–61.

Schwartz, R. K. 1975. *Nature and genesis of some washover deposits.* Tech. Mem. U.S. Army Corps. Engrs Coastal Engng Res. Centre 61, 98 pp.

Schweller, W. J. and L. D. Kulm 1978. Depositional patterns and channelised sedimentation in active E. Pacific trenches. In Stanley & Kelling (1978), 311–24.

Sclater, J. G., R. N. Anderson and M. L. Bell 1971. Elevation of ridges and evolution of the central eastern Pacific. *J. Geophys. Res.* **76**, 7888–915.

Sclater, J. G., S. Hellinger and C. Tapscott 1977. the paleobathymetry of the Atlantic Ocean from the Jurassic to the present. *J. Geol.* **85**, 509–52.

Scoffin, T. P. 1970. The trapping and binding of subtidal carbonate sediments by marine vegetation in Bimini Lagoon, Bahamas. *J. Sed. Petrol.* **40**, 249–73.

Scott, A. C. 1979. The ecology of Coal Measures floras from N. Britain. *Proc. Geol. Ass.* **90**, 97–116.

Sedimentation Seminar 1981. Comparison of methods of size analysis for sands of the Amazon–Solimes rivers, Brazil and Peru. *Sedimentology* **28**, 123–8.

Seed, H. B. and K. L. Lee 1966. Liquefaction of saturated sands during cyclic loading. *J. Soil Mech. Found. Div., A.S.C.E.,* 92, 105–34.

Segonzac, Dunoyer de 1970. The transformation of clay minerals during diagenesis and low-grade metamorphism: a review. *Sedimentology* **15**, 281–346.

Seilacher, A. 1967. Bathymetry of trace fossils. *Mar. Geol.* **5**, 413–28.

Selley, R. C. 1976. *Introduction to sedimentology.* London: Academic Press.

Sellwood, B. W. 1968. The genesis of some sideritic beds in the Yorkshire Lias (England) *J. Sed. Petrol.* **38**, 854–8.

Sellwood, B. W. 1972. Tidal flat sedimentation in the Lower Jurassic of Bornholm, Denmark. *Palaeogeogr. Palaeoclimat. Palaeoecol.* **11**, 93–106.

Sellwood, B. W. and W. S. McKerrow 1973. Depositional environments in the lower part of the Great Oölite Group of Oxfordshire and North Gloucestershire. *Proc. Geol. Ass.* **85**, 189–210.

Shapiro, A. H. 1961. *Shape and flow: the fluid dynamics of drag.* New York: Doubleday (London: Heinemann).

Sharma, G. D., A. S. Naidu and D. W. Hood 1972. Bristol Bay: a model contemporary graded shelf. *Bull. AAPG* **56**, 2000–12.

Sharp, R. P. 1963. Wind ripples. *J. Geol.* **71**, 617–36.

Shaw, A. B. 1964. *Time in stratigraphy.* New York: McGraw-Hill.

Shaw, J. and J. Archer 1979. Deglaciation and glaciolacustrine sedimentation conditions. Okanagan Valley, British Columbia, Canada. In *Moraines and varves,* C. Schluchter (ed.), 347–56. Rotterdam: Balkema.

Shearman, D. J. 1966. Origin of marine evaporites by diagenesis *Trans Inst. Min. Metall.* **75B**, 208–15.

Shearman, D. J. 1970. Recent halite rock, Baja California, Mexico. *Trans Inst. Min. Metall.* **79B**, 155–62.

Shearman, D. J. and J. G. C. M. Fuller 1969. Anhydrite diagenesis, calcitisation and organic laminites, Winnipegosis Formation, M Devonian, Saskatchewan. *Bull. Can. Petrolm Geol.* **17**, 496–525.

Shearman, D. J., J. Twyman and M. Z. Karimi 1970. The genesis and diagenesis of oölites. *Proc. Geol. Assoc.* **81**, 561–75.

Sheen, S. J. 1964. *Turbulence over a sand ripple.* M. Engng thesis, Univ. Auckland.

Shepard, F. P. 1979. Currents in submarine canyons and other types of sea valleys. *SEPM Spec. Publ.,* no. 27, 85–94.

Shepard, F. P. and D. L. Inman 1950. Nearshore circulation. *Proc. 1st Conf. Coast. Engng,* 50–9. Berkeley, Ca: Council on Wave Research.

Shideler, G. L. 1978. A sediment-dispersal model for the South Texas continental shelf, NW Gulf of Mexico. *Mar. Geol.* **26**, 284–313.

Shinn, E. A. 1969. Submarine lithification of Holocene carbonate sediments in the Persian Gulf. *Sedimentology* **12**, 109–44.

Shinn, E. A., R. M. Lloyd and R. N. Ginsburg 1969. Anatomy of a modern carbonate tidal flat, Andros Island, Bahamas. *J. Sed. Petrol.* **39**, 1202–28.

Simons, D. B., E. V. Richardson and C. F. Nordin 1965. Sedimentary structures generated by flow in alluvial channels. In *Primary sedimentary structures and their hydrodynamic interpretation*, G. V. Middleton (ed.), 34–52. SEPM Spec. Pubn, no. 12.

Simpson, J. E. 1972. Effects of the lower boundary on the head of a gravity current. *J. Fluid Mech.* **53**, 759–68.

Sly, P. G. 1978. Sedimentary processes in lakes. In Lerman (1978), 166–200.

Smale, D. 1978. Silicretes and associated silica diagenesis in southern Africa and Australia. In Langford-Smith (1978), 261–80.

Smalley, I. J. 1966. Formation of quartz sand. *Nature* **211**, 476–9.

Smalley, I. J. 1971. Nature of quickclays. *Nature* **231**, 310.

Smalley, I. J. and C. F. Moon 1973. High voltage electron microscopy of fine quartz particles. *Sedimentology* **20**, 317–22.

Smith, A. G. and J. C. Briden 1977. *Mesozoic and Cenzoic palaeocontinental maps*. Cambridge: Cambridge Univ. Press.

Smith, D. B. 1973. The origin of the Permian Middle and Upper Potash deposits of Yorkshire, England: an alternative hypothesis. *Proc. Yorks. Geol. Soc.* **39**, 327–46.

Smith, D. B. 1974. Sedimentation of Upper Artesia (Guadalupian) cyclic shelf deposits of northern Guadalope Mountains, New Mexico. *Bull. AAPG* **58**, 1699–730.

Smith, D. B. and A. Crosby 1979. The regional and stratigraphical context of Zechstein 3 and 4 potash deposits in the British sector of the southern North Sea and the adjoining land areas. *Econ. Geol.* **74**, 397–408.

Smith, N. D. 1971. Transverse bars and braiding in the Lower Platte River, Nebraska. *Bull. Geol Soc. Am.* **82**, 3407–20.

Smith, N. D. 1974. Sedimentology and bar formation in the Upper Kicking Horse River, a braided outwash stream. *J. Geol.* **82**, 205–24.

Sneed, E. D. and R. L. Folk 1958. Pebbles in the lower Colorado River, Texas: a study in particle morphogenesis. *J. Geol.* **66**, 114–50.

Sommer, F. 1978. Diagenesis of Jurassic sandstones in the Viking graben. *J. Geol Soc. Lond.* **135**, 63–8.

Southard, J. B. 1971. Representation of bed configurations in depth-velocity-size diagrams. *J. Sed. Petrol.* **41**, 903–15.

Southard, J. B., L. A. Boguchwal and R. D. Romea 1980. Test of scale modelling of sediment transport in steady unidirectional flow. *Earth Surf. Proc.* **5**, 17–24.

Sorby, H. C. 1856. On the origin of the Cleveland Hill ironstone. *Proc. Yorks. Geol Soc.* **3**, 457–61.

Sorby, H. C. 1879. The structure and origin of limestones. *Proc. Geol Soc. Lond.* **35**, 56–94.

Spearing, D. R. 1976. Upper Cretaceous Shannon Sandstones: an offshore, shallow-marine sand body. *Wyoming Geol. Ass. Guidebook 28th Field Conf.*, 65–72.

Spears, D. A. 1976. The fissility of some Carboniferous shales. *Sedimentology* **23**, 721–6.

Spears, D. A. and R. Kanaris-Sotirious 1979. A geochemical and mineralogical investigation of some British and other European tonsteins. *Sedimentology* **26**, 407–25.

Spencer, A. M. 1971. *Late Precambrian glaciation in Scotland*. Mem. Geol Soc. Lond., no. 6.

Stanley, D. J. and J. J. P. Swift (eds) 1976. *Marine sediment transport and environmental management*. New York: Wiley.

Stanley, D. J. and G. Kelling (eds) 1978. *Sedimentation in submarine canyons, fans and trenches*. Stroudsburg, Pa: Dowden, Hutchinson & Ross.

Staub, J. R. and A. D. Cohen 1979. The Snuggedy Swamp of S. Carolina: a back-barrier estuarine coal-forming environment. *J. Sed. Petrol.* **49**, 133–44.

Steel, R. J. 1974. New Red Sandstone floodplain and piedmont sedimentation in the Hebridean province, Scotland. *J. Sed. Petrol.* **44**, 336–57.

Steel, R. J. 1976. Devonian basins of western Norway – sedimentary response to tectonism and varying tectonic contrast. *Tectonophysics* **36**, 207–24.

Steel, R. J. and A. C. Wilson 1975. Sedimentation and tectonism (?Permo-Triassic) on the margin of the North Minch Basin, Lewis. *J. Geol Soc. Lond.* **131**, 183–202.

Steel R. J. and S. M. Aasheim 1978. Alluvial sand deposition in a rapidly subsiding basin (Devonian, Norway). In Miall (1978), 385–412.

Steinen, R. P. 1978. On the diagenesis of lime mud: scanning electron microscopic observations of subsurface material from Barbados, WI. *J. Sed. Petrol.* **48**, 1139–48.

Stewart, F. H. 1949. The petrology of the evaporites of the Eskdale No. 2 boring, East Yorkshire. Part 1: The lower evaporite bed. *Min. Mag.* **28**, 621–5.

Stockman, K. W., R. N. Ginsburg and E. A. Shinn 1967. The production of lime mud by algae in South Florida. *J. Sed. Petrol.* **37**, 633–48.

Stoddart, D. R. 1971. World erosion and sedimentation. In *Introduction to fluvial processes*, R. J. Chorley (ed.), 8–29. London: Methuen.

Stokes, W. L. 1968. Multiple parallel-truncation bedding planes – a feature of wind deposited sandstone formations. *J. Sed. Petrol.* **38**, 510–15.

Stommel, H. 1948. The westward intensification of wind-driven ocean currents. *Trans Am. Geophys. Union* **29**, 202–6.

Stommel, H. 1957. The abyssal circulation. *Deep-sea Res.* **4**, 149–84.

Stow, D. A. V. and J. P. B. Lovell 1979. Contourites: their recognition in modern and ancient sediments. *Earth Sci. Rev.* **14**, 251–91.

Straaten, L. M. J. U. Van and P. H. Keunen 1957. Accumulation of fine grained sediments in the Dutch Wadden Sea. *Geol. Mijn.* **19**, 320–54.

Strakhov, N. M. 1967, *Principles of lithogenesis, 1*. Edinburgh: Oliver and Boyd.

Stride, A. H. 1963. Current swept floors near the southern half of Great Britain. *Q. J. Geol Soc. Lond.* **119**, 175–99.

Sturm, M. and A. Matter 1978. Turbidites and varves in Lake Brienz (Switzerland): deposition of clastic detritus by density currents. In Matter & Tucker (1978), 145–66.

Suess, E. and D. Futterer 1972. Aragonitic oöids: experimental precipitation from seawater in the presence of humic acid. *Sedimentology* **19**, 129–39.

Surdam, R. C. and J. R. Boles 1979. Diagenesis of volcanic sandstones. In Scholle & Schluger (1979), 227–242.

Surdam, R. C. and K. O. Stanley 1980. Effects of changes in drainage-basin boundaries on sedimentation in Eocene Lakes Gosiute and Uinta of Wyoming, Utah and Colorado. *Geology* **8**, 135–9.

Sverdrup, H. U., M. W. Johnson and R. H. Fleming 1942. *The oceans: their physics, chemistry and general biology*. New York: Prentice-Hall.

Swift, D. J. P. 1972. Implications of sediment dispersal from bottom current measurements; some specific problems in understanding bottom sediment distribution and dispersal on the continental shelf: a discussion of two papers. In *Shelf sediment transport: process and pattern*, D. J. P. Swift, D. B. Duane & O. H. Pilkey (eds.), 363–71. Stroudsburg, Pa: Dowden, Hutchinson & Ross.

Swift, D. J. P. 1974. Continental shelf sedimentation. In *The geology of continental margins*, C. A. Burk and C. L. Drake (eds), 117–35. Berlin: Springer.

Swift, D. J. P. 1976. Continental shelf sedimentation. In Stanley & Swift (1976), 311–50.

Swift, D. J. P., D. J. Stanley and J. R. Curray 1971. Relict sediment on continental shelves: a reconsideration *J. Geol.* **79**, 327–46.

Swift, D. J. P., D. B. Duane and O. H. Pilkey (eds), 1972. *Shelf sediment transport: process and pattern.* Stroudsburg, Pa: Dowden, Hutchinson & Ross.

Swift, D. J. P., D. B. Duane and T. F. McKinney 1973. Ridge and swale topography of the Middle Atlantic Bight, North America: secular response to the Holocene hydraulic regime. *Mar. Geol.* **15**, 227–47.

Takahishi, T. 1975. Carbonate chemistry of seawater and the calcite compensation depth in the oceans. *Cushman Found. Foram. Res. Spec. Pubn*, no. 13, 11–26.

Talbot, M. R. 1973. Major sedimentary cycles in the Corallian Beds. *Palaeogeog., Palaeoclim., Palaeoecol.* **14**, 293–317.

Talbot, M. R. 1980. Environmental responses to climatic change in the West African Sahel over the past 20 000 years. In *The Sahara and the Nile*, M. A. J. Williams & H. Faure (eds), 37–62. Rotterdam: Balkema.

Talbot, M. R. and M. A. J. Williams 1978. Erosion of fixed dunes in the Sahel, central Niger. *Earth Surf. Proc.* **3**, 107–13.

Talbot, M. R. and M. A. S. Williams 1979. Cyclic alluvial fan sedimentation on the flanks of fixed dunes, Janjari, Central Niger. *Catena* **6**, 43–62.

Taylor, G. and I. E. Smith 1975. The genesis of sub-basaltic silcretes from the Monaro, New South Wales. *J. Geol Soc. Austral.* **22**, 377–85.

Teichmuller, M. and R. Teichmuller 1968. Canozoic and Mesozoic coal deposits of Germany. In *Coal and coal bearing strata*, D. Murchison & T. S. Westoll (eds), 347–77. Edinburgh: Oliver & Boyd.

Thesiger, W. 1964. *Arabian sands*. London: Longman (Penguin edn, 1974).

Thiede, J. and T. H. van Andel 1977. The paleoenvironment of anaerobic sediments in the late Mesozoic South Atlantic Ocean. *Earth Plan. Sci. Lett.* **33**, 301–9.

Thierstein, H. R. and W. H. Berger 1978. Injection events in ocean history. *Nature* **276**, 461–6.

Thomas, B. 1938. *Arabia Felix: across the empty quarter of Arabia*. London: Readers Union.

Thorstenson, D. C., F. T. Mackenzie and B. L. Ristvet 1972. Experimental vadose and phreatic cementation of skeletal carbonate sand. *J. Sed. Petrol.* **42**, 162–7.

Thunell, R. C., D. F. Williams and J. P. Kennett 1977. Late Quaternary palaeoclimatology, stratigraphy and sapropel history in eastern Mediterranean deep-sea sediments. *Mar. Micropal.* **2**, 371–88.

Till, R. 1974. *Statistical methods for the Earth scientist.* London: Macmillan.

Till, R. 1978. Arid shorelines and evaporites. In Reading (1978), 178–206.

Tissot, B. P. and D. H. Welte 1978. *Petroleum formation and occurrence*. Berlin: Springer.

Townsend, A. A. 1976. *The structure of turbulent shear flow*. Cambridge: Cambridge Univ. Press.

Townson, W. G. 1975. Lithostratigraphy and deposition of the type Portlandian. *J. Geol Soc. Lond.* **131**, 619–38.

Tricker, R. A. R. 1964. *Bores, breakers, waves and wakes*. London: Mills & Boon (New York: Elsevier).

Tritton, D. J. 1977. *Physical fluid dynamics*. London: Van Nostrand Reinhold.

Tucker, M. E. 1973. Sedimentology and diagenesis of Devonian pelagic limestones (Cephalopodan kalk) and associated sediments of the Rheno-Hercynian Geosyncline, West Germany. *Neues Jb, Geol. Palaont. Abh.* **142**, 320–50.

Tucker, M. E. 1974. Sedimentology of Palaeozoic pelagic limestones: the Devonian Griotte (S. France) and Cephalopodan kalk (Germany). In Hsü & Jenkyns (1974), 71–92.

Tucholke, B. E. 1975. Sediment distribution and deposition by the western boundary undercurrent: the Great Antilles Outer Ridge. *J. Geol.* **83**, 177–207.

Turner, P. 1980. *Continental red beds*. Amsterdam: Elsevier.

Valloni, R. and J. B. Maynard 1981. Detrital modes of recent deep-sea sands and their relation to tectonic setting: a first approximation. *Sedimentology* **28**, 75–84.

Valyashko, M. G. 1972. Playa lakes – a necessary stage in the development of a salt-bearing basin. In *Geology of saline deposits*, G. Richter-Bernberg (ed.), 41–51. Paris: Unesco.

Veen, F. R. van 1977. *Prediction of permeability trends for water injection in a channel-type reservoir, Lake Maracaibo, Venezuela.* Society of Petroleum Engineers, American Institute Mining Metallurgy & Petroleum Engineering 6703, 1–4.

Visher, G. S. 1969. Grain size distributions and depositional processes. *J. Sed. Petrol.* **39**, 1074–106.

Wahlstom, E. E. 1948. Pre-Fountain and Recent weathering on Flagstaff Mountain near Boulder, Colorado. *Bull. Geol Soc. Am.* **59**, 1173–90.

Walker, R. G. 1966. Shale Grit and Grindslow Shales: transition from turbidite to shallow water sediments in the Upper Carboniferous of northern England. *J. Sed. Petrol.* **36**, 90–114.

Walker, R. G. 1978a. Facies and facies models: general introduction. In Walker (1978b), 1–8.

Walker, R. G. (ed.) 1978b. *Facies models* Toronto: Geol. Ass. Canada.

Walker, R. G. and E. Mutti 1973. Turbidite facies and facies associations. In *Turbidites and deep water sedimentation*, 119–57. SEPM Short Course, Anaheim.

Walker, T. R. 1976. Diagenetic origin of continental red beds. In *The continental Permian of central, west and south Europe*, H. Falke (ed.), 240–82. Dordrecht: Reidel.

Walker, T. R. and J. C. Harms 1972. Eolian origin of flagstone beds, Lyons Sandstone (Permian) type area, Boulder County, Colorado. *Mountain Geol.* **9**, 279–88.

Walker, T. R., B. Waugh and A. J. Crone 1978. Diagenesis in first-cycle desert alluvium of Cenozoic age, southwestern United States and northwestern Mexico. *Bull. Geol Soc. Am.* **89**, 19–32.

Walter, M. R. (ed.) 1976. *Stromatolites*. Amsterdam: Elsevier.

Wanless, H. R. and J. R. Cannon 1966. Late Paleozoic glaciation. *Earth Sci. Rev.* **1**, 247–86.

Wanless, H. R., J. R. Baroffio, J. C. Gamble, J. C. Horne, D. R. Orlopp, A. Rocha-Campos, J. E. Souter, P. C. Trescott, R. S. Vail and C. R. Wright 1970. Late Palaeozoic deltas in the central and eastern United States. In Morgan (1970), 215–45.

Wasson, R. J. 1977. Late-glacial alluvial fan sedimentation in the Lower Derwent Valley, Tasmania. *Sedimentology* **24**, 781–99.

Watkins, D. J. and L. M. Kraft 1978. Stability of continental shelf and slope off Louisiana and Texas: geotechnical aspects. In Bouma, Moore & Coleman (1978), 267–86.

Watkins, J. S., L. Montodert and P. W. Dickerson (eds) 1979. *Geological and geophysical investigations of continental margins*. Mem. AAPG, no. 29.

Watson, G. S. 1966. The statistics of orientation data. *J. Geol.* **74**, 786–97.

Watts, N. L. 1980. Quaternary pedogenic calcretes from the Kalahari (southern Africa), mineralogy, genesis and diagenesis. *Sedimentology* **27**, 661–86.

Waugh, B. 1970a. Formation of quartz overgrowths in the Penrith Sandstone (L. Permian) of NW England as revealed by scanning electron microscopy. *Sedimentology* **14**, 309–20.

Waugh, B. 1970b. Petrology, provenance and silica diagenesis of the Penrith Sandstone (Lower Permian) of NW England. *J. Sed. Petrol.* **40**, 1226–40.

Weber, K. J. and E. Daukoru 1975. Petroleum geology of the Niger delta. *Proc. 9th World Petrolm Cong. Tokyo* **2**, 209–21. London: Applied Science.

Wellendorf, W. and D. Krinsley 1980. The relation between the crystallography of quartz and upturned aeolian cleavage plates. *Sedimentology* **27**, 447–54.

West, I. M. 1964. Evaporite diagenesis in the lower Purbeck beds of Dorset. *Proc. Yorks. Geol Soc.* **34**, 315–30.

West, I. M. 1975. Evaporites and associated sediments of the basal Purbeck Formation (U. Jurassic) of Dorset. *Proc. Geol. Ass.* **86**, 205–25.

White, D. A. 1980. Assessing oil and gas plays in facies-cycle wedges. *Bull. AAPG* **64**, 1158–78.

White, D. E. 1965. *Fluids in subsurface environments*. AAPG Mem. no. 4.

Wiley, M. (ed.) 1976. *Estuarine processes*, vols 1 & 2. Chichester: Wiley.

Wilkinson, B. H. and E. Landing 1978. 'Eggshell diagenesis' and primary radial fabric in calcite oöids. *J. Sed. Petrol.* **48**, 1129–38.

Williams, G. P. 1970. *Flume width and water depth effects in sediment transport experiments*. USGS Prof. Pap., no. 562-H.

Williams, P. B. and P. H. Kemp 1971. Initiation of ripples on flat sediment beds. *J. Hydraul. Div. A.S.C.E.* **97**, 505–22.

Williams, P. F. and B. R. Rust 1969. The sedimentology of a braided river. *J. Sed. Petrol.* **39**, 649–79.

Wilson, H. H. 1977. 'Frozen-in' hydrocarbon accumulations or diagenetic traps – exploration targets. *Bull. AAPG* **61**, 483–91.

Wilson, I. G. 1971. Desert sand flow basins and a model for the development of ergs. *Geog. J.* **137**, 180–99.

Wilson, I. G. 1972a. Aeolian bedforms – their development and origins. *Sedimentology* **19**, 173–210.

Wilson, I. G. 1972b. Universal discontinuities in bedforms produced by the wind. *J. Sed. Petrol.* **42**, 667–9.

Wilson, I. G. 1973. Ergs. *Sed. Geol.* **10**, 77–106.

Wilson, J. B. 1967. Palaeoecological studies on shell beds and associated sediments in the Solway Firth. *Scott. J. Geol.* **3**, 329–71.

Wilson, J. L. 1975. *Carbonate facies in geologic history*. Berlin: Springer.

Wilson, L. 1973. Variations in mean annual sediment yield as a function of mean annual precipitation. *Am. J. Sci.* **273**, 335–49.

Wilson, M. D. and E. D. Pittman 1977. Authigenic clays in sandstones: recognition and influence on reservoir properties and palaeoenvironmental analysis. *J. Sed. Petrol.* **47**, 3–31.

Wolman, M. G. and J. P. Miller 1960. Magnitude and frequency of forces in geomorphic processes. *J. Geol.* **68**, 54–74.

Wood, G. V. and M. J. Wolfe 1969. Sabkha cycles in the Arab/Darb Formation off the Trucial Coast of Arabia. *Sedimentology* **12**, 165–91.

Woods, P. J. and R. G. Brown 1975. Carbonate sedimentation in an arid zone tidal flat, Nilemash Embayment, Shark Bay, Western Australia. In Ginsburg (1975), 223–33.

Woods, P. J. E. 1979. The geology of the Boulby mine. *Econ. Geol.* **74**, 409–18.

Wright, A. E. and F. Moseley (eds) 1975. Ice ages: ancient and modern. *Geol J.* Spec. Issue, no. 6.

Wright, L. D. 1977. Sediment transport and deposition at river mouths: a synthesis. *Bull. Geol Soc. Am.* **88**, 857–68.

Wright, L. D. and J. M. Coleman 1973. Variations in morphology of major river deltas as functions of ocean wave and river discharge regimes. *Bull. AAPG* **57**, 370–98.

Yalin, M. S. 1977. *Mechanics of sediment transport*, 2nd edn. Oxford: Pergamon.

Young, F. G. and G. E. Reinson 1975. Sedimentology of Blood Reserve and adjacent formations (Upper Cretaceous), St Mary River, S. Alberta. In *Guidebook to selected sedimentary environments in SW Alberta, Canada*, M. S. Shawa (ed.), 10–20. *Can. Soc. Petrolm Geol.*

Yuretich, R. F. 1979. Modern sediments and sedimentary processes in Lake Rudolf (Lake Turkana) eastern Rift Valley, Kenya. *Sedimentology* **26**, 313–32.

Zenger, D. H. 1972. Significance of supratidal dolomitisation in the geologic record. *Bull. Geol Soc. Am.* **83**, 1–12.

Zenger, D. H. and J. B. Dunham 1980. Concepts and models of dolomitization – an introduction. In Zenger *et al.* (1980), 1–9.

Zenger, D. H., J. B. Dunham and R. L. Ethington (eds) 1980. *Concepts and models of dolomitization*. SEPM Spec. Pubn, no. 28. Tulsa.

Ziegler, P. A. 1975. North Sea Basin history in the tectonic framework of NW Europe. In *Petroleum and the continental shelf of NW Europe*, A. W. Woodland (ed.), 131–49. London: Applied Science.

Index

Page numbers are given in roman type. The numbers of text figures are given in italic type. Table and plate numbers are given as '*Table 2.2*' and '*Plate 1*'. Text section numbers are given in bold type.

abrasion **4e**
Abu Dhabi 304, *23.5*, *23.6*, *23.20*
abyssal plains **25d**, *25.2*
accretionary wedge 246
Achanarras Limestone 158
active margins **25a**, **e**
adhesion ripples 135
advection 192, 204
aeolian
  bounding surfaces *13.3*, *13.4*
  dust 250
  facies **13c**, **d**
aeolianite *Plate 8*, *29.5*
Agulhas current 234
air flow **8e**, **13**
Airy waves **8c**, *18.2*
aklé dunes 99–100
Aleutians trench *25.12*
algae **109**, 216, 222, 285, 291
algal (see also stromatolites)
  blooms 156
  borings *2.9*
  coatings 26
  mats 199. 212
  ooze 159
  scum 220
  tufa 217
alkali feldspar *Tables 1.1–3*, **28**
alkalinity 18
allochems 27
alluvial fans **14**, 159
  plains 313
  ridge 145
Amazon submarine fan *Table 25.1*
amino-acids 293
amphibole *Tables 1.1–3*, 275
amorphous lumps 26
amphidromic point *18.11*, *18.12*, *199*
Andros Island 215, 216, 217, *23.10–12*
angle
  of initial yield 76, 111
  of residual shear 76
anhydrite 167, 213, 299, **3a**, **23i**, **30**, *3.3*, *3.4*, *30.4*, *30.5*, *Table 3.1*
ankerite 17, 271, 281, 296
anoxic conditions 233, **26c**
Antarctic Bottom Water 235
Antarctic convergence 234
Antarctic Intermediate Water 234
anthracite *Table 31.1*
antidunes *8.9*, 92
aphanitic texture 304
aquathermal pressuring 266
Arab Formation 318
Arabian Gulf 212, 217, 221, 298
aragonite 1, 106, 156, 170, 214, 217, 220, 224, **2g**, **29**
arenites **28h**
arkoses **28h**, 276
Athabaska glacier *17.3*
atoll *23.24*
augite *Tables 1.1*, *1.2*
Australia 128, 133
authigenic minerals **28**, **29**, **30**
autosuspension **App. 7.2**, 80
avalanches 86, 93, *7.2*
avulsion 146, 186

back-arc basins 245
bacterial processes 262, 270–2
Bagnold Number 73
Bahamas 1, 18, 20, 25, 27, 217, 223, 224, 225, *2.3*, **23c**
barchan dunes 99
barrier(s) **21**, 186, 313, *21.10*
  inlet *21.16*
  reef *23.24*
bars 93, 143, 146, 147, 167
base-surge 93
bays **23d**
beach(es) 183, **21**
  berms 196
beachrock 26, 289, *29.6*, *Plate 8*
Beatton River *15.6*
bedforms **8**, **9**, 62, 63
  hierarchies 89, 97, *12.9*
  lag **8f**, *8.26*
  phase diagrams 85, 93, *8.1*, *8.13*, *8.14*, *8.17*
  theory 94, **App. 8.1**
bedload **6d**, **6f**, *6.6*
Bengal submarine fan *Table 25.1*
bentonites 272
Bering shelf 206
Bernouilli
  effect 68, 72
  equation 50, 61, **App. 5.1**
Bimini 225, *23.18*
'birds-eyes' 291
birdsfoot 185
biotite 5, 276, *Tables 1.1*, *1.2*
bioturbation 109, 151
bittern 304
bituminous coal *Table 31.1*
bituminous limestone *Table 28.3*
Black Sea 158, 253, *26.6*, *26.7*
black shale 233, 239, 253, 315, *Tables 28.2*, *28.3*
Blake Plateau 225, 235
boils 58
Bonaire Lagoon 300
boring 109, 291
Bouma logs 120
  sequence 81
boundary layers **5**, *5.7–10*
bounding planes 135
brachiopods 21
Brahmaputra River 147, *15.11*, *15.12*
braiding **15b**, 140, *15.9*, *15.10*
brown coal *Table 31.1*
build-ups 219, 221, 224–6, **23g**, *23.23*, *23.24*
burst 92, 100, **5i**, *5.17–20*

Caithness Flagstone Group 158
calcareous algae 25
calcite **2**, 156, 158, 220, 224, 225, 271, 276, 281, **29**, *2.5*
calcium carbonate **2**, **29**
  world distribution *2.1*
calcrete 289, *29.5*
caliche 287
Callianassid burrows 216
Cantabrian Mountains 141
carbon isotopes 271, **27d**, *27.4*, *27.5*
carbonate
  banks **23a**, *23.2*, *23.3*
  compensation depth 231, 238, 250, 251–2, **24c**, *24.3*, *26.3*
  platforms *23.2*, *23.3*, **23a**
  ramps *23.2*, *23.3*, **23a**
carnallite 31, 304, *Table 3.1*
Carnegie Ridge 256
catastrophes 125
cement fabrics
  dripstone 287
  grain contact *29.4*
  isopachous 291, *29.4*, *29.6*
  meniscus 291
  microstalactitic 291, *29.4*
  radiaxial fibrous 293, *29.9*, *29.10*
  syntaxial 296, *2.8*
cementation **27–29**
Central Australian desert 135
chalcedony 289, 308
chalk 158, 253, 255–6, 296
chamosite 309
Chandeleur Islands 190, 197
channel morphology *15.1*
  sinuosity *15.2*
  sedimentation **15**
  width 142
charcoal 314
cheniers **21**
chert 239, **30b**, *Table 3.2*
chevron marks 105
Chèzy coefficient 50
chicken-mesh texture 213, 304
chlorite 13, 242, 273, 280, *28.7*
chute bar 150
chute-and-pool 92

cirque glacier *17.6*
classification of sedimentary rocks **28d**, **h**, **29g**
clay 42, **9**, **11a**
  cement *28.7*, **28**
  infiltration 275
  lumps 114
  minerals *1.9*, *1.10*, **28b–f**, 250
  pellicles 278
  stone 273
cleavage 273
climbing ripple cross lamination 115, 164, *8.7*
coal 141, 190, **31b**, **c**, *14.8*
  rank 311, *31.2–4*, *Table 31.1*
coasts **Part 6**
coccoliths 250, 253, 256, 296
cohesion 67, **9a**
collision breeding 27
colloids 103
colluvium 140
compaction 106, 124, 316, **27f**, *27.8*, *27.11*, *28.2*
complex ions 16
concretions 271–2, 272, *28.3*
Congaree River 147
connate water 266
'connectedness' 151
continental environments **Part 5**
continental rises **25b**, *25.1*, *25.3*
continental slopes **25b**, 237
contourites 235, 238–9, 242
convolute laminae *11.2*
Coorong Lagoon 297, 300
coral 25, 226
  reefs 292
Coriolis Force 231–2, 234
cotidal line 178
crevasse *15.7*, 314
  deltas 185
  splays 144, 147, 314
crinoids 21, 226
cross lamination **8a**, **c**, 92, 150, 207
  cosets **8a**
  sets **8a**, 147
  stratification **8a**, 112, 150

Danikil Depression 254
Danube delta *Table 19.2*
d'Arcy–Weisbach equation 50
Dead Sea 20, 156
Death Valley *14.7*
debris flows 115, 140, 225, 247, **7a**, **c**, **f**, **24f**, *7.3*, *7.4*, *24.11*
decarbonatisation 281
decarboxylation 272, 281, *Tables 28.1*, *28.2*
decompaction 265
dedolomitisation 301, *29.8*
Deep Sea Drilling Project 231, 254
deflation 135
Delmarva peninsula 197
deltaic cycles *19.12*
deltas **19**, 115, 313
  abandonment 186, 190
  cycles 186
  Gilbert-type 182
density 4, *Table 5.1*
  currents **7**, 156, *7.6–8*
deposition rate 42, 129, **12d**, *12.6*
deserts **8e**, **13**
desiccation cracks **11f**, *11.6*
dewatering pipes 112
diachronism **12e**
diagenesis **Part 8**
  carbonates **29**
  clastic **28**
  diagenetic realms **27h**
diapirs **11d**
diatoms 31, 158, 250, 252
dickite 273
diffusion 192
dilitancy 76, 111
Diplocraterion yoyo 109, *10.3*
dish and pillar 112, *11.2*
dispersion 103
dispersive clay fabric *11.1*
dispersive pressure **App. 7.1**, 74, 76, 77
distal 81
distributary mouth bar 183
dolomite 17, 20, 30, 106, 158, 271, 281, 318, **29f**
domal dunes 100
Donjek River 146, *15.9*
downward dipping sets *8.3*, *15.12*
draa 97, 100, *8.21*
drag
  force 68, *6.3*
  form 51
  particle 51
dripstone 287
dropstones 164, *17.12*
drumlin 164
du Boys equation 50
dunes 88–9, **86**, 93, **8e**, **App. 8.1**, 112, 143, 147
  aeolian **13b**
duricrusts 289

Ebro delta *Table 19.2*
echinoderms 25
eddy viscosity 48, 55
Eh **1c**, 4, **27e**, *1.2*
Ekman spiral 231, *24.1*, *24.2*
  transport **24d**, 32, 235
Ekofisk 318
Eleuthera Bank 291
energy budget in fluids **5d**
endolithic algae 26
Endrick River 147, 151
enfacial junctions 260, 289, 293
enterolithic layers 304
environments of deposition *Table 12.1*
epilimnion 156
ergs **13**, *13.1*, *13.5*
escape burrows 109
esker 164, *17.10*
estuaries **20**
eustasy 126
euxinic 4
evaporites 252–4, **3d**, **23b**, **i**
  diagenesis **30**
evaporative pumping 303
Exuma Sound 220

facies analysis **12–26**
  associations 120
  principles of **12**
  transitions *Table 12.2*
fall velocity **6b**, *6.1*
fanhead entrenchment *14.4*
fans
  alluvial **14**
  submarine **25c**
fault scarp 140
feldspar 3, 5, 8, 11, 272, 276, 279–80
felted lath texture 304
fenestrae 212, 217, 291
fermentation 262, 271–3, *Tables 28.1*, *28.2*
ferroan calcite **2b**, **29d**
ferroan dolomite **2b**, 296
fibroradiate texture 304
fissility 275
flagstone 92
flame structures 114, *11.2*
Flandrian transgression 202, 207
flasers 95, 194, 199–200, *8.18*
flocculation 42, 103, 111, 192, 270, 275, *11.1*
Florida 1, 18, 215, 217, 222–3, *23.14*
  current 225
  swamp 313
flow
  laminar **5e**, **5g**
  non-uniform 49
  plug 54
  rapid 53
  reattachment 104
  regime 93
  separation 86, 93, 104, 150, **5j**, **App. 8.1**, *5.21–4*
  steady 49
  till *17.8*
  tranquil 53
  turbulent **5e**, **h**, **i**
  uniform 49
  unsteady 49
  visualisation **5a**, **i**
fluid
  Bingham 48, 77
  density 47
  flow **5**
  Newtonian 48, 53, 54
  Non-Newtonian 48, 54
  viscosity 47–8
fluidisation 78
fluorapatite 32
flute marks 103, *9.1–4*
foraminifera 25, 239, 250, 252, 293, 296
forearc basins **25e**
  region *25.11*, *25.14*
formation water 261, 266, 300, **27f**, **29d**, *27.1*
forsterite 7
fractures 318
Fragum hamelini 218
free energy 7

frequency curve 35
freshwater composition *Table 3.1*
friction coefficients 93–4
frost weathering 8
Froude Number 93, 183, **5f**
fulvic acids 20

Gallup Sandstone 201
Ganges–Brahmaputra abyssal cone 242
  delta 188
garnet 12
Garwoodia 25, *2.8*
gas (natural) **31**
gastropods 106
geomorphic threshold 139
geopetal structure 292
geostatic pressure gradient *27.2*
Girvanella 25
glacial cycles 128
glaciers **17**
glass
  volcanic 279
glauconite 32, 256, 291, 309
goethite 308–9
grading 39, 77, 97, **8d**
grain
  abrasion **4e**
  aggregates **4g**
  collisions 72
  concentration **4g**, **App. 8.1**
  creep 99
  diameter **4a**
  fabric **4h**
  flows **7a**, **b**, **f**, **App. 7.1**, **7.2**
  form **4f**
  mass **4a**
  packing 42
  populations **4c**
  shape **4f**
  size **4**
  stresses **6e**
  suturing 268
Grand Banks *24.12*, *24.13*
Grand Banks Earthquake 238
grapestones **2k**, 26
gravity flows **7**, *7.1*, *7.9*, 224
Great Barrier Reef 217
Great Plains 144
Great Salt Lake 26
Great Sand Dunes 135
Great Valley *25.14*
Green River Formation 159, 253, 319, *16.6*, *16.7*
greywackes **28h**
grooves 103
groundwater mixing 299
growth faults **11e**, 317, *11.2*, *31.9*
Gulf of Mexico 114, 206
Gulf Stream 234
gutter marks 104
gypsum **3a**, 135, 157, 212–14, 278, 299, 304–5, **23i**, **30**, *Tables 3.1*, *3.3*, *30.4*, *30.5*

haematite 263–4, 276, 308–9, **30c**
Halimeda 25, 224
halite **3a**, 135, 156–7, 227, 254, 278, **23i**, **30**, *Table 3.1*
halmyrolysis 259
halokinesis 114
hardgrounds 26, 109, 225, 291–2, 296, *29.6*, *29.8*
Hatteras abyssal plain 245, *25.9*
heavy minerals 12, *1.8*
helical flow 143, **App. 15.1**, *15.3*
helical vortices 205
Heligoland Bight 206
histogram 35
Holocene, peculiarities of **12g**
homopycnal 182
Honduras 217
hopper crystals 158
hornblende 5, *Tables 1.1*, *1.2*
Hornelen basin 151
humic acids 20, 311
  coal 311
huminite 311
hurricanes 220
Hutchison embayment 214
Hydrobia 109, *10.3*
hydrocarbons **31**
hydrogen bubbles 58, 70, *5.16–18*
hydrogen sulphide 271
hydrolysis 4
hydrostatic pressure gradient *27.2*
  stress 268
hypersaline oceans **26d**
hypolimnion 156
hypopycnal 183

ice **17**
  bergs 161, 164
  shelf 164
illite 5, 8, 12–13, 242, 270–3, 276, 279–81, 296, *28.7*
illitisation 281
imbrication 43, 167, **66**
inertinite 311
interconnectedness 318
interdistributary bay 183, 185
interflows 156–7, 164
internal sediment 222
intersection point 140
intertidal flats *Plate 6*, **21**
intraclasts 26, 213, *2.9*
intrastratal solution **28e**, **f**
ion activity products *Tables 2.2*, *3.1*
  pairs 16
ionic strength 16
ionisation constant (k) 18
Iran 114
iron 4, 263
  diagenesis 263–4, **30c**
  minerals **30c**
  sulphides 271
isostasy *12.4*, *12.5*
isotopes **27d**

jets 182, *19.2–4*
Joulters Cay 220
  Shoal *23.19*
Jura Quartzite 207, *22.12*

kaolinite 5, 8, 12, 103, 273, 276, 279, 281, 313, *28.7*
Karman–Prandtl equation 57, 72, **App. 5.4**
karst 218, 223
kerogen 274, 315
Kimmeridge Clay 272, 319
kinematic viscosity 48
kinetic energy 49
Kosi River 147, 151, *15.13*
Kuro Shio 234

lag (bedform) **8f**
lag deposits 40, 135
lagoons **21**, **23d**
Laguna Madre 27
Lake Brienz 156–7
Lake Chad 135
Lake Kivu 158
Lake Ontario *16.1*
Lake Rudolf *16.1*
Lake Turkana 158
Lake Zürich 156, 158, *16.3*
lakes **16**, 115
laminar (viscous) flow **5g**, 161
laminations 95, **8d**
  stromatolitic 106
laminites 158
lateral accretion 151, 194, 199, 217
  bars 142
  deposition 143
lateritisation 308
Laurentian abyssal cone 242, 245
layered halite rock 158
law of the wall 66
lenticular lamination *8.18*, *8.19*
levees 147
lift 62, 68, 72, *6.3*
lignin 311
lime muds 289
limestone **29**
  classification *Tables 29.1*, *29.2*
  diagenesis **29**
limonite 276, 308–9
limpid dolomite 300
liptinite 311
lithic fragments 11, 279
lithification 109, **27–30**
lithofacies 120
Lithothamnion 25
liquefied flows **7a**, **d**, **f**, **11a–c**, *7.5*, *11.3*
Llandovery transgression 209
load casts 114, 227, *11.2*
lodgement 162
loess 133
Lofer Cycles 217
log hyperbolic 37
log normal 37
'loose boundary' hydraulics 67, 173–6, 183, 241, *18.6*
lumps *2.9*
lunettes 135
lutecite 308
lysocline 29, 232

macerals 311

macrocell structure 304
macrotidal 178, **21**
magnetic anomalies 238
magnetite 263–4
manganese diagenesis **30d**
 crusts 255–6
 nodules 250
mangroves 217
Manlius Facies 217
marine diagenetic realm **27h**, **29c**
Markov processes 123
marl *Table 28.3*
marshes **23c**
mass flow 227
matrix 78, 284
maturity 13
mean (grain size) 38
mean annual discharge 142
meander belt 144, 145, 151
 loop 147, 150–1, *15.18*
 wavelength 142
meandering **15b**, **App. 15.1**
median (grain size) 38
Mediterranean 293
meltwater 162
Mesaverde Group 201
Mesotidal 178, **21**
Messinian evaporites **26d**, 227, 254, *26.8*
metamorphic reactions 273, 279, *Tables 28.1*, *28.2*
meteoric diagenetic realm **27h**, **29b**
methane 266, 271, 312, **31d**, **e**
micrite **29**
 envelopes **2h**, **g**, *2.9*, 285
microspar 289
microtidal 178, **21**
Mid-Atlantic Channel 245
Mid-Atlantic ridge 251
Middle Atlantic Bight *22.11*
mid-ocean ridge 15, *26.3–5*
migration (hydrocarbon) **31e**
Millstone Grit 190
Mississippi
 abyssal cone 242
 delta 185–6, 197, *12.8*, *19.5*, *19.8–11*, *Table 19.2*
 river 115, 142, 150, *15.17*
 submarine fan *25.8*, *Table 25.1*
mixed-layer clays 13, 272, 276, 280
mixing length 57
 models *29.15*, *29.16*, *30.7*
 theory 291
mixtite 167
mode (grain size) 38
molecular viscosity 47–8
molluscs 21
momentum flux 172
montmorillonite 5, 8, 13, 103, 158, 247, 266, 272, 276, 280
morainic complex *17.9*
Morrison Formation 151
mouth bar 185, *19.2–5*
mud
 classification *Table 28.3*
 cracks 217
 curls 115
 diagenesis **28b**, **c**
 rocks **28d**
 waves 242
Muddy Sandstone 200
multistorey sandstones *15.19*
muscovite 8, 273, *Tables 1.1–3*

nacrite 273
Nankai trench *25.13*
natural gas reservoir 136
neomorphism 286, 293, 296
nepheloid layer 236, *24.10*
Nernst equation 262–3
Niger
 cone 242
 delta 186, 188, *19.13*, *Table 19.2*
 river 115, 186
nodules 225, 291–3
North Atlantic Deep Water 235
North Sea 114, 136

oceanic currents
 drift 231–2
 gradient 231–2, 234
 surface **24d**
 thermohaline (contour) 233, 235, 239, 242, 252
oceanic environments **Part 7**
 currents **24d**, **e**
 facies **26b**
 processes **24**
 water *Table 27.1*
ocean ridges 126
offshore bars 196, 198, *21.8*, *21.9*
oil **31**, 266
 shale 159, 200, 253, 266, **31h**, *Table 28.3*
Old Red Sandstone 158
olivine *Tables 1.1–3*, 276
Omo River *16.1*
oöliths **2j**, 20, 199, 285, 294–5, 309, **23a**, *2.9*
ooze
 calcareous 15, *26a*
 siliceous *26a*
opal-A 32, 250–1, **30b**, *3.5*, *Table 3.2*
ophiolite 254–5
orbital paths *18.3*
Orcadian Lake *16.5*
Oregon shelf 204, 206, *22.4*
organic diagenesis **31**
Ortonella 25
outwash 161, 164, *17.13*
overbank floods 185
overflow 164
overgrowths
 silica 278
 syntaxial 278, *2.8*
overpressuring 465–6, *27.9*, *27.10*, *27.12*
overturned cross laminae **11c**, *11.3*
oxidation **1c**, 21, *Table 28.1*
oxygen isotopes 239, 253, *27.3*, *27.5*, **27d**
 minimum layer 231, 252–3

packing *4.5*
palaeoceanography **24g**
palaeocurrents **12f**, **App. 12.1**
palaeohydraulics **App. 15.2**
palaeoslope 128
Papua delta 188
particle paths 49
passive margins **25a**, **b**
P.D.B. standard **27d**
peat 145, **31c**, *Table 31.1*
pelagic 15, **2b**
peloids **2i**, 192, 213, 216–7, 250, 285, 287, *2.8*
Penicillus 1, 25, *Plate 1*
permeability 42
Persian Gulf 20, 27, 114, 291
petrography **27c**
pH 4, 103, **1c**, **27e**
phi scale 35
phosphate 20, 32, 158, 233, 239, 250, 252, 256, 291, **3c**
phreatic diagenetic realm **27h**
physical weathering **1f**
pisoliths 287, 289, *29.5*
pitted pebbles 268
plagioclase 275, *Tables 1.1–3*
planar laminations 92
plane bed **App. 8.1**
 lower stage 93
 upper stage 89, **8b**, 92–4
plate tectonics **1i**, **12h**, **25**
platform margin slopes **23h**
platforms **23g**, **h**
Platte River 147
playa 135–6, 156, 159
plug zones *7.2*, *7.3*
plumes 204
point bar 109, 115, 143, 147, 150, 151, 194, 199–200, *15.3*, *15.14–16*, *21.14*
polyhalite *Table 3.1*
pore pressure 111
porosity 42, 264, *4.6*, *27.7*, *Table 28.1*
Port Askaig Tillite 167
potential energy 49
power 74, 93
Prandtl 54, 57
preservation potential 21, 92, 106, 122
pressure
 energy 50
 fluid **5d**
 geostatic 265, **27b**
 hydrostatic 265, **27b**
 solution 295, 296, **27g**
primary current lineation 89, 92, *8.9*
probability 35, 37
pro-delta 186
productivity 32
proglacial lakes 164
progradation 122, 125
protodolomite 298
proximal 81
pseudopleochroism 293
pteropods 28, 250, 293
pyrite 264, 271, 273, 309, 313
pyroxene 275, *Tables 1.1–3*

quartz 5, 8, 11, 41–2, 111, 274, 276, *Tables 1.1–1.2*

overgrowths *28.6*
undulosity *1.6*
quartzine 308
quick condition 111
clay 111–12
sand 112

radiation stress 172
radiolaria 31, 239, 250, 252, 255
Rayleigh–Taylor instability 114
reactivation surfaces 89
red beds **28e**
clay 250, *Table 28.2*
pigment 309
Red Sea 254, 293
reduction **1c**
reefs **23g**, **h** 318, *23.22*, *23.24*
reflectance 311
regelation ice 162
layer *175*
regression **12e**, *Table 12.3*
relict shelfs 202, 221
Reynolds Number 68, 69, 75, **5e**, **App. 5.2**
Reynolds stresses 58, 60, 62, *5.12*, *5.20*
Rheno-Hercynian 256
rhourds 100
ridge-and-runnel 198
rip cells 173, *21.9*
rip currents 173–4, 241, *18.6*, *18.8*
ripples
aeolian (ballistic) **8e**, 135, *8.21–23*
climbing 88
counterflow 89
current **8a**, **b**, **App. 8.1**, 143, *8.2–6*, *8.12*, *8.16*
granule **8e**
indices 86, 95
rolling grain 94
vortex 94
wave current 198
wave formed 159, *8.15*, *8.16*
river channels **15**
flood facies 115
floodplains **15**
plains **15**
River Omo 158
Riverine Plain 144
rock fragments 5
Roda Sandstone 208, *22.14*
rolling **6d**
Rosin's Law 37
rotational faults *11.5*
slides **11e**
Rotliegendes **13d**
rough boundary *5.14*
roughness 56–7, 60, **App. 8.1**
roundness **4f**, *4.4*
rudists 222

sabkha 135, 170, 227, 278, 298–9, 303, **23b**, *23.5–8*, *30.1–3*, *31.11*
Sahara 97, 128, 133, *13.2*
St Lawrence River 242
Saline Valley *16.4*
salt diapirs 304, **11d**
marsh 193, 199, *Plate 6*
weathering 9
wedge 183, 192–3
saltation **6d**, *6.6*, *6.8*
sand ribbons 204–5, *22.7*
sheet (low angle) 135
volcanoes 112
waves 89, 147, 204–5
sandstone 3, **28h**
cementation **28**, *28.4*, *28.8*, *28.10*
classification **28h**, *28.14*
diagenesis **28e–h**
San Francisco delta *Table 19.2*
sapropel 253, 311, *Table 28.3*
Saskatchewan glacier *17.2*
River 147, *15.10*
saturation depth 28
savannas 133
Scalby Formation 151
Schithothrix 106, 217
scour lag *21.6*
scree 140
scroll bars 150
Scytonema 217
seagrass 219
seawater
alkalinity *2.2*
composition *Table 2.1*
second generation cement 295
secondary currents **5j**
flows 205
porosity **28g**, 318, *28.11*, *28.13*
sediment ridges 241–2, *25.4*
transport **6**, 174
yields *1.3*
sedimentary basins **12h**
seiches 155
seif dunes 100
Senegal delta *Table 19.2*
sericite 273
settling lag *21.6*
shards 272
Shark Bay 212, 218–19, 222, *23.15–17*
shear stress 69, 94, **5**, **6**, *5.4*
sheet floods 140–1
shelf(s) **Part 6**
currents *Table 22.1*
relict **22b**, *22.2*
tide dominated **22b**, **c**
weather dominated **22b**, **c**
Shield's diagram 69–70, *6.4*
shoal retreat complexes 206
ridges 207
shorelines
carbonate/evaporite **23**
clastic **21**
shrinkage cracks 159
siderite 158, 264, 272, 273, 281, 309
sieve deposits 140
silcrete 278, 308, *30.5*
silica
biogenic **3b**
cycle *3.2*, *Table 3.2*
diagenesis **30b**
rim cement 280
silicoflagellates 250
Simpson Desert 100
skeletal carbonate 217, *2.6–8*
grains 213
skewness 38–9, *4.2*, *4.3*, *Table 4.5*
skin friction lines 49, 62–3, 86, *8.5*
slumping **11e**, 225, 247, **24f**
slurry 78
smectite 272, 279–80, 296
smooth boundary *5.14*
SMOW standard **27d**
Snuggedy Swamp 314
soft sediment deformation 151, **11**
Sohm abyssal plain 245, *25.10*
soil 140, 145
solubility products *Table 2.2*, *3.1*
solution fronts 286, 293
Solway Firth 194
sorting 37–8, *4.3*, *Table 4.5*
source rocks **31**
sourcelands **1h**, **i**
Southern Uplands (Scotland) 249
South Esk River *15.14*
sphericity **4f**
spherulites 293
spillovers **23e**, *23.18*
spirit levels 222
spit 183
Spitsbergen 153
sponges 222, 226
spreite 109, *10.4*
stable isotopes **27d**, **29**
staining **App. 2a**
standard deviation 38
Sternberg's Law 40
Stokes Law **6b**, **App 6.1**
storm *22.13*
surge 206
washover 195
streaklines 49, **5c**, *5.3*
streaks **5i**, 89, 92, 103, **App. 8.1**, *5.17*, *5.19*
stress **6e**
strike–slip faults 241
stromatolites **10a**, 167, 212–13, 217, 300, *10.1*, *10.2*, *23.6*, *23.8*, *23.9*
strontium 17, 286
'structure grumeleuse' 296
stylolites 268
subaqueous shrinkage cracks 115
subduction complex 247, 249
subglacial tunnel *17.10*
submarine canyon 237, 241
fans 190, 239, 241, 247, **25c**, *25.5–8*
subsidence **12d**
subsurface (burial) diagenetic realm **27h**, **29d**
pressure **27b**
temperature **27c**
sulphate reduction 271, 273, 294, *Tables 28.1*, *28.2*
sulphur 305
superficial oöids 26
suspension **6d**, 60
Sussex Sandstone 209
suturing 281
swale 144, 150
swamp 313

sweeps 89, 92, **5i**
sylvite 31, 304, *Table 3.1*
syneresis cracks 112, **11f**, *11.6*

tar sands **31g**
Taylor–Görtler vortices 63
tepees 287
terminal riverine cones 151
Tethys 255, *26.10*
Thalassia 218, 220, 222
thalweg 142, 147
thermal stratification **16b**
thermocline 155–7, 250
thermodynamics 7
thixotropy 48, 112
threshold
  diagrams *6.4*, *6.5*
  for grain movement **6c**
tidal channels 188, 194, 197, 199, 217, 314, *21.13*
  current ridges 188, 194, 206, 220, *22.9*, *22.10*
  currents 178, 203–4, *18.13*
  deltas 195, 199, **23e**
  flats 109, 193, **23b**, **c**, *21.12*
  inlets 186, 197–9
  resonance 178
  transport paths *22.3*
  wave **18c**
till **17**, 40, 167, *17.7*
tool marks **96**
tourmaline 12
trace fossils **10b**
transgression **12e**, *Table 12.3*
transport stage 70
trenches **25e**
triple boundaries 260
trona 159–60
Troodos Massif 254
Trucial Coast 217, 221

turbidites **7f**, 115, 157, 225, 227, 235, 242–7, 253
turbidity
  currents 245–7, 251, 284, **24f**
  flows 115, **7a**, **e**, **f**
  maximum 192, *20.2*
turbulent intensity 60, **App. 8.1**, *5.20*
  flow **5h**, **i**
  stresses 86

Udden–Wentworth scale 35
umbers 255
undercompaction 265
underflows 156–7, 164
uplift **12d**
upper phase plane beds *8.9*
upward-transition probability matrix *Table 12.2*
upwelling 32, 233, 235, 250, *24.7*

vadose diagenesis 276, 285, 289, **27h**
varves 31, 157–8, 164, 253, 275, *3.4*
vector mean 127
  statistics **App. 12.1**
velocity (flow)
  fluctuations 53, 72
  free stream 49, 53
  mean 51
  profiles *5.10*, **App. 5.3**
vertical accretion 144
viscosity 4, **8b**, *Table 5.1*
viscous sublayer 55–8, 92, 103
vitrinite 311
volcanic arc 247
vortex 62–3

Wabash 131, 147, *15.5*
wackes **28h**
wadi 136
Walther's Law 122, *12.2*

washovers 198, 216, **6d**
water
  composition **2c**
  escape pipes *11.2*
  in rock weathering **1b**
  structure of *1.1*
  table 135
  waves **8c**, **18b**, **App. 18.1**
Waulsortian 222
wave(s) **8c**, **18b**, **App. 18.1**
  breaking *18.4*
  energy 172
  group velocity 172
  height 173
  orbitals 172, *8.15*
  period 173
  power 172, 174, 183
  reflection *18.5*
  refraction 173, *18.5*
  set-up 173, *18.7*
weatherability series 5
weathering **1b–f**
weathering potentials index 6
  profile *1.3*
West Atlantic shelf 207
Western Boundary Current *24.9*
Western Boundary Undercurrent 235, 241–2
Westwater Canyon Member 151
wet fan **14a**, 147
whitings 18, 222
Wilcox Group 281
wüstenquartz 250

yardangs 134
Yoredale cycles 189
Yucatan 221

Zechstein 31, 304, *13.6*
zeolites 279